HANDBOOKS IN OPERATIONS RESEARCH
AND MANAGEMENT SCIENCE
VOLUME 8

Handbooks in Operations Research and Management Science

Volume 8

ELSEVIER
Amsterdam – Lausanne – New York – Oxford – Shannon – Tokyo

Network
Routing

Edited by

M.O. Ball
University of Maryland

T.L. Magnanti
Massachusetts Institute of Technology

C.L. Monma
Bellcore

G.L. Nemhauser
Georgia Institute of Technology

1995
ELSEVIER
Amsterdam – Lausanne – New York – Oxford – Shannon – Tokyo

ELSEVIER SCIENCE B.V.
Radarweg 29, 1043 NX Amsterdam, The Netherlands

For information on all Elsevier publications
visit our website at books.elsevier.com

ISBN: 9780444821416

Transferred to digital printing 2007
Printed and bound by CPI Antony Rowe, Eastbourne

Preface

Whenever we use a telephone, shop at our neighborhood foodstore or mall, read our mail or fly for business or for pleasure, we are the beneficiaries of some system that has routed messages, goods or people from one place to another. Indeed, many great advances in the evolution of civilization (the telephone, trains, airplanes) represent quantum leaps in our ability to transport information, goods or services.

The papers in this volume consider a general area of study known as network routing. The underlying problems are conceptually simple, yet mathematically complex and challenging. How can we best route material or people from one place to another? Or, how can we best design a system (e.g., locate facilities) to provide services and goods as efficiently and equitably as possible?

The problems encountered in answering these questions often have an underlying combinatorial structure, for example, either we dispatch a vehicle or we don't, or we use one particular route or another. The problems also typically have an underlying network structure (a communication or transportation network). In addition, models for these problems often are very large with hundreds or thousands of constraints and variables.

Over the past three decades, researchers have made enormous progress in developing theory, models and solution methods for addressing these problems. Indeed, the advances in this general problem domain have been one of the major success stories of operations research and applied mathematics. With contributions by many individuals who have fuelled this success, this volume attempts simultaneously to describe the exciting challenges of the field of network routing and the enormous progress in modeling and solving these problems.

The first five chapters deal with several topics whose development has primarily been motivated by transportation applications, particularly those involving routing and scheduling issues. Chapter 1 by Fisher treats the "core" vehicle routing problem. While the traveling salesman problem can be interpreted as the problem of finding a best route for a single vehicle, the vehicle routing problem requires the determination of optimal routes for a fleet of vehicles, each with capacity limitations. In many cases, vehicle routing applications involve one of many possible time constraints. Chapter 2 by Desrosiers, Solomon and Soumis treats this class of problems. Frequently, practitioners need to solve routing problems repeatedly with changing, uncertain requirements. Such environments lead to the type of stochastic and dynamic problems that Powell, Jaillet and Odoni cover in Chapter 3. In Chapter 4, Federgruen and Simchi-Levi address the worst case and asymptotic analysis of routing algorithms. Analyses of this type have been useful not only in comparing algorithms, but also in investigating the overall design of distribution systems that involve inventory and other considerations. The classical vehicle routing problem associates demands with network nodes. Arc routing problems arise when service requirements are associated with network arcs. These problems, which trace their history back to the "first problem in graph theory", the Königsburg bridge problem, are treated in Chapter 5 by Assad and Golden.

The second set of chapters deal with a diverse set of routing related issues which arise in several different application domains. Network equilibrium problems, which are addressed in Chapter 6 by Florian and Hearn, arise when several users compete for the limited resources of a network. Practitioners use this class of models for a wide range of problems, including the prediction of traffic patterns within an urban transportation system and the traffic flows within a telecommunications network. Chapter 7 by Labbé, Peters and Thisse treats location problems defined on networks. Networks serve as a very rich framework for modeling location problems and for developing efficient location algorithms. Chapter 8 by Möhring, Wagner and Wagner treats a new, but very significant application area for network optimization, the design of VLSI networks. Chapter 9 by Sharkey takes a broad look at economic equilibrium models arising in a network setting. Included are topics from the economic literature as well as closely related topics from non-cooperative game theory.

A companion volume in the Handbook series, entitled "Network Models", treats basic network models such as minimum cost flows, matching and the traveling salesman problem, as well as, several complex network topics, not directly related to routing, such as network design and network reliability.

An examination of the chapters in this volume reveals network routing to be an area of particular dynamism. The majority of the material in several of the chapters was developed over the past ten years. The intellectual advances represented by the recent work in this area, the ability of the models studied to solve practical problems and advances in computing technology indicate that this area can have a tremendous impact on research and practice in the coming years. Our hope is that this volume will help network routing achieve its full potential.

Michael Ball
Tom Magnanti
Clyde Monma
George Nemhauser

Contents

M.O. Ball et al., Eds., *Handbooks in OR & MS, Vol. 8*

Chapter 1

Vehicle Routing

Marshall Fisher

Operations and Information Management Department, The Wharton School, University of Pennsylvania, Philadelphia, PA 19104, U.S.A.

1. Introduction

Transportation comprises a significant fraction of the economy of most developed nations. For example, a National Council of Physical Distribution Study [1978] estimates that transportation accounts for 15% of the U.S. gross national product. This economic importance has motivated both private companies and academic researchers to vigorously pursue the use of operations research and management science to improve the efficiency of transportation.

Various modes of transportation exist including air, rail, ship and motor vehicle. The research on transportation has focused on different issues in each mode. In air, the efficient scheduling of airline crews [Lavoi, Minous & Odier, 1988] has received primary attention while in rail, use of large-scale real-time computer control systems has dominated the research agenda (e.g. Harker 1990). The literature on both motor vehicles (trucks, school buses and general passenger buses) and ships has focused on a common problem — the efficient use of a fleet of vehicles that must make a number of stops to pick up and/or deliver passengers or products. The problem requires one to specify which customers should be delivered by each vehicle and in what order so as to minimize total cost subject to a variety of constraints such as vehicle capacity and delivery time restrictions. This chapter provides an introduction to research on this problem.

We'll confine ourselves to the most popular model of the vehicle routing problem, in which a fleet of identical vehicles makes deliveries to customers from a central depot. The model is defined by the following parameters.

K = number of vehicles in the fleet,

n = number of customers to which a delivery must be made. Customers are indexed from 1 to n and index 0 denotes the central depot,

b = capacity (e.g., weight or volume) of each vehicle,

a_i = size of the delivery to customer i, measured in the same units as vehicle capacity,

c_{ij} = cost of direct travel between points i and j (we assume $c_{ij} \geq 0$ and $c_{ij} = c_{ji}$ for all ij).

The vehicle routing problem is to determine K vehicle routes, where a route is a tour that begins at the depot, traverses a subset of the customers in a specified sequence and returns to the depot. Each customer must be assigned to exactly one of the K vehicle routes and the total size of deliveries for customers assigned to each vehicle must not exceed the vehicle capacity b. The routes should be chosen to minimize total travel cost.

This chapter will present the most important algorithms that have been developed for this model. However, before plunging into the mathematical development, it will be helpful to consider some examples of vehicle routing problems and various practical issues that arise in the use of vehicle routing models. Vehicle routing problems are all around us in the sense that many consumer products such as soft drinks, beer, bread, snack foods, gasoline and pharmaceuticals are delivered to retail outlets by fleets of trucks whose operation fits the vehicle routing model. Other examples of vehicle routing problems include the delivery of liquified industrial gases and the collection of milk from farms for transportation to a processing center. Many companies have reported successful implementation of mathematical algorithms to optimize trucking operations including Air Products and Chemicals [Bell, Dalberto, Fisher, Greenfield, Jaikumar, Kedia, Mack & Prutzman, 1983], Chevron [Brown & Graves, 1981], DuPont [Fisher, Greenfield, Jaikumar & Lester, 1982], Edward Don and Company [Walter & Zielinski, 1983; Zielinski, 1985], Exxon [Collins & Clavey, 1986], Kraft, Inc., [Evans & Norback, 1985], North American Van Lines [Powell, Sheffi, Nickerson, Butterbaugh & Atherton, 1986], Southland Corporation [Belardo, Duchessi & Seagle, 1985], and a number of soft drink distributors [Golden & Wasil, 1987].

Real vehicle routing problems usually include complications beyond the basic model considered here. Typical complications include the following.

1. Travel costs can be asymmetric so that $c_{ij} \neq c_{ji}$ in general.

2. The characteristics of the vehicles can introduce a variety of constraints beyond the simple vehicle capacity constraints. The vehicle fleet can be heterogeneous, with each vehicle k having a distinct capacity b_k. There can be multiple capacity constraints, arising, for example, if there are both weight and volume restrictions. Sometimes vehicles are divided into compartments for storage of different products (e.g., different blends of gasoline), which further complicates vehicle capacity constraints. Customer/vehicle compatibility constraints may restrict the set of customers that a vehicle can feasibly service. A vehicle may be capable of making more than one trip within a planning horizon. Finally, vehicles may both deliver and pick up product. This complicates vehicle capacity constraints and may also impose precedence constraints on the sequencing of stops on a route if products are picked up at one point for delivery later in the route.

3. The total time duration of a route may be constrained.

4. The time of delivery to customer i may be constrained to fall within a designated 'time window' or windows. Time window constraints arise frequently in practice and usually require fundamental changes in the algorithm used. For this

reason time constrained routing problems are given special attention in Chapter 2 of this Volume.

5. There may be multiple depots with each vehicle in the fleet assigned to a particular depot.

6. Precedence constraints can impose a partial ordering on the customer delivery sequence. A common example of a precedence constraint requires some customers to be the first or last stop on a route.

7. In fleet planning problems, the number of vehicles used is a variable rather than a constant.

8. Delivery to some customers may be optional provided we incur a specified penalty cost for not delivering. This situation can arise when any customer not delivered on the company fleet can be serviced by an outside carrier at a known cost.

9. Period routing problems arise in the distribution of products such as soft drinks, snack foods, beer and bread. In these applications the distributing firm is interested in developing a set of daily routes for some T day period so that each customer receives delivery at a designated frequency. For example, the period might be a 5-day week and the delivery frequency for a specific customer could be twice per week.

10. Inventory routing problems arise in the distribution of liquid products such as industrial gases or gasoline. In these problems, each customer has an inventory of the product, and the distributor must determine the timing and amount of deliveries so that the customer does not run out of product.

11. The objective function in real problems can be quite complex, including terms dependent on the distance travelled, the number of vehicles used, the time duration of routes (as with overtime pay for drivers) and penalties for not delivering to all customers.

The algorithms described in this chapter can usually be extended to accommodate most of these variations, although this is generally more challenging for the exact algorithms.

The data required for an application includes vehicle fleet characteristics (the number of vehicles and their capacity), customer order information (the size of the delivery to each customer), and geographic data (the travel cost and time between any two points). Travel costs and times are by far the most difficult data to obtain. Vehicle characteristics are straightforward to assemble because the amount of information is small, and customer order information can usually be obtained from a company's order entry system. On the other hand, it's rare to find a company that maintains travel costs and distances in computerized form. Usually, when a company is considering introducing an optimization model for vehicle routing, this data exists in the head of a dispatcher responsible for manual routing. The dispatcher knows roughly which customers are close to each other, and will consult a map to augment this informal knowledge as necessary.

Two approaches have been taken to providing geographic information. The first approach is easy to implement and thus preferred if it is accurate enough. It consists of assigning coordinates to each customer and assuming that the travel distance between customers is the Euclidean distance between their coordinate pairs. Cost is taken to be proportional to distance. The distance is often scaled up by a factor to compensate for roads that deviate from a straight-line path between customers. The degree of approximation can be improved further by defining the distance between customers to be the length of a direct path that does not cross various polygons defined to represent rivers, lakes, mountain ranges and other geographic barriers to travel. Finding such a path has been the subject of a number of papers, e.g. Viegas & Hansen [1985].

The second approach obtains travel distances by applying a shortest route algorithm to a computerized network model of the road system. A number of consulting firms and other organizations have compiled road network models. For example, the Swedish Postal Service has developed a network model of all roads in Sweden.

Vehicle routing algorithms can be applied in one of two modes: variable routing or fixed routing. In a variable routing system, an algorithm is used with actual customer delivery requirements to develop routes for the next planning horizon. For example, in a daily variable routing system, customers might be allowed to place orders up to one day prior to delivery time. At the end of a given day, we know precisely which customers require delivery tomorrow and can apply a vehicle routing algorithm to optimize routes for those customers.

Fixed routing is applicable when customer demands are sufficiently stable to allow use of the same routes repeatedly. In a fixed route application, the vehicle routing problem is solved using average demand data. For example, the delivery of bread to grocery stores requires a delivery to most stores every day. While the volume of bread delivered may vary somewhat due to factors such as holidays or days of the week, the set of customers requiring delivery doesn't change so fixed routes are generally used. Companies use fixed routes to avoid the effort required in frequently resolving the vehicle routing problem and to provide stability. For example, it's usually desirable for the same driver to deliver a particular customer repeatedly.

Theory and practice have always been tightly interwoven in vehicle routing research. From the theoretical perspective, the vehicle routing problem contains as special cases some of the most popular models in combinatorial optimization, including the traveling salesman problem and the generalized assignment problem. The many successful implementations of vehicle routing algorithms cited earlier testify to the practical importance of the topic.

The connection between theory and practice dates to the first paper on vehicle routing [Dantzig & Ramser, 1959] which describes both a practical problem concerned with delivering gasoline to gas stations by the Atlantic Refining Company and provides the first formulation of the general vehicle routing problem. Dantzig and Ramser also suggest a solution procedure motivated by their particular example. Soon after, Clarke & Wright [1964] presented a greedy heuristic that improved on the Dantzig–Ramser method. They too were motivated by a practical problem — the delivery of consumer nondurable products by the Cooperative Wholesale Society Limited in the Midlands of England.

This early work introduced what might be called the first generation of vehicle routing research which relied on greedy methods and various local improvement heuristics. A number of commercial vehicle routing systems developed in the 1970's incorporated Clarke and Wright's heuristic, most notably IBM's VSPX system. Literally hundreds of companies tried these commercial systems, but with no reported successes. The generation 1 methodology created in the 60's and 70's simply lacked the sophistication required to solve complex, real problems. Companies that tried these methods frequently reported obtaining solutions inferior to those produced by their existing manual system and noted an inability to handle all of the complex constraints in their real problems.

Success with vehicle routing in the real world had to await the second generation of research which began in the mid to late 70's when a number of researchers began to apply the machinery of mathematical programming to the vehicle routing problem. These efforts employed various mathematical programs that approximated the vehicle routing problem. Although the mathematical programs were solved to optimality, the overall procedures were heuristic since portions of the vehicle routing problem had to be modelled inexactly to achieve tractability. Examples include using a generalized assignment problem to assign customers to vehicles and a set partitioning model to select vehicle routes from a list of candidate routes.

The next major shift in emphasis in vehicle routing research has not yet emerged in a definitive enough form to characterize the third generation of vehicle routing research, but we can try to predict the form this generation will take by looking at the deficiencies of the current technology. What are the major limitations of the generation 2 algorithms? One response is a variation on the old question "If you're so smart, why aren't you rich?" If these algorithms are so effective, why aren't they used by more companies? Certainly the list of implementations given at the beginning of this introduction is impressive, but they barely scratch the surface of the tens of thousands of companies that own truck fleets and make up the 15% of the economy devoted to transportation. What inhibits these other firms from using the algorithms that have been developed?

The answer seems to be a lack of robustness in currently implemented algorithms that makes them hard to transfer from one company to another. In talking with researchers involved in implementing vehicle routing algorithms, one hears a common story. A heuristic algorithm is selected which appears reasonable for a particular situation. Under testing, the algorithm may work well in most cases, but occasionally produces obviously unreasonable results. The heuristic is then 'patched up' to fix the troublesome cases, leading to an algorithm with growing complexity and computational requirements. After considerable effort, a procedure is created that works well for the situation at hand, but one is left with the disquieting feeling that what has been produced is extremely sensitive to nuances in the data and will not perform well when transferred to other environments. It's not uncommon that a heuristic developed for a particular geographic region of a company's operation will perform poorly in another region served by the same company. What's needed are more robust tools for vehicle routing.

As the need for robustness has become more clear during the last decade, fortunately the resources for achieving robustness have also grown. Rapidly decreasing computation costs are pushing the tradeoff between computation time and solution quality in the direction of higher quality solutions. The accuracy of data on the cost of travel between customers has been greatly improved by the creation of road network databases. Finally, the base of fundamental research on which to draw has greatly expanded, including optimization research on related models like the traveling salesman problem and new approaches to heuristic problem solving growing out of the artificial intelligence community.

One could imagine various approaches to achieving robustness. The simplest might be to provide an interactive interface for vehicle routing algorithms based on a graphic display that would allow a human dispatcher to manually correct difficulties with solutions. A more complex approach could draw on past research in artificial intelligence. For example, we could use an expert system to capture the expertise used by an operations research analyst in developing and tuning a vehicle routing algorithm for a particular application. The expert system could have available to it all existing vehicle routing algorithms so as to be able to select an appropriate algorithm for a particular application. Test data could be supplied to the expert system with which to evaluate the performance of candidate algorithms and tune algorithm parameters as appropriate.

My own prediction is that optimization algorithms offer the best promise for achieving robustness. Although optimization has not been considered a practical approach for real problems in the past, rapidly decreasing computation costs and promising new research are causing a reevaluation of this assumption. In a practical application, an optimization algorithm need not be run to full optimality but can be stopped as soon as an acceptable solution has been obtained. As such, these algorithms offer all the computational tractability advantages of heuristics, but they also allow a user to control the tradeoff between solution quality and computational cost.

Finally, significant research has been conducted on the closely related traveling salesman problem. This provides a base of theoretical research on which to draw for vehicle routing optimization. A number of successful vehicle routing optimization algorithms are adaptations of traveling salesman algorithms. The dramatic increase in the size of traveling salesman problems solvable to optimality also suggests what could be accomplished for vehicle routing with a concerted research effort.

The three remaining sections of this paper correspond to the three generations of vehicle routing research and present key examples of simple heuristics, math-programming based heuristics and two recent approaches that are emerging as the focal points of generation 3: optimization algorithms and various heuristic approaches based on new results in artificial intelligence. Other survey's of vehicle routing which one may wish to consult include Bodin, Golden, Assad & Ball [1983], Bodin [1991], Christofides [1985], Golden & Assad [1986], Laporte & Nobert [1987], and Magnanti [1981].

2. Generation one — simple heuristics

For convenience in defining vehicle routing heuristics we introduce the graph $G = (N, A)$ with node set $N = \{0, 1, \ldots, n\}$ indexing the depot (node 0) and customers 1 to n and arc set $A = N \times N$ corresponding to all links between points in the problem. The length of arc ij is c_{ij}.

The simple heuristics developed during generation 1 can be grouped into three categories: route building heuristics, route improvement heuristics, and two-phase methods. Route building heuristics select arcs sequentially until a feasible solution has been created. Arcs are chosen based on some cost minimization criterion subject to the restriction that the selection does not create a violation of vehicle capacity constraints. Route improvement heuristics begin with a set of arcs $S \subseteq A$ that constitute a feasible schedule and seek an interchange of a set $S_1 \subset S$ with a set $S_2 \subset A - S$ that reduces cost while maintaining feasibility. Two-phase methods first assign customers to vehicles without specifying the sequence in which they are to be visited. In phase 2 sequenced routes are obtained for each vehicle using some traveling salesman problem algorithm. This section describes the most important examples of these three types of heuristics.

The Clarke & Wright [1964] method is the best known route building heuristic. Clarke and Wright begin with an infeasible solution in which every customer is supplied individually by a separate vehicle. By combining any two of these single customer routes, we would use one less vehicle and also reduce cost. Recall that 0 indexes the central depot. The cost of serving customers i and j individually by two vehicles is $c_{0i} + c_{i0} + c_{0j} + c_{j0}$ while the cost of one vehicle serving i and j on the same route is $c_{0i} + c_{ij} + c_{j0}$. Thus, combining i and j results in a cost savings of $s_{ij} = c_{i0} + c_{0j} - c_{ij}$.

Clarke and Wright select the arc ij with maximum s_{ij} subject to the requirement that the combined route is feasible. Customers i and j are now regarded as a single macro customer. A customer l can be linked to the macro customer at cost $\min\{c_{li}, c_{lj}\}$. With this convention, the route combining operation can be applied iteratively. In combining routes, we can simultaneously form partial routes for all vehicles or sequentially add customers to a given route until the vehicle is fully loaded. The latter is called the sequential Clarke and Wright method.

There have been many modifications to the basic Clarke and Wright method. Gaskell [1967] and Yellow [1970] independently introduced the concept of a modified savings given by $s_{ij} - \theta c_{ij}$ where θ is a scalar parameter. By varying θ, one can place greater or less emphasis on the cost of travel between two nodes relative to their distance from the central depot. This parameter can be altered and different solutions obtained. The best of these is then chosen. Golden, Magnanti & Nguyen [1977] have used computer science techniques to substantially reduce the running time of Clarke and Wright.

Altinkemer & Gavish [1991] have recently achieved significantly improved computational results with an implementation of Clarke-Wright in which several pairs of routes are combined on a single iteration. The choice of which pairs of routes to combine is determined by solving a maximum matching problem in which the nodes correspond to existing partial routes and the weight on an edge joining two nodes is the savings that would result from joining the two partial routes corresponding to the nodes.

Lin [1965] and Lin & Kernighan [1973] demonstrated the effectiveness of local improvement for the traveling salesman problem. They introduced the term k-optimal for a traveling salesman solution that can not be improved by deleting k or fewer arcs and replacing them feasibly with different arcs. In extensive computational testing they showed that a 3-optimal solution can be computed quickly and closely approximates the optimal value. Christofides & Eilon [1969] adapted the Lin and Kernighan method for vehicle routing by employing a reformulation of the vehicle routing problem without capacity constraints (called a K-traveling salesman problem) into a traveling salesman problem. A solution to the K-traveling salesman problem consists of K tours beginning and ending at the depot. This problem can be converted to a traveling salesman problem with $n + K$ cities by constructing K copies of the depot. Each copy of the depot is joined to customer i by an arc of length c_{0i}. Arcs joining copies of the depot have infinite length. It's easy to see that a solution to this $n + K$ city traveling salesman problem is also a solution to the K traveling salesman problem. Christofides and Eilon used this mapping to apply the Lin and Kernighan 3-optimal interchange procedure to feasible vehicle routing solutions, with the added requirement that any arc exchange made cannot destroy feasibility of the vehicle capacity constraints. Russell [1977] improved on the Christofides–Eilon procedure by selectively considering some exchanges involving more than 3 arcs. More recent research on local improvement methods is reported in Savelsberg [1985] and Thompson [1988].

Gillett & Miller's [1974] 'Sweep' algorithm is the simplest and best known two phase method. This method is directly applicable only to planar problems in which customers are located at points in the plane and c_{ij} is the Euclidean distance between points i and j. In phase 1, customers are represented in a polar coordinate system with the origin at the depot. A customer is chosen at random and the ray from the origin through the customer is 'swept' either clockwise or counterclockwise. Customers are assigned to a given vehicle as they are swept, until further assignment of customers would exceed the capacity of that vehicle. Then a new vehicle is selected and the sweep continues, with assignments now being made to the new vehicle. This process continues until all customers have been assigned to a vehicle. In phase 2, the customers assigned to each route are sequenced using some traveling salesman algorithm. Other two phase heuristics have been suggested by Christofides, Mingozzi & Toth [1979] and by Tyagi [1968].

We note that none of the simple heuristics described here place much emphasis on the vehicle capacity constraints. While these constraints are checked for violation as necessary, they have no other influence on the choices made in forming a solution. For this reason, these simple heuristics can easily terminate with an infeasible or poor solution if capacity constraints are tight.

We conclude this section with a computational comparison of four of the heuristics we have described. Table 1 summarizes problem characteristics and results.

Table 1
Computational comparison of heuristics

Problem	n	K	$\dfrac{\sum_{i=1}^{n} a_i}{Kb}$	Clarke & Wright		Sweep		Christofides & Eilon		Russell	
				Cost	CPU[a]	Cost	CPU[a]	Cost	CPU[b]	Cost	CPU[c]
1	50	5	0.97	585	0.8	532	12.2	556	120	524	15
2	75	10	0.97	900	1.7	874	24.3	876	240	854	244.8
3	100	8	0.93	886	2.4	851	65.1	863	600	833	100
4	150	12	0.94	1204	6.6	1079	142.0	d	d	d	d
5	199	17	0.95	1540	11.0	1389	252.2	d	d	d	d
6	100	10	0.91	831	2.4	937	50.8	d	d	d	d

[a] CDC 6600 seconds.
[b] IBM 7090 seconds.
[c] IBM 370/168 seconds.
[d] No results for this problem.

All test problems are planar in the sense that the given data are coordinates of customers and c_{ij} equals the Euclidean distance between customers i and j. The customer coordinates for problems 1–3 are randomly generated and are given in Christofides & Eilon [1969]. Problem 4 was produced by adding the customers of problems 1 and 3 with the depot and vehicle capacities as in problem 3. Problem 5 was produced by adding the customers of problem 4 with the first 49 customers of problem 2. The data from problem 6 is given in Christofides, Mingozzi & Toth [1979]. This problem was designed to resemble real problems by grouping customer locations into clusters. These six problems are 'classics' in that virtually every researcher proposing heuristics or optimization algorithms has tested their procedure on these problems. Hence they are useful in providing a common base of comparison across a wide range of algorithms.

The value in the column headed K gives the number of vehicles used. Except for problem 5, $K = \lceil \sum_{i=1}^{n} a_i/b \rceil$, the smallest number of vehicles that will admit a feasible solution. For problem 5, $\lceil \sum_{i=1}^{n} a_i/b \rceil = 16$, but using $K = 16$ results in vehicle capacity utilization of 99.9%. This is such a tightly constrained problem that, with a couple of exceptions that will be noted when we discuss Generation 3 research, all researchers have used $K = 17$. The ratio of total demand to total vehicle capacity is listed in Table 1 to provide a measure of the tightness of vehicle capacity constraints.

Results for the first two heuristics in Table 1 are taken from Christofides, Mingozzi & Toth [1979], for the third from Christofides & Eilon [1969] and for the fourth from Russell [1977]. The solution values for Sweep are the best of n executions with different starting customers and the solution times are the sums of the times for all of these runs. It should also be noted that the parallel implementation of Clarke–Wright based on maximum matchings as developed by Altinkemer & Gavish [1991] achieved a value of 1351 on problem 5, although they do not report the computation time required to find the solution.

One qualification needs to be made for computational results reported throughout this chapter on these six problems. Since c_{ij} is the Euclidean distance between two points, it will in general be a real number. Nonetheless, some researchers have rounded c_{ij} to the nearest integer, or in some cases even truncated to the next smaller integer. Clearly, either of these changes can have a significant effect on the cost of a particular solution. For example, it is known that the optimal cost for problem 1 with all c_{ij} rounded to the nearest integer is 521, while the best feasible solution found to date for this problem with real c_{ij} is 524.61.

This complicates the comparison of heuristics tested on these problems. Because researchers have not been explicit in reporting how they were computing c_{ij}, apparent differences between two heuristics can be the result of nothing more than different rounding conventions for the costs. Also, even researchers who used real c_{ij} often report the resulting objective function as an integer either rounding or truncating downwards. For example, the value of 524 reported by Russell for problem 1 is probably the best found real solution of 524.61. In reporting results in the rest of this chapter, I will not try to second guess the rounding assumptions employed by various researchers, but will simply reproduce their results as reported in the published literature. Recently, researchers have become more careful in declaring their rounding assumptions, so for the generation 3 results I shall generally be able to identify the rounding assumption that was used.

3. Generation two — mathematical programming based heuristics

Mathematical programming based heuristics are very different in character from the simple heuristics described in the preceding section. They solve to optimality some mathematical programming approximation of the vehicle routing problem. We describe in this section two well known examples of this approach which have produced computational results that are usually superior to the simple heuristics described in the last section. The first uses the generalized assignment problem and the second the set partitioning problem to approximate the vehicle routing problem. Haimovich & Rinnooy Kan [1985], Altinkemer & Gavish [1987] and Bramel & Simchi-Levi [1992] use other advanced concepts from mathematical programming to develop asymptotically optimal heuristics for vehicle routing.

Fisher & Jaikumar [1981] solved a generalized assignment problem approximation of the vehicle routing problem to obtain an assignment of customers to vehicles. The customers assigned to each vehicle can then be sequenced using any traveling salesman algorithm. To motivate the generalized assignment problem approximation, first note that the vehicle routing problem can be represented exactly by the following *nonlinear* generalized assignment problem. Defining

$$y_{ik} = \begin{cases} 1, & \text{if point } i \text{ (customer or depot) is visited by vehicle } k \\ 0, & \text{otherwise} \end{cases}$$

and $y_k = (y_{0k}, \ldots, y_{nk})$, the vehicle routing problem is defined as

$$\min \sum_k f(y_k)$$

such that

$$\sum_i a_i y_{ik} \le b, \qquad k = 1, \ldots, K$$

$$\sum_k y_{ik} = \begin{cases} K, & i = 0 \\ 1, & i = 1, \ldots, n \end{cases}$$

$$y_{ik} = 0 \text{ or } 1, \qquad i = 0, \ldots, n, \ k = 1, \ldots, K$$

where $f(y_k)$ is the cost of an optimal traveling salesman problem tour of the points in $N(y_k) = \{i \,|\, y_{ik} = 1\}$. Defining

$$x_{ijk} = \begin{cases} 1, & \text{if vehicle } k \text{ travels directly from point } i \text{ to point } j \\ 0, & \text{otherwise} \end{cases}$$

the function $f(y_k)$ can be defined mathematically by

$$f(y_k) = \min \sum_{ij} c_{ij} x_{ijk}$$

such that

$$\sum_i x_{ijk} = y_{jk}, \qquad j = 0, \ldots, n$$

$$\sum_j x_{ijk} = y_{ik}, \qquad i = 0, \ldots, n$$

$$\sum_{ij \in S \times S} x_{ijk} \le |S| - 1, \qquad S \subseteq N(y_k), \ 2 \le |S| \le n$$

$$x_{ijk} = 0 \text{ or } 1, \qquad i = 0, \ldots, n, \ j = 0, \ldots, n$$

Of course, this definition doesn't help us in computations since we lack a closed form expression for $f(y_k)$. The generalized assignment heuristic replaces $f(y_k)$ with a linear approximation $\sum_i d_{ik} y_{ik}$ and solves the resulting linear generalized assignment problem to obtain an assignment of customers to vehicles. To obtain the linear approximation, first specify K 'seed' customers $i_1, \ldots, i_K$ that are assigned one to each vehicle. Without loss of generality, assume customer i_k is assigned to vehicle k for $k = 1, \ldots, K$. Then set the coefficient d_{ik} to the cost of inserting customer i into the route on which vehicle k travels from the depot directly to customer i_k and back. Specifically, $d_{ik} = c_{0i} + c_{ii_k} - c_{0i_k}$. Clearly, the seed customers define the general direction in which each vehicle will travel and the generalized assignment problem completes the assignment of customers to routes given this general framework.

Many algorithms exist for optimal solution of the generalized assignment problem with a linear objective function, such as the Lagrangean relaxation algorithm described in Fisher, Jaikumar & Van Wassenhove [1986]. One can also solve the generalized assignment problem heuristically by assigning customers sequentially to the vehicle with minimum d_{ik} in which they fit. Commonly used sequences for processing customers include decreasing order of distance from the depot or decreasing order of the difference between the assignment cost for best and second best vehicles.

Seed customers can be chosen by a variety of heuristics. Fisher & Jaikumar ([1981] suggest the following procedure for the planar case. They actually determine K seed points $w_1, \ldots, w_K$ in the plane rather than K seed customers. These points are used exactly like seed customers in determining d_{ik}.

To determine $w_1, \ldots, w_K$, the plane is partitioned into K cones with origin at the depot corresponding to the K vehicles. To determine these cones, first partition the plane into n smaller cones, one for each customer. The infinite half ray forming the boundary between two customer cones is positioned to bisect the angle formed by half rays through the two customers. Associate the weight a_i with customer cone i and define $\bar{b} = \sum_{i=1} a_i / K$.

Each vehicle cone is then formed from a contiguous group of customer cones or fractions of customer cones. The weight of the group is required to equal $\bar{b}$. A fraction of a customer cone i contributes the same fraction of a_i to the total group weight. The point w_k is located along the ray bisecting the kth cone. The distance of w_k from the origin is fixed so that the weight included inside the arc through w_k is $0.75\,\bar{b}$. This weight is defined to equal the sum of the a_i for all customers inside the arc plus a fraction of a_i for the customer just outside the arc. This fraction is $A/(A + B)$, where A is the distance to the arc from the customer just inside the arc and B is the distance to the arc from the customer just outside the arc.

In the more general case, when planarity is not assumed, seeds are frequently chosen with the following simple rule. Choose the first seed s_1 to be a customer farthest from the depot. If k seeds have been chosen, choose s_{k+1} to solve

$$\max_i \min\, \{c_{i0}, \min_{j=1,\cdots,k} c_{is_j}\}$$

Bramel & Simchi-Levi [1992] have provided an elegant approach to seed selection by expanding the generalized assignment model to involve both choice of which customers will be seeds and assignment of customers to seeds.

Let

$$z_j = \begin{cases} 1, & \text{customer } j \text{ is chosen to be a seed} \\ 0, & \text{otherwise} \end{cases}$$

$$y_{ij} = \begin{cases} 1, & \text{customer } i \text{ is assigned to a route on which customer } j \text{ is a seed} \\ 0, & \text{otherwise} \end{cases}$$

Define $d_{ij} = c_{0i} + c_{ij} - c_{0j}$ and $v_j = 2c_{0j}$. Then an assignment of customers to routes is determined by solving

$$\min \sum_{ij} d_{ij} y_{ij} + \sum_j v_j z_j$$

$$\sum_j z_j = K$$

$$\sum_i a_i y_{ij} \leq b, \quad j = 1, \cdots, n$$

$$\sum_j y_{ij} = 1, \qquad i = 1, \cdots, n$$

$$y_{ij} \le z_j, \qquad \text{all } ij$$

$$y_{ij} = 0 \text{ or } 1, \qquad \text{all } aj$$

$$z_j = 0 \text{ or } 1, \qquad j = 1, \cdots, n$$

The model is called a capacitated concentrator location problem; Bramel & Simchi-Levi [1992] give an optimization algorithm for it based on a Lagrangian relaxation in which the constraints $\sum_j y_{ij} = 1$ are dualized for all i.

The generalized assignment heuristic can be adapted to accommodate various side constraints. Examples include time windows [Nygard, Greenberg, Bolkan & Swenson, 1988; Koskosidis, Powell & Solomon, 1989] and split deliveries [Brenninger-Goethe, 1989].

The set partitioning heuristic begins by enumerating a number of candidate vehicle routes. A candidate route is defined by a set $S \subseteq \{1, \ldots, n\}$ of customers to be delivered by a single vehicle and a delivery sequence for these customers. We index the candidate routes by j and define the following parameters.

$$c_j = \text{the cost of candidate route } j$$

$$a_{ij} = \begin{cases} 1, & \text{if customer } i \text{ is included on route } j \\ 0, & \text{otherwise} \end{cases}$$

$$J = \text{number of candidate routes generated}$$

Letting

$$y_j = \begin{cases} 1, & \text{if candidate route } j \text{ is used} \\ 0, & \text{otherwise} \end{cases}$$

the vehicle routing problem can then be approximated by the following set partitioning problem.

$$\min \sum_{j=1}^{J} c_j y_j$$

$$\sum_{j=1}^{J} y_j = K$$

$$\sum_{j=1}^{J} \sum_{k=1}^{K} a_{ij} y_j = 1, \qquad i = 1, \ldots, n$$

$$y_j = 0 \text{ or } 1, \qquad j = 1, \ldots, J$$

There are many effective optimization algorithms for set partitioning. For example, the Lagrangian relaxation algorithm in Fisher & Kedia [1990] has solved problems as large as $n = 200$ and $J = 10,000$.

The set partitioning approach will find an optimal solution if the candidate route list contains all feasible routes. In most situations, this would result in a set partitioning problem too large to be solvable, so one instead heuristically generates a candidate route list designed to be short enough to allow solution of the set partitioning problem yet long enough to result in an optimal or near optimal solution to the vehicle routing problem. We can usually do this if the number of feasible routes is small, which can happen if the a_i are large relative to b so that the number of customers delivered by a vehicle is small (on the order of 1–4 deliveries per truck) or if there are additional restrictions (such as time window constraints) that limit the number of feasible routes. The set partitioning approach is guaranteed to find a feasible solution if one initially applies any of the heuristics described in this chapter to obtain a feasible solution and includes the routes of this solution in the candidate list.

The set partitioning approach can easily accommodate certain kinds of complex constraints. For example, any feasibility condition for an individual vehicle route can usually be incorporated in the heuristic for generating candidate routes. Similarly, this approach can handle costs which are a complicated function of route characteristics since all one needs to do is evaluate this function for given routes.

The set partitioning approach was originally proposed by Balinski & Quandt [1964], although not implemented for large scale problems until much later. Successful applications in truck scheduling include Foster & Ryan [1976] and Fleuren [1988] for time window constrained problems and in ship scheduling Fisher & Rosenwein [1989]. Callen, Jarvis & Ratliff [1981] developed an interactive implementation of this approach in which a user (aided by a color graphics interface) specifies additional candidate routes during the solution process. Agarwal, Mathur & Salkin [1989] and Desrochers, Desrosiers & Solomon [1990] (for time window constrained problems) report exact algorithms based on a set partitioning model and column generation. Bramel & Simchi-Levi [1993] provide theoretical results on the effectiveness of the set partitioning/column generation approach.

Table 2 summarizes on the 6 problems introduced in section 2 the computational performance of the Fisher & Jaikumar [1981] generalized assignment heuristic, Bramel & Simchi-Levi [1992] extended generalized assignment heuristic, and the set partitioning heuristic as implemented by Foster & Ryan [1976].

Table 2
Computational comparison of mathematical programming based heuristics

Problem	n	Fisher & Jaikumar		Bramel & Simchi-Levi		Foster & Ryan	
		Cost	CPU[a]	Cost	CPU[b]	Cost	CPU[c]
1	50	524	9.33	524.6	68	523	2.6
2	75	857	11.95	848.2	406	864	6.8
3	100	833	17.7	832.9	890	825	16.9
4	150	1014	33.6	1088.6	2552	d	d
5	199	1420	40.1	1461.2	4142	d	d
6	100	824	6.1	826.1	400	d	d

[a] DEC 10 seconds.
[b] RS600 Model 550 seconds.
[c] IBM 370/168 seconds.
[d] No results for this problem.

In these computational experiments, the generalized assignment problem was solved using the Lagrangian relaxation algorithm described in Fisher, Jaikumar & Van Wassenhove [1986] and the traveling salesman problem for the customers assigned to each vehicle were solved optimally using an algorithm similar to the one reported in Miliotis [1976].

4. Generation three — what are the new frontiers?

4.1. Artificial intelligence

Artificial intelligence techniques have been applied to vehicle routing in two ways. First, expert systems are being developed to assist a user faced with a particular application in choosing an appropriate algorithm and tuning the parameters of that algorithm to the problem at hand. Second, successful new algorithms have recently been developed using artificial intelligence search techniques like simulated annealing and tabu search.

The first approach begins with the observation that there are many types of vehicle routing problems and also many different algorithms have been developed, so that a typical problem faced by a vehicle routing analyst is to match an appropriate algorithm with a particular application. Relevant information about the application includes not only the formal definition of the problem but characteristics of the data such as average number of stops per vehicle or tightness of the vehicle capacity constraints. To illustrate the problem of choosing an appropriate algorithm, consider two of the algorithms presented in the last section — the generalized assignment and set partitioning algorithms. Each of these algorithms works best on problems with particular data characteristics. For example, the set partitioning algorithm would be particularly appropriate for a problem that had an average of 1.5 customers per vehicles because the small number of customers per vehicle would imply that the number of ways customers could be combined to form vehicle routes would also be small. The generalized assignment algorithm would not be a good choice for such a problem since the focus of this method is to efficiently assign remaining customers once a single seed customer has been assigned to each route. With such a small number of customers per route, choosing the seed customers would almost completely specify the routes and the contribution made through solution of the generalized assignment problem would be minimal. On the other hand, the generalized assignment method fits particularly well on problems that have very tight vehicle capacity constraints since the vehicle capacity constraints play a prominent role when the generalized assignment problem is solved.

Each of these algorithms also have parameters whose specification should depend on the data characteristics of the particular problem being solved. For example, the set partitioning algorithm requires rules for generating the candidate route list and the generalized assignment algorithm requires rules for setting seed customers, as well as a choice of whether the generalized assignment problem will be solved optimally or heuristically, and if the latter, a specification of the heuristic to be used.

Several researchers have studied the question of choosing and tuning an appropriate algorithm for a particular vehicle routing application. Desrochers, Lenstra & Savelsbergh [1990] have developed a scheme for classifying vehicle routing problems which is a helpful first step in guiding a user in the selection of a method that is well suited for his specific situation. Kadaba, Nygard & Juell [1991] have applied various artificial intelligence techniques such as neural networks and genetic algorithms to the choice of seed customers in the generalized assignment heuristic.

Finally, Potvin, Lapalme & Rousseau [1989] have developed an interactive graphic computer system that allows an expert user to formulate and test various versions of vehicle routing algorithms within a broad class. The system works within the framework of a generalized assignment heuristic in which the generalized assignment problem is solved heuristically. The graphical interface allows the user to either manually perform or provide rules for performing the functions of choosing seed customers, adding customers to routes in the heuristic solution of the generalized assignment problem and improving the delivery sequence of a set of customers assigned to a particular vehicle.

All of this research constitutes a very promising beginning towards the goal of developing an expert system for automatically choosing and tuning algorithms for a particular application. It's hard to compare the computational performance of this approach with other algorithms, since no researcher has tested a procedure of this type on standard test problems such as the six problems we have been using in this chapter. Indeed, given that the goal of this approach is adaptability to a wide range of applications, testing on a small number of standard problems would not make sense.

Significantly improved results for the six standard test problems have been obtained by adapting artificial intelligence search procedures to vehicle routing. A generic search procedure for the vehicle routing problem begins with a starting solution S and a rule for constructing a neighborhood $N(S)$ of alternative solutions that are near to S in some sense. The search procedure then chooses a new solution from $N(S)$ and iterates until some stopping criterion is reached.

Traditional local improvement methods choose a solution from $N(S)$ with a strictly improved objective function value and stop when $N(S)$ contains no such improving solution. Tabu search and annealing allow the selection of nonimproving solutions under certain conditions to be defined below.

An initial solution for a search algorithm can be obtained by applying any of the heuristics defined in this chapter. Usually the neighborhood $N(S)$ is defined to be all solutions obtained by exchanging n_1 customers on a given route with n_2 customers on another route. Typically, $n_1 \leq 1$ and $n_2 \leq 1$, so the exchanges permitted consist of moving a single customer from one route to another route or exchanging two customers on different routes.

Different approaches can be used to compute the change in cost resulting from the addition or deletion of a customer on a route. Ideally, one would evaluate this change by optimally solving a traveling salesman problem over the depot plus the customers on the route before and after the addition or deletion of the new customer. However, this approach can be computationally expensive so a less accurate but faster method is generally used. When a customer is deleted, the cost of the new route is computed without changing the sequence of the customers that remain on the route. Similarly, the customer added to a route is inserted between two existing customers without changing the sequence of the original customers on the route. The choice of where to insert the new customer is made to minimize the increase in cost.

With a traditional local improvement method that only accepts solutions that strictly improve the objective function, one can choose whether to accept the first solution in $N(S)$ that improves the objective function or to examine all solutions in $N(S)$ and choose the one producing the greatest reduction in the objective function.

Tabu search chooses the best solution contained in $N(S)$ that does not violate certain restrictions designed to prevent the algorithm from cycling. Typically, these restrictions prevent for the next t iterations a movement of customers that would 'undo' a previous movement. For example, if customer c is moved from route A to some other route on iteration k, then on iterations $k + 1$ through $k + t$ we prohibit any exchange that would move customer c back to route A. Similarly, if on iteration k customer c_1 on route A and c_2 on route B are exchanged, then on iterations $k + 1$ through $k + t$ we prohibit any movement that would return customer c_1 to route A and customer c_2 to route B. Tabu search stops after some fixed iteration limit has been reached.

Simulated annealing searches the neighborhood of $N(S)$ in a defined order. Let Δ denote the increase in object value (over the current incumbent solution S) for some $S' \epsilon N(S)$. Then S' is accepted as the new incumbent solution either if $\Delta \leq 0$ or $\Delta > 0$ and $e^{-\Delta/T} \geq \theta$ where θ is a uniform random parameter $0 < \theta < 1$ and T is a control parameter of the search called the 'temperature.' Typically, T is gradually lowered during the search procedure, steadily reducing the probability that a nonimproving solution will be accepted. If a complete search of $N(S)$ fails to produce a new incumbent, the value of T is raised by some amount and the search repeated. The simulated annealing algorithm stops when a fixed consecutive number of searches of $N(S)$ fails to produce a new incumbent solution.

Three authors have constructed successful implementations of either tabu search or simulated annealing. Gendreau, Hertz & Laporte [1993] applied tabu search using neighborhoods consisting of all solutions that could be constructed by moving a single customer from one route to another. Osman [1993] applied tabu search and simulated annealing using a neighborhood consisting of all solutions that could be constructed by moving one customer from a given route to another or exchanging two customers on different routes. Taillard [1993] applied tabu search with the same neighborhood definition as Osman. Taillard also penalized moves that had been performed frequently by adding a penalty term to the cost improvement for the move. The penalty term was proportional to the number of times the move had been performed thus far. Taillard also designed a parallel implementation of tabu search suitable for large problems. In his parallel

Table 3
Computational comparison of artificial intelligence search algorithms

Problem	Tabu search					Simulated annealing		
	Gendreau, Hertz & Laporte		Osman		Taillard	Osman		
	Cost	CPU[a]	Cost	CPU[b]	Cost	Cost	CPU[b]	
1	524.61	6	524	2	524.61	528	3	
2	835.32	53.8	844	3	835.26	538	107	
3	826.14	18.4	835	26	826.14	829	156	
4	1031.07	58.8	1044	59	1028.42	1058	84	
5	1311.35	90.9	1334	54	1298.79	1378	39	
6	819.56	16	819	15	819.56	826	11	

[a] Silicon Graphics 36MHZ workstation minutes.
[b] VAX 8600 minutes.

implementation, the customers are divided into groups and tabu search is initially performed within these groups.

The results achieved by these various implementations of tabu search and simulated annealing for the six standard problems considered in this chapter are shown in Table 3.

All computations are with real c_{ij}. Taillard does not report CPU times. The solutions to problem 5 obtained by Osman have $K = 16$ rather than $K = 17$ as used by other researchers.

4.2. Optimization algorithms

This section presents four optimization algorithms for vehicle routing based on polyhedral combinatorics, a matching relaxation, a shortest path relaxation and a K-tree relaxation. Other research on optimization is reported in Christofides, Mingozzi & Toth [1981b] and Lucena [1986].

The first approach closely parallels the highly successful use of polyhedral combinatorics developed for the traveling salesman problem by Chvatal [1973], Grotschel [1980], Grotschel & Padberg [1979], Padberg & Hong [1980] and Grotschel & Pulleyblank [1986]. The book by Lawler, Lenstra, Rinnooy Kan & Shmoys [1985] contains an excellent review of this research. This research relies on the following formulation of the n city symmetric TSP, where x_{ij} is a 0–1 variable equal to 1 if an only if the salesman travels directly between cities i and j.

$$\min \sum_{i<j} c_{ij}x_{ij} \tag{1}$$

$$\sum_{j<i} x_{ij} + \sum_{j>i} x_{ji} = 2, \quad i = 1, \dots, n \tag{2}$$

$$\sum_{\substack{ij \in S \times S \\ i<j}} x_{ij} \leq |S| - 1, \quad \text{for all } S \subset \{1, \dots, n\}, \ 2 \leq |S| \leq n - 1 \tag{3}$$

$$x_{ij} = 0 \text{ or } 1, \qquad 1 \leq i < j \leq n \qquad (4)$$

As discussed in Laporte, Nobert & Desrochers [1985], this formulation can be adapted to the vehicle routing problem by adding variables and a constraint like (2) to model the depot and by changing the right-hand side of (3) to impose the vehicle capacity constraints. We present the resulting formulation below. The variable x_{ij} now represents the number of vehicles traveling directly between points i and j and $V(S) = \lceil \sum_{i \in S} a_i / b \rceil$ where $\lceil y \rceil$ denotes the smallest integer not less than y. This value for $V(S)$ can be replaced by a tighter lower bound on the number of vehicles needed for delivery to the customers in S if one is available.

$$\min \sum_{i<j} c_{ij} x_{ij} \tag{1'}$$

$$\sum_{j<i} x_{ij} + \sum_{j>i} x_{ji} = 2, \qquad i = 1, \ldots, n \tag{2}$$

$$\sum_{\substack{ij \in S \times S \\ i<j}} x_{ij} \leq |S| - V(S), \qquad \text{for all } S \subset \{1, \ldots, n\}, \ 2 \leq |S| \leq n - 1 \tag{3'}$$

$$x_{ij} = 0 \text{ or } 1, \qquad 1 \leq i < j \leq n \tag{4}$$

$$\sum_{j} x_{0j} = 2K \tag{5}$$

$$x_{0j} = 0, \ 1 \text{ or } 2, \qquad j = 1, \ldots, n. \tag{6}$$

The case $x_{0j} = 2$ corresponds to a route containing only customer j. If single customer routes cannot occur, we can require $x_{0j} = 0$ or 1.

Laporte, Nobert & Desrochers [1985] applied a general purpose integer programming algorithm to this formulation. They added capacity constraints (3') to the formulation as they were violated since these constraints are too numerous to specify a priori.

In gauging the computational difficulty of a vehicle routing problem, we must consider the number of customers, number of vehicles and tightness of the vehicle capacity constraints as measured by average vehicle utilization $(\sum_{i=1} a_i)/Kb$. Laporte and coworkers were able to optimally solve randomly generated problems with 50 to 60 customers, 1 or 2 vehicles and average vehicle utilization to 0.74.

The TSP formulation (1)–(4) can be strengthened by the addition of other valid inequalities, such as the comb inequalities developed by Chvatal [1973] and generalized by Grotschel & Padberg [1979]. Laporte & Nobert [1987] and Laporte & Bourjolly [1984] generalized these comb inequalities to the following valid constraints for the vehicle routing problem. Let $W_l \subseteq \{1, \ldots, n\}, l = 0, \ldots, k$ denote sets satisfying

$$|W_l - W_0| \geq 1 \qquad (l = 1, \ldots, k)$$
$$|W_l \cap W_0| \geq 1 \qquad (l = 1, \ldots, k)$$
$$|W_l \cap W_{l'}| = 1 \qquad (1 \leq l, < l' \leq k).$$

Then the following comb inequality holds for every feasible vehicle routing solution, where, as before, $V(S)$ denotes any lower bound on the number of vehicles required to deliver the customers in set S.

$$\sum_{l=0}^{k} \sum_{i,j \in W_l} x_{ij} \leq \sum_{l=0}^{k} |W_l| -$$
$$- \left\lceil \frac{1}{2} \sum_{l=1}^{k} [V(W_l) + V(W_l - W_0) + V(W_l \cap W_0)] \right\rceil$$

Using comb inequalities and formulation (1')–(6), Cornuejols & Harche [1989] solved to optimality the first problem reported in Table 1 with c_{ij} rounded to integer values. The size of this problem (50 customers, 5 vehicles) and the tightness of the vehicle capacity constraints ($\sum_{i=1}^{n} a_i/Kb = 0.97$) makes this a notable accomplishment. Campos, Corberan & Mota [1991] and Araque, Kudva, Morin & Pekny [1994] have specialized the polyhedral approach to problems with $a_i = 1$ for all i.

Miller [1993] has developed an optimization algorithm that dualizes constraints (2) in formulation (1')–(6) to obtain a maximum b-matching problem that is solved using the algorithm in Miller & Pekny [1994]. This algorithm has solved to optimality the first problem reported in Table 1 with c_{ij} rounded to integral values.

Christofides, Mingozzi & Toth [1981a] compute bounds on the vehicle routing problem using a shortest path calculation. They define a graph with nodes (i, q) for $i = 0, 1, \ldots, n$ and $q = 0, 1, \ldots, b$ and with an arc of length c_{ij} joining each pair of nodes (i, q) and $(j, q + a_j)$. Letting $l(i, q)$ denote the length of a shortest path from node (o, o) to node (i, q), $F(i, q) = l(i, q) + c_{i0}$ is a lower bound on the cost of delivering q units on a vehicle route in which customer i appears last. This route is feasible except that some customers may receive more than one delivery.

Define

$v(i, q, k) = $ minimum cost of k routes carrying load q with different last customers chosen from $\{1, \ldots, i\}$.

We can compute $v(i, q, k)$ with the following recursion

$$v(i, q, k) = \min\{v(i - 1, q, k); \min_{q'}[v(i - 1, q - q', k - 1) + F(i, q')]\}$$

Then a lower bound on the vehicle routing problem is given by $v(n, \sum_{i=1} a_i, K)$. This bound corresponds to a set of K routes that constitute a feasible solution except that some customers may not be delivered exactly once. The bound can be strengthened by introducing Lagrangian penalties on these violated customer delivery constraints.

Kolen, Rinnooy Kan & Trienekens [1987] generalized this algorithm to accommodate time window constraints. Christofides [1985] reports that an algorithm based on state space relaxation solved to optimality a problem with 53 customers and 8 vehicles. In addition to the vehicle capacity constraints, there were constraints on driving time and a few loose customer time window constraints.

Fisher [1990] describes an optimization algorithm in which lower bounds are obtained from a relaxation based on a generalization of spanning trees called K-trees. Given a graph with $n + 1$ nodes, a K-tree is defined to be a set of $n + K$ edges that span the graph. The vehicle routing problem is modeled as a minimum cost degree constrained K-tree problem with side constraints. The side constraints are then dualized to obtain a Lagrangian relaxation.

Let $N = \{1, \ldots, n\}$, $N_0 = N \cup \{0\}$ and x_{ij} denote a $0 - 1$ variable equal to 1 if edge (i, j) is selected in a solution. Since the edge (i, j) is undirected, it will simplify notation to adopt the convention that the subscripts ij on x_{ij} are an unordered pair, so x_{ij} and x_{ji} denote the same variable. Then x_{ij} is defined for the $[n(n + 1)]/2$ unordered pairs in $N_0 \times N_0$. Also define

$$x = (x_{01}, x_{02}, \ldots x_{0n}, x_{12}, \ldots, x_{n-1,n})$$

$$X = \left\{ x \mid x = \text{0–1 and defines a } K\text{-tree satisfying} \sum_{i=1}^{n} x_{0i} = 2K \right\}.$$

For $S \subseteq N$, let $\bar{S} = N_0 - S$, $a(S) = \sum_{i \in S} a_i$ and $r(S) = \lceil [a(S)]/b \rceil$ where $\lceil y \rceil$ denotes the smallest integer not less than y. For $S \subseteq N_0$, let $E(S)$ denote the edge set of a complete, undirected graph on the node set S, i.e., $E(S)$ is the set of all unordered pairs $ij, i \in S, j \in S, i \neq j$. Then the VRP can be formulated as

$$Z^* = \min_{x \in X} \sum_{ij \in E(N_0)} c_{ij} x_{ij} \tag{7}$$

$$\sum_{\substack{j \in N_0 \\ j \neq i}} x_{ij} = 2, \qquad \text{for all } i \in N \tag{8}$$

$$\sum_{i \in S} \sum_{j \in \bar{S}} x_{ij} \geq 2r(S), \qquad \text{for all } S \subseteq N \text{ with } |S| \geq 2 \tag{9}$$

In this formulation, routes with single customers are not allowed. Generally this is not a serious restriction for problems with tight vehicle capacity constraints and no a_i that is large relative to b. Note that customer j cannot appear alone on a route if $(\sum_{i=1}^{n} a_i) - a_j > (K - 1)b$, because using a dedicated vehicle to deliver customer j would leave insufficient vehicle capacity to service the remaining customers. Rearranging terms in this expression, we can see that the prohibition of single customer routes is not constraining if $a_j/b < \sum_{i=1}^{n} a_i/b - (K - 1)$ for all j, a condition that will be satisfied if vehicle capacity constraints are sufficiently tight and no customer is large relative to vehicle capacity. The K-tree approach could be modified to allow for particular customers to be served on single stop routes by including two edges between these customers and the depot in the graph used to compute lower bounds, although, of course, this could affect lower bound strength.

Letting $u_i, i \in N$ and $v_s \geq 0$ for $S \subseteq N, |S| \geq 2$ denote Lagrange multipliers for constraints (2) and (3), we can define the following Lagrangian relaxation of (1)–(3).

$$Z_D(u, v) = \min_{x \in X} \sum_{ij \in E(N_0)} \bar{c}_{ij} x_{ij} + 2 \sum_{i=1}^{n} u_i + 2 \sum_{S \subseteq N} v_S r(S) \tag{10}$$

where $u_0 = 0$, and

$$\bar{c}_{ij} = c_{ij} - u_i - u_j - \sum_{\substack{S \text{ such that } i \in S, j \in \bar{S} \\ \text{or } i \in \bar{S}, j \in S}} v_S \tag{11}$$

It is well known and easy to show that $Z_D(u, v) \leq Z^*$. A polynomial algorithm for (10) is given in Fisher [1991]. To obtain a tight lower bound, the subgradient method is used to approximate an optimal solution to $\max_{u, v \geq 0} Z_D(u, v)$. Since there are $0(2^n)$ constraints in the set (9), it is not feasible to tabulate explicitly all of these constraints prior to computation. Rather, a subset of these constraints is generated dynamically as they are violated. All other constraints have $v_s = 0$ and are ignored. Feasible solutions are obtained from a Lagrangian heuristic applied to Lagrangian solutions to remove infeasibilities.

Constraints (9) also lend themselves to tightening in the following way. The rationale underlying (9) is that $br(S)$ defines the minimum vehicle capacity that must enter and leave S to feasibly carry the customers in set S. However, if $x_{ij} = 1$ in the left hand side of (9) for some customer $j \in \bar{S}$, then customer j is serviced by the same vehicle as at least one customer in set S and its demand subtracts from capacity available for set S. Hence, the total vehicle capacity entering and leaving set S must be at least $a(S)$ plus the customer demand for all customers $j \in \bar{S}$ with $x_{ij} = 1$. This implies

$$\sum_{i \in S} \sum_{j \in \bar{S}} x_{ij} \geq 2 \left\lceil \frac{a(S) + \sum_{\substack{i \in S \\ j \in \bar{S}}} a_j x_{ij}}{b} \right\rceil \tag{11}$$

Constraint (11) is not directly useful, since the right-hand side is a nonlinear function of x, but it can be used to derive other constraints. An example is given below.

For any $S \subseteq N$, let $S' = \{j \; \varepsilon \; \bar{S} | j \geq 1 \text{ and } a_j > b \, r(S) - a(S)\}$

$$e_j = \begin{cases} 0, & j \in S \\ 0, & j \in S' \text{ and } |S'| \leq 2 \\ \dfrac{r(S)}{r(S) + 1}, & j \in S' \text{ and } |S'| > 2 \\ 1, & j \in \bar{S} - S' \end{cases}$$

We can now define the tightened vehicle capacity constraints

$$\sum_{j=0}^{n} e_j \sum_{i \in S} x_{ij} \geq 2r(S) \quad \text{for all} \ S \subseteq N \ \text{with} \ |S| \geq 2 \tag{9'}$$

Computational results using the K-tree approach are presented in Table 4. The first 6 problems are the ones introduced in Table 1, Section 2 of this chapter except that for problem 5, $K = \lceil \sum_{i=1}^{n} a_i/b \rceil = 16$ here, rather than the value of 17 used in most previous studies. The data in problems 7–12 are taken from real vehicle routing applications. Problems 7–9 are concerned with the delivery of industrial gases in cylinders and are based on data provided by Air Products and Chemicals, Inc. Problems 10 and 12 represent a day of grocery deliveries from the Peterboro and Bramalea, Ontario terminals, respectively, of National Grocers Limited [see Arizza & Karellas, 1983]. Problem 11 is concerned with delivery of tires, batteries and accessories to gasoline service stations and is based on data provided by Exxon.

All problems except problems 7–9 were planar and c_{ij} is the distance between points i and j, computed as a single precision real value. In problem 7–9, integral values for all c_{ij} were defined as part of the input.

The condition $a_j/b < \sum_{i=1}^{n} a_i/b - (K-1)$, $j = 1, \cdots, n$ was satisfied for 8 out of the 12 test problems used (problems 1, 2, 3, 5, 7, 8, 11, 12) so the prohibition of single customer routes was not constraining since these problems have no feasible solution with single customer route. For two other problems (problems 9 and 10), a few customers violated this condition, but the resulting single route solutions were easily shown to be nonoptimal using the lower bounding procedure presented here and a negligible amount of computation time. While it has not been formally established that the optimal solutions to problems 4 and 6 do not contain single customer routes, it seems unlikely since deleting the largest customer and a single vehicle would leave a very tightly constrained problem.

These computations suggest that real problems are easier for optimization than randomly generated problems. For example, compare real problems 10 and 11 together with the 'realistic' problem 6 with the random problems 1–3 of similar size. For problems 6, 10 and 11, the lower bounds computed with no more than 2000 subgradient iterations were 99.1% of the optimum, on average, while for problems 1–3, the lower bounds were 94% of the optimum, on average and required 3,000 subgradient iterations.

The difference between uniform random and real problems is illustrated in Figures 1 and 2, which show depot and customer locations for problems 2 and 11, respectively. The rather even distribution of customers in Figure 1 is typical of problems 1–5, just as the grouping of customers into clusters is typical of problems 6–12. This clustering seems to play a role akin to sparsity in linear programming in providing a structure that can be exploited in computations. For example, a group of customers clustered together tends to act as one customer for vehicle capacity constraints (9); they are either all in a set S defining a constraint or none of them is.

Table 4
Computational results with the K-tree algorithm

Problem	n	K	$\dfrac{\sum_{i=1}^{n} a_i}{Kb}$	Upper bounds*		Lower bound	$\dfrac{\text{Lower}}{\text{Upper}}$	Subgradient iterations	Computation time Apollo domain 3000 minutes
				Best previously known feasible solution	Optimal or improved solution found by the K-tree algorithm				
1	50	5	0.97	524.61[1]	–	507.09	0.97	3000	95.75
2	75	10	0.97	635.26[1]	–	755.50	0.90	3000	183.97
3	100	8	0.91	826.14[1]	–	785.86	0.95	3000	307.95
4	150	12	0.93	1028.42[1]	–	932.68	0.91	3000	682.41
5	199	16	0.999	1334.55[2]	–	1096.72	0.82	3000	1186.00
6	100	10	0.91	819.56[1]	819.56[4]	817.77	0.998	2000	259.64
7	25	3	0.92	3104[3]	3070[4]	3070	1.00	769	11.23
8	29	4	0.94	5830[3]	5829[4]	5829	1.00	2444	53.33
9	36	4	0.76	5032[3]	4961[4]	4961	1.00	320	4.87
10	44	4	0.90	723.54[3]	723.54[4]	720.76	0.996	2000	49.74
11	71	4	0.96	244.92[3]	241.97[4]	237.76	0.98	2000	105.03
12	134	7	0.95	1216.66[3]	1163.60[4]	1133.73	0.97	2000	253.84

* Sources for upper bounds: [1] Taillard [1992].
[2] Osman [1993].
[3] Algorithm in Bramel & Simch-Levi [1992]. These problems were run and the results communicated to me by Julien Bramel and David Simchi-Levi.
[4] First Lagrangian heuristic described in Fisher [1990].

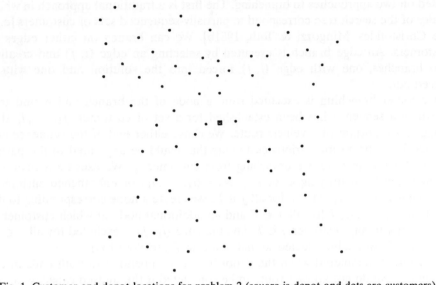

Fig. 1. Customer and depot locations for problem 2 (square is depot and dots are customers).

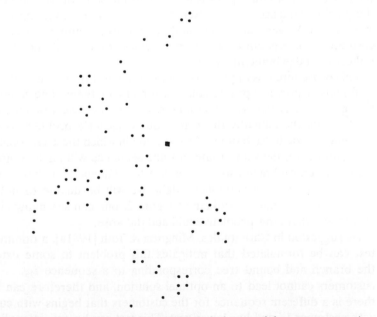

Fig. 2. Customer and depot locations for problem 11 (square is depot and dots are customers).

The K-tree algorithm found proven optimal solutions to problems 7–9 without branching and to problems 6, 10 and 11 using a branch and bound algorithm based on two approaches to branching. The first is a traditional approach in which nodes of the search tree correspond to partially-sequenced sets of customers [e.g., see Christofides, Mingozzi & Toth, 1981b]. We can branch on either edges or customers. An edge branch is executed by selecting an edge (i, j) and creating two branches, one with edge (i, j) forced into the solution and one with it forced out.

Customer branching is executed from a node of the branch and bound tree at which a sequence has been established for a set of customers $i_1, \ldots, i_k$ that comprises a portion of a vehicle route. We chose either end of the sequence and branch by enumerating various customers that could be appended to the partial route. Assume that we are branching from customer i_k. We execute a customer branch step by identifying a set $T \subseteq N_0 - \{i_1, \ldots, i_k\}$ of unbranched customers satisfying $a\{i_1, \ldots, i_k, j\} \leq b$ for all $j \in T$. We create a node corresponding to the sequence $i_1, \ldots, i_k, j$ for all $j \in T$ and an additional node at which customer i_k cannot link to any customer $j \in T$, i.e., the edge (i_k, j) is excluded for all $j \in T$. Normally, T would be selected so that points in T are close to i_k.

A route is completed when the depot has been appended to both ends in the sequence. We do not allow a route to be completed if the unused capacity on the vehicle is so great that the remaining vehicles have insufficient capacity to deliver the remaining customer orders.

This procedure begins with an edge branch on an edge (i_1, i_2). At the node of the search tree where (i_1, i_2) is forced into the solution, a customer branch using the partial sequence i_1, i_2 is executed. At the other node, another edge branch is executed. In general, at any node of the tree defining a sequenced subset of customers corresponding to a partial vehicle route, customer branching is used. Otherwise, edge branching is used.

This procedure was applied to several of the test problems, but was unsuccessful in finding a proven optimal solution for any of them. The major problem with this approach is that the decisions resolved when we branch are quite minor. To illustrate the difficulty this can create, consider a problem with a cluster of k customers close to each other. Any solution in which these customers are delivered contiguously on the same route in some sequence will have about the same cost. Hence, when we branch so as to resolve the sequence for these customers, unless the lower bound is exceptionally tight, we will be unable to fathom any of the $0(k!)$ nodes generated. Looking at Figure 2, one can see many clusters of 4 to 5 customers where this problem could and did arise.

As suggested in Christofides, Mingozzi & Toth [1981a], a dominance fathoming test can be formulated that mitigates this problem to some extent. A node of the branch and bound tree corresponding to a sequence $i_1, \ldots, i_k$ for a set of customers cannot lead to an optimal solution, and therefore can be fathomed, if there is a different sequence for the customers that begins with customer i_1, ends with customer i_k, and has lower cost. This test can be operationalized by applying the Lin & Kernighan [1973] 3-opt rule to the customer sequence specified at a

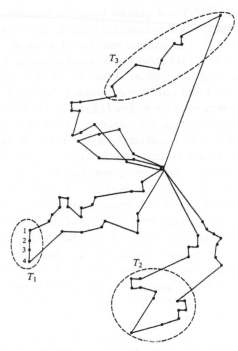

Fig. 3. Examples of branch clusters for problem 11.

node. If the 3-opt rule finds an improved sequence, then the node is fathomed. This dominance test improved performance somewhat, but not enough to allow optimal solution of any of the test cases.

This experience suggests that one might obtain a better branching procedure by identifying macro properties of an optimal solution whose violation would have a sufficiently large impact on cost to allow fathoming. Figure 3 shows the optimal solution for problem 11. One property that stands out in this optimal solution is the presence of clusters of customers delivered contiguously on the same route which are close to each other and far from the remaining customers in the problem. Three examples are encircled in Figure 3 and labelled as T_1, T_2, and T_3. Requiring the customers in any of these three clusters to be delivered on two different routes would appear to have a significant impact on cost.

This observation was used to develop a new branching rule. Let $I(T)$ denote the incidence of edges on the node set T in a solution graph, i.e., $I(T) = \sum_{i \in T} \sum_{j \in \bar{T}} x_{ij}$. In the solution depicted in Figure 3, $I(T_k) = 2$, $k = 1, 2, 3$. Note that in a feasible solution the incidence on any customer set must be an even integer not less than $2\lceil (a(T))/b \rceil$. Branching occurs by selecting any set $T \subseteq N$ and creating two nodes corresponding to $I(T) = 2\lceil (a(T))/b \rceil$ and $I(T) \geq 2\lceil (a(T))/b \rceil + 2$. The constraint $I(T) \geq 4\lceil (a(T))/b \rceil$ is of the same form as constraints (9') and hence easy to incorporate within the Lagrangian relaxation. Branching in this way for the sets T_1, T_2 and T_3 produces a lower bound at the node corresponding to $I(T_k) \geq 4$ greater than the feasible value of 244.92, the best known feasible value prior to applying branch and bound, thus establishing that $I(T_k) = 2, k = 1, 2, 3$ in an optimal solution to problem 11.

The constraint $I(T) = 2\lceil(a(T))/b\rceil$ can be used to derive additional restrictions that tighten the Lagrangian problem. First of all, for any $j \notin T$ such that $a_j + a(T) > b$, we can force out of the solution the edges (i, j) for all $i \in T$. Second, if there is another set S for which $I(S) = 2$, $S \cap T = \phi$ and $a(S \cup T) > b$, we can force out of the solution the edges (i, j) for all $i \in T, j \in S$. Third, if $\lceil(a(T))/b\rceil = 1$ and $b - a(T)$ is sufficiently small, it may be feasible to enumerate all combinations of customers that can fit in the remaining space $b - a(T)$ and branch by generating a node corresponding to each combination. Fourth, if $\lceil(a(T))/b\rceil = 2$, we can select a subset $S \subset T$ of a few large customers for which it is computationally feasible to enumerate all partitions of S into two sets corresponding to the customers assigned to each of the 2 vehicles that must deliver the customers in T. We can then branch on the choice of a partition.

Finally, we describe a simple dominance test that can sometimes establish an optimal sequence for the customers in a set T with $I(T) = 2$. By way of illustration, consider set T_1 in Figure 3. The customers in this set are indexed from 1 to 4. Because $I(T_1) = 2$, there will be precisely two customers in T_1 joined to customers outside of T_1. There are $\binom{4}{2} = 6$ choices for this pair of customers. Index these pairs $(i_k, j_k), k = 1, \ldots, 6$, and assume $(i_1, j_1) = (1, 4)$. For any pair (i_k, j_k), the path through the remaining two customers must minimize cost and can be determined by enumeration. For example, it is apparent that for (i_1, j_1), the optimal path is $(1, 2, 3, 4)$. Let C_k denote the cost of the optimal path from i_k to j_k, e.g., $C_1 = c_{12} + c_{23} + c_{34}$.

We call a pair (i_k, j_k) dominated if, for each pair i, j, $i \neq j$, $i \in N_0 - T$, $j \in N_0 - T$, there exists $k^* \neq k$ such that

$$C_{k^*} + \min(c_{ii_{k^*}} + c_{jj_{k^*}}, \; c_{ji_{k^*}} + c_{ij_{k^*}}) \leq C_k + \min(c_{ii_k} + c_{jj_k}, \; c_{ji_k} + c_{ij_k})$$

$$(12)$$

It is clear that a path through T joining a dominated pair can be ignored as a sequence for the customers in T, since it could be replaced in any feasible solution by a different sequence (namely, the shortest path through T joining i_{k^*} and j_{k^*}) without increasing cost. Returning to our example, for the set T_1, all pairs $k = 2, \ldots, 6$ can be shown by direct computation of (13) to be dominated by $k^* = 1$. Hence, we can fix the sequence of customers in T_1 to $1, 2, 3, 4$. For the set T_3, there are four sequences that dominate all others. In this event, we can branch by selecting one of the sequences.

In computational work, the dominance test was applied to each $k \in T$ by computing (7) for all possible ij and each possible k^*. The step of finding the shortest path through T joining a pair of customers in T can be accomplished by a straightforward modification of the dynamic programming algorithm for the traveling salesman problem given by Held & Karp [1962].

The ideas defined above can be combined in many ways to create a branch and bound algorithm, depending on how branch sets are chosen and the order in which the various methods of branching are combined. We describe here the particular algorithm used in Fisher [1990].

We need to identify sets of customers on which to branch. Two types of branch sets were used. One was the set of customers S on a single route or a pair of crossed routes for which average vehicle utilization exceeded 98%. If S was a pair of crossed routes, add to S any customers within the convex hull of $S \cup \{0\}$ that fit (i.e. $a(S) \le 2b$ with the customers added), starting with customers farthest from the depot. Sets like these make good branch sets because the branch $I(S) = 2\lceil(a(S))/b\rceil$ removes many edges from the problem, namely those edges (i, j) for which $i \in S$, $j \in \bar{S}$ and $a(S) + a_j > \lceil(a(S))/b\rceil b$.

The second type of branch sets were clusters of customers $S = \{i_1, \cdots, i_k\}$ delivered contiguously in the order $i_1, \cdots, i_k$ on a single route of a starting feasible solution. S was also required to contain the customer on the route farthest from the depot, to be separable from $\bar{S}$ by a straight line and to have a sufficiently small value of

$$D(S) = \frac{\sum_{j=1}^{k-1} c_{i_j i_{j+1}}}{(k-1) \min_{i \in S, j \in \bar{S}} c_{ij}}.$$

The quantity $D(S)$ is used measure the extent to which the customers in S are close to each other and far from the remaining points $\bar{S}$. It's easy to see that the sets T_1, T_2 and T_3 in Figure 3 satisfy the required properties and have relatively small values of $D(S)$.

We describe the specific steps in the branch and bound procedure in the order they are executed. First branch on any route S with $a(S) = b$ and then on $\lfloor n/10 \rfloor$ of the second type of branch sets described above with the smallest values of $D(S)$ for which we can fathom the node corresponding to $I(S) \ge 4$.

Next, the dominance test described previously is applied to any set S with $I(S) = 2$ and $|S| \le 11$ (the computation time for the dominance test is prohibitive for larger sets). We branch on choice of sequence whenever at most 4 sequences are nondominated. For a set S with $I(S) = 2$ and $b - a(S) > 0$ but sufficiently small that at most one other customer could fit feasibly with S on a single route, we branch by enumerating all feasible completions of the vehicle route containing S.

We then branch on all single routes or pairs of crossed routes with average vehicle utilization exceeding 98%. If S is a pair of routes, at the node corresponding to $I(S) = 4$, we branch on all partitions of $R \subseteq S$ into 2 sets corresponding to the set of customers assigned to each of the 2 routes, where R contains the 11 largest customers in S. A set $T \subset S$ for which $I(T) = 2$ has been imposed can be treated as a single customer for this purpose. Finally, at any node still unfathomed, the traditional branching method described earlier is applied.

This algorithm has solved to optimality problems 6–11. Results are reported in Table 5. Because of round off error, the problems with real c_{ij} were solved to

Table 5
Optimization results for the K-tree algorithm

Problem	Time to find optimal solution* Apollo domain 3000 minutes	Nodes in branch and bound tree
6	591	89
7	11	1
8	64	1
9	6	1
10	342	148
11	948	37

*Time includes the time (reported in the last column of Table 4) to be bound to the root node.

ε optimality (the solution obtained may exceed the true optimum by an additive constant ε) with $\varepsilon = 0.0001$.

Optimization now appears on the verge of becoming a practically useful tool. In the last 5 years, the size problems solvable to optimality has been extended to 50–100 customers. Moreover, the methods developed can be or have been extended to incorporate real world complications like time windows.

References

Agarwal, Y., K. Mathur and H.M. Salkin (1989). A set-partitioning-based exact algorithm for the vehicle routing problem. *Networks* 19, 731–749.

Altinkemer, K., and B. Gavish (1987). Heuristics for unequal weight delivery problems with a fixed error guarantee. *Oper. Res. Lett.* 6, 149–158.

Altinkemer, K., and B. Gavish (1991). Parallel savings based heuristics for the delivery problem. *Oper. Res.* 39, 456–469.

Araque, J.R., G. Kudva, T.L. Morin and J.F. Pekny (1994). A branch-and-cut algorithm for vehicle routing problems. *Ann. Oper. Res.*,

Arizza, A., and P. Karellas (1983). A feasibility study of the ROVER system at National Grocers Co. Ltd., B.A. Thesis of Applied Science, Department of Industrial Engineering, University of Toronto.

Bartholdi, J.J., L.K. Platzman, R.L. Collins and W.H. Warden, (1983) A minimal technology routing system for meals on wheels. *Interfaces* 13, 1–8.

Balinski, M.L., and R.E. Quandt (1964). On an integer program for a delivery program for a delivery problem. *Oper. Res.* 12, 300–304.

Belardo, S., P. Duchessi and J. Seagle (1985). Microcomputer graphics in support of vehicle fleet routing. *Interfaces* 15, 84–92.

Bell, W.J., L. Dalberto, M.L. Fisher, A.J. Greenfield, R. Jaikumar, P. Kedia, R.G. Mack and R.J. Prutzman (1983). Improving the distribution of industrial gases with an on-line computerized routing and scheduling optimizer, *Interfaces* 13, 4–23.

Bodin, L.D. (1991). Twenty years of routing and scheduling, *Oper. Res.* 38, 571–579.

Bodin, L.D., B.L. Golden, A. Assad and M. Ball (1983). Routing and scheduling of vehicles and crews: the state of the art. *Comput. Oper. Res.* 10, 69–211.

Bramel, J., and D. Simchi-Levi (1993). On the effectiveness of set partitioning formulations for the vehicle routing problem, Working Paper, Department of Industrial Engineering and Oper. Res., Columbia University.

Bramel, J., and D. Simchi-Levi (1992). A location based heuristic for general routing problems, Working Paper, Department of Industrial Engineering and Oper. Res., Columbia University.

Brenninger-Goethe, M. (1989). Two vehicle routing problems — Mathematical programming approaches, Linkoping Studies in Science and Technology. Dissertations 200. Department of Mathematics, Linkoping University, Linkoping.

Brown, G., and G. Graves (1981). Real-time dispatch of petroleum tank trucks, *Manage. Sci.* 27, 19–31.

Campos, V., A. Corberan and E. Mota (1991). Polyhedral results for a vehicle routing problem. *Eur. J. Oper. Res.* 52, 75–85.

Christofides, N. (1985). Vehicle routing, in: E.L. Lawler, J.K. Lenstra, A.H.G. Rinnooy Kan and D.B. Shmoys (eds.), *The Traveling Salesman Problem: A Guided Tour of Combinatorial Optimization*, Wiley, Chichester.

Christofides, N., and S. Eilon (1969). An algorithm for the vehicle dispatching problems, *Oper. Res. Q.* 20, 309–318.

Christofides, N., A. Mingozzi and P. Toth (1979). The vehicle routing problem, in: N. Christofides, A. Mingozzi, P. Toth and C. Sandi (eds.). *Combinatorial Optimization*, Wiley, Chichester, pp. 318–338.

Christofides, N., A. Mingozzi and P. Toth (1981a). Space state relaxation procedures for the computation of bounds to routing problems. *Networks* 11, 145–164.

Christofides, N., A. Mingozzi and P. Toth (1981b). Exact algorithms for the vehicle routing problem, based on spanning tree and shortest path relaxations, *Mathematical Programming* 20, 255–282.

Chvatal, V. (1973). Edmonds polytopes and weakly Hamiltonian graphs. *Math. Program.* 5, 29–40.

Clarke, G., and J. Wright (1964). Scheduling of vehicles from a central depot to a number of delivery points. *Oper. Res.* 12, 568–581.

Collins, D.H.G., and W.M. Clavey (1986). The optimization of Route scheduling: a case study. *Proc. Int. Trans. Conf.*, Institute of Industrial Engineers.

Cornuejols, G., and F. Harche (1989). Polyhedral study of the capacitated vehicle routing problem, Management Science Research Report No.553, Carnegie Mellon University, Pillsburgh, PA.

Cullen, F., J. Jarvis and D. Ratliff (1981). Set partitioning based heuristics for interactive routing. *Networks* 11, 125–144.

Dantzig, G.B., and J.H. Ramser (1959). The truck dispatching problem, *Manage. Sci.* 6, 80–91.

Desrochers, J. Desrosiers and M. Solomon (1992). A new optimization algorithm for the vehicle routing problem with time windows. *Oper. Res.* 40, 342–354.

Desrochers, M., J.K. Lenstra and M.W.P. Savelsbergh (1990). A classification scheme for vehicle routing and scheduling problems. *J. Oper. Res.* 46, 322–332.

Evans, S., and J. Norback (1985). The impact of a decision-support system for vehicle routing in a food service supply situation. *J. Oper. Res. Soc.* 36, 467–472.

Fisher, M.L. (1991). A polynomial algorithm for the degree constrained K-tree problem, Working Paper 91-07-06, Department of Decision Sciences, University of Pennsylvania, to appaear.

Fisher, M. L. (1990). Optimal solution of vehicle routing problems using minimum K-trees, Working Paper, Department of Decision Sciences, University of Pennsylvania, to appear.

Fisher, M.L. (1982). Greenfield, A.J., Jaikumar, R. and Lester, J.T., III, A computerized vehicle routing application. *Interfaces* 12, 42–52.

Fisher, M.L., and R. Jaikumar (1981). A generalized assignment heuristic for vehicle routing, *Networks* 11, 109–124.

Fisher, M.L. (1986). Jaikumar, R. and Van Wassenhove, L., A multiplier adjustment method for the generalized assignment problem. *Manage. Sci.* 32, 1095–1103.

Fisher, M.L., and P. Kedia (1990). Optimal solution of set covering/partitioning problems using dual heuristics. *Manage. Sci.* 36, 674–688.

Fisher, M.L., and M.B. Rosenwein (1989). An interactive optimization system for bulk cargo ship scheduling. *Nav. Res. Logist.* 36, 27–42.

Foster, B.A., and D.M. Ryan (1976). An integer programming approach to the vehcle scheduling problem. *Oper. Res.* 27, 367–384.

Fleuren, H. (1988). A computational study of the set partitioning approach for vehicle routing and scheduling problems, Ph.D. Dissertation, University of Twente, Enschede.

Gaskell, T.J. (1967). Basis for vehicle fleet scheduling, *Oper. Res. Q.* 18, pp. 281.

Gillett, B. and Miller, L. (1974). A heuristic algorithm for the vehicle dispatch problem. *Oper. Res.* 22, 240–349.

Gendreau, M., A. Hertz and G. Laporte (1993). A tabu search heuristic for the vehicle routing problem, Centre de Recherche sur les Transports, Publication 777, Université de Montreal.

Golden, B.L., and A.A. Assad (1986). Perspectives on vehicle routing: exciting new developments. *Oper. Res.* 14, 803–810.

Golden, B.L., T.L. Magnanti and H.Q. Nguyen (1977). Implementing vehicle routing algorithms. *Networks* 7, 113–148.

Golden, B., and E. Wasil (1987). Computerized vehicle routing in the soft drink industry. *Oper. Res.* 35, 6–17.

Grotschel, M. (1980). On the symmetric traveling salesman problem: solution of a 120-city problem. *Math. Program. Studies* 12, 61–77.

Grotschel, M., and M.W. Padberg (1979). On the symmetric traveling salesman problem: I–II. *Math. Program.* 16, 265–280 and 281–302.

Grotschel, M., and W. Pulleyblank (1986). Clique tree inequalities and the traveling salesman problem. *Math. Oper. Res.* 11, 537–569.

Harker, P.T. (1990). The use of advanced train control system in scheduling and operating railroads: models, algorithms and applications, in: *Transportation Research Record* 1263, TRB, National Research Council, Washington, D.C., pp. 101–110.

Haimovich, M., and A.H.G. Rinnooy Kan (1985). Bounds and heuristics for capacitated routing problems. *Math. Oper. Res.* 10, 527–542.

Held, M., and R.M. Karp (1962). A dynamic programming approach to sequencing problems. *J. Soc. Industr. Appl. Math.* 10, 196–210.

Kadaba, N., K.E. Nygard and P.L. Juell (1991). Integration of adaptive machine learning and knowledge-based systems for routing and scheduling applications. *Expert Systems Appl.* 2, 15–27.

Kolen, A.W.J., A.H.J. Rinnooy Kan and H.W.J.M. Trienekens, (1987). Vehicle routing with time windows. *Oper. Res.* 35, 266–273.

Koskosidis, Y.A., W.B. Powell and M.M. Solomon (1989). An optimization based heuristic for vehicle routing and scheduling with time window constraints, No. TMS-88-09-1, The Institute for Transportation Systems, The City College of CUNY, New York, NY.

Laporte, G., and J.-M. Bourjolly (1984). Some further results on *k*-star constraints and comb inequalities. *Cahiers du GERAD, G-84, G-84-17, Ecole des Hautes Etudes Commerciales de Montreal.*

Laporte, G., and Y. Nobert (1987). Exact algorithms for the vehicle routing problem. *Ann. Discr. Math.* 31, 147–184.

Laporte, G., Y. Nobert and M. Desrochers (1985). Optimal routing under capacity and distance restrictions. *Oper. Res.* 33, 1050–1073.

Lavoi, S., M. Minoux and E. Odier (1988). A new approach for crew pairing problems by column generation with an application to air transportation. *Eur. J. Oper. Res.* 35, 45–58.

Lawler, E.L., J.K. Lenstra, A.H.G. Rinnooy Kan and D.B. Shmoys, eds. (1985). *The Traveling Salesman Problem: A Guided Tour of Combinatorial Optimization*, Wiley, Chichester.

Lin, S. (1965). Computer solution of the traveling salesman problem. *Bell Systems Tech. J.* 44, 2245–2269.

Lin, S., and B. Kernighan (1973). An effective heuristic algorithm for the traveling salesman problem. *Oper. Res.* 21, 498–516.

Lucena, A.P. (1986). Exact solution approaches for the vehicle routing problem, Ph.D. Thesis, Department of Management Science, Imperial College of Science and Technology, University of London.

Magnanti, T. (1981). Combinatorial optimization and vehicle fleet planning: perspectives and prospects. *Networks* 11, 179–214.

Miliotis, P. (1976). Integer programming approaches to the traveling salesman problem. *Math. Program.* 10, 367–378.

Miller, D. (1993). A matching based exact algorithm for capacitated vehicle routing paper, Working Paper, Central Research & Development Department, E.I. du Pont de Nemours and Company, Wilmington, DE.

Miller, D.L., and J.F. Pekny (1994). A staged primal–dual algorithm for perfect *b*-matching with edge capacities. *ORSA J. Comp.*.

National Council of Physical Distribution Management Study, (1978). Measuring productivity in physical distribution, prepared by A.T. Kearney, Inc., Chicago, IL.

Nygard, K.E., P. Greenberg, W.E. Bolkan and E.J. Swenson, (1988). Generalized assignment methods for the deadline vehicle routing problem, in: B.L. Golden and A.A. Assad (eds.), *Vehicle Routing: Methods and Studies*, Elsevier Science Publishers B.V., North-Holland.

Nygard, K.E., and P.L. Juell (1991). Integration of adaptive machine learning and knowledge-based systems for routing and scheduling applications. *Expert Systems Appl.* 2, 15–27.

Osman, I.H. (1993). Metastrategy simulated annealing and tabu search algorithms for the vehicle routing problem. *Ann. Oper. Res.* 41, 421–451.

Padberg, M.W. and Hong, S. (1980). On the symmetric traveling salesman problem: a computational study. *Math. Program. Studies* 12, 78–107.

Potvin, J.-Y., G. Lapalme and J.-M. Rousseau (1989). ALTO: A computer system for the design of vehicle routing algorithms. *Comput. Oper. Res.* 16, 451–470.

Powell, W.B., Y. Sheffi, K.S. Nickerson, K. Butterbaugh and S. Atherton (1988) Maximizing profits for North American Van Lines' Truckload Division: a new framework for pricing and operations. *Interfaces* 18, 21–41.

Russell, R.A. (1977). An effective heuristic for the *M*-tour traveling salesman problem with some side conditions. *Oper. Res.* 25, 5217–524.

Savelsberg, M. (1985). Local search in routing problems with time windows, *Oper. Res.* 4, 285–305.

Taillard, E. (1993). Parallel iterative search methods for vehicle routing problems, *Networks* 23,

Thompson, P.M. (1988). Local search algorithms for vehicle routing and other combinatorial problems. Submitted in partial fulfillment of the requirements of the degree of Doctor of Philosophy in Operations Reserach at the Massachusetts Institute of Technology.

Toth, P. (1984). Heuristic algorithms for the vehicle routing problems, presented at the Workshop on Routing Problems, Hamburg,

Tyagi, M. (1968). A practical method for the truck dispatching problem, *J. Oper. Res. Soc. Jpn.* 10, 76–92.

Viegas, J., and P. Hansen (1985). Finding shortest paths in the plane in the presence of barriers to travel (for any 1_p-norm). *Eur. J. Oper. Res.* 20, 373–381.

Walter, D.F., and E. Zielinski (1983). Computerized vehicle routing package–management issues. Talk delivered at the 1983 Annual Meeting of the National Council of Physical Distribution Management, New Orleans.

Zielinski, E.L. (1985). Implementing a computerized vehicle routing system: A case study. *Auerbach Rep. 1*, 24, 1–14.

Yellow, P. (1970). A computational modification to the savings method of vehicle scheduling. *Oper. Res. Q.* 21, 281–283.

M.O. Ball et al., Eds., *Handbooks in OR & MS, Vol. 8*

Chapter 2

Time Constrained Routing and Scheduling

Jacques Desrosiers, Yvan Dumas *

GERAD and École des Hautes Études Commerciales, Montréal, Qué. H3T 1V6, Canada

Marius M. Solomon

Northeastern University and GERAD, Boston, MA 02115, U.S.A.

François Soumis

GERAD and École Polytechnique de Montréal, Montréal, Qué. H3C 3A7, Canada

À la mémoire de Martin Desrochers, décédé prématurément le 3 juin 1991 à l'âge de 33 ans, pour sa contribution exceptionnelle au succès de notre équipe.

1. Introduction

Over the last two decades, economic phenomena such as the oil crisis of the early '70s, the deregulation of the U.S. airline and trucking industries in the '80s, and Europe '92 have highlighted transportation as an area where large productivity improvements could be achieved. This interest in transportation has also been fueled by other important developments such as the continuing reduction in the labor cost of products time-based competition, and the constant breakthroughs in computer technology. In turn, operation researchers have found a renewed interest in routing and scheduling problems. These problems generally involve the assignment of vehicles to trips such that the assignment cost and the corresponding routing cost is minimized.

During this period, the temporal aspect of these intrinsically spatial problems has also become increasingly important as manufacturing, service and transportation companies have tried to not only cut their logistics costs, but also compete on service differentiation. Not surprisingly, we have then witnessed the development of a fast growing body of research focused on time constrained routing and scheduling. The time dimension has been incorporated in these problems in the form of customer-imposed time window constraints. These constraints restrict the start of service at a customer to begin later than or at a prespecified earliest time and earlier than or at a prespecified deadline. When each time window consists of

* Now Visiting Professor of R&D at Ad Opt Technologies Inc., Montréal, Qué., Canada.

a single value, a fixed schedule problem is obtained. When the time windows are treated as hard constraints, a vehicle is not permitted to arrive at a node after the latest time to begin service. However, if a vehicle arrives too early at a customer, it is permitted to wait until the customer is ready for the beginning of service. A much less studied, but nevertheless important, variant of the problem incorporates soft time windows which can be violated at a cost.

In general, researchers have considered a homogeneous vehicle fleet mix. Yet, as the field has matured, the increased level of sophistication of the methods developed has permitted the treatment of heterogeneous fleet mixes. The fleet size has been assumed either fixed a priori or free, i.e., to be determined simultaneously with the best set of routes and schedules. However, many methods can actually optimize the fleet size in the process.

The costs involved in time-constrained routing and scheduling consist of fixed vehicle utilization costs and variable routing and scheduling costs. These latter costs include distance and travel time costs, waiting time costs, and loading/unloading time costs. In addition, inconvenience costs are associated with the transport of individuals. For problem variants involving the assignment of personnel to tasks, corresponding costs are defined.

The first articles to mention problems with temporal constraints date back to Dantzig & Fulkerson [1954], Ford & Fulkerson [1962], Appelgren [1969, 1971], Levin [1971], Madsen [1976], and Orloff [1976]. Since then, a flurry of activity has been directed at the classical routing model, many of its realistic variants, important generalizations including the temporal aspect, and a variety of practical applications. This research has been reviewed in several insightful surveys written by Magnanti [1981], Bodin, Golden, Assad & Ball [1983], Desrochers, Lenstra, Savelsbergh & Soumis [1988], Solomon & Desrosiers [1988], and Halse [1992].

Time constrained routing and scheduling problems are encountered in a variety of industrial and service sector applications, ranging from logistics and transportation systems to material handling systems in manufacturing. Fixed schedule problems appear in several domains such as airline and rail transportation. The shortest path problem with time windows has been successfully utilized as a subproblem for the multiple traveling salesman problem used to construct school-bus routes. The traveling salesman problem with time windows has applications in single and multiple vehicle problems. The vehicle routing problem with time windows has many industrial applications including those where dock availability is a bottleneck such as for distribution centers. Pick-up and delivery problems with time windows have been applied to the transportation of the handicapped and elderly persons. Fixed schedule problems with resource constraints have found a fertile area in urban-bus driver scheduling and airline crew scheduling environments. In addition to the hard time window applications just described, soft time windows constraints have been applied to the transport of individuals in dial-a-ride problems.

The goal of this paper is to describe the significant advances made in time-constrained routing and scheduling. In terms of solution methodology capable of solving realistic size problems, this field has seen a natural progression from ad-hoc methods to simple heuristics, to optimization-based heuristics and recently optimal algorithms. The paper provides an extensive overview of the algorithms developed and focuses on optimization methods. Helped by continuously better insights into problem structures with time windows and rapid advances in computer technology, these methods are becoming a viable tool for solving practical size problems. While we have tried to be as comprehensive as possible, we have undoubtedly omitted a few papers. We apologize in advance for any such occurrence.

The paper is organized as follows. Section 2 addresses fixed schedule problems and develops in detail the Dantzig–Wolfe decomposition/column generation approach which will then be applied to many of the other problem types. Next, Section 3 treats the traveling salesman problem with time windows. In Section 4, we examine time and resource constrained shortest path problems which constitute important modules of many of the algorithms to be described later. Section 5 explores the vehicle routing problem with time windows and several important problem variants including the multiple traveling salesman problem. Then, Section 6 treats the pick-up and delivery problem with time windows. In particular, the dial-a-ride problem with time windows is treated in this section. Section 7 examines an unified framework for fleet and crew scheduling problems. Finally, our conclusions and perspectives on future research are presented in Section 8.

2. Fixed schedule problems

The problem treated in this section considers a set of tasks (or trips) characterized by an origin, a destination, a duration and a fixed starting time. The tasks are given a priori and are composed of one or several activities. Between two successive tasks there is an inter-task transit time period of a certain duration and cost. The problem to be solved consists of forming a schedule, i.e., sequences of tasks where each task is performed exactly once. These sequences are sometimes called routes.

This type of problem is encountered in several fields, such as airline, rail, school-bus and urban transportation. In airline transportation, the trips are the flight legs: they are described by a city pair, an aircraft type and a schedule. Connections are possible only between two flight segments where the first ends and the second begins at the same airport, respectively. In rail transportation, this problem is found in the assignment of engines to train cars. A trip is determined by a train to haul at a given time. Connections are possible between the trains arriving at a station and those leaving the same station or other stations nearby. Several locomotives can be assigned to the same train. In school-bus transportation (the morning problem), a trip consists of visiting a sequence of stops, where school children are picked-up, and one or several schools, where these children are delivered. Connections are possible between the trips belonging to the same geographical region. In urban transportation, buses must be assigned to trips and connections are possible on the city network, in conformance to certain rules.

Each application differentiates itself by the structure of the network, its size and its periodicity (e.g., monthly, weekly or daily). Dantzig & Fulkerson [1954] were the first to formulate a fixed schedule problem as a minimum cost flow circulation problem. This section presents several refinements brought since then to the formulation of the problem as well as to the underlying network. We will be particularly interested in the construction of routes for vehicles to be assigned to fixed-schedule trips. The application to the assignment of personnel to tasks is immediate. In what follows we assume that all the vehicles are identical and that they all originate from a unique depot. It is therefore possible to solve the problems addressed in polynomial time [Lenstra & Rinnooy Kan, 1981]. The problem involving several depots is treated at the end of this section. More complex models involving flexible schedules, different vehicle types, and other generalizations, are treated in the subsequent sections.

2.1. The single depot vehicle scheduling problem

Notation. To describe the single depot vehicle scheduling problem (SDVSP), consider a set of n trips $\{T_1, T_2, \ldots, T_n\}$, where trip T_i has a given duration and starts at time a_i, $i = 1, 2, \ldots, n$. Consider also a single depot where v vehicles are stationed. Let the set of nodes, $N = \{1, 2, \ldots, n\}$ represent the set of trips, and node $\{n + 1\}$ represent the single depot. Let also t_{ij} denote the travel time between the ending point of trip T_i and the starting point of trip T_j. For simplicity we will assume that t_{ij} includes the duration of the trip T_i. An ordered pair of trips (T_i, T_j) is said to be compatible if it satisfies the relation $a_i + t_{ij} \leq a_j$, for all $i, j \in N$. If two trips are to be covered by the same vehicle, additional restrictions may be imposed for compatibility. Denote the set of compatible trips by I, a subset of $N \times N$. Then define $A = I \cup (\{n + 1\} \times N) \cup (N \times \{n + 1\})$ to be the set of arcs. Let c_{ij}, $(i, j) \in I$ be the cost incurred if a vehicle performs trip T_j immediately after trip T_i. This cost is usually a function of the distance between the ending point of trip T_i and the starting point of trip T_j. It can also include the fixed cost incurred to cover one of the trips, say trip T_i. Let next $c_{n+1,j}$ be the cost incurred if a vehicle undertakes trip T_j, $j = 1, 2, \ldots, n$ as the first trip after leaving the depot. Similarly, the cost $c_{j,n+1}$ is incurred when trip T_j is the last trip before the vehicle returns to the depot. This cost can also include the fixed cost incurred to cover trip T_j. Finally, let c be the fixed cost associated with the use of a vehicle. In general, the above costs are nonnegative. The problem consists of finding an assignment of vehicles to trips in such a way that:

- each trip is covered exactly once by a vehicle;
- each vehicle used in the solution leaves from the depot, covers a sequence of pairwise compatible trips and returns to the depot at the end of the route;
- the number of vehicles leaving from the depot does not exceed v (this number may be fixed a priori at v, minimized during the solution process, or free);
- the sum of the costs of the routes traveled by the vehicles used in the solution is minimized.

To formally describe the SDVSP, define the graph $G = (V, A)$, where $V = N \cup \{n + 1\}$ is the set of nodes and A is the set of arcs. This directed graph is acyclic on a cylinder. Note first that graph (N, I) is already acyclic. By making duplicates of the depot node, o and d for the origin and the destination of any

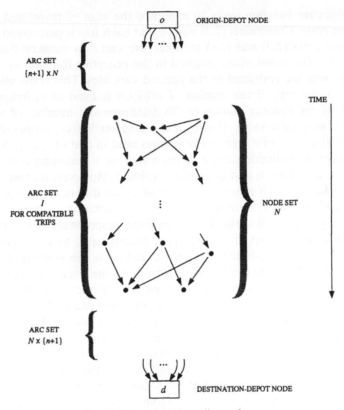

Fig. 1. The unfolded acyclic graph.

feasible route, respectively, graph G can be unfolded to become acyclic in the plane (Figure 1). If however, depot nodes o and d are merged, graph G is no longer acyclic, except on a cylinder. Let the variables X_{ij} be the binary flow on arc $(i, j) \in A$: $X_{ij} = 1$, if a vehicle covers trips T_j after trip T_i, and $X_{ij} = 0$, otherwise.

Formulation. The SDVSP can be formulated as a pure network flow problem:

$$\text{Minimize} \sum_{(i,j) \in A} c_{ij} X_{ij} + \sum_{j \in N} c X_{n+1,j} \tag{2.1}$$

subject to

$$\sum_{j \in V} X_{ij} = 1, \qquad \forall i \in N \tag{2.2}$$

$$\sum_{j \in N} X_{n+1,j} \leq v, \tag{2.3}$$

$$\sum_{i \in V} X_{ij} - \sum_{i \in V} X_{ji} = 0, \qquad \forall j \in V \tag{2.4}$$

$$X_{ij} \geq 0, \qquad \forall (i, j) \in A. \tag{2.5}$$

The objective function seeks to minimize the sum of travel and fixed vehicle utilization costs. Constraints (2.2) ensure that each trip is performed exactly once, while constraints (2.3) and (2.4) are fleet size and flow conservation constraints, respectively. The summations involved in the objective function as well as in the constraint sets are restricted to the defined variables. This convention is utilized throughout the text. If the number of vehicles is fixed at v, inequality (2.3) is replaced by an equality constraint. To minimize the number of vehicles, c is assigned a very large value. Finally, if this number is free, simply set $c = 0$. For simplicity, the value of c is generally incorporated in that of $c_{n+1,j}$, $\forall j \in N$.

The above formulation can be solved either as a minimum cost network flow problem or a minimum cost circulation problem. However, it must be noted that the networks on which algorithms are applied are different from that defined by the graph G. On these networks, each node representing a trip is split into two nodes, a trip-origin and a trip-destination node, respectively. These two nodes are then linked by an arc. A lower and upper bound equal to 1 is imposed on these n trip arcs. These transformations ensure that each trip is covered exactly once. The cost of such an arc can be set to zero or to the fixed cost incurred to cover that trip. Since each trip has to be covered exactly once, the sum of the costs of all the trips is a constant. Note that inter-trip costs must be defined accordingly. Similarly the depot node is split into an origin-depot node and a destination-depot node. These two nodes are then linked by a directed arc with zero cost. If this arc is directed from the origin to the destinaton depot it represents the slack variable in constraints (2.3). Given a supply of v at the origin-depot and an equal demand at the destination-depot, the capacity of this arc is $v - 1$, since it needs at least one vehicle to cover the trips. From the above description it follows that formulation (2.1)–(2.5) can be solved as a minimum cost network flow problem. If however the arc has the reverse orientation, i.e., from the destination-depot node to the origin-depot node, then the flow on this arc is restricted to be less than or equal to v, and a minimum cost circulation problem is derived. Each network is comprised of $2n + 2$ nodes and $|A| + n + 1$ arcs. Strongly polynomial time algorithms can be used to solve these problems [see Ahuja, Magnanti & Orlin, 1989]. In what follows, we examine several specialized cases of network representation, corresponding algorithms and cost structures.

Transportation/assignment network. Further analysis of the formulation (2.1)–(2.5) also reveals the structure of a transportation problem on a bipartite graph with $n + 1$ nodes on each side (Figure 2). The $n + 1$ nodes on the left are the origin-depot node o and the n trip-destination nodes. The $n + 1$ nodes on the right are the destination-depot d and the n trip-origin nodes. The supply and the demand equal v for the depot node, and equal 1 for all the trips of N. The arcs are those of A together with the additional zero cost arc (o, d) symbolizing the unused vehicles or the slack variable in (2.3). Note however that this transportation problem has a structure which very closely resembles an assignment problem and can be solved with a specialized algorithm developed by Paixão & Branco [1987]. Model (2.1)–(2.5) can be directly transformed into an assignment problem with $n + v$ rows

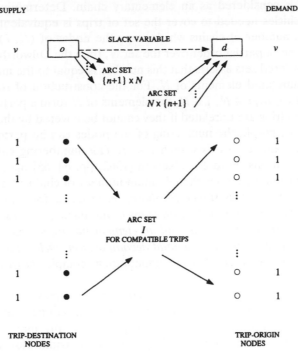

Fig. 2. A transportation problem structure.

and $n + v$ columns by replacing the depot node $(n + 1)$ with v identical nodes $(n + 1), \ldots, (n + v)$, and by setting the cost of the arcs connecting these nodes equal to zero (if the number of vehicles is to be minimized or if it is free) or equal to ∞ (if exactly v vehicles must be used), and by setting $c_{n+k,j} = c + c_{n+1,j}$, for $k = 1, \ldots, v$ and $j \in N$.

The elimination of all the depot nodes is equally possible if the minimization of the number of vehicles (large c) is sought, or if this number is free ($c = 0$). This leads to an assignment problem formulation [Orloff, 1976]. The arcs $(i, j) \in I$ are assigned the cost c_{ij}. Otherwise, if $(i, j) \in N \times N - I$, such an arc is defined nevertheless and it is assigned a cost of $c_{i,n+1} + c + c_{n+1,j}$. This cost is the sum of the cost $c_{i,n+1}$ of returning from the end of trip i to the depot, the cost c of utilizing a new vehicle, and the cost $c_{n+1,j}$ of going from the depot to the beginning of trip j. This formulation is however not suitable if the number of vehicles is fixed a priori since the additional constraint needed to fix this number destroys the assignment structure.

Minimal fleet size. Given that the fixed cost associated with a vehicle is often very high, an important objective is to determine the minimum fleet size to meet the schedule of trips. In this case, the solution could be developed on the acyclic graph (N, I). A chain in this graph is a set of trips of N and inter-trips of I where a single trip is considered as an elementary chain. Determining the minimum number of vehicles needed to cover the set of trips is equivalent to determining the minimum number of chains which cover the nodes of (N, I). If the relation $(i, j) \in I$ defines a partial order over the set N, then the Dilworth theorem [1950] on partially ordered sets asserts that this number is equal to the maximum number of mutually unrelated elements of N. For the construction of routes, under the relation $a_i + t_{ij} \leq a_j, i \in N, j \in N$, the elements of N form a partially ordered set. Therefore, two trips are unrelated if they cannot be covered by the same vehicle.

In any acyclic graph, the numbering of the nodes can be performed in such a way that if the arc (i, j) exists, then $i < j$. This numbering can be performed in $O(|I|^2)$ operations. Ford & Fulkerson [1962, pp. 64–65] have shown that the following algorithm generates the minimum number of chains: "*Assuming that the set has been numbered as stipulated above, select an undominated element (e.g., element 1), then proceed to its first (in terms of numbering) successor j, then to the first successor k of j, and so on until an element having no successor is reached. This traces out one chain of a minimal decomposition. Delete the elements of this chain and repeat the process.*" The computational complexity of this algorithm is $O(n)$.

Under the relation $(i, j) \in I$, the elements of N may not form a partially ordered set. For example, if long duration inter-trip connections are forbidden, the transitivity relations are not necessarily satisfied any more. Nevertheless, a minimal decomposition can be found by using the maximal flow algorithm [Ford & Fulkerson, 1962, and Ahuja, Magnanti & Orlin, 1989] which remains valid for acyclic graphs. The minimal fleet size covering all the trips is then given by $n - \sum_{(i,j) \in I} X_{ij}$. That is, the utilization of a number of inter-trips reduces the number of vehicles by the same amount. In the special case where the travel times t_{ij} are independent of j, they can be added to the service time for each trip T_i and the problem of finding the minimum number of vehicles reduces to the fixed job schedule problem [see Gertsbakh & Stern, 1978, and also Fischetti, Martello & Toth, 1987, 1989].

Network flow formulation (revisited). The graphs used in the preceding sections enable the utilization of various specialized algorithms. Depending on the case, examples include the minimum cost flow, the minimum cost circulation, the transportation and the assignment algorithms, respectively. In addition, each application can also lead to the construction of a very specific and generally much smaller network. The time-space network of a fixed schedule problem involving only one depot and a single type of vehicle or individual is characterized by arcs corresponding to activities, specific to the context of each application. Furthermore, the quantity flowing on each arc can be constrained depending on the nature of the arc. Examples of such arcs include the case of a trip to be covered (in this instance both the lower and the upper bounds are equal to one), of potential trips (here only the upper bound is equal to one), of diverse activities, such as periods of waiting, idle time, pauses for meals, and of travel.

The network type formulation which at the same time is the most general and the most simple is that where all the activities are on the arcs. In this case, each activity has an origin node and a destination node. For a graph defined by a set of nodes V and a set of arcs A, the structure of the costs c_{ij}, $(i, j) \in A$, and of the lower and upper bounds, ℓ_{ij} and u_{ij}, respectively, imposed on the variables X_{ij}, $(i, j) \in A$, completely characterizes a given application. Several examples are presented in the following section. The mathematical formulation of this network flow problem is then the following:

$$\text{Minimize} \sum_{(i,j)\in A} c_{ij} X_{ij} \tag{2.6}$$

subject to

$$\sum_{i\in V} X_{ij} - \sum_{i\in V} X_{ji} = 0, \quad \forall j \in V \tag{2.7}$$

$$\ell_{ij} \leq X_{ij} \leq u_{ij}, \qquad \forall (i, j) \in A. \tag{2.8}$$

Relations (2.7) are the flow conservation equations at each node, while constraints (2.8) give the bounds on the X_{ij} variables. This is in fact a minimum cost circulation problem. The oriented graph (V, A) is no longer acyclic. We can however still consider that it is acyclic on a cylinder. This type of formulation allows the consideration of problems where a repetitive cyclic schedule is desired, that is where the initial and final conditions are identical but yet to be determined. The above formulation generalizes formulation (2.1)–(2.5). Any instance of (2.1)–(2.5) can be reformulated as (2.6)–(2.8) by appropriately defining the costs, setting the bounds ℓ_{ij} and u_{ij} to the value one on the trip-arcs, as well as imposing an upper bound for v on the returning arc assuring the circulation in the network and limiting the number of vehicles available. The advantages of this formulation will be clearly seen in the following section.

2.2. Time-space networks in applications

The present section illustrates on several examples drawn from Desrosiers, Soumis & Desrochers [1982] how each application can give rise to the construction of a very specific and generally much smaller network. These networks are especially important in the more complex models: called several hundred times in relaxation or decomposition schemes, they permit a considerable acceleration of the solution time.

Aircraft fleet assignment. Consider a set of n flight legs. For each airport, a node is defined for each point in time where a flight leg starts or terminates. Next, for each flight leg define an arc connecting its departure node at one airport to its arrival node at a different airport. Finally, at each airport, ground arcs describing waiting are added to connect the nodes which are successive in time. The flight connection arcs exist only between the nodes at the same airport: such arcs describe a waiting time period and not a physical movement as is the case in road transportation.

When the initial conditions on the number of aircraft stationed at each airport are known, the network includes, for each airport, an origin-depot node which is connected to the node indicating the first departure. In the same manner, the node indicating the last arrival at each airport is connected with the destination-depot node. The final conditions are obtained by fixing the lower and upper bounds at the required value. If the initial and final conditions are imposed, in particular if they are the same as in a periodic schedule, ferries (empty flight legs) might be used to ensure feasibility of the problem.

Let us emphasize that the network constructed does not correspond to a multi-depot problem. In fact, even if we require that the number of aircraft available at each airport at the beginning of the planning period should equal that at the end of this period, it is however not required that the aircraft be the same. The multi-depot problems are formulated in the same manner as the problems with several types of vehicles by means of multi-commodity flow problems introduced in Section 2.3.

Consider next the case where the number of aircraft available at each airport at the beginning of the planning period is unknown and it is necessary that this number should be identical to that at the end of this period. To obtain a periodic solution, the depot nodes are eliminated and, at each airport, one arc, called a periodic arc, is directed from the last arrival node of the period considered to the first departure node of that period (Figure 3). The derived graph is acyclic on a cylinder and the problem to be solved becomes a minimum cost circulation problem. In this network, a lower and upper bound of value one is imposed on the flow of arcs corresponding to the flight legs that necessarily must be covered. In the case of potential flight legs, the lower bound is zero. On the waiting time arcs, it can frequently happen that the flows are larger than one, that is, several aircraft waiting at the same time at an airport. The cost structure is defined in terms of gains (or losses) associated with a necessary or potential flight leg. The number of aircraft is obtained by summing the total flow on all the arcs of the network at a given point in time. The size of this network is $2n$ nodes and $3n$ arcs, the degree of each node being 3 (an arc depicting a flight leg and two arcs depicting waiting time). Note that the size of the arc set is only $O(n)$ as compared to the potential $O(n^2)$ size.

The size of this network can be reduced by grouping together some nodes corresponding to the same airport (Figure 4). At each airport, the nodes are examined in chronological order and a new group is started each time a flight arrival node is encountered, provided that in the current group there already exists a flight departure node. This aggregation rule ensures that a flight leaving an airport utilizes an aircraft already arrived at this airport.

To appreciate the size of the networks involved, consider the case of Air Canada described in Soumis, Ferland & Rousseau [1980]. It consists of 325 flight legs and 300 potential flight legs. These give rise to an original time-space network which comprises 1250 nodes and 1875 arcs. The aggregation scheme groups together the nodes, on the average by sets of three, hence reducing the size of the network to 420 nodes and 1050 arcs.

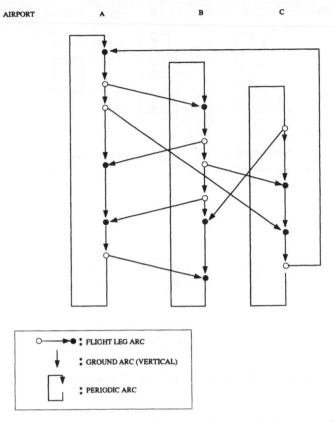

Fig. 3. Time-space network for the periodic aircraft assignment problem.

Locomotive assignment. Similar network transformations as above arise in rail transportation where engines are waiting at the same station. An additional distinctive aspect is to be noted: depending on the engine type, an origin-destination arc defining the trip of a train can have a lower and an upper bound different from the value one. On one hand, train cars may need to be hauled by several engines, while on the other hand it is possible to haul supplementary locomotives on that train to satisfy the demand in a neighboring city.

School-bus assignment. Consider the problem of assigning buses to school trips for the case involving students returning from school to their homes. The problem is to determine routes that start at the depot and cover a set of trips, each of which starts at a fixed time. There are no capacity constraints since each trip satisfies these by definition and vehicles moving between trips are empty. The essential feature of this application resides with the fact that each school is the departure point of a set of trips. Hence, the network contains only the n_S school-nodes which represent the origins of the trips and the n destination-nodes of the trips.

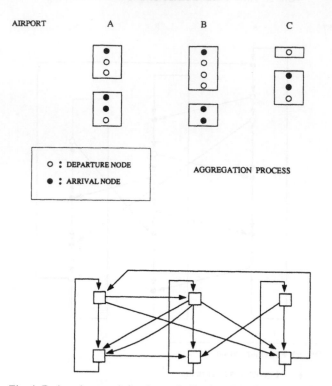

Fig. 4. Reduced network for the periodic aircraft assignment problem.

The inter-trip arcs are defined between the destinations of the trips and the schools. Therefore, this network representation reduces the number of nodes of the school-bus network from $2n$ to $n + n_S$, while the number of arcs is reduced by approximatively a factor n_S/n. For a school-bus network of 273 trips and 70 schools, the number of inter-trip arcs is cut from 17,000 to less than 3000.

Urban-bus assignment. For this problem, the tasks are also bus trips. The inter-trip arcs correspond to two types of bus movements: 1) a bus going directly from a trip-destination node to a trip-origin node and incurring a bus/driver cost of travel and waiting, or 2) a bus returning to the depot for a driver rest period between two trips and incurring only a travel cost. The inter-trip arc chosen is the least costly of the two types. In the applications considered by Bökinge & Hasselström [1980] and by Soumis, Desrosiers & Desrochers [1985], the number of inter-trip arcs corresponding to buses returning to the depot is of the order of the square of the number of trips. In practice these arcs could correspond to returns to the depot after the morning peak hour and departures from the depot for the afternoon peak hour. A reduction in the number of arcs of this type is realized by creating new depot nodes, one for each potential bus departure and arrival time at the depot (Figure 5). Given n trips to be covered, $2n$ depot nodes are hence created, with only $2n$ waiting time

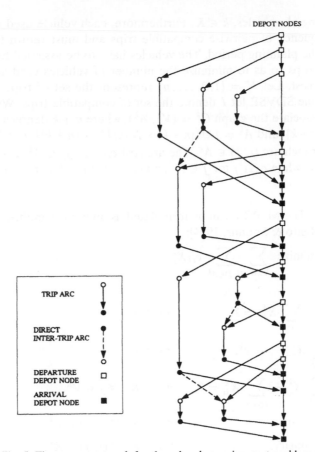

DEPOT NODES

TRIP ARC	
DIRECT INTER-TRIP ARC	
DEPARTURE DEPOT NODE	□
ARRIVAL DEPOT NODE	■

Fig. 5. Time-space network for the urban bus assignment problem.

arcs. In an example with $n = 128$ trips, the size of the original network comprises $2n + 2 = 258$ nodes and $3n + |I| = 7085$ arcs, where the arcs belonging to I satisfy $a_i + t_{ij} \leq a_j$, $i, j \in N$. The transformed network comprises $4n = 512$ nodes and only 1360 arcs ($5n$ arcs + direct inter-trip arcs, i.e., $640 + 720$ arcs). An aggregation at the level of the $2n$ depot nodes identical to the one presented for aircraft fleet assignment can also be performed, reducing again the size of this network.

Note finally that the time-space network model for the single depot problem can be extended to a multi-commodity flow model for the multiple depot problem in a straightforward manner [Lamatsch, 1992]. An optimal solution approach to this problem is presented next.

2.3. The multiple depot vehicle scheduling problem

Notation. The multiple depot vehicle scheduling problem (MDVSP) is similar to the single depot version, except that now there is a set of depots K, the kth depot housing v^k vehicles, $k \in K$. Furthermore, each vehicle used in the solution covers a sequence of pairwise compatible trips and must return to its depot at the end of the planning period. The vehicles have to be assigned to a set of fixed time-tabled trips so as to minimize the number of vehicles used and the overall operational cost. Let $N = \{1, 2, \ldots, n\}$ represent the set of trips. As previously defined for the SDVSP, let I denote the set of compatible trips. With each depot $k \in K$, we associate the graph $G^k = (V^k, A^k)$, where $n + k$ denotes the kth depot, $V^k = N \cup \{n+k\}$, and $A^k = I \cup (\{n+k\} \times N) \cup (N \times \{n+k\})$. Let X_{ij}^k be the flow of type k through arc $(i, j) \in A^k$. The arc cost c_{ij}, $(i, j) \in A^k$, is independent of k if $(i, j) \in I$, while $c_{n+k,j}$ for $j \in N$, and $c_{i,n+k}$ for $i \in N$, are usually dependent on k.

Formulation. The MDVSP can be formulated as an integer multi-commodity flow problem [Ribeiro & Soumis, 1994]:

$$\text{Minimize} \sum_{k \in K} \sum_{(i,j) \in A^k} c_{ij} X_{ij}^k \tag{2.9}$$

subject to

$$\sum_{k \in K} \sum_{j \in V^k} X_{ij}^k = 1, \qquad \forall i \in N \tag{2.10}$$

$$\sum_{j \in N} X_{n+k,j}^k \leq v^k, \qquad \forall k \in K \tag{2.11}$$

$$\sum_{i \in V^k} X_{ij}^k - \sum_{i \in V^k} X_{ji}^k = 0, \qquad \forall k \in K, \ \forall j \in V^k \tag{2.12}$$

$$X_{ij}^k \geq 0, \qquad \forall k \in K, \ \forall (i, j) \in A^k \tag{2.13}$$

$$X_{ij}^k \text{ binary}, \qquad \forall k \in K, \ \forall (i, j) \in A^k. \tag{2.14}$$

The objective function (2.9) seeks to minimize the total cost. Constraints (2.10) ensure that each trip is performed exactly once, while constraints (2.11) and (2.12.) are fleet size and flow conservation constraints, respectively. The MDVSP has been shown to be $\mathcal{NP}$-hard when $|K| \geq 2$ [Bertossi, Carraresi & Gallo, 1987]. As discussed in the previous section, the case $|K| = 1$ is solvable in polynomial time as a minimum cost network flow problem. When the costs $c_{n+k,j}$ and $c_{j,n+k}$ are independent of the depot k, $k \in K$, the problem reduces to the single depot case. In particular, the multi-depot minimum fleet size problem reduces to the single depot case. In fact, one has to only define the cost structure as follows: $c_{ij} = 0$ for all $(i, j) \in I$, $c_{n+k,j} = 1$ and $c_{j,n+k} = 0$, for all trips $j \in N$ and all depot $k \in K$. In general, solving the minimum fleet size problem is also a way to provide a feasible solution (if one exists) to the MDVSP in polynomial time.

Literature. Heuristic algorithms have been proposed by Bodin, Rosenfield & Kydes [1978], Ceder & Stern [1981], Smith & Wren [1981], El-Azm [1985], and Bertossi, Carraresi & Gallo [1987]. Surveys on the subject can be found in Bodin & Golden [1981], Wren [1981], Bodin, Golden, Assad & Ball [1983], and Carraresi

& Gallo [1984]. Two new heuristics have also been published very recently by Lamatsch [1992] and Mesquita & Paixão [1992]. Both are based on Lagrangean relaxations of integer programming formulations.

Lamatsch [1992] has proposed a multi-commodity flow model equivalent to the formulation (2.9)–(2.14). The solution approach he used requires the sum over k of constraint set (2.12), yielding an additional set of constraints

$$\sum_{k \in K} \sum_{i \in V^k} X_{ij}^k - \sum_{k \in K} \sum_{i \in V^k} X_{ji}^k = 0, \quad \forall j \in N \cup \bigcup_{k \in K} \{n + k\}. \quad (2.12a)$$

The problem obtained from the relaxation of the multi-commodity flow conservation constraints (2.12) in the objective function may then be regarded as a single-commodity flow problem with new flow variables X_{ij} defined as $\sum_{k \in K} X_{ij}^k$. The resulting subproblem possesses the integrality property (see Section 5.5), so that the best bound obtained using this Lagrangean relaxation is also equal to the linear relaxation value, i.e., as good as the bound provided by the Dantzig–Wolfe decomposition described below. For large problems however, the computation time reported is extremely high, as convergence is very slow. For this reason, a strategy is developed to modify the Lagrangean relaxation solution in order to obtain a good heuristic solution for the original multi-commodity flow problem without going in a branch-and-bound exploration tree. Computational results are reported for three small real-world problems of a company running up to 250 trips from two depots.

The most recent polynomial time heuristic algorithm, which also always guarantees the use of the minimum number of vehicles, is due to Dell'Amico, Fischetti & Toth [1990]. These authors report extensive computational results on test problems involving up to 1000 trips and 10 depots, showing that it outperforms other heuristics from the literature. Two exact algorithms have also been proposed. The first [Carpaneto, Dell'Amico, Fischetti & Toth, 1989] is a branch-and-bound algorithm based on the computation of lower bounds by an additive scheme (see Fischetti & Toth [1989] for the theoretical exposition). The average integrality gap is relatively small, on the order of 0.9% on a set of randomly generated test problems, involving up to 70 trips and 3 depots. The second exact method [Ribeiro & Soumis, 1994] is also a branch-and-bound algorithm. It is based on the solution of the linear relaxation of formulation (2.9)–(2.14) by a decomposition scheme (Dantzig and Wolfe 1960). Computational results are reported for problems generated similarly to the ones above but of size up to four to five times larger (up to $n = 100$ and $|K| = 10$, and $n = 250$ and $|K| = 6$). On these problems the average gap is only 0.0008%.

To explain the gap difference between the two methods, let Z_{IP} denote the optimal value for an integer programming problem. Denote by Z_{LP} the value of its linear relaxation. Briefly, the additive bounding procedure applied to the multidepot vehicle scheduling problem uses a combination of an assignment bound and shortest path bounds. Let also Z_{ADD} be the value of the lower bound provided by this additive bounding procedure. Finally, let Z_{DW} be the value of the lower bound obtained by the Dantzig–Wolfe decomposition presented next. Then the following result is obtained by Ribeiro & Soumis [1994]:

$$Z_{ADD} \leq Z_{LP} = Z_{DW} \leq Z_{IP}. \tag{2.15}$$

This relation states that the value provided by the additive bounding procedure can be below that of the linear relaxation of the problem while the decomposition scheme used provides a bound which is equal to the LP bound. In practice, the value Z_{ADD} is very good. In theory the worst case ratio can be arbitrarily bad. Even under the symmetric cost assumption for the underlying travel network, the worst case ratio is greater or equal to 50% and this bound is tight.

Theoretical aspects. The remaining part of this section exposes the decomposition principle applied to the MDVSP. The theory is based on Minkowski's theorem [Lasdon, 1970; Nemhauser & Wolsey, 1988]. Let $\mathcal{X} = \{X \in \mathbb{R}^n_+ | AX \leq b\}$ be a nonempty polyhedron defined by a finite set of constraints and lying within the nonnegative orthant of real numbers. A point $x_p \in \mathcal{X}$ is defined to be an extreme point of $\mathcal{X}$ if there do not exist $X^1, X^2 \in \mathcal{X}$, $X^1 \neq X^2$, such that $x_p = 1/2X^1 + 1/2X^2$. If $\mathcal{X}(0) = \{X \in \mathbb{R}^n_+ | AX \leq 0\} \neq \{0\}$, then $X \in \mathcal{X}(0) \setminus \{0\}$ is called a ray of $\mathcal{X}$. A point $x_p \in \mathcal{X}$ is defined to be an extreme ray of $\mathcal{X}$ if there do not exist rays $X^1, X^2 \in \mathcal{X}(0) \setminus \{0\}$, $X^1 \neq \lambda X^2$, for any constant $\lambda > 0$, such that $x_p = 1/2X^1 + 1/2X^2$.

Then, a polyhedron $\mathcal{X}$ has a finite number of extreme points and extreme rays, and any point $X \in \mathcal{X}$ can be written as a convex combination of extreme points of $\mathcal{X}$ plus a nonnegative linear combination of extreme rays of $\mathcal{X}$, i.e.,

$$X = \sum_p x_p \theta_p,$$

where

$$\sum_p \theta_p \delta_p = 1, \quad \theta_p \geq 0,$$

and

$$\delta_p = \begin{cases} 1, & \text{if } x_p \text{ is an extreme point of } \mathcal{X} \\ 0, & \text{if } x_p \text{ is an extreme ray of } \mathcal{X}. \end{cases}$$

A Dantzig–Wolfe decomposition. The decomposition scheme is applied to the linear relaxation of the MDVSP obtained by dropping the integrality requirements (2.14). The master problem is given by (2.9)–(2.11). The subproblem involves a modified objective function $\sum_{k \in K} \sum_{(i,j) \in A^k} \bar{c}_{ij} X_{ij}$, where the coefficients $\bar{c}_{ij}$ will be defined later. The subproblem set of constraints is defined by the flow conservation equations (2.12) over the nonnegative X^k_{ij} variables (2.13).

The subproblem. The subproblem constraints define a homogeneous system. Therefore the only extreme point of the subproblem is the null solution and any positive solution can be expressed as a nonnegative linear combination of the extreme rays. Hence, the convexity constraint will not appear in the decomposition scheme. Let $(x^k_{ijp}, k \in K, (i,j) \in A^k)$, $p \in \Omega$, be the set of extreme rays of the subproblem set of constraints. It will be shown later that these extreme rays may be defined by paths starting from some depot, covering a number of trips and returning to the same depot. To characterize an extreme ray it is then sufficient to send only one unit of flow on the corresponding path. Hence, the constant x_{ijp}^k takes only binary values.

Express next any solution X_{ij}^k to the master problem as a nonnegative linear combination of the extreme rays, i.e.,

$$X_{ij}^k = \sum_{p \in \Omega} x_{ijp}^k \theta_p, \quad \forall k \in K, \ \forall (i, j) \in A^k$$

$$\theta_p \geq 0, \qquad\qquad \forall p \in \Omega.$$

The master problem. Making the substitution into (2.9)–(2.11) and rearranging the summation order gives the revised formulation:

$$\text{Minimize} \sum_{p \in \Omega} \left(\sum_{k \in K} \sum_{(i,j) \in A^k} c_{ij} x_{ijp}^k \right) \theta_p \qquad (2.16)$$

subject to

$$\sum_{p \in \Omega} \left(\sum_{k \in K} \sum_{j \in V^k} x_{ijp}^k \right) \theta_p = 1, \quad \forall i \in N \qquad (2.17)$$

$$\sum_{p \in \Omega} \left(\sum_{j \in N} x_{n+k,jp}^k \right) \theta_p \leq v^k, \quad \forall k \in K \qquad (2.18)$$

$$\theta_p \geq 0, \qquad\qquad \forall p \in \Omega. \qquad (2.19)$$

Let the cost coefficients of the objective function be denoted by c_p, $p \in \Omega$. The costs c_p are the sum of the costs of the arcs used on the path. Therefore,

$$c_p = \sum_{k \in K} \sum_{(i,j) \in A^k} c_{ij} x_{ijp}^k. \qquad (2.20)$$

Since each node is visited at most once by a path, define

$$a_{ip} = \sum_{k \in K} \sum_{j \in V^k} x_{ijp}^k, \qquad (2.21)$$

a binary constant taking value 1 if path p visits node i. Finally define

$$b_p^k = \sum_{j \in N} x_{n+k,jp}^k, \qquad (2.22)$$

another binary constant taking the value 1 only if path p starts from depot $k \in K$. The resulting master problem is then given by:

$$\text{Minimize} \sum_{p \in \Omega} c_p \theta_p \qquad (2.23)$$

subject to

$$\sum_{p \in \Omega} a_{ip} \theta_p = 1, \quad \forall i \in N \qquad (2.24)$$

$$\sum_{p\in\Omega} b_p^k \theta_p \le v^k, \qquad\qquad \forall\, k \in K \tag{2.25}$$

$$\theta_p \ge 0, \qquad\qquad \forall\, p \in \Omega. \tag{2.26}$$

Linear relaxation solution. The above formulation defines the linear relaxation of a set partitioning problem with availability constraints for each depot. Given an optimal solution to the restricted master problem on a subset of variables $\Omega' \subset \Omega$, let α_i, $i \in N$ and β_k, $k \in K$, be the dual variables associated with the covering of the trips $i \in N$ and the use of a vehicle from a depot $k \in K$, respectively. A new minimum marginal cost column is generated by solving:

$$\underset{p\in\Omega-\Omega'}{\text{Minimize}}\; c_p - \sum_{i\in N} \alpha_i a_{ip} - \sum_{k\in K} \beta_k b_p^k. \tag{2.27}$$

The constants c_p, a_{ip} and b_p^k are unknown on $\Omega - \Omega'$. If a negative marginal cost column exists, it is added to the current restricted master problem. Since there is no negative marginal cost columns on the subset Ω' (the current solution is optimal for the restricted master), optimization (2.27) can be performed on Ω. Constants c_p, a_{ip} and b_p^k can be represented by equations (2.20)–(2.23) which define them, expressed in terms of the original decision variables X_{ij}^k. These variables must satisfy the subproblem constraints (2.12) and (2.13). Hence, minimization (2.27) can be reformulated as:

$$\text{Minimize}\quad \sum_{k\in K}\sum_{(i,j)\in A^k} c_{ij} X_{ij}^k$$

$$- \sum_{i\in N} \alpha_i \left(\sum_{k\in K}\sum_{j\in V^k} X_{ij}^k\right) - \sum_{k\in K} \beta_k \left(\sum_{j\in N} X_{n+k,j}^k\right) \tag{2.28}$$

subject to   (2.12)–(2.13).

The set of arc A^k, $k \in K$, can be partitioned the following way: (i, j), with $i \in N$ and $j \in V^k$, or $(n + k, j)$, with $k \in K$ and $j \in N$. Hence the objective function can be rewritten as

$$\sum_{k\in K}\sum_{i\in N}\sum_{j\in V^k}(c_{ij} - \alpha_i)X_{ij}^k - \sum_{k\in K}\sum_{j\in N}(c_{n+k,j} - \beta_k)X_{n+k,j}^k. \tag{2.29}$$

It follows that the marginal cost of an arc $(i, j) \in A^k$, $k \in K$ is given by:

$$\bar{c}_{ij} = \begin{cases} c_{ij} - \alpha_i, & \text{if } i \in N \\ c_{ij} - \beta_k, & \text{if } i = n+k, \quad k \in K. \end{cases} \tag{2.30}$$

Then, minimization (2.27) simply becomes:

$$\text{Minimize}\quad \sum_{k\in K}\sum_{(i,j)\in A^k} \bar{c}_{ij} X_{ij}^k \tag{2.31}$$

subject to $(2.12) - (2.13)$.

If a negative marginal cost solution exists, then, by the flow conservation equations, it is defined by a number of negative marginal cost paths. We can choose one path or any number of paths from any number of different depots. The way to characterize an extreme ray is to choose only one path. Then, one unit of flow is sent on it to define the extreme ray. Note that the subproblem is separable by depot, and that for each depot, the problem solution amounts to a shortest path, on an acyclic graph, which admits an easy solution. The algorithmic complexity is only $O(n^2)$ for a n-node problem [Fulkerson, 1972].

The solution process entails alternating between the restricted master problem and the subproblem. The procedure is started with an artificial set of columns for the restricted master problem which forms the initial identity basis. Each time the restricted master problem is solved on a subset of columns, the dual variables are updated and used for the subproblem solution. Then, to accelerate the process, each time the generating phase is called, every depot may provide several path columns. The column generation procedure terminates when no more negative marginal cost columns exist for any depot. The process then gives the value Z_{LP}, i.e., the solution value of problem (2.9)–(2.13), which is the linear relaxation of the MDVSP.

Optimal integer solution. To obtain an optimal integer solution to the original problem (2.9)–(2.14), cuts and branch-and-bound strategies are used, if necessary. An integer solution is found if constraints (2.14), i.e., X_{ij}^k binary, $\forall k \in K$, $\forall (i, j) \in A^k$, are all satisfied. Initially, one needs to check if all θ_p variables take binary values. This is a sufficient condition since, in this case, the solution decomposes into a set of disjoint paths. Otherwise, one has to evaluate if

$$X_{ij}^k = \sum_{p \in \Omega} x_{ijp}^k \theta_p \text{ binary}, \quad \forall k \in K, \ \forall (i, j) \in A^k \tag{2.32}$$

are all satisfied. This is accomplished by using only the fractional θ_p variables in the optimal basis, i.e., $p \in \Omega$ such that $0 < \theta_p < 1$.

If the solution is fractional, the value Z_{LP} can be rounded up to the next integer, if all coefficients c_{ij} are integer. In other words, the following constraint is added to the linear relaxation (2.9)–(2.13):

$$\sum_{k \in K} \sum_{(i,j) \in A^k} c_{ij} X_{ij}^k \geq \lceil Z_{LP} \rceil. \tag{2.33}$$

Branching can be performed on the number of vehicles used at a depot, i.e., on $\sum_{j \in N} X_{n+k,j}^k$, for a given k. It may also be done on a specific binary flow variables X_{ij}^k, or on a subset of flow variables like $\sum_{k \in K} X_{ij}^k$. Branching decisions to set these variables at zero or at one define new arc sets for each subproblem. If new constraints are added to the linear relaxation (2.9)–(2.13), the decomposition process can be applied as described before so that new constraints appear in the master problem definition. Otherwise, branching decisions are easily transferred to the subproblem structure (elimination of arcs, dual variables) so that only feasible negative marginal cost columns are generated. The column generation scheme is then applied at each node of the branch-and-bound tree to obtain the overall optimal solution. Note that while solving (2.23)–(2.26) by a simplex algorithm, if a feasible integer solution is found, it is recorded and whenever a better solution is encountered, the previous best solution is updated. The experimental behavior of the algorithm is described in more details in Ribeiro & Soumis [1994].

Additional details on the solution of integer programs using Dantzig–Wolfe decomposition are provided in Sections 5, 6 and 7. We can nevertheless offer here some insights into branching schemes for integer programs solved by decomposition methods, including column generation. If the formulation (2.9)–(2.14) is solved directly, branching decisions will naturally be taken on the original variables [see for example Forbes, Holt & Watts, 1991]. This is also true should the problem be solved by Lagrangean relaxation where no new variables are defined. Given that Lagrangean relaxation and Dantzig–Wolfe decomposition are equivalent dual and primal methods, respectively [see Magnanti, Shapiro & Wagner, 1976], this suggests that in a decomposition approach, branching should be performed on the original variables, even if new variables have been introduced for extreme points or rays.

3. The traveling salesman problem with time windows

The traveling salesman problem (TSP) with time windows (TSPTW) consists of finding the minimum cost path to be traveled by a vehicle, starting and returning at the same depot, which must visit a set of nodes exactly once. The service at a node must begin within the time window defined by the earliest time and the latest time when the start of service is allowed at that node. A vehicle is not permitted to arrive at a node after the latest time to begin service. However, if a vehicle arrives too early at a node, it is permitted to wait until the node is ready for the beginning of service. The TSPTW is a time-constrained vehicle routing problem involving only one uncapacitated vehicle.

This type of problem is encountered in a variety of industrial and service sector applications. Examples include bank deliveries, postal deliveries and school-bus routing and scheduling. The TSPTW constitutes an important element of more complex vehicle routing problems. In particular, for *cluster-first, route second* approaches to such problems, if the clustering phase is performed heuristically, the a posteriori optimization of the TSPTW in each cluster is necessary. This type of problem has also notable applications in manufacturing. In automated manufacturing systems, for example, an automated guided vehicle must visit a set of flexible machines. To achieve full machine utilization, the automated guided vehicle must visit each machine within the time window determined by the processing time of the job currently being machined.

3.1. Problem formulation

Notation. Consider a set of nodes $N = \{1, \ldots, n\}$ to be visited and the additional single depot. Next, duplicate the depot into an origin-depot o and a destination-depot d. Let $V = N \cup \{o, d\}$. Next, associate with each node $i \in V$ a time window $[a_i, b_i]$. The constant a_o is the earliest starting time of the vehicle from the depot, while b_d is the latest arrival time of the vehicle to the depot. Denote by A the set of arcs and associate with each arc a nonnegative duration t_{ij} and a cost c_{ij}. We assume next that the service time at node i is included in the time value t_{ij}, for all $i \in N \cup \{o\}$. Finally, an arc (i, j) is defined in the set A if the following condition holds:

$$a_i + t_{ij} \leq b_j, \qquad i \in N \cup \{o\}, \quad j \in N \cup \{d\}, \quad i \neq j. \tag{3.1}$$

Formulation. The mathematical programming formulation given below involves two types of variables: the set of binary flow variables $X = (X_{ij}, (i, j) \in A)$ and the set of time variables $T = (T_i, i \in V)$. The variable X_{ij} is equal to 1, if arc (i, j) is used in the optimal tour, and is set at 0 otherwise. Given that waiting at a node before the specified time window is permitted, the variable T_i specifies the start of service at node i, $i \in N \cup \{o\}$, while the variable T_d gives the arrival time at the destination-depot node.

The minimal cost tour starting within the time interval $[a_o, b_o]$ at the origin-depot node, visiting exactly once all the nodes of N within their time windows, and finishing at the destination-depot node before the time limit b_d, can be formulated as follows:

$$\text{Minimize} \quad \sum_{(i,j) \in A} c_{ij} X_{ij} \tag{3.2}$$

subject to

$$\sum_{j \in N \cup \{d\}} X_{ij} = 1, \qquad\qquad \forall\, i \in N \tag{3.3}$$

$$\sum_{j \in N} X_{o,j} = 1, \tag{3.4}$$

$$\sum_{i \in N \cup \{o\}} X_{ij} - \sum_{i \in N \cup \{d\}} X_{ji} = 0, \quad \forall\, j \in N \tag{3.5}$$

$$\sum_{i \in N} X_{i,d} = 1, \tag{3.6}$$

$$X_{ij}(T_i + t_{ij} - T_j) \leq 0, \qquad \forall\, (i, j) \in A \tag{3.7}$$

$$a_i \leq T_i \leq b_i, \qquad\qquad \forall\, i \in V \tag{3.8}$$

$$X_{ij} \geq 0, \qquad\qquad \forall\, (i, j) \in A \tag{3.9}$$

$$X_{ij} \text{ binary,} \qquad\qquad \forall\, (i, j) \in A. \tag{3.10}$$

This formulation closely replicates the fixed schedule formulation introduced in Section 2. It is a nonlinear program with $O((n + 2)^2)$ variables and $O((n + 2)^2)$ constraints, where the objective function (3.22) represents the total cost. The subproblem defined by (3.2)–(3.5) and (3.9) is a minimum cost flow problem (in fact an assignment problem with $(n + 2)$ rows and $(n + 2)$ columns). Constraints (3.7)–(3.8) ensure feasibility of the time schedule.

Subtours are eliminated by constraints (3.7). Using a big M constant, these constraints can be linearized and rewritten as follows [Desrosiers, Pelletier & Soumis, 1983; Solomon, 1983]:

$$T_i + t_{ij} - T_j \leq M(1 - X_{ij}), \qquad \forall (i, j) \in A. \tag{3.7a}$$

They represent a generalization of the classical TSP subtour elimination constraints proposed by Miller, Tucker & Zemlin [1960]. Note that the large constant M can be replaced by $M_{ij} = \max\{b_i + t_{ij} - a_j, 0\}$, $(i, j) \in A$. When $b_i + t_{ij} \leq a_j$, these constraints are always satisfied for all values of T_i, T_j and X_{ij}. Then, the formulation requires the presence of constraints (3.7) or (3.7a) only for arcs $(i, j) \in A$, such that $M_{ij} > 0$.

Nonlinear integrality property. The nonlinear formulation (3.2)–(3.10) has also an interesting property. If the problem is feasible, integrality requirements (3.10) can be eliminated from the above formulation [Desrosiers, Soumis & Desrochers, 1984]. To show that this can be done, let (X^*, T^*) be an optimal solution. If X^* is not integer, define $A^* = \{(i, j) \in A | X_{ij}^* > 0\}$. Now fix T^*, and consider the minimum cost assignment problem on the subnetwork obtained by retaining only the arcs of A^* for which there exists an integer optimal solution $\overline{X}$. It is then easy to verify that $(\overline{X}, T^*)$ is an optimal integer solution to (3.2)–(3.9). Note however that the linearized formulation containing (3.7a) does not necessarily provide an integer solution. Hence, constraints (3.10) must be retained in the linearized formulation.

Literature. Research on the TSPTW has been scant. Savelsbergh [1985] has proved that even finding a feasible solution to the TSPTW is an $\mathcal{NP}$-complete problem. Therefore, the author has proposed heuristic algorithms based on the k-interchange concept [Lin, 1965; Lin & Kernighan, 1973]. This involves the replacement of k arcs currently on the TSP tour with k other arcs. For the TSP, the processing of a single k-interchange can be performed in constant time, for any k. This is because one has to only examine the savings derived from such an interchange. For the TSPTW, however, the treatment of a k-interchange could require $O(n)$ time. Such an interchange may shift the time to start the service at a node, which in turn could alter the start of service at all the other nodes on the tour. Therefore, in addition to its profitability one has to also test the feasibility of an interchange. Savelsbergh [1985] shows that for 2-interchanges and the restricted 3-interchanges proposed by Or [1976], the processing of a single interchange can still be executed in constant time for the objective of minimizing the total travel time. Savelsbergh [1988, 1989] has extended this result to the case of minimizing the total schedule time. Extensive research has been conducted on heuristics for more complex routing problems and this will be discussed in Sections 5 and 6.

Despite the difficulty of the TSPTW, several other authors have focused on exact algorithms to minimize the *total cost* or the *total schedule time*. Objective (3.2) describes the former case; minimizing the value of $T_d - T_o$ is appropriate for the latter case. For this objective, it is also easy to define waiting time variables $W_i \geq 0$, $i \in N \cup \{d\}$ at each node to be serviced and the destination-depot node, introduce them in constraints (3.7) or (3.7a) and replace the inequality sign by an equality sign, and rewrite the objective function as:

$$\text{Minimize } T_d - T_o = \sum_{(i,j) \in A} t_{ij} X_{ij} + \sum_{i \in N \cup \{d\}} W_i \qquad (3.2a)$$

Under special conditions, Baker [1983] introduces another model for the minimization of the total schedule time. His model as well as the solution process are discussed in Section 3.5.

The formulation (3.2)–(3.10) or the model with the objective function (3.2) replaced by (3.2a) can be used in a branch-and-bound scheme to determine an optimal time window constrained tour. For example, one can relax constraints (3.7) and (3.8) and use the lower bound provided by the assignment problem solution. Branching on the X_{ij} variables as do Carpaneto & Toth [1980] for the asymmetric TSP may provide a very good algorithm. Considering the linearized version, another lower bound is given by the solution of the linear relaxation. This bound is at least as good as the assignment bound, yet much more time-consuming to calculate. If the graph is dense, it is expected to be very close to the assignment bound (see Section 5.2).

Many other relaxations or decompositions can be designed using the previous formulations. In any search tree, branching on the flow variables X_{ij} $(i, j) \in A$ is easily implemented. However when the most important component of a specific TSPTW application involves the time window restrictions, branching on the time variables T_i is definitely much more appropriate. This is also true for any time-constrained vehicle routing problem solved by branch-and-bound.

3.2. Time window reduction and arc elimination

The complexity of the algorithms proposed for time window constrained routing and scheduling problems is a function of the width of the time windows. To improve the efficiency of these algorithms, the time windows' width can be reduced a priori by applying several rules described in Cyrus [1988] and in Desrochers, Desrosiers & Solomon [1992]. These rules, which are valid for both objective functions (i.e., the minimization of the total cost or the total schedule time), are presented in Figure 6.

At each node, an attempt is made to reduce the time window width by testing conditions (3.11)–(3.14) in order. Arcs which become infeasible are removed from the arc set A. After examining all the nodes, a new iteration is started. In practice, after two or three iterations no further reductions are possible.

Arc elimination for the TSPTW. Additional computational efficiency can be gained by reducing the set of arcs for the TSPTW by applying the rules developed by Langevin, Desrochers, Desrosiers & Soumis [1993]. To present these conditions, let EAT(i, j) be the earliest arrival time at node j from node i. This can be

LOWER BOUND ADJUSTMENT DERIVED FROM THE EARLIEST ARRIVAL TIME
AT NODE k FROM ITS PREDECESSORS, FOR $k \in N \cup \{d\}$:

$$a_k := \max\{a_k, \min_{(i,k) \in A} \{a_i + t_{ik}\}\} \tag{3.11}$$

UPPER BOUND ADJUSTMENT DERIVED FROM THE LATEST ARRIVAL TIME
AT NODE k FROM ITS PREDECESSORS, FOR $k \in N \cup \{d\}$:

$$b_k := \min\{b_k, \max\{a_k, \max_{(i,k) \in A} \{b_i + t_{ik}\}\}\}; \tag{3.12}$$

LOWER BOUND ADJUSTMENT DERIVED FROM THE EARLIEST DEPARTURE TIME
FROM NODE k TO ITS SUCCESSORS, FOR $k \in N \cup \{o\}$:

$$a_k := \max\{a_k, \min\{b_k, \min_{(k,j) \in A} \{a_j - t_{kj}\}\}\}; \tag{3.13}$$

UPPER BOUND ADJUSTMENT DERIVED FROM THE LATEST DEPARTURE TIME
FROM NODE k TO ITS SUCCESSORS, FOR $k \in N \cup \{o\}$:

$$b_k := \min\{b_k, \max_{(k,j) \in A} \{b_j - t_{kj}\}\}. \tag{3.14}$$

Fig. 6. Time window reduction.

computed by solving a shortest time path problem (including waiting times) starting at time a_i and satisfying the time window constraints from node i to node j. This can be calculated using dynamic programming (see Section 4).

For arc (i, j) to exist it should be possible to visit node k either before node i or after node j, for all $k \in N$. Therefore, arc (i, j) can be eliminated if $\exists k \in N$ such that

$$\text{EAT}(k, i) > b_i \quad \text{and} \quad \text{EAT}(j, k) > b_k. \tag{3.15}$$

Note that when the depot is node j (node i), only the former (latter) condition needs to be tested. Rule (3.15) can even be strengthened. Let $\text{LDT}(i, j)$ be the latest departure time from i such that the time to begin service at j is feasible. This can also be computed by solving a shortest time path problem (including waiting times) by using the reverse arc directions and starting at time b_j at node j.

Let $k \sim i$ represent nodes k and i linked by a path, or the single arc (k, i) in the degenerate case. Then, arc (i, j) can be eliminated if there exists a node k such that the two sequences $k \sim i \sim j$ and $i \sim j \sim k$ are not feasible, i.e., if $\exists k \in N$ such that

$$\text{EAT}(k, i) > \text{LDT}(i, j) \quad \text{and} \quad \text{EAT}(i, j) > \text{LDT}(j, k). \tag{3.16}$$

Time window reduction for the TSPTW. Gélinas [1993] suggests specific time window reductions to be performed for the TSPTW. Let BEFORE(k) be the set of all nodes that must necessarily be visited before k. Therefore, BEFORE(k) = $\{i \in N | \text{EAT}(k, i) > b_i\}$. Similarly define AFTER(k) = $\{j \in N | \text{EAT}(j, k) > b_k\}$, the set of all nodes that must necessarily be visited after k. Given these sets, the adjustment relations (3.11) and (3.14) can be strengthened. Another lower bound adjustment can then be derived from the earliest arrival time at node k from nodes only in BEFORE(k), for $k \in N \cup \{d\}$:

$$a_k := \max\{a_k, \max_{i \in \text{BEFORE}(k)} \{\text{EAT}(i, k)\}\}. \tag{3.11a}$$

Symmetrically, another upper bound adjustment can be derived from the latest departure time from node k to nodes in AFTER(k), for $k \in N \cup \{o\}$:

$$b_k := \min\{b_k, \min_{j \in \text{AFTER}(k)} \{\text{LDT}(k, j)\}\}. \tag{3.14a}$$

For the TSPTW, preprocessing involves time window reduction and arc elimination procedures until no further modifications are possible. Finally, note that the values $\text{EAT}(i, j)$ and $\text{LDT}(i, j)$ can be overestimated and underestimated, respectively, by using unconstrained shortest time paths. This is much faster and it gives satisfactory results in practice. For the rest of this section we assume that A represents the set of arcs after time window reduction and arc elimination.

The next sections focus on three optimal branch-and-bound approaches (Sections 3.3, 3.4 and 3.6) and on a fourth approach (Section 3.5) which solves the problem directly. The first two methods solve the problem of minimizing the total schedule time [Baker, 1983; Christofides, Mingozzi & Toth, 1981b]. The last two approaches are suitable for both types of objective function. The third algorithm is a very powerful dynamic programming algorithm [Dumas, Desrosiers, Gélinas & Solomon, 1991]. The fourth method is presented not for its numerical performance, but rather for its original formulation involving only $O(n^2)$ variables and $O(n)$ constraints in the linear relaxation [Langevin, Desrochers, Desrosiers & Soumis, 1993].

3.3. A time constrained critical path approach

In this subsection, it is assumed that the time matrix t_{ij}, $i, j \in V$ is symmetric and that the triangle inequality holds. Under these conditions, Baker [1983] presents a branch-and-bound algorithm for the following time oriented formulation of the TSPTW:

$$\text{Minimize} \quad T_d - T_o \tag{3.17}$$

subject to

$$T_i \geq T_o + t_{o,i}, \quad \forall i \in N \tag{3.18}$$

$$|T_i - T_j| \geq t_{ij}, \quad 1 \leq i < j \leq n \tag{3.19}$$

$$T_d - T_i \geq t_{i,d}, \quad \forall i \in N \tag{3.20}$$

$$a_i \leq T_i \leq b_i, \quad \forall i \in V. \tag{3.21}$$

The solution process also imposes a third condition which specifies that the tour starts at a fixed time value, i.e., $a_o = b_o = T_o$. As discussed later, allowing $b_o - a_o > 0$ implies an additional degree of complexity.

The approach involves constructing a related acyclic network. This network has the same set of nodes V. The arcs in the network are constructed as follows. Given any two nodes i and j, if the time window constraints impose a precedence relationship between these two nodes, say $T_i < T_j$ if $a_j + t_{ji} > b_i$, then this relationship is represented by an arc. The constraints (3.19) which do not create precedence relationships are relaxed.

In order to minimize the total schedule time $T_d - a_o$, the critical path, starting at time a_o from the origin-depot node and returning to the destination-depot node at the end of the tour, is computed on the related acyclic network as a longest path with time window constraints. Given that the starting time of the critical path is fixed, the computational complexity of such an algorithm is $O(n^2)$ for a n-node network (see Section 4.5). This critical time path provides a lower bound in the search tree. Branching creates two subproblems corresponding to the case where arrival at node i precedes the arrival at node j, and vice versa. Therefore the branching decisions are taken on time variables. The best solution gives the minimal value of $T_d - a_o$. The optimal tour can be found on the corresponding acyclic network of the search tree. In this case, a maximum cardinality path labeling convention must be adopted for the longest path algorithm to resolve ties in the triangle inequality. Otherwise, paths of equal schedule time to the critical path (the optimal tour) but containing less than $n + 1$ arcs would be allowed. Baker [1983] reports good performance for the algorithm on 50-node problems where only a small percentage of time windows overlap.

Note. Baker [1983] presented the above algorithm to find the minimum schedule time for the TSPTW under several conditions including the fixed starting time of the tour. If waiting at the origin-depot node is feasible within the time window $[a_o, b_o]$ for $b_o > a_o$, then a specific critical path length must be calculated for each possible starting time value, i.e., for $D_o = b_o - a_o + 1$ values in the case of integer values, for all the time parameters. The optimal critical path length is then given by the smallest value. For a n-node problem, the computational complexity of the algorithm becomes $O(n^2 D_o)$.

An alternative way to solve the critical path problem with time windows when the starting time is flexible at the origin-depot node is to use a two-dimensional labeling procedure. Such labeling procedures for the shortest path problem with time windows and dominance criteria which permit label elimination at a node are discussed in the next section. The first dimension of this two-dimensional labeling represents the start of service at a node within the time interval. The second dimension represents the negative of the elapsed schedule time (the duration). Additional comments on the critical path problem with time windows are given in Section 4.5.

3.4. State-space relaxation

State-space relaxation is a general relaxation procedure proposed by Christofides, Mingozzi & Toth [1981b] for a number of routing problems. The motivation for this methodology stems from the fact that very few combinatorial optimization problems can be solved by dynamic programming alone due to the dimensionality of their state-space. To overcome this difficulty, the number of states is reduced by mapping the state-space associated with a given dynamic programming recursion to a smaller cardinality space. This mapping, denoted by g, must associate to every transition from a state S_1 to a state S_2 in the original state space, a transition $g(S_1)$ to $g(S_2)$ in the new state-space. To be effective, the function g must give rise to a transformed recursion over the relaxed state-space which can be computed in polynomial time. Furthermore, this relaxation must generate a good lower bound for the original problem. This lower bound could then be used in a branch-and-bound algorithm for the solution of the original problem.

We present this approach in the context of the minimization of the total schedule time for the TSPTW. Assume that the time-constrained path starts at the fixed time value a_o. Define $F(S, i)$ as the shortest time it takes for a feasible path starting at node o, passing through every node of $S \subseteq N$ exactly once, to end at node $i \in S$. Note that the optimization of the total arc cost would involve an additional dimension to account for the arrival time at a node (see Section 3.5). The function $F(S, i)$ can be computed by solving the following recurrence equations:

$$F(S, j) = \min_{(i,j)\in A} \left\{ F(S - \{j\}, i) + t_{ij} \mid i \in S - \{j\} \right\} \quad \forall S \subseteq N, j \in S.$$
(3.22)

The recursion formula is initialized by

$$F(\{j\}, j) = \begin{cases} \max \{a_j, a_o + t_{o,j}\} & \text{if } (o, j) \in A, \\ \infty, & \text{otherwise.} \end{cases}$$
(3.23)

The optimal solution to the TSPTW is given by:

$$\min_{j \in N} \left\{ F(N, j) + t_{j,d} \right\}.$$
(3.24)

Note that equation (3.22) is valid if $a_j \leq F(S, j) \leq b_j$. If however $F(S, j) < a_j$, then $F(S, j) = a_j$; if $F(S, j) > b_j$, $F(S, j) = \infty$. Formulas (3.22)–(3.24) define a shortest path algorithm on a state graph whose nodes are the states (S, i) and whose arcs represent transitions from one state to another. This algorithm is a forward dynamic programming algorithm where at step s, $s = 1, \ldots, n + 1$, a path of length s is generated. The states (S, i) of cost $F(S, i)$ are defined as follows: S is an unordered set of visited nodes and i is the last visited node, $i \in S$.

Christofides, Mingozzi & Toth [1981b] suggest several alternatives for the mapping g. We present here the shortest q-path relaxation, i.e., $g(S) = q = \sum_{i \in S} q_i$, where $q_i \geq 1$ is an integer associated with node $i \in N$; then $g(S - \{j\}) = g(S) - q_j$. Define $Q = \sum_{i \in N} q_i$. Hence the transformed recursion equations are:

$$F(q, j) = \min_{(i, j) \in A} \left\{ F(q - q_j, i) + t_{ij} \mid q - q_j \geq q_i \right\},$$

$$q \in \{1, \ldots, Q\}, j \in N. \qquad (3.25)$$

Recursions (3.25) hold if $a_j \leq F(q, j) \leq b_j$. Otherwise, if $F(q, j) < a_j$, then $F(q, j) = a_j$ and if $F(q, j) > b_j$, then $F(q, j) = \infty$. The recursion formula is initialized by

$$F(q, j) = \begin{cases} \max\{a_j, a_o + t_{o,j}\}, & \text{if } (o, j) \in A \text{ and } q = q_j, \\ \infty, & \text{otherwise,} \\ & \text{for } q \in \{1, \ldots, Q\}, j \in N. \end{cases} \qquad (3.26)$$

The lower bound is given by:

$$\min_{j \in N} \left\{ F(Q, j) + t_{j,d} \right\}. \qquad (3.27)$$

The complexity of the bounding procedure for a n-node problem is $O(n^2 Q)$.

This lower bound can be increased by applying node penalties. This method consists of associating Lagrangean multipliers with the travel times of arcs incident to nodes that are not visited, or visited more than once. These multipliers decrease the former travel times and increase the latter. They are adjusted through subgradient optimization. The lower bound can also be increased by modifying the state-space through the application of subgradient optimization to the weights q_i. Branching on flow variables, Christofides, Mingozzi & Toth [1981b] report solving 50-node problems with moderately tight time windows. The reader may also refer to Friis [1985] for a Pascal code of this algorithm.

The same methodology has been applied to more complex routing problems and it will be discussed in Section 5. State-space relaxation parallels constraint aggregation and Lagrangean relaxation for integer programming. Constraints in integer programming formulations appear as state variables in dynamic programming recursions and hence constraint aggregation and relaxation correspond to state-space relaxation.

3.5. A dynamic programming method

In the previous section, we have illustrated the use of transformed recursions derived from relaxing the state-space to obtain lower bounds which are then embedded in a tree search approach. However, when enough constraints are present, dynamic programming can be used directly to solve realistic size problems. The method presented in this section is based on the paper by Dumas, Desrosiers, Gélinas & Solomon [1991] which takes advantage of the time window constraints. These constraints permit the reduction of the state-space dimension and the number of state transitions which is achieved in a preprocessing stage as well as while the algorithm is being run. The dynamic programming method presented here aims at minimizing the total arc cost. This is a more difficult optimization than when the minimum schedule time is sought. As in the previous methods, it is also assumed that the tour starts at time a_o. If the starting time at node o is within a time window $[a_o, b_o]$, the optimal tour can still be found by starting at time a_o, since waiting time does not increase the objective function.

To present this method, let $F(S, i, t)$ be the least cost of a path originating at node o, visiting every node of $S \subseteq N$ exactly once, ending at node $i \in S$, and ready to service node i at time t or later. Note the difference between the cost $F(S, i, t)$ of a path, and the time t when service can start at node i. Given $S \subseteq N$ and $i \in S$, this creates two-dimensional labels $(t, F(S, i, t))$, $a_i \leq t \leq b_i$, of the time and cost components since there is no relation of total order in $\mathbb{R}^2$. The function $F(S, i, t)$ can be obtained by solving the recursions:

$$F(S, j, t) = \min_{(i,j) \in A} \left\{ F(S - \{j\}, i, t') + c_{ij} \mid i \in S - \{j\}, \right.$$
$$\left. t' \leq t - t_{ij}, a_i \leq t' \leq b_i \right\}$$

$$\text{for all } S \subseteq N, j \in S \text{ and } a_j \leq t \leq b_j. \tag{3.28}$$

Equation (3.28) holds true for $a_j \leq t \leq b_j$. If instead $t < a_j$, then $F(S, j, t) = F(S, j, a_j)$, while if $t > b_j$, then $F(S, j, t) = \infty$. Equation (3.28) is initialized by

$$F(\{j\}, j, \max\{a_j, a_o + t_{o,j}\}) = \begin{cases} c_{o,j}, & \text{if } (o, j) \in A, \\ \infty, & \text{otherwise.} \end{cases} \tag{3.29}$$

The optimal solution to the TSPTW is given by:

$$\min_{(i,d) \in A} \min_{a_i \leq t \leq b_i} \left\{ F(N, i, t) + c_{i,d} \mid i \in N, t \leq b_d - t_{i,d} \right\}. \tag{3.30}$$

We present now the tests used to reduce the state-space dimension and the number of transitions. The first test deals with the situation where several paths that visit set S and end at i have different times to begin service and costs.

Test 1. For two different states, (S, i, t^1) and (S, i, t^2), if $t^1 \leq t^2$ and $F(S, i, t^1) \leq F(S, i, t^2)$, then state (S, i, t^1) dominates state (S, i, t^2).

After the application of this test only the Pareto optimal states are further considered. A number of applications of dominance between states as a function of time and cost values can be found in the literature [see for example Desrochers & Soumis, 1988a, b; Kolen, Rinnooy Kan & Trienekens, 1987]. Incidentally, this test has been successfully applied in algorithms for the shortest path problem with time windows to be presented in Section 4. If integer data are used to define the time windows and the traveling times, the number of states (S, i, t), $a_i \leq t \leq b_i$ is countable in t. Then, $F(S, i, t)$ is stepwise decreasing as a function of t over the interval $[a_i, b_i]$. Therefore, after applying Test 1, the two-dimensional labels $\big(t, F(S, i, t)\big)$ of (S, i) can be ordered in a list by increasing time and decreasing cost value. Let FIRST(S, i) be the time value of the first label of that list.

An optimal solution for the TSPTW can be thought of as a set of n nodes, ordered according to the time to begin service at each node. Partial orderings of the nodes are also dictated when precedence relationships are present. The state-space dimension and the number of transitions can be greatly reduced by examining whether partial paths satisfy these orderings. The tests developed for this purpose detect when a current partial path cannot be farther extended, thereby eliminating the corresponding states and transitions.

To present these tests, recall that LDT(i, j) is the latest departure time from i such that the time to begin service at j is feasible. Recall also that BEFORE(i) is the set of nodes k that must necessarily be visited before i. Finally, a state (S, i, t) admits a feasible extension towards j, i.e., the state $(S \cup \{j\}, j, \max\{a_j, t + t_{ij}\})$ can be created, if $t + t_{ij} \leq b_j$. We now present several post-feasibility tests. The first test was introduced by Desrosiers, Dumas & Soumis [1986] for the single vehicle pick-up and delivery problem.

Test 2. Given the states (S, i, t), for all $a_i \leq t \leq b_i$, if the smallest time value to begin service at node i is greater than the latest feasible departure time towards a certain node j, $j \notin S$, i.e., FIRST$(S, i) > \min_{j \notin S}$ LDT(i, j), then the states (S, i, t), for all $a_i \leq t \leq b_i$, do not admit feasible extensions towards any node.

This global test eliminates the states (S, i, t) for all t by only examining the earliest time to begin service at i. If there exists a node which cannot feasibly be reached from node i even when service at node i begins as early as possible, then there is no reason to consider the states (S, i, t) any further. The next test asserts that all the nodes in the set BEFORE(j) must necessarily be visited before a given node j.

Test 3. Given the states (S, i, t), for $a_i \leq t \leq b_i$, and given node j, $j \notin S$, $(i, j) \in A$, if BEFORE$(j) \not\subset S$, then no feasible extensions exist towards j.

Test 4. Given the state (S, i, t), for a fixed t, $a_i \leq t \leq b_i$, and a node j, $j \notin S$, $(i, j) \in A$, assume that node j is such that it admits a feasible extension of (S, i, t) towards j, i.e., $t \leq$ LDT(i, j). If there exists a node $k, k \notin S \cup \{j\}$ such that $t + t_{ij} >$ LDT(j, k), i.e., the additional extension from j to k does not satisfy the time window constraint at node k, then j cannot succeed i for (S, i, t). In this case the states (S, i, t'), for all $t' \geq t$, are not extended towards j.

This fourth test is implemented by choosing node k only among the immediate successors of i.

The time complexity of this algorithm is clearly exponential. Nevertheless Dumas, Desrosiers, Gélinas & Solomon [1991] report having optimally solved large scale problems with fairly wide, overlapping time windows using integer data for all parameters. When the density of the points inside the chosen geographical region increased with problem size, problems with up to 200 nodes were solved. As expected, the CPU time increases with the width of the time windows. However, for small time window widths, the behavior of the algorithm is less than exponential. When the density of the nodes in the geographical region was kept constant as the problem size was increased, the algorithm was capable of solving problems with up to 800 nodes. For these problems, the CPU time increased linearly with problem size.

Several extensions of the approach presented here can be readily seen. First, as can be observed from Test 3, the algorithm is already adapted to account for additional precedence constraints. Second, as forward dynamic programming is used to solve the TSPTW, much more general cost functions could be utilized. These include step functions, nonlinear functions, and functions capturing the additional cost for waiting before service can begin at a node. In fact, as long as the cost function is a nondecreasing function of time, the algorithm and the dominance process can be adapted. Finally, when there are several nodes at the same physical location but with different time windows, the algorithm can be accelerated by removing a priori a set of arcs between these nodes. This is the *same location criterion* used in Dumas, Desrosiers & Soumis [1991].

3.6. A two-commodity flow formulation

Finke, Claus & Gunn [1984] have introduced a two-commodity network flow formulation for the classical TSP. In their formulation, one unit of the first commodity, flow Y, is delivered at each node while one unit of the second commodity, flow Z, is picked-up at each node. Hence the total combined amount of the two flows at each node is always the same. Lucena [1986] has considered a similar formulation with various demands d_i at each node $i \in N$. Langevin, Desrochers, Desrosiers & Soumis [1993] have extended the approach to the TSPTW. Figure 7 depicts several two-commodity flow representations for traveling salesman tours, each one being described by the sequence $o \rightarrow i_1 \rightarrow i_2 \ldots i_n \rightarrow d$.

In time constrained traveling salesman problems, time represents each of the two commodities. Along any feasible traveling salesman path, flow Y_{ij}, $(i, j) \in A$ is a decreasing flow at each node, while flow Z_{ij}, $(i, j) \in A$ increases at each node. The total combined amount of flow is always the same at each node and equals the upper bound value b_d at the destination-depot node. Recall that W_i, $i \in N \cup \{d\}$ is the waiting time before the start of service of node i. The two-commodity formulation of the TSPTW is given by:

$$\text{Minimize} \quad \sum_{(i,j)\in A} c_{ij}(Y_{ij} + Z_{ij})/b_d + \sum_{i \in N \cup \{d\}} W_i \qquad (3.31)$$

subject to

$$\sum_{j \in N \cup \{d\}} (Y_{ij} + Z_{ij}) = b_d, \qquad \forall i \in N \qquad (3.32)$$

$$\sum_{i \in N \cup \{o\}} (Y_{ij} + Z_{ij}) = b_d, \qquad \forall j \in N \qquad (3.33)$$

$$\sum_{j \in N} (Y_{o,j} + Z_{o,j}) = b_d, \qquad (3.34)$$

$$\sum_{i \in N} (Y_{i,d} + Z_{i,d}) = b_d, \qquad (3.35)$$

$$T_o = \sum_{j \in N} Z_{o,j}, \qquad (3.36)$$

$$T_i = \sum_{j \in N \cup \{d\}} Z_{ij}, \qquad \forall i \in N \qquad (3.37)$$

$$T_i = \sum_{j \in N \cup \{o\}} (Z_{ji} + t_{ji}(Y_{ji} + Z_{ji})/b_d) + W_i, \qquad \forall i \in N \qquad (3.38)$$

$$T_d = \sum_{i \in N} (Z_{i,d} + t_{i,d}(Y_{i,d} + Z_{i,d})/b_d) + W_d, \qquad (3.39)$$

$$a_i \le T_i \le b_i, \qquad \forall i \in V \qquad (3.40)$$

$$Y_{ij} \ge 0, \ Z_{ij} \ge 0, \qquad \forall (i, j) \in A \qquad (3.41)$$

$$W_i \ge 0, \qquad \forall i \in N \cup \{d\} \qquad (3.42)$$

$$(Y_{ij} + Z_{ij})/b_d \ \text{binary}, \qquad \forall (i, j) \in A. \qquad (3.43)$$

The objective function (3.31) minimizes the sum of the costs over the arcs and the waiting time costs. If the waiting time costs do not appear in the objective function, then variables $W_i, i \in N \cup \{d\}$ can be considered simply as surplus variables and thus removed. At each node, the sum of the Y and Z flows in constraints (3.32)–(3.35) is always equal to b_d. Relation (3.36) describes the starting time variable at the origin-depot node, while the times to begin service at the other nodes are given in (3.37)–(3.39). For example, relation (3.38) states that at node $i, i \in N$, the total amount of flow on arcs entering node i is increased by the sum of the amount of time needed to travel on those arcs and the waiting time incurred before node i is ready for service. Since $(Y_{ij} + Z_{ij})/b_d$ are restricted to binary values by (3.43), relation (3.38) states that in any feasible integer solution, the time to begin service at node i (evaluated by flow Z) is given by the sum of the time at the previous node, say j, the travel time t_{ji} to reach node i, and the waiting time W_i incurred before the lower bound a_i. Time window constraints appear in (3.40), while nonnegativity requirements for all variables are given in relations (3.41) and (3.42). The binary variables $(Y_{ij} + Z_{ij})/b_d$, $(i, j) \in A$ defined by (3.43) are the flow variables X_{ij} of the previous formulations while time variables $T_i, i \in N \cup \{o, d\}$ have been defined in (3.36)–(3.39).

This model is very general and as can be seen, it already accounts for a flexible starting time at the origin-depot node. If on each arc, the cost is identical to the

Finke, Claus and Gunn (1984)

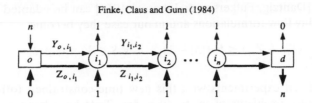

Lucena (1986)

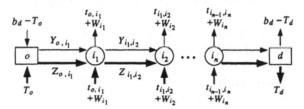

Langevin, Desrochers, Desrosiers and Soumis (1993)

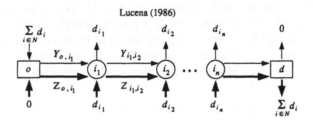

Fig. 7. Two-commodity flow illustrations.

time needed to travel on the arc, i.e., $c_{ij} = t_{ij}$, $(i, j) \in A$, and if $c_i = 1, i \in N \cup \{d\}$, then the objective function minimizes the total schedule time of the optimal tour while respecting the time windows. The formulation (3.31)–(3.43) for the TSPTW is valid only if the fixed parameter b_d is sufficiently large not to eliminate integer solutions. If there exists an integer solution, the upper bound value may be taken as $b_d = \max_{(i,d) \in A} \{b_i + t_{i,d}\}$. Relaxing the integrality conditions (3.43), the linear relaxation formulation of the TSPTW involves $O(n^2)$ variables and only $O(n)$ constraints. The lower bound obtained from the linear relaxation is at least as good as the assignment bound [given by the solution of (3.31)–(3.35) and (3.41)–(3.42)]. It can be demonstrated that it improves as the value of b_d decreases, as long as the TSPTW instance stays feasible.

It should be noted that even though the subtour elimination constraints are verified for any feasible integer solution for the TSPTW, they are not necessarily verified by the linear relaxation of the previous formulation. It is then possible to improve the lower bound obtained using the LP relaxation of the two-commodity flow formulation by detecting violated subtour elimination constraints and by adding them to the LP relaxation. The classical subtour elimination constraints for the TSP [Dantzig, Fulkerson & Johnson, 1954] can be adapted to the various two-commodity flow formulations and in our case they become

$$\sum_{i,j \in S} (Y_{ij} + Z_{ij}) \le b_d(|S| - 1), \qquad \forall S \subset V, \ |S| \ge 2. \tag{3.44}$$

Computational experiments with this new time-constrained formulation show that it can solve problems of up to 40 nodes. Tests have been conducted using a fixed starting time value, i.e., $T_o = a_o = b_o$. Further research might possibly introduce improvements and extensions of the Miller–Tucker–Zemlin subtour elimination constraints [Desrochers & Laporte, 1991]. Despite their relative weakness, it is straightforward to identify which of these constraints are violated. Furthermore they are valid for any problem of the branch-and-bound tree. Computational tests should show the viability of this approach.

3.7. Special routing structures and alternative objective functions

Related research has examined the TSP with precedence constraints, special routing structures and alternative objective functions. The TSP with precedence constraints will be addressed in Section 6. We next examine the other two areas.

The shoreline network. A routing structure which has recently received attention is the shoreline network. Following the work of Psaraftis, Solomon, Magnanti & Kim [1990] we define an ordered set of points $N = \{1, 2, \ldots, n\}$ to be located on the shoreline if the interpoint travel distances $c_{ij}, i \in N, j \in N$ satisfy the following conditions: For all $1 \le i \le k \le j \le n$,

$$c_{ii} = 0, \tag{3.45}$$

$$c_{ij} = c_{ji}, \tag{3.46}$$

$$c_{ij} \ge c_{ik}, \tag{3.47}$$

$$c_{ij} \ge c_{kj}, \tag{3.48}$$

$$c_{ij} \le c_{ik} + c_{kj}. \tag{3.49}$$

Note that a shoreline network is the triangle-inequality metric with two additional restrictions, (3.47) and (3.48). Note further that it is not necessary for a path along the shoreline to be convex, i.e., lie on the boundary of a convex region. A further restriction of the shoreline problem is the straight-line case where condition (3.49) is strengthened to $c_{ij} = c_{ik} + c_{kj}$. Shoreline problems arise in transportation environments such as the routing and scheduling of cargo ships. In addition, the straight-line case occurs in a wide range of applications including rail, rivers, and highways.

In the path version of the shoreline TSPTW, a vehicle starting at time zero from point 1 and traveling at unit speed must visit once all the points within their respective time windows. In the tour version, the vehicle must return to point 1 after visiting all the other points exactly once.

The computational complexity of the shoreline TSPTW remains open at this time. It should however be noted that the shoreline TSP is polynomially solvable, while the Euclidean TSP is $\mathcal{NP}$-complete [Papadimitriou, 1977]. In this light, Psaraftis, Solomon, Magnanti & Kim [1990] propose heuristic algorithms for its solution and derive data-dependent worst-case performance ratios for these heuristics. The authors also illustrate the performance of these algorithms on a real-world shoreline. Kim [1985] has performed a cursory analysis of other problem variants involving alternative objective functions, different types of time window constraints and multi-vehicles.

Research has also been carried out for the straight-line case. We discuss here only the path variant of the problem which has proved to be more challenging than its tour counterpart. Psaraftis, Solomon, Magnanti & Kim [1990] have examined the version of the problem where only earliest pick-up time-constraints are present. The authors develop an $O(n^2)$ optimal dynamic programming algorithm which minimizes the total schedule time, or equivalently, the maximum completion time. When service times are present, the problem was later shown to be strongly $\mathcal{NP}$-complete by Tsitsiklis [1992]. This author has also constructed an $O(n^2)$ dynamic programming algorithm for the version of the problem involving solely deadlines. Furthermore, he has shown that when general time windows are present the problem is strongly $\mathcal{NP}$-complete.

Alternative objective functions. A related problem is the traveling repairman problem (TRP) with time windows (TRPTW). The objective is to minimize the sum of completion times, i.e., $\sum T_i$, or equivalently, the sum of flowtimes, i.e., $\sum (T_i - a_i)$. Afrati, Cosmadakis, Papadimitriou, Papageorgiou and Papakonstantinou [1986] have shown that the TRP can be solved in $O(n^2)$ time. They have further shown that when deadlines are imposed, the problem becomes $\mathcal{NP}$-complete but can be solved by a pseudo-polynomial algorithm. Recently, Tsitsiklis [1992] proved that the TRPTW is strongly $\mathcal{NP}$-complete. The author also examines the TSPTW and the TRPTW when the number of nodes is bounded by a constant. We refer the reader to the original paper for details on the different complexity results and open problems. If the order in which the nodes are visited is fixed in advance, the TRPTW reduces to optimizing the schedule for a fixed route. We address this problem in Section 6.3.

Research has also been recently undertaken on the objective of minimizing the number of late deliveries. This is an important objective encountered in several practical applications. One example is encountered in technologically advanced manufacturing systems where an automated guided vehicle must deliver parts to a set of machines such that the number of idle machines is minimized. Another example involves express pizza delivery systems. Gendreau, Laporte & Solomon [1992] develop a specialized enumerative algorithm capable of optimally solving problems involving up to 100 customers.

Finally, the time dependent TSP, i.e., the TSP where the travel time between every pair of nodes depends not only on the distance between the nodes but also on the time of day, has been studied by Malandraki & Daskin [1989]. The authors propose a cutting plane heuristic.

4. Constrained shortest path problems

The shortest path problem with time windows consists of finding the least cost route between two specified nodes in a graph whose nodes can only be visited within a specified time interval. This problem has found its first application as a subproblem in the construction of school-bus routes in the early '80s. Since then it has appeared as a subproblem in many time constrained routing and scheduling problems to be described in the following sections. The shortest path problem with time windows is a relaxation of the traveling salesman problem with Time Windows obtained by relaxing the requirement that each node must be visited.

4.1. The shortest path problem with time windows

Notation. Let $G = (V, A)$ be a graph where A is the set of arcs and V is the set of nodes $N \cup \{o, d\}$, where N consists of nodes that can be visited from an origin o to a destination d. A time window $[a_i, b_i]$, $i \in V$ is associated with each node. A path in the graph G is defined by a sequence of nodes $i_0, i_1, \ldots, i_H$, such that each arc (i_{k-1}, i_k) belongs to A. All paths start at time a_o from node $i_0 = o$ and finish at node $i_H = d$ no later than b_d. A path is elementary if it contains no cycles. Each arc $(i, j) \in A$ has a positive or negative cost c_{ij} and a positive duration t_{ij}. We assume next that the service time at node i is included in the time value t_{ij}, for all $i \in N$. An arc (i, j) is defined in the set A only if it is feasible, i.e., if it respects the condition: $a_i + t_{ij} \leq b_j$. This problem was first introduced in Desrosiers, Pelletier & Soumis [1983] and next used in Desrosiers, Soumis & Desrochers [1984] as a subproblem for the multiple traveling salesman problem with time windows.

Formulation. The mathematical programming formulation involves two types of variables: flow variables X_{ij}, $(i, j) \in A$, and time variables T_i, $i \in V$. Define variable X_{ij} to be the flow on arc $(i, j) \in A$, and variable T_i to be the start of service at node i, $i \in V$. Waiting time is allowed before the start of service at each node. Using this notation, the elementary (or acyclic) shortest path problem with time windows (ESPPTW) may be formulated as follows:

$$\text{Minimize} \sum_{(i,j)\in A} c_{ij} X_{ij} \tag{4.1}$$

subject to

$$\sum_{j\in V} X_{ij} - \sum_{j\in V} X_{ji} = \begin{cases} +1, & i = o \\ 0, & \forall i \in N \\ -1, & i = d \end{cases} \tag{4.2}$$

$$X_{ij} \geq 0, \qquad\qquad \forall (i, j) \in A \tag{4.3}$$

$$X_{ij}(T_i + t_{ij} - T_j) \leq 0, \qquad \forall (i, j) \in A \tag{4.4}$$

$$a_i \leq T_i \leq b_i, \qquad\qquad \forall i \in V. \tag{4.5}$$

The objective function (4.1) seeks to minimize the total travel cost. Constraints (4.2)–(4.3) define flow conditions on the graph G, time windows appear in (4.5), and compatibility requirements between flow and time variables are given in (4.4).

Nonlinear integrality property. This nonlinear formulation has an appealing characteristic: as in the case of the TSPTW, it can be shown that if the problem is feasible, then there exists an optimal integer solution. To prove this, note that constraints (4.4) indicate that if $X_{ij} > 0$, then $T_i + t_{ij} \leq T_j$. If the flow values are fractional, the optimal solution of value Z^* is composed of several paths of cost c_p, each of which has a positive flow θ_p, i.e., $Z^* = \sum_p c_p \theta_p$, where $\sum_p \theta_p = 1$. Assign a unit flow to the arcs of the minimum cost path c_{min}: then, this path satisfies the time constraints and also constitutes an optimal integer solution since:

$$Z^* \leq c_{min} = \sum_p c_{min}\theta_p \leq \sum_p c_p\theta_p = Z^*.$$

Literature. The shortest path with additional linear constraints is a related problem which has been discussed by a number of authors, e.g., Joksch [1966], Minoux [1975], Hansen [1980], Aneja, Aggarwal & Nair [1983], Jaffe [1984] and Martins [1984]. The proposed algorithms use Lagrangean relaxation, branch-and-bound and dynamic programming. These algorithms are designed, however, for problems with only a few additional linear constraints, while the ESPPTW involves many such constraints.

Dynamic programming algorithms. The ESPPTW can be solved by dynamic programming. To introduce this approach, define $F(S, i, t)$ as the minimum cost of the path going from node o to node i, $i \in N \cup \{d\}$, visiting all nodes in set $S \subseteq N \cup \{d\}$ only once, and servicing node i at time t or later. The cost $F(S, i, t)$ can be computed by solving the following recurrence equations:

$$F(\phi, o, a_o) = 0 \tag{4.6}$$

$$F(S, j, t) = \min_{(i,j) \in A} \left\{ F(S - \{j\}, i, t') + c_{ij} \mid i \in S - \{j\}, \right.$$
$$\left. t' \leq t - t_{ij}, a_i \leq t' \leq b_i \right\},$$

$$\text{for all } S \subseteq N \cup \{d\}, \quad j \in S \quad \text{and} \quad a_j \leq t \leq b_j. \tag{4.7}$$

The optimal solution is given by

$$\min_{S \subseteq N \cup \{d\}} \min_{a_d \leq t \leq b_d} F(S, d, t). \tag{4.8}$$

Note that equation (4.7) is valid only if $a_j \leq t \leq b_j$. If however $t < a_j$, then $F(S, j, t) = F(S, j, a_j)$, and if $t > b_j$, then $F(S, j, t) = \infty$. The ESPPTW is $\mathcal{NP}$-hard in the strong sense as can be shown by reduction from the problem denoted *sequencing within intervals* [Dror, 1994; see also Garey & Johnson, 1979]. Thus, the proposed dynamic programming algorithm has an exponential complexity and no pseudo-polynomial algorithm is known for this problem.

An easier problem to solve is obtained from the ESPPTW by relaxing the elementary path requirement. This relaxed problem called the shortest path problem with time windows (SPPTW) admits paths with cycles when some c_{ij} are negative. However, the time window constraints and the positive arc durations t_{ij} guarantee the finiteness all of paths. The SPPTW can also be solved by dynamic programming. To present this methodology, define $F(i, t)$ as the minimum cost of the path going from node o to node i, $i \in N \cup \{d\}$, and servicing node i at the latest at time t. The shortest path from o to d can be computed by solving the recursions:

$$F(o, a_o) = 0 \tag{4.9}$$

$$F(j, t) = \min_{(i,j) \in A} \left\{ F(i, t') + c_{ij} \mid i \in N \cup \{d\} - \{j\}, \right.$$
$$\left. t' \leq t - t_{ij}, a_i \leq t' \leq b_i \right\},$$

$$\text{for all } j \in N \cup \{d\} \text{ and } a_j \leq t \leq b_j. \tag{4.10}$$

The optimal solution is given by

$$\min_{a_d \leq t \leq b_d} F(d, t). \tag{4.11}$$

Again, equation (4.10) is valid only if $a_j \leq t \leq b_j$. If however $t < a_j$, then $F(j, t) = F(j, a_j)$ and if $t > b_j$, then $F(j, t) = \infty$. The SPPTW includes the knapsack problem as a special case and therefore is also $\mathcal{NP}$-hard. However, there are pseudo-polynomial algorithms to solve it. These algorithms will be discussed in the next subsection.

4.2. Dynamic programming algorithms for the SPPTW

Solving the SPPTW by dynamic programming is equivalent to solving an unconstrained shortest path problem on an expanded acyclic graph. Given that the time window parameters are integer values, the maximum number of nodes in the expanded acyclic graph is equal to $D = \sum_{i \in V} (b_i - a_i + 1)$. This requires large amounts of computer time and memory when the set of nodes is large and especially when the time windows are wide. Nevertheless, we present here two efficient algorithms to solve the SPPTW.

First, introduce the concept of a *label* which permits to solve the problem on the graph $G = (V, A)$, rather than on the acyclic graph with D nodes. With each path $\mathcal{P}_i$ from the origin o to the node i satisfying the time windows is associated a two-dimensional (time, cost) label corresponding to the start of service at node i and the cost of the path $\mathcal{P}_i$, respectively. At node i, these labels will be denoted by

$$(T_i^k, C_i^k), \qquad i \in V, k \geq 1$$

to indicate the characteristics of the kth path from o to i. The indices k and i may be dropped when the context is unambiguous. These labels are calculated iteratively along the path $\mathcal{P}_i = (i_0, i_1, i_2, \ldots, i_H)$ as:

$$(T_{i_0}, C_{i_0}) = (a_o, 0)$$

$$(T_{i_h}, C_{i_h}) = \left(\max\{a_{i_h}, T_{i_{h-1}} + t_{i_{h-1}, i_h}\}, C_{i_{h-1}} + c_{i_{h-1}, i_h}\right), \quad h = 1, \ldots, H.$$

where $i_0 = o$ and $i_H = i$

Definition. Let (T_i^1, C_i^1) and (T_i^2, C_i^2) be two different labels for two paths from o to i. The first label dominates the second, i.e., $(T_i^1, C_i^1) \prec (T_i^2, C_i^2)$ if and only if $(T_i^2, C_i^2) - (T_i^1, C_i^1) \geq (0, 0)$.

Definition. A label (T_i, C_i) at a given node i is said to be efficient if no other labels at i dominate it. A path from o to i is said to be efficient if the corresponding label is efficient.

This dominance relation is not a total ordering and therefore does not allow all paths to be ordered. However, this relation defines a partial order on the labels and therefore it does allow us to conclude that the efficient path $\mathcal{P}_i$ is the shortest path from node o to node i such that service at node i starts at the latest at time T_i. This implies the possibility of several efficient paths for each node. The importance of the dominance relation stems from allowing the cost of the feasible path to be defined as a decreasing step function of the time at which the service begins.

Define now Q_i, $i \in V$, to be the set of labels of node i and let $\mathrm{EFF}(Q_i)$ denote the set of efficient labels among the set of labels Q_i of node i. The set of efficient labels at each node can be calculated by dynamic programming. The shortest path from o to d satisfying the time window constraints is obtained directly from the set $\mathrm{EFF}(Q_d)$: it is represented by the least cost label.

Since arc costs c_{ij} might be negative, we first present a *label correcting* algorithm which is an adaptation of the Ford–Bellman–Moore algorithm [1956, 1958 and 1959, respectively] for the classical shortest path problem. We next present a *label setting* algorithm, which is a generalization of the Dijkstra's algorithm [1959] on graphs with nonnegative arc costs.

A label correcting algorithm [Desrosiers, Pelletier & Soumis, 1983]. Let $\Gamma(i) = \{j | (i, j) \in A\}$ be the set of successors of node i, and $Q_i = \bigcup_k \{(T_i^k, C_i^k)\}$ be the set of labels of node $i \in V$. A basic operation in shortest path algorithms is the *treatment of a label* (T_i^k, C_i^k). It consists of creating new labels at nodes $j \in \Gamma(i)$ by adding arcs (i, j) to the path from o to i associated with label (T_i^k, C_i^k). The new label for a given $j \in \Gamma(i)$ is:

$$f_{ij}(T_i^k, C_i^k) = \begin{cases} (\max\{a_j, T_i^k + t_{ij}\}, C_i^k + c_{ij}), & \text{if } T_i^k + t_{ij} \leq b_j \\ \phi, & \text{otherwise.} \end{cases} \quad (4.12)$$

The *treatment of node i* is the treatment of all labels in Q_i. The set of new labels at each node $j \in \Gamma(i)$ is $\bigcup_k f_{ij}(T_i^k, C_i^k)$. Since the definition of f_{ij} allows waiting at node j, the new labels will not all be efficient, as the time values T_j^k may pile up at the value a_j. In fact $T_j^k = a_j$, for all k such that $T_i^k + t_{ij} \leq a_j$. Furthermore, some new labels may dominate or may be dominated by some labels already in Q_j. Hence, the new set of efficient labels at node j is given by $\text{EFF}(\bigcup_k f_{ij}(T_i^k, C_i^k) \cup Q_j)$. The algorithm to be presented next uses a list $\mathcal{L}$ of nodes which consists of the nodes which have to be treated after improvement or modification of their labels. This algorithm can be described as follows:

A label correcting algorithm for SPPTW

Step 0. Initialization
$$Q_o = \{(T_o^1 = a_o, C_o^1 = 0)\};$$
$$Q_i = \{(T_i^1 = a_i, C_i^1 = \infty)\} \quad \forall i \in V \setminus \{o\}; \quad \mathcal{L} = \{o\}.$$

Step 1. Treatment of node i
 Choose a node $i \in \mathcal{L}$;
 For all $j \in \Gamma(i)$ do:
 $$Q_j' = \text{EFF}(\bigcup_k f_{ij}(T_i^k, C_i^k) \cup Q_j),$$
 If $Q_j' \neq Q_j$ then $Q_j = Q_j'$ and $\mathcal{L} := \mathcal{L} \cup \{j\}$.

Step 2. Reduction of $\mathcal{L}$
 $\mathcal{L} := \mathcal{L} \setminus \{i\}$;
 If $\mathcal{L} = \phi$ then STOP, otherwise return to *Step 1*.

The labels of set Q_i of a node i are placed in a list in increasing time order. Next, let Δ_i, a subset of Q_i, be the set which only contains the labels which have been modified since i appeared in $\mathcal{L}$. Labels in Δ_i are the only labels to generate modifications at node j when coming from node i. The set Q_j is then obtained by a sequential update of set Q_j based only on Δ_i.

The complexity of this algorithm depends on the rule for selecting the node to be treated next. FIFO or LIFO strategies can produce an exponential worst-case complexity. A polynomial complexity of order $O(D^3)$ can however be obtained by utilizing the expanded acyclic graph. Desrochers & Soumis [1988a] have implemented a $\mathcal{L} - 2$ queue method [Pape, 1980; Pallottino, 1984]. In short, this method uses a double queue to represent the list $\mathcal{L}$. The first time a node is inserted in $\mathcal{L}$, it is added at the tail of the second queue. Subsequent insertions of this node occur at the tail of the first queue. The nodes to be treated are withdrawn from the head of the first queue. If the first queue is empty, the second queue becomes the first, and the second queue becomes empty.

In an acyclic graph the nodes can be numbered in such a way that if $(i, j) \in A$, then $i < j$. This numbering of the nodes ensures that each node is inserted in list $\mathcal{L}$ once and only once. In a graph with time windows where cycles exist, the nodes can be numbered in chronological order, for example, by start time. If the time windows are sufficiently narrow, this numbering ensures that the relation $(i, j) \in A \Rightarrow i < j$ is satisfied by the majority of the arcs. Therefore, the treatment of the nodes from the first queue in this order significantly reduces reinsertions and computation time (see the computational experiments in Desrosiers, Pelletier & Soumis [1983]).

The label correcting algorithm solves problems with 500 nodes, 50,000 arcs and 100 discrete time units per window in a matter of a few seconds. This corresponds to solving a classical shortest path problem in an expanded network with 50,500 nodes and 5,050,000 arcs.

A label setting algorithm [Desrochers & Soumis, 1988a]. This method is a generalization of Dijkstra's algorithm to the SPPTW. This permanent labeling algorithm can be used even if the SPPTW contains negative costs and negative cycles. The procedure is based on a new treatment order of labels.

In classical shortest path problems, the notions of *node* and *label* are closely linked: one label is associated to each node. The order of treatment is thus defined simultaneously for nodes and labels. In the case of time constrained shortest path problems with an objective function which minimizes the travel cost, a set of labels is associated with each node, thus the two notions are distinct. The label correcting algorithm previously presented did not fully exploit this distinction. The treatment order was defined on the *nodes* and all the labels associated with a node were treated. The computational aspect was improved by the use of subsets Δ_i, $i \in V$, and by the chronological numbering of the nodes by start time.

In contrast, the permanent labeling algorithm treats sequentially the *labels* in increasing order of time. The positive duration of the arcs insures that once a label has been treated it is impossible to improve it any further. The algorithm does not necessitate the use of list $\mathcal{L}$. Instead, the method uses sets P_i of permanent labels at node i. Each set P_i contains the labels of node i which has been previously treated. The algorithm consists of the following steps:

A label setting algorithm for SPPTW

Step 0. Initialization
$$Q_o = \{(T_o^1 = a_o, C_o^1 = 0)\}; \quad Q_i = \emptyset, \quad \forall i \in N \cup \{d\};$$
$$P_i = \emptyset, \quad \forall i \in V.$$
Step 1. Selection of the next label to be treated
    Choose a label (T_i^k, C_i^k) with minimal T_i^k from $\bigcup_{i \in V}(Q_i \setminus P_i)$;
    If $\bigcup_{i \in V}(Q_i \setminus P_i) = \emptyset$ then STOP.
Step 2. Treatment of label (T_i^k, C_i^k)
    For all $j \in \Gamma(i)$,
$$Q_j := \text{EFF}\big(f_{ij}(T_i^k, C_i^k) \cup Q_j\big);$$
$$P_i := P_i \cup \big\{(T_i^k, C_i^k)\big\};$$
    Return to *Step 1*.

The complexity of Step 1 can be reduced by using the concept of a *bucket* introduced by Denardo & Fox [1979]. For classical shortest path problems, a bucket is a set of nodes whose label costs lie within a specified interval. For time constrained shortest path, a bucket contains those labels whose time lies within a specified interval. The kth bucket consists of the set of labels with a time label in the interval $[kw, (k + 1)w)$, where the width of the bucket, w, is fixed at $w = \min_{(i,j) \in A} t_{ij}$. The search for the earliest untreated time label is thus replaced by the search for an element of the earliest untreated bucket. The optimality of the algorithm with buckets can be easily proved. Its worst-case complexity is of order of $O(D^2)$. The algorithm solves problems with up to 2,500 nodes and 250,000 arcs in only a few seconds.

4.3. The shortest path problem with resource constraints

The shortest path problem with resource constraints (SPPRC) was introduced by Desrochers [1986] as a multi-dimensional generalization of the SPPTW. The SPPRC appears as a subproblem in the Dantzig–Wolfe decomposition/column generation and in the Lagrangean relaxation approaches used to solve vehicle routing problems (Section 5), vehicle fleet planning problems, bus driver scheduling problems, airline crew scheduling problems and many other time constrained routing and scheduling problems (Section 7).

Notation. The problem is defined on a graph $G = (V, A)$ where A is the set of arcs and V is the set of nodes $N \cup \{o, d\}$, where N consists of the nodes that can be visited from an origin o to a destination d. A path in the graph G is defined as a sequence of nodes $i_0, i_1, \ldots, i_H$, such that each arc (i_{h-1}, i_h) belongs to A. All paths start at the origin ($i_0 = o$) and end at the destination ($i_H = d$). A cost c_{ij} is associated with each arc $(i, j) \in A$. The cost of a path is defined as the sum of the costs of the arcs of the path. The paths need not be elementary, i.e., the same node is allowed to be visited more than once (i_h can be equal to $i_{h'}$, $h \neq h'$).

Let R denote the set of resources used to model a complex situation. The travel time t_{ij} on each arc (i, j) is replaced by the consumption of t_{ij}^r units of resource r, for all $r \in R$. The time interval constraint $[a_i, b_i]$ on node i is replaced by $|R|$ interval constraints of the form $[a_i^r, b_i^r]$, $r \in R$, restricting the amounts of resources used by the path to reach node i. The resource consumptions accumulate along a path. Define then T_i^r as the amount of the rth resource used to reach node i using a path from o to i. A path reaching node i and using less than a_i^r units of the rth resource becomes feasible by wasting some units of the rth resource. However, a path reaching node i and using more than b_i^r units of the rth resource is infeasible.

An arc (i, j) is infeasible if it is not possible to visit node j after node i while respecting the feasibility windows for both nodes i and j. All infeasible arcs are then excluded from the set A. Conversely, all arcs $(i, j) \in A$ respect the following conditions:

$$a_i^r + t_{ij}^r \leq b_j^r, \quad \forall r \in R.$$

The SPPRC is then defined as follows: find a minimal cost feasible path from the origin o to the destination d, i.e., a sequence of nodes $(i_0, i_1, \ldots, i_H)$ and an associated resource consumption, $T_{i_h}^r$, $0 \leq h \leq H$, for each resource such that:

(1) $(i_{h-1}, i_h) \in A$, for $1 \leq h \leq H$, with $i_0 = o$ and $i_H = d$;

(2) $T_{i_{h-1}}^r + t_{i_{h-1}, i_h}^r \leq T_{i_h}^r$, for $1 \leq h \leq H$ and $r \in R$;

(3) the resource constraints are respected at all the visited nodes: $a_{i_h}^r \leq T_{i_h}^r \leq b_{i_h}^r$, for $0 \leq h \leq H$ and $r \in R$;

(4)] the total cost ($\sum_{h=1}^{H} c_{i_{h-1}, i_h}$) of this path is the minimum among all feasible paths.

Formulation. The mathematical formulation (4.1)–(4.5), introduced for the elementary shortest path problem with time windows can easily be extended to the SPPRC. Let X_{ij}, $(i, j) \in A$, be the flow variables, and T_i^r, $i \in V$, $r \in R$, be the resource variables. Then, the formulation is:

$$\text{Minimize} \sum_{(i,j) \in A} c_{ij} X_{ij} \tag{4.13}$$

subject to

$$\sum_{j \in V} X_{ij} - \sum_{j \in V} X_{ji} = \begin{cases} +1, & i = o \\ 0, & \forall i \in N \\ -1, & i = d \end{cases} \tag{4.14}$$

$$X_{ij} \geq 0, \qquad \forall (i, j) \in A \tag{4.15}$$

$$X_{ij}(T_i^r + t_{ij}^r - T_j^r) \leq 0, \qquad \forall (i, j) \in A, \forall r \in R \tag{4.16}$$

$$a_i^r \leq T_i^r \leq b_i^r, \qquad \forall i \in V, \forall r \in R. \tag{4.17}$$

The above formulation contains only elementary shortest paths in the solution space. If the graph G is acyclic, like in many practical fleet assignment and crew scheduling applications, the dynamic programming described next is optimal. Otherwise, the solution space may include paths with finite cycles.

4.4. A dynamic programming algorithm for the SPPRC

Solving the SPPRC by dynamic programming may be done by solving an unconstrained shortest path on an acyclic graph. Given that resource parameters are integer values, the maximum number of nodes of the resulting acyclic graph is equal to $\sum_{i \in V} \prod_{r \in R} (b_i^r - a_i^r + 1)$. This may require a large amount of computer time and memory. Note that an interval reduction process can be performed a priori, similar to the one described for the TSPTW in Section 3.1. Next, we use again the concept of labels which permits to solve the problem on the original graph $G = (V, A)$.

With each path $\mathcal{P}_i$ from the origin o to node i is associated a state denoted T_i^R, corresponding to the quantity of each resource $(T_i^1, T_i^2, \ldots, T_i^{|R|})$ used by the path, and a cost value C_i, which is a function of the state. For a given node i, the set of feasible states is $\{(T_i^1, T_i^2, \ldots, T_i^{|R|}) | a_i^r \leq T_i^r \leq b_i^r, \text{ for all } r \in R\}$. To use a simpler notation, for each path we define a label (T_i^R, C_i). The label represents both the state value and the cost of a path $\mathcal{P}_i$. The definitions of dominance, efficient label and efficient path can be generalized to this multi-dimension problem.

Definition. Let (T_i^{1R}, C_i^1) and (T_i^{2R}, C_i^2) be two distinct labels. Then, the first label dominates the second, i.e., $(T_i^{1R}, C_i^1) \prec (T_i^{2R}, C_i^2)$ if and only if $C_i^1 \leq C_i^2$, $T_i^{1r} \leq T_i^{2r}$ for all $r \in R$.

Definition. A label (T_i^R, C_i) at a given node i is said to be efficient if no other label at i dominates it. A path $\mathcal{P}_i$ from o to i is said to be efficient if its label is efficient.

A pulling dynamic programming algorithm. Similarly to the SPPTW, the SPPRC can be solved by a dynamic programming algorithm which preserves only nondominated labels. For the SPPTW, the basic step of the label correcting algorithm and the permanent labeling algorithm is the *reaching* process. This consists of starting with a label associated with a path $\mathcal{P}_i$, and for all $j \in \Gamma(i)$, determining the labels corresponding to feasible paths $\mathcal{P}_j$ which extend the path $\mathcal{P}_i$. For each node i, these two algorithms create new labels for nodes $j \in \Gamma(i)$; in other words, labels can be created for node j every time one of its predecessors is treated. The labels at each node j must therefore be updated several times. This is costly, because the process is more complex than a sequential update, especially if there are a significant number of resource constraints.

The efficiency of the dynamic programming algorithm can be increased by using a *pulling* process. This process was introduced by Desrochers & Soumis [1988b] in a reoptimization algorithm for the SPPTW and in Desrochers [1986] for the SPPRC. For a given node j, the *pulling* process consists of creating labels at this node for all feasible labels (T_i^R, C_i) associated with paths $\mathcal{P}_i$, obtained by extending all these paths for which the addition of arc $(i, j) \in A$ allows arrival at node j in a feasible state T_j^R. A new label $(T_j^R, C_i + c_{ij})$, with T_j^R given by

$$\left(\max\{a_j^r, T_i^r + t_{ij}^r\}, \forall r \in R\right) \tag{4.18}$$

is created at node j if $T_i^r + t_{ij}^r \leq b_i^r$ for all $r \in R$. As the new labels are created at a single node, this algorithm requires the updating of labels at only one node, in contrast with the reaching approach requiring updating at several nodes.

A label setting algorithm can be developed if, for each arc (i, j), at least one resource consumption or cost takes positive values, i.e., $(t_{ij}^1, t_{ij}^2, \ldots, t_{ij}^{|R|}, c_{ij}) \not\leq 0$, for all $(i, j) \in A$. To simplify the notation, we will use a slightly stronger hypothesis:

$$\text{for all } (i, j) \in A, \quad t_{ij}^1 \geq w > 0. \tag{4.19}$$

A simple algorithm is then obtained. To present it, the following notation is needed. For each node $i \in V$, let Q_i be the set of labels and P_i the subset of permanent labels. The set P_i is characterized by using a variable bound π_i

on the consumption of the first resource: P_i is the subset of labels such that $a_i^1 \leq T_i^1 \leq \pi_i \leq b_i^1$. The algorithm consists of the following steps:

A label setting algorithm for SPPRC

Step 0. Initialization
$$Q_o = P_o = \{(a_o^1, \ldots, a_o^{|R|}, 0)\}; \qquad \pi_o = b_o^1;$$
$$Q_i = P_i = \phi, \ \pi_i = a_i^1, \quad \forall i \in N \cup \{d\}.$$

Step 1. Selection of node j
If $\pi_i = b_i^1 \quad \forall i \in N \cup \{d\}$, then STOP.
Otherwise select $j \in \arg\min_{i \in N \cup \{d\}} \{\pi_i | \pi_i < b_i^1\}$;

Step 2. Pulling at node j
$Q_j(\pi_j)$ is the set of labels created such that $\pi_j \leq T_j^1 \leq \min\{b_j, \pi_j + w\}$.
$Q_j := Q_j \cup Q_j(\pi_j)$;
$P_j := \text{EFF}(Q_j \cup P_j)$;
$\pi_j = \min\{b_j, \pi_j + w\}$;
Return to *Step 1*.

The new labels created in set P_j at Step 2 are permanent labels. The multi-dimensional procedure EFF(·) used to select the subset of efficient labels is complicated and it needs specialized data structures to obtain a pseudo-polynomial algorithm of low complexity. An efficient algorithm using the pulling process was designed by Desrochers [1986]. The author describes a primal-dual approach which permits to identify permanent and nonpermanent labels using the bucket concept. Problems with 1,000 nodes, 50,000 arcs, and 5 resource constraints at each node were solved in less than 1 minute. Such problems necessitate the creation of about 6,000,000 states.

4.5. Extensions

The first special case of the SPPTW is the problem where the cost of a path is the path duration, including the waiting time. In this case, for any label (T_i, C_i) at node i, $C_i = T_i - a_o$. The two dimensional labels can be replaced by one dimensional labels. Since a relationship of total order exists between the one dimensional labels, it is sufficient to keep only one nondominated label at each node. Hence, the classical shortest path algorithms can be adapted to the context of time windows: if $T_i > b_i$, the label is eliminated; if $T_i < a_i$, the label is set to the value a_i. If the graph is acyclic or if arc durations are nonnegative, an $O(n^2)$ label setting algorithm can be used. Otherwise, an $O(n^3)$ label correcting algorithm must be utilized.

A second special case of the SPPTW is the longest time path problem (the critical path problem) with time windows. This is obtained by minimizing the negative value of the duration. In this case, for any label (T_i, C_i) at node i, $C_i = -(T_i - a_o)$. Hence, there is no dominance relation between the labels of a node. Note however that the pseudo-polynomial algorithms can still be used. If the graph is acyclic, the implementation is facilitated by treating nodes in topological order. In the special case where time windows impose a precedence relationship between the nodes, an $O(n^2)$ can be used. This fact was exploited by Baker [1983] in his approach for the TSPTW described in Section 3.

The algorithms presented can also be adapted to solve the 2-cycle free SPPTW, a relaxation of ESPPTW, yet stronger than SPPTW. A 2-cycle free path is a path without cycles of the form $(j \rightarrow i \rightarrow j)$. For the classical shortest path problem, the dynamic programming algorithm can be adapted for 2-cycle elimination by keeping the best and the second best labels at each node [Houck, Picard, Queyranne & Vemuganti, 1980; and Christofides, Mingozzi & Toth, 1981a]. The modification appears in the treatment of a node i for which $j \in \Gamma(i)$, and the best label is obtained by using the arc (j, i). The creation of a path with the cycle $(j \rightarrow i \rightarrow j)$ is avoided by adding the arc (i, j) to the second best label. It creates the best 2-cycle free path arriving at node j via node i. Algorithms for 2-cycle free SPPTW are easily obtained by doubling the number of labels in the SPPTW algorithms [Kolen, Rinnooy Kan & Trienekens, 1987, and Desrochers, Desrosiers & Solomon, 1992].

Some generalizations of the SPPTW involving linear costs on the waiting time at each node can also be solved. Consider first a linear penalty function $c_j(T)$, with a nonnegative slope within the time window $a_j \leq T \leq b_j$, for each node j. In this case, the least cost value of a given partial path at node j is the one with the earliest arrival time. Provided that the global cost function is updated correctly, i.e., c_{ij} is replaced by $c_{ij} + c_j(T_j)$ in the computations, no other modifications to the algorithm are required. If all the penalty functions are defined using nonpositive slopes, the shortest path problem should be solved on the mirror network, i.e., the direction of all arcs should be reversed and the time windows should be updated according to a certain constant parameter. Such an application is presented in Dumas, Salomon, Solomon and Van Wassenhove [1993] in the manufacturing context. When positive and negative slope penalty functions appear simultaneously, the algorithmic difficulties increase so much that this extension will not be discussed here [Ioachim, Gélinas, Desrosiers & Soumis, 1994].

Additional extensions of the SPPRC concern the cost function and the resource functions. Since the shortest path problem with resource constraints is solved using forward dynamic programming, nonlinear cost functions can easily be used. Desrochers, Gilbert, Sauvé & Soumis [1992] provide examples of piecewise linear and step functions utilized for the urban crew scheduling problem (see also Sections 6 and 7). The essential condition on a valid cost function is that it must share some nondecreasing properties (see Desaulniers, Desrosiers, Ioachim, Solomon & Soumis [1994] for a proof). Finally, as the cost can be considered as a resource, nonlinear nondecreasing resource functions can also be used to model complex situations.

5. The vehicle routing problem with time windows

The vehicle routing problem with time windows (VRPTW) consists of designing a set of minimum cost routes, originating and terminating at a central depot, for a fleet of vehicles which services a set of customers with known demands. The customers must be assigned exactly once to vehicles such that the vehicle capacities are not exceeded. The service at a customer must begin within the time window defined by the earliest time and the latest time when the customer permits the start of service.

Time windows can be hard or soft. In the hard time window case, if a vehicle arrives too early at a customer, it is permitted to wait until the customer is ready to begin service. However, a vehicle is not permitted to arrive at a node after the latest time to begin service. In contrast, in the soft time window case, the time windows can be violated at a cost. We are going to focus on the hard time window variant where most of the research effort has been directed.

The VRPTW is a generalization of the vehicle routing problem (VRP) involving the added complexity of time windows. The VRP and a variety of its practical applications have been the subject of a wide body of research [see Magnanti, 1981; Bodin, Golden, Assad & Ball, 1983; Laporte & Nobert, 1987; Laporte, 1992, for survey papers]. A similar flurry of activity is currently being experienced in the VRPTW area. Figures 8 and 9 illustrate the temporal and spatial interplay of VRPTW routes which in many cases do not exhibit the geographical cohesiveness of VRP routes.

The special case in which the vehicle capacities are not binding is called the multiple traveling salesman problem with time windows (*m*-TSPTW). It is also a direct extension of the single depot vehicle scheduling problem, a fixed schedule problem introduced in Section 2.1. The *m*-TSPTW involves the determination of routes that start at a single depot and cover a set of trips, each of which starts

Fig. 8. Léa's perception of the VRPTW.

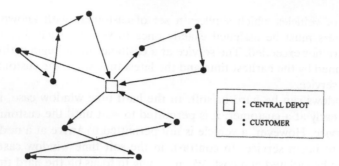

Fig. 9. Temporal and spatial interplay of VRPTW routes.

within a time window. Note that trips are considered as customers. In addition, there are no capacity constraints, since each trip satisfies these by definition, and vehicles traveling between trips are empty. An additional complexity encountered in some VRPTW consists of precedence relationships among certain customers. An example is the backhauling problem with time windows which involves both pick-ups and deliveries.

Time windows arise naturally in problems faced by business organizations which work on flexible time schedules. Specific examples of problems with hard time windows include bank deliveries, postal deliveries, industrial refuse collection and school-bus routing and scheduling. Among the soft time window problems, dial-a-ride problems constitute an important example.

Since the VRP is $\mathcal{NP}$-hard, by restriction, the VRPTW is also $\mathcal{NP}$-hard. In fact, even finding a feasible solution to the VRPTW when the number of vehicles is fixed a priori, is itself a $\mathcal{NP}$-complete problem. This is a corollary of the result derived by Savelsbergh [1985] for the case of one uncapacitated vehicle, i.e., the TSPTW, mentioned in Section 3. Therefore, many algorithms have been developed under the assumption that the number of vehicles used is free. This implies the simultaneous determination of the vehicle fleet size and the optimal set of routes and schedules for this fleet. While most methods have assumed a single depot and a homogeneous fleet, they admit easy extensions. Such extensions with specific initial and final conditions for each vehicle of the fleet are explicitly given here in the formulation provided in Section 5.1. This generalized model and the solution approach are therefore well suited to be used in a planification phase as well as in an operational mode.

Literature. Given the inherent computational challenge of this problem class, the early work on the VRPTW has consisted of case studies [Pullen & Webb, 1967; Knight & Hofer, 1968; Madsen, 1976]. One line of later research has been directed at the development and analysis of heuristics capable of solving realistic size problems. A second line has been the design of effective optimal approaches for easier problem variants such as the m-TSPTW [Desrosiers, Soumis & Desrochers, 1984]. Recently, substantial progress has been made in removing computational barriers for optimal VRPTW algorithms [Kolen, Rinnooy Kan & Trienekens, 1987; Desrochers Desrosiers & Solomon, 1992; Halse, 1992]. We present these approaches later in this section.

Heuristic approaches. A number of route construction and route improvement procedures have been proposed. Route construction algorithms build a feasible solution by inserting at every iteration one unrouted customer into a current partial route. Sequential methods construct one route at a time, while parallel methods build several routes simultaneously. Insertion criteria based on maximum savings, minimum additional distance and time, and nearest neighbor concepts have been proposed. Iterative improvement methods start from a feasible solution and seek to improve it through a sequence of local modifications. They terminate when no more improvements are possible, generally at a local optimum. Branch exchange improvement methods implement local modifications by interchanging k arcs in the current solution with k arcs presently not used. A k-interchange is performed if and only if it is feasible and it generates an improved solution. Such exchanges can be performed within and/or between routes. These procedures terminate with a k-optimal solution, i.e., one that cannot be improved through any k-interchange.

Solomon [1987] was the first to generalize a number of VRP route construction heuristics to the VRPTW. His computational results show the effectiveness of a two-phase sequential algorithm based on insertion criteria. First, the best feasible insertion position for each unrouted customer is chosen with respect to the minimum additional distance and time required. Second, the customer to insert is selected using a measure of the maximum savings derived. Insertion based algorithms have also been successfully utilized in routing environments where additional precedence constraints exist, such as the dial-a-ride problem which will be discussed in Section 6. Nevertheless, the worst-case ratio behavior of these approximation methods and of k-interchange heuristics on n-customer problems was proven to be at least of order n [Solomon, 1986].

Route improvement procedures have been developed by Russell [1977], Cook & Russell [1978], and Baker & Schaffer [1986] for the VRPTW, and by Potvin, Lapalme & Rousseau [1989] specifically for the m-TSPTW. These methods are extensions to the VRPTW of the k-interchange heuristic of Lin & Kernighan [1973] and of the 2-opt and 3-opt branch exchange procedures of Lin [1965]. While effective, these methods suffer from the major drawback of a very large processing time requirement. To alleviate this problem, Solomon, Baker & Schaffer [1988] have proposed a very efficient within route improvement algorithm based on the OR-opt procedure [Or, 1976]. The algorithm restricts the branch exchanges examined to only those where one, two, or three adjacent customers are inserted in a later position on the route, between two currently consecutive customers. It also incorporates the lexicographic ordering for processing suggested by Savelsbergh [1985] and uses push-forward and backward shifts in customer arrival times for feasibility checks. The authors report that the algorithm obtained solutions which were within 1% of those derived by 3-opt and required only 26% of its processing time.

A competitive algorithm based on a parallel insertion heuristic and the OR-opt method has been proposed by Potvin & Rousseau [1993]. Another competitive approach and a similar OR-opt implementation has been suggested by Thompson & Psaraftis [1989]. Their method is based on the concept of cyclic k-transfers which involves transferring k demands from route I^j to route $I^{\delta(j)}$ for each j and a fixed integer k. The set of routes $\{I^r\}$, $r = 1, \ldots, m$ constitutes a feasible solution and δ is a cyclic permutation of a subset of $\{1, \ldots, m\}$. In particular, when δ has fixed cardinality $\mathcal{C}$, we obtain a $\mathcal{C}$-cyclic k-transfer. By allowing k dummy demands on each route, demand transfers can be performed among permutations rather than cyclic permutations of routes. Due to the complexity of the cyclic transfer neighborhood search, this is performed heuristically. In particular, similarly to OR-opt, only sets of demands belonging to adjacent customers are considered for transfer. Several variants of the generic algorithm are developed. One that proved effective consists of a mix of modules involving 2- and 3-cyclic 1-transfers, some of which consider dummy demands. The method has also been applied to problems to be discussed later in Section 6. Finally, Potvin, Kervahut & Rousseau [1992] have introduced a new Tabu search heuristic which appears to be quite good at improving Solomon's initial solutions.

Recently, Kontoravdis & Bard [1992] have developed a parallel heuristic and lower bounds for the fleet size. The approach is a greedy randomized adaptive search procedure (GRASP) which combines a greedy heuristic and randomization to construct a feasible solution. Local search is then used to improve upon this solution. During this phase, each route is considered for elimination, with routes having fewer customer examined first. The authors have also developed three different lower bounds. The first stems from the underlying Bin Packing structure created by the capacity constraints. The second is derived from the maximum clique of the associated incompatibility graph. An arc in this graph corresponds to a pair of customers which cannot be on the same route due to capacity or time window violations. The third also considers a Bin Packing problem generated by the time window constraints. Here, bin capacity is the length of the scheduling horizon, while the items are of size equal to either the time needed to go from a customer to its closest neighbor or the depot, or the time from the depot to the first customer on each route. The authors report computational results which indicate that on random and semi-clustered problems, GRASP performed better than Solomon's heuristic. GRASP performed at least as well as Potvin and Rousseau's heuristic on all data sets and was superior for problems with a short scheduling horizon. The lower bounds have shown an encouraging performance as, for 14 problems from Solomon's data set, the gap between the best lower bound and the heuristic was zero.

Research on problems involving soft time windows has been scant. The objective function considered is to minimize a linear combination of total vehicle operating cost and total customer penalty due to missing any of the time windows. Sexton & Choi [1985] present a heuristic Benders decomposition algorithm for the single vehicle pick-up and delivery problem with soft time windows. It is a two-phase routing and scheduling procedure similar to the one developed in the dial-a-ride context. Koskosidis, Powell & Solomon [1992] have developed an iterative optimization-based heuristic which extends the generalized assignment heuristic of Fisher & Jaikumar [1981]. The problem is decomposed into an assignment/clustering phase and a series of TSP with soft time windows components. At each iteration, the assignment of customers to vehicles is solved as a capacitated clustering problem. The routes and schedules are obtained by solving a TSP subproblem for each vehicle. These solutions also provide the improved approximate clustering costs to be used at the next iteration.

5.1. Problem formulation

Notation. Let $N = \{1, \ldots, n\}$ be the set of customers and K, indexed by k, be the set of available vehicles to be routed and scheduled. Consider the graphs $G^k = (V^k, A^k)$, for all $k \in K$, each of them consisting of a set V^k of nodes and a set A^k of arcs. The set V^k consists of $N \cup \{o(k), d(k)\}$, where $o(k)$ and $d(k)$ represent respectively the origin-depot and the destination-depot of vehicle k, $k \in K$ (or any initial and final locations). The set A^k contains all feasible arcs, which is a subset of $V^k \times V^k$.

In the pick-up version, for each customer $i \in N$, there is a known demand p_i to be picked-up and a time window $[a_i, b_i]$ within which the customer permits the start of service. Define now a load parameter $\ell_i = p_i$, $i \in N$. For each vehicle $k \in K$, assume an initial load value $\ell_{o(k)}$, a time window $[a_{o(k)}, b_{o(k)}]$ at the starting location place as well as another time window at the destination location defined by $[a_{d(k)}, b_{d(k)}]$. Without loss of generality we assume that all the parameters are integer. Note that any VRPTW involving only deliveries can be reformulated as an equivalent pick-up problem by letting $\ell_i = d_i$, for all $i \in N$.

For each arc $(i, j) \in A^k$, $k \in K$, there is a cost c_{ij}^k and a travel time t_{ij}^k. As in previous sections, we assume that the service time at node i is included in the travel time t_{ij}^k, for all i. All the customers must be assigned to at most v vehicles, $v \leq |K|$, such that the capacity Q^k of each vehicle used is not exceeded. Note that an arc $(i, j) \in A^k$, $k \in K$ can be eliminated by temporal constraints if $a_i + t_{ij}^k > b_j$, by capacity constraints if $\ell_i + \ell_j > Q^k$, or by other considerations.

The mathematical programming formulation to be introduced next involves three types of variables: flow variables X_{ij}^k, $(i, j) \in A^k$, $k \in K$, equal to 1 if arc (i, j) is used by vehicle k, and 0 otherwise; time variables T_i^k, $i \in V^k$, $k \in K$ specifying the start of service at node i; and load variables L_i^k, $i \in V^k$, $k \in K$, specifying the load of the vehicle k just after servicing node i.

Formulation. The problem of finding the minimal cost set of routes satisfying the VRPTW constraints can then be formulated as follows:

$$\text{Minimize} \sum_{k \in K} \sum_{(i,j) \in A^k} c_{ij}^k X_{ij}^k \qquad (5.1)$$

subject to

$$\sum_{k \in K} \sum_{j \in N \cup \{d(k)\}} X_{ij}^k = 1, \qquad \forall i \in N \tag{5.2}$$

$$\sum_{k \in K} \sum_{j \in N} X_{o(k),j}^k \leq v, \tag{5.3}$$

$$\sum_{j \in N \cup \{d(k)\}} X_{o(k),j}^k = 1, \qquad \forall k \in K \tag{5.4}$$

$$\sum_{i \in N \cup \{o(k)\}} X_{ij}^k - \sum_{i \in N \cup \{d(k)\}} X_{ji}^k = 0, \qquad \forall k \in K, \ \forall j \in V^k \setminus \{o(k), d(k)\} \tag{5.5}$$

$$\sum_{i \in N \cup \{o(k)\}} X_{i,d(k)}^k = 1, \qquad \forall k \in K \tag{5.6}$$

$$X_{ij}^k (T_i^k + t_{ij}^k - T_j^k) \leq 0, \qquad \forall k \in K, \ \forall (i,j) \in A^k \tag{5.7}$$

$$a_i \leq T_i^k \leq b_i, \qquad \forall k \in K, \ \forall i \in V^k \tag{5.8}$$

$$X_{ij}^k (L_i^k + \ell_j - L_j^k) \leq 0, \qquad \forall k \in K, \ \forall (i,j) \in A^k \tag{5.9}$$

$$\ell_i \leq L_i^k \leq Q^k, \qquad \forall k \in K, \ \forall i \in N \cup \{d(k)\} \tag{5.10}$$

$$L_{o(k)}^k = \ell_{o(k)}, \qquad \forall k \in K \tag{5.11}$$

$$X_{ij}^k \geq 0, \qquad \forall k \in K, \ \forall (i,j) \in A^k \tag{5.12}$$

$$X_{ij}^k \text{ binary}, \qquad \forall k \in K, \ \forall (i,j) \in A^k. \tag{5.13}$$

This is a nonlinear formulation where the objective function (5.1) represents the total cost. It is possible to include a fixed charge c of using a vehicle by adding it to all $c_{o(k),j}^k$, $j \in N$. To minimize the number of vehicles, c is assigned a very large value, while it is set to zero if this number is free. To fix the number of vehicles used at exactly v, inequality (5.3) is replaced by an equality. Additional restrictions on the fleet composition can easily be formulated and incorporated in the above model. Constraints (5.2) impose that each customer be assigned exactly once to a vehicle route. Constraints (5.4)–(5.6) describe the flow on the path that vehicle k will use. Constraints (5.7) and (5.8) ensure feasibility of the time schedule while constraints (5.9)–(5.11) guarantee feasibility of the loads. Binary conditions on the flow variables are given in (5.13).

In the presence of the binary conditions on the flow variables, constraints (5.7) and (5.9) can be linearized and rewritten as:

$$T_i^k + t_{ij}^k - T_j^k \leq \left(1 - X_{ij}^k\right) M_{ij}^k, \qquad \forall k \in K, \ \forall (i,j) \in A^k \tag{5.7a}$$

$$L_i^k + \ell_j - L_j^k \leq \left(1 - X_{ij}^k\right) Q^k, \qquad \forall k \in K, \ \forall (i,j) \in A^k \tag{5.9a}$$

where M_{ij}^k are large constants. Note that one can replace the large constant M_{ij}^k in constraints (5.7a) by the value $\max\{b_i + t_{ij}^k - a_j, 0\}$, $(i,j) \in A^k$, and only require constraints (5.7) or (5.7a) for arcs $(i, j) \in A^k$, such that $M_{ij}^k > 0$; when $b_i + t_{ij}^k \leq a_j$, these constraints are satisfied for all values of T_i^k, T_j^k and X_{ij}^k.

This very general formulation can account for a homogeneous as well as a heterogeneous fleet of vehicles, the single and multiple-depot cases, and even for optimization situations requiring specific initial conditions for each vehicle. This is a practical model useful in an operational mode. The definition of a single time window per customer can also be relaxed to include multiple time windows. This may necessitate changing the objective function to account for preferred service times. Multiple time windows have primarily been examined in the context of the multi-period vehicle routing problem where the time windows are full days. Each customer must receive a specified number of visits within the planning horizon. This problem has been surveyed by Solomon & Desrosiers [1988]. Incidentally, this generalization can be treated by Dantzig–Wolfe decomposition, State-Space and Lagrangean relaxation schemes which use time window constrained shortest paths as sub-structures.

We consider next two special cases of the VRPTW.

The multiple traveling salesman problem with time windows. Eliminating constraints (5.9)–(5.11) yields an m-TSPTW formulation. This is a VRPTW involving uncapacitated vehicles. It is also a natural extension of the fixed schedule problem which occurs when the time windows consist of a single value, i.e., the arrival time at each customer is prespecified. Aircraft, ship, engine, school bus and urban bus scheduling have proven to be very fruitful grounds for dealing with time window constraints. To our knowledge, Appelgren [1969] was the first author to use a Dantzig–Wolfe decomposition/column generation algorithm to solve a ship scheduling problem with time window constraints. The resulting master problem is a set partitioning problem, while the subproblem consists of a shortest path problem over an expanded network arising from the discretization of the time windows. Optimal integer solutions are not guaranteed in this first paper, nor in the second one where the author wrote: *"There are fundamental difficulties in combining these integer programming methods with the Dantzig–Wolfe decomposition, since the constraints generated in the master program have to be taken into account in the solution of the subprograms."* [Appelgren, 1971]. As reported in Section 2, these difficulties are all removed if, first, master program solutions are transferred back to the original formulation on which the decomposition method is applied, and second, all branching decisions are taken on this original structure. The decomposition process can then be reapplied on the modified structure. Additional details are provided in following sections.

Another approach is that of Levin [1971] who presents and directly solves an integer programming formulation including discretized time windows for the minimization of aircraft fleet size. The same integer programming formulation has been later used by Swersey & Ballard [1983] to solve school-bus scheduling problems. The authors have been successful in manually adjusting the few fractional variables obtained, without increasing the number of bus fleet size.

Other heuristics have been proposed for school-bus scheduling by Orloff [1976], Bodin & Berman [1979], Graham & Nuttle [1986], and Desrosiers, Ferland, Rousseau, Lapalme & Chapleau [1986], and for aircraft scheduling by Martin [1981]. Exact algorithms for the m-TSPTW are presented in the following subsections.

The simple backhauling problem with time windows. The vehicle routing problem with backhauls under time window constraints is an extension of the VRPTW. Linehaul customers are located at sites which are to receive a quantity of goods from the depot, while backhaul customers are located at sites which have to send a quantity of goods back to the depot. In the simple backhaul problem, all deliveries must be made before any pick-ups occur. Heuristics have been designed only for the version without time windows at customers, and most of them are adaptations of insertion methods created for the VRP [Deif & Bodin, 1984; Golden & Assad, 1984; Casco, Golden & Wasil, 1988; Goetschalckx & Jacobs-Blecha, 1989]. To our knowledge, only one mathematical optimization approach has been successfully implemented by Yano, Chan, Ritcher, Cutler, Murty & McGettigan [1987]. It is based on a set covering formulation and a priori enumeration of a restricted set of feasible routes.

Let us now modify formulation (5.1)–(5.13) to account for the simple backhauling problem. For this, define $N = N^D \cup N^P$ to be the set of customers N, divided into a set of delivery locations N^D, and a set of pick-up locations N^P. The delivery demands are given by $\ell_i = d_i$, $i \in N^D$, while the pick-up demands are fixed to $\ell_i = p_i$, $i \in N^P$. The network structure is also partitioned in two main parts: the delivery level (which needs to be addressed first), and the pick-up level. The only difference with formulation (5.1)–(5.13) is the replacement of (5.10) by

$$\ell_i \le L_i^k \le Q^k, \qquad \forall k \in K, \quad \forall i \in N^D \tag{5.10a}$$

$$Q^k + \ell_j \le L_i^k \le 2Q^k, \qquad \forall k \in K, \quad \forall i \in N^P \cup \{d(k)\}. \tag{5.10b}$$

At the delivery level, constraints (5.10) and (5.10a) are identical. Due to the capacity window constraints, at any customer i of N^D, the total quantity L_i^k delivered since the start is at most equal to Q^k. At the end of the delivery portion, any vehicle k should be empty, or equivalently, it should allow Q^k new units of capacity. Crossing from the delivery level to the pick-up level on an arc $(i, j) \in A^k$, $i \in N^D$ and $j \in N^P$, the load L_j^k is set to the lower bound of the capacity window at customer j, i.e., at the value

$$L_j^k = \max\{L_i^k + \ell_j, \ Q^k + \ell_j\} = Q^k + \ell_j = Q^k + p_j, \quad i \in N^D, \ j \in N^P.$$

Constraints (5.9) and (5.10b) describe such conditions. As stated before, waiting until the permitted start of service is allowed, that is, if customer j follows customer i, and the arrival time $T_i^k + t_{ij}$ at customer j is less than a_j^k on vehicle k, the service only starts at time $T_j^k = a_j = \max\{T_i^k + t_{ij}, a_j\}$. The vehicle load variables act in a similar fashion.

Note finally that (5.9) need not be defined for arcs crossing from N^D to N^P since they are always satisfied. The exact algorithms designed for the VRPTW can then be utilized for the backhauling case where all the deliveries are performed before the start of pick-ups [Gélinas, Desrochers, Desrosiers & Solomon, 1992]. More complex algorithms are however necessary when the problem involves pick-up or delivery requests performed in any order as well as for the problem which involves simultaneous pick-up and delivery requests at the same customer. Halse [1992] describes some exact and approximate algorithms for these cases.

All the optimization algorithms for the VRPTW or its special cases, such as the m-TSPTW and the backhauling problem, employ branch-and-bound enumeration trees. In the next subsections, we will review several approaches used to compute lower bounds. To improve the efficiency of these approaches, the time windows' width can be reduced using the conditions presented in Section 3.2.

5.2. Network and linear programming relaxations

The network relaxation was analyzed for the m-TSPTW with a single depot and a homogeneous fleet of vehicles. It is obtained by dropping constraints (5.9)–(5.11) and then relaxing the scheduling and time window constraints (5.7)–(5.8). If $a_i = b_i$, for all $i \in N$, then the problem reduces to the fixed schedule problem and the network relaxation produces an optimal solution. The quality of the bound deteriorates with an increasing width of the time windows.

The network relaxation bound is often of poor quality as there is usually a gap on the number of vehicles. Two branching rules have been proposed: branching on flow variables and branching by partitioning the time windows. Using the former branching rule in the case of very tight time windows, Soumis, Desrosiers & Desrochers [1985] have optimally solved urban-bus scheduling problems involving up to 150 trips. However, for school-bus scheduling problems where the time windows become wider, the branch-and-bound tree grows too rapidly in size for the approach to be practical [Desrosiers, Soumis, Desrochers & Sauvé, 1986]. Hence, the authors used the latter branching rule which proved successful in handling the wider time intervals. Nevertheless, it was concluded that the network relaxation approach was inferior to the Dantzig–Wolfe decomposition scheme which will be presented in the next section.

The linear relaxation bound is obtained by replacing constraints (5.7) and (5.9) by their linearized versions (5.7a) and (5.9a) and by discarding the binary requirements (5.13). The resulting linear program is the above network flow problem with the additional time and capacity constraints. Following the work of Desrosiers, Sauvé & Soumis [1988], when setting the time variables at the center of the time window, i.e., $T_i^k = (a_i + b_i)/2$, the load variables similarly at $L_i^k = (\ell_i + Q^k)/2$, $i \in N, k \in K$, constraints (5.7a) and (5.9a) are satisfied if $X_{ij}^k \leq 1/2$, for all $(i, j) \in A^k$. It is thus relatively easy to obtain a fractional optimal linear programming solution to problem (5.1)–(5.13) for which the time and capacity constraints are inactive. In most cases, this bound is thus no better than the network relaxation bound.

5.3. Dantzig–Wolfe decomposition/column generation

The method to be presented next has sometimes been referred to as a Dantzig–Wolfe decomposition or a column generation technique. As reported previously, early work with this approach dates back to Appelgren [1969, 1971] for a vessel scheduling problem with time window constraints. A modified version of the subproblem structure as well as the first exact branch-and-bound scheme was originally applied to the m-TSPTW [Desrosiers, Soumis & Desrochers, 1984], and subsequently to many vehicle routing and crew scheduling problems (see Sections 2, 6 and 7], including the VRPTW [Desrochers, Desrosiers & Solomon, 1992].

Theoretical aspects. In the presence of binary constraints on the flow variables, the nonlinear problem (5.1)–(5.13) can be linearized to provide an integer problem denoted by IP. The classical Dantzig–Wolfe decomposition for linear programming can thus be applied to the linear relaxation LP of that integer problem. The decomposition scheme involves a master problem MP and a subproblem SP. If the binary requirements are kept in the subproblem definition and if the subproblem possesses the integrality property, i.e., the optimal value of SP, when solved as a linear program, is not altered by dropping the binary constraints, then $Z_{LP} = Z_{MP}$. If however SP does not possess the integrality property, then SP can be solved as an integer problem. Theoretically, the optimization can be done by solving a linear program defined on the convex hull of SP, or it can be done directly, as an integer problem, by some other means. Solving such a SP as an integer problem allows for a possible partial reduction of the integrality gap between Z_{IP} and Z_{LP}, that is

$$Z_{LP} \leq Z_{MP} \leq Z_{IP}.$$

To obtain an optimal integer solution to the original problem IP, branching decisions can be taken on the original variables of IP, and the Dantzig–Wolfe decomposition process embedded into a branch-and-bound enumeration tree. These features are exploited in the following Dantzig–Wolfe decomposition applied to the VRPTW.

A Dantzig–Wolfe decomposition. The decomposition presented here parallels the scheme introduced in Section 2.3 for the multiple depot vehicle scheduling problem. The master problem is given by (5.1)–(5.3), i.e., the objective function (5.1), the covering of each customer $i \in N$ exactly once (5.2), and the constraint on the total number of vehicles (5.3). The subproblem involves a modified objective function $\sum_{k \in K} \sum_{(i,j) \in A^k} \bar{c}_{ij}^k X_{ij}^k$, where the coefficients $\bar{c}_{ij}^k$ will be defined later, and constraint set (5.4)–(5.13). That is, the path constraints (5.4)–(5.6), constraints (5.7) and the time windows (5.8), which together insure the feasibility of the time schedule, also constraints (5.9), the capacity intervals (5.10) along with the initial load conditions (5.11), which guarantee the feasibility of the load, and finally the binary requirements (5.13) on the flow variables $X_{ij}^k, k \in K, (i, j) \in A^k$.

The subproblem. The first observation to make on the subproblem structure is that it decomposes into $|K|$ disjoint subproblems, one for each vehicle. The second observation characterizes each subproblem as an elementary shortest path problem with time and capacity constraints, its solution being obtained on a bounded polyhedron. Consequently, the flow variables X_{ij}^k can be expressed as a nonnegative convex combination of the paths generated from the corresponding subproblem. Hence, the usual convexity constraints will appear in the master problem definition. We will show later in this section how they can be removed in some special cases, one of them being the homogeneous fleet and single depot case.

In Section 2.3, the subproblem for the MDVSP can be viewed as a classical shortest path problem defined on an acyclic network. This problem possesses the integrality property, i.e., a linear programming algorithm applied on the problem gives rise to an integer solution. Therefore, solving the shortest path problem using a dynamic programming algorithm is equivalent to using a linear programming algorithm, and thus the bound obtained from the Dantzig–Wolfe decomposition is equal to the optimal value of the linear relaxation of the MDVSP. In the case of the VRPTW, the linearized subproblem together with the integrality requirements is an elementary shortest path problem with time windows and capacity constraints. This subproblem does not possess the integrality property. Consequently, solving it as a nonlinear integer program permits a reduction of the integrality gap between the optimal solution of the linearized version of formulation (5.1)–(5.12) and the optimal integer VRPTW solution to (5.1)–(5.13).

As mentioned in Section 4.1, the elementary shortest path problem with resource constraints is a $\mathcal{NP}$-hard problem for which no polynomial or pseudo-polynomial algorithms are known. It was also shown in Section 4.2 that when nonelementary path solutions for the subproblem are allowed, i.e., solutions where a path may contain cycles of finite duration and/or load due to the presence of time windows and capacity intervals, pseudo-polynomial algorithms have been developed for its solution.

The introduction of these paths containing cycles augments the size of the set of admissible columns generated for the master problem. The lower bound on the VRPTW solution obtained by solving the coordinating master problem may decrease, yet it can be slightly improved if the constrained shortest path problem is solved using a 2-cycle elimination procedure within the solution process [Houck, Picard, Queyranne & Vemuganti, 1980; Kolen, Rinnooy Kan & Trienekens, 1987; Desrochers, Desrosiers & Solomon, 1992]. However, the covering constraints (5.2) of the master problem ensure that each customer is visited exactly once. Hence, the supplementary columns containing cycles cannot appear in any integer solution of the master problem, whether optimal or not, and they are automatically constrained to take a zero value during the branch-and-bound process.

The master problem. Let Ω_B^k be the set of feasible paths of subproblem $k, k \in K$. Each $p \in \Omega_B^k$ corresponds to an elementary path which can be described using *binary* values $(x_{ijp}^k, (i, j) \in A^k)$. Any solution X_{ij}^k to the master problem can be expressed as a nonnegative convex combination of a finite number of elementary paths, i.e.,

$$X_{ij}^k = \sum_{p \in \Omega_B^k} x_{ijp}^k \theta_p^k, \qquad \forall k \in K, \ \forall (i, j) \in A^k$$

$$\sum_{p \in \Omega_B^k} \theta_p^k = 1, \qquad \forall k \in K$$

$$\theta_p^k \geq 0, \qquad \forall k \in K, \ \forall p \in \Omega_B^k.$$

Since the solution process of the constrained shortest path subproblem may generate paths containing finite cycles, such a path p can be described using *integer* flow values x_{ijp}^k, $(i, j) \in A^k$. Let Ω_I^k be the set of the additional paths and define $\Omega^k = \Omega_B^k \cup \Omega_I^k$. Any solution X_{ij}^k to the master problem can still be expressed as a nonnegative convex combination of paths, i.e.,

$$X_{ij}^k = \sum_{p \in \Omega^k} x_{ijp}^k \theta_p^k, \qquad \forall k \in K, \ \forall (i, j) \in A^k$$

$$\sum_{p \in \Omega^k} \theta_p^k = 1, \qquad \forall k \in K$$

$$\theta_p^k \geq 0, \qquad \forall k \in K, \ \forall p \in \Omega^k.$$

Next, define the parameters c_p^k, a_p^k and b_p^k the following way:

$$c_p^k = \sum_{(i, j) \in A^k} c_{ij}^k x_{ijp}^k, \qquad \forall k \in K, \ \forall p \in \Omega^k$$

$$a_{ip}^k = \sum_{j \in N \cup \{d(k)\}} x_{ijp}^k, \qquad \forall k \in K, \ \forall i \in N, \ \forall p \in \Omega^k$$

$$b_p^k = \sum_{j \in N} x_{o(k), jp}^k, \qquad \forall k \in K, \ \forall p \in \Omega^k.$$

We make these substitutions into (5.1)–(5.3) and rearrange the summation order. These substitutions together with the convex combination constraints gives the revised formulation of the master problem:

$$\text{Minimize} \sum_{k \in K} \sum_{p \in \Omega^k} c_p^k \theta_p^k \tag{5.14}$$

subject to

$$\sum_{k \in K} \sum_{p \in \Omega^k} a_{ip}^k \theta_p^k = 1, \qquad \forall i \in N \tag{5.15}$$

$$\sum_{k \in K} \sum_{p \in \Omega^k} b_p^k \theta_p^k \leq v, \tag{5.16}$$

$$\sum_{p \in \Omega^k} \theta_p^k = 1, \qquad \forall k \in K \tag{5.17}$$

$$\theta_p^k \geq 0, \qquad \forall k \in K, \ \forall p \in \Omega^k. \tag{5.18}$$

The coefficient $c_p^k, k \in K, p \in \Omega^k$ is the cost of the path $p \in \Omega^k$. Coefficient a_{ip}^k, $k \in K, i \in N, p \in \Omega^k$ is a constant taking a nonnegative integer value: it indicates the number of times customer i is visited by the vehicle k on path p. Coefficient $b_p^k, k \in K, p \in \Omega^k$, takes only binary values: value 1 if vehicle k visits at least one customer in path p, and 0 otherwise. In the latter case, path $p \in \Omega^k$ only uses the arc $(o(k), d(k))$. In (5.17), the coefficient of θ_p^k is equal to 1, for all $k \in K$ and $p \in \Omega^k$. In fact, this constraint corresponds to (5.4), or equivalently to (5.6), in the original formulation, i.e.,

$$\sum_{j \in N \cup \{d(k)\}} X_{o(k),j}^k = \sum_{i \in N \cup \{o(k)\}} X_{i,d(k)}^k = 1.$$

The mathematical formulation (5.14)–(5.18) is then the linear relaxation of a set partitioning type problem with an additional constraint on the total number of vehicles and a set of convex combination constraints.

In the case of a single depot and a homogeneous fleet of vehicles, together with the same initial conditions for all vehicles, the sets $\Omega^k, k \in K$ are all identical, i.e., $\Omega = \Omega^k$, for all $k \in K$. The subproblem solution involves only a single network $G = (V, A) = (V^k, A^k)$, for all $k \in K$. In this case, the convex combination constraints (5.17) can be aggregated. First, let $\theta_p = \sum_{k \in K} \theta_p^k$; then $\sum_{p \in \Omega} \theta_p = |K|$ and index k can then be removed from (5.14)–(5.16) and (5.18). Since the aggregated constraint counts the total number of used and unused vehicles, it is redundant when compared to constraint (5.16) limiting the number of vehicles used to $v, v \leq |K|$. The resulting formulation given below becomes the classical linear relaxation of the set partitioning formulation with the additional restriction on the number of vehicles routed and scheduled:

$$\text{Minimize} \sum_{p \in \Omega} c_p \theta_p \tag{5.19}$$

subject to

$$\sum_{p \in \Omega} a_{ip} \theta_p = 1, \qquad \forall i \in N \tag{5.20}$$

$$\sum_{p \in \Omega} \theta_p \leq v, \tag{5.21}$$

$$\theta_p \geq 0, \qquad \forall p \in \Omega. \tag{5.22}$$

In the presence of multiple depots and a heterogeneous fleet size, similar aggregations can be performed as long as the conditions are identical for all vehicles in the same group. One constraint is retained for each group and it describes the number of available vehicles within that group. The route assignment to a specific vehicle number within a group can be done a posteriori, i.e., after the solution is obtained.

Linear relaxation solution. The solution process consists of two levels. At the first level, the coordinating master problem, a linear program, is optimized using the current columns. At the second level, the subproblems are solved to find minimum marginal cost columns.

If a path with negative marginal cost is found, the corresponding column is added to the known ones in the master problem and the solution process returns to the first level. Otherwise the current solution is optimal for the master problem and it provides a lower bound on the optimal integer solution of the constrained multi-commodity flow model.

Solving the master problem (5.14)–(5.18) over the current set of columns $\Omega' \subset \bigcup_{k \in K} \Omega^k$ by using the simplex method gives the dual variables α_i, $i \in N$, β and γ^k, $k \in K$ associated with constraints (5.15)–(5.17), respectively, necessary for the solution of the subproblem. The marginal cost $\bar{c}_p^k$ of path p for vehicle $k \in K$ is given by:

$$\bar{c}_p^k = c_p^k - \sum_{i \in N} \alpha_i a_{ip}^k - \beta b_p^k - \gamma^k \tag{5.23}$$

$$= \sum_{(i,j) \in A^k} c_{ij}^k x_{ijp}^k - \sum_{i \in N} \alpha_i \left(\sum_{j \in N \cup \{d(k)\}} x_{ijp}^k \right)$$

$$- \beta \left(\sum_{j \in N} x_{o(k),jp}^k \right) - \gamma^k \left(\sum_{j \in N \cup \{d(k)\}} x_{o(k),j}^k \right). \tag{5.24}$$

The arc set A^k, $k \in K$, can be partitioned the following way: $i \in N$ and $j \in N \cup \{d(k)\}$, or $i = o(k)$ and $j \in N$, or $i = o(k)$ and $j = d(k)$. Hence, the marginal cost $\bar{c}_p^k$ can be rewritten as

$$\sum_{i \in N} \sum_{j \in N \cup \{d(k)\}} (c_{ij}^k - \alpha_i) x_{ijp}^k$$

$$+ \sum_{j \in N} (c_{o(k),j}^k - \beta - \gamma^k) x_{o(k),jp}^k + (c_{o(k),d(k)}^k - \gamma^k) x_{o(k),d(k),p}^k \tag{5.25}$$

Therefore, we can define the marginal cost $\bar{c}_{ij}^k$, $(i,j) \in A^k$, $k \in K$, of an arc as:

$$\bar{c}_{ij}^k = \begin{cases} c_{ij}^k - \alpha_i & \text{if } i \in N \\ c_{ij}^k - \beta - \gamma^k & \text{if } i = o(k) \text{ and } j \in N \\ c_{ij}^k - \gamma^k & \text{if } i = o(k) \text{ and } j = d(k). \end{cases} \tag{5.26}$$

Finding the minimum marginal cost column over the set Ω^k results in the following optimization problem:

$$\text{Minimize} \quad \sum_{(i,j) \in A^k} \bar{c}_{ij}^k X_{ij}^k$$

subject to   (5.4)–(5.13).

This optimization problem corresponds to an elementary shortest path problem with time and capacity windows. We have described solution strategies for relaxed versions of this problem in Section 4.5.

Solving the master problem, i.e., the linear program (5.14)–(5.18), is accelerated by generating several columns simultaneously, since the one-time solution of a subproblem by dynamic programming not only produces the minimum marginal cost column, but also many other columns of negative marginal cost.

Optimal integer solution. To obtain an optimal integer solution to the original multi-commodity flow formulation (5.1)–(5.13), one can try to find an integer solution to the master problem. This is a node-path formulation of the VRPTW equivalent to the original arc-node formulation. As pointed out by Appelgren [1969, 1971] in the context of vessel scheduling, as well as by Chvátal [1983, pp. 197–198] in the context of the cutting-stock problem, it is not an easy task to find an optimal integer solution to a problem solved using a column generation scheme. All the cuts imposed along with the branching decisions taken must be compatible with the decomposition scheme, i.e., with the master and the subproblem structures. A branch-and-bound tree must be explored, where additional columns might be generated in each branch. The main difficulty stems from the fact that fixing a fractional optimal basic variable at zero results in the regeneration of the corresponding column. This can be avoided if the subproblem is allowed to generate second, third, ..., nth best solutions [see Hansen, Jaumard & Poggi de Aragão, 1991; Maculan, Michelon & Plateau, 1992].

The simplest approach to obtain an optimal integer solution is however to take all the decisions on the original formulation (5.1)–(5.13). Cuts and branching decisions can be taken on the number of vehicles used, on one or several binary flow variables, on time variables, or on load variables. The decomposition scheme is then reapplied on each branch. Local decisions, i.e., decisions which concern only a single path, such as fixing a flow variable at 0 or at 1, or dividing the time window of a time variable, are carried out directly on the subproblem networks without changing the shortest path solution approach. Global decisions, i.e., those which concern more than one path, such as an integer cut on the total cost, or cuts or branching decisions on the number of vehicles used of each type at each depot, stay in the master problem definition. In other words, cuts and branch-and-bound decisions are taken on the multi-commodity flow model using flow and resource variables. These decisions are thereafter transferred to the adequate structure, i.e., the master problem or the subproblem network parameters.

Computational results. This approach proved to be very successful on a variety of practical applications. In school bus fleet planning (the problem of assigning buses to trips), which is a m-TSPTW, problems with up to 151 trips were solved in Desrosiers, Soumis & Desrochers [1984], while the algorithm has obtained the optimal solution to a real world 279-trip problem [Girard, 1990]. These problems involved a single depot and a homogeneous fleet. In the context of the transportation of elderly and handicapped persons, a m-TSPTW model was also used as an approximate model after a grouping phase of the initial requests. The Dantzig–Wolfe decomposition/column generation algorithm was then used to solve problems of more than 300 trips [Dumas, Desrosiers & Soumis, 1989a, see also Section 6.3]. In this case, more than thirty different initial conditions on the vehicles were treated, i.e., more than thirty different subproblems.

For the single depot VRPTW involving a homogeneous fleet, the algorithm found optimal solutions to a number of 100-customer problems [Desrochers, Desrosiers & Solomon, 1992], mainly using branching strategies based on the value of the flow variables. This decomposition scheme has also been applied to the backhauling problem to produce optimal solutions to randomly generated problems with up to 100 customers [Gélinas, Desrochers, Desrosiers & Solomon, 1992]. In this case, the branching decisions were taken on the time variables, a strategy which reduced the size of the search tree as well as the computational time by a factor of two.

An empirically observed property is that the optimal integer solution value of the set partitioning problem resulting from the use of a Dantzig–Wolfe decomposition or a column generation scheme is very close to its linear programming relaxation. Bramel & Simshi-Levi [1993] showed that the relative gap between fractional and integer solutions becomes arbitrarily small as the number of customers increases. This makes the branch-and-bound process more efficient.

The approach presented here has also been used on more complex problems. Those involving multiple depot sites and a heterogeneous vehicle fleet were solved using many subproblem structures, one for each depot/vehicle type [see Dumas, Desrosiers & Soumis, 1989a; Haouari, Dejax & Desrochers, 1991]. By an adequate design of the network, the approach can also be generalized to include multiple time window constraints per customer, to determine coffee breaks or meal locations within flexible time periods, and to deal with many other practical restrictions [Sansó, Desrochers, Desrosiers, Dumas & Soumis, 1990]. For much more complex models involving more than two resources and solved using the Dantzig–Wolfe decomposition/column generation approach, the reader is referred to Section 7 on fleet planning, crew scheduling and crew rostering problems.

5.4. State-space relaxation and integer programming relaxation

Kolen, Rinnooy Kan & Trienekens [1987] were the first to describe an optimization approach and to present computational results for the VRPTW, for the case of a single depot housing a homogeneous fleet. Their method was partially inspired by the work of Christofides, Mingozzi & Toth [1981a] on the classical VRP. The algorithm involves a branch-and-bound tree, where decisions are taken on fixing a flow variable at 0 or 1, and embeds state-space relaxation, which is an adaptation of the method described previously in Section 3.4 for the TSPTW.

Each node in the search tree corresponds to a set of fixed routes starting and finishing at the depot, or to at most one partial route starting at the depot, and to a set of customers that are forbidden to be inserted next on the partial route. In the branching step, a customer who does not already belong to one of the fixed routes or the partial route and is not forbidden to belong to the partial route is selected. This creates a binary decision on the corresponding flow variable consisting of whether or not to extend the partial route. When there is no current partial route, the decision is whether or not to initiate a new route with the selected customer as the first customer on the route.

The state-space relaxation method then calculates a lower bound on all feasible extensions of the incomplete solution. It relaxes the condition stating that each customer not yet on a route must be served exactly once. The least cost extension of the current incomplete solution to a set of routes is however computed so that the total load on all these routes is equal to $\sum_{i \in N} \ell_i$, and so that each route has a different last customer, i.e., the one visited just prior to the return to the depot.

A state-space relaxation. At the first node of the search tree, the states of the dynamic programming algorithm are of the form STATE(i, ℓ, k), $i \in N$, $0 \le \ell \le \sum_{i \in N} \ell_i$, $0 \le k \le v$ where each directed path from STATE$(0, 0, 0)$ to STATE(i, ℓ, k) corresponds to a set of k routes with total load ℓ, and with different last customers, each one belonging to $\{1, \ldots, i\}$. The transition cost from STATE(i, ℓ, k) to STATE$(i + 1, \ell, k)$ is zero: this is the case when customer $i + 1$ is not covered by the k routes. The transition cost from STATE(i, ℓ, k) to STATE$(i + 1, \ell + \ell', k + 1)$ is computed using the value $F(i + 1, \ell')$ of a shortest path problem with time windows originating at the depot node and having node $i + 1$ as the last customer, and with total load on that path equal to ℓ'. The lower bound at the root node is given by

$$\min_{1 \le k \le v} \text{STATE}\left(n, \sum_{i \in N} \ell_i, k\right). \tag{5.27}$$

At the other nodes of the branch-and-bound tree, if there is no partial route currently fixed, then a set of routes is already fixed and the problem to be solved is identical to the root node problem, except that it is of a smaller dimension. In the case where a partial route is fixed, exactly one of the routes appearing in the lower bound must be an extension of the partial route. This implies that new states must be added to the original state space. These are defined in a similar manner and the reader is referred to the original paper for details.

Improvement of the lower bound is obtained using a time constrained shortest path algorithm with 2-cycle elimination and node penalties for uncovered customers. These penalties are obtained heuristically using subgradient optimization.

An integer programming relaxation. This lower bound is closely related to the set partitioning type formulation (5.19)–(5.22) given in the case of a single depot and a homogeneous vehicle fleet. Rather than relaxing the binary conditions on the variables θ_p, $p \in \Omega$, Kolen, Rinnooy Kan & Trienekens [1987] relax the constraints (5.20) which require that each customer must be visited exactly once, and replace them by two types of constraints: a weighted row aggregation constraint on the one hand, and a different last customer constraint on the other hand. Define f_{ip}, $i \in N$, $p \in \Omega$ to be a binary constant taking value 1, if customer i is the last customer visited on route p, and 0, otherwise. Then, the integer programming relaxation is the following:

$$\text{Minimize} \quad \sum_{p \in \Omega} c_p \theta_p \tag{5.28}$$

subject to

$$\sum_{p \in \Omega} \left(\sum_{i \in N} \ell_i a_{ip} \right) \theta_p = \sum_{i \in N} \ell_i, \tag{5.29}$$

$$\sum_{p \in \Omega} f_{ip} \theta_p \leq 1, \qquad\qquad \forall i \in N \tag{5.30}$$

$$\sum_{p \in \Omega} \theta_p \leq v, \tag{5.31}$$

$$\theta_p \quad \text{binary}, \qquad\qquad \forall p \in \Omega. \tag{5.32}$$

Problem (5.28)–(5.32) is then solved to optimality, directly using dynamic programming at the first node of the tree. It is evident that the set Ω of feasible routes can be dramatically reduced, since only the least cost route among all those having the same load and last customer is retained in the state-space relaxation procedure. The optimal *integer* solution to (5.28)–(5.32) provides a lower bound on the VRPTW optimal solution cost. This solution might not be feasible since some customer may not be visited exactly once. Note that there is no dominance relation between the lower bounds provided by this state-space relaxation and by the linear programming relaxation of the set partitioning type formulation, respectively.

At the other nodes of the search tree, if there is no partial route currently fixed, the same formulation (5.28)–(5.32) is solved, except that some θ_p variables are already fixed at 1, while some others are removed because of a capacity constraint, or because some customers must be forbidden. In the case of a node in the search tree with a partial route already fixed, define e_p, $p \in \Omega$ a binary constant taking value 1, if route p is an extension of that partial route, and 0, otherwise. Then, the following supplementary constraint must be added to formulation (5.28)–(5.32):

$$\sum_{p \in \Omega} e_p \theta_p = 1. \tag{5.33}$$

This constraint claims that exactly one of the routes appearing in this branch must be an extension of the partial route.

The reader is also referred to Bianco, Mingozzi, Ricciardelli & Spadoni [1989] who use a similar integer programming relaxation solved by dynamic programming in the context of urban crew scheduling. Such a problem is further addressed in Section 7.

Computational results. Optimal solutions are reported for problems with up to 15 customers. In our point of view, the limitations of this approach are essentially due to the size of the state space, even if this has been reduced by the use of a state-space relaxation procedure. Furthermore, this methodology cannot easily incorporate new constraints, multi-vehicle types or multi-depot problems, except by using an enlarged state space.

5.5. Lagrangean relaxation methods

Lagrangean relaxation can be applied to various VRPTW formulations in many ways. On the one hand, the difficult constraints (time window and capacity constraints) can be relaxed so that the resulting Lagrangean subproblem is easy to solve. On the other hand, part of the pure network flow constraints may be relaxed, thus retaining the complicating constraints in the Lagrangean subproblem. We will describe such approaches after we first present some theoretical aspects of Lagrangean relaxation.

Theoretical aspects. Let IP denote the integer linear programming problem

$$Z_{IP} = \text{minimize} \quad \sum_j c_j X_j$$

$$\text{subject to} \quad \sum_j a_{ij} X_j = b_i \; (\text{or} \; \geq b_i), \; \forall i$$

$$X_j \in \mathcal{X}, \text{ integer}, \qquad \forall j,$$

where $\mathcal{X}$ is a feasible region defined by a set of linear constraints. Then, problem LP is defined as the linear relaxation of problem IP obtained by dropping the integrality conditions and Z_{LP} is its optimal value. Let also subproblem $L(\lambda)$ be the Lagrangean relaxation of IP relative to $\sum_j a_{ij} X_{ij} = b_i$ (or $\geq b_i$), $\forall i$, with a vector of Lagrangean multipliers $\lambda = (\lambda_i)$, $\forall i$, unrestricted in sign (or $\lambda_i \geq 0$, $\forall i$), i.e.:

$$Z_{L(\lambda)} = \text{minimize} \quad \sum_j c_j X_j + \sum_i \lambda_i \left(b_i - \sum_j a_{ij} X_j \right)$$

$$\text{subject to} \quad X_j \in \mathcal{X}, \text{ integer}, \qquad \forall j.$$

Given that the Lagrangean dual problem is $Z_L = \max_\lambda Z_{L(\lambda)}$, it is well known [Geoffrion, 1974] that

$$Z_{LP} \leq Z_L \leq Z_{IP}.$$

Finally, if the problem IP is feasible and the subproblem $L(\lambda)$ possesses the integrality property, i.e., the optimal value of $L(\lambda)$ when solved as a linear program is not altered by dropping the integrality constraints, then $Z_L = Z_{LP}$. This is a fundamental result which states that the lower bound Z_L on Z_{IP} provided by a Lagrangean relaxation approach is no better than the linear relaxation bound Z_{LP}, if the subproblem $L(\lambda)$ possesses the integrality property. If the integrality gap $Z_{IP} - Z_{LP}$ is small, the integrality property is a *good* property in the sense that subproblem $L(\lambda)$ is a linear problem easy to solve, and Z_L provides a good lower bound on Z_{IP}. On the other hand, if the integrality gap $Z_{IP} - Z_{LP}$ is large, the integrality property is a *bad* property. To obtain satisfactorily good lower bounds with Lagrangean relaxations in this case, integer Lagrangean subproblems $L(\lambda)$ without the integrality property must be solved, so that part of the integrality gap is explored in solving $L(\lambda)$. This approach is effective if the

Lagrangean problems can be easily solved by exploiting their special structure. Examples include knapsack problems, shortest path problems with time window and resource constrained shortest path problems.

There are several possible approaches for finding the value Z_L, including subgradient optimization, dual ascent methods, and Dantzig–Wolfe decomposition. In the latter case, if one defines the master problem by retaining the constraints $\sum_j a_{ij} X_j = b_i$, (or $\leq b_i$) $\forall i$ and the objective function, the resulting subproblem is identical to the subproblem $L(\lambda)$. Lagrangean relaxation and Dantzig–Wolfe decomposition yield the same lower bound on Z_{IP} as long as the subproblem is the same for both. These two approaches are respectively dual and primal methods to obtain the same bound [see Magnanti, Shapiro & Wagner, 1976; Magnanti, 1981; Desrosiers, Sauvé & Soumis, 1988; Nemhauser & Wolsey, 1988].

Lagrangean relaxation of the time and capacity constraints. Since constraints (5.7) and (5.9) can be linearized [see (5.7a) and (5.9a)], they can be relaxed in the objective function (5.1) with the properly defined multipliers. The resulting Lagrangean subproblem then becomes a pure network flow problem, which possesses the integrality property. The best Lagrangean bound is thus equal to the linear programming bound described in Section 5.2. This approach is analyzed in Desrosiers, Sauvé & Soumis [1988] for the m-TSPTW. It was never tested numerically since it is known that the integrality gap in this case is generally too large to be explored by branch-and-bound.

Lagrangean relaxations based on shortest path problems with resource constraints. Relaxing the constraints (5.2) on visiting each customer once, and the constraints (5.3) on the number of available vehicles, results in a set of Lagrangean subproblems, one for each specific vehicle. Given the set of multipliers $\alpha = (\alpha_i, \ i \in N)$, which are unrestricted in sign, and the multiplier $\beta \geq 0$, subproblem $L^k(\alpha, \beta)$ is defined as follows:

$$\text{Minimize} \sum_{(i,j) \in A^k} c_{ij}^k X_{ij}^k + \sum_{i \in N} \alpha_i \Big(1 - \sum_{j \in N \cup \{d(k)\}} X_{ij}^k\Big) + \beta\Big(v - \sum_{j \in N} X_{o(k),j}^k\Big)$$

subject to (5.3)–(5.13).

The above subproblem is a shortest path problem with time window and capacity constraints. This approach has been tested in Desrosiers, Sauvé & Soumis [1988] using classical and augmented Lagrangean relaxation methods for the m-TSPTW. It has also been used by Fisher, Jörnsten & Madsen [1992] for the VRPTW in conjunction with variable splitting [Guignard & Kim, 1987]. This involved a semi-assignment problem, defined using constraint set (5.2) and solved by inspection, and a set of shortest path problems with time and capacity constraints, one for each available vehicle. In this case, variable splitting does not allow any improvement of the Lagrangean lower bound since the semi-assignment problem possesses the integrality property.

Variable splitting allows for various relaxation schemes, each one exploiting different solvable structures. In addition to the above scheme, Fisher, Jörnsten &

Madsen [1992] and Fisher [1994] also used one based on a K-tree structure, where K is the set of available vehicles. An additional approach has been considered by Jörnsten, Madsen and Sørensen [1986] who used the generalized assignment problem (which accounts for the vehicle capacity constraints) and a shortest path with only time window constraints. The reader is referred to the original papers and to Halse [1992] for further details.

Computational results. For the m-TSPTW, the various Lagrangean relaxation schemes have not been computationally competitive with the Dantzig–Wolfe decomposition/column generation approach described in Section 5.3. In the context of school bus fleet planning, Lagrangean relaxation was used by Desrosiers, Sauvé & Soumis [1988] only to estimate the optimal fleet size. The best approach found was the augmented Lagrangean relaxation method (which uses Frank–Wolfe decomposition and the SPPTW). This approach was far better than the classical Lagrangean relaxation which uses subgradient optimization to adjust the multipliers. Problems with up to 223 nodes, where each node represented a trip, were solved in the context of school bus fleet assignment.

For the VRPTW, Lagrangean relaxation schemes have been used to minimize the total distance traveled for a fixed fleet size. Using the K-tree structure, Fisher, Jörnsten & Madsen [1992] report having solved a number of 25- and 50-customer problems and two 100-customer problems from the benchmark problem sets developed by Solomon [1983]. This method did not perform as well as the Dantzig–Wolfe decomposition/column generation approach. It also did not prove competitive with the other variable splitting method proposed by the same authors and further developed and tested by Halse [1992]. The two approaches were better than the K-tree method even on problems involving geographically clustered customers, where the latter approach should perform at its best.

The Dantzig–Wolfe decomposition/column generation algorithm and the variable splitting approach have exhibited similar computational results. The former method proved slightly superior in solution quality to the latter as it solved several more problems. In addition, the former method is much more efficient in terms of computational time. For example, the largest problem solved to optimality, consisting of 105 customers and 11 identical vehicles, has been solved by Halse [1992] in 33295 seconds (more than 9 hours) on a HP 9000/835 computer system and by Desrosiers & Gélinas [1993] in 142 seconds on a HP 9000/755 (the HP 9000/755 is 8–10 times faster than the HP 9000/835).

There are several reasons which explain the performance of the Dantzig–Wolfe decomposition/column generation approach relative to that of the various Lagrangean relaxations on vehicle routing problems with time and capacity constraints. First, the simplex algorithm used to solve the master problem in the Dantzig–Wolfe decomposition makes use of a lot of information (e.g., all the added columns) and rapidly adjusts the multipliers, while subgradient optimization is dependent on only the previous iteration. Second, this method is primal and allows for the use of many heuristics to rapidly converge. For example, a small portion of the network may be used to increase the speed of convergence while the whole network is used only for final optimality tests. If such a reduction of the network size is used in the dual approach, the intermediate lower bounds are not valid. Third, the primal method provides more information to better design a successful branch-and-bound to explore the integrality gap. For example, branching decisions taken on the flow variables, on the time variables, or on global constraints were designed and implemented in the Dantzig–Wolfe decomposition approach. Most of these can be transferred to Lagrangean approaches, thus improving their performance. This is indeed the direction taken by Kohl & Madsen [1993] who report very promising results using bundle methods [see Lemaréchal, 1989]. They were able to dramatically reduce the CPU time required to solve a number of problems.

6. Pick-up and delivery problems with time windows

The pick-up and delivery problem with time windows (PDPTW) involves the satisfaction of a set of transportation requests by a heterogeneous vehicle fleet housed at several depots. A transportation request consists of picking-up a certain number of customers at a predetermined pick-up location during a departure time interval and transporting them to a predetermined delivery location to be reached during an arrival time interval. The departure and arrival time windows are based on desired pick-up or delivery time requests specified by the customers. Loading or unloading times are incurred at each vehicle stop.

Depending on the context, the problem consists of minimizing two or three objectives, in a hierarchical fashion, subject to a variety of constraints. The first objective involves the minimization of the number of vehicles or the total vehicle fixed costs required to satisfy the transportation requests. For this minimum value, the second objective addresses the minimization of the total distance or travel time. These two objectives are sufficient to model the problem of transportation of goods. However, for the transportation of persons, a third objective is necessary. This minimizes the inconvenience created by pick-ups or deliveries performed either sooner or later than desired by the customers. This latter PDPTW context is called dial-a-ride.

The PDPTW involves a multitude of constraints: *visiting* constraints which ensure that each pick-up and delivery stop is visited exactly once; *time window* constraints to be satisfied at each stop; *capacity* constraints on the vehicles (this can even be a multi-dimensional parameter, as in the case of transportation for the handicapped, where vehicles have two capacity limits - one on the number of wheelchairs and the other for the ambulatory passengers); *depot* constraints guaranteeing that a vehicle returns to the same depot from which it has started after each service run; *coupling* constraints stating that for a given request the same vehicle must visit the pick-up and the delivery stops; *precedence* constraints imposing that each customer must be picked-up before it is dropped-off; finally, *resource* constraints on the availability of drivers and vehicle types. The PDPTW is a generalization of the VRPTW. In the pick-up version, the VRPTW is the particular case of the PDPTW where the destinations are all a common depot. In the delivery version, the VRPTW is the particular case of the PDPTW where the origins are all a common depot.

Several cases of the pick-up and delivery problem will not be treated here. One case is that of full-loads which can be modeled as a m-TSPTW. A second case is that involving simple backhauling, i.e., where all the deliveries take place before all the pick-ups. This can be modeled as a VRPTW problem with one resource for the time and one for the vehicle capacity, and has been treated in Section 5. In the general backhauling problem, where all goods to be delivered (pick-up) are routed from (to) the depot, and where goods may be picked-up anytime there is sufficient space in the vehicle, the capacity constraints require two resources. The first is the maximum load on the vehicle up to the present, while the second is the actual load on the vehicle at the last customer visited. Note that the former resource is a nondecreasing function, hence it can be used to forbid subtours even when time windows are absent [Halse, 1992].

We will examine separately the one-vehicle case in Section 6.1 and the multi-vehicle case in Section 6.2. In addition we will present the related problem of schedule optimization to minimize the customer inconvenience for a fixed route in Section 6.3.

6.1. The single vehicle pick-up and delivery problem

The dial-a-ride context constitutes the core of research on pick-up and delivery problems. The early work is due to Wilson and coworkers [Wilson, Sussman, Wang & Higonnet, 1971; Wilson & Weissberg, 1976; Wilson & Colvin, 1977] who sought real-time solutions. The single vehicle pick-up and delivery problem (1-PDPTW) is a constrained TSPTW, where additional restrictions consist of capacity and precedence constraints. Psaraftis [1980, 1983] has exploited this in the development of the first algorithms to minimize the total customer inconvenience in the 1-PDPTW where time window constraints are defined by maximum position shifts. His forward and respectively backward dynamic programming algorithms have a $O(n^2 3^n)$ time complexity, where n is the number of requests. Consequently they were limited at that time to solving very small size problems (at most 10 requests).

The same objective function is examined by Sexton & Bodin [1985a, b] in the presence of one-sided time windows (i.e., specified maximum delivery times or minimum pick-up times, respectively). The overall problem is decoupled into a coordinating routing master problem, formulated as an integer program, and a scheduling subproblem for a fixed route, which admits a linear programming formulation. A heuristic version of Benders' decomposition is then used for solution, where only the scheduling subproblem is solved optimally. The optimal solution to the subproblem can be obtained very efficiently by a network flow algorithm (see Section 6.3). The efficiency of the scheduling algorithm makes the overall procedure computationally tractable. This allows the solution of medium-sized real problems ranging from 7 to 20 customers in an average of 18 seconds of UNIVAC 1100/81A CPU time. Sexton & Choi [1986] use a similar methodology to minimize a linear combination of total vehicle operating time and total customer penalty due to missing any of the time windows for the single vehicle pick-up and delivery problem with soft time windows.

Recently, Van der Bruggen, Lenstra & Schuur [1993] have developed a heuristic algorithm based on arc-exchange procedures to minimize the route duration. Tested on real-life problems, their method is shown to produce near-optimal solutions in a reasonable amount of computation time. They have also developed an alternative algorithm based on simulated annealing which finds high quality solutions in a relatively large CPU time.

An optimal algorithm for minimizing the total distance traveled has been designed by Desrosiers, Dumas & Soumis [1986]. The notation and the mathematical formulation used, as well as some algorithmic details are presented next.

Notation. Consider a set of n requests. Associate to the pick-up location of request i a node i, and to his delivery location a node $n + i$. Note that different nodes may correspond to the same physical location. Let $N^P = \{1, \ldots, n\}$ be the set of pick-up nodes and $N^D = \{n + 1, \ldots, 2n\}$ be the set of delivery nodes, and define $N = N^P \cup N^D$. The set of nodes of the network is $V = N \cup \{o, d\}$, where o is associated with the origin-depot location while d is associated with the destination-depot location.

Request i demands that p_i units (goods or customers) be transported from node $i \in N^P$ to the corresponding node $n + i \in N^D$. Let $\ell_i = p_i$, for $i \in N^P$, and $\ell_{n+i} = -p_i$, for $n + i \in N^D$. The capacity of the single vehicle is given by Q. For each pair of distinct nodes $i, j \in V$, t_{ij} represents the travel time, which includes the service time at node i, and c_{ij} denotes the travel cost. Let $[a_i, b_i]$ denote the time window associated to node i, $i \in N$. Retain in the set A only the arcs $(i, j), i, j \in V$, which satisfy a priori capacity and time constraints, and other restrictions such as those based on precedence, and the same location. Note that a preliminary step in the determination of the admissible arcs is a time window reduction process as described in Section 3.

Three types of variables are used in the formulation: binary flow variables X_{ij}, $(i, j) \in A$, time variables T_i, $i \in V$, and load variables L_i, $i \in V$. The flow variable X_{ij}, $(i, j) \in A$ equals 1, if the vehicle travels from node i to node j, and 0, otherwise. The time variable T_i, $i \in V$, represents the time at which service begins at node i. The load variable L_i, $i \in V$, gives the total load on the vehicle just after the service is completed at node i. It is assumed that the vehicle departs empty from the depot at a given fixed time, i.e., $L_o = 0$ and $a_o = b_o = T_o$.

Formulation. A nonlinear mathematical formulation of the single vehicle pick-up and delivery problem with time windows is as follows:

$$\text{Minimize} \quad \sum_{(i,j) \in A} c_{ij} X_{ij} \qquad (6.1)$$

subject to

$$\sum_{j \in N \cup \{d\}} X_{ij} = 1, \qquad \forall i \in N \qquad (6.2)$$

$$\sum_{j \in N^P} X_{o,j} = 1, \qquad (6.3)$$

$$\sum_{i \in N \cup \{o\}} X_{ij} - \sum_{i \in N \cup \{d\}} X_{ji} = 0, \qquad \forall j \in N \qquad (6.4)$$

$$\sum_{i \in N^D} X_{i,d} = 1, \qquad (6.5)$$

$$X_{ij}(T_i + t_{ij} - T_j) \leq 0, \qquad \forall (i, j) \in A \qquad (6.6)$$

$$a_i \leq T_i \leq b_i, \qquad \forall i \in V \qquad (6.7)$$

$$X_{ij}(L_i + \ell_j - L_j) = 0, \qquad \forall (i, j) \in A \qquad (6.8)$$

$$\ell_i \leq L_i \leq Q, \qquad \forall i \in N^P \qquad (6.9)$$

$$0 \leq L_{n+i} \leq Q - \ell_i, \qquad \forall n + i \in N^D \qquad (6.10)$$

$$L_o = 0, \qquad (6.11)$$

$$T_i + t_{i,n+i} \leq T_{n+i}, \qquad \forall i \in N^P \qquad (6.12)$$

$$X_{ij} \text{ binary}, \qquad \forall (i, j) \in A. \qquad (6.13)$$

This formulation includes a $(2n + 1) \times (2n + 1)$ assignment problem structure (6.1)–(6.5) where each node is visited only once, time window constraints (6.7), capacity intervals at pick-up nodes (6.9) and at delivery nodes (6.10), and constraints on the initial load of the vehicle at the origin-depot node (6.11). Constraints (6.6) describe the relation between the flow variables and the time variables. They involve an inequality sign since waiting time is permitted before the start of service at a node. These constraints prevent subtour formation. Constraints (6.8) describe the relation between the flow variables and the vehicle load at each node. These constraints do not prevent subtour formation since the vehicle load may increase or decrease. Finally, constraints (6.12) impose the precedence relations on the associate pick-up and delivery nodes.

The objective function is not as general as those addressing user inconvenience issues. Nevertheless, such issues can be incorporated by defining the time windows appropriately. Furthermore, since the solution strategy is based on a forward dynamic programming approach, the linear objective function can easily be replaced by a more general nonlinear function. For example, let $c(L_i) > 0$, $i \in V$, denote a nondecreasing function of the total load transported on the vehicle, just after the service is completed at node i. This function acts as a penalty factor on the travel cost and constraint (6.1) can be replaced by:

$$\text{Minimize} \sum_{(i,j) \in A} c(L_i) c_{ij} X_{ij}. \qquad (6.1a)$$

A dynamic programming solution approach. Desrosiers, Dumas & Soumis [1986] have developed an optimal forward dynamic programming algorithm which also capitalizes on the fact that the single vehicle PDPTW has a TSPTW structure with additional capacity and precedence constraints. The method is based on reductions of the state space and of the number of state transitions which are performed both a priori and during the execution of the algorithm, similar to the ones developed for the TSPTW algorithm discussed in Section 3.5.

The states are defined as follows: the state (S, i) is defined if there exists a feasible path which starts at node o, visits all the nodes in $S \subseteq N$ and ends at node $i \in S$. A path is feasible if it satisfies vehicle capacity, precedence and time window constraints. State (S, i) is said to be post-feasible if there is a path starting at i, visiting all the nodes not belonging to S and satisfying the above constraints. Several criteria are proposed which eliminate all nonfeasible states and a number of not post-feasible states. An additional criterion involves the case of several customers situated at the same location. The authors show that when several customers have the same physical location, some arcs can a priori be eliminated without restricting the optimality of the solution. This criterion allows the a priori imposition of a servicing order for the customers.

For each state (S, i), two-dimensional labels are defined for each path from node o to node i. A two-dimensional label represents the arrival time of the path at node i and the total distance traveled on the path, respectively. Similarly to the dynamic programming algorithm for the TSPTW, a label is nondominated if it is a minimal element in the set of labels of state (S, i) under the natural partial order over both time and cost. Several of the above criteria for state elimination are implemented through the use of labels. All the criteria developed lead to the elimination of a large number of states and state transitions. This is the main reason why the algorithm proved successful. It obtained the optimal solution to problems containing up to 40 customers (80 nodes) in less than 6 seconds on a CYBER 173.

An application of the 1-PDPTW. This type of problem has seen an interesting application to the discharge of larvicide in rivers to combat the growth of the black fly larvae. The larvae are the root cause of millions of cases of onchocerciasis (river blindness) that occur annually in Africa and hence is the object of large-scale larvicide control programs. This multi-period 1-PDPTW involves the discharge by helicopters of up to 5 types of larvicide subject to many types of constraints. The actual pick-up points as well as refueling locations are chosen from a larger set during the solution process. Chalifour, Solomon, Boisvert & Desrosiers [1992] cluster sets of points where the same larvicide is to be used into small groups (mini-clusters) and use an insertion heuristic to solve the problem.

6.2. The multiple vehicle pick-up and delivery problem

The general multiple vehicle pick-up and delivery problem with time windows (m-PDPTW) has only recently received attention. The only optimal algorithm has been proposed in the context of goods transportation by Dumas, Desrosiers & Soumis [1991]. This algorithm is based on a Dantzig–Wolfe decomposition approach embedded into a branch-and-bound search tree. The master problem results in the linear relaxation of a set partitioning type model, while feasible routes or columns are generated by a subproblem modeled as a shortest path problem with coupling, precedence, time window and capacity constraints.

Notation. The notation used is a direct extension of the one presented in the previous section. Let K, indexed by k, be the set of vehicles to be routed and scheduled. Then (V^k, A^k) is the network associated with a specific vehicle k. The set $V^k = N \cup \{o(k), d(k)\}$ is the set of nodes with $o(k)$ and $d(k)$ denoting the origin-depot and the destination-depot of vehicle k, respectively; the set A^k contains all feasible arcs, a subset of $V^k \times V^k$. The cost and time elements, the vehicle capacities and the three types of variables are defined adequately. Note that the index k serves to define a specific vehicle and a depot location, and more generally, the initial vehicle conditions of a heterogeneous fleet of vehicles. It is assumed that each vehicle k departs empty from its origin-depot, at a specified time value given by $T^k_{o(k)} = a_{o(k)} = b_{o(k)}$.

Formulation. We are now in a position to present a mathematical formulation of the multiple depot multiple vehicle types pick-up and delivery problem with time windows:

$$\text{Minimize} \sum_{k \in K} \sum_{(i,j) \in A^k} c^k(L^k_i) c^k_{ij} X^k_{ij} \tag{6.14}$$

subject to

$$\sum_{k \in K} \sum_{j \in N \cup \{d(k)\}} X^k_{ij} = 1, \qquad \forall i \in N^P \tag{6.15}$$

$$\sum_{j \in N^P \cup \{d(k)\}} X^k_{o(k),j} = 1, \qquad \forall k \in K \tag{6.16}$$

$$\sum_{i \in N \cup \{o(k)\}} X^k_{ij} - \sum_{i \in N \cup \{d(k)\}} X^k_{ji} = 0, \qquad \forall k \in K, \ \forall j \in N \tag{6.17}$$

$$\sum_{i \in N^D \cup \{o(k)\}} X^k_{i,d(k)} = 1, \qquad \forall k \in K \tag{6.18}$$

$$X^k_{ij}(T^k_i + t^k_{ij} - T^k_j) \le 0, \qquad \forall k \in K, \ \forall (i, j) \in A^k \tag{6.19}$$

$$a_i \le T^k_i \le b_i, \qquad \forall k \in K, \ \forall i \in V^k \tag{6.20}$$

$$X^k_{ij}(L^k_i + \ell_j - L^k_j) = 0, \qquad \forall k \in K, \ \forall (i, j) \in A^k \tag{6.21}$$

$$\ell_i \le L^k_i \le Q^k, \qquad \forall k \in K, \ \forall i \in N^P \tag{6.22}$$

$$0 \le L^k_{n+i} \le Q^k - \ell_i, \qquad \forall k \in K, \ \forall n + i \in N^D \tag{6.23}$$

$$L^k_{o(k)} = 0, \qquad \forall k \in K \tag{6.24}$$

$$T^k_i + t^k_{i,n+i} \le T^k_{n+i}, \qquad \forall k \in K, \ \forall i \in N^P \tag{6.25}$$

$$\sum_{j \in N} X_{ij}^k - \sum_{j \in N} X_{j,n+i}^k = 0, \qquad\qquad \forall k \in K, \forall i \in N^{\mathrm{P}} \qquad (6.26)$$

$$X_{ij}^k \geq 0, \qquad\qquad\qquad\qquad \forall k \in K, \forall (i,j) \in A^k \qquad (6.27)$$

$$X_{ij}^k \text{ binary}, \qquad\qquad\qquad \forall k \in K, \forall (i,j) \in A^k. \qquad (6.28)$$

The nonlinear objective function minimizes the total travel cost. Note also that the fixed cost of a vehicle k is usually incorporated into the values $c_{o(k),j}^k$, for all $j \in N^{\mathrm{P}}$. Constraints (6.15)–(6.18) and (6.28) form a multi-commodity flow problem. Next, constraints (6.19) describe the compatibility requirements between routes and schedules, while (6.20) are the time window constraints. Constraints (6.21) express the compatibility requirements between routes and vehicle loads, while (6.22) and (6.23) are the capacity intervals at pick-up and delivery nodes, respectively. These intervals are dependant of the vehicle used to satisfy a given request. Next, constraints (6.24) impose the initial load condition on each vehicle. Constraints (6.25), which are also vehicle dependent, are the precedence constraints forcing node i to be visited before node $n+i$, $i \in N^{\mathrm{P}}$. Along with the coupling constraints (6.26) they ensure that the same vehicle k visits both nodes i and $n+i$, $i \in N^{\mathrm{P}}$. Finally, the formulation involves nonnegativity constraints on flow variables (6.27) and binary requirements (6.28).

A Dantzig–Wolfe decomposition. The master problem, a linear program, retains the constraints (6.14)–(6.15), while the subproblem consists of a modified cost function and constraints (6.16)–(6.28). Note that the subproblem is separable by vehicle, i.e., one subproblem for each vehicle $k \in K$. As in the VRPTW case, the master problem will become the linear relaxation of a set partitioning type problem, while the subproblems are constrained shortest path problems.

Given the dual variables provided by the master problem, the arc costs are modified accordingly for a given vehicle k, $k \in K$ (see Section 5.3). An integer solution to the constrained shortest path problem is obtained by a forward dynamic programming in a manner similar to the algorithm presented for the single-vehicle PDPTW. The efficiency of the algorithm is derived from powerful label elimination criteria which greatly reduce the state space and the number of state transitions. These criteria generalize some of the corresponding criteria introduced for the single-vehicle version. For each path p starting at an origin-depot and ending at node i, denoted $\mathcal{P}_i^p$, labels of the form (set of visited nodes, time, cost) are defined. The three-dimensional label (S_i^p, T_i^p, Z_i^p) represents the set of nodes visited on the path p, the time the service starts at node i on path p and the total cost of path p, respectively. The information contained in the first component serves to determine the vehicle load at node i, i.e., $\sum_{j \in S_i^p} \ell_j$. A path is admissible, if it respects at each node the time window, priority, and capacity constraints. The coupling constraints must also be respected when the last visited node i is a destination-depot. Enumeration of all the feasible paths is impossible in practice. To reduce the number of labels, we associate with each admissible path $\mathcal{P}_i^p$ a label $\left(R(S_i^p), T_i^p, Z_i^p\right)$, where $R(S_i^p) \subseteq N^{\mathrm{P}}$ is the subset of S_i^p containing only the pick-up nodes which have been visited, but whose corresponding delivery nodes have not yet been visited. The information provided by the first component is still sufficient to calculate the vehicle load at node i. However, the resulting optimal shortest path may contain a cycle, i.e., a path which satisfies the same *request* twice while respecting all the constraints. Since a request cycle must satisfy a sequence of at least 5 nodes (e.g., $i \to n+i \to j(\neq i) \to i \to n+i$, as arc $(n+i, i)$ never exists), such a cycle is not common in practice when time windows are small in comparison with the travel times. An admissible label associated with path $\mathcal{P}_i^p$, which cannot be extended from node i to a destination-depot due to time window or coupling constraints violations, is called a non-post-feasible label. Many of these non-post-feasible labels are eliminated during the solution process. In addition, labels are eliminated based on a dominance criterion, which extends the similar criterion for the single-vehicle case. A stronger dominance criterion is possible if the costs satisfy the triangle inequality.

Each time a subproblem is solved, if there exits a negative marginal cost path, it is added to the restricted set partitioning type formulation which is in turn solved by the simplex method. When no more negative marginal cost columns can be found from any subproblem, the solution to the restricted master problem constitutes the optimal solution to the master problem.

An integer solution to the original problem (6.14)–(6.28) is then obtained by exploring a branch-and-bound tree. The branching strategy directly exploits the pick-up sequence of the requests. This limits the size of the branch-and-bound tree when compared to other potential strategies. Order variables O_{ij}, $i, j \in N^P$ are set at 1 if pick-up j is the next pick-up after pick-up i, and set at 0 otherwise. To transfer the information provided by the order variables to the subproblem, a fourth component is added to each label in the shortest path algorithm. This component represents the last pick-up node visited. This additional component does not make the subproblem much harder to solve since the information it provides is closely related to the information provided by the first component.

The algorithm has been successful in solving two real life problems of sizes 19 and 30 requests, respectively, taken from Guinet [1984], as well as for randomly generated problems involving up to 55 requests with tight vehicle capacity constraints [Dumas, Desrosiers & Soumis, 1991]. Both real life problems were solved to optimality, without any branching. For all the other problems, the number of vehicles was always minimal after the addition of a cut at the first node, while the gap on travel cost ranged from 0.6% to 3.2%. Very good solutions were obtained within the exploration of a few nodes in the branch-and-bound tree, typically less than 10. The algorithm is appropriate for the PDPTW when the problem solutions require an average of up to 5 requests per vehicle, i.e., 10 pick-up and delivery stops. However it becomes ineffective in the dial-a-ride context due to the actual size of practical problems [over 3,000 requests] and the large number of requests per vehicle. Such problems have been solved by heuristic methods.

Heuristics for the dial-a-ride problem with time windows. As in the single vehicle version, the research on the multi-vehicle PDPTW has been concentrated on the dial-a-ride context. A parallel insertion procedure was developed by Roy, Rousseau, Lapalme & Ferland [1984a, b]. This algorithm simultaneously constructs routes for all vehicles starting at the beginning of the day by using proximity criteria. It can also insert new requests into a set of existing routes with the possibility of initializing new routes as needed.

A similar approach for minimizing a weighted combination of customer disutility and system costs was employed by Jaw, Odoni, Psaraftis & Wilson [1986]. Simulated problems involving 250 customers and 10-14 vehicles took about 20 seconds each. A real problem involving about 2,617 customers and 20 vehicles (a much larger size than ever attempted before) was successfully solved in 12 minutes of VAX 11/750 CPU time. A comparison with a previous approach developed by the authors is provided in Psaraftis [1986]. An improved version of the algorithm was implemented by Madsen, Ravn & Rygaard [1993] for a dynamic dial-a-ride problem with time windows. This real world problem with 300 requests and 24 vehicles was solved in less than 10 seconds on a HP 9000/720.

A traditional *cluster first, route second* approach was proposed by Bodin & Sexton [1986]. The set of requests is partitioned into vehicle clusters and a single vehicle dial-a-ride problem solved for each cluster. The single-vehicle problems are solved by using the methodology described in Sexton & Bodin [1985a, b]. To further reduce total user inconvenience, requests are then relocated one at a time in vehicles different than that where they were originally assigned.

Dumas, Desrosiers & Soumis [1989a] improve upon the above methodology by moving a part of the clustering problem into the routing problem. In the classical *cluster first, route second* philosophy, clusters consisting of the set of customers assigned to each vehicle are formed, and routing is performed separately in each cluster. However, it is very difficult to construct a good set of clusters, especially without relying on routing information. Hence, the authors have devised heuristic algorithms which group together customers who efficiently can be served by the same vehicle route segment to form mini-clusters. A 2,000-request problem is typically reduced to an approximately 800-task problem, where each mini-cluster defines a task to perform. This strategy moves part of the clustering process into the routing phase. The new routing problem, which constructs routes corresponding to drivers' workdays by stringing together the mini-clusters, is less difficult to solve than the original, since it is of smaller size and certain constraints have partly (in the case of time window and capacity restrictions] or totally (in the case of coupling and precedence constraints) been satisfied. In fact, the mini-clustering problem corresponds to a m-PDPTW while the routing problem is a m-TSPTW with multiple depots and a heterogeneous vehicle fleet for which a solution strategy has been described in Section 5.3. Figure 10 taken from Ioachim, Desrosiers, Dumas & Solomon [1991] illustrates this two-phase solution approach.

The global routing solution, i.e., the m-TSPTW solution, can be approximated to permit the insertion of additional requests during the operation day. Such insertions are facilitated when periods of free time can be joined into longer noninterrupted periods. A nonlinear objective function which accomplishes this

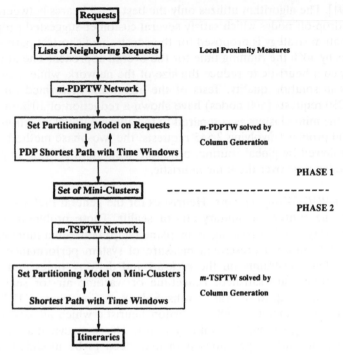

Fig. 10. Mini-clustering/route construction process for dial-a-ride problems.

while also minimizing the operational costs is given by:

$$\text{Maximize} \quad \sum_{k \in K} \sum_{(i,j) \in A^k} (T_j^k - T_i^k - t_{ij}^k)^\alpha X_{ij}^k \qquad (6.29)$$

where $\alpha > 1$ is a constant, $T_j^k - T_i^k - t_{ij}^k$ is the free time between nodes i and j, and the nodes are either mini-clusters or depot pull-outs or pull-ins. Since the total schedule time is the sum of the total travel time and total free time, minimizing the first component of the sum is equivalent to maximizing the second when $\alpha = 1$. This approach can easily solve problems with up to 200 customers and 85 mini-clusters in less than 5 minutes on a CYBER 173. To solve larger problems, the day is partitioned into time slices and the algorithm is applied several times. A real-world application involving 880 customers and 282 mini-clusters has been solved in 22 minutes on the same computer. Recently, the algorithm has been utilized to solve problems involving up to 3,586 requests. It takes only one hour of CPU time on a HP–730 workstation to obtain a solution.

While Dumas, Desrosiers & Soumis [1989a] use a sequential insertion heuristic to create the mini-clusters, Desrosiers, Dumas, Soumis, Taillefer & Villeneuve [1991] perform the insertions in parallel. Finally, Ioachim, Desrosiers, Dumas & Solomon [1991] utilize mathematical optimization techniques to globally define a set of mini-clusters. These are generated by solving a m-PDPTW by column generation using an enhanced version of the approach of Dumas, Desrosiers &

Soumis [1991]. The algorithm utilizes only the best feasible arcs between the pick-up and the drop-off nodes which satisfy several customer-suggested properties. A new initialization strategy is proposed for the constrained shortest path algorithm. This reduces by 40% the running time for the overall approach. The authors have also developed a heuristic to reduce the size of the network, while incurring only small losses in solution quality. Tests of the algorithm performed on problems with up to 250 requests (500 nodes) have shown a reduction of 10% in the travel time within the mini-clusters as compared to the above parallel insertion heuristic. For an actual problem involving 2,545 requests, the two-phase method, i.e., mini-clustering followed by global routing, produced a substantial 5.9% improvement in total traveling time over the same heuristic.

Additional m-PDPTW applications. Heuristics for the general PDPTW have been proposed in the context of military air- or sealift. These problems involve the movement of cargo and/or passengers by plane (ship) between a number of bases (ports) which optimize a prescribed measure of system performance such that the time window constraints on the pick-up and delivery times and the other problem constraints are satisfied. Peacetime or wartime air- or sealift involve planning horizons of up to several months. Solanki & Southworth [1991] present an application of Solomon's [1987] insertion heuristic which adjusts an existing schedule execution planning for military airlift. This is a dynamic environment where the existing airlift operations plan may need to be modified daily. The authors report having solved a problem involving 42 aircrafts and 10 airfields for seven days in less than two minutes on a SUN 4 workstation.

Another heuristic, the airlift planning algorithm (APA) was devised by Rappoport, Levi, Golden & Toussaint [1992] for a much longer planning horizon of up to 90 days. To simplify the problem, the assignment of payload to aircraft is treated before the routes and schedules are tested for feasibility. The concept of *best fit* drives the former problem. The latter problem is solved using shortest paths. To obtain feasibility, the APA may necessitate several iterations through the two phases. Even then however, due to the complexity of the many types of constraints, feasibility is not guaranteed. For a 90-day scheduling horizon problem, involving 13 bases and 285 planes, APA was able to move 95.9% of cargo (61.2% on time) and 99.9% of passengers (92.3% on time) in 696 seconds on a SUN SPARC 1 workstation. The algorithm can be used to initialize the Solanki & Southworth [1991] method. It has also been enhanced by a preprocessor that sequences the requirements to be moved more intelligently. This involves the solution of a transportation problem and is described in Rappoport, Levy, Toussaint & Golden [1992].

The sealift problem has been examined by Psaraftis, Orlin, Bienstock & Thompson [1985]. The authors propose a heuristic approach to be used in military emergencies. The heuristic is initialized with a one-to-one assignment of some cargoes to some ships. The scheduling horizon is then partitioned into time slices, numbered from the earliest to the latest. At iteration k, the heuristic temporarily assigns the cargoes whose earliest pick-up times fall within the kth time slice.

A transportation problem where each arc cost is a measure of the utility of the corresponding cargo/ship assignment is used for this purpose. Certain cargoes are then permanently assigned, allowing for cargo splitting if necessary. The horizon is then rolled to the next time slice. The authors have examined scenarios involving 7 ships which must pick-up and deliver 20 cargoes over a scheduling horizon of 98 days divided into bi-weekly intervals. This rolling horizon heuristic has been used to provide good initial solutions to be improved by cyclic transfer algorithms (see Thompson & Psaraftis, 1989). These have proved very successful in decreasing total tardiness.

While the above approaches permit each ship to carry many cargoes at the same time, Fisher & Rosenwein [1989] consider the pick-up and delivery of bulk cargoes. Their optimization algorithm is based on research conducted earlier for a similar truck scheduling problem, which is presented in Fisher, Greenfield, Jaikumar & Kedia [1982], and Bell, Dalberto, Fisher, Greenfield, Jaikumar, Kedia, Mack & Prutzman [1983]. The first phase of the approach is the generation of a list of candidate schedules for each ship. All, or a limited number of schedules, can be generated based on an input parameter that controls the amount of unloaded travel permitted. The schedule selection phase is a set packing problem that is solved with the branch-and-bound procedure described in Fisher & Kedia [1990], where bounds are obtained through Lagrangean relaxation. Fisher & Rosenwein [1989] report having optimally solved a realistic problem involving 28 cargoes to be lifted by a fleet of 17 ships in 50 seconds of VAX 8600 CPU time. Finally, the transport of a single bulk product between production and consumption ports is being addressed by Christiansen [1993]. The load to be picked-up or discharged is either fixed or variable leading to an inventory–PDPTW model.

6.3. Schedule optimization for a fixed route

Any solution approach to the VRPTW or the PDPTW must examine three aspects of the problem: the clustering of the nodes (or the assignment of customers to vehicles), the routing of the vehicles and the development of each vehicle's schedule. Approaches differ in terms of the simultaneous or hierarchical treatment of these issues. In all cases, however, the schedule associated with the route determined for each vehicle must be optimized. As time windows generally are fairly narrow in these problems, the service times obtained in the travel cost minimization part of the problem will provide a certain quality of service. To maximize the service quality, a final explicit optimization can be performed with respect to the set of desired service times at nodes. Hence, beginning service at a node at a less desirable time incurs a penalty or inconvenience cost.

Notation and formulation. Given a fixed route which visits, in sequence, nodes $1, 2, \ldots, n$, let N be the set of these nodes. Let also $t_{i,i+1}$, $i \in N \setminus \{n\}$ be the travel time from node i to node $i + 1$. For each node $i \in N$, define the variable T_i as the time service begins at node i. Define also the convex function $f_i(T_i)$ as the inconvenience function. Then, the schedule optimization problem for a fixed route is:

$$\text{Minimize} \sum_{i \in N} f_i(T_i) \tag{6.30}$$

subject to

$$T_i + t_{i,i+1} \le T_{i+1}, \quad \forall i \in N \setminus \{n\} \tag{6.31}$$
$$a_i \le T_i \le b_i, \quad \forall i \in N. \tag{6.32}$$

We seek to minimize the total inconvenience such that the schedule is compatible with the fixed route (6.31) and the service times fall within the time windows (6.32).

Literature. Three approaches to this problem have been presented to date (see Figure 11). In the first approach, Sexton & Bodin [1985a] consider time windows of the form $(-\infty, b]$ and a linear decreasing inconvenience function $f_i(T_i)$ such that $f_i(b_i) = 0$. In a related study, Sexton & Choi [1986] consider a piecewise linear function with $f_i(T_i) = 0$, if $T_i \in [a_i, b_i]$, $f_i(T_i)$ decreasing for $T_i \in (-\infty, a_i)$ and increasing for $T_i \in (b_i, +\infty)$. The two proposed algorithms have $O(n)$ complexity.

The third and most general approach to date is that of Dumas, Soumis & Desrosiers [1990] who consider a general convex function on the time window $[a_i, b_i]$. The authors propose a dual approach based on relaxing constraints (6.31). The violated constraints are then reintroduced one by one until they are all satisfied. The complexity of the algorithm is of the order of n unidimensional minimizations. For time windows, the use of convex inconvenience functions is more interesting than the use of linear functions since it allows solutions at the interior of the feasible region. In addition, the model admits two simple enhancements. The first consists of adding a linear cost to waiting time. This is

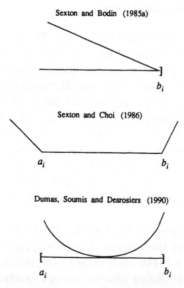

Fig. 11. Examples of inconvenience functions.

of particular interest when the waiting time causes an inconvenience. A second generalization entails discretizing the service time variables. These extensions do not imply any modifications to the algorithm. An adaptation of the algorithm has also found an application in statistics for the estimation of a step function in a nonlinear regression model [Dumas, Desrosiers & Soumis, 1989b]. This occurs when the dependent variable of a regression model is discrete while the optimization determines where the steps take place. In the above mentioned application, the algorithm is used to estimate a travel time function in an urban context where the function values are only multiples of 5 minutes. In that case, decision variables are linked by ratio constraints, i.e., $X_i \leq R_{i,i+1} X_{i+1}$, where $R_{i,i+1} > 0$ for all $i \in N \setminus \{n\}$.

Two special cases of the schedule optimization for a fixed route are worth noting. The first involves a quadratic inconvenience function, for which the authors proved that the complexity is linear in the number of elementary operations. The second deals with a linear inconvenience function. In this case, the optimization is even easier as the dual can be modeled as a minimum cost flow problem, a result previously obtained by Sexton & Bodin [1985a]. The linear case is presented next.

The linear case. Let us rewrite problem (6.30)–(6.32) with linear inconvenience functions:

$$\text{Minimize} \sum_{i \in N} \alpha_i T_i + \beta_i \tag{6.33}$$

subject to

$$T_i - T_{i+1} \leq -t_{i,i+1}, \quad \forall i \in N \setminus \{n\} \quad [-X_{i,i+1}] \tag{6.34}$$

$$-T_i \leq -a_i, \quad \forall i \in N \quad [-X_{n+1,i}] \tag{6.35}$$

$$T_i \leq b_i, \quad \forall i \in N \quad [-X_{i,n+1}] \tag{6.36}$$

The variables between brackets indicate the corresponding dual variables. Ignoring the constants β_i, $i \in N$, the dual problem can be formulated as follows:

$$\text{Minimize} -\sum_{i \in N \setminus \{n\}} t_{i,i+1} X_{i,i+1} - \sum_{i \in N} a_i X_{n+1,i} + \sum_{i \in N} b_i X_{i,n+1} \tag{6.37}$$

subject to

$$\begin{aligned} X_{1,2} + X_{1,n+1} &\quad - X_{n+1,1} = -\alpha_1, \\ X_{i,i+1} + X_{i,n+1} - X_{i-1,i} &\quad - X_{n+1,i} = -\alpha_i, \quad \forall i \in N \setminus \{1, n\} \\ + X_{n,n+1} - X_{n-1,n} &\quad - X_{n+1,n} = -\alpha_n, \end{aligned} \tag{6.38}$$

$$X_{i,i+1} \geq 0, \quad \forall i \in N \setminus \{n\} \tag{6.39}$$

$$X_{n+1,i} \geq 0, \quad X_{i,n+1} \geq 0, \quad \forall i \in N. \tag{6.40}$$

The dual is a transportation problem with $n + 1$ nodes and $3n - 1$ arcs. One node is associated with each variable T_i, $i \in N$, with a demand of α_i units, and an artificial node $n + 1$, with a supply $\alpha_{n+1} = \sum_{i \in N} \alpha_i$. This formulation has also been efficiently treated by Sexton & Bodin [1985a] using an $O(n)$ algorithm. However, the number of arcs can further be reduced a priori to only $2n - 1$ arcs.

Note first that the linear function $f_i(T_i) = \alpha_i T_i + \beta_i$, $a_i \leq T_i \leq b_i$, reaches its minimum at a_i, if $\alpha_i > 0$, and at b_i, otherwise. Assume next that the time windows have been reduced to eliminate the portions which are never used. This can be done recursively by means of the following simple relations:

$$a_{i+1} := \max\{a_{i+1}, a_i + t_{i,i+1}\}, \qquad i = 1, \ldots, n-1 \tag{6.41}$$

$$b_i \quad := \min\{b_i, b_{i+1} - t_{i,i+1}\}, \qquad i = n-1, \ldots, 1. \tag{6.42}$$

The complementary slackness conditions help to show that $X_{n+1,i} X_{i,n+1} = 0$, $\forall i \in N$, i.e., that at least one of the two variables is zero. Dumas, Soumis & Desrosiers [1990] then prove that the variables set at zero can in fact be specified:

$$\begin{aligned} X_{n+1,i} = 0, & \qquad \forall i \in \{i \,|\, \alpha_i < 0\} \\ X_{i,n+1} = 0, & \qquad \forall i \in \{i \,|\, \alpha_i \geq 0\}. \end{aligned} \tag{6.43}$$

The dual variables associated with the network nodes are the variables T_i of the original problem (the additional variable T_{n+1} is set at zero). The solution process starts with a basis consisting of variables $X_{i,n+1}$, if $\alpha_i < 0$, and $X_{n+1,i}$, if $\alpha_i \geq 0$, $i \in N$, and uses as entry criterion, the variable $X_{i,i+1}$ of smallest index i with negative marginal cost, i.e., $-t_{i,i+1} - T_i + T_{i+1} < 0$. The entry of one of these arcs into the basis corresponds to setting at equality the ith constraint in (6.34). The solution approach decides sequentially which constraint is set to equality, and previous decisions are not reconsidered. The algorithm takes at most $n-1$ iterations. Hence it is of linear complexity.

7. A unified framework for fleet and crew scheduling problems

Network models and algorithms have been used for many years in the field of vehicle fleet planning and crew scheduling problems. For example, the survey paper by Carraresi & Gallo [1984] reviews the network-based optimization algorithms and their use in the context of Urban Mass Transit vehicle routing and crew scheduling applications. In the same context, many other papers also appear in the very recent proceedings of the workshop on computer-aided transit scheduling [Desrochers & Rousseau, 1992].

In Section 2, we have considered a number of fixed schedule problems. The simplest ones were formulated and solved using classical network models and algorithms. On the other hand, the multi-depot vehicle scheduling problem was formulated as a multi-commodity network flow problem. The best known optimal solution approach uses a Dantzig–Wolfe decomposition/column generation scheme embedded in a branch-and-bound enumeration tree. Its subproblem/column generator is as a classical shortest path problem on an acyclic network.

Section 4 has introduced the shortest path problem with time window constraints (SPPTW), its generalized version which involves multi-resource constraints (SPPRC) and their corresponding solution approaches based on dynamic programming algorithms. In the vehicle routing applications of Section 5, one and two resource constrained shortest path algorithms were again used as sub-problems or feasible route generators in a Dantzig–Wolfe decomposition/column generation optimal solution approach to the m-TSPTW and to the VRPTW. This approach has been presented for multiple depots and heterogeneous vehicle fleet cases. Yet it can also be extended to the simple backhauling problem, to multiple time window intervals at each customer, to models which incorporate rest and meal periods, and most other realistic situations. Even more sophisticated shortest path algorithms, with additional precedence and coupling constraints, have been introduced in Section 6 to design an optimal approach for pick-up and delivery routing problems. At that time, we have also introduced nonlinear nondecreasing objective functions, (6.14) for example, which have been optimized using dynamic programming algorithms in the subproblems. All the previous problem definitions have involved at most two resources: elapsed time and cumulative load; these resources came from time window and capacity restrictions.

In this section, we consider vehicle fleet planning and crew scheduling problems which involve many resource constraints. These constraints relate mostly to maintenance schedules in vehicle fleet planning, and worker collective agreement regulations in crew scheduling. There are many important application domains for fixed or variable schedule problems with resource constraints: vehicle fleet assignment for buses, aircraft, ships and locomotives; crew pairing construction and monthly crew workload assignment for urban mass transit, airline, naval and rail companies. They all share the fact that feasible solutions correspond to paths on properly designed multi-commodity networks.

In the airline context, the crew costs represent a large share of these companies' total costs and any improvement translates into sizable savings. For example, at Northwest Airlines, the total crew costs represented \$1050 millions in 1989. Thus an improvement of 5% in the solution cost would amount to annual savings of more than \$50 millions [Barutt & Hull, 1990]. This explains the existence of several specialized software packages for solving these problems.

We will focus this section on an optimal solution approach for two fixed schedule pairing construction problems: the urban bus driver scheduling problem, and the classical airline crew scheduling problem. Nevertheless, the multi-commodity network flow model with resource constraints, as well as the Dantzig–Wolfe decomposition/column generation solution approach described below, have also been applied to aircraft fleet planning (from daily to multiple month schedules), to train driver pairing construction, and also to monthly crew workload assignment problems, again in the airline context. Furthermore, the model and methodology is by no means restricted to fixed schedule problems as time window restrictions involve only a supplementary resource. This has been recently applied to aircraft fleet planning in conjunction with a passenger demand model on a set of European and North-American test problems where benefits obtained went up to 20% on some scenarios [Desaulniers, Desrosiers, Solomon & Soumis, 1994].

7.1. A multi-commodity network flow model with resource constraints

Notation. Let F, indexed by f, represent the set of operational tasks to be performed exactly once. For example, F may represent the set of operational flight legs in an airline context. Define K as the set of commodities. With each commodity $k \in K$, associate the graph $G^k = (V^k, A^k)$, where V^k is the set of nodes and $A^k \subseteq V^k \times V^k$ is the set of arcs. Let $o(k)$ and $d(k)$, $k \in K$, denote the origin-depot and the destination-depot nodes, respectively. Let X_{ij}^k be the flow of type k through arc $(i, j) \in A^k$.

A given commodity k, $k \in K$, corresponds to a single vehicle in vehicle routing problems, to a specific crew in crew pairing problems, and to individual workers in monthly workload assignment problems. Graphs (V^k, A^k), for all $k \in K$, are oriented time-space networks, where nodes correspond to real locations at different times, while arcs represent for example a vehicle, a crew or a worker activity such as briefing, debriefing, an operational task, or a deadhead task, a rest period, a meal, a connection, or a partial or a full workday, or, finally, a partial or even a full schedule. Arcs may represent single or multiple activities. Define $A_f^k \subseteq A^k, k \in K, f \in F$, the subset of arcs on graph G^k such that each of these arcs contains an operational task. Description of some specialized time-space networks can be found in Section 2.2, and in Barnhart, Johnson, Anbil & Hatay [1991], Desrochers, Gilbert, Sauvé & Soumis [1992], Desrosiers, Dumas, Desrochers, Soumis, Sansó & Trudeau [1991], Lamatsch [1992], and Lavoie, Minoux & Odier [1988], to name a few.

Define R as the set of resources. The intervals $[a_i^{kr}, b_i^{kr}]$, $k \in K$, $i \in V^k$, $r \in R$, restrict the amounts of resources used to reach node $i \in V^k$, and t_{ij}^{kr} is the number of units of resource $r \in R$ used on arc $(i, j) \in A^k$. The resource variables are given by $T_i^{kr}, k \in K, i \in V^k, r \in R$. Assume that initial conditions are given by $T_{o(k)}^{kr} = a_{o(k)}^{kr} = b_{o(k)}^{kr}$, for all $r \in R$, at the origin-node $o(k)$, for all $k \in K$.

Finally let M, indexed by m, be a set of global constraints. Global constraints do not involve just a single commodity path, but many or even all commodities. In a simplified model, for a given $m \in M$, let $B_m^k \subseteq A^k, k \in K$, be the subset of arcs on graph (V^k, A^k) included in the constraint definition. Specific coefficients $b_{mij}^k, m \in M$ are assigned to arcs $(i, j) \in B_m^k, k \in K$. Let $\underline{b}_m$ and $\overline{b}_m$, $m \in M$ be the lower and upper bound values on a global constraint m.

Note that an arc $(i, j) \in A^k, k \in K$ is infeasible if it is not possible to go from node i to node j while respecting the resource intervals at both nodes. All infeasible arcs are then excluded from the sets A^k, for all $k \in K$. Conversely, all arcs $(i, j) \in A^k$ respect the following conditions:

$$a_i^{kr} + t_{ij}^{kr} \leq b_j^{kr}, \quad \forall k \in K, \quad \forall r \in \dot{R}.$$

Formulation. The mathematical formulation given below makes use of binary network flow variables $\mathbf{X}^k = (X_{ij}^k, (i, j) \in A^k), k \in K$, and resource variables $\mathbf{T}^k = (T_i^{kr}, r \in R, i \in V^k), k \in K$.

Vehicle fleet planning and crew scheduling problems refer to a large class of optimization problems in which vehicles and crews must be assigned to fixed time-tabled tasks in such a way that:

- a properly defined cost function $\sum_{k \in K} c^k(\mathbf{X}^k, \mathbf{T}^k)$ is minimized;
- each task in F is carried out once;
- a given set of global and local constraints are satisfied.

The multi-commodity network flow model with resource constraints is:

$$\text{Minimize} \quad \sum_{k \in K} c^k(\mathbf{X}^k, \mathbf{T}^k) \tag{7.1}$$

subject to

$$\sum_{k \in K} \sum_{(i,j) \in A_f^k} X_{ij}^k = 1, \qquad \forall f \in F \tag{7.2}$$

$$\underline{b}_m \le \sum_{k \in K} \sum_{(i,j) \in B_m^k} b_{mij}^k X_{ij}^k \le \bar{b}_m, \quad \forall m \in M \tag{7.3}$$

$$\sum_{j:(o(k),j) \in A^k} X_{o(k),j}^k = 1, \qquad \forall k \in K \tag{7.4}$$

$$\sum_{i:(i,j) \in A^k} X_{ij}^k - \sum_{i:(j,i) \in A^k} X_{ji}^k = 0, \quad \forall k \in K, \ \forall j \in V^k \setminus \{o(k), d(k)\} \tag{7.5}$$

$$\sum_{i:(i,d(k)) \in A^k} X_{i,d(k)}^k = 1, \qquad \forall k \in K \tag{7.6}$$

$$X_{ij}^k \left(T_i^{kr} + t_{ij}^{kr} - T_j^{kr} \right) \le 0, \qquad \forall k \in K, \ \forall r \in R, \ \forall (i,j) \in A^k \tag{7.7}$$

$$a_i^{kr} \le T_i^{kr} \le b_i^{kr}, \qquad \forall k \in K, \ \forall i \in V^k, \ \forall r \in R \tag{7.8}$$

$$X_{ij}^k \ge 0, \qquad \forall k \in K, \ \forall (i,j) \in A^k \tag{7.9}$$

$$X_{ij}^k \ \text{binary}, \qquad \forall k \in K, \ \forall (i,j) \in A^k. \tag{7.10}$$

The above formulation generalizes the previous ones given in Sections 2 and 5. The cost function (7.1) may be linear, such as $\sum_{k \in K} \sum_{(i,j) \in A^k} c_{ij}^k X_{ij}^k$, or be a nondecreasing nonlinear function using the operators $\sum$, min and max. Relations (7.2) require that each operational task be performed once. Relations (7.3) describe the global set of constraints, where coefficient values $\underline{b}_m$, $\bar{b}_m$ and b_{mij}^k may be restricted to integer values. Special cases of (7.3) are the fleet size constraints (2.3), (2.11) and (5.3) of the previous sections. This type of constraints can also represent global constraints on the number of full time and part time bus drivers in a feasible schedule, or airline base constraints related to the number of available crews. Constraint sets (7.4)–(7.10) describe the solution space of a shortest path problem with resource constraints introduced in Section 4.

Other types of local constraints, such as constraint sets (6.26)–(6.27) for the precedence and coupling requirements already encountered in pick-up and delivery problems, can be introduced in the proposed model. The dynamic programming solution approach must therefore be adapted accordingly to these new restrictions.

The previous model is easily generalized in two additional ways. First, the cost component can be regarded as a resource which is cumulated along the path structure of a subproblem. Since this kind of subproblem is solved using a forward dynamic programming algorithm, some nonlinear cost functions are easily treated. The basic condition on a valid cost function is that it must share some nondecreasing properties [Desrochers, Gilbert, Sauvé & Soumis, 1992; Desaulniers, Desrosiers, Ioachim, Solomon & Soumis, 1994]. Therefore nonlinear resource functions can be used to model very complex conditions. The second extension concerns the set of global constraints which can include constraints written in terms of the resource variables. Such restrictions may appear, for example, in aircraft routing problems in a situation where start times of flights are flexible within time intervals and, additionally, there must be a delay of at least 45 minutes between consecutive flight legs with the same origin-destination pair. Another example is the imposition of separate time windows on school starting and ending times which imply an adjustment of the time windows assigned to the arrival and starting times of bus-trips [Ferland & Fortin, 1989]. These restrictions, which may involve more than a single path, therefore appear as global constraints.

A Dantzig–Wolfe decomposition. Following the same principles described in Sections 2.3, 5.3 and 6.2, the decomposition scheme is applied to (7.1)–(7.10), i.e., the master problem includes relations (7.1)–(7.3), while the column generators or subproblems come from (7.4)–(7.10) and a modified cost function described below.

Each column $p \in \Omega^k$ of the master problem can be represented by a path satisfying constraint sets (7.4)–(7.10). This path p is in turn described by binary flow values x_{ijp}^k, $(i, j) \in A^k$, $k \in K$. Any solution X_{ij}^k to the master problem can be expressed as a nonnegative convex combination of a finite number of paths, i.e.,

$$X_{ij}^k = \sum_{p \in \Omega^k} x_{ijp}^k \theta_p^k, \quad \forall k \in K, \, \forall (i, j) \in A^k$$

$$\sum_{p \in \Omega^k} \theta_p^k = 1, \quad \forall k \in K$$

$$\theta_p^k \geq 0, \quad \forall k \in K, \, \forall p \in \Omega^k.$$

Define next the constants a_{fp}^k and b_{mp}^k:

$$a_{fp}^k = \sum_{(i,j) \in A_f^k} x_{ijp}^k, \quad \forall k \in K, \forall f \in F, \, \forall p \in \Omega^k, \tag{7.11}$$

$$b_{mp}^k = \sum_{(i,j) \in B_m^k} b_{mij}^k x_{ijp}^k, \quad \forall k \in K, \, \forall m \in M, \, \forall p \in \Omega^k. \tag{7.12}$$

The binary constant a_{fp}^k takes value 1, if path $p \in \Omega^k$, $k \in K$ covers task $f \in F$ and 0, otherwise, while the constant b_{mp}^k takes integer values. Given the flow values describing a path $p \in \Omega^k$, $k \in K$, i.e.,

$$\mathbf{x}_p^k = (x_{ijp}^k, \ (i, j) \in A^k), \qquad k \in K, \ p \in \Omega^k,$$

and the resource values along that path, i.e.,

$$\mathbf{t}_p^k = (t_{ip}^{kr}, \ i \in V^k, \ r \in R), \qquad k \in K, \ p \in \Omega^k,$$

the cost of such a path p is evaluated as

$$c_p^k = c(\mathbf{x}_p^k, \mathbf{t}_p^k), \qquad \forall k \in K, \ \forall p \in \Omega^k. \tag{7.13}$$

Making the substitutions for X_{ij}^k in (7.1)–(7.3) gives a linear program which is close to the linear relaxation of a set partitioning type formulation:

$$\text{Minimize} \sum_{k \in K} \sum_{p \in \Omega^k} c_p^k \theta_p^k \tag{7.14}$$

subject to

$$\sum_{k \in K} \sum_{p \in \Omega^k} a_{fp}^k \theta_p^k = 1, \qquad \forall f \in F \tag{7.15}$$

$$\underline{b}_m \le \sum_{k \in K} \sum_{p \in \Omega^k} b_{mp}^k \theta_p^k \le \overline{b}_m, \qquad \forall m \in M \tag{7.16}$$

$$\sum_{p \in \Omega^k} \theta_p^k = 1, \qquad \forall k \in K \tag{7.17}$$

$$\theta_p^k \ge 0, \qquad \forall k \in K, \ \forall p \in \Omega^k. \tag{7.18}$$

As previously presented in Section 5.3, column coefficients of the master problem formulation are generated by solving resource constrained shortest path problems (see Section 4 for their solution through dynamic programming algorithms). Given the dual variables α_f, $f \in F$, β_m, $m \in M$, and γ^k, $k \in K$, associated respectively with (7.15), (7.16) and (7.17) in the linear program above, the marginal cost $\overline{c}_p^k$, $k \in K$, $p \in \Omega^k$ of a path is given by

$$\overline{c}_p^k = c^k(\mathbf{x}_p^k, \mathbf{t}_p^k) - \sum_{f \in F} \alpha_f a_{fp}^k - \sum_{m \in M} \beta_m b_{mp}^k - \gamma^k. \tag{7.19}$$

Considering the coefficient definitions (7.11) and (7.12), the dual variables must be properly assigned to the adequate arcs in the set A^k. The marginal cost of a path is thus defined from all the arc parameters, i.e., resources consumed, arc costs, and dual variables. In the case of more complex models involving nonlinearities, the marginal costs are computed using the $\sum$, min and max operators. The nonlinearities introduced by the min and max operators are treated by using a multi-dimensional linear cost during the paths' construction, and by applying min and max operators once the paths are completed.

Optimal integer solution. The Dantzig–Wolfe decomposition/column generation scheme is embedded in a branch-and-bound search tree. Integer solutions are related to the original formulation (7.1)–(7.10). Branch-and-bound strategies can then be applied to the flow variables X_{ij}^k, $(i, j) \in A^k$, $k \in K$, to the resource variables T_i^{kr}, $i \in V^k$, $k \in K$, $r \in R$, and to any sets of constraints added to the set M. Every decision taken on the original formulation (7.1)–(7.10) can be transferred, either to the master problem or to the subproblems, and thus it is easily usable in the proposed Dantzig–Wolfe decomposition/column generation approach for integer programs.

7.2. The bus driver scheduling problem

The bus driver scheduling problem consists of covering every part of the bus schedule by bus drivers at minimum cost. The bus schedule defines the vehicle blocks, each vehicle block (or simply block) being a bus trip starting at the depot and going back to the depot. Along such a block, there are relief points where there can be a change of driver. The portion of a block which falls between two consecutive relief points, and which is thus always served by a single driver, is a task. A piece of work consists of several consecutive tasks in a block to be performed by a single driver.

The internal regulations of a bus company, and the collective agreement between the drivers' union and the transit operators define all of the details relevant to duty (workday) and schedule feasibility. A duty consists of one or more pieces of work executed by the same driver. In cases where a duty is composed of more than one piece of work, breaks or unworked periods are inserted between the pieces. The feasibility of a duty depends both on the length of the pieces of work and on the length of the breaks. It is also restricted by constraints such as limits on the number of pieces of work in a duty, the total worked time, the paid time, or the total spread. The collective agreement may also classify the duties into types and define constraints on the global manpower schedule.

Literature. Three main types of algorithms have been developed to solve the urban transit crew scheduling problem and later implemented: the runcutting method, HASTUS, and set partitioning methods. These methods and others have been described in detail in the proceedings of the last four workshops on computer-aided transit scheduling [Wren, 1981; Rousseau, 1985; Daduna & Wren, 1988; Desrochers & Rousseau, 1992] and in the survey of Carraresi & Gallo [1984]. Here we restrict our attention to optimal approaches: the set partitioning method and its variants. The reader is however referred to branch-and-bound algorithms for the exact solution of particular cases of the bus driver scheduling problem arising only when the *spread time constraint* or the *working time constraint* are imposed. These are given in Fischetti, Martello & Toth [1987] and [1989], respectively.

The set partitioning method uses a set partitioning type problem (see (7.14)–(7.18) above) to choose a set of feasible duties (columns) which cover all the tasks (rows) in such a way that a minimal cost schedule is obtained. The variable θ_p^k takes the value 1, if duty $p \in \Omega^k$, $k \in K$ is chosen, and 0, otherwise, where the cost of duty $p \in \Omega^k$ is c_p^k, $k \in K$. The partitioning constraints (7.15) guarantee that each task $f \in F$ will be covered by one duty p, a_{fp}^k being equal to 1, if the duty $p \in \Omega^k$, $k \in K$ covers the task f, and 0, otherwise. The additional constraints (7.16) restrict the proportion of duties of a given type in the schedule. Convex combination constraints (7.17) are aggregated whenever initial conditions are identical for a k-group of indices, and these aggregated constraints are integrated to the set of global constraints (7.16). Except for very small size problems, it is impossible to enumerate all the feasible columns, so that heuristic rules are used for the elimination of improbable relief points and less efficient workdays [Heurgon, 1975; Manington & Wren, 1975; Ryan & Foster, 1981; Mitra & Darby-Dowman, 1985; Wren, Smith & Miller, 1985]. The reader is also referred to Falkner & Ryan [1992] for a recent successful implementation of a set partitioning approach for Bus Crew Scheduling in Christchurch, New Zealand.

The Dantzig–Wolfe decomposition/column generation approach can be used to avoid the explicit use of the extremely large number of columns in the set partitioning type formulation of the crew scheduling problem. It permits to implicitly consider all the feasible columns using a subproblem generator. This subproblem consists of finding the minimal marginal cost feasible duty for a driver. An optimal implementation of this approach was developed by Desrochers & Soumis [1989]. The subproblems are shortest path problems with resource constraints, where the resource constraints insure that each feasible path corresponds to a feasible workday. Details on how to model the subproblem are presented in Desrochers, Gilbert, Sauvé & Soumis [1992]. The paper describes the design of the time-space network, the resources to account for the number of pieces of work and the spread, and the modelling of more complicated situations, such as overtime and spread premiums, and constraints on the duration of a half-day or on workday types.

Computational results. Results on real life problems involving 167 and 235 tasks and requiring 3 and 5 resources are reported in Desrochers & Soumis [1989]. Integer solutions with a gap of 0.05% and 0.07% with respect to the linear programming lower bounds were obtained by branch-and-bound. The commercial software package CREW-OPT based on this approach has been integrated into the HASTUS System [Hamer & Séguin, 1992] and is in operation in the cities of Lyon, Toulouse and Vienna. The main advantage of the Dantzig–Wolfe decomposition/column generation solution approach is that it always produces very high quality feasible schedules which do not require any manual adjustments. This column generation method has also been used in a successful benchmark for train driver assignment with up to 15.3% improvement over manual solutions at East Japan Rail. Tested problems ranged from 384 to 779 tasks.

7.3. The airline crew pairing construction problem

In recent years, several surveys have been published on the classical airline crew scheduling problem, i.e., the crew pairing construction problem [Etschmaier & Mathaisel, 1985; Gershkoff, 1989; Barutt & Hull, 1990]. The subject is also devoted a chapter in the book by Teodorović [1988]. They express the renewed current interest in this problem.

Let us first describe some terms used in the airline industry. A flight leg (or leg) is the portion of a flight between a take-off and the subsequent landing. A duty starts with a briefing period, which is then followed by several legs separated by connections and it ends with a debriefing period. A deadhead leg is a leg where a crew member is assigned to travel as a passenger. It is used to transfer a crew member to a city where there is an undersupply, or to return a crew member to its base. A pairing is a trip from the base and back to the base, composed of one or more duties separated by rest periods.

The purpose of the classical airline crew scheduling problem is to construct a set of minimal cost crew pairings covering all the legs and satisfying all the rules in the workers convention and in the air traffic regulations. Some constraints, such as minimum and maximum connection time, maximum time flown, and maximum time in operation relate to duty construction. Other constraints, such as minimum and maximum rest period, the 8 in 24 rule which forbids to allocate more than 8 hours of flight time in 24 to a pilot without extra compensation, restrict the pairing construction. There can also be global constraints on the crew schedule to insure a proper allocation of the pairings to the bases.

One of the most important aspects of the crew pairing construction problem is cost evaluation. The costs can be divided into three components: crew costs, accommodation costs, and penalty costs. The crew costs include the basic crew salaries, and the various guarantees on minimum flight time per duty, on the duty duration and on the pairing duration. The deadhead costs are also included in the crew costs, while the post-pairing cost may either be evaluated or not. The accommodation costs include the per diem or meal allowance, and the hotel and transportation costs incurred every time a crew is spending a rest period away from its base. The penalty costs are used mainly to penalize undesirable but legal pairings, or to increase the robustness of the crew schedule. For example, there often is a penalty on aircraft change during a duty, to reduce the consequences of a potential plane delay on the whole system.

Literature: The airline crew pairing construction problem can be formulated as a set partitioning type problem similarly to the bus crew scheduling problem. In air transportation, rows correspond to flight legs and the columns correspond to feasible pairings [see Arabeyre, Fearnley, Steiger & Teather, 1969]. Some interesting results on problems with relatively few feasible pairings have been obtained by Marsten, Muller & Killion [1979], Marsten & Shepardson [1981] and Ryan & Garner [1985]. Several recent optimization systems based on the a priori generation of a subset of pairings are described in Baker, Bodin & Fisher

[1985], Gershkoff [1989], Anbil, Gelman, Patty & Tanga [1991], and Graves, McBride, Gershkoff, Anderson & Mahidhara [1993]. These systems all make use of the following three main components: a generator, an optimizer and a local improvement module.

Hoffman & Padberg [1993] have presented a branch-and-cut approach to optimally solving set partitioning problems and set partitioning problems with base constraints. The branch-and-cut solver is applied to problems for which feasible pairings had been given a priori. It generates cutting planes based on the underlying structure of the polytope and incorporates these cuts into a tree-search algorithm that uses automatic reformulation procedures, heuristics and linear programming technology to assist in the solution.

Wedelin [1993] presents an approximation algorithm for 0–1 integer programming problems, which has been used as a solver for a set covering formulation with constraints. The algorithm, originally developed in a probabilistic framework, can be interpreted in this context as a simple ascent algorithm for the dual LP-relaxation, along with an approximation scheme which manipulates the reduced costs to produce integer solutions. This method is used in a feedback process in conjunction with a heuristic pairing generator. Computational tests indicate that this method solves faster than other LP-software packages, such as CPLEX [CPLEX, 1992], large problems up to 2×10^6 pairings while providing solution quality similar.

Problems with a huge number of feasible pairings have been solved by column generation. A heuristic implementation of the column generation approach was developed by Crainic & Rousseau [1987]. The new negative marginal cost columns were generated using a partial enumeration method. An integer solution was produced using a heuristic single branch enumeration method. Another implementation of the column generation approach was developed by Lavoie, Minoux & Odier [1988]. The columns were generated by solving a classical shortest path problem in an expanded network which integrates all the constraints. The reduced cost of a pairing was evaluated by a linear approximation. This method produced significant savings (over 5%) at Air France compared to previous solutions on medium and long haul problems. The problems on which it was applied involved up to 1100 flight legs. The same approach was recently used on long haul problems by Barnhart, Johnson, Anbil & Hatay [1991]. Three of the previous authors, Barnhart, Hatay & Johnson [1992], also developed a new methodology to improve crew pairing solutions through the efficient selection of deadhead flights from other airline companies. Their methodology uses the dual information determined in solving the linear programming relaxation of the crew pairing problem to build arrival and departure cost profiles at each station. These profiles provide a mechanism to price-out potential deadhead flights. Preliminary numerical results indicate significant reductions in crew costs.

An exact algorithm based on Dantzig–Wolfe decomposition/column generation is presented in Desrosiers, Dumas, Desrochers, Soumis, Sansó & Trudeau [1991]. The minimal marginal cost columns are generated with resource constrained shortest path subproblems. Many subproblem networks have been considered, where flight legs or duties have appeared either as nodes or as arcs. Networks representing duties as nodes or as arcs insure feasibility relatively to duty restrictions by network construction. Networks representing flight legs on nodes or on arcs are smaller but need more resources. A detailed description of the design of the networks and the resources, as well as some complex cost functions are given in Desaulniers, Desrosiers, Ioachim, Solomon & Soumis [1994].

Computational results. Very recent results presented in Desrosiers, Dumas, Solomon & Soumis [1993] show the strength of the approach used to solve optimally model (7.1)–(7.10) [see also Desrosiers, Dumas, Desrochers, Soumis, Sansó & Trudeau, 1991]. A 6-resource model was applied to a regional carrier's data set. The weekly schedule obtained for a 986 leg problem resulted in a 5.3% improvement over the sum of the costs of the optimal solutions obtained separately for each of the five fleet involved. On seven medium haul test problems from Air France, of sizes 154, 280, 342, 392, 477 legs (April'90 data set), 566 and 701 legs (January'92 data set) respectively, a 4-resource model allowed an average improvement of 5.9% over previous solutions.

Finally, on two test problems from a North American carrier involving 313 and 282 legs, respectively, and also over 2,000 deadhead legs, a 5-resource model resulted in savings of respectively 4.0% and 9.6% over other computerized solutions. If the objective considered is the excess crew cost (also called pay-and-credit), than the savings are of 31.5% and of 37.4%, respectively. In both test problems, the number of pairings generated was less then 20,000. This represents a very small fraction of the total number of feasible pairings, which was estimated at 190×10^{12}, i.e. 190 million of millions pairings in the case of the 282 leg problem.

The multi-commodity flow model with resource constraints solved using a Dantzig–Wolfe decomposition/column generation approach embedded into a branch-and-bound search tree has proved to be a very strong mathematical tool. Eight of the fifteen test problems mentioned above have been solved to optimality, while the optimality gap for the seven others was always within 1%. For example, the 9.6% improvement on the above 282 leg problem is almost optimal, the integrality gap being of $68 over the LP lower bound of $206,651. This primal approach, which can implicitly consider an extraordinarily huge number of pairings, is the only one which can also provide the optimality gap value on a feasible schedule solution.

During the writing of this survey, an optimization system based on this approach was used to solve test problems from several airline companies. The larger test problems involved more than 4,000 flight legs and they were all solved on workstations. For these, we have designed a solution procedure using time slice decomposition with overlapping time periods, and hence, we had to apply the algorithm several times. This procedure is similar to the one mentioned in the previous section and used for large scale dial-a-ride problems. On each of the test problems, this system has obtained the best optimized solutions compared to any other system. This approach has proved usable in practice and it has opened the way for optimal and near optimal solutions to many other applications in crew scheduling and vehicle fleet planning problems.

8. Conclusions and perspectives

In this paper we have tried to illustrate the spectacular growth of the field of time constrained routing and scheduling during the past fifteen years. We have provided an extensive overview of the models and algorithms developed, focusing on optimization methods. While heuristics will remain a viable tool for very large-scale and/or complex problems, optimization methods are becoming effective for solving practical size problems. The size of problems solved by such methods increases constantly. This new generation of optimal algorithms blends the effectiveness of advanced optimization methods, tailored to take advantage of special problem structures, with the efficiency of advanced computer science methods, and the computing power of workstations. Several models and algorithms developed over this period have already proved themselves in a number of successful implementations, while challenging new applications are being attacked every year.

For complex problems with fixed or flexible schedules, optimal algorithms based on Dantzig–Wolfe decomposition/column generation schemes have emerged as the most powerful solution methodologies. The equivalent dual schemes based on Lagrangean relaxation are currently showing very encouraging results. Dynamic programming has proved as one of the fundamental building blocks of these approaches. Furthermore, for simpler and sufficiently constrained problem structures, it constitutes a straightforward method for obtaining an optimal solution to realistic size problems. In addition, it is also flexible in that it allows the utilization of a variety of objective functions. All the above methods are also very flexible from the perspective that they can be transformed into optimization-based heuristics. The success of these optimal methodologies are an expression of the maturity level reached in this field. They constitute a solid platform for future research developments.

The work on mathematical models amenable to be solved by optimal methods in different time constrained areas has led to a first unified framework. This is a multi-commodity network flow model with additional resource constraints. Dantzig–Wolfe decomposition can be applied to it to provide the classical set partitioning formulation which in turn is solved by column generation. Resource variables help to manage complex nonlinear cost functions and difficult local constraints (e.g., time windows, vehicle capacity, and union rules]. Forward dynamic programming algorithms can handle these resources in shortest path computations. These developments have eventually provided insight into the design of the branch-and-bound search tree: all decisions must relate to the original multi-commodity model. This methodology can be extended far beyond the routing and scheduling field. It can be applied to develop optimal approaches to integer programs solved by column generation, an open problem since the early '60s. While the framework has been used in many applications, it can still be generalized. One way could be through new global constraints on resources. This could be complemented by more complexly constrained shortest paths.

A more specific research direction concerns the VRPTW area. We believe that further advances will be generated by promising research on polyhedral structures. For example, the linear programming bound in the Dantzig–Wolfe decomposition approach to the VRPTW could be enhanced by means of cuts. These have been instrumental for the solution of loosely constrained problems, while at the other extreme, dynamic programming embedded in decomposition approaches has solved sufficiently constrained problems. Combining these methodologies could lead to methods capable of solving moderately tight problems. The interface with artificial intelligence could provide means for selecting an appropriate method for a given problem structure. Problem characteristics such as routing structure, time window tightness and distribution, vehicle capacity, length of the scheduling horizon and the number of customers per route are important knowledge acquisition tools that could be embedded in expert systems for the VRPTW.

Future research should also continue to address computational efficiency issues, as optimal methods are being used for increasingly larger problems. We expect that more powerful algorithms will come from the interface with computer science. Specifically, new data structures and parallel implementations could lead the way.

As companies are making a substantial effort toward the computer integration of their operations, much more knowledge should be transferred between logistics and manufacturing. In particular, the application of time constrained routing models and algorithms to manufacturing is an important research direction that has barely been tapped.

The increased capability of optimal methods has lead to some very interesting applications. We expect this trend to continue. There are a number of strategies that run across the successful implementations. One set involves preserving as much realism as possible in the mathematical model. This translates into having a global perspective on the real world problem at hand. Decomposition techniques can then partition the overall problem into models to be solved by powerful algorithms. Feedback mechanisms between these models allow the return to the overall problem. In addition, the underlying networks need to be close to the real life settings. Over the last decade we have witnessed the use of increasingly realistic network structures. This was and will continue to be a product of the augmenting strength of optimal solution techniques. Nevertheless, one may need to choose among different network representations. A related idea is the use of realistic objective functions. In the current customer oriented business environment, lateness oriented objectives are such examples for routing problems. A second set involves simplification strategies. They will continue to play a central role in lifting the ceiling of solvable problem size. Mini-clustering is a good example of such a strategy.

The improved efficiency of optimal algorithms has substantially cut the planning lead time. By being run closer to the actual decision time, the algorithms can use more accurate data, and hence produce more accurate results. This is leading to a wider acceptance of these planning tools in practice. Other important research directions that we hope to see developing in the next few years, such as sensitivity analysis and reoptimization capabilities, could also significantly increase their acceptance. The recent work on soft time windows is a small step in this direction.

In addition to the further advances for optimization methods, there are a number of other key factors that are needed for the widespread utilization of these methodologies in practice. These are as valid for the routing and scheduling field as they are for any O. R. area. First, O. R. implementations should continue to integrate the new developments in mathematics, computer science and technology. Better information systems, data bases and computer graphics should partner optimization methods. So should powerful computers such as workstations and parallel machines. Teams are naturally the next ingredient. They should include researchers, programmers and business people. To be successful, the researchers on a team should rediscover the interdisciplinary nature of O. R. Too often, their narrow specialization has hindered the implementation process. As computer implementation strategies are an integral component of bringing to light the power of an optimization method, a team should also take advantage of the new generation of programmers. They have been around computers since early childhood and have been trained in computer science. Hence, they can harness the power of computers much better than their predecessors. Business people should be another element of such teams. They have the know-how necessary to sell O. R. solutions to interested companies. In addition they can serve the users' need to talk to nonmathematical people. Overall, each team member needs to also be a generalist in order to facilitate communication among the team members. A final factor is the necessary cooperation between O. R. people (teams) and users. An important element of this cooperation is the integration of the O. R. solutions within the current business practices of the company implementing the methodology.

We hope that this paper has offered a comprehensive view of this rapidly moving field. While it has reached a certain level of maturity, many important problems remain open. We can only hope that this paper has steered sufficient interest that many of its readers will embark on or continue their research in this field. The growth of the field can only be furthered by challenging competition among different research teams.

Acknowledgements

During the writing of this survey we have benefitted from numerous discussions with our friends and colleagues including Pierre Hansen, Brigitte Jaumard, André Langevin, Gilbert Laporte, Celso Ribeiro, Brunilde Sansó and Ben Smith. We are particularly grateful to Ed Baker, Oli Madsen and Alexander Rinnooy Kan who have taken the time to read an earlier version of this manuscript and have provided us with very helpful suggestions. We wish to especially thank Paolo Toth for his extensive and insightful comments. Mike Ball and an anonymous referee have made additional comments that have improved the manuscript. Special thanks are due to Guy Desaulniers, Irina Ioachim and Sylvie Gélinas who have also been instrumental in going over the paper with a fine comb. We are very grateful to Francine Benoît and Nicole Paradis at GERAD who have spent innumerable hours typing and editing the many versions of the manuscript.

This text has been used at the NorFA Research Course in Narvik, Norway, June 21–28, 1993. We would like to thank the organizers, Min Hwei Chuang, Per Olov Lindberg and Oli Madsen for providing us with a vehicle for testing this manuscript in the classroom. We have been involved in fruitful discussions with Susan Powell, Erik Anderson, Marielle Christiansen, Eric Gélinas, Niklas Kohl, Michael Lind, Andreas Nöu, Dag Wedelin and others during this Summer Course.

This work could have never been accomplished without the support and understanding of our families and special friends. We are most grateful to them.

Jacques Desrosiers, Yvan Dumas and François Soumis were partially supported by the Natural Sciences and Engineering Research Council of Canada grants. Marius Solomon was partially supported by the Patrick F. and Helen C. Walsh Research Professorship.

References

Afrati, F., S. Cosmadakis, C. Papadimitriou, G. Papageorgiou and N. Papakostantinou (1986). The complexity of the traveling repairman problem. *Theor. Inf. Appl. 20*, 79–87.

Ahuja, R.K., T.L. Magnanti and J.B. Orlin (1989). Network flows, in: G.L. Nemhauser, A.H.G. Rinnooy Kan and M.J. Todol (eds.), *Handbooks in Operations Research and Management Science, Vol. 1: Optimization*, North-Holland, Amsterdam, pp. 211–369.

Anbil, R., E. Gelman, B. Patty and R. Tanga (1991). Recent advances in crew pairing optimization at American airlines, *Interfaces* 21, 62–74.

Aneja, Y.P., V. Aggarwal and K.P.K. Nair (1983). Shortest chain subject to side constraints. *Networks* 13, 295–302.

Appelgren, L.H. (1969). A column generation algorithm for a ship scheduling problem. *Transp. Sci.* 3, 53–68.

Appelgren, L.H. (1971). Integer programming methods for a vessel scheduling problem. *Transp. Sci.* 5, 64–78.

Arabeyre, J.P., J. Fearnley, C. Steiger and W. Teather (1969). The airline crew scheduling problem: a survey, *Transp. Sci.* 3, 140–163.

Baker, E. (1983). An exact algorithm for the time constrained traveling salesman problem. *Oper. Res.* 31, 938–945.

Baker, E., and J. Schaffer (1986). Computational experience with branch exchange heuristics for vehicle routing problems with time window constraints. *Am. J. Math. Manage. Sci.* 6, 261–300.

Baker, E.K., L.D. Bodin and M. Fisher (1985). The development of a heuristic set covering based system for air crew scheduling. *Transp. Policy Decision Making* 3, 95–110.

Barnhart, C., E. Johnson, R. Anbil and L. Hatay (1991). A column generation technique for the long-Haul crew assignment problem, Industrial and Systems Engineering Reports Series, COC-91-01, Georgia Institute of Technology, Atlanta, Ga.

Barnhart, C., L. Hatay and E.L. Johnson (1992). Deadhead selection for the long-Haul crew pairing problem, Working Paper COC 92-01, School of Industrial and Systems Engineering, Georgia Tech., Atlanta, Ga., 34 pp.

Barutt, J., and T. Hull (1990). Airline crew scheduling: supercomputers and algorithms. *SIAM News*,

23(6), 1, 20–22.

Bell, W., L. Dalberto, M. Fisher, A. Greenfield, R. Jaikumar, P. Kedia, R. Mack and P. Prutzman (1983). Improving the distribution of industrial gases with an online computerized routing and scheduling optimizer. *Interfaces* 13, 4–23.

Bellman, R.E. (1958). On a routing problem. *Q. Appl. Math.* 16, 87–90.

Bertossi, A.A., P. Carraresi and G. Gallo (1987). On some matching problems arising in vehicle scheduling models. *Networks* 17, 271–281.

Bianco, L., A. Mingozzi, S. Ricciardelli and M. Spadoni (1989). *Algorithms for the crew scheduling problem based on the set partitioning formulation*, Istituto di Analisi dei Sistemi ed Informatica del CNR, R. 280, Roma.

Blais, J.-Y., J. Lamont and J.-M. Rousseau (1990). The HASTUS vehicle and manpower scheduling system at the Société de Transport de la Communauté Urbaine de Montréal. *Interfaces* 20, 26–42.

Bodin, L., and L. Berman (1979). Routing and scheduling of school buses by computer. *Transp. Sci.* 13, 113–129.

Bodin, L., and B. Golden (1981). Classification in vehicle routing and scheduling. *Networks* 11, 97–108.

Bodin, L., and T. Sexton (1986). The multi-vehicle subscriber dial-a-ride problem. *TIMS Studies Manage. Sci.* 26, 73–86.

Bodin, L., D. Rosenfield and A. Kydes (1978). UCOST: A Micro approach to a transit planning problem. *J. Urban Anal.* 5, 46–69.

Bodin, L., B. Golden, A. Assad and M. Ball (1983). Routing and scheduling of vehicles and crews: the state of the art. *Comput. Oper. Res.* 10, 62–212.

Bökinge, U., and D. Hasselström (1980). Improved vehicle scheduling in public transport through systematic changes in time-table. *Eur. J. Oper. Res.* 5, 388–395.

Bramel, J., and D. Simshi-Levi (1993). On the effectiveness of set partitioning formulations for the vehicle routing problem. Working Paper, Graduate School of Business, Columbia University.

Carpaneto, D., M. Dell'amico, M. Fischetti and P. Toth (1989). A branch and bound algorithm for the multiple vehicle scheduling problem. *Networks* 19, 531–548.

Carpaneto, G., and P. Toth (1980). Some new branching and bounding criteria for the asymmetric travelling salesman problem. *Manage. Sci.* 26, 736–743.

Carraresi, P., and G. Gallo (1984). Network models for vehicle and crew scheduling. *Eur. J. Oper. Res.* 16, 139–151.

Casco, D., B. Golden and E. Wasil (1988). Vehicle routing with backhauls: models, algorithms and case studies, in: B.L. Golden and A.A. Assad (eds.), *Vehicle Routing: Methods and Studies*, North-Holland, Amsterdam, pp. 127–147.

Ceder, A., and H.I. Stern (1981). Deficit function bus scheduling with deadheading trip insertions for fleet size reduction. *Transp. Sci.* 15, 338–363.

Chalifour, A., M.M. Solomon, J. Boisvert and J. Desrosiers (1992). An application of vehicle-routing methodology to large-scale larvicide control programs. *Interfaces* 22, 88–99.

Christiansen, M. (1993). Ship routing and scheduling (preliminary version). Ph.D. Dissertation, Department of Economics, The Norwegian Institute of Technology, Trondheim.

Christofides, N., A. Mingozzi and P. Toth (1981a). Exact algorithms for the vehicle routing problem based on the spanning tree and shortest path relaxations. *Math. Program.* 20, 255–282.

Christofides, N., A. Mingozzi and P. Toth (1981b). State-space relaxation procedures for the computation of bounds to routing problems. *Networks* 11, 145–164.

Chvátal, V. (1983). *Linear Programming*, Freeman, New York, N.Y., 478 pp.

Cook, T., and R. Russell (1978). A simulation and statistical analysis of stochastic vehicle routing with timing constraints. *Decision Sci.* 9, 673–687.

CPLEX Reference Manual (1992). Using the CPLEX Callable Library and CPLEX Mixed Integer Library, CPLEX Optimization, Inc., Incline Village, Nev.

Crainic, T.G., and J.-M. Rousseau (1987). The column generation principle and the airline crew scheduling problem. *INFOR* 25(2), 136–151.

Cyrus, J.P. (1988). The vehicle scheduling problem: models, complexity and algorithms. Ph.D. Dissertation, Technical University of Nova Scotia, Halifax, N.S.

Daduna, J.R., and A. Wren, eds. (1988). *Computer-Aided Transit Scheduling*. Lecture Notes in Economic and Mathematical Systems 308, Springer Verlag, Berlin, Heidelberg.

Dantzig, G., and D. Fulkerson (1954). Minimizing the number of tankers to meet a fixed schedule. *Nav. Res. Logistics Q.* 1, 217–222.

Dantzig, G.B., and P. Wolfe (1960). The decomposition algorithm for linear programming. *Oper. Res.* 8, 101–111.

Dantzig, G., D. Fulkerson and S. Johnson (1954). Solution of large-scale traveling salesman problem. *Oper. Res.* 2, 393–410.

Deif, I., and L. Bodin (1984). Extension of the Clarke and Wright algorithm for solving the vehicle routing problem with backhauling, in: A.E. Kidder, (ed.), *Conference on Computer Software Uses in Transportation and Logistics Management*, Babson Park, Mass., pp. 75–96.

Dell'Amico, M., M. Fischetti and P. Toth (1990). Heuristic algorithms for the multiple depot vehicle scheduling problem (to appear).

Denardo, E.V., and B.L. Fox (1979). Shortest-route methods: 1. Reaching, pruning and buckets. *Oper. Res.* 27, 161–186.

Desaulniers, G., J. Desrosiers, I. Ioachim, M.M. Solomon and F. Soumis (1994). A Unified Framework for Deterministic Time Constrained vehicle routing and crew scheduling problems. Cahiers du GERAD G-94-46, École des Hautes Études Commerciales, Montréal.

Desaulniers, G., J. Desrosiers, M.M. Solomon and F. Soumis (1994). Daily aircraft routing and scheduling. Cahiers du GERAD G-94-21, École des Hautes Études Commerciales, Montréal.

Desrochers, M. (1986). La fabrication d'horaires de travail pour les conducteurs d'autobus par une méthode de génération de colonnes. Ph.D. Dissertation, University of Montreal, Montreal (in French).

Desrochers, M., and G. Laporte (1991). Improvements and extensions to the Miller–Tucker–Zemlin subtour elimination constraints. *Oper. Res. Lett.* 10, 27–36.

Desrochers, M., J.K. Lenstra, M.W.P. Savelsbergh and F. Soumis (1988). Vehicle routing with time windows: optimization and approximation, in: B. Golden and A.A. Assad (eds.), *Vehicle Routing: Methods and Studies*, North-Holland, Amsterdam, pp. 65–84.

Desrochers, M., and J.-M. Rousseau, eds. (1992). *Computer-Aided Transit Scheduling*. Lecture Notes in Economics and Mathematical Systems 386, Springer Verlag, Berlin Heidelberg.

Desrochers, M., and F. Soumis (1988a). A generalized permanent labeling algorithm for the shortest path problem with time windows. *INFOR* 26, 191–212.

Desrochers, M., and F. Soumis (1988b). A reoptimization algorithm for the shortest path problem with time windows. *Eur. J. Oper. Res.* 35, 242–254.

Desrochers, M., and F. Soumis (1989). A column generation approach to the urban transit crew scheduling problem. *Transp. Sci.* 23, 1–13.

Desrochers, M., J. Desrosiers and M.M. Solomon (1992). A new optimization algorithm for the vehicle routing problem with time windows. *Oper. Res.* 40, 342–354.

Desrochers, M., J. Gilbert, M. Sauvé and F. Soumis (1992). CREW-OPT: Subproblem modeling in a column generation approach to urban crew scheduling, in: M. Desrochers and J.-M. Rousseau (eds.), *Computer-Aided Transit Scheduling*, Lecture Notes in Economics and Mathematical Systems 386, Springer Verlag, Berlin Heidelberg, pp. 395–406.

Desrosiers, J., and É. Gélinas (1993). NorFA Research Course, Narvik, Norway, June 21–28.

Desrosiers, J., F. Soumis and M. Desrochers (1982). Routes sur un réseau espace-temps. *C.R. Congr. ASAC, Rech. Opér.* 3, 28–44 (in French).

Desrosiers, J., P. Pelletier and F. Soumis (1983). Plus court chemin avec contraintes d'horaires, *RAIRO* 17, 357–377 (in French).

Desrosiers, J., F. Soumis and M. Desrochers (1984). Routing with time windows by column generation. *Networks* 14, 545–565.

Desrosiers, J., Y. Dumas and F. Soumis (1986). A dynamic programming solution of the large-scale single-vehicle dial-a-ride problem with time windows. *Am. J. Math. Manage. Sci.* 6, 301–325.

Desrosiers, J., J.-A. Ferland, J.-M. Rousseau, G. Lapalme and L. Chapleau (1986). TRANSCOL: A multi-period school bus routing and scheduling system. *TIMS Studies Manage. Sci.* 22, 47–71.

Desrosiers, J., F. Soumis, M. Desrochers and M. Sauvé (1986). Methods for routing with time windows. *Eur. J. Oper. Res.* 23, 235–245.

Desrosiers, J., M. Sauvé and F. Soumis (1988). Lagrangian relaxation methods for solving the minimum fleet size multiple traveling salesman problem with time windows. *Manage. Sci.* 34, 1005–1022.

Desrosiers, J., Y. Dumas, M. Desrochers, F. Soumis, B. Sansó and P. Trudeau (1991). A breakthrough in airline crew scheduling. *Proc. 26th Annu. Meet. of the Canadian Transportation Research Forum*, Quebec City, pp. 464–478.

Desrosiers, J., Y. Dumas, F. Soumis, S. Taillefer and D. Villeneuve (1991). *An algorithm for mini-clustering in handicapped transport*. Cahiers du GERAD G-91-02, École des Hautes Études Commerciales, Montréal.

Dijkstra, E. (1959). A note on two problems in connection with graphs. *Numer. Math.* 1, 269–271.

Dilworth, R. (1950). A decomposition theorem for partially ordered sets. *Ann. Math.* 51, 161–166.

Dror, M. (1994). Note on the complexity of the shortest path models for column generation in the VRPTW. *Oper. Res.* 42, 977–978.

Dumas, Y., J. Desrosiers and F. Soumis (1989a). Large scale multi-vehicle dial-a-ride problems. Cahiers du GERAD G-89-30, École des Hautes Études Commerciales, Montréal.

Dumas, Y., J. Desrosiers and F. Soumis (1989b). Minimisation d'une fonction convexe séparable avec contraintes de rapport entre les variables. *RAIRO* 23, 305–317 (in French).

Dumas, Y., F. Soumis and J. Desrosiers (1990). Optimizing the schedule for a fixed vehicle path with convex inconvenience costs. *Transp. Sci.* 24, 145–152.

Dumas, Y., J. Desrosiers and F. Soumis (1991). The pick-up and delivery problem with time windows. *Eur. J. Oper. Res.* 54, 7–22.

Dumas, Y., M. Salomon, M.M. Solomon and L.N. Van Wassenhove (1993). Discrete lotsizing and scheduling with sequence dependent setup times and setup costs, Working Paper, INSEAD, Fontainebleau.

Dumas, Y., J. Desrosiers, E. Gélinas and M.M. Solomon (1995). An optimal algorithm for the traveling salesman problem with time windows. *Oper. Res.* (to appear).

El-Azm, A. (1985). The minimum fleet size problem and its applications to bus scheduling, in: J.-M. Rousseau (ed.), *Computer Scheduling of Public Transport 2*, North-Holland, Amsterdam, pp. 493–512.

Etschmaier, M.M., and D.F.X. Mathaisel (1985). Airline scheduling: an overview. *Transp. Sci.* 19, 127–138.

Falkner, J.C., and D.M. Ryan (1992). EXPRESS: Set partitioning for bus crew scheduling in Christchurch, in: M. Desrochers and J.-M. Rousseau (eds.), *Computer-Aided Transit Scheduling*, Lecture Notes in Economics and Mathematical Systems 386, Springer Verlag, Berlin, Heidelberg, pp. 359–378.

Ferland, J.A., and L. Fortin (1989). Vehicle routing with sliding time-windows. *Eur. J. Oper. Res.* 38, 213–226.

Finke, G., A. Claus and E. Gunn (1984). A two-commodity network flow approach to the traveling salesman problem. *Congr. Numer.* 41, 167–178.

Fischetti, M., and P. Toth (1989). An additive bounding procedure for combinatorial optimization problems. *Oper. Res.* 37, 319–328.

Fischetti, M., S. Martello and P. Toth (1987). The fixed job schedule problem with spread-time constraints. *Oper. Res.* 35, 849–858.

Fischetti, M., S. Martello and P. Toth (1989). The fixed job schedule problem with working-time constraints. *Oper. Res.* 37, 395–403.

Fisher, M. (1994). Optimal solution of vehicle routing problems using minimum k-trees. *Oper. Res.* 37, 319–328.

Fisher, M., and R. Jaikumar (1981). A generalized assignment heuristic for vehicle routing. *Networks* 11, 109–124.

Fisher, M., and P. Kedia (1990). Optimal solution of set covering/partitioning problems using dual heuristics. *Manage. Sci.* 36, 674–688.

Fisher, M.L., and M.B. Rosenwein (1989). An interactive optimization system for bulk-cargo ship scheduling, *Nav. Res. Logistics* 36, 27–42.

Fisher, M., A. Greenfield, R. Jaikumar and P. Kedia (1982). Real-time scheduling of bulk delivery fleet: practical application of Lagrangian relaxation. WP 82-10-11, Department of Decision Sciences, University of Pennsylvania, Philadelphia, Pa.

Fisher, M., K.O. Jörnsten and O.B.G. Madsen (1992). Vehicle routing with time windows, Research Report 4C/1991, IMSOR, The Technical University of Denmark, Lyngby.

Forbes, M.A., J.N. Holt and A.M. Watts (1991). Exact solution of locomotive scheduling problems. *J. Oper. Res. Soc.* 42, 825–831.

Ford, L.R. (1956). Network Flow Theory. Report P-923, The Rand Corporation, Santa Monica, Calif.

Ford, L., and D.R. Fulkerson (1962). *Flows in Networks.* Princeton University Press, Princeton, N.J.

Friis, J. (1985). Rutelægning med tidsvinduer: metoder og algoritmer, Master Thesis, Research Report 11/85, IMSOR, The Technical University of Denmark, Lyngby (in Danish).

Fulkerson D.R. (1972). Flow networks and combinatorial operations research, in: A.M. Geoffrion (ed.), *Perspective on Optimisation*, Addison-Wesley, Reading, Mass., pp. 139–171.

Garey, M.R., and D.S. Johnson (1979). *Computer and Intractability: a Guide to the Theory of NP-completeness.* Freeman, San Francisco, Calif.

Gélinas, S. (1993). Private communication.

Gélinas, S., M. Desrochers, J. Desrosiers and M.M. Solomon (1992). A new branching strategy for time constrained routing problems with application to backhauling. Cahiers du GERAD G-92-13, École des Hautes Études Commerciales, Montréal [Revised April, 1994].

Gendreau, M., G. Laporte and M.M. Solomon (1992). Single- vehicle routing and scheduling to minimize the number of missed deadlines (forthcoming).

Geoffrion, A.M. (1974). Lagrangian relaxation and its uses in linear programming. *Math. Program. Study* 2, 82–114.

Gershkoff, I. (1989). Optimizing flight crew schedules. *Interfaces* 19, 29–43.

Gertsbakh, I., and H. Stern (1978). Minimal resources for fixed and variable job schedules. *Oper. Res.* 26, 68–85.

Girard, P. (1990). Conversion de données et comparaison des résultats de TRANSCOL et GENCOL. Département de Génie Électrique et Informatique, Ecole Polytechnique de Montréal.

Goetschalckx, M., and C. Jacobs-Blecha (1989). The vehicle routing problem with backhauls. *Eur. J. Oper. Res.* 42, 39–51.

Golden, B., and A. Assad (1984). The simple backhaul problem. Internal Memo, Distribution Systems Technologies Inc., Columbia, Md.

Graham, D., and H. Nuttle (1986). A comparison of heuristics for a school bus scheduling problem. *Transp. Res.* 20B, 175–182.

Graves, G.W., R.D. McBride, I. Gershkoff, D. Andersion and D. Mahidhara (1993). Flight crew scheduling, *Manage. Sci.* 39, 736–745.

Guignard, M., and S. Kim (1987). Lagrangean decomposition: a model yielding stronger Lagrangean bounds. *Math. Program.* 39, 215–228.

Guinet, A. (1984). Le système T.I.R.: un système d'établissement de tournées industrielles routières, Ph.D. dissertation, Université Claude Bernard, Lyon.

Halse, K. (1992). Modeling and solving complex vehicle routing problems, Ph.D. Dissertation no. 60, IMSOR, Technical University of Denmark, Lyngby, 372 pp.

Hamer, N., and L. Seguin (1992). The HASTUS system: new algorithms and modules for the 90s, in: M. Desrochers and J.- M. Rousseau (eds.), *Computer-Aided Transit Scheduling,* Lecture Notes in Economics and Mathematical Systems 386, Springer Verlag, Berlin, Heidelberg, pp. 17–29.

Hansen, P. (1980). Bicriterion path problem, in: G. Fandel and T. Gal (eds.), *Lecture Notes in Economics and Mathematical Systems 177,* Springer-Verlag, Heidelberg, pp. 109–127.

Hansen, P., B. Jaumard and M. Poggi de Aragão (1991). Un algorithme primal de programmation linéaire généralisée pour les programmes mixtes. *C.R. Acad. Sci.Paris* 313, 557–560.

Haouari, M., P. Dejax and M. Desrochers (1991). Modelling and solving complex vehicle routing problems using column generation, Working Paper, LEIS, École Centrale, Paris.

Heurgon, E. (1972) Un problème de recouvrement: l'habillage des horaires d'une ligne d'autobus. *RAIRO* 6, 13–29 (in French).

Heurgon, E. (1975). Preparing duty rosters for bus routes by computer, in: L. Bodin and D. Bergmann (eds.), *Preprints, Workshop on Automated Techniques for Scheduling of Vehicle Operators for Urban Public Transportation Services*, Chicago, Ill.

Hoffman, K.L., and M. Padberg (1993). Solving airline crew scheduling problems by branch-and-cut. *Manage. Sci.* 39, 657–682.

Houck, D.J. Jr., J.-C. Picard, M. Queyranne and R.R. Vemuganti (1980). The travelling salesman problem as a constrained shortest path problem: theory and computational experience. *Opsearch* 17, 93–109.

Ioachim, I., S. Gélinas, J. Desrosiers and F. Soumis (1994). A dynamic programming algorithm for the shortest path problem with time windows and linear node costs. Cahiers du GERAD G-94-24, École des Hautes Études Commerciales, Montréal.

Ioachim, I., J. Desrosiers, Y. Dumas, M.M. Solomon and D. Villeneuve (1995). A request clustering algorithm for door-to-door handicapped transportation. *Transp. Sci.* 29, 63–78.

Jaffe, J.M. (1984). Algorithms for finding paths with multiple constraints. *Networks* 14, 95–116.

Jaw, J., A. Odoni, H. Psaraftis and N. Wilson (1986). A heuristic algorithm for the multi-vehicle advance-request dial-a-ride problem with time windows. *Transp. Res.* 20B, 243–257.

Joksch, H.C. (1966). The shortest route problem with constraints. *J. Math. Anal. Appl. 14,*, 191–197.

Jörnsten, K., O.B.G. Madsen and B. Sørensen (1986). Exact solution of the vehicle routing and scheduling problem with time windows by variable splitting, Research Report 5/86, IMSOR, The Technical University of Denmark, Lyngby.

Kim, T. (1985). *Solution algorithms for sealift routing and scheduling problems,* Ph.D. Dissertation, M.I.T., Boston, Mass.

Knight, K., and J. Hofer (1968). Vehicle scheduling with timed and connected calls: a case study. *Oper. Res. Q.* 19, 299–310.

Kohl, N., and O.B.G. Madsen (1993). An optimization algorithm for the vehicle routing problem with time windows based on Lagrangean relaxation, Research Report 17/1993, IMSOR, The Technical University of Denmark, Lyngby.

Kolen, A.W.J., A.H.G. Rinnooy Kan and H.W.J.M. Trienekens (1987). Vehicle routing with time windows. *Oper. Res.* 35, 266–273.

Kontoravdis, G., and J.F. Bard (1992). A GRASP for the vehicle routing problem with time windows, Working Paper, revised October 1993, Department of Mechanical Engineering, The University of Texas, Austin, Texas.

Koskosidis, Y.A., W.B. Powell and M.M. Solomon (1992). An optimization based heuristic for vehicle routing and scheduling with soft time window constraints. *Transp. Sci.* 26, 69–85.

Lamatsch, A. (1992). An approach to vehicle scheduling with depot capacity constraints, in: M. Desrochers and J.-M. Rousseau (eds.), *Computer-Aided Transit Scheduling*, Lecture Notes in Economics and Mathematical Systems 386, Springer Verlag, Berlin, Heidelberg, pp. 181–195.

Langevin, A., M. Desrochers, J. Desrosiers and F. Soumis (1993). A two-commodity flow formulation for the traveling salesman and makespan problems with time windows. *Networks* 23, 631–640.

Laporte, G. (1992). The vehicle routing problem: an overview of exact and approximate algorithms. *Eur. J. Oper. Res.* 59, 345–358.

Laporte, G., and Y. Nobert (1987). Exact algorithms for the vehicle routing problem. *Ann. Discr. Math.* 31, 147–184.

Lasdon, L.S. (1970). *Optimization Theory for Large Systems.* MacMillan, New York, N.Y.

Lavoie, S., M. Minoux and E. Odier (1988). A new approach of crew pairing problems by column generation and application to air transport. *Eur. J. Oper. Res.* 35, 45–58.

Lemaréchal, C. (1989). Non differentiable optimization, in: G.L. Nemhauser, A.H.G. Rinnooy Kan and M.J. Todol (eds.), *Handbooks in Operations Research and Management Science, Vol. 1: Optimization*, North-Holland, Amsterdam, pp. 529–572.

Lenstra, J.K., and A.H.G. Rinnooy Kan (1981). Complexity of vehicle routing and scheduling problems. *Networks* 11, 221–228.

Levin, A. (1971). Scheduling and fleet routing models for transportation systems. *Transp. Sci.* 5, 232–255.

Lin, S. (1965). Computer solutions of the traveling salesman problem. *Bell System Tech. J.* 44, 2245–2269.

Lin, S., and B. Kernighan (1973). An effective heuristic algorithm for the traveling salesman problem. *Oper. Res.* 21, 498–516.

Lucena, A. (1986). Exact solution approaches for the vehicle routing problem. Ph.D. Thesis, Department of Management Science, Imperial College of Science and Technology, University of London.

Maculan, N., P. Michelon and G. Plateau (1992). Column- generation in linear programming with bounding variable constraints and its application in integer programming, Working Paper ES-268/92, Federal University of Rio de Janeiro, P.O. Box 68511, 21945 Rio de Janeiro, 13 pp.

Madsen, O.B.G. (1976). Optimal scheduling of trucks — a routing problem with tight due times for delivery, in: H. Strobel, R. Genser and M. Etschmaier (eds.), *Optimization Applied to Transportation Systems*, International Institute for Applied System Analysis, CP-77-7, Laxenburgh, pp. 126–136.

Madsen, O., H. Ravn and J.R. Rygaard (1993). REBUS, a system for dynamic vehicle routing for the Copenhagen Fire Fighting Company, Research Report 2/1993, IMSOR, the Technical University of Denmark, Lyngby.

Magnanti, T.L. (1981). Combinatorial optimization and vehicle fleet planning: perspectives and prospects. *Networks* 11, 179–214.

Magnanti, T.L., J.F. Shapiro and M.H. Wagner (1976). Generalized linear programming solves the dual. *Manage. Sci.* 22, 1195–1203.

Malandraki, C., and M. Daskin (1989). A cutting plane heuristic algorithm for the time dependent traveling salesman problem, Working Paper, Department of Civil Engineering, Northwestern University, Evanston, Ill.

Manington, P., and A. Wren (1975). Experiences with bus scheduling algorithm which saves vehicles, in: L. Bodin and D. Bergmann (eds.), *Preprints, Workshop on Automated Techniques for Scheduling of Vehicle Operations for Urban Public Transportation Services*, Chicago, Ill.

Marsten, R.E., and F. Shepardson (1981). Exact solution of crew scheduling problems using the set partitioning model: recent successful applications. *Networks* 11, 165–177.

Marsten, R.E., M.R. Muller and C.L. Killion (1979). Crew planning at flying tiger: a successful application of integer programming. *Manage. Sci.* 25, 1175–1183.

Martin, G. (1981). Aircraft scheduling considered as an N- task, M-parallel machine problem with start-times and deadlines. *INFOR* 19, 152–161.

Martins, E.Q.V. (1984). On a multicriteria shortest path problem. *Eur. J. Oper. Res.* 16, 236–245.

Mesquita, M., and J. Paixão (1992). Multiple depot vehicle scheduling problem: a new heuristic based on quasi-assignment algorithms, in: M. Desrochers and J.-M. Rousseau (eds.), *Computer-Aided Transit Scheduling*, Lecture Notes in Economics and Mathematical Systems 386, Springer Verlag, Berlin, Heidelberg, pp. 181–195.

Miller, C., A. Tucker and R. Zemlin (1960). Integer programming formulations and traveling salesman problems. *J.A.C.M.* 7, 326–329.

Minoux, M. (1975). Plus court chemin avec contraintes: algorithmes et applications. *Ann. Télécom.* 30, 1–12 (in French).

Mitra, G., and K. Darby-Dowman (1983). CRU-SCHED: A computer based bus crew scheduling system using integer programming, in: J.-M. Rousseau (ed.), *Computer Scheduling of Public Transport 2*, North-Holland, Amsterdam, pp. 223–232.

Moore, E.F. (1959). The shortest path trough a maze, *Proc. Int. Symp. on the Theory of Switching,*

Part II, April 2–5, 1957, Harvard University Press, Cambridge, Mass.

Nemhauser, G.L., and L.A. Wolsey (1988). *Integer and Combinatorial Optimisation*. John Wiley & Sons, New York, N.Y.

Or, I. (1976). *Travelling salesman-type combinatorial problems and their relation to the logistics of blood banking*, Ph.D. Dissertation, Department of Industrial Engineering and Management Sciences, Northwestern University, Evanston, Ill.

Orloff, C. (1976). Route constrained fleet scheduling. *Transp. Sci.* 10, 149–168.

Paixão, J., and I.M. Branco (1987). A quasi-assignment algorithm for bus Scheduling. *Networks* 17, 249–270.

Pallottino, S. (1984). Shortest path methods: complexity, interrelations and new propositions. *Networks* 14, 257–267.

Papadimitriou, C. (1977). The Euclidean traveling salesman problem is NP-complete. *Theor. Comput. Sci.* 4, 237–244.

Pape, U. (1980). Algorithm 562: shortest path lengths. *ACM Trans. Math. Software* 6, 450–455.

Potvin, J.-Y., and J.-M. Rousseau (1993). A parallel route building algorithm for the vehicle routing and scheduling problem with time windows. *Eur. J. Oper. Res.* 66, 331–340.

Potvin, J.-Y., G. Lapalme and J.-M. Rousseau (1989). A generalized K-opt exchange procedure for the MTSP. *INFOR* 27, 474–481.

Potvin, J.-Y., T. Kervahut and J.-M. Rousseau (1992). A Tabu search heuristic for the vehicle routing problem with time windows. Publication 855, Centre de Recherche sur les Transports, Université de Montréal.

Psaraftis, H. (1980). A dynamic programming solution to the single-vehicle, many-to-many, immediate request dial-a-ride problem. *Transp. Sci.* 14, 130–154.

Psaraftis, H. (1983). An exact algorithm for the single-vehicle many-to-many dial-a-ride problem with time windows. *Transp. Sci.* 17, 351–357.

Psaraftis, H. (1986). Scheduling large-scale advance-request dial-a-ride systems. *Am. J. Math. Manage. Sci.* 6, 327–367.

Psaraftis, H.N., J.B. Orlin, D. Bienstock and P.M. Thompson (1985). Analysis and solution algorithms of sealift routing and scheduling problems: Final report, Working Paper 1700-85, M.I.T., Sloan School of Management, Boston, Mass.

Psaraftis, H., M.M. Solomon, T.L. Magnanti and T. Kim (1990). Routing and scheduling on the shoreline with release times. *Manage. Sci.* 36, 212–223.

Pullen, H., and M. Webb (1967). A computer application to a transport scheduling problem. *Comput. J.* 10, 10–13.

Rappoport, H.K., L.S. Levy, B.L. Golden and K. Toussaint (1992). A planning heuristic for military airlift. *Interfaces* 22, 73–87.

Rappoport, H.K., L.S. Levy, K. Toussaint and B.L. Golden (1992). A transportation problem formulation for the MAC airlift planning problem, Working Paper, University of Maryland, College Park, Md.

Ribeiro, C., and F. Soumis (1994). A column generation approach to the multiple depot vehicle scheduling problem, *Oper. Res.* 42, 41–52.

Rousseau, J.-M., ed. (1985). *Computer Scheduling of Public Transport 2*. North-Holland, Amsterdam.

Roy, S., J.-M. Rousseau, G. Lapalme and J.A. Ferland (1984a). Routing and scheduling for the transportation of disabled persons — the algorithm, Working Paper TP 5596E, Transport Canada, Transport Development Center, Montreal.

Roy, S., J.-M. Rousseau, G. Lapalme and J.A. Ferland (1984b). Routing and scheduling for the transportation of disabled persons — the tests, Working Paper TP 5598E, Transport Canada, Transport Development Center, Montreal.

Russell, R. (1977). An effective heuristic for the M-tour traveling salesman problem with some side conditions. *Oper. Res.* 25, 517–524.

Ryan, D.M., and B.A. Foster (1981). An integer programming approach to scheduling, in: A. Wren (ed.), *Computer Scheduling of Public Transport Urban Passenger Vehicle and Crew Scheduling*, North-Holland, Amsterdam, pp. 269–280.

Ryan, D.M., and K.M. Garner (1985). The solution of air-crew scheduling problems for Air New Zealand, *Proc. 21st Annu. Conf. of O.R.S.N.Z.*, pp. 42–48.

Sansó, B., M. Desrochers, J. Desrosiers, Y. Dumas and F. Soumis (1990). Modeling and solving routing and scheduling problems: GENCOL User Guide, GERAD, 5255 Decelles, Montréal.

Savelsbergh, M.W.P. (1985). Local search in routing problems with time windows. *Ann. Oper. Res.* 4, 285–305.

Savelsbergh, M.W.P. (1988). Computer aided routing, Ph.D. Dissertation, Centre for Mathematics and Computer Science, Amsterdam.

Savelsbergh, M.W.P. (1989). The vehicle routing problem with time windows: minimizing route duration, Working Paper, Centre for Mathematics and Computer Science, Amsterdam.

Sexton, T., and L. Bodin (1985a). Optimizing single vehicle many-to-many operations with desired delivery times: I. Scheduling. *Transp. Sci.* 19, 378–410.

Sexton, T., and L. Bodin (1985b). Optimizing single vehicle many-to-many operations with desired delivery times: II. Routing. *Transp. Sci.* 19, 411–435.

Sexton, T., and Y. Choi (1986). Pickup and delivery of partial loads with time windows. *Am. J. Math. Manage. Sci.* 6, 369–398.

Smith, B., and A. Wren (1981). VAMPIRES and TASC: Two successfully applied bus scheduling programs, in: A. Wren (ed.), *Computer Scheduling of Public Transport: Urban Passenger Vehicle and Crew Scheduling*, North-Holland, Amsterdam, pp. 97–124.

Smith, B.M., and A. Wren (1988). A bus crew scheduling system using a set covering formulation. *Transp. Res.* 22A, 97–108.

Solanki, R.S., and F. Southworth (1991). An execution planning algorithm for military airlift. *Interfaces* 21, 121–131.

Solomon, M.M. (1983). The vehicle routing and scheduling problem with time window constraints: models and algorithms. Ph.D. Dissertation, Department of Decision Sciences, University of Pennsylvania, Philadelphia, Pa.

Solomon, M.M. (1986). On the worst-case performance of some heuristics for the vehicle routing and scheduling problem with time window constraints. *Networks* 16, 161–174.

Solomon, M.M. (1987). Algorithms for the vehicle routing and scheduling problem with time window constraints. *Oper. Res.* 35, 254–265.

Solomon, M.M., and J. Desrosiers (1988). Time window constrained routing and scheduling problems. *Transp. Sci.* 22, 1–13.

Solomon, M.M., E. Baker and J. Schaffer (1988). Vehicle routing and scheduling problems with time window constraints: efficient implementations of solution improvement procedures, in: B.L. Golden and A.A. Assad (eds.), *Vehicle Routing: Methods and Studies*, North-Holland, Amsterdam, pp. 85–106.

Soumis, F., J. Ferland and J.-M. Rousseau (1980). A large scale model for airline fleet planning and scheduling problem. *Transp. Res.* 14B, 191–201.

Soumis, F., J. Desrosiers and M. Desrochers (1985). Optimal urban bus routing with scheduling flexibilities, in: P. Thoft Christensen (ed.), *Lecture Notes in Control and Information Sciences 59*, Springer-Verlag, Berlin, Heidelberg, pp. 155–165.

Swersey, A., and W. Ballard (1983). Scheduling school buses. *Manage. Sci.* 30, 844–853.

Teodorović, D. (1988). *Airlines Operations Research*. Gordon and Breach, New York, N.Y.

Thompson, P.M., and H.N. Psaraftis (1989). Cyclic transfer algorithms for multi-vehicle routing and scheduling problems, Working Paper 89-008, Leavey School of Business and Administration, Santa Clara University, Calif.

Tsitsiklis, J. (1992). Special cases of the traveling salesman and repairman problems with time windows. *Networks* 22, 263–282.

Van der Bruggen, L.J.J., J.K. Lenstra and P.C. Schuur (1993). Variable-depth search for the single-vehicle pickup and delivery problem with time windows, *Transp. Sci.* 27, 298–311.

Wedelin, D. (1993). Efficient algorithms for probabilistic inference, combinatorial optimization and the discovery of causal structure from data, Ph.D. Dissertation, Department of Computer Sciences, Chalmers University of Technology, Göteborg.

Wilson, H., and N. Colvin (1977). Computer control of the rochester dial-a-ride system, Report R77-31, Dept. of Civil Engineering, M.I.T., Boston, Mass.

Wilson, H., and H. Weissberg (1976). Advanced dial-a-ride algorithms research project: Final report, Report R76-20, Dept. of Civil Engineering, M.I.T., Boston, Mass.

Wilson, H., J. Sussman, H. Wang and B. Higonnet (1971). Scheduling algorithms for dial-a-ride systems, Urban Systems Laboratory Report USL TR-70-13, M.I.T., Boston, Mass.

Wren, A., ed. (1981). *Computer Scheduling of Public Transport Urban Passenger Vehicle and Crew Scheduling*. North- Holland, Amsterdam.

Wren, A., B.M. Smith and A.J. Miller (1985). Complementary approaches to crew scheduling, in: J.-M. Rousseau (ed.), *Computer Scheduling of Public Transport 2*, North-Holland, Amsterdam, 263–278.

Yano, C., T. Chan, L. Richter, T. Cutler, K. Murty and D. McGettigan (1987). Vehicle routing at quality stores. *Interfaces* 17, 52–63.

M.O. Ball et al., Eds., *Handbooks in OR & MS, Vol. 8*

Chapter 3

Stochastic and Dynamic Networks and Routing

Warren B. Powell

Department of Civil Engineering and Operations Research, Program in Statistics & Operations Research, Princeton University, Princeton, NJ 08544, U.S.A.

Patrick Jaillet

MSIS Department, The University of Texas at Austin, U.S.A, and Laboratoire de Mathématiques et Modélisation, ENPC, France

Amedeo Odoni

Operations Research Center, Massachusetts Institute of Technology, Cambridge, MA 02159, U.S.A.

1. Introduction

The field of logistics is becoming increasingly dominated by the need for technologies that support real-time decision making. Using recent advances in information technologies, decisions concerning the routing and scheduling of drivers and vehicles, management of vehicle inventories, and the design of service offerings, can be made in a real-time environment with information that is constantly changing. Dynamic and stochastic models are playing an increasingly important role in such a setting: by definition, one is forced to make decisions before all the information one would wish to have becomes available and then modify these decisions as new information is received. The most common form of stochasticity arises as a result of uncertainty concerning some aspect of demand (level of demand, location, timing, etc.) but many other forms may be present, as well (length of travel times, resource availability, service breakdowns, etc.). As a rule, once a set of decisions has been made and some action has been taken, the decision-makers have the opportunity to observe an outcome of (some of) the uncertain events and must then respond to these events. The inherent dynamism of this type of operation often introduces important analytical complications: initial decisions may be greatly affected by how well the decision-makers — and the system they manage — are equipped to respond to subsequent random events.

Dynamic and stochastic models undoubtedly represent the 'wave of the future.' Their increasing prominence is driven by technology: the explosive growth in the availability of real-time information about transportation and logistics systems is turning the focus of operations researchers away from the traditional static planning models. Carriers and shippers are avidly seeking the benefits afforded by the ability to rapidly and continually 'reconfigure' the operation of a transportation system to improve service or reduce cost. In addition, even when it comes to strategic planning, methodological developments over the last few years have made possible the development of models that capture better the uncertainty and the dynamic characteristics associated with the transportation and logistics environment. Clearly such models often constitute far more accurate representations of reality than some of their deterministic and static forebears.

Model applications in this general area can, in fact, be divided into two broad classes along the strategic vs. tactical lines just mentioned. The first class of models deals with a priori optimization, or, in more technical terms, two-stage programming with recourse. A good example is the location and sizing of a warehouse. Typically, a decision on the warehouses location and capacity must be made on the basis of a probabilistic description of customer demands. Once the facility is built, the characteristics of the 'market' become better known and decisions can then be made about how to serve demands using a fleet of vehicles. However, this service may be unnecessarily costly or inefficient, if the selected location or size of the warehouse turn out to have been poor, in the first place. Note that, in such problems, the time from when the first set of decisions is made to when the random outcomes become (partially) known, is typically fairly large.

The second set of applications involves multistage problems where the process of making decisions and observing outcomes occurs on a continuous, rolling basis. A set of vehicles may be routed over the course of the day while demands are continuously being called in. With each new demand, the vehicle tours may be redesigned, not only to accommodate actual demands, but anticipated demands as well. Another example arises in dynamic fleet management, where empty vehicles are repositioned to anticipate future demands. Such decisions are made daily, as shipper demands are continuously being received at a dispatching center.

The two classes of problems, then, are both stochastic, since we have to anticipate the future, and dynamic, since decisions are made over time. However, the study of the properties of stochastic two stage problems (a priori optimization) can often be done without explicitly handling the dynamics of the problem — in essence, through a static model. By contrast, it is very common to solve stochastic, multistage problems using deterministic models. We may approximate the future using a deterministic model, updating the information on a rolling basis as random events become known. As a result, our presentation spans (i) static, stochastic models, (ii) deterministic dynamic models, and (iii) multistage stochastic models. We feel that a unified presentation of all three perspectives provides valuable insights into the formulation and solution of this difficult and important class of problems.

One question that comes up with surprising frequency is: what constitutes a dynamic model? To answer this, we must first distinguish between a problem, a model, and the application of a model. A problem is dynamic if one or more of its parameters is a function of time. This includes such problems as vehicle routing with time windows or with time-varying travel times. Note that two types of dynamic problems are covered here. The first type, which we call problems with dynamic data, are characterized by information that are constantly changing.

Dynamic data might include real-time customer demands, traffic conditions, or driver statuses. The second type is problems with time-dependent data which is known in advance. In this category, we would include problems such as vehicle routing with time windows where all the information is known in advance, but where this data is a known function of time. Other examples of time-dependent data might be customer demands or travel times, if we can assume them to be known functions of time.

Similarly, a model is dynamic if it incorporates explicitly the interaction of activities over time. The simplest dynamic model is a dynamic network, a construct widely used in routing and scheduling problems. It is useful, however, to distinguish between deterministic, dynamic models, and stochastic models which explicitly capture the *staging* of decisions and the realization of random variables. For example, it is not unusual to solve deterministic, dynamic models without recognizing in any way the dynamic structure of the problem. By contrast, stochastic, dynamic models require specific steps to be taken in the design of a solution strategy.

Finally, we have a dynamic application if a model is solved repeatedly as new information is received. Dynamic applications of models place tremendous demands on access to real-time data and on the performance of algorithms. Typically, it is necessary to update information, optimize and return results in a matter of minutes or even seconds.

A useful illustration of these concepts arises in the problem of routing a vehicle over a congested transportation network. Clearly, traffic conditions are changing over time, and hence the problem is dynamic. We might select an optimal route using a static model, if we chose not to represent dynamic conditions explicitly within the model. (For example, we could find the optimal route by simply minimizing average travel times, thus working with one particular static representation of the network.) We might also solve the static model repeatedly as new information became known, giving us a dynamic application of a static model. Yet a third alternative might be to develop a model that would consider explicitly the anticipated dynamic changes in traffic conditions and apply it once at the outset of the period of interest. In other words we would select one route and then stick to it even as conditions change. This would constitute a static application of a dynamic model. Finally, had we chosen to solve the dynamic model repeatedly as new information became known, we would have a dynamic application of a dynamic model. This example, then, indicates how one can have a static or dynamic application of a static or dynamic model.

Stochastic, dynamic models represent a rich area of research. Important dimensions of the problem that require attention include:

1. Definition of specific application areas.
2. Development of tractable mathematical models, with particular emphasis on optimization under uncertainty.
3. Development of efficient formulations and solution algorithms.
4. Evaluation of alternative models.
5. Integration of real-time optimization models and on-line databases.
6. Implementation and user acceptance of real-time decision support systems.

In this chapter, we cover aspects of the first four issues only; the last two are beyond the scope of our presentation. The following list is only indicative of the enormous variety of problems that might be addressed using dynamic and stochastic models:

1. Production and inventory planning: Each week (or month) it is necessary to develop a production, inventory and transportation plan to meet anticipated demands over a specified time horizon of perhaps several months or one year).

2. Vehicle routing: This involves the dynamic routing of a fleet of vehicles to meet real-time customer demands as they are called in. Problems in this class can arise in a real-time setting, where these decisions must be made in response to actual conditions, or in a planning environment, where future demands must be forecast and are subject to considerable uncertainty.

3. Design of a service network: A carrier must design a package of scheduled transportation services which must then be communicated to the market and modified dynamically in response to the market's feedback.

4. Repositioning of empty vehicles to anticipate future demands: When too many vehicles accumulate in a single city or region, it may be necessary to move excess vehicles to adjacent areas where better opportunities for generating revenue may exist.

5. Booking strategies: In static models, it is common to minimize costs subject to the constraint of satisfying all demands. In a dynamic context, it is usually not possible to satisfy all the demands all the time. It may be necessary to turn down customer requests that cannot be satisfied or to try to postpone certain services. Dynamic booking controls the commitment of a carrier to handle incoming demands and allows the carrier to schedule when they will be served.

6. Costing and pricing: Costing and pricing activities in transportation require understanding the interactions among different markets over time. For example, the profit from a particular vehicle movement may depend on the scheduling and timing of other movements around the network. Costing and pricing issues can arise in both a real-time setting (requiring a 'spot' evaluation) as well as in the longer run, e.g., evaluating a new contract which might affect existing conditions on a network.

7. Determination of level-of-service for specific loads/shipments: It is possible to provide different levels of service to different customers, by design. Carriers must regularly make decisions about how much effort to put into a particular shipment to ensure adequate service. With proper planning, it should be possible to set appropriate expectations when a load is accepted for service.

8. Tactical sales planning and load solicitation: In some types of transportation, it is possible to use sales and marketing adjustments to help control the size of demand served on a day to day basis. In truckload trucking, for instance, it is possible to anticipate the total number of trucks that will be available in a given region one to three days in advance. Then, the telemarketing department of the carrier can call shippers and solicit enough freight in each region to help balance the number of loads with the available capacity.

9. Fleet mix specification: This is the standard problem of determining how many of each type of vehicle to include in the fleet of a carrier.

10. Facility planning and design. Facility location and sizing decisions are customarily viewed as static problems which are solved using static models. Quite often, however, there are important advantages in examining such problems in a dynamic context. Examples include determining the size and location over time of plants and warehouses in a logistics network, as well as the location of terminals and driver domiciles for a carrier.

Because of the length of the chapter it is useful to provide a 'roadmap' though it for the reader. It consists of four major parts:

Part I   A priori optimization
          Section 2: A priori (two-stage) stochastic models
Part II   Dynamic models in logistics
          Section 3: Modelling issues for dynamic problems
          Section 4: Dynamic models in transportation and logistics
          Section 8: Stochastic programming models in networks and routing
Part III Dynamic networks
          Section 5: Algorithms for deterministic, dynamic networks
          Section 6: Infinite horizon network models
          Section 7: Stochastic programming for networks
          Section 9: Approximations for networks with random arc capacities
Part IV   Model evaluation
          Section 10: Evaluating dynamic models

Part I is dedicated to a priori optimization in routing, covering shortest paths, traveling salesman-type problems and vehicle routing. These problems arise when decisions must be made before random outcomes (typically customer demands) are known.

Part II covers dynamic models of problems arising in transportation and logistics, and includes a discussion of important modeling issues, as well as a summary of dynamic models for a number of key problem areas. Section 8, which covers stochastic, dynamic models, is presented near the end of the chapter, following a review of fundamental concepts in stochastic programming.

Part III focuses specifically on dynamic networks, which provide an important foundation for addressing many problems in logistics planning. Section 5 presents algorithms that have been specialized for dynamic networks. Section 6 discusses results for solving infinite networks, including both exact results for stationary infinite networks, and model truncation techniques. Section 7 presents basic results and concepts from the field of stochastic programming, oriented toward their application to network problems. This discussion provides a general framework for formulating and solving stochastic, dynamic network problems. That framework is used to present two stochastic programming models in Section 8. Section 9 then covers a series of results that have been derived for the special case of dynamic networks with random arc capacities. Links with random upper bounds can be used to model random demands for many problems in logistics and distribution.

This section summarizes work that has leveraged this special structure to obtain much stronger results than have been possible in other, more difficult problem areas such as dynamic vehicle routing.

Finally, Part IV addresses the special question of evaluating stochastic, dynamic models. The issue of evaluation is an important one and distinguishes sharply static from dynamic models. In static models, the choice of objective function is usually fairly obvious. That same objective function also provides the yardstick for evaluating the quality of the solution. In dynamic models, by contrast, the objective function used to make decisions on a rolling horizon basis may often have little to do with the measures developed to evaluate the overall quality of a solution.

We have sought to present our material in a way that might enhance two perspectives, namely the formulation and solution of models and the analysis of operating strategies. Our goal is to provide a structured presentation of alternative modeling and algorithmic frameworks that may be useful to both practitioners and researchers.

While every effort has been made to present a reasonably complete picture, this chapter is not intended to be a comprehensive survey of all applications of dynamic networks. Early bibliographies that the reader may wish to consult include those of Bradley [1975] and of Golden and Magnanti [1988]. Assad [1987] provides an extensive review of the literature on multi-commodity flows. A thorough treatment of network models and algorithms is provided in Ahuja, Magnanti & Orlin [1992]. Finally, of particular value is the extensive review of dynamic networks given in Aronson [1989]. This review provides a summary of specialized algorithms for dynamic networks, as well as a fine discussion of the planning horizon literature as it relates to dynamic networks. Aronson's survey also includes work drawn from other applications and a number of papers that involve static networks. In contrast, our presentation is restricted to dynamic models and related methodologies, focusing mostly on applications in logistics. In addition, while we consider deterministic models in depth, we also put considerably greater emphasis on stochastic models than did Aronson's review. We also cover a wider range of issues associated with model formulation, and the interaction between model formulation and the design of algorithms.

2. A priori (two-stage) stochastic models

2.1. Introduction and problem definitions

This section is concerned with a specific family of routing problems on networks. Most of these problems have been addressed only recently, and their common characteristic is the explicit inclusion of probabilistic elements in the problem definitions, as will be explained below. For this reason we shall refer to them as stochastic routing problems (SRPs). In the sense that these problems are addressed in a non-dynamic environment, they are also designated as 'static' — to distinguish them from the dynamic and stochastic routing problems (DSRPs) which will be the subject of the next section.

The objective here is to offer a broad overview that emphasizes the fundamental concepts and provides an up-to-date reference list for the numerous developments which have been published in this area during the last few years. The topics that will be reviewed deal with:

(i) the concept of 'robust' a priori solutions to SRPs that perform well on average;

(ii) the performance of the a priori optimization approach relative to the strategy of 're-optimization', i.e., the strategy of solving optimally every potential instance of the original problem;

(iii) the computational complexity and the combinatorial properties of a priori optimization;

(iv) exact and heuristic algorithms, together with theoretical worst-case and average-case bounds for the heuristics;

(v) the asymptotic behavior of a priori optimization and of re-optimization strategies;

(vi) available computational experience.

This subsection will introduce all these ideas, along with formulations of the several problems to be discussed in more detail later. The presentation shall draw heavily on the general framework, which was first established in Bertsimas, Jaillet & Odoni [1990].

There are several motivations for investigating the effect of including probabilistic elements in routing problems. Among them two are of particular importance. The first is the desire to define and analyze models which are more appropriate for those real-world problems in which randomness is not only present but a major concern, as well. There is a plethora of important and interesting applications of SRPs, especially in the context of strategic planning for collection and distribution services, communication and transportation systems, job scheduling, organizational structures, etc. For such applications, the probabilistic nature of the models makes them particularly attractive as mathematical abstractions of real-world systems.

The second motivation is interest in investigating the robustness (with respect to optimality) of optimal solutions to deterministic routing problems, when the instances for which these problems have been solved, are modified. In our case, we confine the investigation to problems on networks and the perturbation of a problem's instance is simulated by the presence or absence of subsets of the network's set of nodes. Such considerations are particularly important for $\mathcal{NP}$-hard problems, for which the effort to get an optimal solution is important, if not prohibitive.

We next discuss one of the central themes of the conceptual approach adopted in addressing and characterizing SRPs, namely the idea of a priori optimization. In many applications, one finds that, after solving a given instance of a route optimization problem, it becomes necessary to solve repeatedly many other instances of the same problem. These other instances are usually just variations of the instance solved originally. Yet, they may be sufficiently different from that original instance to necessitate every time a re-consideration of the entire problem on the part of the analyst.

The most obvious approach in dealing with such cases is to attempt to solve optimally (or near-optimally with a good heuristic) every potential instance of the original problem. We call this approach the 're-optimization strategy' and denote it with the Greek letter Σ. The approach, however, suffers from several disadvantages. For example, if the routing problem considered is $\mathcal{NP}$-hard, one might have to solve exponentially many instances of a hard problem. Moreover, in many applications it is necessary to find a solution to each new instance quickly, but one might not have the required computing or other resources for doing so.

As an alternative, we shall also investigate here a different strategy. Rather than re-optimizing every potential instance, we wish to find an a priori solution to the original problem and then update in a simple way this a priori solution to answer each particular instance/variation. Clearly, the natural questions to ask are: What is the measure of 'effectiveness' of such an a priori solution? Once such a measure has been defined, how does one find the best a priori solution? And, how does one update the a priori solution for each particular problem instance?

The above discussion is general, in the sense that it applies to any combinatorial optimization problem. In order to address these questions concretely, we restrict our attention to a class of network routing problems. Consider then a complete graph $G = (V, E)$ on n nodes on which a routing problem is defined (for example the traveling salesman problem). If every possible subset of the node set V may or may not be present on any given instance of the optimization problem (for example, on any given day, the traveling salesman may have to visit only a subset S of the nodes in V), then there are 2^n possible instances of the problem — all the possible subsets of V. Suppose instance S has probability $p(S)$ of occurring. Given a method $\mathcal{U}$ for updating an a priori solution f to the 'full-scale' optimization problem on the original graph $G(V, E)$, $\mathcal{U}$ will then produce for problem instance S, a feasible solution $t_f(S)$ with value ('cost') $L_f(S)$. (In the case of the TSP, $t_f(S)$ would be a tour through the subset S of nodes and $L_f(S)$ the length of that tour.) Then, given that we have already selected the updating method $\mathcal{U}$, the natural choice for the a priori solution f is to select f so as to minimize the expected cost

$$E[L_f] = \sum_{S \subseteq V} p(S) L_f(S), \tag{1}$$

with the summation being over all subsets of V. In other words, we would like to minimize the 'weighted average' over all problem instances of the values $L_f(S)$ obtained by applying the updating method $\mathcal{U}$ to the a priori solution f.

This choice of a measure of effectiveness for the a priori solution f that we seek, namely the expected cost (1), gives a reasonable answer to our first question. But what properties should the updating method $\mathcal{U}$ have? The most desirable property of $\mathcal{U}$ would be for $L_f(S)$ to be 'close' to the value of the optimal solution $L_{\text{OPT}}(S)$, for every instance S. A less restrictive and more global property is to require $E[L_f]$ to be 'close' to the expected cost $E[\Sigma]$, over all problem instances, of the re-optimization strategy:

$$E[\Sigma] = \sum_{S \subseteq V} p(S) L_{\mathrm{OPT}}(S). \tag{2}$$

In addition, $\mathcal{U}$ must be able to update efficiently the solution from one problem instance to the next.

In the following definitions of the updating methods $\mathcal{U}$, the choices of $\mathcal{U}$ may initially seem arbitrary. But these choices will turn out to be natural ones. First, for every choice of $\mathcal{U}$ we are proposing, the updating of the solution to a particular instance S can be done very easily. Moreover, these updating methods are well suited for applications.

After this general discussion of the rationale behind the definitions which follow, we describe informally the problems we are considering.

The probabilistic traveling salesman problem

The probabilistic traveling salesman problem (PTSP) is perhaps the most fundamental stochastic routing problem that can be defined. It is essentially a traveling salesman problem (TSP), in which the number of points to be visited in each problem instance is a random variable.

Consider a problem of routing through a set of n known points. On any given instance of the problem only a subset S consisting of $|S| = k$ out of n points $(0 \leq k \leq n)$ must be visited. Suppose that the probability that instance S occurs is $p(S)$. As mentioned above, ideally we might like to re-optimize the tour for every instance, but in many cases we may not have the resources to do so or, even if we had them, re-optimization might turn out to be too time consuming. Instead, we wish to find a priori a tour through all n points. On any given instance of the problem, the k points present will then be visited in the same order as they appear in the a priori tour (see Figure 1 for an illustration). The problem of finding such an a priori tour which is of minimum length in the expected value sense is defined as the PTSP. The updating method $\mathcal{U}$ for the PTSP is therefore to visit the points

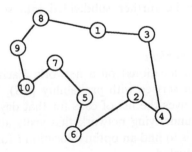

A priori tour through 10 points.

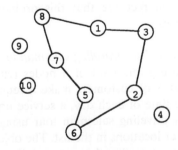

The tour when points 4, 9, and 10 need not be visited.

Fig. 1. The PTSP methodology.

on every problem instance in the same order as in the a priori tour, i.e. we simply skip those points which are not present in that problem instance.

The expectation is computed over all possible instances of the problem, i.e. over all subsets of the vertex set $V = \{1, 2, \ldots n\}$. That is, given an a priori tour τ, if problem instance $S(\subseteq V)$ will occur with probability $p(S)$ and will require covering a total distance $L_\tau(S)$ to visit the subset S of customers, that problem instance will receive a weight of $p(S)L_\tau(S)$ in the computation of the expected length. If we denote the length of the tour τ by L_τ (a random variable), then our problem is to find an a priori tour through all n potential customers, which minimizes the quantity

$$E[L_\tau] = \sum_{S \subseteq V} p(S)L_\tau(S), \tag{3}$$

with the summation being over all subsets of V.

The probabilistic vehicle routing problem

Consider a standard VRP but with demands which are probabilistic in nature rather than deterministic. The problem is then to determine a fixed set of routes of minimal expected total length, which corresponds to the expected total length of the fixed set of routes plus the expected value of extra travel distance that might be required. The extra distance will be due to the possibility that demand on one or more routes may occasionally exceed the capacity of a vehicle and force it to go back to the depot before continuing on its route.

The following two solution-updating methods can be defined: Under method $\mathcal{U}_a$ the vehicle visits all the points in the same fixed order as under the a priori tour, but serves only customers requiring service during that particular problem instance. The total expected distance traveled corresponds to the fixed length of the a priori tour plus the expected value of the additional distance that must be covered whenever the demand on the route exceeds vehicle capacity. Method $\mathcal{U}_b$ is defined similarly to $\mathcal{U}_a$ with the sole difference that customers with no demand on a particular instance of the vehicle tour are simply skipped. An example of the PVRP under both updating methods can be seen in Figure 2. In Section 2.3, we will in fact see that this problem can be further subdivided into several categories.

The probabilistic traveling salesman location problem

We are given a set of n nodes (customer locations) on a network. Each day a subset S of customers make a request for service with probability $p(S)$. By a specific time of each day, a service unit receives the list of calls for that day and starts a traveling salesman tour using the underlying network that visits all the customer locations in the list. The objective is to find an optimal location i for the service unit, so that the expected distance traveled

$$E[\Sigma_{\text{TSLP}}(i)] = \sum_{S \subseteq V} p(S)L_{\text{OPT}}(S \cup i) \tag{4}$$

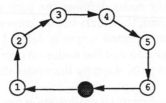

(i) A prior route through 6 customers (each with a demand of 0 or 1 unit) by a vehicle of capacity 2.

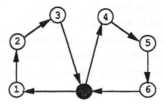 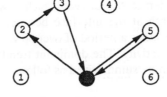

Nonskipping strategy (method a).   Skipping strategy (method b).

(ii) The two strategies when only customers 2, 3, and 5 have a positive demand.

Fig. 2. The PVRP methodology.

is minimized. This problem is called the traveling salesman location problem (TSLP).

The difficulty of having to compute the optimal tour for every instance can be overcome by using an a priori tour τ_p and then follow the PTSP updating method $\mathcal{U}$ described before, i.e., skip customer locations with no demand. The problem is then to find a node and an a priori tour to minimize the expected distance traveled using the PTSP approach, i.e. to minimize

$$h(i, \tau) \triangleq \sum_{S \subseteq V} p(S) L_\tau(S \cup i). \tag{5}$$

The problem of finding simultaneously an optimal location and an optimum a priori tour is called the probabilistic traveling salesman location problem (PTSLP).

The probabilistic shortest path problem

The probabilistic shortest path problem (PSPP) can be described as follows: consider the problem of finding a shortest path between a node source s and a node sink t in a complete network having a length associated with each arc. On any given instance of the problem only a subset among intermediate nodes can be used to go from s to t, the subset being chosen according to a given probability law. We wish to construct an a priori path such that, on any given instance of the problem, the sequence of nodes defining the path is preserved but only the permissible nodes are traversed, the others being skipped. The problem of finding an a priori path of minimum expected length is defined as a PSPP. The updating method $\mathcal{U}$ for the PSPP is again to visit the nodes on every problem instance in the same order as in the a priori path, skipping the nodes that cannot be used in that problem instance.

During the last decade combinatorial optimization has undoubtedly been one of the fastest growing and most exciting areas in mathematical programming. Needless to say, the related scientific literature has been expanding at a very rapid pace. Examples of particular relevance to this chapter are the three excellent review volumes on the traveling salesman problem [Lawler, Lenstra & Shmoys, 1985], on routing and scheduling [Bodin, Golden, Assad & Ball, 1983], and on vehicle routing [Golden & Assad, 1988], each of which offers several hundreds of references.

Research at the interface between probability theory and combinatorial optimization spans a period of over 30 years and in recent years has been at the center of much activity. The dominant trends of this interplay which are relevant to this paper can be summarized as follows.

Probabilistic analysis of combinatorial optimization problems in the Euclidean plane. Research in this area was initiated by the pioneering paper of Beardwood, Halton & Hammersley [1959]. After a period of more than 15 years and motivated by the significant advances in theoretical computer science, Karp [1977] used their main result to propose a partitioning heuristic, which constitutes an ϵ-approximation algorithm for the TSP in the Euclidean plane.

In the last decade, the asymptotic properties of many combinatorial optimization problems in the Euclidean plane have been investigated. The most general analysis in this direction is due to Steele [1981a], who developed the theory of subadditive Euclidean functionals to obtain very sharp limit theorems for a broad class of combinatorial optimization problems. Rates of convergence for general functionals have also recently been addressed by Jaillet [1992a] (see also Rhee & Talagrand [1988] for the TSP).

Probabilistic analysis on problems with random lengths. In the last decade there have been numerous papers dealing with the behavior of combinatorial optimization problems when the costs involved are taken from a probability distribution. Interest in this area intensified after the pioneering paper of Karp [1979] on the TSP and the attempts to explain probabilistically the success of the simplex method for linear programming. Of particular relevance to this chapter are papers on stochastic (and static) versions of the shortest path problem in which arc lengths are random [see Frank, 1969; Andreatta, Ricaldone & Romeo, 1985; Frieze & Grimmet, 1985; Hassin & Zemel, 1985; Andreatta & Romeo, 1988; and Bertsekas & Tsitsiklis, 1991].

Stochastic routing problems. In contrast to their deterministic counterparts, the professional literature on SRPs to date is very sparse. Jaillet [1985] introduced the PTSP, examined some of its combinatorial properties and proved asymptotic theorems in the plane. A summary of these results as well as a discussion on the applications of the PTSP and the PVRP are contained in Jaillet & Odoni [1988]. Bertsimas [1988] introduced the framework of a priori optimization and further studied the PTSP, PVRP as well as other problems.

Except for an isolated result in the 1970's [Tillman, 1969], VRPs with stochastic elements in their definitions have received attention only recently. Stewart & Golden [1983], Dror & Trudeau [1986], Dror, Laporte & Trudeau, [1989]; and Laporte & Louveau [1990] use techniques from stochastic programming to solve optimally small problems and find bounds for them. The definitions of these problems are different from the ones we are considering in this section.

The traveling salesman location problem has been considered by Eilon, Watson-Gandy & Christofides [1971] and Burness & White [1976], where heuristic approaches are proposed. Recently in a series of papers, Berman & Simchi-Levi [1986, 1988a, b] and Simchi-Levi & Berman [1988] solved the problem on a tree network and proposed a heuristic of relative worst error $1/2$ for the general network case as well as for the Euclidean and the rectilinear metric.

The probabilistic shortest path problem was introduced in Jaillet [1988] in which a branch and bound scheme was proposed. A thorough study of the complexity as well as of some special polynomial cases of this problem is contained in Jaillet [1992b].

A final remark has to do with the relationship between network reliability theory and the class of SRPs we are considering. In network reliability theory [see for example Colbourn, 1987] the nodes are usually assumed to be always reliable and the questions addressed concern the existence of paths among pairs of nodes. In case of the SRPs the type of questions we are addressing as well as the motivation for their definition are different.

As noted earlier, SRPs could prove highly useful in many application contexts in which the explicit consideration of randomness is essential. For instance, the PTSP arises in practice whenever a company, on any given day, is faced with the problem of collections (deliveries) from (to) a random subset of some known global set of customers in an area and does not wish to or, simply, cannot redesign the tours from scratch every day. Examples in this category include a 'hot meals' delivery system described by Bartholdi, Platzman, Collins, Lee & Warden [1983], routing of forklifts in a cargo terminal or in a warehouse and, interestingly, the daily delivery of mail to homes and businesses by postal carriers everywhere. In fact it was this last application that led to the initial formulation of the PTSP in Jaillet [1985]. Jaillet & Odoni [1988] describe in considerable detail an application in a strategic planning context in which a package distribution company has decided to begin service in a particular area. After carrying out a market survey and identifying a set of potential major customers who during any single time period have a significant probability of requiring a visit, the company wishes to estimate the resources necessary to serve these customers. The PTSP then provides a model for computing approximately the expected amount of travel that will be required per time period and, by implication, the number of vehicles, drivers, etc.

In a non-routing context, PTSP models can also be of interest in many situations in which an ordering of entities of any type has to be found and that sequence has to be preserved even when some of the entities may be absent. One such example can be given from the area of job-shop scheduling: Consider the problem of loading n jobs on a machine at which a changeover cost is incurred whenever a new job is loaded. With any given ordering of the n jobs on the machine, we can then associate a total changeover cost. Any given ordering of the n jobs may also impose specific long-term requirements on the job-shop, such as a set of tasks to be performed before and after the processing of the jobs on the machine. These requirements may often be difficult to modify on a daily basis so that, if on a given day some jobs need not be processed, the relative ordering previously specified for the remaining jobs is nonetheless left unmodified. The PTSP is again relevant in analyzing such situations.

PVRPs are of course 'constrained' cases of PTSPs and thus arise in the same collection and distribution contexts as PTSPs, whenever the vehicle capacity Q becomes a practically significant issue. The capacity Q may be expressed in terms of a maximum allowable vehicle load, maximum number of stops, maximum distance per tour or some other physical or statutory limitation. For instance, in the case of the delivery of cash by a bank to a set of automatic teller machines spatially distributed throughout a city, Q might be the upper bound on the amount of money that a vehicle might carry for safety reasons. The uncertainty in this problem is due to the fact that each machine may or may not require a visit during any given time period, depending on the amount of money it dispenses. Similar applications of the PVRP can be found in most problems that combine inventory and routing considerations.

TSLPs and PTSLPs arise similarly in the complex but also very common contexts in which facility location, routing and, possibly, inventory-related decisions must be made simultaneously. Note the difference between these problems and the classical 'median' (or 'minisum') and 'center' (or 'minimax') problems in facility location theory. In the case of (P)TSLPs, once a facility is located, demands are visited through tours; therefore, the facility location problem must be 'central' relative to the *ensemble* of the demand points, as ordered by the (yet unknown) tour through all of them. By contrast, in the classical problems the facility (or facilities) must be located by considering distances to *individual* demand points, thus making the problem more tractable.

The generic PSPP, as stated, can be of interest in many applications. First consider a network in which arcs represent streets of a city, and nodes are intersections, and suppose we want to go from an origin s to a destination t along this network. The length of an arc (i, j) is defined to be the time needed to go from i to j, a value which may vary greatly. Usually, one either uses the means of the traveling times on each arc and solves a deterministic shortest path problem, or one considers the travel times as random variables and tries to solve one of many possible problems such as: finding the path with maximum probability of being the shortest, or finding the path among the shortest paths which has minimum variance, etc. (see Andreatta, Ricaldone & Romeo [1985] for a good discussion of these formulations). However, if the traffic is close to saturation, a slight perturbation of the flow can create a traffic blocking condition. In that case, we have a critical situation in which the travel time does not vary much anymore (being constantly high because of congestion) except when a blocking condition is faced (a street unexpectedly blocked for some reason, an intersection gridlocked with conflicting flows of vehicles; etc.). Depending on the frequencies of these blockings one might then select longer but less risky paths to travel from one point to another. Let us see how the PSPP offers an analytical way to choose among paths with such uncertainties. First, one can model a risky street (i, j) by adding an artificial 'probabilistic' node k that can or cannot be traversed with certain probabilities: in the first case the length of the street would be its normal travel time, in the second case it would be much higher. In the case of a risky intersection the model is even simpler: we simply model the intersection as a 'probabilistic' node. Other applications will be discussed in Section 2.5.

We now turn to a more thorough investigation of each of these problems. To keep the length of the presentation within reasonable limits, detailed derivations have been omitted. All but the more important theorem proofs are only sketchily outlined, with appropriate references given for interested readers.

2.2. The probabilistic traveling salesman problem

As indicated in the introduction, the PTSP is the most fundamental of the SRPs. For this reason, the PTSP will be used in this section to illustrate how the entire set of issues identified earlier can be addressed for a specific SRP.

2.2.1. The expected length of a given PTSP tour τ

Let us consider a PTSP defined on a given complete graph $G = (V, E)$, $|V| = n$, with a cost $d : E \rightarrow R$ and a vector $(p_1, \ldots, p_n)$ of the probabilities of presence of the vertices. To facilitate the derivation of analytic results and without loss of generality, it will be assumed henceforth that the vertices are mutually independent with regard to probability of presence.

Consider now the tour $\tau = (1, 2, \ldots, n, 1)$ and let $E[L_\tau]$ be the expected length of τ. Note that in this case the tour length L_τ can take 2^n different values, the same as the number of instances involving 'present' and 'absent' vertices. For each such instance, we would require $O(n)$ additions to compute L_τ. Thus were we to use an enumeration approach, the computational effort to compute $E[L_\tau]$ would be $O(n2^n)$ for any given tour τ. Fortunately, a much more efficient approach exists.

Theorem 1.

$$
E[L_\tau] = \sum_{i=1}^{n} \sum_{j=i+1}^{n} d(i, j) p_i p_j \prod_{k=i+1}^{j-1} (1 - p_k) +
$$

$$
+ \sum_{j=1}^{n} \sum_{i=1}^{j-1} d(j, i) p_i p_j \prod_{k=j+1}^{n} (1 - p_k) \prod_{k=1}^{i-1} (1 - p_k). \tag{6}
$$

Sketch of the proof. This result follows directly from the following argument: the instances of τ for which $d(i, j)$ makes a contribution to $E[L_\tau]$ are those in which the vertices i and j are both present, while the vertices $i + 1, \ldots, j - 1$ are absent (and are thus skipped in traversing the tour). □

Theorem 1 shows that $E[L_\tau]$ can be computed in $O(n^2)$ time under very general conditions. Jaillet [1985, 1991a] gives a generalization of Theorem 1 for the case in which the independence assumption does not hold and also discusses a number of variations and extensions of the theorem.

2.2.2. Asymptotic comparison of re-optimization and a priori optimization

We turn next to the issue of characterizing and comparing the asymptotic behavior of the re-optimization and the a priori strategies for the PTSP, if the locations of the points are uniformly and independently distributed in the Euclidean plane. This comparison is important in order to assess the promise and potential usefulness of the a priori strategies. We will be quite informal; the interested reader can consult Jaillet [1992c] for a detailed and rigorous treatment of these issues, as well as for generalizations.

Let $X^{(n)} = (X_1, \ldots, X_n)$ be n points uniformly and independently distributed in the unit square. Let L_{TSP}^n be the length of the TSP defined on $X^{(n)}$.

Let $E[\Sigma_{TSP}^n]$ be the expectation of the TSP solutions obtained under the re-optimization strategy defined on $X^{(n)}$.

Let $E[L_{PTSP}^n]$ be the expectation of the a priori strategy, i.e. the expected length of the optimal a priori solution to the PTSP defined on $X^{(n)}$.

It is well known [see Beardwood, Halton & Hammersley, 1959] that we can characterize very sharply the solutions to the deterministic TSP.

Theorem 2. *With probability 1*

$$\lim_{n \to \infty} \frac{L_{TSP}^n}{\sqrt{n}} = \beta_{TSP}. \tag{7}$$

This almost sure convergence was later strengthened by Steele [1981b] to include complete convergence, i.e.,

Theorem 3.

$$\forall \varepsilon > 0, \sum_n \mathbb{P}\left(\left| \frac{L_{TSP}^n}{\sqrt{n}} - \beta_{TSP} \right| > \varepsilon \right) < +\infty. \tag{8}$$

We now characterize the expectation of the re-optimization strategy for the PTSP assuming that each of the n points is present with the same constant probability p, which is called the coverage probability. We remark that in the following theorem the expectation is taken over all the possible 2^n instances of the problem and the probability 1 statement refers to the random locations of the points.

Theorem 4 *With probability 1*

$$\lim_{n\to\infty} \frac{E[\Sigma^n_{\text{TSP}}]}{\sqrt{n}} = \beta_{\text{TSP}}\sqrt{p}. \tag{9}$$

Sketch of the proof. The intuitive idea in the proof is that the principal contribution to $E[\Sigma^n]$ comes from the sets S with $|S|$ close to np. The reason is that the number of points present is given by a Binomial random variable with parameters n, p and hence is almost surely asymptotically equivalent to np. In this range of $|S|$ we can apply Theorem 3 to obtain Theorem 4. Note that in Jaillet [1992c], an explanation is given on why the complete convergence of Theorem 3 is crucial for this kind of argument. $\square$

Intuitively Theorem 4 means that solutions under the re-optimization strategy behave asymptotically similarly to those of the corresponding deterministic TSP but on np rather than n points. We next characterize asymptotically the a priori optimization strategy.

Theorem 5. *With probability 1*

$$\lim_{n\to\infty} \frac{E[L^n_{\text{PTSP}}]}{\sqrt{n}} = \beta_{\text{PTSP}}(p). \tag{10}$$

Sketch of the proof. We first prove that the PTSP belongs to the class of subadditive Euclidean functionals whose asymptotic behavior has been characterized by Steele [1981a]. Their value is almost surely asymptotic to $c\sqrt{n}$, where c depends on the functional. For a detailed proof the reader is again referred to Jaillet [1992c]. $\square$

Comparing Theorems 4 and 5 we can observe that the a priori and re-optimization strategies have similar asymptotic behaviors almost surely. Both theorems prove the existence of a constant but without determining the value of the constant analytically; in fact, for most similar asymptotic results, the respective limiting constants are unknown and only bounds or experimental estimations have been established [see Avram & Bertsimas [1992] and Jaillet [1993] for an important exception concerning the minimum spanning tree problem]. In fact the current best known result on the relationship between $\beta_{\text{PTSP}}(p)$ and $\beta_{\text{TSP}}\sqrt{p}$ was obtained in Jaillet [1985] and is reproduced here:

$$\tfrac{5}{8}\sqrt{p} \leq \beta_{\text{TSP}}\sqrt{p} \leq \beta_{\text{PTSP}}(p) \leq \min(\beta_{\text{TSP}}, 0.9204\sqrt{p}). \tag{11}$$

On the other hand extensive experimental work by Johnson (1989) suggests that $\beta_{\text{TSP}} \approx 0.72$. Note that it is tempting to conjecture that $\beta_{\text{TSP}}\sqrt{p} = \beta_{\text{PTSP}}(p)$, but no correct proof exists of this result (contrary to the erroneous claim in Bertsimas, Jaillet & Odoni [1990]). Yet, the a priori strategy does seem to behave (asymptotically) equally well on average with the re-optimization strategy on Euclidean problems.

2.2.3. *The complexity of a priori optimization*

Having shown that, in terms of performance, a priori strategies are attractive compared with re-optimization strategies (at least for the PTSP) we now turn to the question of how difficult it is to find the optimal a priori solution from a computational complexity perspective.

We first introduce the decision version of a PTSP. Given a complete graph $G = (V, E)$, $|V| = n$, a cost $d : E \to R$, a vector $(p_1, \ldots, p_n)$ of the probabilities of presence of the vertices and a bound B, does there exist a PTSP tour f such that $E[L_f] \leq B$?

We can then characterize the complexity of the a priori strategy as follows:

Theorem 6. *The decision version of the PTSP is $\mathcal{NP}$-complete.*

Sketch of the proof. We only need to show membership in $\mathcal{NP}$, since the PTSP is a generalization of a well known $\mathcal{NP}$-complete problem [see Garey & Johnson, 1979]. Membership in $\mathcal{NP}$ is seen to hold, since, given a solution f, we can compute $E[L_f]$ in $O(n^2)$ as we have shown in Theorem 1. $\square$

Thus, although we can compute efficiently the expected length of any given a priori solution to a PTSP, it is still $\mathcal{NP}$-hard to find an optimal a priori solution.

2.2.4. *Theoretical approximations to optimal a priori solutions*

In the previous section we found that it is still $\mathcal{NP}$-hard to obtain optimal a priori solutions to the PTSP. In this section we address the question of approximating the optimal a priori solution with polynomial time heuristics, whose worst case behavior we can characterize.

The first natural question to address is how heuristic approaches to the deterministic problem perform when applied to the corresponding probabilistic problem. For example, what is the performance of the well-known Christofides heuristic for the TSP [see Larson & Odoni, 1981] if applied to the PTSP? In order to find useful bounds for the routing problems (PTSP) we assume below that the triangle inequality holds. We can then prove the following:

Theorem 7. *Let L_{TSP} be the length of the optimal solution to the deterministic TSP and let L_H be the length of a heuristic solution to the same problem. Let p be the coverage probability and $E[L_{\text{PTSP}}]$ the expected length of the optimum a priori solution to the corresponding PTSP. If the heuristic has the property that $L_H/L_{\text{TSP}} \leq c$, then $E[L_H]/E[L_{\text{PTSP}}] \leq c/p$.*

Sketch of the proof. Using the triangle inequality, we know that, for any tour f, $E[L_f] \leq L_f$. In addition, we show that $E[L_{\text{PTSP}}] \geq pL_{\text{PTSP}}$. Combining these inequalities the result follows. $\square$

Theorem 7 suggests that if the coverage probability is large then constant guarantee heuristics for the deterministic problem still behave well for the corresponding probabilistic problem. But if $p \to 0$ the bound is not informative and indeed one can find examples with $p \to 0$, $np \to \infty$ for which $E[L_{\text{TSP}}]/E[L_{\text{PTSP}}] \to \infty$, that is, even if $c = 1$, the optimal deterministic solution is an arbitrarily bad approximation to the optimal a priori solution. (For a number of interesting examples see Jaillet [1985, 1991a].) As an indication of the rate at which the ratio $E[L_{\text{TSP}}]/E[L_{\text{PTSP}}]$ tends to infinity, Bertsimas [1988] proves the following:

Theorem 8. *For the PTSP with triangle inequality*

$$\frac{E[L_{\text{TSP}}]}{E[L_{\text{PTSP}}]} = O(\sqrt{n}). \tag{12}$$

We next investigate the existence of constant guarantee heuristics. We restrict our attention to Euclidean problems and examine the spacefilling curve heuristic, first proposed by Platzman & Bartholdi [1989] for the Euclidean TSP. The spacefilling curve heuristic can be described as follows:

1. Given the n coordinates (x_i, y_i) of the points in the plane compute the number $f(x_i, y_i)$ for each point. The function $f : R^2 \to R$ is called the Sierpinski curve (for details on the computation of $f(x, y)$ see Bartholdi & Platzman [1982]).

2. Sort the numbers $f(x_i, y_i)$ and visit the corresponding initial points (x_i, y_i) in that order, producing a tour SF.

The key property of the spacefilling curve heuristic that makes its analysis for the PTSP possible is the following: Consider an instance S of the problem. Suppose the spacefilling curve heuristic produces a tour $SF(S)$ if we run the heuristic on the instance S. Consider now the tour SF produced by the heuristic on the original instance of the problem, i.e. when all points are present. What is the tour that the PTSP strategy would produce in instance S if the a priori tour is SF?

The answer is precisely $SF(S)$, because sorting has the property of preserving the order in which the points in S will be visited by the spacefilling curve, which is exactly the property of the PTSP strategy as well. Based on this critical observation we can then analyze the spacefilling curve heuristic.

Theorem 9. *For the Euclidean PTSP the spacefilling curve heuristic produces a tour SF with the property*

$$\frac{E[L_{\text{SF}}]}{E[L_{\text{PTSP}}]} \le \frac{E[L_{\text{SF}}]}{E[\Sigma_{\text{TSP}}]} = O(\log n). \tag{13}$$

Sketch of the proof. In Platzman & Bartholdi [1989] it is proven that the length of the spacefilling curve heuristic satisfies:

$$\frac{L_{\text{SF}}}{L_{\text{TSP}}} = O(\log n). \tag{14}$$

Consider an instance S of the problem. If the spacefilling curve heuristic is applied to the instance S, it will similarly produce a tour $SF(S)$ with length

$$\frac{L_{\text{SF}}(S)}{L_{\text{TSP}}(S)} = O(\log |S|) = O(\log n). \tag{15}$$

But since $SF(S)$ is the tour produced by the PTSP strategy at instance S then

$$\frac{E[L_{SF}]}{E[\Sigma_{TSP}]} = \frac{\sum_{S \subseteq V} p(S) L_{SF}(S)}{\sum_{S \subseteq V} p(S) L_{TSP}(S)} \leq \frac{\sum_{S \subseteq V} p(S) O(\log n) L_{TSP}(S)}{\sum_{S \subseteq V} p(S) L_{TSP}(S)} = O(\log n).$$

$\square$

Note that this result does not depend on the probabilities of points being present. It holds even if there are dependencies on the presence of the points. Observe also that the spacefilling curve heuristic ignores the probabilistic nature of the problem but surprisingly produces a tour which is globally (in every instance) close to the optimal.

As a corollary to Theorem 9 we can compare the PTSP and the re-optimization strategies from a worst-case perspective. For the Euclidean PTSP, since $E[L_{PTSP}] \leq E[L_{SF}]$,

$$\frac{E[L_{PTSP}]}{E[\Sigma_{TSP}]} = O(\log n). \tag{16}$$

Platzman & Bartholdi [1989] conjecture that the spacefilling curve heuristic is a constant-guarantee heuristic. Unfortunately, Bertsimas & Grigni [1989] showed this conjecture to be false, and the existence of a constant guarantee heuristic for the Euclidean PTSP remains open.

2.3. The probabilistic vehicle routing problem

We shall now look at generalizations of the PTSP, which in Section 2 were introduced as the PVRP. Specifically, we still consider demands which are probabilistic in nature and our problem is to determine a priori routes of minimal expected length for vehicles with finite capacity. The complications introduced by the finite capacity of the vehicles is a major point of interest. A first problem is to consider a single vehicle and design a giant *a priori* vehicle tour through all the demand points. While covering this tour the vehicle may run out of capacity and, in such an event, it will have to return to the depot — for instance, in order to deposit the load it has picked up at the points it has already visited. Thus, the expected tour length to be minimized must also include any additional distance traveled to and from the depot whenever the vehicle reaches its capacity. There is, of course, an alternative interpretation under which the very same problem can be viewed as a multi-vehicle PVRP. This can be seen best if one sets p_i, the probability of visiting point i, equal to 1 for all i. Then the approach just described is identical to one of the two standard approaches to multi-vehicle deterministic VRPs, namely 'route first, cluster second'. Under this interpretation, the returns of the vehicle to the depot result in multiple tours, so that we are dealing with multiple-VRP tours as solutions to the overall problem. However, in the general case when some of the p_i are strictly less than 1, some criterion or criteria must be used in order to break up the giant a priori tour into clusters of customers — with each cluster served by a different vehicle.

In order to consider all these aspects in a more specific way let us now define four generalized versions of the PTSP that can be classified as PVRPs.

The capacitated probabilistic traveling salesman problem. Assume that each point x_i, requiring a visit with a probability p_i, independently of all others, has a unit demand, and that the salesman (vehicle) has a capacity q. We wish to find a priori a tour through all n points. On any given instance, the subset of points present will then be visited in the *same order* as they appear in the a priori tour. Moreover if the demand on the route exceeds the capacity of the vehicle, the salesman has to go back to the depot before continuing on his route. The problem of finding such a tour of minimum expected total length (the expected length of the tour in the PTSP sense plus the expected extra distance due to overloading) is defined as a capacitated PTSP. This problem is the 'probabilistic vehicle routing problem under updating method $\mathcal{U}_b$' as described in Section 2.1.

The m-probabilistic traveling salesmen problem (m-PTSP). Consider the problem of routing through a set of n points starting from and ending at a depot. On any given instance of the problem, only a random subset of points (each unit-demand point x_i being present with a probability p_i, independently of the others) has to be visited. We wish to find, a priori, m subtours, each starting from and ending at the depot, such that each point is included in exactly one tour, and of minimum total expected length under the skipping strategy (method $\mathcal{U}_b$). For this problem we have m vehicles with no capacity limits.

The capacitated m-probabilistic traveling salesmen problem. This capacitated vehicle problem is a natural combination of the capacitated PTSP and of the m-PTSP.

The general probabilistic vehicle routing problem. This is the same as the capacitated m-PTSP except that the demand of each customer x_i is no longer modeled by a Bernoulli random variable with parameter p_i, but rather by a more general random variable. Note that, for this problem, the two updating methods $\mathcal{U}_a$ and $\mathcal{U}_b$ of Section 2.1 are identical if the demand of each customer is strictly positive with probability one.

Several variations of this last problem have been referred to as stochastic vehicle routing problems in the existing literature [see for example Stewart & Golden, 1983; Dror & Trudeau, 1986; Dror, Laporte & Trudeau, 1989; Laporte & Louveau, 1990]. There is, however, an important difference between the approach typically adopted in these references and the one described here: These other stochastic vehicle routing problems are formulated by using techniques from stochastic programming (i.e., chance-constrained optimization, or stochastic programming with recourse) that allow one to transform these problems into deterministic VRPs, and then either solve them optimally or obtain bounds for them. One major consequence of these approaches is that it is necessary to introduce additional parameters (in the form of performance criteria) whose choice is at the analysts' discretion and may be related to routing costs only indirectly. Examples of such parameters might be: 'the probability of a vehicle having to return to the depot more than once, while serving its cluster of customers, should not exceed δ, $0 < \delta < 1$' for the chance-constrained formulation; or, 'if the vehicle reaches capacity while serving its customers a penalty α is incurred' for stochastic programming with recourse. It should be emphasized that the routing strategies described below constitute in fact, a form of stochastic programming with recourse (the recourse is to go back to the depot whenever vehicle capacity is reached), but in our case the cost of the recourse simply corresponds to extra travel distance.

We now summarize our main results on the m-PTSP, capacitated-PTSP, and capacitated m-PTSP. For details, the reader is referred to Jaillet [1987], Jaillet & Odoni [1988], Bertsimas [1988], Jaillet [1991b], and Bertsimas [1992]. To facilitate the presentation we will restrict ourselves to the case of uniform coverage probability p. We begin with some basic definitions: Let x_0 denote the location of the depot, while $x = (x_1, x_2, \ldots)$ represents an arbitrary infinite sequence of points in $\mathbf{R}^2$; $x^{(n)} = (x_1, x_2, \ldots, x_n)$ are the first n points of x. If the position of the points is random, the sequence is denoted by upper-case letters, i.e., $X = (X_1, X_2, \ldots)$. For each point x_i its distance to the depot will be written as d_i and the average distance $(\sum_{i=1}^{n} d_i)/n$ as $\bar{d}$. Associated with x is a sequence of i.i.d. Bernoulli random variables with parameter p describing the presence or absence of the points. $E[L_{PTSP}(x^{(n)})]$, $E[L_{m\text{-}PTSP}(x^{(n)})]$, $E[L_{CPTSP}(x^{(n)})]$, and $E[L_{m\text{-}CPTSP}(x^{(n)})]$ will represent respectively the optimal-solution expected lengths, through $x^{(n)}$, of the PTSP, the m-PTSP, the capacitated PTSP, and the capacitated m-PTSP. Finally, for any i and j, we will use $s(x_i, x_j)$ to denote the 'savings' quantity $d(x_i, x_0) + d(x_0, x_j) - d(x_i, x_j)$.

2.3.1. Expected length of a priori solutions

Theorem 10. *The expected length, $E[L_{\tau}^c]$ of the capacitated PTSP for a given tour $(x_0, x_1, x_2, \ldots, x_n, x_0)$ through $x^{(n)}$ is given by*

$$E[L_{CPTSP}(x^{(n)})] = E[L_{PTSP}(x^{(n)})] + \sum_{i=q}^{n-1} \sum_{j=i+1}^{n} \gamma_{ij} s(x_{\sigma(i)}, x_{\sigma(j)}), \qquad (17)$$

where $E[L_{PTSP}(x^{(n)})]$ is the expected length of the PTSP tour through $x^{(n)}$, and where

$$\gamma_{ij} = \sum_{k=1}^{\lfloor i/q \rfloor} \binom{i-1}{kq-1} p^{kq+1} (1-p)^{j-kq-1}. \qquad (18)$$

Sketch of the proof. The expected length of the capacitated PTSP is the sum of two terms: The expected length of the PTSP, plus the expected value of extra travel distance that might be required due to limited capacity. For the first term, the reader is referred to Jaillet [1985, 1991a] for a detailed expression (i.e., a generalization of Theorem 1 which includes the presence of a depot). For the second term, note that γ_{ij} is the probability that the vehicle reaches capacity at point x_i and that the next point present along the tour is x_j (in that case the extra distance is simply $s(x_i, x_{i+r+1})$). $\quad\square$

Expressions (17) and (18) show that the objective function of the capacitated PTSP can also be obtained in $O(n^2)$ steps. The expected lengths for the m-PTSP and the capacitated m-PTSP are easily obtained by applying Theorems 1 and 10, respectively, to each of the m subtours separately.

The following theorem establishes some relationships between the values of the optimal solutions to the various types of PVRPs. For each n, we sort the distances from the depot in a non-decreasing order $(d_{(1)}, d_{(2)}, \ldots, d_{(n)})$ (i.e., $d_{(j)} \leq d_{(j+1)}$ for $j \in [1, \ldots, n-1]$).

Theorem 11. *The optimal-solution expected lengths of the m-PTSP, capacitated PTSP, and capacitated m-PTSP are related to the expected length of the optimal PTSP as follows.*

(i) m-PTSP:

$$\max\left\{ E[L_{\text{PTSP}}(x^{(n)})]; 2p\left(\sum_{j=1}^{m-1} d_{(j)} + d_{(n)}\right)\right\} \leq$$

$$\leq E[L_{m\text{-PTSP}}(x^{(n)})] \leq E[L_{\text{PTSP}}(x^{(n)})] + 2p\sum_{j=1}^{m-1} d_{(j)}. \tag{19}$$

(ii) Capacitated PTSP:

$$\max\left\{ E[L_{\text{PTSP}}(x^{(n)})]; \frac{2p}{q}n\overline{d}\right\} \leq\leq E[L_{\text{CPTSP}}(x^{(n)})] \leq$$

$$\leq E[L_{\text{PTSP}}(x^{(n)})] + 2p(1-(1-p)^{n-q})\sum_{j=q}^{n-1} d_{(j+1)}. \tag{20}$$

(iii) Capacitated m-PTSP:

$$E[L_{m\text{-CPTSP}}(x^{(n)})] \geq \max\left\{ E[L_{\text{PTSP}}(x^{(n)})]; 2p\sum_{j=1}^{m-1} d_{(j)} + \frac{2p}{q}\sum_{j=m}^{n} d_{(j)}\right\}, \tag{21}$$

and

$$E[L_{m\text{-CPTSP}}(x^{(n)})] \leq E[L_{\text{PTSP}}(x^{(n)})] +$$

$$+ 2p\sum_{j=1}^{m-1} d_{(j)} + 2p(1-(1-p)^{n-q-m+1})\sum_{j=q}^{n-m} d_{(j+m)}. \tag{22}$$

Sketch of the proof. (i) A feasible solution to the m-PTSP can consist of having each of $m-1$ vehicles visit a single point each and the last vehicle go through the remaining $n-m+1$ points in an optimal PTSP manner. The upper bound on $E[L_{m\text{-PTSP}}(x^{(n)})]$ then follows from the fact that the PTSP functional is monotone and that the expected length of the tour a vehicle which visits a single point, say x_j, is $2pd_j$; if we take the single points to be the $m - 1$ closest to the depot, we obtain the best upper bound of this form. The first lower bound on $E[L_{m\text{-PTSP}}(x^{(n)})]$ is obvious, while the second one follows from the fact that, in an optimal solution to the m-PTSP, the expected length of each one of the m subtours is greater than the expected distance needed to visit the farthest point on the subtour. This, in turn, is greater than the bound given in (19).

(ii) A feasible solution to the capacitated PTSP is given by an optimal PTSP tour. To obtain the upper bound on $E[L_{\text{CPTSP}}(x^{(n)})]$ we then have to bound the expected length of the extra distance to the depot for an optimal PTSP tour (see Jaillet 1991b] for details). The first lower bound on $E[L_{\text{CPTSP}}(x^{(n)})]$ is obvious. The second one follows from the fact that, on each instance of the problem, if the set V of points present is of cardinality k, the vehicle will cover $\lceil k/q \rceil$ subtours, each consisting of less than q points. The length of such a subtour will then be greater or equal to $2(\sum_{i \in \text{subtour}} d_i)/q$. The lower bound in this instance is then $2(\sum_{i \in V} d_i)/q$. By summing over all possible instances (weighted by their probability of presence) we finally obtain the desired result.

(iii) The bounds for $E[L_{m\text{-CPTSP}}(x^{(n)})]$ are obtained by combining the previous arguments on $E[L_{m\text{-PTSP}}(x^{(n)})]$ and $E[L_{\text{CPTSP}}(x^{(n)})]$. $\square$

We also note that there are no clear relations between $E[L_{m\text{-CPTSP}}(x^{(n)})]$ and $E[L_{\text{CPTSP}}(x^{(n)})]$. In addition, although $E[L_{\text{PTSP}}(x^{(n)})]$ and $E[L_{m\text{-PTSP}}(x^{(n)})]$ are monotone functionals, this is in general not true for $E[L_{\text{CPTSP}}(x^{(n)})]$ and $E[L_{m\text{-CPTSP}}(x^{(n)})]$.

2.3.2. Asymptotic results

The main objective of this asymptotic analysis is to obtain *strong* limit laws for PVRPs similar to the PTSP results given in Section 2.2.2. In order to be consistent with Section 2.2.2, we will write, hereafter, for all the expected lengths, $E[L^n_-]$ for $E[L_-(X^{(n)})]$, whenever $X = (X_1, X_2, \ldots)$ is a sequence of points independently and uniformly distributed over $[0, 1]^2$.

The m-PTSP

Let us assume that m is a non-decreasing (possibly constant) function of n, and let us consider the asymptotic behavior of $E[L^n_{m\text{-PTSP}}]$. We have the following theorem.

Theorem 12. *Consider an infinite sequence of points independently and uniformly distributed over $[0, 1]^2$ and let p be the coverage probability for each point. For any position of the depot within a finite distance μ of the unit square (the depot can possibly be inside the square, i.e. $\mu = 0$) we have:*
(i) If $m = o(\sqrt{n})$, or if $m = \Theta(\sqrt{n})$ and $\mu = 0$, then

$$\lim_{n \to \infty} \frac{E[L^n_{m\text{-PTSP}}]}{\sqrt{n}} = \beta_{\text{PTSP}}(p) \ (a.s.). \tag{23}$$

(ii) *If $m = \Omega(\sqrt{n})$ and $m = o(n)$, then*

$$\lim_{n\to\infty} \frac{E[L^n_{m\text{-PTSP}}]}{m} = 2p\mu \; (a.s.). \tag{24}$$

(iii) *If $\lim_{n\to\infty} m/n = \alpha$, $0 < \alpha \le 1$, then*

$$\lim_{n\to\infty} \frac{E[L^n_{m\text{-PTSP}}]}{m} = \frac{2p}{\alpha} \int_0^\alpha F_D^{-1}(x)\,dx \; (a.s.). \tag{25}$$

where $F_D^{-1}(x)$ is the generalized inverse of the cumulative distribution function of D, the random variable representing the distance from a random point of the unit square to the depot.

In order to prove this theorem we need a result which, because of its independent interest, we present in a general setting. The reader is referred to Jaillet [1991b] for details.

Lemma 1. *Let $(Y_i)_i$ be a sequence of i.i.d. absolutely continuous (with no singular part) real random variables with bounded support $[a, b]$. For each n, let $(Y_{(i,n)})_{1\le i\le n}$ be the corresponding order statistics (i.e., $Y_{(i,n)} \le Y_{(i+1,n)}$ for $i \in [1..n-1]$). Let s be a non-decreasing integer valued function of n, $s \le n$. Finally let F_Y be the common cumulative distribution function of the Y_i's. We then have:*

(i) If $s = o(n)$, then

$$\lim_{n\to\infty} \frac{1}{s} \sum_{i=1}^s Y_{(i,n)} = a \; (a.s.). \tag{26}$$

(ii) If $\lim_{n\to\infty} s/n = \alpha$, $0 < \alpha \le 1$ then

$$\lim_{n\to\infty} \frac{1}{s} \sum_{i=1}^s Y_{(i,n)} = \frac{1}{\alpha} \int_0^\alpha F_Y^{-1}(x)\,dx \; (a.s.). \tag{27}$$

Sketch of the proof of Theorem 12. For case (i) we have from (19)

$$E[L^n_{\text{PTSP}}] \le E[L^n_{m\text{-PTSP}}] \le E[L^n_{\text{PTSP}}] + 2p\frac{\sum_{j=1}^{m-1} D_{(j)}}{m-1}(m-1). \tag{28}$$

If $m = o(\sqrt{n})$, the result follows from the fact that the D_i's are bounded and from Theorem 5. If $m = \Theta(\sqrt{n})$ and $\mu = 0$, the result follows from Lemma 1 part (i) and from Theorem 5.

For case (ii) and case (iii) we have from (19)

$$2p\frac{\sum_{j=1}^{m-1} D_{(j)} + D_{(n)}}{m} \le \frac{E[L^n_{m\text{-PTSP}}]}{m}, \tag{29}$$

and

$$\frac{E[L^n_{m\text{-PTSP}}]}{m} \leq \frac{(E[L^n_{\text{PTSP}}]/\sqrt{n})\sqrt{n}}{m} + 2p\,\frac{\sum_{j=1}^{m-1} D_{(j)}}{m}. \tag{30}$$

The results follow respectively from Lemma 1 part (i) and part (ii) and from Theorem 5. □

Note that in the case where $\lim_{n\to\infty} m/\sqrt{n} = \delta$ and $\mu \neq 0$, we have, with probability one, from Theorem 5, Lemma 1 and (19):

$$\max\{\beta_{\text{PTSP}}(p); 2p\mu\} \leq \liminf_{n\to\infty} \frac{E[L^n_{m\text{-PTSP}}]}{\sqrt{n}}$$

$$\leq \limsup_{n\to\infty} \frac{E[L^n_{m\text{-PTSP}}]}{\sqrt{n}} \leq \beta_{\text{PTSP}}(p) + 2p\mu. \tag{31}$$

The capacitated versions

Consider an infinite sequence of points independently and uniformly distributed over $[0, 1]^2$ and let p be the coverage probability for each point. For any position of the depot within a finite distance μ of the unit square (including possibly inside the square, i.e. $\mu = 0$) we have the following bounds.

$$\frac{2p}{q}E[D] \leq \liminf_{n\to\infty} \frac{E[L^n_{\text{CPTSP}}]}{n} \leq \limsup_{n\to\infty} \frac{E[L^n_{\text{CPTSP}}]}{n} \leq 2pE[D] \text{ (a.s.).} \tag{32}$$

These inequalities follow directly from Theorem 5, Theorem 11, and the strong law of large numbers.

The asymptotic properties of these capacitated problems will be characterized more sharply in the next subsection in conjunction with algorithmic procedures.

2.3.3. Algorithmic considerations

All PVRPs that are being considered in this section are obviously $\mathcal{NP}$-hard. One is therefore justified in looking for approximations to the optimal solutions.

Our goal is to present some examples of worst-case and probabilistic analyses of heuristics for the PTSP, m-PTSP, capacitated PTSP, and capacitated m-PTSP, drawing on results presented earlier.

The PTSP

As mentioned in Section 2.2 several 'good' heuristics have been proposed for the PTSP but none of them has been so far amenable to a theoretical analysis which would prove that the heuristic provides a 'constant guarantee' (in the worst-case sense). Thus, unlike the case of the TSP, the existence of a constant-guarantee heuristic for the PTSP is still an open problem. As we will now see this is an important question since most of the worst-case analyses of the PVRPs depend heavily on the analysis of the PTSP.

For the analysis of the PVRPs we will nevertheless assume that it is possible to obtain optimal or near-optimal solutions to the PTSP. Any progress on the PTSP will automatically translate into progress on the more complicated probabilistic vehicle routing problems.

The m-PTSP

In contrast to the relationship between the TSP and the m-TSP, the m-PTSP cannot be transformed into an equivalent PTSP. Nevertheless the heuristic which consists of assigning the first $m - 1$ vehicles to the $m - 1$ closest points to the depot, and the last vehicle to the remaining points can be analyzed mathematically with the help of the previous sections. Indeed, if $H(x^{(n)})$ is the value given by this heuristic we have from Theorem 11:

$$\frac{H(x^{(n)})}{E[L_{m\text{-PTSP}}(x^{(n)})]} \leq \frac{E[L_{\text{PTSP}}(x^{(n)})] + 2p \sum_{j=1}^{m-1} d_{(j)}}{\max\left\{E[L_{\text{PTSP}}(x^{(n)})]; 2p \sum_{j=1}^{m-1} d_{(j)}\right\}} \leq 2. \tag{33}$$

If instead of having an optimal PTSP tour we have an alternative tour τ such that $E[L_\tau(x^{(n)})] \leq \kappa E[L_{\text{PTSP}}(x^{(n)})]$ then the new heuristic, say $\mathcal{H}_{alt}$, is such that

$$\frac{H_{alt}(x^{(n)})}{E[L_{m\text{-PTSP}}(x^{(n)})]} \leq \frac{\kappa E[L_{\text{PTSP}}(x^{(n)})] + 2p \sum_{j=1}^{m-1} d_{(j)}}{\max\left\{E[L_{\text{PTSP}}(x^{(n)})]; 2p \sum_{j=1}^{m-1} d_{(j)}\right\}} \leq \kappa + 1. \tag{34}$$

Also for any relation between m and n, we can show, from Theorem 12, that such a heuristic is asymptotically optimal with probability one for points uniformly and independently distributed over the unit square.

The capacitated PTSP

If we denote by $H^c(x^{(n)})$ the value of the heuristic, say $\mathcal{H}^c$, which consists of simply using the optimal PTSP tour for this problem, we have, from Theorem 11:

$$\frac{H^c(x^{(n)})}{E[L_{\text{CPTSP}}(x^{(n)})]} \leq$$

$$\leq \frac{E[L_{\text{PTSP}}(x^{(n)})] + 2p(1 - (1-p)^{n-q}) \sum_{j=q}^{n-1} d_{(j+1)}}{\max\left\{E[L_{\text{PTSP}}(x^{(n)})]; (2p/q)n\bar{d}\right\}} \leq q + 1, \tag{35}$$

which is not very good especially when q is big. We can also see that (35) does not allow us to derive an asymptotically optimal algorithm. Let us analyze an improvement of this algorithm, introduced in Bertsimas [1988] for a simpler version of this problem. This heuristic, based on ideas contained in Haimovich, Rinnooy Kan & Stougie [1988], is the following:

Cyclic heuristic

Step 1. Find an optimal PTSP tour $t_1 = (x_0, x_{\sigma^*(1)}, x_{\sigma^*(2)}, \ldots, x_{\sigma^*(n)}, x_0)$ and consider the tours $t_k^* = (x_0, x_{\sigma^*(k)}, x_{\sigma^*(k+1)}, \ldots, x_{\sigma^*(n)}, x_{\sigma^*(1)}, \ldots, x_{\sigma^*(k-1)}, x_0)$ for $k \in [2..n]$.

Step 2. Compute the objective values $E_{t_k}^c$ for $k \in [1..n]$.

Step 3. Take the tour with the minimum objective value.

Let $IH^c(x^{(n)})$ be the value of this improved heuristic. We then have the following [see Jaillet, 1991b]:

Theorem 13.

$$IH^c(x^{(n)}) \leq \left(1 - \frac{1}{n}\right) E[L_{\text{PTSP}}(x^{(n)})] + 2\left(1 + \frac{np}{q}\right)\overline{d}. \tag{36}$$

In conclusion, from Theorems 11 and 13 we have the following result:

$$\frac{IH^c(x^{(n)})}{E[L_{\text{CPTSP}}(x^{(n)})]} \leq$$

$$\leq \frac{(1 - 1/n)E[L_{\text{PTSP}}(x^{(n)})] + 2(1 + np/q)\overline{d}}{\max\{E[L_{\text{PTSP}}(x^{(n)})]; \ (2p/q)n\overline{d}\}} \leq 2 + \frac{1}{n}\left(\frac{q}{p} - 1\right). \tag{37}$$

When n goes to infinity, we then have a constant-guarantee (independent of q) heuristic. If instead of having an optimal PTSP tour, we have a tour τ such that $E[L_\tau(x^{(n)})] \leq \kappa E[L_{\text{PTSP}}(x^{(n)})]$ then this alternative heuristic gives the following bound:

$$\frac{IH_{alt}^c(x^{(n)})}{E[L_{\text{CPTSP}}(x^{(n)})]} \leq \kappa + 1 + \frac{1}{n}\left(\frac{q}{p} - 1\right). \tag{38}$$

Moreover, with the help of the cyclic heuristic, one can get the asymptotic behavior of the capacitated PTSP. Indeed, the following theorem, obtained from Theorem 13 and equation (32), shows that the cyclic heuristic is asymptotically optimal with probability one.

Theorem 14. *Consider an infinite sequence of points independently and uniformly distributed over $[0, 1]^2$ and let p be the coverage probability for each point. For any position of the depot within a finite distance μ of the unit square (including possibly inside the square, i.e. $\mu = 0$) we have:*

$$\lim_{n\to\infty} \frac{E[L_{\text{CPTSP}}^n]}{n} = \lim_{n\to\infty} \frac{IH^c(x^{(n)})}{n} = \frac{2p}{q}E[D](a.s.). \tag{39}$$

The capacitated m-PTSP

Let $I H_m^c(x^{(n)})$ be the value of the heuristic, here called the m-cyclic heuristic, which consists of assigning the first $m-1$ vehicles to the $m-1$ closest points to the depot, and using the cyclic heuristic on the remaining $n-m+1$ points. Then, letting $n_1 = n - m + 1$, we have, from the same arguments as before, the following bound:

$$\frac{I H_m^c(x^{(n)})}{E[L_{m\text{-CPTSP}}(x^{(n)})]} \leq 2 + \frac{1}{n_1}\left(\frac{q}{p} - 1\right). \tag{40}$$

One can finally show the asymptotic optimality of the m-cyclic heuristic [see Jaillet, 1991b]

Theorem 15. *Consider an infinite sequence of points independently and uniformly distributed over $[0, 1]^2$ and let p be the coverage probability for each point. For any position of the depot within a finite distance μ of the unit square (including possibly inside the square, i.e. $\mu = 0$) we have:*

(i) If $m = o(n)$, then

$$\lim_{n\to\infty} \frac{E[L_{m\text{-CPTSP}}^n]}{n} = \lim_{n\to\infty} \frac{I H_m^c(X^{(n)})}{n} = \frac{2p}{q} E[D] \ (a.s.). \tag{41}$$

(ii) If $\lim_{n\to\infty} m/n = \alpha$, $0 < \alpha \leq 1$, then

$$\lim_{n\to\infty} \frac{E[L_{m\text{-CPTSP}}^n]}{n} = \lim_{n\to\infty} \frac{I H_m^c(X^{(n)})}{n}$$

$$= 2p(1 - \frac{1}{q}) \int_0^\alpha F_D^{-1}(x)\, dx + \frac{2p}{q} E[D] \ (a.s.). \tag{42}$$

2.3.4. More general probabilistic vehicle routing problems

We have discussed here cases of unit demand at each demand point. It is in fact possible to consider the case of unequal demands in much the same way as in Haimovich, Rinnooy Kan & Stougie [1988; see for example Bertsimas, 1992]. Following Jaillet [1987], one can also consider more general probability distributions for the demand at each point, such as binomial distributions, and re-derive most of the previous results. Finally in our multi-vehicle models, we did not consider the importance of balanced routes. If this is an issue, modifications of several heuristics proposed in Haimovich, Rinnoy Kan & Stougie [1988] could be analyzed successfully for a probabilistic environment, as well.

2.4. The probabilistic traveling salesman location problem

2.4.1. Algorithmic issues

The traveling salesman location problem (TSLP) and its generalization, the probabilistic traveling salesman location problem (PTSLP), were defined in Section 2.1. Note that the difference between the two problems is that in the TSLP it is assumed that an a priori tour through the n probabilistic nodes of the network where potential demands reside is already given at the outset. The only issue is to locate a facility that will minimize the expected length of any particular instance of this a priori tour, if demands are always visited in the same order in which they appear on the a priori tour. By contrast, in the PTSLP both the facility location *and* the a priori tour must be found.

It is interesting that, until the recent work of Berman & Simchi-Levi [1986, 1988a, b] problems that involve simultaneously location and routing considerations were considered practically intractable by locational theorists [see, e.g., Burness & White, 1976] at the same time when other facility location problems on networks, such as 'median' and 'center' problems [Mirchandani & Francis, 1990], were being studied extensively.

Berman & Simchi-Levi [1986] showed initially that the TSLP can be solved as a discrete combinatorial optimization problem:

Lemma 2. *At least one optimal facility location for the TSLP is a node.*

Sketch of the proof. The proof of this lemma is easy and follows the line of the classical Hakimi-like proofs of node-optimality for median location problems [Hakimi, 1964]. Lemma 2 can be readily extended to the PTSLP. $\square$

It follows that the TSLP on trees can be solved efficiently:

Theorem 16. *The TSLP on a tree network can be solved in $O(n)$ time.*

Sketch of the proof. The proof of Theorem 16 is based on showing that a 'majority rule' [Goldman, 1974] holds for the TSLP on trees. Specifically, suppose we identify any edge (i, j) of the tree and suppose we 'cut' this edge. The tree is then subdivided into two subtrees, one containing node i and the other node j. It can then be shown that the optimal facility location must be in the 'majority' subtree, i.e., the subtree which contains the set of nodes whose total probability of presence, summed over all possible instances of the problem, is the larger of the two. Once again, Theorem 16 can be readily extended to the PTSLP. $\square$

For a general network, we note first that both the TSLP and the PTSLP are, in general $\mathcal{NP}$-hard problems. Berman & Simchi-Levi [1988b] have therefore proposed a surprisingly simple, constant-guarantee heuristic for the case in which calls from demand nodes occur independently, with the probability of a call from node i being equal to p_i. Let $d(i, j)$ denote the shortest distance between nodes i and j. Then the heuristic can be described as follows:

1. For each node i on the network do the following:
   (a) Sort all the n nodes j_1, j_2, j_n such that $d(i, j_1) \leq \cdots \leq d(i, j_n)$. [Note that $j_1 = i$.]
   (b) Then compute $f(i) = \sum_{k=2}^{n} d(i, j_k) p_{j_k} \prod_{m=1}^{k-1} (1 - p_{j_m})$.
2. Select the node i with the minimum value of $f(i)$ as the solution to the TSLP.

Given a complete minimum distance matrix, Step 1 above takes $O(n \log n)$ time and thus the entire algorithm takes $O(n^2 \log n)$ time.

In Berman & Simchi-Levi [1988b] it is shown that this heuristic has a relative worst-case performance guarantee of 1/2, i.e., it can be at most 50% off from the optimal TSLP solution. However, it was shown later by Bertsimas that this constant-guarantee bound can be improved to

$$\tfrac{1}{2}(1 - p_{i^*}) \leq \tfrac{1}{2}(1 - \min_i p_i) \tag{43}$$

where i^* is the true optimal location for the TSLP. (For an elegant proof of this result see Bertsimas [1989].) For example, for the special case in which, for all i, $p_i = p$, $p = 1/2$ implies a relative worst-case error of 25% instead of 50%.

No constant-guarantee heuristic is available, on the other hand for the PTSLP — a more difficult problem. Bertsimas [1989] has characterized the worst case performance of a nearest neighbor location heuristic and of a spacefilling curve location heuristic and showed that both provide a worst-case performance bound which is $O(\log n)$.

The discerning reader may also have already observed that any node i with $p_i = 1$ is an optimal location for both the TSLP and the PTSLP. (If a node is to be visited on every instance of the problem anyway, we are assured that there will be no extra-distance penalty, if the facility is also located on that node!) This observation, however, is true only if the distance matrix satisfies the triangular inequality.

Finally, it should be noted that only recently have some more formal attempts been made — by using large-scale mathematical programming formulations — to obtain optimal or heuristic solutions to stochastic problems, like the TSLP and the PTSLP, which combine locational and routing considerations [see Laporte, 1988, and Laporte, Louveau & Mercure, 1989].

2.4.2. Asymptotic results

Sharp asymptotic performance results have also been obtained for the TSLP and the PTSLP. For example, it is easy to show, drawing on Theorems 2 and 4, that for $p_i = p$ and demands uniformly distributed in the unit square, we have [Berman & Simchi-Levi, 1988b]:

$$\lim_{n \to \infty} \frac{E[\Sigma^n_{\text{TSLP}}]}{\sqrt{n}} = \beta_{\text{TSP}} \sqrt{p}, \tag{44}$$

where $E[\Sigma^n_{\text{TSLP}}]$ is the expected length of the TSLP solution obtained under the re-optimization strategy. The limit in (44) is identical to that of Theorem 4, a result which is not surprising if one considers the fact that, in the limit, the location of the TSLP facility is not important when the number of potential demands to be visited is arbitrarily large. Interestingly Bertsimas [1989] has shown that the asymptotic behavior of the TSLP heuristic described earlier in this section is exactly the same as in (44). Thus the heuristic not only provides a constant-guarantee performance but is also asymptotically optimal.

For the PTSLP, by analogy to Theorem 5, we have [Bertsimas, 1989]:

$$\lim_{n\to\infty} \frac{E[\Sigma^n_{\text{PTSLP}}]}{\sqrt{n}} = \beta_{\text{PTSP}}(p). \tag{45}$$

2.4.3. A different class of stochastic facility location problems

To conclude this section, we note that an extensive literature also exists on a different set of stochastic facility location problems which, however do not involve any routing considerations. These problems are characterized by demands which appear at random times and random locations on a network and are served by stationary facilities or traveling vehicles that can serve only one demand at a time. Coupled with service times whose duration is a random variable, these conditions generate queueing phenomena. Thus, in this set of problems 'optimal facility (or vehicle) locations' are typically defined to be those that minimize the expected total time that elapses between the occurrence of a demand and the completion of service to that demand. Note that this total time includes any time spent waiting for a facility or vehicle to become available. Comprehensive reviews of this class of facility location problems are provided in Odoni [1987] and in Batta, Berman, Chiu, Larson & Odoni [1990].

2.5. The probabilistic shortest path problem

This section will be drawn to a large extent from Jaillet [1992b]; it will give non-technical review of the main results obtained on this problem.

2.5.1. Notation and assumptions

Let $G = (N, A, d)$ denote a complete, loopless, directed, weighted graph where N is the node set, A is the set of arcs joining the nodes of N, and d is a function $A \mapsto R$. We consider a node source s, a node sink t and paths from s to t. A path c will be given by the sequence of nodes defining it, i.e., $c = (s, n_1, n_2, \ldots, n_k, t)$. The set of nodes N is partitioned into two subsets N_1 and N_2:

N_1 is the set of nodes which never experience failure ('black' nodes), of cardinality $|N_1| = m$ ($m \geq 2$ since s and t belong to N_1).

N_2 is the set of nodes with possible failure ('white' nodes), of cardinality $|N_2| = n$.

We are given a probability law $\mathbb{P}$ on Ω, the power set of N_2 (an instance $\omega \in \Omega$ defines the subset of white nodes with no failure on this particular instance). We *restrict* $\mathbb{P}$ to be such that all outcomes of equal cardinality have the same probability of occurring, i.e.:

$$\forall \omega_1 \in \Omega, \ \forall \omega_2 \in \Omega, \ |\omega_1| = |\omega_2| \Rightarrow \mathbb{P}(\omega_1) = \mathbb{P}(\omega_2). \tag{46}$$

If W is the random variable that represents the number of white nodes with no failure we have :

$$\Pr(W = |\omega|) = \binom{n}{|\omega|} \mathbb{P}(\omega). \tag{47}$$

Hence, our probabilistic models can be specified equivalently by either giving the probability $\mathbb{P}$ or the probability distribution of W. Note also that the restriction imposed on $\mathbb{P}$ implies that, given $W = k$, the k nodes are selected uniformly at random among the set of n nodes; any probability $\mathbb{P}$ satisfying (46) will then be said to be *node-invariant*.

One important specific example (hereafter named $\mathbb{P}_1$) is

$$\mathbb{P}_1(\omega) = p^{|\omega|}(1 - p)^{n-|\omega|}, \tag{48}$$

which corresponds to the case in which every white node has a probability p of being present, independently of the others; we then speak informally of a Bernoulli process with parameter p.

For a given a priori path c between s and t, the length L_c covered in traversing the nodes without failure on each instance of the problem is a random variable. The general PSPP can then be stated as follows.

Problem PSPP. Given a network $G = (N, A, d)$ with node source s and node sink t and a probability $\mathbb{P}$ find an a priori path c of minimum expected length, $E[L_c]$.

2.5.2. The expected length of a given path

The most general result, based on an extension of results obtained for the PTSP is the following:

Theorem 17. *Given a node-invariant PSPP and a path* $c = (0, 1, \ldots, k, k+1)$ *from* s *to* t *(by convention* $0 \equiv s$ *and* $k+1 \equiv t$*) we have:*

$$E[L_c] = \sum_{r=0}^{k-2} \alpha_r A_c^{(r)} + \sum_{r=0}^{k-1} \beta_r B_c^{(r)} + \gamma_k C_c^{(k)}, \tag{49}$$

where:

$$A_c^{(r)} = \sum_{i=1}^{k-1-r} d_c(i, i+r+1),$$

$$B_c^{(r)} = d_c(0, r+1) + d_c(k-r, k+1),$$

$$C_c^{(k)} = d_c(0, k+1),$$

$$\alpha_r = \sum_{j=r}^{n-2} \left[\binom{n-2-r}{j-r} \Big/ \binom{n}{j} \right] \Pr(W = n - j), \text{ for all } r \in [0..k-2]$$

$$\beta_r = \sum_{j=r}^{n-1} \left[\binom{n-1-r}{j-r} \Big/ \binom{n}{j} \right] \Pr(W = n - j), \text{ for all } r \in [0..k-1]$$

$$\gamma_k = \sum_{j=k}^{n} \left[\binom{n-k}{j-k} \Big/ \binom{n}{j} \right] \Pr(W = n - j),$$

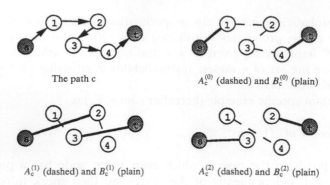

The path c $A_c^{(0)}$ (dashed) and $B_c^{(0)}$ (plain)

$A_c^{(1)}$ (dashed) and $B_c^{(1)}$ (plain) $A_c^{(2)}$ (dashed) and $B_c^{(2)}$ (plain)

Fig. 3. A path and the arcs representing some A_c's and B_c's.

with the following convention: $d_c(i, i + r + 1) = \sum_{e=0}^{s} d(b_e, b_{e+1})$ where $b_0 \equiv i$, $b_{s+1} \equiv i + r + 1$, and $(b_1, \ldots, b_s)$ is the sequence of black nodes drawn from $(i + 1, \ldots, i + r)$.

Sketch of proof. On any given instance of the problem, the arc $(i, i + r + 1)$ is in the resulting path if and only if the nodes i and $i + r + 1$ are working, and the nodes $i + 1, \ldots, i + r$ are not working. The definitions of A_c's, B_c's, and C_c^k are then based on a regrouping of arcs with equal probabilities; this is illustrated in Figure 3 on a simple example. $\square$

Note that when $k = 1$ or when $k = 0$, equation (49) gives respectively $E[L_c] = \beta_0 B_c^{(0)} + \gamma_1 C_c^{(1)}$ and $E[L_c] = \gamma_0 C_c^{(0)}$. The closed form expression (49) computes the expected length of a path from s to t through k intermediate nodes in $O(k^2)$ elementary operations (for a general, *node-invariant* $\mathbb{P}$ and assuming that the α_r's have been previously computed). Thus, Theorem 17 also shows that the recognition version of this problem belongs to the class $\mathcal{NP}$. Finally, for the case of a Bernoulli process $\mathbb{P}_1$, $\alpha_r = p^2(1 - p)^r$, $\beta_r = p(1 - p)^r$, and $\gamma_k = (1 - p)^k$.

2.5.3. The complexity of the PSPP and its relationship to the SPP

The SPP is a special case of the PSPP in which all nodes are black; it is then natural to investigate the possible links between the two problems. We show in this section that the two problems are fundamentally different so that the PSPP requires entirely new solution procedures. The following results are easily proved and confirm our earlier observations regarding the drastic changes that randomness induces into well-known combinatorial problems.

1. The PSPP is $\mathcal{NP}$-hard. Indeed consider the special Bernoulli case in which we have only two black nodes (s and t). The expected length of a path containing k ($k \leq n$) white nodes depends on $d(s, t)$ via a weight equal to $(1 - p)^k$. Since $(1 - p)^k \geq (1 - p)^n$, if we take d(s,t) arbitrarily large (everything else being equal), we can force the potential candidate paths for the corresponding PSPP

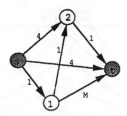

Fig. 4. The optimal deterministic shortest path can be bad.

to go through all the n white nodes. But this last problem is the probabilistic Hamiltonian path problem which is $\mathcal{NP}$-hard (being equivalent to the PTSP, see Jaillet [1985]).

2. Suppose that the distances do not satisfy the triangular inequality, since otherwise the optimal PSPP path would simply be the arc (s, t). Under that condition it is easy to construct examples in which an optimal SPP path is *arbitrarily* bad for the corresponding PSPP. See for example Figure 4 in which all arcs not shown are of length 4 and in which $M > 4$. The optimal SPP path $(s, 1, 2, t)$ of length 3 has an expected length depending on M, the length of arc $(1, t)$, which is traversed when node 1 works and node 2 fails; the expected length of $(s, 1, 2, t)$ can then be made arbitrarily large as compared to the expected length of path (s, t) (of value 4).

3. The principle of optimality (which helps solve the SPP) cannot be applied here. The main reason is that the *expected* length of a path is not an additive functional (as opposed to the length of a path), in the sense that in general $E[L_{c_1 \oplus c_2}] \neq E[L_{c_1}] + E[L_{c_2}]$ where $c_1 \oplus c_2 = (i_1, \ldots, i_k)$ stands for the concatenation of the two paths $c_1 = (i_1, \ldots, i_j)$ and $c_2 = (i_j, \ldots, i_k)$.

Based on Result 1, a polynomial time algorithm for the PSPP seems out of reach; from Result 2, the optimal SPP path cannot be considered as a good candidate for approximating the corresponding PSPP; and from Result 3, one has to be careful about utilizing classical SPP algorithms (see for example Papadimitriou & Steiglitz [1982] for the PSPP).

A practical consequence of these results is the necessity to develop entirely new solution procedures. As mentioned earlier, a branch and bound scheme was proposed in Jaillet [1988]. In Jaillet [1992b], it has been shown that, if the number of probabilistic ('white') nodes is, at any time, bounded by a constant, the PSPP can still be solved in polynomial time. Since this is the first instance in this chapter in which a large subclass of problems are solvable by polynomial time procedures, let us briefly review the detailed arguments in the next section.

2.5.4. *Polynomial procedures for special cases of the problem*

We now consider some special cases of the PSPP for which we are able to give polynomial time procedures. Let us first consider the simplest of these cases and show how this can be done. The other cases will be straightforward extensions of the main idea developed here.

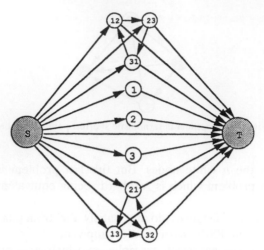

Fig. 5. The auxiliary network when $N_2 = \{1, 2, 3\}$.

A simple special case. Let us assume that $m = 2$, i.e., the only black nodes are s and t. Let us also consider a probability $\mathbb{P}$ such that

$$\Pr(W \leq n - 2) = 0 \text{ and } \Pr(W = n - 1) > 0, \tag{50}$$

i.e., either all white nodes are working or only one of them has a failure.

Let us then construct an auxiliary network (V, E, φ) as follows (see Figure 5 for an illustration): The set of nodes is $V = (N_2 \otimes N_2) \cup N_1 \cup N_2$ of cardinality $|V| = n^2 + 2$ (where the notation $A \otimes A$ stands for $(A \times A) \setminus diag(A)$); the set of arcs E and the arc-length function φ are defined by:

- Arc (s, t) of length $\varphi(s, t) = d(s, t)$.
- Arcs (s, v) for all $v = i$ in N_2 of length $\varphi(s, i) = \beta_0 d(s, i) + (\gamma_1/2)d(s, t)$.
- Arcs (s, v) for all $v = (i, j)$ in $N_2 \otimes N_2$ of length $\varphi(s, (i, j)) = \beta_0 d(s, i) + (\alpha_0/2)d(i, j) + \beta_1 d(s, j)$.
- Arcs (v, t) for all $v = i$ in N_2 of length $\varphi(i, t) = \beta_0 d(i, t) + (\gamma_1/2)d(s, t)$.
- Arcs (v, t) for all $v = (i, j)$ in $N_2 \otimes N_2$ of length $\varphi((i, j), t) = \beta_0 d(j, t) + (\alpha_0/2)d(i, j) + \beta_1 d(i, t)$.
- Arcs (v, w) for all $v = (i, j), w = (j, k), i \neq j \neq k$ of length $\varphi((i, j), (j, k)) = (\alpha_0/2)(d(i, j) + d(j, k)) + \alpha_1 d(i, k)$.

The cardinality of E is then $|E| = n(n - 1)(n - 2) + 1 + 2(n(n - 1) + n) = n^3 - n^2 + 2n + 1$.

One can now give the fundamental relationship between the two networks:

Lemma 3. *There is a one to one correspondence between the set of paths from s to t in (N, A, d) and the set of paths from s to t in (V, E, φ). Moreover the expected length of each path from s to t in (N, A, d) is equal to the length of a corresponding path in (V, E, φ).*

We are now in a position to give our main result in this simple situation:

Theorem 18. *Given a node-invariant PSPP with $m = 2$ and with a probability $\mathbb{P}$ such that $\Pr(W \leq n - 2) = 0$ and $\Pr(W = n - 1) > 0$, one can find an optimal PSPP path between s and t in time $O(n^3)$.*

Theorem 18 follows directly from Lemma 3. Indeed, from Lemma 3, one can solve such a PSPP by finding the shortest path in the auxiliary network (V, E, φ). This can be done by a careful implementation of the Dijkstra algorithm in time $O(|E| + |V| \log |V|)$ [see Driscoll, Gabow, Shraiman & Tarjan, 1988, or Fredman & Tarjan, 1987].

Generalizations

We can extend the idea of Theorem 18 in several directions.

We can first consider cases in which several white nodes can fail at the same time. For example let us look at the case in which $\Pr(W \leq n - 3) = 0$. The set of nodes V will be augmented to include all triplets (i, j, k) with $i \neq j \neq k$, i.e. of $n(n - 1)(n - 2)$ nodes. The set of arcs E and the function φ will remain the same except that we add arcs of the types $(s, (i, j, k))$, $((i, j, k), t)$, and $((i, j, k), (j, k, l))$ (with $l \neq i$) and that we delete arcs of the type $((i, j), (j, k))$. Now the PSPP can be solved in $O(n^4)$.

Another example of a generalization is to consider more than two black nodes: the set V would be augmented to include $m - 2$ additional nodes, and we would have to add arcs of the types $(i, v), (v, i), ((i, j), v)$, and $(v, (i, j))$ for all v in $N_1 \setminus \{s, t\}$ and $i, j, i \neq j$ in N_2, and arcs of the types (s, v), (v, t), (v, w), and (w, v) for all $v \neq w$ in $N_1 \setminus \{s, t\}$.

The most general result is obtained by considering a combination of the two previous extensions:

Theorem 19. *Given a node-invariant PSPP with m black nodes ($m \geq 2$) and a probability $\mathbb{P}$ such that $\Pr(W \leq n - k - 1) = 0$ and $\Pr(W = n - k) > 0$, one can find an optimal PSPP path between s and t in time $O(mn^{k+1} + n^{k+2} + m^2)$.*

We note that similar techniques can be applied to special versions of other problems such as the PTSP (i.e., for some special cases, a PTSP can be transformed into a TSP). Finally these results can be used to obtain good heuristics to the PSPP or PTSP when the probability distribution of W — the total number of nodes that are present, is such that $\Pr(W \leq n - k) = \epsilon$ with ϵ very small (for example for a Bernoulli case with p very close to 1). One can also analyze exactly the quality of these heuristics using the general framework given in Jaillet [1985, pp. 163–169].

2.5.5. Applications

In Section 2.1. we have seen one application of the PSPP. We now summarize a few other applications. In the context of flying operations, nodes s and t can be airports and the other nodes can represent geographical areas (mountains, countries, etc.) that can or cannot be flown over by aircraft going from s to t (for example, because of weather conditions, unexpected military restrictions, etc.). The modifications of the route the plane has to take, because of such unexpected restrictions, can be costly, if not anticipated beforehand. One would prefer to consider explicitly these uncertainties in the model in order to find routes of minimum expected costs. The PSPP model is appropriate for such a situation.

More important, consider the following class of problems: suppose one has to go from a 'city' s to another city t on a probabilistic network, possibly by passing through other cities in which one receives some 'revenue'. With the objective of minimizing the expected total net cost, this problem can be modeled as a PSPP: The underlying network is a complete graph built on all cities of interest, and the length of an arc (i, j) is the net cost of traveling between city i and city j (minimum cost of traveling from i to j minus half the total revenue in the cities, $c_{ij} - (r_i + r_j)/2$).

In the context of network reliability, the PSPP can be interpreted as providing a local (hence, easily implementable) strategy of handling node failures (by skipping them) which is optimal in a global sense (minimization of the expected cost). This can be useful either for describing operating strategies on unreliable networks, or for having a polynomial-time computable estimate of 'connecting cost versus node failures' for such networks.

Finally, it is worth mentioning that if, for a specific application, the physical underlying network is not complete, the PSPP model can still be used by first transforming the network into a complete one. In that case a crucial condition for making the network complete with respect to the PSPP is the following: Given the selected a priori path $(s, n_1, n_2, \ldots, n_k, t)$, if nodes n_{i+1} through n_{i+r} fail, then they are skipped by either taking the arc (n_i, n_{i+r+1}) if it exists in the original network, or by taking an alternative path from n_i to n_{i+r+1} *using only nodes whose probability of failure is equal to zero*. This latter condition is to insure that we do not define a 'random' length for this missing edge, i.e., an edge whose length would depend on the presence or not of 'probabilistic' nodes. We illustrate this procedure in Figure 6 for the case of a simple symmetric network with three 'probabilistic' nodes (2, 4, and 5), and four 'deterministic' nodes (s, t, 1, and 3). In order to insure that our procedure respects the previous crucial condition, we split it into three main steps: In Step 1, we consider missing arcs between 'deterministic' nodes; in Step 2, we consider in turn each 'probabilistic' node, and missing arcs between it and 'deterministic' nodes; and in Step 3, we consider in turn all missing arcs between pairs of 'probabilistic' nodes.

Transforming a network into a complete one

1. First, delete all 'probabilistic' nodes and their adjacent arcs, and call the remaining network the *backbone B*. Make B complete by using shortest paths. In our example, edges (s, t), $(1, 3)$, and $(1, t)$ are thus added. Note that if B is

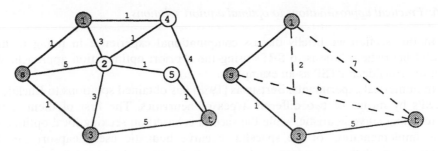

(i) The original graph. Nodes 2, 4, and 5 may fail with a probability p.

(ii) The backbone B with its initial arcs (plain lines). B is then made complete by adding arcs (dashed lines).

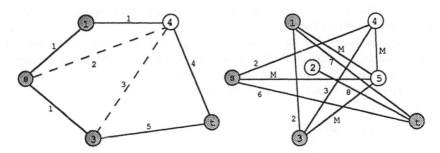

(iii) The complete backbone B with node 4 and its initial arcs adjacent to B (plain lines). The arcs $(s, 4)$ and $(3, 4)$ are then added (dashed lines).

(iv) The final set of all added weighted arcs.

Fig. 6. An original network and the addition of missing weighted arcs.

disconnected, every arc added between disconnected components should then be given a large weight, say M (this case is not shown in our example).

2. Then, consider a 'probabilistic' node, say i, and add it (together with adjacent arcs) to B. From i, add all necessary arcs with weight equal to that of the shortest paths that do not use t as an intermediate node. Then, remove i from B and repeat the procedure for the other 'probabilistic' nodes. In our example, the edges $(2, t)$, then $(s, 4)$, $(3, 4)$, and finally $(s, 5)$, $(1, 5)$, $(3, 5)$ are thus added. Again, if the 'probabilistic' node is not initially directly connected to B by a node other than t, each added arc will then be given the large value M (in our example this is the case for node 5).

3. Finally, consider two non-adjacent 'probabilistic' nodes, say i and j, and add them (with adjacent arcs) to B. Add arc (i, j) with weight equal to the shortest path between i and j that does not use t as an intermediate node. Then remove the pair i, j, and repeat the procedure for the other pairs of non-adjacent 'probabilistic' nodes (in our example, edge $(4, 5)$ is thus added).

Note that this entire procedure might require (see above) the choice of a single large value M.

2.6. Practical approximations to optimal a priori solutions

In this section we briefly discuss computational experience in trying to find useful heuristic solutions to SRPs using the a priori optimization approach. We use the Euclidean PTSP as an example.

In numerical experiments, Bertsimas [1988] has obtained solutions to Euclidean PTSPs by means of two different types of heuristics. The first of them is the spacefilling curve heuristic, while the second is based on seeking local optimality. The implementation of the spacefilling curve heuristic uses heapsort for the sorting part of the procedure, and thus requires only $O(n \log n)$ time to find a nearly optimal tour SF. Interestingly, this is even faster than the computation of the expected length of that tour, $E[L_{SF}]$, which requires $O(n^2)$ time. Since the computed tour SF is independent of the probabilities p_i, the spacefilling curve heuristic can be used when these probabilities are not all the same, or even when they are not accurately known.

For problems involving equal probabilities $p_i = p$, and not more than a few hundred nodes, considerable success was achieved with two separate iterative improvement algorithms based on the idea of local optimality. Given a tour τ and a set $S(\tau)$ of tours which are minor modifications of τ, the tour τ is said to be locally optimal if

$$E[L_\tau] \leq \min_{\tau' \in S(\tau)} E[L_{\tau'}]. \tag{51}$$

The iterative improvement algorithm works by choosing an initial tour τ_0, then testing to see if τ_0 is locally optimal. If a better tour τ_1 is found, it then replaces τ_0 and is itself tested. Since there are only a finite number of possible tours, this procedure must eventually converge to a locally optimal tour τ_* — which, it is hoped, will be a nearly-optimal solution to the problem.

Lin [1965] used an iterative improvement algorithm for the TSP based on what he called the λ-opt local neighborhood. For a given tour τ consisting of n links between nodes, the neighborhood $S_\lambda(\tau)$ consists of those tours which differ from τ by no more than λ links. For $\lambda = 2$ this is the set of tours which can be obtained by reversing a section of τ; for $\lambda = 3$ it is the set of tours obtainable by removing a section of τ and inserting it, with or without a reversal, at another place in the tour. Both the 2-opt and 3-opt TSP algorithms were implemented, since when p is greater than about 0.5 the TSP solutions provide useful starting points for our more general PTSP routines.

Unlike the TSP case, the expected length $E[L_\tau]$ in the PTSP sense depends on all $(n^2 - n)/2$ independent elements of the distance matrix. We cannot, therefore, speak of some links leaving and others entering the tour; rather, it is only the weight given to each of the $d(i, j)$ by equation (6) which changes. We can still use Lin's λ-opt neighborhoods, but the computation of the changes in expected length becomes considerably more complicated. It takes $O(n^2)$ time to calculate the change in expected length from τ to an arbitrary tour in $S_2(\tau)$, so it would seem at first that testing for even 2-p-optimality (referred to heretofore as '2-p-opt') would take $O(n^4)$ time. We can, however, reduce this to $O(n^2)$ if we examine the tours in the proper sequence and maintain certain auxiliary arrays of information as the computation proceeds.

Another neighborhood tried by Bertsimas [1988] consists of moving a single node to another point in the tour, rather than reversing an entire section. The corresponding neighborhood, which we call the 1-shift neighborhood, has roughly twice as many members as S_2, it is a subset of S_4, and yields much better results than S_2 in our experiments.

The best general approach seems to be to first use the spacefilling curve algorithm, followed by 3-opt if p is fairly large, and then finish by applying 1-shift. The threshold point below which 3-opt ceases to be helpful is uncertain and probably depends strongly on the specifics of the problem. For a detailed description of the numerical results the reader is referred to Bertsimas 1988].

3. Modelling issues for dynamic problems

Dynamic problems pose a rich set of modeling issues which must be resolved as a model is developed. In this section, we review these issues to provide a foundation for the technical presentations on model formulation and solution. First, Section 3.1 reviews some general modeling issues that must be addressed in the formulation of a model, and discusses some important choices that have to be made in the process of formulating a dynamic model. Next, Section 3.2 provides a general taxonomy of dynamic models covering different types of models that arise in different applications. Section 3.3 discusses two key issues that arise in the formulation of dynamic models, and Section 3.4 lists alternative objective functions that can be used for dynamic optimization problems, including a review of some basic concepts and terminology from the planning horizon literature. Finally, Section 3.5 summarizes major solution approaches that have been used.

3.1. General issues

In contrast with static models, dynamic models offer a rich and difficult set of issues that must be resolved before we even understand the basic framework in which we are working. Key decisions that must be made in the choice of the model include:

Deterministic vs. stochastic — Deterministic models are fundamentally different than stochastic models, and pose different algorithmic challenges, as well as different answers to questions such as the choice of the proper planning horizon.

Myopic vs. dynamic — Most operational models are basically myopic, optimizing based on what is known in the near future. While the problem is dynamic, the models are basically static, and forecasted future activities are incorporated in at best a very heuristic way. A dynamic model actually attempts to capture future activities.

Choice of objective function — Infinite horizon problems must be approximated to handle forecasting uncertainties and ensure boundedness. A standard choice uses discounted or undiscounted costs over a finite horizon.

The planning horizon — Dynamic models are typically solved over a finite planning horizon. The modeler must choose an appropriate length for the planning horizon, which will in turn effect the type of forecasting that is required as well as the basic formulation of the model.

Spatial and temporal aggregation — The preponderance of models proposed to date use a discrete space, discrete time formulation. This requires choices about the level of temporal and, in some cases, spatial aggregation.

There are six broad research issues that arise in the design and solution of large-scale dynamic network models.

1. *Developing accurate models for optimization under uncertainty.* There are a variety of techniques for optimizing problems under uncertainty. Applied in a standard manner, these techniques are often hopelessly intractable for problems of realistic size. Approximations need to be proposed and tested on the basis of accuracy and computational efficiency. There are a variety of approaches for developing approximations, and a considerable amount of research is needed to determine the best approach.

2. *Identifying 'efficient' formulations.* There is often more than one formulation of the same problem, and some formulations lead more naturally to efficient solution algorithms. Using linear, nonlinear and stochastic networks, we show how alternative formulations can significantly impact algorithmic efficiency.

3. *Design of efficient solution algorithms.* Large stochastic, dynamic network models are intractably difficult. While significant progress may be made in the formulation of the model, continued progress in the development of solution algorithms specifically designed for dynamic networks will likely prove extremely valuable. We review a number of algorithms that have been specially developed for linear, nonlinear and stochastic, dynamic networks.

4. *Planning horizons and truncation errors.* The replacement of infinite horizon models with finite approximations inevitably produces errors. We need to evaluate truncation errors and develop methods for reducing these errors without significantly increasing the size of the model.

5. *Errors due to spatial and temporal aggregation.* It is often necessary to discretize both time and space. Relatively little research has addressed the errors produced by aggregation.

6. *Evaluating a stochastic, dynamic model.* Determining the quality of the solution from a stochastic, dynamic model is a difficult question. Tight bounds are not available, and exact solutions are not possible for problems of realistic size. Extensive rolling horizon simulations are needed to compare two models empirically, raising the difficult problem of experimental design and evaluation. We need better bounds to evaluate the quality of the solutions we obtain, and better experimental methodologies for testing and comparing approximations.

Stochastic formulations of dynamic models are proposed in Sections 7 (in particular, Section 8.2) and 9. Most of the work in stochastic models has focused on the dynamic vehicle allocation problem, but there has been some work on stochastic vehicle routing [Dror, Laporte & Trudeau, 1989; Stewart & Golden, 1983]. The investigation of alternative formulations is reviewed in Sections 5.4

and 5.5. The development of efficient solution algorithms is covered in depth in Section 5 for deterministic networks and Sections 7 and 9. Issues regarding planning horizons are covered in Sections 3.4.3, 6.2.3 and, from a different perspective, 10.4. Finally, the problem of evaluating stochastic, dynamic models (in a rolling horizon framework) is covered broadly in Section 10.

3.2. A taxonomy of dynamic models

Dynamic models come in several forms depending on the nature of the application. These include:

1. *Fixed (finite) horizon* — these models have a natural, finite horizon that most typically occurs because of the bounds of a work shift. In other cases, management might be specifically designing a weekly schedule, producing a fixed horizon.

2. *Rolling (infinite) horizon* — infinite horizon models arise when there is a continuum of decisions being made which impact the future. Examples include dynamic traffic assignment and dynamic fleet management. These problems generally require some form of model truncation.

3. *Periodic* — periodic models arise in a planning context when there is a natural period to the data, such as time of day and/or day of week. These models often arise in the context of inventory routing problems, and involve the development of routine schedules that can be run on a day to day basis.

4. *Dynamic equilibrium* — dynamic equilibrium problems arise in a (long term) planning context for fleet management problems, where a dynamic network model is being used to capture flows over time, but where the model must be solved in such a way that the inventories of vehicles in each city in the first time period are the same as those in the last time period. A typical application of this model would be in fleet sizing, where we wish to simulate the flows of vehicles moving loaded and empty over time. However, we want our model to capture a typical week or month, and we do not want the model to start out with all the fleet capacity in one city and end up with it in another city.

3.3. Formulation issues

Two issues of special importance in the formulation of dynamic models are a) the choice of stochastic versus deterministic formulations, and b) the use of a simultaneous versus recursive model structure. The first issue addresses the basic representation of the problem itself, while the second is a mathematical choice of two problem formulations.

3.3.1. Stochastic versus dynamic models

Uncertainty arises in dynamic models from five exogenous sources:

1. Uncertainty in demand forecasts.
2. Uncertainty in forecasts of external supplies of equipment and drivers.
3. Randomness in the performance of the network (e.g. weather).

4. Randomness in the management and operation of the network in future time periods.

5. Errors in the data provided to the model.

Errors in demand forecasts can often be quantified and modeled in a formal way. Demand forecasts are generally derived from time series models, and past patterns of errors can provide a measure of future errors. Uncertainty in forecasts of external supplies of equipment or drivers arises most often as a result of uncertainty in travel times for movements that originated at an earlier time. Randomness in the performance of the network might refer to uncertainty in travel times (due to weather or congestion) or network capacity. One effect of randomness in network performance is randomness in downstream vehicle supplies. For example, a container sent to a shipper may be held for a random number of days before being released back to the carrier. The fourth source of error is much more difficult to quantify. Planning at some time t requires anticipating how the system will be managed or operated in later time periods. These anticipated actions will not always be implemented as planned, producing a different source of error. Finally, all models must acknowledge the presence of errors in the input data. Such errors are difficult to model and should, as a rule, be viewed as a management problem. However, we cannot always eliminate them, and our models must acknowledge that these errors exist.

Deterministic models are most widely used in practice due to their ease of formulation and the availability of existing solution algorithms. There are, however, several reasons to support the use of stochastic models, including:

1. Deterministic models do not (in general) exhibit a natural planning horizon, and truncation errors can in practice be significant. The result is that deterministic models are often very large, and may be difficult to solve, preventing their use in an operational setting.

2. Deterministic models can exhibit 'nervousness' as a result of sensitivity to changes in forecasted information. Recommendations can change unnecessarily with modifications to forecasted information, increasing costs and reducing confidence in the model.

3. A deterministic model is an approximation, and can produce inferior results, sometimes significantly so, compared to models that handle uncertainty explicitly.

Depending on the choice of formulation, incorporating uncertainty can make a model either hopelessly intractable, or compact and easy to solve.

3.3.2. Simultaneous vs. recursive formulations

A key theme in development of dynamic network models is the role of model formulation. Within the chapter, the presentation of different solution approaches has been organized more along the lines of deterministic and stochastic models and formulations. However, it is often the case that we can choose among multiple formulations for the same model. In Section 5.4 we show how a nonlinear, dynamic network can be solved much more efficiently by using a special flow splitting formulation. In Section 5.5 we show how dynamic multicommodity networks can be solved using decomposition by using formulations that mitigate the effects of degeneracy.

In this section, we review a separate issue. Consider a generic (linear) dynamic network flow problem with the general form:

$$\min_{x_t} \sum_{t=0}^{P} c^T x_t \tag{52}$$

subject to

$$A_t x_t - \sum_{t'=0}^{t-1} B_{t',t} x_{t'} = R_t \quad t = 1, \ldots, P \tag{53}$$

where $B_{t',t}$ is the node-arc incidence matrix giving the elements of $x_{t'}$ that send flow from nodes in period t' to period t, and A_t gives the elements of x_t that send flow from period t to periods $t' \geq t$. We may often simplify the problem by assuming that $B_{t',t} = 0$ for $t' < t - 1$.

We refer to this form of the model as the *simultaneous* formulation, since we consider the optimization of flows in all time periods simultaneously. Alternatively, we can formulate the problem using a state variable. Let S_t be the flow entering each node in period t from periods $t' < t$, defined by:

$$S_t = \sum_{t'=0}^{t-1} B_{t',t} x_{t'} \tag{54}$$

Now rewrite (52) as follows:

$$\min_{x_0, S_1} c^T x_0 + Q_1(S_1) \tag{55}$$

where the function $Q_1(S_1)$ is defined recursively using:

$$Q_t(S_t) = \min_{x_t} c^T x_t + Q_{t+1}(S_{t+1}) \tag{56}$$

subject to

$$A_t x_t = R_t + S_t \tag{57}$$
$$S_{t+1} - B_{t,t+1} x_t = S_t \tag{58}$$

We refer to the formulation (55)–(58) as the *recursive* formulation. The simultaneous formulation is most commonly associated with deterministic models, while the recursive form is almost required by any stochastic model. In stochastic models, $Q_t(S_t)$ plays the role of the expected recourse function or value function. However, in Section 5.4 we see that the recursive formulation is of tremendous value for solving deterministic nonlinear, dynamic networks. At the same time, the most widely used approach for solving stochastic optimization problems, called scenario aggregation (see Section 7.4), uses the simultaneous formulation. The recursive form seems better suited to dynamic models, but at this time both approaches are being used in the research literature.

3.4. Objective functions

The travelling salesman problem is often cited as the archetypal example of a 'hard' combinatorial optimization problem. However, in one important respect, this problem is significantly easier than the dynamic models presented in this chapter — the objective function is well defined. While the determination of the optimal solution may be difficult, the comparison of two heuristics is generally straightforward. When solving a dynamic model under uncertainty, the choice of the objective function is not obvious. This issue has been studied extensively in the research literature, primarily in the context of inventory planning and dynamic lot sizing.

We briefly discuss three issues in the formulation of the objective function. Section 3.4.1 presents alternative criteria that can be used in dynamic models. Section 3.4.2 reviews methods for formulating bounded objective functions for infinite horizon problems. Finally, Section 3.4.3 provides a summary of basic terminology from the planning horizon literature for solving finite approximations of infinite horizon problems.

3.4.1. Cost criteria for dynamic models

In static models, minimizing cost, subject to constraints of servicing all the demand, is the standard criterion for evaluating the quality of a solution. In dynamic models, the choice of measurement is not as obvious. Depending on the application, we might consider:

1. *Costs*. Of course, minimizing costs (over a finite or infinite horizon) remains a standard criterion for many applications.

2. *Profits*. Unlike static models where we usually service all the demand, it is invariably the case that dynamic models must consider the possibility of not servicing all the demand. In a real-time setting, it is simply not always possible to guarantee that you will service all the demand all the time. The possibility of refusing service arises in some applications, implying that operating profits (revenue minus cost) is more relevant. In other settings, all demands are handled, but with varying degrees in the quality of the service provided, suggesting that penalties be introduced to capture the effect of poor service.

3. *Minimum covering*. If we want the fewest number of vehicles or crews required to handle a set of demands, then we want a minimum flow solution subject to covering constraints [Dantzig & Fulkerson, 1954].

4. *Maximum throughput*. We may wish to design a network that maximizes the total flow through the network, thereby maximizing the total demand served or service rendered [Orlin, 1983].

5. *Average time in network*. Traffic assignment problems are often primarily interested in the time required to push flow through the network. We may minimize average time in the network, which is fairly easy, or we may wish to minimize a nonlinear function of travel time, which is generally much harder.

6. *Maximum clearance time*. Evacuation problems typically focus on the latest time at which flow remains in the network. The minimum time (or 'bottleneck

time') transportation problem seeks to minimize the latest time that all demands are satisfied [see Szwarc, 1966, 1971a; Hammer, 1969, 1971].

3.4.2. Objective functions for infinite horizon optimization

Let x_t be the set of actions in period t, c_t be the cost vector and S_t be the state of the system given $\{S_{t-1}, x_{t-1}\}$. Let $\mathcal{X}_t(S_t)$ be a set of actions that we may take in period t, let $g_t(S_t, x_t)$ be the costs incurred in period t, and let G_t be the total optimal costs over periods $\{0, \ldots, t\}$. Three different optimality criteria have been suggested in the literature to produce bounded objective functions over infinite horizons:

Infinite discounted programming [Derman, 1970]. For a given discount factor $\{0 \le \alpha < 1\}$ we solve:

$$G^* = \inf_{x \in \mathcal{X}} \sum_{t=0}^{\infty} \alpha^t g_t(S_t, x_t)$$

If g_t is bounded, then $\alpha < 1$ guarantees that G^* exists.

Average reward criterion [Derman, 1970]. For stationary problems with homogeneous costs and demands, it may make sense to minimize the average costs per period:

$$G^* = \inf_{x \in \mathcal{X}} \lim_{T \to \infty} \frac{1}{T} \sum_{t=0}^{T} g_t(S_t, x_t)$$

Relative gain optimization [Howard, 1971]. The first two formulations represent methods for avoiding unbounded objective functions. An alternative approach that can be used on the context of discrete dynamic programs minimizes the expected relative cost, where the cost of each state is measured relative to a base state which is defined as zero. Let π_{it} be the probability of being in state i at time t (given initial state S_0) and let S_t be the set of accessible states at time t given S_0. For a given state $i \in S_t$, the expected reward in period t is given by:

$$v_{ti} = \min_{x_t \in \mathcal{X}_t} C_t(x_t, i) \tag{59}$$

where $C_t(x_t, i)$ is the cost of being in state i and taking action x_t. The total expected reward in period t is then:

$$V_t = \sum_{i \in S_t} \pi_{it} v_{it} \tag{60}$$

If the underlying problem is a homogeneous Markov reward process, then it is well known that [Bhat, 1984; Heyman & Sobel, 1984]:

$$\sum_{t=0}^{T} v_{it} = v_{i0} + gT \quad \text{as } T \to \infty \tag{61}$$

where v_{i0} is a state dependent constant and g is a system dependent constant. If we let state 1 be a base state, then we can define new values:

$$\bar{v}_{it} = v_{it} - v_{1t} \tag{62}$$

In the limit:

$$\lim_{T \to \infty} \sum_{t=0}^{T} \bar{v}_{it} = v_{i0} - v_{01} \tag{63}$$

which is bounded. Thus, by using relative gains, we may solve an infinite *undiscounted* optimization problem.

Remarks. The average reward criterion works only for purely stationary models, since, in the limit, no weight is put on activities in the near future. This criterion was used in Orlin [1983], [1984b] in the study of stationary dynamic networks. The relative gains model is intuitively appealing since it captures the actual tradeoff between competing activities and produces a bounded objective function without requiring the use of artificial discounting. This method was the basis of a heuristic presented in Powell [1987] for the dynamic fleet management problem for truckload motor carriers. A Markov reward model was used to develop relative salvage values to capture the value of additional vehicles in one region over another.

The discounted infinite horizon formulation is the most standard basic model. The textbook motivation for the discount factor is accounting for the time value of money. In most operational models, however, the effective planning horizon is too short for this effect to have any impact. Instead, discount factors are used to heuristically account for uncertainty in the data. For example, if the time period is one week, we might use a discount factor $\alpha = 0.3$, equivalent to a *weekly* interest rate of 70%! This approach is simple and intuitively appealing, but is a clearly heuristic approach to handling forecasting uncertainties. For example, this method discounts all activities in period t by α^t, whereas in practice there may be very different degrees of uncertainty associated with different activities in the same time period. For example, containers dispatched from Chicago to Tokyo will arrive in Tokyo with a very high probability in about two weeks, and yet the demands for containers in Tokyo two weeks from now may be quite uncertain.

3.4.3. Planning horizons

The most common approach to approximating infinite horizon problems is to solve a finite, or truncated, problem. Let $P < \infty$ be a specified *planning horizon*. Then, for any α, $\{0 \leq \alpha \leq 1\}$ we solve:

$$G^*(P) = \inf_{x \in \mathcal{X}} \sum_{t=0}^{P} \alpha^t g_t(S_t, x_t)$$

This method, which may be used with discounted or undiscounted costs, is the most standard approach used in practice. Although a discount factor is not required to ensure boundedness, it is often used in practice as a heuristic approach for accounting for uncertainty. Planning horizon methods possess the intuitive appeal of working in a manner similar to the way people work [Morton, 1981]. The idea is that, for sufficiently large P, the quality of the solution in the first period, x_0, will be sufficiently good for practical purposes.

This raises the problem of choosing an appropriate planning horizon, and determining the properties of the solution provided by a finite approximation. In this section, we briefly review some of the basic concepts and terminology that has been developed within the planning horizon literature. First, the term *planning horizon* is used fairly loosely to describe how far into the future the model extends in order to make decisions now. The term is not well defined, because there are different ways that forecasted data can be used in a dynamic model. For example, *salvage values* [Grinold & Hopkins, 1973] can be used to capture the value of extra flows into a node at the end of the planning horizon. These salvage values approximate activities beyond the formal end of the planning horizon.

Excellent discussions of the planning horizon literature can be found in Bean & Smith [1984], Sethi & Bhaskaran [1985], Bhaskaran & Sethi [1987], Bes & Sethi [1988] and Morton [1979]. In this literature, general concepts of planning horizons have been replaced with more precise notions of *forecast horizons* and *decision horizons*. A forecast horizon (or exact forecast horizon) τ^f is a period such that planning horizons $P \geq \tau^f$ produce the same solutions for periods $\{1, \ldots, \tau^d\}$, where $\tau^d \leq \tau^f$ is called a decision horizon. τ^f is called a *near forecast horizon* if the first period solution 'closely approximates' the infinite horizon optimal solution. A *weak forecast horizon* exists when conditions on future data must be imposed to produce a forecast horizon.

There is an extensive literature on planning horizons, most of it motivated by dynamic lot size and inventory problems. One of the earliest results on forecast horizons is given by Wagner & Whitin [1958] for the dynamic lot size problem, which can be formulated as a dynamic network with concave costs. Wagner–Whitin show how a forecast horizon can accelerate the solution algorithm, as well as identify how far into the future forecasted data can impact first period decisions. Comprehensive surveys of forecast/decision horizons are given by Bhaskaran & Sethi [1987], Morton [1979] and Bensoussan, Crouhy & Proth [1983]. Bes & Sethi [1988] provide a formal treatment of forecast/decision horizons for stochastic, dynamic problems.

Research in this area tends to fall along two general lines. The first depends on the identification of *regeneration points* which produce forecast horizons by creating points in time when the system effectively restarts itself. Morton [1979] provides a general framework for the use of regeneration points in undiscounted, infinite horizon problems. The use of regeneration points has proved particularly effective in the solution of deterministic problems with concave costs [Zangwill, 1969; Lundin & Morton, 1975] which often arise in lot size models and production planning. The second line of investigation has looked at convex, stochastic problems which tend to exhibit a convergence of state vectors in future periods as a result of the ergodic structure of the underlying stochastic process [see, for example, Morton & Wecker, 1977]. A closely related line of investigation looks at deterministic convex problems with discounted costs [Bean & Smith, 1984].

In contrast with the extensive research literature on inventory and production problems, planning horizon results for dynamic networks (with the notable exception of concave cost networks) are much more limited. This is even more true of the types of dynamic networks that arise in dynamic fleet management. Aronson & Chen [1986] present a specialization of network simplex for dynamic networks, drawing on related work for staircase linear programs [Aronson & Thompson, 1984; Aronson & Chen, 1985]. This algorithm is oriented toward dynamic production planning problems which consist of a static transportation problem in each period, where inventory arcs that carry excess supply from one period to the next. Aronson & Chen note that the simplex basis tends to exhibit breaks which create regeneration points which can be used to identify decision horizons (see Section 5.2). However, a break in the basis for a P-period problem is not guaranteed to remain as the planning horizon is increased. For this reason, Aronson & Thompson [1984] introduce the notion of an *empirical* (or *computational*) decision horizon, which is a decision horizon that is likely to hold with a high probability. Aronson & Chen [1989] summarize empirical studies of decision horizons in dynamic networks for production planning applications.

3.5. Solution approaches

We divide solution strategies into seven broad areas, five of which deal explicitly with stochastic problems. These are:

1. *Deterministic model/linear programming.* We assume that all deterministic formulations can be formulated in some way as a linear program. Of course, some of these linear programs are exceptionally large, and specialized algorithms have been devised to handle both the side constraints [Lasdon & Terjung, 1971; Assad, 1987; Kennington & Helgason, 1980; Farvolden, Powell & Lustig, 1993] and the dynamic structure [Ho & Manne, 1974; Aronson & Chen, 1986].

2. *Chance constrained programming* [Charnes & Cooper, 1959]. For the case of random demands, it may be useful to specify the fraction of demand that should be satisfied, representing a level of service constraint (see Section 9.1.3).

3. *Stochastic programming/scenario aggregation.* Stochastic programs can be approximated by optimizing decisions over a specified set of *scenarios*. The solution for each scenario must be aggregated to produce a single recommended action in the first time period (see Section 7.4).

4. *Stochastic gradient methods.* These methods approximate expected future costs by taking a sample of future activities and then calculating a sample gradient, which can then be used to optimize decisions in the first time period. Several techniques are available for using sample information within an optimization procedure (see Section 7.5).

5. *Approximations of stochastic programs.* This class of methods refers to techniques that replace the original stochastic program with an approximation that

simplifies the problem in such a way that it can be solved with classical techniques. These methods are heuristic and must be evaluated experimentally. The simplest approximation that is widely used in practice is stochastic programming with simple recourse, but other methods can be used (see Section 9).

6. *Markov decision processes*. While MDP's have not been successfully applied to the solution of large scale, stochastic networks, there is an extremely rich theoretical literature on MDP's with considerably more attention given to the study of the properties of optimal solutions, the errors due to model truncation and the convergence of optimal policies.

7. *Optimal control*. Optimal control theory is the only approach that explicitly considers problems in continuous time. Research in this class of problems is limited.

4. Dynamic models in transportation and logistics

We consider the following sequence of models:
* dynamic shortest paths;
* dynamic traffic assignment (system optimization);
* dynamic production and inventory planning;
* dynamic vehicle allocation;
* dynamic assignment;
* dynamic vehicle routing;
* dynamic service network design;
* dynamic facility planning.

The first two problems address the task of routing flows over a network. The first considers a single vehicle or traveler, while the dynamic traffic assignment problem considers many vehicles moving between multiple origins and destinations.

Next we introduce the classical dynamic production and inventory planning problem. Strictly speaking, this problem area is outside of the scope of this chapter [see Shapiro, 1993]. However, there is a substantial literature on dynamic networks that is motivated by these applications, and many of the concepts regarding dynamic models can be traced to results derived in the context of production and inventory planning.

The next three areas deal with routing of vehicles, in order of difficulty. The dynamic vehicle allocation problem is typically approached as a (possibly pure) dynamic network. Vehicles are modeled as aggregate flows over a network, with very little ability to enforce driver work rules. The dynamic assignment problem considers the assignment of drivers to individual tasks (e.g. moving loads from one point to another) on a rolling basis. Finally, the dynamic vehicle routing problem addresses the problem of clustering customers into loads.

The last two problems are the hardest from a combinatoric perspective. We review the limited literature on these problems, but do not cover them in any depth in this chapter.

The presentation begins in Section 4.1 with a short discussion of basic conventions in notation and vocabulary. The rest of the section presents a sample of specific models, in order of increasing complexity.

4.1. Nomenclature and conventions

A dynamic network normally begins with a physical, spatial network that is then defined over time. We assume that movements occur between a set of cities C, which can represent a city, terminal, region of the country, a single shipper or trailer pool, port, or any other physical facility.

Our problem is generally to move goods (generally shipments, but also passengers, equipment or other objects) between pairs of cities. In general, goods moving from an *origin city r* to a *destination city s*, and might be designated D^{rs}. However, for vehicle routing problems, the pickup or delivery location may be fixed at a warehouse, and therefore might be referenced by a single index (e.g. D^r). A city pair (r, s) is usually referred to as a *market* or, in freight terms, a *traffic lane*. In dynamic problems, market demands are defined over time by D^{rs}_t, where t refers to the time the freight enters the network. An important dimension of dynamic flow problems is the nature in which demands are called into the system, known as the *booking process*. The mathematical specification of the booking process also reveals the structure of forecasting uncertainties for future demands. In truckload trucking, rail and container movements, an individual demand usually fills the entire container. In this presentation, we use the word *load* to refer to shipments that occupy an entire vehicle, while *shipment* is used specifically for customer orders that require in-vehicle consolidation.

Vehicles moving between an origin i and destination j often move through a series of intermediate transshipment points. We refer to this itinerary as a *route*, made up of individual transportation segments called *legs*. The timing of the movement of vehicles over a route is referred to as a *schedule*.

In our view of the world, goods move in containers (trailers, boxcars) pulled by vehicles (tractors, locomotives, ships) whose movements are determined by transportation schedules and managed by drivers. We let x and y refer to the flows of loaded and empty containers, respectively. The flow of loaded containers on leg (i, j) in time t is denoted x_{ijt}. The flows of loaded vehicles are similarly denoted v_t, and the movement of empty vehicles is denoted u_t. The flow of goods is referred to using w.

We try to use a consistent vocabulary and notation throughout, but in some instances there is an established literature that has evolved in the context of a particular problem. For example, in freight transportation (such as trucking or rail) we may refer to the dynamic vehicle allocation problem, but we are really referring to the containers (trailers, boxcars) instead of the individual tractors or locomotives. In the dynamic traffic assignment problem, we consider only the movement of goods (no containers or vehicles). In truckload trucking, we ignore the goods, since the movement of a truckload shipment from r to s is equivalent to the movement of the trailer (and hence we focus on the movement of trailers, and simply distinguish between loaded and empty trailers).

4.2. The dynamic shortest path problem

Section 2.5 provided results for finding a priori shortest paths over a stochastic network. In this section, we consider deterministic and stochastic shortest path problems which involve time varying data.

4.2.1. Deterministic shortest paths

The deterministic dynamic shortest path problem was first introduced by Cooke & Halsey [1966]. They assume that travel times are multiples of δ for some $\delta > 0$. Also, they assume that these travel times are known at times $t_0, t_0 + \delta, t_0 + 2\delta, \ldots$ Now let $f_i(t)$ be the minimum travel time from node i to the destination node d given that you are visiting node i at time t. Cooke and Halsey then introduce the functional:

$$f_i(t) = \begin{cases} \min_{j \neq i} [c_{ij}(t) + f_j(t + c_{ij}(t))] & \forall i \neq d \\ 0 & i = d \end{cases}$$

which can be solved using dynamic programming. This is equivalent to solving a shortest path problem over a time expanded network, where each node in the dynamic network corresponds to a point in space-time in the original network. Dreyfus [1969] simplified this approach by adapting Dijkstra's algorithm, whereby labels v_i are maintained for each node i. At each iteration, the node i with the minimum v_i is made permanent. The labels v_i are updated using:

$$v_j = \min[v_j, v_i + c_{ij}(v_i)] \quad \forall j$$

More recently, Kaufman & Smith [1992] show that Dreyfus' method will solve the deterministic dynamic shortest path problem only if it satisfies a consistency condition given by:

$$s + c_{ij}(s) \leq t + c_{ij}(t) \quad \forall s, t, s \leq t$$

This assumption is known as the overtaking condition in traffic problems. It effectively assumes that if car i enters a link later than car j, then it will also leave the link later. In other words, we do not allow car i to overtake car j on the link.

A variation of the dynamic shortest path problem allows cars to wait at intermediate nodes (sometimes referred to as parking). This was first studied by Halpern [1977], who showed that if the overtaking condition is not imposed, then optimal paths may involve cycling (where cars effectively travel around the block waiting for more favorable conditions). He proposed *parking* at nodes so that cars could wait at a node (instead of cycling) until a more favorable time to traverse the node.

Orda & Rom [1990] consider three types of dynamic networks. The first allows unlimited waiting at any node; the second allows waiting only at the origin node, and the third where no waiting is allowed anywhere. They show that if the departure time from the source node is unrestricted, then a shortest path can be found that is simple and achieves a delay as short as the most unrestricted path. They also give examples that show that under the assumption of no waiting (and arbitrary link travel time functions) can produce networks with no finite optimal path.

4.2.2. *Stochastic shortest paths*

Shortest path problems over stochastic, dynamic networks come in several forms depending on the motivating application. Bertsekas & Tsitsiklis [1991] consider a very general stochastic shortest path problem where at each node i we choose a probability distribution over the set of outbound arcs so that the expected cost to the destination is a minimum. The problem is formulated as a Markov decision process and used to investigate the properties of an optimal policy. Standard methods are proposed for finding the optimal policy.

Andreatta & Romeo [1988] consider a network where individual arcs may or may not be present. A vehicle arriving to a node i may then learn that arc (i, j) is no longer present, and must then adopt a recourse strategy to develop a new path. Each arc in the network has a known probability of being blocked, but they also assume that once an arc (i, j) is found to be blocked, no other arc emanating from i can be blocked. Andreatta [1987] shows that this problem is harder than $\mathcal{NP}$, and also shows that the cost of path derived from a deterministic approximation of the network can be arbitrarily bad relative to the optimal stochastic path.

Psaraftis & Tsitsiklis [1990] investigate a problem motivated by the routing of ships over the sea subject to weather delays. The assumption is that at specific 'nodes' the ship may move immediately over one or more arcs, or wait until more favorable conditions arise. Two variations of this problem arise because we can assume that the ship will wait and then choose an outbound arc, or choose an outbound arc first and then wait for better conditions.

Assume that costs are known functions of variables that evolve according to a Markov process. More precisely, the cost of traversing each arc (i, j) of the graph is a known function $f_{ij}(e_i)$ of the state e_i of a variable at node i at the time the vehicle departs from node i on its way to node j. The variables are mutually independent, and each is governed by a finite-state Markov process, with discrete time transitions. The actual state e_i of the variable at node i is known by the vehicle only when it is at this node. If the vehicle reaches node i and then decides to immediately go to node j, it incurs a cost $f_{ij}(e_i)$. The vehicle may also want to wait at node i in anticipation of a more favorable state at i. It can wait as much as it wants, but at a cost of C per state transition. The problem is then to decide on a policy that minimizes the expected total cost of a traversal, say from source node 1 to destination node n.

The main results obtained by Psaraftis & Tsitsiklis [1990] are the following: (1) Focusing on an individual arc of the network, the authors develop and analyze different policies for traversing that arc. Recognizing that this problem can be formulated as an 'infinite-horizon, total cost, stochastic dynamic programming problem' [see Bertsekas, 1987], they investigate two methods, the *successive approximation* procedure and the *policy iteration* procedure, and prove the computational superiority of the latter. They also consider linear programming in order to solve this infinite horizon dynamic programming problem, and mention some simple case for which closed form solutions are obtainable. (2) Considering the entire network, they then consider a suboptimal policy in which the direction decision precedes the 'go/no go' decision, the latter being solved as before. Finally they develop an optimal policy in which the previous two decisions are reversed, and they show that it is also better in terms of computational effort.

4.3. The dynamic traffic assignment problem

The dynamic traffic assignment problem addresses the problem of routing goods over a network from origin (representing the origin location as well as the time the flow entered the network) to the destination (the time of arrival to the destination is an output of the model). In between, the flow will move over one or more links which may be capacitated or have nonlinear costs. The problem is in routing flows between multiple origins to multiple destinations, and handling the competition for limited capacity over a network of transportation services.

In this section, we consider the pure traffic assignment problem. We also focus only on freight problems which can be formulated as system optimization (global cost minimization) problems, and exclude passenger applications, such as the dynamic traffic assignment problem for auto traffic, where passenger behavior is typically modeled as an equilibrium problem. For research on dynamic traffic equilibrium problems, see Friesz, Luque, Tobin & Wie [1995], Friesz, Bernstein, Smith, Tobin & Wie [1993], Ran, Boyce & LeBlanc [1992], and the references cited there.

We consider three versions of the dynamic traffic assignment problem. Section 4.3.1 considers the capacitated, linear traffic assignment problem in discrete time, which arises when routing freight flows over a fixed network of transportation services. Next, Section 4.3.2 describes the nonlinear dynamic traffic assignment, which models congestion effects without hard capacity constraints. Both of these formulations are presented in a discrete time framework. Finally, 4.3.3 describes evacuation models, which represent an important special case of the dynamic traffic assignment problem.

4.3.1. The capacitated version

To formulate the dynamic traffic assignment problem, we distinguish between flows of shipments, 'containers' and scheduled transportation services (vehicles). The dynamic 'vehicle' allocation problem focuses on the flows of containers. Here, we focus on the flows of shipments over a scheduled network of transportation services. Define:

D_t^{rs} = total flow of shipments moving from origin r to destination s, originating in period t.

$h^{rs}(\tau)$ = service penalty assessed on shipments moving from r to s delivered in τ time periods.

P_t^{rs} = set of feasible space-time paths joining r and s, departing at time t, moving over a dynamic network.

τ_{pt}^{rs} = travel time for path $p \in P_t^{rs}$.

$$\sigma_{ij}^{p,rs}(t_1, t_2) = \begin{cases} 1 & \text{if link } (i, j) \text{ is on path } p \text{ joining } r \text{ and } s, \text{ departing in} \\ & \text{period } t_1 \text{ and moving over link } (i, j) \text{ departing in period } t_2, \\ 0 & \text{otherwise.} \end{cases}$$

w_{pt}^{rs} = total flow on path p joining r and s, departing in period t.

w_{ijt} = total flow on link (i, j) departing in period p.

v_{ijt} = total transportation capacity available between i and j, departing at time t, determined by the routing of vehicles.

c_{ij}^w = cost per unit to move a shipment over link (i, j).

The flow of vehicles is given by v_{ijt}, which is assumed to be expressed in the units of the flow w_{ijt}. Note that we are not concerned with how the capacity v_{ijt} is obtained (this is considered later). We have also adopted a path flow formulation, since recent experimental work has shown that this approach significantly outperforms more classical link-based formulations (see Section 5.5.1).

It is important to model transfer activities, where shipments must be unloaded, sorted and reloaded. These activities, as well as their costs and constraints, are easily modeled using standard network techniques (for example, splitting nodes to model transfer costs). Interestingly, while there is a cost for handling shipments at transshipment points, the cost per shipment c_{ij}^w for a transportation link is typically zero, since transportation costs are normally associated with the flows of vehicles reflected by v_{ijt}. Interestingly, existing dynamic traffic assignment models have not really taken advantage of this.

The traffic assignment problem is most easily visualized using the network in Figure 7. Service penalties are handled using a fairly standard device. The flow D_t^{rs} which originates at time t is assumed to exit through a supersink for destination s. The costs on links into the supersink reflect service penalties, where the model is significantly simplified if the service penalty function, $h^{rs}(\tau)$, depends only on the destination and is linear in τ. A mathematical statement of what we refer to as the dynamic traffic assignment problem is given by:

$$F_w(v) = \min_{w_{pt}^{rs}} \sum_{t=0}^{P} \left[(c^w)^T \omega_t + \sum_{r \in C} \sum_{s \in C} \sum_{p \in P_t^{rs}} w_{pt}^{rs} h_p^{rs}(\tau_{pt}^{rs}) \right] \tag{64}$$

subject to, for $t = 0, \ldots, P$

$$\sum_{p \in P_t^{rs}} w_{pt}^{rs} = D_t^{rs} \qquad\qquad\qquad\qquad r, s \in C$$

$$w_{pt}^{rs} \geq 0 \qquad\qquad\qquad\qquad r, s \in C, \ p \in P_t^{rs}$$

$$w_{ijt} - \sum_{t' \leq t} \sum_{r \in C} \sum_{s \in C} \sum_{p \in P_t^{rs}} w_p^{rs}(t') \sigma_{ij}^{p,rs}(t', t) = 0 \quad i, j \in C$$

$$w_{ijt} \leq v_{ijt} \qquad\qquad\qquad\qquad i, j \in C \tag{65}$$

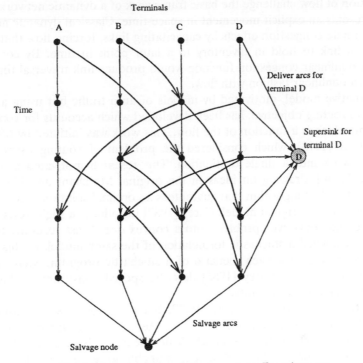

Fig. 7. Space-time network for dynamic traffic assignment.

$F_w(v)$ is the optimum shipment routing costs given a set of transportation services v_t.

From an algorithmic perspective, the complicating feature is the bundle constraint (65). The development of algorithms for this problem is discussed in Section 5.5.

Problem (64) focuses on moving a shipment through a network of transportation services, with transfers through intermediate transshipment points for sorting. The same problem arises when moving a loaded container (vehicle) through the network. In rail and container shipping, the shipment and loaded railcar or container are equivalent. In LTL trucking, the problem of moving a trailer loaded with shipments from origin to destination is also a type of traffic assignment problem. In this problem, however, the movements are constrained not by available transportation capacity, v_{ijt}, but rather by the availability of drivers governed by work rules.

4.3.2. The nonlinear version

The model above does not account for the possibility of congestion on the links of the network. By this, we refer explicitly to a process whereby travel times on a link increase as a nonlinear function of the flow on the link. Travel times that are a function of flow challenge the basic framework of a dynamic network, where a link represents an explicit movement in space-time. Classical dynamic networks can approximate congestion effects by capacitating links, forcing flow that cannot move over a link to hold in inventory to a later point in time. By contrast, a link with a nonlinear congestion function would produce link traversal times that increase in a nonlinear fashion with flow.

An alternative model, motivated by models of auto traffic but using a system optimization routing objective, has been developed which accounts for congestion on each link which is a function of the flow. This work was initiated by Merchant & Nemhauser [1978], which considered the problem of routing users over a dynamic network into a single destination. This research motivated a series of papers. Ho [1980], drawing on ideas in the original Merchant and Nemhauser paper, introduced an algorithm that guarantees an optimal solution to a piecewise linear version of the original model. Carey [1987] introduces a slight revision that transformed the nonconvex problem into a convex one. Most recently, Birge & Ho [1987] introduced a stochastic formulation of the same model. In this model, Ho's results are used for each scenario of a stochastic program. Merchant and Nemhauser [1978b] and Carey [1986] describe special constraint qualifications necessary to guarantee optimality.

We use the formulation in Carey [1987], but adopt our own notation from above. The basic formulation introduced by Merchant and Nemhauser models congestion by restricting the rate at which flow leaves a link, a rate that varies as a nonlinear function of flow. Flow that is unable to leave the link in a given time period is assumed to hold until the next time period. As a result, we have to divide the 'flow' on a link into three categories: flow entering a link (in a given time period), flow leaving a link, and flow that is held in inventory from one time period to the next. Keeping with our earlier notation, we use ω_t to denote flows, and define:

w_{ijt}^e = flow *entering* link (i, j) at the beginning of time period t.

w_{ijt}^h = flow *held in inventory* on link (i, j) at the beginning of time period t, which has been held over from the previous time period.

w_{ijt}^d = flow *departing* from link (i, j) during period t.

$g_{ij}(w_{ijt}^h)$ = maximum allowable outflow from link (i, j) in period t.

$c_{ij}(w_{ijt}^h)$ = total cost incurred from holding w_{ijt}^h units of flow on link (i, j) in period t.

R_{it} = exogenous flow entering the network at node i in period t (with a common destination).

We assume that all values are assigned initial values for $t = 0$. The dynamic traffic assignment with congestion can be stated as follows:

$$\min_{w^e, w^h, w^d} \sum_{t=0}^{P} \sum_{i,j} c_{ij}(w_{ijt}^h) \tag{66}$$

subject to, for $t = 0, \ldots, P$

$$(w_{ijt}^h + w_{ijt}^e - w_{ijt}^d) - w_{ij,h,t+1} = 0 \quad i, j \in C$$

$$\sum_{k \in C} w_{jkt}^e - \sum_{i \in C} w_{ijt}^d = R_{jt} \qquad j \in C$$

$$w_{ijt}^d - g_{ij}(w_{ijt}^h) \leq 0 \qquad i, j \in C \qquad (67)$$

$$w_{ijt}^e, w_{ijt}^h, w_{ijt}^d \geq 0 \qquad i, j \in C$$

The important characteristic of this formulation is the exit functions $g_{ij}(w_{ijt}^h)$ which restrict the amount of flow allowed to leave the link in a given time period. Note that we do not restrict the problem if we require the exit functions to satisfy $g(w) \leq w$. If $g(w) = w$, then the problem is unconstrained and we have an instance of an uncapacitated traffic assignment problem. Normally, congestion reduces the efficiency of the network, resulting in situations where $g(w) < w$. It is reasonable to expect that $g(w)$ is concave in w, which implies that constraint (67) is convex (that is, the set of values w_{ijt}^h that satisfy (67) is a convex set). In their original paper, Merchant and Nemhauser assume that (67) holds with equality, which is not convex, and demonstrate that the resulting problem can have local optima. They investigate a form of the model where the exit functions $g_{ij}(w_{ijt}^h)$ are replaced with piecewise linear approximations, and introduce the following cost function assumption (CFA):

(a) $c_{ij}(w_{ijt}^h)$ is nonnegative, convex in w_{ijt}^h and nondecreasing in t.

(b) Let (i, j) and (k, l) represent two distinct arcs on a path to the destination in time period t, with (k, l) closer to the destination. Then $\partial c_{ij}(u, t)/\partial u \geq \partial c_{ij}(v, t)/\partial v$ for all u and v.

(c) The inequality in (b) is a strict inequality when $t = P$.

Merchant and Nemhauser consider the case where the nonlinear functions $g_{ij}(w_{ijt}^h)$ are replaced with piecewise linear approximations. For this case, they show that CFA guarantees that the global optimal solution can be expressed as a basic solution of a linear program. Ho [1980] exploits these properties and presents a method for finding the optimal solution by solving at most $P + 1$ linear programs (where P is the length of the planning horizon). Birge & Ho [1987] extend this methodology to handle uncertainty in the input flows R_t.

Carey [1987] introduces the formulation we use above where the exit functions are used as inequalities as in (67). The difference $g_{ij}(w_{ijt}^h) - w_{ijt}^h$, which gives the extent to which the actual flow departures fall below the capacity of the link, is referred to as a *flow control*, with the interpretation that a master controller might want the actual flow departure to be below the capacity of the link (which is consistent with a system optimization formulation). This raises the question of the conditions that would produce positive flow controls. For this question, Carey introduces the following exit function assumption (EFA):

(a) $0 \leq g_{ij}(w_t) \leq w_{ijt}$;

(b) $0 \leq \partial g_{ij}(w_t)/\partial w_t < 1$;

(c) $g_{ij}(0) = 0$.

Carey shows that under EFA, and a weaker form of CFA, that constraint (67) will always be binding. The CFA conditions are needed to ensure that the model will try to push as much flow as possible into the supersink before the end of the planning horizon. Condition (b) eliminates conditions that encourage holding flow in inventory, while condition (c) is a tie-breaking condition to force flow out of the network before the final period.

It should be noted that as stated, this model does not incorporate travel times. If there is no congestion, then all flow moves through the network in the same time period. The model can be modified to incorporate travel times through suitable changes to constraint (67), although a modification to EFA may be necessary to ensure zero flow controls.

This line of research has focused on problems with flows into a single destination, allowing a single commodity formulation. Carey argues that multiple destination problems can be transformed into single destination problems by adding a supersink. Appropriate use of constraints on links into the supersink can guarantee that the correct amount of flow exits the network from a particular destination, but such a formulation is unable to guarantee conservation of flow when commodity flows are specified in an origin/destination format.

4.3.3. Evacuation models

A special case of the traffic assignment problem is one that minimizes the time required to move all flow from source to sink. Chalmet, Francis & Saunders [1982] presents a network model to solve a building evacuation problem. Jarvis & Ratliff [1982] show that if the cost of exiting in period t is c_t, and $c_1 \leq c_2 \leq \ldots \leq c_P$, then the optimal solution a) minimizes the average transit time, b) maximizes the total number of evacuees in the interval $(1, t)$ for $1 \leq t \leq P$, and c) minimizes the time at which the network is cleared. Choi, Hamacher & Tufekci [1988] present algorithms to handle building evacuation problems with side constraints. Finally, Hamacher & Tufeki [1987] develop additional properties of flows for evacuation problems.

A separate line of research has considered stochastic evacuation problems. Karbowicz and MacGregor Smith [1984] present a k-shortest path algorithm for handling random events when evacuating a stochastic networks. MacGregor Smith & Towsley [1981] describe a queueing network model for building evacuations. More recently, MacGregor Smith [1993] presents a method for multiobjective stochastic evacuation problems. In this model, efficient routes are identified which are pareto optimal in terms of total distance travelled and total evacuation time. The goal of the process is to identify a set of evacuation routes for each origin. The algorithm proceeds by identifying *noninferior* (NI) paths, and then testing these paths using a queueing network simulator. This simulator measures the level of congestion resulting from a set of paths. After each iteration, new paths are generated and then reevaluated using the queueing network simulator.

Evacuation problems are important in traffic assignment problems for freight. Here, the problem is routing freight over a dynamic network of transportation services. In contrast with classical routing problems, we are interested primarily in how long it takes to move shipments over the network, as opposed to minimizing costs (which are determined by the flows of vehicles, rather than the flows of goods). Evacuation problems are characterized by an initial supply of flow into the network, whereas traffic assignment problems have a continuous flow of goods that must be moved to their destination.

4.4. Production and inventory planning

Production and inventory planning is a special type of routing problem, but the field is important since it is home to many of the basic concepts in dynamic models. Section 4.4.1 provides a brief summary of some of the literature in this area, with special emphasis on dynamic network models of production and inventory planning. Then, Section 4.4.2 presents a dynamic network model that illustrates a dynamic inventory network model, and summarizes some key developments in this context.

4.4.1. Survey and applications

Production and inventory planning is one of the most widely studied topics in the operations research literature. These problems have often been formulated using dynamic networks, and hence represents an important step in the historical development of dynamic models. We do not attempt to review this substantial literature, but instead refer the reader to reviews such as Zahorik, Thomas & Trigeiro [1984] and Bahl, Ritzman & Gupta [1987], or any of the popular textbooks in the field. Important applications of dynamic networks for production, inventory and transportation planning include Glover, Jones, Karney, Klingman & Mote [1979] and Klingman & Mote [1982]. These applications demonstrate the effectiveness of linear network models for aggregate production planning. Our motivation for reviewing this literature stems from the observation that the problem of planning the movements of large fleets of vehicles can be viewed in the context of multilocation inventory problems. While classical inventory and distribution problems focus on getting goods from plants to customers, dynamic fleet management can be viewed as moving empty vehicles and containers from surplus areas to deficit areas.

Dynamic production and inventory planning problems are often formulated using dynamic transportation problems. An early introduction to this problem is given in Szwarc [1971b], and an excellent review of the literature on dynamic transportation problems is given in Bookbinder & Sethi [1980]. Aronson [1989] reviews in depth algorithms for dynamic transportation problems motivated by production planning and scheduling, in particular, the forward network simplex algorithm introduced by Aronson & Chen [1986]. This work, which is reviewed further in Section 5, shows that solution algorithms can be accelerated by taking advantage of the dynamic properties of the network. Bowman [1956] showed that the linear production planning problem over time could be solved as a single transportation problem, assuming no constraints on inventories. Bellmore, Ecklof & Nemhauser [1969] reformulate this problem as a dynamic transshipment network, which is actually a dynamic transportation problem, and describe a modification of the Busacker & Gowen [1961] algorithm for dynamic networks (this algorithm solves a sequence of least-cost flow augmenting path problems). A number of authors have also considered a minimum time formulation of the problem [Szwarc, 1966; Hammer, 1969, 1971; Tapiero & Soliman, 1972; and Tapiero, 1975].

The research on stochastic multilocation inventory problems is much more limited. For the most part, the literature on inventory planning problems has focused on the analysis of policies with good performance under uncertainty with stationary demand patterns. Stochastic dynamic models quickly become computationally intractable. Shapiro [1993] reviews some of the literature on stochastic production and distribution models. An important contribution to this literature is a series of papers by Karmarkar who uses a stochastic programming framework to study distribution strategies as well as developing bounds and approximations for the objective function [see Karmarkar, 1979, 1981a, b, 1987]. Karmarkar [1981b] shows that the form of an optimal policy for a multiperiod, multilocation inventory problem is the same as for one-period, which is to say that a base-stock policy is optimal. In this policy, if the inventory falls below a base-stock y, then it is optimal to bring the inventory up to the level y. Karmarkar [1987] then develops bounds and approximations for the expected recourse function for this class of problems.

4.4.2. Models

In logistics applications, the determination of how much to ship, when and where, is typically determined by the solution of a production and inventory planning problem. The classical production and inventory planning problem is illustrated in Figure 8. Here, we have modeled demands as upper bounds on links from each time period into a supersink. Thus, it is possible not to satisfy all the demand, presumably at a penalty.

A more general form of this basic problem is the multiregion, dynamic production and inventory planning problem, which can be stated mathematically as:

Decision variables

w_{ijt} = flow of goods from city i to city j, in period t.

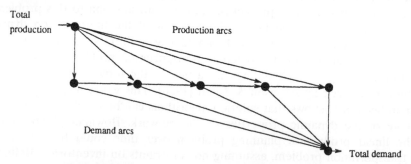

Fig. 8. for production and inventory planning at a single location.

$w_t \quad = \{\ldots, w_{ijt}, \ldots\}.$

$W_{it} \quad$ = net surplus or deficit of goods at city i in time t, where W_{it} is positive if i is a surplus point, and negative if a deficit.

$B_{t_1,t_2} \quad$ = node-arc incidence matrix where the rows correspond to cities $i \in C$ at time $t = t_2$, and columns correspond to flows between cities (i, t_1) and (j, t_2). An element equals minus one if flow departing from i at time t_1 arrives in j at time t_2, and zero otherwise, for $t_1 \leq t_2$.

$A_t \quad$ = node-arc incidence matrix where the rows correspond to cities i at time t, and columns correspond to flows between cities i at time t and cities j at time $t' \geq t$. An element is minus one if the arc is entering node (i, t) and minus one if it is emanating from the node.

$$\min_{w_t} \sum_{t=0}^{P} \alpha^t c^T w_t \tag{68}$$

subject to, for $t = 0, \ldots, P$

$$
\begin{aligned}
A_0 w_0 & & = W_0 \\
B_{0,1} w_0 + A_1 w_1 & & = W_1 \\
B_{0,2} w_0 + B_{1,2} w_1 + A_2 w_2 & & = W_2 \\
B_{0,3} w_0 + B_{1,3} w_1 + B_{2,3} w_2 + A_3 w_3 & & = W_3 \\
\vdots & \ddots & \vdots \\
B_{0,t} w_0 + B_{1,t} w_1 + B_{2,t} w_2 + & \cdots + A_t w_t & = W_t
\end{aligned}
\tag{69}
$$

Equations (68)–(69) describe a general, dynamic production planning problem. If a time period is relatively long, such as a month, then we generally have that $B_{t_1,t_2} = 0$ for $i \neq j$. In this case, we have an instance of a dynamic inventory network as shown in Figure 9, with the specific form of a dynamic transportation problem. A good example of a practical application of this network is given in Glover, Jones, Karney, Klingman & Mote [1979] and Klingman & Mote [1982].

The solution of dynamic inventory networks of this type has been investigated in depth by Aronson & Chen [1986]. This work applies the principles of forward linear programming, described in Aronson & Thompson [1984] and Aronson [1980], and in particular the review article Aronson [1989]. These methods are based on the principles of decision horizons which tend to arise in practice in this class of models (see Aronson & Chen [1985, 1989].

4.5. The dynamic vehicle allocation problem

The dynamic vehicle allocation problem arises in a variety of settings which involve the management of fleets of vehicles over time. Below review applications in trucking, rail, container shipping, and air. Related applications can be found in the management of taxi cabs and rental fleets (especially the so-called 'one-way' rentals of trucks). Following this, we present some simple models for the dynamic vehicle allocation problem.

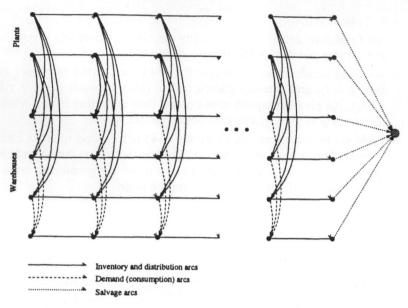

Fig. 9. Dynamic network for production and inventory planning.

4.5.1. Applications

Trucking. For truckload trucking, the primary problem addressed in the research literature has been the dynamic vehicle allocation problem for managing large fleets of trucks. Powell, Sheffi & Thiriez [1984] present a nonlinear dynamic network model for accounting for uncertainties in forecasted demands. Powell [1986] refines this model to allow for stochastic vehicle inventories. This model extends [Powell, Sheffi & Thiriez, 1984] to allow for the possibility that trucks that are not needed will remain in inventory, and extends the model in Jordan & Turnquist [1983] by tracking both loaded and empty movements. Powell [1987] further extends this model by providing the most realistic model of future vehicle trajectories. Whereas prior research makes very restrictive assumptions about how a truck may be used under uncertainty, Powell [1987] provides for a very general model of truck dispatching under uncertainty. A more formal mathematical model is provided in Powell [1988], where several different models are formulated for the same problem. This paper provides the first formal model of the dynamic vehicle allocation problem as a dynamic network with random arc capacities. Frantzeskakis & Powell [1990] introduce a new heuristic for solving multistage dynamic networks with random arc capacities, motivated by truckload motor carriers. These results have been further extended in Powell & Cheung [1992]. A real application of these results is given in Powell, Sheffi, Nickerson, Butterbaugh & Atherton [1988].

Rail. As one of the oldest modes of transportation, railroads have been a popular subject for operations research. The main challenge has been managing large fleets of rail cars. Feeney [1957], Leddon & Wrathall [1967], Gorenstein, Poley & White [1971] and Herren [1973, 1977] represent a number of early examples of efforts to optimize fleets of rail cars. Misra [1972] formulates the problem as a linear program, while White [1972] presents a dynamic transshipment network over a finite planning horizon. White & Bomberault [1969] use the dynamic structure of the network to develop a specialized algorithm, one of the earliest efforts to specialize an algorithm for a dynamic network. Mendiratta [1981] and Mendiratta & Turnquist [1982] present inventory models for managing empty cars, taking into account the decentralized nature of the decision-making process. Jordan & Turnquist [1983] present the first stochastic model of the empty car management problem. Ratcliffe, Vinod & Sparrow [1984] use a simulation model of empty freight cars. Glickman & Sherali [1985] address the problem of pooled fleets of empty cars, recognizing that railroads share fleets of cars. Shan [1985] uses a dynamic, multicommodity network flow model to handle multiple car types, using resource directive decomposition to solve the resulting network. Chih [1986] extends this model to handle multiple railroads. Despite the increasing sophistication of these models, there is little real evidence that they are being adopted and used by the railroads. By contrast, much simpler myopic models, such as the transportation formulations given in Turnquist [1986] and Turnquist & Markowicz [1989], are seeing wider acceptance.

Other researchers are going beyond the problem of managing empty cars. Haghani [1989] presents a combined model for train makeup and empty car repositioning, representing one of the earliest efforts to address both the flows of loaded and empty cars. Chih, Hornung, Rothenberg & Kornhauser [1990] consider the problem of managing locomotives. Smith & Sheffi [1990] present a locomotive distribution model that accounts for uncertainty in the need for locomotives, using a simple recourse strategy to handle the effects of uncertainty. Kraay, Harker & Chen [1989] address the problem of dynamically managing the movement of trains over a rail line, which requires optimizing the use of sidings to allow for train passings.

Container shipping. Ocean applications of dynamic models arise when planning the movement of ocean vessels, and the optimization of fleets of containers over a global logistics network. Dantzig & Fulkerson [1954] provide one of the earliest applications of optimization over a dynamic network to minimize the number of tankers required to meet a given schedule. Other efforts to optimize the movement of vessels include Brown, Graves & Ronen [1987], Psaraftis, Orlin, Bienstock & Thompson [1985], Fisher & Rosenwein [1989], and Perakis & Papadakis [1989]. Ermoliev, Krivets & Petukhov [1976] and Florez [1986] consider the optimization of containers. Crainic, Gendreau & Dejax [1992] looks at the dynamic management of containers over land within a region near a port.

Air. Dynamic problems in air transportation include the assignment of aircraft to routes, crew scheduling, pricing and booking problems, and the dynamic management of aircraft between airports (the air traffic control problems). Dantzig & Ferguson [1956] use the fleet assignment problem as an early example of linear programming under uncertainty. Magnanti & Simpson [1978] describe a series of dynamic network models with side constraints to handle fleet assignment. The crew scheduling problem for airlines has become a popular area of research. Surveys of this area are given by Arabeyre, Feranley, Steiger & Teather [1969], Marsten & Shepardson [1981], and more recently by Crainic & Rousseau [1987]. The common approach used in this area is based on set partitioning problems to choose from among the best set of possible crew schedules. Methods contrast based on whether a generator is used to generate all possible 'reasonable' schedules, or whether column generation techniques are used. A different approach is suggested by Ball & Roberts [1985] which uses a matching algorithm to sequentially generate possible crew schedules.

The second problem is the dynamic management of aircraft moving between airports, sometimes referred to as the flow management problem in air traffic control. An important reference in this is Andreatta & Romanin-Jacur [1987]. An excellent discussion of models and issues arising in the flow management problem is given in Odoni [1986]. Mulvey & Zenios [1987] give a nonlinear, dynamic network model for routing aircraft. Bielli, Calicchio, Micoletti & Ricciardelli [1982] also formulate the flow control problem as a dynamic network.

4.5.2. Models for the dynamic vehicle allocation problem

An important class of problems that can be solved as pure networks are dynamic fleet management problems involving a single equipment type. Dynamic fleet management refers broadly to the problem of managing a fleet of vehicles (containers) over time to maximize a set of objectives. A component of this problem is the *dynamic vehicle allocation problem* which specifically addresses decisions regarding the repositioning of empty vehicles, as well as load acceptance and rejection. A presentation of alternative models and formulations of the dynamic vehicle allocation problem is given in Powell [1988].

The basic objective of dynamic vehicle allocation is to generate revenues by carrying a set of *loads*, each of which fills an entire vehicle (there is no consolidation). Carrying a load implies a vehicle must move from one city to another, at which point the vehicle becomes empty and must be assigned to a new load. In some instances, more loads will terminate in a particular area than originate, thus requiring excess vehicles to be repositioned empty out of one area and into another in anticipation of future loads. Thus, there are three types of activities for vehicles: moving loaded, moving empty, and holding in inventory (doing nothing). Empty moves may also be divided between moving empty to satisfy an actual demand, and moving empty to satisfy a forecasted demand.

Dynamic fleet management encompasses the dynamic vehicle allocation problem, as well as load solicitation, load evaluation, spot pricing, tactical sales planning, and service planning. Solutions to this broader set of questions depends, however, on the solution of the basic dynamic vehicle allocation problem. We first present the single vehicle version, and then provide a multifleet model.

Single vehicle type. Our basic formulation of the DVA can be expressed easily as a dynamic network. First define:

Decision variables

x_{ijt} = number of vehicles moving loaded from city i to city j, originating in period t.

y_{ijt} = number of vehicles moving empty from city i to city j, originating in period t.

Physical parameters

C = set of cities (points) where loads originate and terminate.

τ_{ij} = travel time (in nonnegative integer periods) from city i to city j (we use the same time for all loaded and empty movements).

r_{ij} = net contribution (revenue minus direct operating costs) generated from moving loaded from i to j (we assume, for notational simplicity, that contribution is not a function of time).

c_{ij} = cost of moving empty from i to j (c_{ii} is the marginal cost of holding vehicles in inventory). As with revenue, we assume that costs are not a function of time; this assumption is easily relaxed.

Activity variables

R_{it} = number of vehicles entering the system for the first time in period t.

D_t^{rs} = market demand, giving the number of loads *available* to be moved from origin r to destination s, originating in period t.

Model parameters

P = length of the planning horizon.

α = discount factor per period.

In our notation, we use a superscript (r, s) to denote transportation *markets* which express the originating and terminating point of a load or shipment. We use a subscripted (i, j) to express transportation movements between cities. In our basic model of the dynamic vehicle allocation problem, which is most directly applicable to truckload trucking, vehicles (trucks) are assumed to move directly from origin to destination. Later, we consider instances of problems where the transportation of a shipment from origin to destination must pass over several transportation links.

The basic dynamic vehicle allocation problem seeks to maximize total (discounted) profits over the planning horizon P, as follows:

$$\max_{x_t, y_t} \sum_{t=0}^{P} (r^T x_t - c^T y_t) \alpha^t \qquad (70)$$

subject to, for $t = 0, \ldots, P$

$$\sum_{j \in C}(x_{ijt} + y_{ijt}) - \sum_{k \in C}\left[x_{ki,(t-\tau_{ki})} + y_{ki,(t-\tau_{ki})}\right] = R_{it} \quad i \in C \quad\quad (71a)$$

$$x_{ijt} \leq D_t^{ij} \quad\quad\quad\quad\quad\quad\quad\quad\quad\quad\quad\quad\quad\quad i, j \in C$$

$$x_{ijt}, y_{ijt} \geq 0 \quad\quad\quad\quad\quad\quad\quad\quad\quad\quad\quad\quad\quad i, j \in C \quad\quad (71b)$$

$$x_t, y_t = 0 \quad\quad\quad\quad\quad\quad\quad\quad\quad\quad\quad\quad\quad\quad t < 0$$

The discount factor α is sometimes included in dynamic models as a heuristic mechanism for accounting for forecasting uncertainties. The choice of planning horizon and the choice of α is discussed in Section 3.4.3.

Equation (71a) enforces flow conservation at the beginning of each time period, where we use the convention of flow out minus flow in. Thus $R_{it} > 0$ when vehicles enter the network for the first time. Equation (71b) is the *demand constraint* where market demands limit the number of loaded movements. Note that there is a single transportation leg (i, j) associated with each market (r, s).

The object of this basic formulation of the DVA is to determine loaded movements x_{ijt}, empty movements, y_{ijt}, and inventory movements, y_{iit}, to maximize total profits over a fixed horizon. This problem also determines which loads should be rejected, given by $D_t^{ij} - x_{ijt}$ (it is common to assess a penalty r_{ij}^p for refusing loads; this results in a modified revenue $\hat{r}_{ij} = r_{ij} + r_{ij}^p$). With a little creativity, this model can be used to handle some of the other dimensions of dynamic fleet management.

There are several important variations of the basic DVA. First, a static form of the problem considers the myopic problem of repositioning vehicles empty now to meet known and forecasted demands within a specific time period (perhaps a week). This model is often used by railroads and container companies, and does not explicitly model the loaded movement. A more general form of this model also considers only the repositioning of empty vehicles, but actually models the holding of vehicles in inventory over time. If empty movements are allowed in future time periods, then this model is specifying not only where to move empty, but when. Another example is the ground holding problem for air traffic control, which consists of determining when to allow an aircraft to depart from an airport, anticipating future demands and landing capacities of other airports. The ground holding problem is an instance of a problem in which we already know *where* the vehicles are going but need to know *when* they should depart.

Multiple vehicle types. The single commodity formulation of the DVA is a powerful framework with practical applications in industry, particular in truckload trucking. At the same time, most applications require consideration of multiple equipment types, where it is possible to substitute some equipment types to handle a given demand. Of course, if there are no substitutions, then the problem can be solved as a sequence of single commodity DVA's. More commonly, some level of substitution is allowed. To formulate this problem, we start with (70)–(71) and introduce the index ℓ to represent flows of equipment type ℓ. So $x_{ijt}(\ell)$ and $y_{ijt}(\ell)$

represent loaded and empty flows of equipment type ℓ, and $D_{ijt}(\ell)$ is the market demand for equipment type ℓ. In addition, define:

$\mathbf{E}$ = set of equipment types.

$D_t^{rs}(\ell)$ = market demand for equipment type ℓ.

$x_{ijt}(\ell, m)$ = flow of equipment type ℓ used to satisfy demand $D_t^{ij}(m)$, $\ell, m \in \mathbf{E}$, $i, j \in C$.

$x_{ijt}(\ell)$ = $\displaystyle\sum_{m \in \mathbf{E}} x_{ijt}(\ell, m)$.

$\sigma_{ijt}(\ell, m)$ = $\begin{cases} 1 & \text{if equipment type } \ell \text{ can be used to service the demand } D_{ijt}(m), \\ 0 & \text{otherwise.} \end{cases}$

$\mathbf{E}^m$ = $\{\ell \mid \sigma^{\ell, m} > 0\}$.

We use $\sigma(\ell, m)$ as an indicator variable. However, we could allow $0 \le \sigma(\ell, m) \le 1$, (where $\sigma(\ell, \ell) = 1$), and give it the interpretation as the fraction of $D_t(m)$ that is realized if it is serviced using equipment type ℓ (allowing lost demand when a less appropriate equipment type is used). The multiple equipment DVA can now be stated as follows:

$$\max_{x_t(\ell), y_t(\ell)} \sum_{\ell \in \mathbf{E}} \sum_{t=0}^{P} \left[r(\ell)^T x_t(\ell) - c(\ell)^T y_t(\ell) \right] \alpha^t \tag{72}$$

subject to, for $t = 0, \ldots, P$

$$\sum_{j \in C} \left(x_{ijt}(\ell) + y_{ijt}(\ell) \right) - \sum_{k \in C} \left(x_{ki,(t-\tau_{ki})}(\ell) + y_{ki,(t-t_{ki})}(\ell) \right) = R_{it}(\ell)$$
$$i \in C, \ \ell \in \mathbf{E} \tag{73}$$

$$\sum_{\ell \in \mathbf{E}^m} x_{ijt}(\ell, m) / \sigma(\ell, m) \le D_{ijt}(m) \qquad i, j \in C, \ m \in \mathbf{E}, \tag{74}$$

$$x_{ijt}(\ell), \ y_{ijt}(\ell) \ge 0 \qquad i, j \in C, \ \ell \in \mathbf{E} \tag{75}$$

Problem (72)–(75) is similar to (70)–(71) with the obvious extensions to account for multiple equipment types. The important difference is that the presence of equipment substitutions transforms the demand constraint (71b) into the bundle constraint (74). This problem is similar in style to the aircraft fleet assignment problem formulated in Simpson & Magnanti [1978]. In this model, *demands* represent scheduled flights, where a particular flight may be satisfied by more than one type of aircraft (if too small an aircraft is used, the conversion factors $\sigma^{\ell, m}$ can be used to model lost passengers). If the problem is not too large (the definition of which seems to change annually) the problem can be solved as a linear program (possibly subject to integrality constraints). Alternatively, decomposition methods can be used to exploit the underlying network structure (as described in depth in the Simpson and Magnanti report). However, if the upper bounds are equal to one, as would occur in airline fleet assignment models, then integrality becomes more important, and different solution approaches are warranted. Jones, Lustig, Farvolden & Powell [1993] investigate the impact of model formulation on the performance of Dantzig-Wolfe decomposition in the context of multifleet assignment problems. This research demonstrated that significant improvements can be realized by using formulations that produce sparse columns in the master problem. Thus, generating a single path in a subproblem is better than generating an entire tree. This effect is particularly pronounced in the context of dynamic networks.

4.6. The dynamic assignment problem

The dynamic assignment problem arises when a set of workers (drivers) must be assigned to a set of tasks over time, responding to new demands as they are called in. The distinguishing characteristic of the dynamic assignment problem over other routing problems is that a worker is never handling more than one task at a time. Different variations of the problem arise in part based on the characteristics of the demands being served. First, tasks may remain at one location, or move from an origin to a destination. Second, we may wish to assign workers to one task at a time, or we may wish to build tours through a series of tasks, typically lasting no longer than a work shift. The tour building requirement tends to arise when tasks are of short duration.

These variations produce the following types of dynamic assignment problems:

1. *The dynamic traveling salesman problem.* A driver must visit a series of points, where the task at each point is relatively short and constant. The problem is to route a driver through a series of points, where demands for a visit are arising dynamically over time.

2. *The dynamic traveling repairman problem.* Tasks occur at one location and are of random length. While it may be possible to develop tours through multiple tasks, often it is necessary to assign workers to one task at a time.

3. *The dynamic driver assignment problem.* This problem arises in truckload trucking, where a driver must be assigned to a load, which has both an origin and a destination. Also, these loads are generally of fairly long duration (one to four days), making it impractical to assign drivers to tours which cover several loads.

4. *The (dynamic) full truckload routing and scheduling problem.* This problem has received very little attention in the literature, but arises when a driver (crew) must be routed over a tour consisting of multiple tasks, and where each task exhibits an origin, destination and service window.

4.6.1. The dynamic traveling salesman problem (DTSP)

Psaraftis [1988] introduces a version of the dynamic travelling salesman problem. Let G be a complete graph of n nodes. Demands for service are independently generated at each node of G according to a Poisson process of parameter λ. These demands are to be serviced by a salesman who takes a (known) time of t_{ij} to travel from node i to node j of G, and spends a (known) time t_0 servicing each demand (on location). If at time $t = 0$ the salesman is at node 1, what should his 'optimal' routing policy be?

'Optimal' here may be defined with respect to a number of different objectives. As in dynamic routing in a communications network [see for example Bertsekas

& Gallager, 1987], there are two main classes of performance measures that are affected by routing decisions: 1) throughput measures (e.g., maximize the average expected number of demands serviced per unit time), and 2) delay measures (e.g., minimize the expected time from the appearance of a demand until its service is completed).

At this point, we are unaware of models for formulating and solving the DTSP explicitly as a dynamic model. Of course, one solution is to find an optimal a priori tour and apply this solution in a dynamic setting. Several other variants of the DTSP can be considered. The graph can be incomplete, symmetric, or Euclidean. The demand generation process can be node specific and may not be Poisson; the service time, t_0, can be a random variable, etc.

A different formulation of the DTSP is the *time-dependent traveling salesman problem*. In this instance, the data is static, but travel times vary as a function of time. Picard & Queyranne [1978] first consider a version of the time-dependent TSP motivated by a machine scheduling problem. Fox, Gavis & Graves [1980] introduce a formulation where travel times between cities are all one period, but where the cost of travel is a function of time. More recently, Malandraki [1989] and Malandraki & Dial [1992b] introduces and develops heuristic solution techniques for the TSP where travel times between cities are a function of time.

4.6.2. The dynamic repairman problem — the uncapacitated case

As a canonical example of a logistics application of the dynamic stochastic vehicle routing problem (DSVRP), consider the following utility repair problem: A utility company (electric power, gas, water, etc) maintains a large, geographically-dispersed network of facilities. The network is subject to failures which occur randomly with respect to both time and location. The company operates a fleet of repair vehicles which are dispatched from a depot in response to reports of local failures. Routing decisions are made on the basis of a real-time log of currently pending failures and possibly some characterization of the process of future failures. Repair vehicles and crews spend a random amount of time servicing each failure before they are free to move on to the next 'call' to be serviced. The utility company wishes to operate its fleet of repair vehicles in a way that minimizes the average downtime due to failures.

The one DSRP that has been investigated in depth to date is the Dynamic Traveling Repairman Problem (DTRP), a Euclidean model of a dynamic VRP. Demands for service arrive according to a renewal process with intensity λ to a connected, bounded Euclidean service region $\mathcal{A}$ of area A. Upon arrival, demands assume an independent and identically distributed (i.i.d.) location in $\mathcal{A}$ according to a continuous probability density function $f(x)$ defined over $\mathcal{A}$. Demands are serviced by m identical vehicles that travel at constant speed v. At each location, vehicles spend some time s in on-site service that is i.i.d. with finite first and second moments denoted by $\bar{s}$ and $\bar{s}^2$ respectively.

A policy for routing the vehicles is called stable if the number of unserved demands in the system is bounded almost surely for all times t. Let M denote the set of stable policies. If a policy is stable, $\rho \equiv (\lambda \bar{s})/m$ is the fraction of time vehicles spend in on-site service. We write T_μ to indicate the system time of a particular policy $\mu \in M$. The DTRP is then defined as the problem of minimizing T_μ. We let T^* denote the optimal value of system time.

Bertsimas and van Ryzin [1991] have analyzed the DTRP for the case in which arrivals are Poisson, demands are uniformly distributed and the entire region is served by a single uncapacitated vehicle.

In the case of light traffic it can be shown that a policy based on locating the server at the median, $x*$, of $\mathcal{A}$ and serving demands in a first-come, first-served (FCFS) order, returning to the median after each service is optimal. This result provides an elegant connection between a dynamic stochastic routing problem (this version of the DTRP) and a static counterpart, the PTSLP (Section 2.4): the optimal location, $x*$, for the static problem is also the location where the server should be located in the dynamic problem and the server-operating policy under the PTSLP is also the optimal operating policy for the DTRP when the traffic is light! The optimal expected system time, $T*$, in this case satisfies

$$T^* \to \frac{E\big[\|X - x^*\|\big]}{v} + \bar{s} \quad \text{as } \lambda \to 0. \tag{76}$$

In the case of heavy traffic, for this single, uncapacitated vehicle problem, policies exist that have finite system times T_μ for all $\rho < 1$. (Recall that ρ is the fraction of time the vehicle spends in on-site service.) This is surprising in that the condition is independent of the service region size and shape; it is also the mildest stability restriction one could hope for. It can be shown that there exists a constant γ such that

$$T^* \geq \gamma^2 \frac{\lambda A}{v^2(1-\rho)^2} - \frac{\bar{s}(1-2\rho)}{2\rho}. \tag{77}$$

Note that this bound grows like $(1 - \rho)^{-2}$ as $\rho \to 1$. Thus, though the stability condition is similar to that of a traditional queue, the system time increases much more rapidly as congestion increases. [In Bertsimas & van Ryzin [1993b] the value $\gamma = 2/(3\sqrt{\pi}) \approx 0.376$ is derived.]

Several interesting operating policies μ that have finite system times, T_μ, for all $\rho < 1$ are investigated in Bertsimas and van Ryzin [1991].

(i) *The partition policy.* This is a policy that applies to the case where $\mathcal{A}$ is a square. It consists of dividing the region into n equal subregions which are served sequentially such that each subregion is adjacent to the next subregion in the sequence except, possibly, for the last one. Within each subregion, demands are serviced in FCFS order until no more demands are left. Then the vehicle moves on to the next subregion in the sequence. The number of subregions n must be optimized in each problem instance.

(ii) *The traveling salesman policy.* As demands arrive arrange them into sets of size n. When all n demands have arrived, consider this the 'arrival of a set'. Service sets in FCFS order by forming a TSP tour on the set of demands. The size of the TSP sets n must be optimized in each problem instance.

(iii) *The space filling curve policy.* Visit demands in the order in which they are encountered in continuous sweeps of the space-filling curve (applies to the case where $\mathcal{A}$ is a square).

(iv) *The nearest neighbor policy.* After each service completion, serve next the demand that is closest to the current location of the vehicle.

It is remarkable that all these diverse operating policies have the same type of asymptotic behavior as the lower bound noted in (77) above, namely, as , $\rho \to 1$,

$$T^* \sim \gamma_\mu^2 \frac{\lambda A}{v^2(1-\rho)^2}, \tag{78}$$

where the constant γ_μ depends on the policy μ. Hence, by comparing this bound to the lower bound in (77), it can be seen that the ratio $\frac{T_\mu}{T^*}$ is bounded by a constant as $\rho \to 1$. Thus all the policies provide a constant factor guarantee in heavy traffic.

The provably best policy of the four listed above is the Traveling Salesman (TS) policy. For a version of this policy, it can be shown that $\gamma_\mu = \beta_{\text{TSP}}/\sqrt{2} \approx 0.51$, where β_{TSP} is the Euclidean TSP constant. Relative to the value noted earlier, this gives a best provable guarantee of

$$\lim_{\rho \to 1} \frac{T_\mu}{T^*} \le \frac{\beta_{\text{TSP}}^2}{2\gamma^2} \approx 1.8. \tag{79}$$

In fact, Bertsimas and van Ryzin [1991] conjecture that the TS policy is optimal in heavy traffic and that the factor of 1.8 is due to slack in the lower bound (77). If true , this conjecture implies, once again a strong connection between a DSRP and a static counterpart, in this case the PTSP on the Euclidean plane.

The TS policy for the DTRP can be extended to a mixed objective involving both waiting time and travel cost. By increasing the size n of the sets that are formed, travel distance per demand can be reduced at the expense of increasing the mean system time. Indeed, one can show that to minimize system time, one should essentially maximize the amount of travel per demand served. Thus, travel cost and system time are conflicting objectives that can be balanced by sizing routes in an appropriate way.

4.6.3. The dynamic driver assignment problem

The driver assignment problem arises in applications where at any point in time, the problem is to assign a driver to a single task. This problem arises in truckload trucking, where a load might take one or more days to complete. As a result, dispatchers assign a driver to a load, and wait for the load to be delivered before planning the next assignment.

In its simplest terms, the driver assignment problem can be formulated as a simple network assignment problem, with arcs connecting driver nodes to load nodes, as illustrated in Figure 10. The cost of the arc from a driver node to the load node would represent the cost of moving empty from the driver's current location to the location of the load. In commercial applications, models such

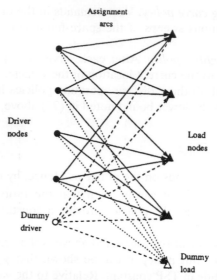

Fig. 10. Static driver assignment network.

as these are run in real-time, responding to changes about the set of drivers and loads. This is especially valuable for carriers which use satellite tracking and mobile communication units to obtain real-time information about the status of each driver.

The model has two attractive features: First, it is extremely simple and can be easily optimized using well-established network optimization codes. Given an optimal solution, the network can be reoptimized following a change in the data in a fraction of the data. Second, the model is extremely flexible, since other, nonquantifiable issues, can be handled by adding various bonuses and penalties to the assignment arcs. For example, one issue that arises with some frequency in the truckload industry is the problem of getting drivers home. Loads go from one city to the next, and there is no guarantee that a driver sent, say, from his home in Atlanta to Chicago will then be sent back home to Atlanta. His next load might be to Dallas. After about two weeks, the carrier might want to try to get the driver home. This can be accomplished in a heuristic fashion by adding a bonus on arcs out of a particular driver to loads which return to points near his home. The size of the bonus can be chosen by asking management how many additional miles they are willing to run a driver to get him home. Needless to say, it is very easy to design a rule-based system for generating these bonuses, taking into account a variety of issues.

One limitation of the basic assignment model is that it is unable to account for the future impacts of decisions made now. For example, if there are too many drivers in one part of the country, it is unable to recommend that drivers be repositioned empty to another region. Also, if too many loads have been booked, it cannot determine the most profitable loads to accept, and which should be 'rejected.' Powell [1991] presents such a model, that combines the basic assignment model with a stochastic, dynamic network model that incorporates forecasted loads. The resulting model is a pure network that can be solved in real-time.

4.7. Dynamic vehicle routing problems

Dynamic vehicle routing arises when we must route vehicles to serve a series of customers with data that is changing over time. The problem differs from the dynamic assignment problem in that multiple customer orders can be consolidated onto the same vehicle, subject to vehicle capacity constraints. Thus, we add the element of clustering orders into tours which satisfy vehicle capacity constraints.

Our discussion of dynamic vehicle routing considers three classes of applications.

The first, introduced in Section 4.7.1 is the capacitated version of the dynamic traveling repairman problem, considered earlier. The second, described in Section 4.7.2, comprises deterministic dynamic vehicle routing problems which exhibit *time-dependent* data, which is all known in advance. Thus we may know a week's worth of demands, and we need to develop a route and schedule for each vehicle that serves these demands over the course of the week. Third, Section 4.7.3 summarizes the literature on stochastic vehicle routing. This literature focuses primarily on a priori (or two-stage) models, but these can form the basis for a stochastic, dynamic *model* that can be used to solve dynamic *problems*. Finally, Section 4.7.4 describes specific applications of vehicle routing models in a dynamic setting, where models were solved in real time. The section on stochastic vehicle routing only discusses the literature. Later, Section 8.1, presents specific models in greater detail. The reason for this delay is that we want to present some of the key concepts in stochastic programming in Section 7.

4.7.1. The dynamic traveling repairman problem - capacitated case

We first visited the uncapacitated dynamic traveling repairman problem as a type of dynamic assignment problem. The capacitated version of this problem is, in effect, the dynamic vehicle routing problem, where customers must be clustered into tours which satisfy vehicle capacity constraints.

Bertsimas and van Ryzin [1993a] examine the case where the region $\mathcal{A}$ is now served by a homogeneous fleet of m vehicles operating out of m depots (whose locations need not be distinct, and where each vehicle is restricted to visiting at most q customers before returning to the depot. A lower bound for the minimum expected system time analogous to that in (77) is obtained. It is also shown that when the vehicle capacity q is very large, policies with the same constant factor performance guarantees as in the single-server case can be constructed by simply partitioning $\mathcal{A}$ into m equal subregions and serving each one independently using a single-server policy. For q finite, constant factor guarantee policies can still be constructed. The best of these policies is based on modifying the Traveling Salesman policy (cf. above) using the tour partitioning heuristic of Haimovich,

Rinnooy Kan & Stougie [1988]. Analytical expressions are given for the constant factor performance guarantees and for stability conditions under this policy. As might be expected, when the vehicle capacity q is finite, stability is no longer independent of the geometry of the service region, unlike the uncapacitated vehicle case.

All the results reviewed so far apply to the case of uniformly distributed demand and Poisson arrivals. In Bertsimas and van Ryzin [1993b] most of these results are extended to the case where demands are distributed in $\mathcal{A}$ according to a general continuous density $f(x)$ and arrivals occur according to a general renewal process. These generalizations require quite different analytical approaches and proof techniques. An additional interesting extension is that results are obtained for operating policies which are 'spatially fair' (i.e., provide the same mean system time for all locations within $\mathcal{A}$) as well as for policies which are 'spatially discriminatory' (i.e., provide different mean system time s to different regions of $\mathcal{A}$).

The DTRP and its variations is clearly a rich problem which, in a dynamic context can play a role analogous to that of the TSP — for deterministic routing problems — or of the PTSP — for stochastic but static problems. It is interesting that the analysis of the DTRP yields simple expressions for the system time that provide insights into the effects of: traffic intensity; on-site service characteristics; number, speed and capacity of vehicles employed; service region size; the distribution of customer locations; and fairness of service constraints.

A recurring observation in the work of Bertsimas and van Ryzin is that static vehicle routing methods, when properly adapted, can yield optimal or near-optimal policies for DSRPs. This is an encouraging result on two levels. On a theoretical level, it suggests that there is indeed a connection between the properties of the static and the dynamic problems. On a practical level, the implication is that exact algorithms, heuristics and insights that have been developed over the years regarding static VRPs and SRPs are relevant to DSRPs and, in fact, form the basis for effective analysis and implementation of DSRP systems.

4.7.2. Deterministic dynamic vehicle routing models

In this section, we consider dynamic *models* of vehicle routing problems, focusing on a single snap-shot of data. These models can be used in real-time to solve dynamic problems, or to develop vehicle schedules over the course of the day using data that is (assumed) known in advance.

Deterministic models of dynamic vehicle routing problems arise in two settings: time-dependent demands, and time-dependent travel times. Time-dependent demands arise generally when demands must be satisfied subject to specific time constraints. Time-dependent travel times usually arise in models which are trying to capture time-of-day congestion effects.

Time-dependent demands. One case of time-dependent demands arises when demands must be satisfied within a specified time window. However, a richer version of this problem arises in the context of the *inventory routing* problem, where the determination of customer demands is based on the need to maintain customer inventories. While a customer might require a delivery no later than day t, a carrier (for example, a company providing fuel oil) might decide to deliver on day $t' < t$ if a truck will be in the area. The basic tradeoff is between routing and inventory holding costs.

Hausman & Gilmour [1967] provide what is likely the first explicit model of vehicle routing which considers the assignment of demands both to days of the week as well as to vehicle tours. The problem is solved using a local search heuristic. More recently, Russell and Gribbin [1991] describe a multiphase approach which uses a sequence of optimization models to first assign customer demands to days of the week, and then build vehicle tours. Their model assumes that each customer must be served with a specific frequency (perhaps twice a week) and the problem is to determine the days of the week to serve the customer which satisfies this frequency. Local improvement heuristics are also used to improve tours. Dror, Ball & Golden [1985] provide a more comprehensive computational comparison of algorithms for inventory routing problems.

Time-dependent travel times. The second class of problems includes time-dependent travel times. We have already reviewed the work on dynamic traveling salesman problems. The only work that we are aware of that deals explicitly with (capacitated) vehicle routing and time-dependent travel times are Malandraki [1989] and Malandraki & Daskin [1992a]. This formulation assumes that each link has a fixed travel time that is a function of the time a vehicle arrives to the link. This creates discontinuities in the travel times from one time period to the next, but vehicles are allowed to wait at a node, which is shown to have the effect of smoothing the travel time function. The objective function is to minimize the total time required to complete all tasks. The decision variable is defined as:

$$x_{ijm} = \begin{cases} 1 & \text{if any vehicle travels directly from node } i \text{ to node } j \\ & \text{starting from } i \text{ during time interval } m \\ 0 & \text{otherwise} \end{cases} \tag{80}$$

In contrast with the classical formulation of the vehicle routing problem, their formulation includes an index that identifies the time interval the vehicle enters the link. Within this time period, the travel time is assumed constant. The intent is to capture major changes in travel times between, for example, peak and off-peak travel. A nearest-neighbor tour construction heuristic is proposed, as is a linear programming-based cutting plane algorithm. Testing is performed on problems with 10 to 25 nodes.

4.7.3. Stochastic vehicle routing models

Stochastic vehicle routing models generally consider the problem of designing vehicle routes with random (forecasted) demands, although there is limited research on the case of stochastic travel times [see Laporte, Louveau & Mercure, 1992]. Routing subject to stochastic demands can be divided into three cases: 1) finite horizon (e.g. routing over a single work shift), 2) infinite/rolling horizon, and 3) periodic. Most vehicle routing problems have a well-defined beginning and end, covering one day's work, where all drivers are expected to return to the home depot at the end of the shift. Rolling horizon problems arise in some situations where drivers are not expected to return home each day (in some cases, drivers may not return home for many days), giving the problem an infinite horizon flavor [see Bell, Dalberto, Fisher, Greenfield, Jaikumar, Kedia, Mack & Prutzman, 1983]. Finally, periodic models typically involve the problem of developing schedules over the course of the week. This includes inventory routing problems, described earlier (see Trudeau & Dror [1992] for a discussion of stochastic inventory routing problems).

The problem of routing vehicles in the face of uncertain demands is an old problem, and was probably one of the motivating applications behind the field of stochastic programming. However, the earliest references of routing vehicles under stochastic demands fall in our presentation under the category of dynamic vehicle allocation problems. The early literature on vehicle routing with stochastic demands considers two-stage problems, where a first-stage decision, such as setting a fleet size or locating a warehouse, is followed in the second stage by the actual routing of the vehicle. In this context, it is necessary to obtain expected routing costs, in a probabilistic sense, given the first stage decision. Some of the first papers which touched on this include Golden & Yee [1979] and Dempster, Fisher, Jansen, Lageweg, Lenstra & Rinnooy Kan [1981], which suggested that these problems could be formulated as multistage, stochastic integer programs. Following on this idea, Spaccamela, Rinnooy Kan & Stougie [1984] show how results from probabilistic analysis of traveling salesman problems [Beardwood, Halton & Hammersley, 1959] can be used to approximate the second stage problem for vehicle routing. The goal of this research, however, is not so much to design the vehicle tour but rather to improve decisions in other areas which depend on routing costs.

The earliest explicit treatment of stochastic demands in the design of vehicle routes appears to be Tillman [1969] who presented a modification of the Clarke-Wright savings algorithm for terminals with Poisson-distributed demands. The first in-depth treatment of this problem is Stewart & Golden [1983] (the references in this paper cite earlier papers by the same authors that appeared in conference proceedings), who formulate the stochastic vehicle routing problem using both a chance-constrained and 'penalty function' approach. The penalty function approach is a form of stochastic programming with *simple recourse*, and subsequent papers differ primarily in the type of simple recourse strategy used. An excellent review of this research is provided in Dror, Laporte & Trudeau [1989]. Dror & Trudeau [1986] present an alternative recourse model which they show to improve on the original models presented in Stewart & Golden [1983].

4.7.4. Dynamic applications in vehicle routing

Interestingly, the limited progress that has been made in developing explicit models of dynamic vehicle routing problems that combine stochastic and dynamic elements has not stopped the implementation of practical tools for these problems. Brown & Graves [1981] provide one of the earliest descriptions of an application of an on-line optimization tool for real-time dispatching of petroleum tank trucks. Also at the same time, Gavish [1981] describes an on-line logistics planning system covering a) selection of leased vehicles, b) assignment of demands to depots, c) choice of return depot following the completion of each trip and d) workload allocation among depots. A series of heuristics were developed to solve these problems, requiring 5 to 10 minutes on a large mainframe of that era.

Bell, Dalberto, Fisher, Greenfield, Jaikumar, Kedia, Mack & Prutzman [1983] describe a similar application of a set partitioning model for routing trucks for Air Products and Chemicals. This model used a fixed planning horizon, using deterministic approximations of forecasted demands. Lagrangian relaxation was used to provide an optimal integer solution to the set partitioning problem. Other, unpublished, examples of real-time uses of optimization for routing and scheduling can also be found. However, the long-term success of these approaches is quite low, and we are unaware of any applications of these advanced techniques in a real-time context which are still operational three years after installation (since these projects typically do not make their way into the research literature, it is quite possible that long term, successful applications have occurred).

An open question in real-time applications is whether advanced optimization methods are really warranted. Bagchi & Nag [1991] describe a completely heuristic, expert systems-based approach for dynamic vehicle scheduling. The Gavish system [Gavish, 1981] uses optimization-based heuristics, and is basically similar in style. As these systems mature, it is expected that a hybrid of optimization models and rule-based expert systems will emerge.

4.8. Dynamic service network design problem

The dynamic service network design problem addresses the task of determining an optimal movement of vehicles, denoted by the vector v_t, which represent transportation services. The function $F_w(v)$ in equation (64) captures the optimal costs of transporting shipments given v. The service network design problem consists of determining the flow of vehicles v_{ijt} moving between two terminals (cities) in time t (arriving at node j in time period $t + \tau_{ij}$). We view v_{ijt} as the set of scheduled transportation services which will move goods. Since vehicles may have to be repositioned empty, we let:

u_{ijt} = number of vehicles moving empty from city i to city j, originating in period t.

V_{it} = net surplus or deficit of vehicles at city i in time period t.

The service network design problem can now be stated as:

$$\min_{v_t, u_t} F_w(v) + \sum_{t=0}^{P} c^T (v_t + u_t) \qquad (81)$$

subject to, for $t = 0, \ldots, P$

$$\sum_{j \in C} (v_{ijt} + u_{ijt}) - \sum_{k \in C} \left[v_{ki,(t-\tau_{ki})} + u_{ki,(t-\tau_{ki})} \right] = V_{it} \quad i \in C \qquad (82)$$

$$v_t, u_t \geq 0 \qquad\qquad\qquad\qquad\qquad\qquad\qquad\qquad\qquad (83)$$

$$v_t, u_t = 0 \qquad\qquad\qquad\qquad\qquad\qquad\qquad\qquad t < 0 \quad (84)$$

Rather than write the problem as a single, large optimization problem which combines traffic assignment along with the routing of loaded and empty vehicles, we have captured the traffic assignment component in the function $F_w(v)$.

Relatively little attention has been given to the dynamic service network design problem, with most prior work focusing on static formulations. Crainic, Ferland & Rousseau [1984] and Crainic, Gendreau & Dejax [1992] propose nonlinear network models for the service network design problem for trucking and rail. Local improvement heuristics are used to search for optimal integer frequencies. Powell & Sheffi [1989] propose an interactive optimization system for less-than-truckload trucking which optimizes the decision of whether to offer transportation service between two points (the frequency of service is handled implicitly). Haghani [1989] proposes a dynamic model of the service network design problem for rail, similar in style to (81). The model is solved using a linear programming package on some small networks. Farvolden & Powell [1991b] define the dynamic service network design problem for less-than-truckload motor carriers, and suggest a subgradient-based local improvement heuristic for deciding where to offer transportation services. The subgradient captures the effect of a change in the vector v_t on the flows of shipments w as well as empty vehicles u_t.

4.9. Dynamic facility planning

Our last topic in algorithms for dynamic networks is that of dynamic facility planning. This encompasses strategic planning of optimal dynamic capacity expansion for networks, and the tactical problem of dynamically repositioning servers and facilities in an operational setting. Examples of problems include adding or dropping plants, warehouses and terminals from a logistics network, serving new markets, or adding link capacity (a problem that arises particularly in communication networks). A number of papers have been written on this topic since the early paper by Erlenkotter [1969]. Reviews of solution approaches are given in Fong [1974], Erlenkotter [1981], Minoux [1987] and Christofides and Brooker [1967]. Zadeh [1974] considers optimal expansion of communication networks, and Fong & Srinivisan [1976] address the combined problem of capacity expansion and shipment planning. Algorithms for multiregion capacity expansion problems are given in Erlenkotter [1975], Fong & Srinivasan [1981a, b, 1986]. Doulliez & Rao [1975] formulate the problem as a specialized shortest path problem, and Laporte & Dejax [1989] show how a combined dynamic location and routing problem can be solved as a vehicle routing problem over an expanded graph. Van Roy & Erlenkotter [1980] suggest a dual-based procedure. Berman [1981] and Berman &

Leblanc [1984] introduce the problem of dynamically repositioning mobile service units over a network. Bean & Smith [1985] address the issue of capacity expansion over an infinite horizon, and Higle, Bean & Smith [1984] consider the problem of capacity expansion under stochastic demands.

5. Algorithms and formulations for deterministic dynamic networks

The special structure of dynamic networks would seem to encourage the development of specialized algorithms. Section 5.1 provides a general introduction to dynamic networks including a historical perspective and some general terminology. Section 5.2 describes a class of inductive algorithms that have been developed for the efficient solution of dynamic networks. Next, Section 5.3 presents a specialization of Dantzig-Wolfe decomposition in a dynamic context. Section 5.4 describes a variable transformation that substantially accelerates the solution of nonlinear dynamic networks. Section 5.5 then raises issues that arise in the solution of dynamic multicommodity network flow problems. Section 5.6 briefly discusses results for dynamic concave cost networks that arise in production planning, and Section 5.7 reviews the literature on optimal control methods for dynamic networks. Finally, Section 5.1.4 notes special measures that can be taken in the generation of dynamic networks.

5.1. An introduction to dynamic networks

Dynamic network models have proved to be an effective modeling framework for a range of planning problems that arise in logistics. These include production and inventory planning, which determines when and where goods should be shipped, as well as the operational problems faced by private fleets and common carriers which must provide the transportation services. Common carrier applications include truckload and less-than-truckload (LTL) trucking, railroads, airlines, and international container operations. These models typically track four types of activities - the movements of vehicles (locomotives, tractors, ships), the scheduling of drivers and crews that guide them, the management of fleets of containers (railcars, trailers, ocean containers), and the flows of goods (shipments/passengers) moving in the containers.

5.1.1. Historical perspective

The use of dynamic networks in an optimization framework was well established in the 1950's. Dantzig & Fulkerson [1954] formulates a tanker scheduling problem using dynamic networks. Ford & Fulkerson [1956] and Gale [1959] derive results for maximum flows over dynamic networks. One of the earliest comprehensive discussions of the use of dynamic networks for common carrier applications is provided by Magnanti & Simpson [1978], which focuses on airline problems. Vemuganti, Oblak & Aggarwal [1989] provides a more recent survey of network models for fleet management, focusing on fleet sizing decisions.

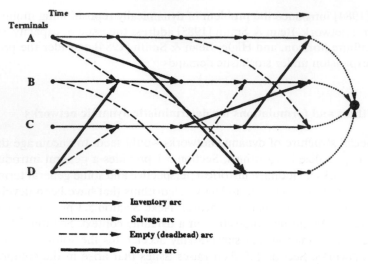

Fig. 11. Dynamic space-time network for airline fleet assignment.

As an illustration, Figure 11 shows a space-time network describing a set of flights over the course of the day. The flows over the network are in units of aircraft. *Revenue* arcs represent (potential) flights moving from one airport at one point in time to another airport at a future point in time. Revenue arcs generally carry an upper bound of one. Other arcs are *empty repositioning* movements and *inventory* arcs (referred to in the airline literature as ground holding arcs). Finally, Figure 11 uses a supersource and supersink, requiring *source* arcs, which carry aircraft from the supersource to the locations of the aircraft at the beginning of the day. Salvage arcs carry aircraft into the supersink at the end of the day. The number of aircraft that are required at each airport at the beginning of the day is an output, rather than an input. However, the network in Figure 11 does not guarantee that the number of aircraft at each airport at the beginning of the day is the same as at the end of the day. For this reason, we need to impose an *equilibrium* condition that specifies that the number of aircraft at each location is the same at the beginning and the end of the day. This equilibrium condition can be posed as a side constraint, or by introducing arcs that loop from the last time period for each city to the first time period for the same city.

5.1.2. Types of dynamic networks

Dynamic networks can be described as variations of two broad flavors. In a *fully dynamic* network, every link moves forward in time from t to some time $t + \tau$, $\tau > 0$. A special case of a fully dynamic network occurs when $\tau = 1$ for all links; we refer to these as *staged networks*. Later, when we deal with dynamic networks under uncertainty, we will have to split time periods into stages to capture the sequence of making decisions, and then observing realizations of random variables. For these problems, it is often convenient to transform networks with links that span more than one time period into networks where all links move forward exactly one time period.

At the other extreme are networks that are *dynamic inventory networks*, which are dynamic sequences of static problems (such as that illustrated in Figure 9). The only dynamic arcs (those moving forward in time) are inventory or holding arcs. Operational problems tend to look more like fully dynamic networks (see Section 4.5), while production planning problems are often modelled using a large time step, where all activities take place within a time period with the exception of inventory holding (illustrated in Section 4.4.2). Hybrid networks arise when a time step is chosen such that some activities move forward in time, while others move within the same time period. Regardless of the time step, however, it is very important that the nodes of the network be defined to ensure that the network is acyclic. For example, links within a time period may always move from plants to customers. If links can move between warehouses within a time period (that is, a link can go from warehouse i to warehouse j and vice versa) then it is important to introduce devices such as splitting each warehouse into inbound and outbound nodes, so that a link from warehouse i to warehouse j actually goes from the inbound node of i to the outbound node of j.

Dynamic networks can also be distinguished by how activities change over time. *Transient networks* exhibit data that, at least in the initial time periods, exhibits no particular pattern, generally reflecting the initial conditions of the problem. On the other extreme are *stationary networks* which exhibit data that is the same from one time period to the next. Finally, *periodic networks* are characterized by data that varies over a fixed cycle (typically a day or a week), which is assumed to be repeated from one cycle to the next.

5.1.3. Graphical representations

One of the most valuable aspects of network models is the ease with which they can be displayed graphically. In contrast with many static networks, dynamic networks pose the problem of dealing with a fourth dimension (time) that produces several strategies for displaying models, with no single strategy appropriate or even preferred for all situations. Four methods that are commonly used include:

1. *Two dimensional figures*. This approach is widely used for displaying small networks. Space is represented in one dimension, and time in the second dimension. Depending on the author, these networks may be displayed with time proceeding left to right, or from top to bottom (occasionally from bottom to top). It is useful for illustrative purposes in papers, but not very practical for viewing real networks.

2. *Three dimensional displays*. Most commonly, this approach consists of a series of two-dimensional maps, stacked from bottom to top to represent time periods. Arcs can move from a point on a map in one time period to a point in different map representing a later time period. On a computer terminal, it is possible to develop true three-dimensional representations of networks which can be manipulated by the user to rotate from one view to the next. These systems

are not widely available, but will come increasingly so as powerful graphics workstations and microcomputers come into wider use.

3. *Simulation.* For practical problems, two or three-dimensional displays of networks can quickly become too cluttered to be of any value. An alternative, then, is to display a dynamic network by literally showing a simulation of activities through space time.

4. *Task schedules.* In this category, we include graphical displays of schedules, such as Gantt charts, which show timing without the spatial movement. These might be used in conjunction with a static (spatial) display of movements.

The goal of graphical displays of dynamic networks is to capture the interaction of activities over space and time. To date, no single approach has emerged as a dominant technique for accomplishing this goal.

5.1.4. Network generators and data handling

A final comment should be made on the structure of network generators for dynamic networks. It is common in dynamic settings for a considerable amount of the data to be essentially static. For example, the distance between two cities does not change over time (although the time to get between two points can change over time, see Malandraki [1989] and Malandraki & Daskin [1992a]). Conventional models do not take advantage of this property, choosing instead to replicate data over many time periods. Klingman & Mote [1982] describe a specialization of a primal network simplex code for a multiperiod production, distribution and inventory planning problem. The code requires that all the parameters of the problem (network structure, costs and bounds) be fixed over time. Using this property, the code is structured to store all this data only once. In addition, each arc is priced over all time periods, with the most favorable arc being used in pivot operations. We are not aware of any work that takes advantage of partial stationarity of data over time, where, for example, demands may vary over time but other activities do not.

5.2. Inductive algorithms for linear networks

The temporal structure of a dynamic network has encouraged the development of specialized algorithms which take advantage of the properties of dynamic networks. White & Bomberault [1969] and White [1972] appears to be the first efforts to design an algorithm explicitly for dynamic networks. The papers outline an inductive procedure which begins with a network with a single node, and successively adds nodes to solve larger problems. Nodes are added to the network in time order. As each node is added, the algorithm looks to push flow from supply nodes to satisfy a demand node.

The methods reviewed above exploit only the dynamic structure of the problem to design specialized algorithms. A different approach exploits *properties* of the underlying *model* to develop efficient algorithms. The most notable example of this is the forward network simplex developed by Aronson & Chen [1986] for multiperiod distribution and inventory problems, for networks with the structure in Figure 9. Their algorithm works by solving a truncated problem over periods $t = 1, \ldots, T - 1$, and then using the solution as an advanced start for period T. The implementation of the network simplex algorithm exploits the fact that when the horizon is expanded to period T, flows in earlier periods may not be affected. Thus the only arcs that are priced out are those in period T, and those in periods $t^o, \ldots, T - 1$, where t^o is the earliest time period affected during the pivoting process following an augmentation. As T increases, t^o tends to increase as well, producing substantial reductions in the number of arcs that are priced. Stanley [1987] introduces a forward convex simplex algorithm for nonlinear dynamic networks.

The performance of the forward network simplex algorithm is due in large part to the properties of the underlying model. As time increases, breaks in the basis tend to occur, connected only by artificial links through the root node. These breaks represent *regeneration* points in the solution, which arise naturally in production and inventory planning problems. The result of these breaks is that pivots in one block of the network do not have an impact on other blocks. As the algorithm progresses forward in time, blocks of the network in early time periods become separate by breaks from later parts of the network. Realizing this, Aronson and Chen modified the simplex algorithm to detect the earliest time period affected by a pivot.

The datasets in Aronson & Chen [1986] use randomly generated supplies and demands in each period which are stationary over time. The model also includes extra 'outsourcing' supplies that allows demands to be satisfied in any period from an outside supplier, but at a higher cost [Aronson, 1990]. The model effectively trades off between inventory holding costs and the outsourcing cost to determine how many periods product should be held in inventory. This tradeoff tends to create frequent periods when no product is held in inventory. The result of these zero inventory periods are breaks in the basis, where pivots in one time period cannot affect decisions in another period with an intervening break. These *regeneration points* [Morton, 1979] create empirical decision horizons and forecast horizons where decisions in early time periods are unaffected by activities in later time periods [Aronson & Thompson, 1984].

The forward network simplex algorithm uses this property to reduce the number of arcs that need to be priced. Alternatively, if we are only interested in the decisions for the initial time periods, we could stop the algorithm as soon as a decision horizon is detected. In the case of dynamic networks, a gap in the basis is referred to as an empirical decision horizon because they are not guaranteed to remain as the problem grows. The term refers to the fact that these breaks seem to hold with a high probability. Exact decision horizons can be obtained for special types of problems. For example, Wagner & Whitin [1958] and Zangwill [1968, 1969], obtain exact decision/forecast horizons for dynamic networks with concave arc costs, and Aronson & Chen [1985] obtain similar results for a special production planning model.

This line of research falls in the class of *forward algorithms* [Morton, 1981] which exploit the presence of *regeneration points* in the model [Lundin & Morton, 1975;

Morton, 1979]. Interestingly, these regeneration points are almost guaranteed *not* to occur in dynamic fleet management. The balancing of supplies and demands in each period that arises in production and inventory planning, resulting in zero inventory arcs, does not occur in fleet management problems.

5.3. A decomposition approach for linear dynamic networks

A broader class of specialized algorithms can be drawn from the literature on dynamic linear programs with the general form:

$$\min \sum_{t=0}^{P} c^T x_t$$

subject to

$$A_0 x_0 = R_0$$
$$B_{t-1} x_{t-1} + A_t x_t = R_t$$
$$x_t \geq 0.$$

One solution approach, presented by Ho & Manne [1974], uses a creative adaptation of Dantzig-Wolfe decomposition. Their approach involves solving the problem iteratively in *cycles*. Each cycle involves solving a sequence of subproblems SP_t, starting for $t = P, P - 1, \ldots, 0$. The unusual aspect of the method is that problem SP_t works simultaneously as a subproblem for periods $t + 1, \ldots, P$, and as a restricted master problem (in the terminology of Dantzig-Wolfe decomposition) for periods $0, 1, 2, \ldots, t - 1$. Let x_τ^k be the kth extreme point solution for period τ, $\tau = 0, \ldots, t$ and let λ_{jt}^k be the weight given to x_t^j in the kth cycle, where $x_t^k = \sum_{j=1}^{k-1} \lambda_{jt}^k x_t^j$ and:

$$\sum_{j=1}^{k-1} \lambda_{jt}^k = 1$$

Let p_t^k be the value of the kth proposal in period t, defined by:

$$p_2^k = c^T x_1^k$$
$$p_t^k = c^T x_{t-1}^k + \sum_{j=1}^{t-1} p_{t-1}^j \lambda_{j,t-1}^k$$

Now define

$$S_t^k = B_{t-1} x_{t-1}^k$$

In network terms, S_t^k is the supply of flow in the kth iteration into period t produced by flows in period $t - 1$. Subproblem $SP(t)$ is now:

$$(SP_t) \quad \min_{x_t, \lambda_{jt}^k} [\sum_{j=1}^{k-1} p_t^j \lambda_{jt}^k] + [c - \pi_{t+1}^k B_t] x_t^k \tag{85}$$

subject to

$$A_t x_t^k + \sum_{j=1}^{k-1} S_t^j \lambda_{jt}^k = R_t \tag{86}$$

$$\sum_{j=1}^{k-1} \lambda_{jt}^k = 1 \tag{87}$$

$$\lambda_{jt}^k, x_t^k \geq 0 \tag{88}$$

The first set of brackets represents a restricted master problem for periods $t = 0, \ldots, t - 1$. The second set of brackets is the relaxed subproblem, where the vector π_{t+1}^k is the dual variables for constraint (71) from subproblem SP_{t+1}. For $t = P$, $\pi_{t+1}^k = 0$. For each cycle through the time periods, the duals π_t^k are improved and an additional extreme point is added to the set, improving the quality of the solution for periods $t = 0, \ldots, t - 1$.

It is unlikely that the Ho and Manne algorithm would ever be directly applied to dynamic networks. It is useful, however, in that it exposes a structure for solving dynamic networks. Problem SP_t is a truncated problem, where the vector π_{t+1} plays the role of a salvage value [Grinold, 1977]. From one perspective, the Ho and Manne procedure can be viewed as a mechanism for finding salvage values in dynamic networks. If the dynamic network were being used on a rolling horizon basis, it may only be necessary for π_{t+1} to be a reasonable approximation, yielding a different interpretation of convergence.

5.4. Flow splitting algorithms for nonlinear, dynamic networks

Classical formulations for dynamic networks consider flows in all time periods simultaneously. The standard form for these models can be stated as:

$$\min \sum_{t=0}^{P} c^T x_t$$

$$A_0 x_0 = R_0$$

$$B_{t-1} x_{t-1} + A_t x_t = R_t \quad t = 1, \ldots, P$$

An alternative view of the same problem is to first introduce a *state* variable S_t defined by:

$$S_t = B_{t-1} x_{t-1}$$

which gives the total flow into each node for period t. We can then replace the decision variables x_{ijt} with:

θ_{ijt} = fraction of flow through node i at time t that should be moved over link (i, j).

Clearly

$$x_{ijt} = S_{it}\theta_{ijt}$$

where

$$\sum_{j \in C} \theta_{ijt} = 1 \tag{89}$$

Thus the problem can be reformulated in terms of the state variables S_{it} and the flow splitting variables θ_{ijt}.

This approach to solving dynamic networks involves both the introduction of a state variable S_{it} and a flow splitting variable θ_{ijt}. In this section, the flow splitting variable, with its simple constraint structure (89), plays the central role in the development of specialized algorithms. However, as we consider more difficult problems later in the presentation, it is the introduction and use of the state variable S_{it} that plays an important role in the more important task of developing effective models. For example, Section 9.2 uses a state variable in conjunction with a more sophisticated class of routing strategies than can be accomplished using a simple flow splitting rule.

Let $c_{ij}(x_{ijt})$ be the total cost of x_{ijt} units of flow in period t where we assume $c_{ij}(x)$ is nonlinear, continuously differentiable and convex. For the traffic assignment problem, let

x^r_{ijt} = flow on link (i, j) at time t headed for destination r.

$$x_{ijt} = \sum_r x^r_{ijt}.$$

The nonlinear dynamic traffic assignment problem can be stated as:

$$\min \sum_{t=0}^{P} c_{ij}(x_{ijt}) \tag{90}$$

such that

$$\sum_{j \in C} \left[x^r_{ijt} - x^r_{ji,(t-\tau_{ji})} \right] = R_{it} \tag{91}$$

$$x^r_{ijt} \geq 0 \tag{92}$$

Gallagher [1978] first used this model to develop a minimum delay routing algorithm for communication networks. The problem with this formulation is that the flow conservation constraints (91) which are so easy to handle using sequential algorithms are not well suited to distributed computation. For this reason, he introduced the flow splitting formulation with

θ^r_{ijt} = fraction of flow through node i at time t with destination r to be routed over link (ij).

In addition he defined the state variable:

$S^r_{it}(\theta)$ = flow through node i at time t with destination r.

where $S_{it}^r(\theta)$ is a nonseparable function of the vector θ. Let

$$x_{ijt} = \sum_r S_{it}^r(\theta)\theta_{ij,t}^r$$

be the total flow on link (i, j), written now as a function of θ.

The nonlinear dynamic assignment problem (90) can now be expressed as follows:

$$\min \sum_{t=0}^{P} c_{ij}(x_{ijt}(\theta)) \tag{93}$$

such that

$$\sum_j \theta_{ij,t}^r = 1 \tag{94}$$

$$\theta_{ij,t}^r \geq 0 \tag{95}$$

In this formulation, the flow conservation constraints (91) are replaced with convexity constraints (94). Problems (90)–(92) are convex, separable (in x_{ijt}) nonlinear programming problems with network constraints. Problems (93)–(95) are convex, *nonseparable*, nonlinear programming problems with constraints that are *separable* by node. For dynamic, acyclic networks, the nonseparable structure of the objective function can be handled easily by defining:

$$C_t(\theta, S_t(\theta)) = \sum_{i,j \in C} c_{ij}(x_{ijt}(\theta)) + C_{t+1}(\theta, S_{t+1}(\theta)) \tag{96}$$

where

$$S_{j,t+1}^r = \sum_{i \in C} S_{i,(t+1-\tau_{ij})}^r \left(\theta_{ij,(t+1-\tau_{ij})}^r\right)$$

Equation (96) allows us to define a simple backward recursion for the derivatives. Let $F(\theta)$ be the total costs over the network, where

$$F(\theta) = C_0(\theta, S_0(\theta))$$

The derivative of $F(\theta)$ with respect to θ_{ijt}^r is then

$$\frac{\partial F(\theta)}{\partial \theta_{ijt}^r} = \frac{\partial C_t(\theta, S_t(\theta))}{\partial \theta_{ijt}^r}$$

$$= \frac{\partial}{\partial \theta_{ijt}^r} \sum_{i,j \in C} c_{ij}(x_{ijt}(\theta)) + \frac{\partial}{\partial \theta_{ijt}^r} C_{t+1}(\theta, S_{t+1}(\theta)) \tag{97}$$

Equation (97) provides a simple backward recursion that is used to calculate two sets of derivatives,

$$\frac{\partial C_t(\theta, S)}{\partial \theta_{ijt}^r}$$

and

$$\frac{\partial C_t(\theta, S)}{\partial S_{it}^r}.$$

Time

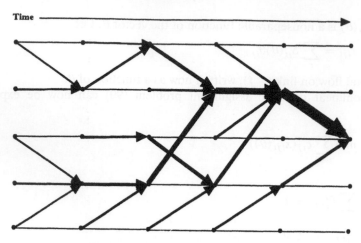

Fig. 12. Funneling in dynamic networks.

This formulation was developed independently by Powell, Sheffi & Thiriez [1984] for a nonlinear formulation of the dynamic vehicle allocation problem. Bertsekas, Gafni & Gallagher [1984] exploited the special structure of the constraint set to develop projection algorithms for a capacitated version of the problem. Bertsekas, Gafni & Gallagher [1984] develop second order algorithms for the same problem. Powell, Berkkam & Lustig [1992] compare the efficiency of the x and θ formulations of the problem, using the specialized algorithms in Genos [Mulvey & Zenios, 1987] to solve the x formulation. These experiments showed that state-of-the-art nonlinear network algorithms could show an almost pathologically poor performance on networks with as few as 10 or 20 time periods. By contrast, even the Frank–Wolfe algorithm worked reasonably well when using the θ formulation, with a gradient projection algorithm performing even better.

The reason for the poor performance of the x formulation, using either Frank–Wolfe or simplicial decomposition, is the nature of an extreme point solution. The solution to a linear subproblem in Frank–Wolfe, which involves routing flows over a shortest path into a supersink for DVA type networks, is illustrated in Figure 12. Following the shortest path, flows from different nodes into the supersink tend to fall into a single path in later time periods, a property we refer to as *funneling*. As a result of this property, which is basic to dynamic networks, extreme point solutions provide especially poor descent directions for first-order search algorithms. By contrast, the θ formulation at every iteration involves solving a subproblem out of every node, thereby providing an adjustment to the allocation of flows out of every single node. The effect is that the solution to the linear subproblem looks more like a dense forest than a narrow path.

5.5. Algorithms for capacitated dynamic multicommodity flow problems

Capacitated, multicommodity flow problems over dynamic networks pose a special challenge that arises in dynamic vehicle routing, crew scheduling, and multifleet assignment problems. As a rule, the principal characteristic of these problems is that they are capacitated multicommodity flow problems first, and dynamic problems second. Most of the research focuses on handling the bundling constraints (also referred to as GUB constraints) without focusing on the special properties of dynamic networks [see, for example, Magnanti & Simpson, 1978; Shan, 1985; Chih, 1986; and the survey by Assad, 1987]. Farvolden, Powell & Lustig [1993] develop a primal partitioning method for the capacitated dynamic traffic assignment problem which takes advantage of an empirical property of the dynamic structure. Flows moving over the network which encounter arcs at capacity tend to 'spill' forward in time, a property that produces a near-triangular working basis. This is exploited in the development of an efficient implementation of the simplex method. Ford & Fulkerson [1958b] and Bellmore & Vemuganti [1973] consider maximizing multicommodity flows over dynamic networks. Bellmore & Vemuganti [1973] show how a stationary solution can be expanded into a dynamic solution (over an infinite dynamic network) to obtain a bound on the optimal solution. Tapiero & Soliman [1972] use optimal control theory to solve multicommodity flow problems in continuous time. Gallagher [1978] and Bertsekas [1979] present methods for nonlinear, multicommodity flow problems over dynamic networks, which is reviewed in Section 5.4.

Below, we review some issues that arise that are specifically related to dynamic networks. First, Section 5.5.1 describes how dynamic problems magnify degeneracy in certain formulations. Next, Section 5.5.2 discusses how certain decomposition strategies are producing efficient solution algorithms for multicommodity flow problems defined over dynamic networks.

5.5.1. Degeneracy in dynamic networks

Multicommodity problems over dynamic networks with moderate to long planning horizons tend to exhibit massive degeneracy that is not found in static problems. This property has been especially pronounced in crew scheduling problems for airlines (one problem generated by American Airlines involves over 12 million tours), and where the interaction of tours over time produces the characteristically high level of degeneracy. The problem is illustrated using the network in Figure 13, where a 16 link network has 256 paths joining origin and destination. Assume that all the links have an upper bound of one, and that we are trying to route four separate drivers over the network. This network can be thought of as having four time periods, with four paths per time period joining a set of *regeneration* nodes. A regeneration node represents a point in the network, which

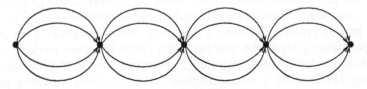

Fig. 13. Degeneracy in dynamic networks.

might often be a driver domicile, where paths might come together. If an n period network has k possible paths in each period, then there will be k^n paths in the network, with only k possible paths with nonzero flow. Furthermore, there are four conservation constraints and 16 upper bound constraints, all of which are binding, implying a basis with rank 20, with 16 basic variables with value zero (assuming an integer solution). Real networks are more complicated, but this simple analysis quickly illustrates the dramatic effect longer planning horizons can have on the potential number of tours.

Researchers are aware of this issue [Desrosiers, 1990], and have developed specialized strategies to help reduce this effect. In airline crew scheduling, one approach that has been used is called graph partitioning. Ball, Bodin & Dial [1983] and Ball & Roberts [1985] use this concept to develop a heuristic tour construction method for crew scheduling in transit and air, respectively. A link in a dynamic network represents a crew handling a flight and is called a duty period. Duty periods are then grouped into pairings, which consists of a small number of duty periods spanning several days, and always beginning and ending at a crew domicile. If each duty period is grouped into a pairing, it is possible to form a new graph, where each node in the network represents a pairing, and links from one pairing to the next represent either a rest period (layover) or empty movement to another domicile. The difficult part of this process is building the pairings, which Ball and Roberts do via a matching-based heuristic. Once this is done, the crew scheduling problem can be solved by finding paths over a greatly reduced graph, where nodes represent pairings. The final part of the procedure then involves performing local switches of duty periods between pairings. The effect of this procedure is to reduce the complexity of the graph, which reduces the number of possible tours and the level of degeneracy.

5.5.2. Decomposition methods

Since dynamic networks exhibit massive numbers of potential paths, it is natural to experiment with decomposition methods based on column generation. It has been conventional wisdom that decomposition methods should be used only when the problem is too large to be solved in-core [see, for example, Ho & Loute, 1983]. However, recent experimental evidence is starting to change this view [Desrosiers, Soumis & Desrochers, 1984; Jones, Lustig, Farvolden & Powell, 1993]. Farvolden, Powell & Lustig [1993] propose a path-based, column generation procedure for solving capacitated, multicommodity traffic assignment problems that arise in freight transportation. Their approach requires solving shortest path subproblems to generate columns, and an LP master problem to determine the best set of paths to use in a current solution. Since every origin-destination pair represents a different commodity, there are a potentially large number of convexity constraints in their master problem, creating a very large linear program. They propose a primal partitioning algorithm that isolates a relatively small working basis. The approach is similar to other algorithms proposed for GUB problems (Lasdon & Terjung [1971] used a similar method for capacitated lot-sizing problems). The master problem has constraints for upper bounds on links, as well as flow conservation constraints over the paths. The standard approach keeps these two types of constraints clustered together in the master problem (that is, all the flow conservation constraints are kept in one set of rows, while the binding upper bound constraints are kept in a different set of rows). In contrast, Farvolden, Powell & Lustig [1993] mixed both types of constraints within the working basis. For dynamic problems, the resulting working basis, with a proper ordering of rows and columns, exhibited a *near triangular* structure which both accelerated the algorithm and increased the degree of integrality in the optimal solution. This near triangular characteristic is an empirical observation, and seems to arise only with dynamic networks.

Jones, Lustig, Farvolden & Powell [1993] extends this idea to explore the effect of problem formulation on the performance of Dantzig-Wolfe decomposition for capacitated multicommodity flow problems. For traffic assignment problems, which involves routing flow from M origins to N destinations, it is possible to formulate the subproblem using N trees, one for each destination, or $M \times N$ paths. Using trees, the master problem involves N convexity constraints, whereas paths requires $M \times N$ convexity constraints. Using modern linear programming technology, Jones, Lustig, Farvolden & Powell [1993] show that working with paths can significantly improve algorithmic efficiency (sometimes by several orders of magnitude). While this work is not specifically tied to dynamic networks, the value of using paths over trees seems to arise only in the context of dynamic networks where degeneracy is much more pronounced. Jones [1992] further experiments with a dynamic decomposition scheme which involves path splitting over networks with long planning horizons (e.g. greater than 10 time periods); instead of using a single path from source to sink over the entire planning horizon, paths are split into two segments, joined in the middle of the planning horizon. Four paths in each segment, joined at the middle, can have the effect of 16 paths over the entire planning horizon. Once again, it is the dynamic structure of the problem that is creating the difficulty.

5.6. Dynamic networks with concave costs

Dynamic networks with concave costs arise frequently in production and inventory planning to capture economies of production and shipping. Wagner & Whitin [1958] formulated the basic production planning problem as a dynamic network with concave costs, and presented a dynamic programming algorithm for its solution. They also give one of the earliest planning horizon results, which has been exploited in the formulation of models as well as the streamlining of algorithms. Zangwill [1968] shows that optimal flows in dynamic networks with concave costs exhibit a tree structure (that is, no cycles), which can be exploited to accelerate dynamic programming algorithms. This result is then used [Zangwill, 1969] to extend Wagner-Whitin's main result to multistage production networks [see also Veinott, 1969; Graves & Orlin, 1985; and Erickson, Monma & Veinott, 1987]. For a thorough mathematical treatment of production planning, see Bensoussan, Crouhy & Proth [1983].

The central result of research on dynamic networks with concave costs is the tree structure of an optimal solution. The implication is that any movement of flow, whether it be production at a plant, shipping from plant to warehouse, or shipping warehouse to customer, is always in an amount that will exactly satisfy the demands for a particular set of time periods. As a result, the problem can be reformulated not in terms of how much flow should be moved, but rather which set of demands should be satisfied. Thus, when we move flow from plant to warehouse in time period 5, it may be to satisfy customer demands in time periods 8, 9 and 10. We can then formulate this problem in terms of deciding which time periods should be satisfied by a particular set of flows. The resulting problem can then be solved in a fairly compact way using dynamic programming. This result has been known for decades, and has not seen wide acceptance. Despite the relative complexity of this model, it still falls short of important real world concerns. High on this list is the ability to deal with many different items simultaneously, subject to bundling constraints, as well as handling uncertainty in demands. Both of these issues tend to destroy the tree structure of the optimal solution that is the foundation of the dynamic programs that have been formulated.

5.7. Optimal control methods for continuous time formulations

The preponderance of models and algorithms for dynamic networks assume a discrete time formulation. While considerably less progress has been made in the area of continuous time formulations, several authors have studied the problem in some depth. Frank [1967] and Frank & El-Bardai [1969] use optimal control theory to optimize flows over communication networks. Tapiero [1971] is the first paper to formulate the dynamic transportation problem using optimal control. Tapiero [1975] solves the single commodity dynamic transportation problem using optimal control; this is extended in Tapiero & Soliman [1972] to the multicommodity case. A nice survey of these results is given in Bookbinder & Sethi [1980]. Ellis & Rishel [1974] consider the airline flow management problem as an optimal control problem, which they solve in discrete time using dynamic programming. D'Ans & Gazis [1976] formulate the management of flow over congested transportation networks using optimal control.

At this point, optimal control methods for network optimization have not been applied in a practical setting. While progress has been made, considerably more work is required in both the formulation and solution of actual problems. In addition, we need to evaluate the value of a continuous time formulation over existing discrete time models that are widely used today.

6. Infinite horizon network models

The previous section presented a number of ways in which algorithms could be designed for dynamic problems. In this section, we address one of the most important problems that arise in dynamic models, which is the challenge of

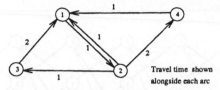

Travel time shown
alongside each arc

Fig. 14. A sample static network.

dealing with infinite planning horizons. The presentation begins in Section 6.1 with stationary dynamic networks, which helps to provide insights into the structure of optimal solutions of infinite networks. Then, Section 6.2 reviews a series of model ending techniques for approximating infinite horizon problems with finite approximations.

6.1. Stationary infinite-horizon networks

One of the earliest results for dynamic networks is given by Ford & Fulkerson [1956, 1958a, b, 1962] who considered the problem of finding maximal dynamic flows from the solution of a static network (see Figure 14). Consider a static graph $\mathcal{G} = \{\mathcal{N}, \mathcal{A}, \tau, l, u\}$ where $\mathcal{N}$ and $\mathcal{A}$ are sets of nodes and (directed) arcs, τ is a vector of (nonnegative) travel times, and l and u are vectors of lower and upper bounds. Assume flow enters through a super source q and leaves through a supersink r, and that $i \in \mathcal{N}$ are transshipment nodes. The problem is to determine the maximum flow that can be pushed from q to r over P time periods (all flow must exit the network by time period P). Let $V(P)$ be the total flow leaving the source node over time periods $1, \ldots, P$. Then the maximum dynamic flow problem for stationary graphs can be restated as:

$$\max V(P)$$

subject to

$$\sum_{t=0}^{P} \sum_{i \in \mathcal{N}} \left[x_{qit} - x_{iq,(t-\tau_{iq})} \right] - V(P) = 0$$

$$\sum_{t=0}^{P} \sum_{i \in \mathcal{N}} \left[x_{ir,(t-\tau_{ir})} - x_{rit} \right] - V(P) = 0$$

$$\sum_{j \in \mathcal{N}} \left[x_{ijt} - x_{ji,(t-\tau_{ji})} \right] = 0 \qquad i \in \mathcal{N}$$

$$x_{ijt} \geq l_{ij} \qquad i, j \in \mathcal{N}$$

$$x_{ijt} \leq u_{ij} \qquad i, j \in \mathcal{N}$$

Figure 15 shows how the original static network can be transformed into a time-expanded, stationary dynamic network. The path from 1, 2, 3, 1 is shown in bold, which spans four time periods. As a result, we can build four distinct paths which cover the same nodes at different points in time.

Time

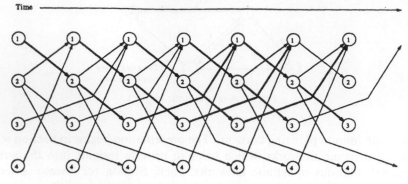

Fig. 15. Stationary, dynamic representation of the static network.

Our interest in this problem is not so much to develop an efficient solution algorithm, but rather to learn about the properties of an optimal solution for both finite and infinite networks. The central result for maximal dynamic network flows, given in Ford & Fulkerson [1958a], is derived from the following intuition. Let $\mathcal{M}$ be a set of directed chains from q to r, and let τ_p be the travel time for path $p \in \mathcal{M}$. We can construct $(P - \tau_p + 1)$ distinct copies of this path, each of which can carry a flow $f_p(t)$, $t = 1, \ldots, (P - \tau_p + 1)$. This is done by starting the first path at time 1; the second path holds in inventory for one time period and then starts at time 2; the last path holds in inventory for $(P - \tau_p)$ time periods and then starts in period $(P - \tau_p + 1)$. Now let f be the vector of path flows. We say that f is *periodic* if $f_p(t) = f_p(t+k)$, for $t^o \leq t \leq t'$, where t^o and t' define boundaries of network boundary conditions. Furthermore, we say that f^* is *stationary* if $k = 1$. Let F^{*d} be the set of all optimal dynamic path flows and let F^{*s} be the set of optimal *stationary* path flows. We can now state the following result:

Theorem 20 [Fulkerson, 1962, Theorem 9.1, p. 149]. $F^{*s} \in F^{*d}$.

Thus, to find the optimal dynamic path flows, we can restrict our attention to stationary path flows, implying $f_p(1) = \ldots = f_p(t) = f_p$. The maximum dynamic flow problem can then be restated as:

$$\max_{f_p} \sum_p (P - \tau_p + 1) f_p \tag{98}$$

Let $v = \sum_p f_p$ be the maximum flow (where $v = V(P)$), and let the link-path incidence relations over the static graph be given by:

$$\delta_p^{ij} = \begin{cases} 1 & \text{if link } (i, j) \text{ is in path } p \\ 0 & \text{otherwise} \end{cases}$$

Clearly

$$\tau_p = \sum_{ij} \delta_p^{ij} \tau_{ij} \tag{99}$$

$$x_{ij} = \sum_p \delta_p^{ij} f_p \qquad (100)$$

Substituting (99) and (100) and the definition of v into (98) gives the following optimization problem over the static graph:

$$\max_{v, x_{ij}} \left[(P+1)\, v - \sum_{ij \in \mathcal{A}} x_{ij} \tau_{ij} \right]$$

subject to a static version of the flow conservation constraints. Thus, given a stationary dynamic network, there exists an optimal solution which is also stationary, and which can be found by solving a static network problem. A set of static flows can then be expanded into a set of stationary flows.

The study of infinite networks leads to understanding the properties of solutions to stationary networks with finite planning horizons. Gale [1959] provides the following result.

Theorem 21. *Let $G(P)$ be the P-period expansion of $\mathcal{G}$, and let $f(P)$ be the optimal dynamic flows for $G(P)$. Then:*

(a) If $f(P)$ is optimal for $G(P)$, it is not *necessarily optimal for $G(P')$ with $P' \le P$.*

(b) There does exist a flow pattern of (P) that is optimal for all $G(P')$, $P' \le P$. In this case, $f(P)$ is termed a universal maximal dynamic flow.

(c) A universal dynamic flow is not necessarily stationary.

Result (c) is surprising, considering that stationary optimal flows exist for all $G(P)$. The implication of this result from a modeling perspective is that a stationary optimal flow pattern $f(P)$ may be unexpectedly sensitive to the choice of planning horizon P. Gale [1959] also finds universal dynamic flows with time varying capacities.

These basic results have been extended by others. Wilkinson [1971] and Minieka [1973] show how the original Ford & Fulkerson [1958a] algorithm can be modified to produce universal dynamic flows. Minieka [1974] provides a further modification of the Ford and Fulkerson algorithm for networks where arcs may be added or dropped from the network in any time period (a special case of the problem considered in Gale [1959]). A conclusion of this research, however, is that if the number of arc changes is 'excessive' then it is better to work directly with the time expanded graph.

An important extension of these results is given by Orlin [1983, 1984a, b]. Orlin [1983] considers the extension of the Ford and Fulkerson result to infinite dynamic networks (with stationary data) where he maximizes the *throughput* of the network instead of the total flow (which would be negative). The throughput $v(t)$ of the network is defined as the total flow in transit in period t (that is, in a dynamic graph, this would be the total flow crossing a line between periods t and $t + 1$). It is shown that for some period t^o, which can be viewed as the end of the initial transient period, that $v(t) = v(t^o) = v$ for all $t \ge t^o$.

Consider now the problem:

$$\max_{x_{ij}} \sum_{ij \in \mathcal{A}} x_{ij}\tau_{ij}$$

subject to

$$\sum_j (x_{ij} - x_{ji}) = 0 \qquad i \in \mathcal{N}$$

$$x_{ij} \geq l_{ij} \qquad\qquad i, j \in \mathcal{N}$$

$$x_{ij} \leq u_{ij} \qquad\qquad i, j \in \mathcal{N}$$

which is a problem defined over the static graph. The optimal static flows x can now be extended into stationary dynamic flows x_t using methods similar in spirit to those developed by Ford and Fulkerson. Decompose x into a set of (static) cycle flows, f_p, where τ_p is the transit time around a cycle. Starting with a cycle at time 1, we can now form an infinite path composed of successive repetitions of the same cycle. We can form τ_p node disjoint copies of this same path, starting at times $1, \ldots, \tau_p$ (the cycle starting at time $t = 1$ returns to the same node by time $\tau_p + 1$ and repeats itself). Let f_{pt}^s denote the stationary path flows created by pushing the static flows f_p along these infinite paths, and let x_t^s be the associated stationary link flows. The major result of Orlin [1983] can be stated as:

Theorem 22. *If the problem is bounded, then:*

(a) The optimum dynamic flow among all stationary flows equals the optimum dynamic flow among all dynamic flows.

(b) The optimum throughput equals the minimum upper capacity of a cut (which is monotone).

(c) If the upper and lower bounds are integral, then the optimum stationary flows may be taken to be integral.

Part (a) of the theorem is the infinite analog of the Ford and Fulkerson result. Part (b) is the dynamic analog of the Ford and Fulkerson max-flow, min-cut theorem.

An important corollary of these results is that they all apply to the *minimum throughput problem*, which is obtained by multiplying all flows and bounds by -1. The minimum throughput problem is applicable to the problem of minimizing the number of vehicles required to serve a predetermined set of schedules that must be repeated in each time period.

A related problem considered in Orlin [1984b] is the problem of minimizing long-run average costs in dynamic networks with convex costs. If $c_{ij}x_{ijt}$ is the cost in period t on link (ij), then we wish to solve

$$\min_{x_{ijt}} \lim_{P \to \infty} \frac{1}{P} \sum_{t=0}^{P} \sum_{i,j \in \mathcal{N}} c_{ij}x_{ijt} \tag{101}$$

An optimal solution can be found for this problem by solving the following static network problem with a side constraint:

$$\min_{x_{ij}} \sum_{i,j \in \mathcal{N}} c_{ij}x_{ij}$$

such that:

$$\sum_j (x_{ij} - x_{ji}) = 0 \quad i \in \mathcal{N} \tag{102}$$

$$\sum_{ij} \tau_{ij} x_{ij} = 0 \tag{103}$$

Equation (103) is called the *throughput constraint* (note that we do not require $x_{ij} \geq 0$). One effect of this constraint is that the optimal solution may be noninteger. Orlin presents an algorithm for rounding this solution to obtain an optimal integer dynamic flow (which is periodic).

6.2. Model ending techniques for infinite horizon networks

Stationary dynamic networks provide intuition into the structure of optimal solutions for infinite, dynamic networks. However, the actual problem being solved is highly specialized. A richer set of problems arise in the solution of rolling horizon problems, which typically exhibit transient data.

A general statement of an infinite dynamic network is given by:

$$\min_{x_t} \sum_{t=0}^{\infty} \alpha^t f_t(x_t) \tag{104}$$

subject to, for $t = 0, 1, 2, \ldots$

$$
\begin{array}{ll}
A_0 x_0 & = R_0 \\
B_{0,1} x_0 + A_1 x_1 & = R_1 \\
B_{0,2} x_0 + B_{1,2} x_1 + A_2 x_2 & = R_2 \\
B_{0,3} x_0 + B_{1,3} x_1 + B_{2,3} x_2 + A_3 x_3 & = R_3 \\
\;\;\vdots \qquad\qquad\qquad \ddots \qquad\qquad \vdots \\
B_{0,t} x_0 + B_{1,t} x_1 + B_{2,t} x_2 + \;\;\cdots\;\; + A_t x_t = R_t
\end{array} \tag{105}
$$

Here, $f_t(x_t)$ is the cost in period t and $\alpha < 1$ is a discount factor. The matrices B_{t_1,t_2}, $t_1 < t_2$, capture the impact of decisions made in period t_1 on period t_2 (if there is no impact, the coefficient is zero). If each constraint corresponds to the flow conservation constraint for a node, then the terms $B_{t_1,t_2} x_{t_1}$ captures the flow into a node in period t_2 from period t_1. The matrices A_t capture the flow out of each node to all future time periods (if some flow goes from period t to period t, then A_t must also capture movements within the time period). The vectors R_t give the surplus/deficit at each node i, where $R_{it} > 0$ represents flow entering the network.

Five methods have been suggested for approximating infinite horizon problems. These include:

1. salvage value;
2. boundary conditions;
3. planning horizon approaches [Morton, 1979];
4. dual equilibrium [Grinold, 1983a];
5. primal equilibrium [Grinold, 1983a].

Below we provide a brief summary of the first four approaches. The primal equilibrium uses concepts similar to the dual equilibrium [see Grinold, 1983a].

6.2.1. Salvage value methods

Salvage value methods involve solving a finite problem and then handling end effects by adding an ad hoc penalty for flow left in the system at the end of the planning horizon. Let S_t be the total flow entering each node in period t from periods $t' < t$, where

$$S_t = \sum_{t'=0}^{t-1} B_{t',t} x_{t'} \tag{106}$$

Let ν be a vector of salvage values, giving the value of a unit of flow in each node in time period $P + 1$. We can now replace equation (104) with:

$$\min_{x_t, S_{P+1}} \sum_{t=0}^{P} \alpha^t f_t(x_t) + \nu^T S_{P+1} \tag{107}$$

subject to constraints (105) for $t = 0, \ldots, P$.

The calculation of ν is ad hoc; it depends on the specific characteristics of the problem, and there is generally not a formal, mathematical basis for it. In inventory problems, this can be a measure of the cost of ending inventories. In dynamic fleet management problems, it can be an estimate of the value of a vehicle at a particular point in the system. In railroad applications, this is sometimes the problem of returning a freight car owned by another railroad (known as a foreign car) back to its home railroad. In any event, we use the term salvage value to refer to any ad hoc penalty that is used to reduce end effects.

6.2.2. Boundary condition methods

Boundary conditions attempt to handle end effects by actually specifying ending inventories or flows, as opposed to simply specifying a penalty. For example, management might specify ending inventories of goods or vehicles in each location.

$$\min_{x_t} \sum_{t=0}^{P} \alpha^t f_t(x_t)$$

subject to constraints (105) for $t = 0, \ldots, P$ and

$$\begin{aligned} S_{P+1} &\geq \bar{S}^l \\ S_{P+1} &\leq \bar{S}^u \end{aligned} \tag{108}$$

where $\bar{S}^l$ and $\bar{S}^u$ represent a specified vector of minimum and maximum final inventories. Equation (108) represents a boundary condition on the final flows. In this example, we are requiring that the supply to each node at the end of the planning horizon be at least $\bar{S}^e$. By constraining the total flow at the end of the planning horizon, we can try to force the model to avoid extreme end of horizon strategies. We could also specify maximum limits, or both minimum and maximum limits. In any event, like salvage values, these limits are generally chosen on an ad hoc basis.

6.2.3. Planning horizon methods

Planning horizon approaches represent a form of model truncation, but where the length of the planning horizon is chosen in such a way to ensure, at least empirically, a good quality solution. This is probably the most widely used approach in practice. Assume that we would like to perform planning over a horizon period of $0, \ldots, P$. To avoid distortions in period P, we solve the model over a longer horizon P', $P' \geq P$. The idea is to choose P' large enough so that the decisions in the interval $0, \ldots, P$ are of 'good' quality. The problem can be viewed as one of solving:

$$\min_{x_t} \sum_{t=0}^{P} c^T x_t + \sum_{t=P+1}^{P'} c^T x_t \qquad (109)$$

subject to flow conservations constraints over $t = 0, \ldots, P, \ldots, P'$. In contrast with the salvage value approach, which uses a linear function to handle end effects, the planning horizon approach simply extends the model from period P to period P'. The extension can be thought of as a nonlinear function of x_P. To see this, let S_{P+1} be the ending flows given $x_0, x_1, \ldots, x_P$, and define $G(S_{P+1})$ as follows:

$$G(S_{P+1}) = \min_{x_{P+1}, \ldots, x_{P'}} \sum_{t=P+1}^{P'} c^T x_t$$

subject to flow conservation constraints over periods $t = P, \ldots, P'$, and given the initial state S_{P+1}. Although the objective function is linear, as a result of the flow conservation constraints, $G(S)$ is a nonlinear (piecewise linear) function of the initial state vector S_{P+1}, which of course is a linear function of the initial flows $x_0, x_1, \ldots, x_P$.

Planning horizon methods are also referred to as *forward* methods for dynamic optimization [Morton, 1981]. The use of forward methods is reviewed in much greater depth in Morton [1979]. The planning horizon literature (see Section 3.4.3) is based on finding forecast horizons P^f such that longer planning horizons $P' > P^f$ will not change the decision in the first P^d periods, $P^d \geq 1$. P^d is then called a decision horizon. Exact forecast/decision horizons have only been found for very special problems, and hence practitioners focus on finding horizons P' that produce 'good' quality solutions for the first time period only.

6.2.4. Dual equilibrium method

The most comprehensive body of research on solving infinite horizon problems is that presented by Grinold [1977, 1983a, b], Hopkins [1971], Grinold & Hopkins [1973] for deterministic problems, and Flam & Wets [1987] for stochastic problems. Grinold [1983a] presents the *dual equilibrium* approach for replacing infinite horizon models with finite approximations. This work was applied by Hughes & Powell [1988] for dynamic networks arising in the dynamic vehicle allocation problem, and has been applied in other contexts as well [see, for example, Murphy & Soyster, 1986].

We start with the model presented by equations (104) and (105). Following Grinold [1983a], we designate the first τ^I time periods as the *transient phase*, representing data that reflects the initial state of the network. For example, the transient phase would capture realizations of forecasted data, which may extend for several days into the future. We let x_0 represent decisions in the first phase, which may represent several time periods. After τ^I, the data is assumed to follow a stationary, periodic pattern with period τ^p. In many applications, the natural period for the stationary phase is one week. If this is the case, then we can simplify the model in equations (104) and (105) by:

$$
\begin{aligned}
B_{t_1,t_2} &= B_{t_2-t_1} \quad t_1 \geq 1, \; t_2 > t_1 \\
A_{t,t} &= A \qquad\quad t \geq 1
\end{aligned}
$$

With this simplification, equation (105) becomes:

$$
\begin{aligned}
A_0 x_0 &&&&&&= R_0 \\
B_{0,1} x_0 + A x_1 &&&&&&= R_1 \\
B_{0,2} x_0 + B_1 x_1 \;\; + A x_2 &&&&&= R_2 \\
B_{0,3} x_0 + B_2 x_1 \;\; + B_1 x_2 \;\; + A x_3 &&&&&= R_3 \\
\vdots \qquad\qquad\qquad\qquad\qquad\ddots\qquad\quad\vdots \\
B_{0,t} x_0 + B_{t-1} x_1 + B_{t-2} x_2 + \quad\cdots\quad + A x_t &= R_t
\end{aligned}
\tag{110}
$$

The dual equilibrium method works by aggregating decisions in the future periods. Costs in period t are weighted by $(1-\alpha)\alpha^{t-1}$ and summed for $t \geq 1$, where $\sum_{t=1}^{\infty}(1-\alpha)\alpha^{t-1} = 1$. The aggregate link flows for the stationary phase would be:

$$
x(\alpha) = (1-\alpha)\sum_{t=1}^{\infty} \alpha^{t-1} x_t
$$

The constraints (110) are aggregated by multiplying the equation for period t by $\alpha^{t-1}(1-\alpha)$ and summing, giving:

$$
(1-\alpha)\left[\sum_{t=1}^{\infty}\alpha^{t-1}B_{0,t}\right]x_0 +
\tag{111}
$$

$$
+ (1-\alpha)\sum_{t=1}^{\infty}\alpha^{t-1}x_t\left\{A+\left[\sum_{t=1}^{\infty}\alpha^{t-1}B_{t-1}\right]\right\} = (1-\alpha)\sum_{t=1}^{\infty}\alpha^{t-1}R_t
$$

Define

$$
R(\alpha) = (1-\alpha)\sum_{t=0}^{\infty}\alpha^{t-1}R_t
$$

$$
B(\alpha) = (1-\alpha)\sum_{t=1}^{\infty}\alpha^{t-1}B_{0,t}
$$

$$
A(\alpha) = A + \sum_{t=1}^{\infty}\alpha^{t-1}B_{t-1}
$$

Substituting these back into (111) gives:

$$
B(\alpha)x_0 + A(\alpha)x(\alpha) = R(\alpha)
$$

Furthermore, if $f_t(x_t)$ is convex then we can write

$$\sum_{t=1}^{\infty} \alpha^t f_t(x_t) = \frac{\alpha}{(1-\alpha)} \left[\sum_{t=1}^{\infty} (1-\alpha)\alpha^{t-1} f_t(x_t) \right]$$

$$\geq \frac{\alpha f(x(\alpha))}{(1-\alpha)} \tag{112}$$

where (112) holds with equality when $f_t(x_t)$ is linear (which is typically the case in transportation). If $f_t(x_t) = c^T x_t$, then

$$\sum_{t=1}^{\infty} \alpha^t c^T x_t = c^T x_0 + \alpha \sum_{t=1}^{\infty} \alpha^{t-1} c^T x_t$$

$$= c^T x_0 + \frac{\alpha c^T x(\alpha)}{(1-\alpha)} \tag{113}$$

The dual equilibrium finite approximation is now given by:

$$\min \ c^T x_0 + \frac{\alpha c^T x(\alpha)}{(1-\alpha)} \tag{114}$$

subject to

$$A_0 x_0 \qquad\qquad\qquad = R_0 \tag{115}$$

$$B(\alpha)x_0 + A(\alpha)x(\alpha) = R(\alpha) \tag{116}$$

$$x_0, x(\alpha) \geq 0 \tag{117}$$

Hughes & Powell [1988] demonstrate this approach for dynamic networks, motivated by the dynamic fleet management problem. The assumption is made in this paper that the transient phase is long enough for all vehicles in transit at $t = 0$ to appear before the beginning of the first stationary phase, implying $R_t = 0, t \geq 1$. Furthermore, the longest travel time is assumed to be shorter than the length of the stationary phase, implying $B_{0,t} = 0, t \geq 2$, and $B_t = 0, t \geq 2$. The important result in Hughes & Powell [1988] is that for a dynamic network, equations (114)–(117) form a generalized network, depicted in Figure 16 for a two city problem with two time periods in each phase. The links that pass from the transient to the stationary phase carry a link multiplier of $1 - \alpha$. Zero cost arcs carry flow from the third set of nodes back to the first set, which feature an arc multiplier of α.

The intuitive foundation of the dual equilibrium method is the assumption that flows in the stationary phases will quickly reach a stationary pattern, implying

$$x \simeq x_2 \simeq \ldots \simeq x \tag{118}$$

Note that since each phase can consist of several time periods, our use of the term stationary is somewhat broader than Ford and Fulkerson. The research on maximal dynamic flows, in particular the work by Orlin [1983, 1984b], suggests that data will *tend* to follow a stationary pattern, supporting the intuition behind (118). At the same time, the presence of initial conditions from the transient phase, as well as the fact that the stationary phases are periodic (as opposed to strictly stationary), suggests that (118) will never be more than an approximation.

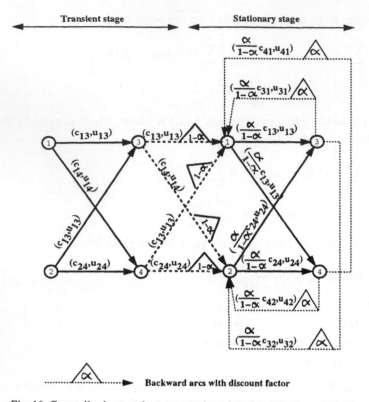

Fig. 16. Generalized network representation of dual equilibrium procedure.

The dual equilibrium method creates flows $x(\alpha)$ in the stationary phase with the same throughput as at the end of the transient phase. As a result, the upper bounds in the aggregate stationary phase are the same as in an individual stationary phase. This occurs since if $x_1 = x_2 = \ldots = x$, then $x(\alpha) = x$. However, the cost coefficients in the stationary phase are all factored by $\alpha/(1 - \alpha)$.

Hughes & Powell [1988] present a somewhat more intuitive approach for aggregating future time periods, referred to as the generalized summation method. In this approach, a more straightforward net present value is used to represent future activities. Thus:

$$x(\alpha) = \sum_{t=1}^{\infty} \alpha^t x_t$$

subject to constraints (105), with the definitions of $R(\alpha)$, $B(\alpha)$ and $A(\alpha)$ modified to use a weight of α^t for period $t, t \geq 1$. This problem is also a generalized network, with the same form as that given in Figure 16. The only changes are (a) the multipliers $(1 - \alpha)$ on links from the transient to stationary phase are replaced with α, (b) cost coefficients in the stationary phase are unfactored, and (c) the upper bounds on links in the stationary phase are factored by $\alpha/(1-\alpha)$. The dual equilibrium and generalized summation methods are effectively equivalent and in numerical experiments give comparable results.

A related result by Grinold & Hopkins [1973] gives conditions under which an infinite dynamic network can be solved exactly as a finite linear program. The main restriction is that the matrices A, B_t and $A(\alpha)$ are *Leontief*. A matrix A is Leontief if and only if there is at most one positive element in each column, and there exists $x \geq 0$ such that $Ax > 0$. These conditions are satisfied if the network is uncapacitated (a condition that would never be satisfied for dynamic fleet management problems). Nonetheless, if this condition were satisfied, we could solve the infinite horizon problem as follows:

First solve:

$$\min \ c^T y$$

such that

$$A(\alpha)y = e \tag{119}$$
$$y \quad \geq 0$$

where e is a vector of ones. Let π^* be the optimal dual variables associated with constraint (119). This problem can be viewed as a kind of discounted infinite shortest path problem for a vehicle entering each node in the aggregate stationary phase. Grinold & Hopkins [1973] give some technical conditions for the existence of π^*. The product $B_1 x_0$ gives the flows into each node of the first stationary phase. The following problem:

$$\min_{x_0} \left[c^T + (\pi^*)^T B_1 \right] x_0$$

will then give the same optimal first phase solution as the original discounted infinite network. Not surprisingly, numerical work reported in Hughes & Powell [1988] indicates that a Leontief model is a poor approximation of the future periods. This result also suggests that linear approximations of future activities will not, in general, be very effective.

There are some significant weaknesses in this approach to the solution of infinite horizon network problems. First is the fact that the dual equilibrium/generalized summation method will return a fractional solution, whereas the solution of a truncated model with any planning horizon [equations (70)–(71)] will return an integer solution (even when a discount factor is used). The second weakness lies in the heuristic basis of a discount factor in a deterministic model. In practice, the discount factor α is not used to determine the time value of money, but rather is an ad hoc approach for capturing uncertainty in the data. As a result, if the length of a stationary phase is one week, a weekly discount factor as low as $\alpha = 0.3$ might be used to reflect forecasting uncertainties.

7. Stochastic programming for networks

The most natural framework for writing stochastic formulations of dynamic fleet management problems is stochastic programming. Since the original articles of Beale [1955] and Dantzig [1955], a substantial literature has developed in the area of solving linear programs under uncertainty. The bulk of this research has been in the context of general linear programs and therefore can in principle be applied to stochastic formulations of all the models described in Section 4. However, as we demonstrate below, classical applications of stochastic programming methods can be extremely difficult to apply to stochastic networks with many random variables.

In this section, we summarize basic results from stochastic programming as a foundation for developing approximations which can be effectively applied to large stochastic networks. The problem is extremely rich, and as a result the presentation here is necessarily brief. Section 7.1 reviews some elementary concepts and notation. Section 7.2 surveys basic solution approaches that are used in stochastic programming, which are divided between methods for approximating the recourse function, scenario aggregation techniques, and finally stochastic gradient techniques. These are then covered in Sections 7.3, 7.4 and 7.5, respectively.

Throughout the presentation, we focus on results that have been specifically applied to, or are especially well adapted to, dynamic networks. This narrow tour through stochastic optimization, then, is intended to highlight emerging lines investigation into stochastic, dynamic networks.

7.1. Basic concepts and notation

Deterministic dynamic network models attempt to simultaneously find an optimal set of decisions x_t over all periods t in a planning horizon, illustrated by the simultaneous optimization problem:

$$\min_{x \in \mathcal{X}} \sum_{t=0}^{P} c^T x_t$$

In stochastic models, a simultaneous optimization formulation does not recognize the sequencing of decisions and realizations. It is important to identify 'what you knew and when you knew it.' To do this, we must define our random variables much more carefully. For the presentation here, we assume the only source of randomness is in the right-hand-side constraints:

ξ_{ijt} = market demand for loads moving from i to j in period t.

ξ_t = $\{\ldots, \xi_{ijt}, \ldots\}$.

We denote by $(\Omega, \mathcal{F}, \mathcal{P})$ the underlying probability space for ξ_t where Ω is a set of elementary outcomes, $\mathcal{F}$ is a set of events, and $\mathcal{P}$ is a probability measure defined over Ω. Let $\omega \in \Omega$ denote an elementary outcome, where $\omega = (\omega_1, \omega_2, \ldots, \omega_t, \ldots, \omega_P)$, $\omega_t \in \Re^m$, denotes the set of elementary outcomes associated with period t. For example, ω_t might be a vector representing the individual decisions of each shipper. We then let $\xi_t(\omega_t)$ be a *realization* of the market demands in period t. If there are n markets, then

$$\xi_t(\omega_t) : \mathfrak{R}^m \to \mathfrak{R}^n$$

represents the aggregation of individual shipper demands to market demands.

For a given outcome ω_t, we need to determine an optimal set of actions $x_t(\omega_t)$. These actions depend on the *history* of the system, which is summarized by all previous outcomes and actions. Let H_t be the history of the process up to and including period t:

$$H_t = \left[(x_0), (\xi_1(\omega_1), x_1(\omega_1)), (\xi_2(\omega_2), x_2(\omega_2)), \ldots, (\xi_t(\omega_t), x_t(\omega_t))\right]$$

The problem is to choose a set of actions x_t given the history H_{t-1} and the outcome ω_t. For this reason, we should write the decision vector as $x_t(H_{t-1}, \omega_t)$ to communicate this dependence on past history and the current set of outcomes. For compactness, we use the notation $x_t(\omega_t)$, to signal the dependence on the current outcome, with the dependence on past history implicit. For additional simplicity, we may use $x_t(\omega)$, where we assume $\omega = \omega_t$ in this context. In practice, with suitable independence assumptions, we can capture past history with a single state vector S_t of greatly reduced dimension. In this case, we assume $x_t(\omega_t)$ is implicitly a function of S_t. The notation $x_t(\omega_t)$ also communicates the property that actions in period t must be made before we know the outcomes $\omega_{t+1}, \omega_{t+2}, \ldots$.

A period t where all data is known, and where decisions must be made prior to realizations of future activities, is referred to as a *stage*. The vast majority of the stochastic programming literature addresses two-stage problems, where decisions must be made in the first, deterministic stage which account for the impacts of future events, which are uncertain. Further, we assume that different actions may be taken in the future once the outcomes of random events become known. This class of problems is referred to as two-stage stochastic programs with recourse. The general two-stage stochastic linear program with recourse can be stated as:

$$\min_{x \in \mathcal{X}_0} c^T x_0 + E_{\omega_t}\left[Q_1(x_0, \omega_1)\right] \tag{120}$$

where

$$\mathcal{X}_0 = \{x \mid A_0 x = R_0, x \geq 0\}.$$

We define

$$Q_1(x_0, \omega_1) = \inf_{y \in \mathcal{X}_1(x_0, \omega_1)} c_1(\omega_1)^T y$$

where

$$\mathcal{X}_1(x_0, \omega_1) = \{y \mid A_1(\omega_1)y = R_1(\omega_1) - B_1(\omega_1)x_0\}$$

In this formulation, we allow $c(\omega)$, $A(\omega)$, $R(\omega)$ and $B(\omega)$ to be stochastic. B is generally referred to as the *technology* matrix while A is the *recourse* matrix. For example, B describes the network which determines how the flows x_0 in period 0 enter period 1 (thus the network is the technology of the problem). $A_1(\omega)$

then determines the set of permissible actions in period 1 (the recourse). If A_1 is deterministic, the problem is said to have *fixed recourse*. If A_1 represents a node-arc incidence matrix, then the problem has *network recourse*. Finally, if we can express A_1 as

$$A_1 = [I, -I]$$

where $y = (y^+, y^-)^T$, $y^+, y^- \in \Re^m$, and $I \in \Re^{m \times m}$ is the identity matrix, then the problem has *simple recourse*. If the problem is feasible for all $x_0 \in \Re^n$, then the problem has *complete recourse*. If $Q_1(x_0, \omega_1) < \infty$ for all $x_0 \in \mathcal{X}_0$ and $\omega \in \Omega$, then the problem is said to have *relatively complete* recourse. For network problems that arise in practice, $\mathcal{X}_t(\omega)$ is nonempty and *compact*, which implies that we can replace the infimum with a simple minimization. $Q_1(x_0, \omega)$ is called the *recourse function*, with the *expected recourse function* given by

$$E[Q_1(x_0)] = \int_{\omega \in \Omega} Q_1(x_0, \omega) \, \mathrm{d}P(\omega) \tag{121}$$

The integration of (121) is computationally intractable for most problems of interest. This implies that not only can we not solve (121) using standard methods, we cannot even calculate the objective function. Thus, the difficulty of working with equation (121) is the heart of what makes stochastic programming a difficult problem.

Dynamic network problems often require looking several time periods into the future. These problems must be formulated as multi-stage stochastic programs to capture the imbedded structure of realizations and decisions. The generic structure of a multistage stochastic program is:

$$\min_{x_0} c^T x_0 + E_{\omega_1} \left\{ \min_{x_1} c^T x_1(\omega_1) + E_{\omega_2} \left\{ \min_{x_2} c^T x_2(\omega_2) + E_{\omega_3} \{\ldots\} \right\} \right\}$$

This formulation exhibits the imbedded minimizations and expectations that are characteristic of multistage stochastic programs. Needless to say, the size and complexity of this problem grows explosively with the number of stages.

7.2. *Solution approaches in stochastic programming*

Methods for solving stochastic programs can be organized into three groups which progress from easy to hard problems:
1. approximating the expected recourse function;
2. scenario methods;
3. stochastic gradient procedures.

The first two methods are based on the seminal papers by Wets on the equivalent convex program [Wets, 1966] and the equivalent deterministic program [Wets, 1974]. The equivalent convex program is the mathematical foundation of the first class of techniques, which seek to replace the expected recourse function with an analytical approximation which allows the first stage problem to be solved using standard optimization techniques. These techniques are only tractable when the underlying problem and the event space exhibit special structure. The second approach, scenario methods, is based on the equivalent deterministic program. This involves solving a large, combined optimization problem for both stages and all possible future scenarios. Finally, the third approach, stochastic gradient methods, uses sampling methods to calculate gradients used in search procedures.

A complete discussion of these approaches is well beyond the scope of this chapter. Excellent reviews can be found in Wets [1982, 1988], Birge & Wets [1986] (approximation schemes for stochastic programs), Vladimirou [1991] (scenario methods), Frantzeskakis [1990] (bounding procedures) and Chong [1991] (stochastic approximation methods). A very small percentage of this work is directed at dynamic networks. At the same time, some of the concepts are easily and directly applied to dynamic networks and should not be ignored simply because the literature has not addressed this special case. Just the same, the scope of our presentation is limited to dynamic networks and related results that readily lend themselves to dynamic networks. Section 7.3 outlines basic results that are used to develop approximations of recourse functions. These ideas are then applied in Section 9 for dynamic networks with random arc capacities. Section 7.4 gives a summary of basic results for scenario methods and their application to stochastic networks. Finally, Section 7.5 introduces stochastic gradient procedures and describes how they can be used to solve more complex problems.

7.3. *Approximating the recourse function*

Consider again the basic stochastic program (120) with recourse function $Q_1(x_0, \omega)$ and expected recourse function:

$$\bar{Q}_1(x_0) = \int_{\omega \in \Omega} Q_1(x_0, \omega) \, dP(\omega) \tag{122}$$

Further, as we defined earlier, let $\mathcal{X}_0$ be the feasible region for the first stage and let $\mathcal{X}_1(\omega)$ be the feasible region for the second stage given x_0 and outcome ω. In general stochastic programs, the issue of feasibility of the second stage is an important problem. For dynamic fleet management problems, however the presence of unbounded inventory arcs guarantees that the second stage exhibits *complete recourse*. Van Slyke & Wets [1969] show that $Q_1(x_0, \xi(\omega))$ is a continuous, convex function in x_0 over $\mathcal{X}_1$ and that $\bar{Q}_1(x_0)$ is also convex in x_0 Wets [1966] shows that (120) can be written as the following *equivalent conve. program*:

$$\min c^T x_0 + \bar{Q}_1(x_0) \tag{123}$$

subject to

$$x_0 \in \mathcal{X}_0$$
$$W x_1 \leq \alpha \tag{124}$$

where the matrix W and vector α are developed from geometric propertie of the feasible region $\mathcal{X}_0 = \bigcap_{\omega \in \Omega} \mathcal{X}_0(\omega)$. For dynamic networks which exhib complete recourse (the second stage is feasible for any x_0), (124) is not needed.

The equivalent convex problem establishes a framework that the recourse function can be viewed as a convex function in x_0, as opposed to a large embedded set of optimization problems given in (161). In practice, $\bar{Q}_1(x_1)$ is impossible to find explicitly. However, we can view (123) as a basis for developing approximations. Birge & Wets [1986] provide an extensive review of approximation schemes for stochastic programs. Some basic strategies that can be used to approximate the recourse function include:

1. primal decomposition methods (such as restricted recourse strategies);
2. dual decomposition methods (Lagrangian relaxation);
3. cutting plane techniques;
4. response surface methods.

Primal decomposition methods impose restrictions on the set of feasible strategies that can be used in the optimization in the second stage [see Frantzeskakis & Powell, 1989a]. The best example of this is simple recourse, where the optimization in the second stage is replaced by a trivial optimization problem whose expectation can be found analytically (see Sections 9.1.2 and 9.2). Dual decomposition methods seek to simplify the calculation of the expected recourse function by relaxing complicating constraints (see Section 9.4). These methods are illustrated in some depth in Section 9 in the context of dynamic networks with random arc capacities.

Response surface methods take samples to estimate a nonlinear approximation of the expected recourse function. This idea was first introduced in Beale, Forest & Taylor [1980] and further investigated in Beale, Dantzig & Watson [1986]. However, there is very little research using this technique.

7.4. Scenario methods

Assume the set Ω consists of the finite set of outcomes $\omega_1, \omega_2, \ldots, \omega_s, \ldots, \omega_L$ with probabilities p_s. Then (120) can be written:

$$\min \; c_0^T x_0 + \sum_{s=1}^{L} c_1^T x_1(\omega_s) p_s \tag{125}$$

subject to

$$
\begin{aligned}
A_0 x_0 & & & = R_0 \\
B_0 x_0 + A_1 x_1(\omega_1) & & & = R_1 \\
B_0 x_0 & + A_1 x_1(\omega_2) & & = R_1 \\
\vdots & & \ddots & \quad \vdots \\
B_0 x_0 & & + A_1 x_1(\omega_L) & = R_1 \\
x_1(\omega_1) & & & \le \xi_1(\omega_1) \\
& x_1(\omega_2) & & \le \xi_1(\omega_2) \\
& & & \quad \vdots \\
& & x_1(\omega_L) & \le \xi_1(\omega_L)
\end{aligned}
\tag{126}
$$

An alternative formulation of the same problem is the *full variable split* form of the problem. In this version, we introduce the variable

$x_0(\omega_s)$ = actions in the first period given scenario s in the second stage.

We then replace x_0 with $x_0(\omega_s)$ in equation (126). Of course, we cannot have a set of actions in the first stage that depends on the outcome of the second stage. The goal of stochastic programming is to choose a single set of actions for the first stage that minimizes expected costs over both stages for all possible scenarios. As a result we need to introduce the *nonanticipativity* constraint:

$$x_0(\omega_s) - x_0 = 0 \tag{127}$$

This form of the nonanticipativity constraint has the same form as (126) with the variable x_0 in each constraint.

Problem (125) is a large optimization problem, where both the first and second stage problems are networks. However, while there is a different solution $x_1(\omega_s)$ for each scenario ω_s in the second stage, there is only a single solution x_0 in the first stage. Constraints (126) represent the flow conservation constraints for each scenario. Since x_0 appears in the flow conservation constraint for each scenario, the problem is no longer a network, and the resulting linear program is very large.

Much more problematically, the number of scenarios will in practice be exponentially large. In typical applications, $|\Omega|$ may easily be on the order of 10^{10} to 10^{1000} (assuming discrete random variables). The standard approach to solving (125) is to replace the set of outcomes Ω with a much smaller sample that is randomly chosen from the original population. In practice, a sample with as few as 100 observations, or smaller, can still be computationally difficult. A number of strategies have been developed to solve this problem, drawing from the field of large scale optimization:

1. techniques for staircase linear programs [see, for example, Wets, 1966];

2. L-shaped decomposition [Van Slyke & Wets, 1969];

3. interior point algorithms for linear programs [see Lustig, Mulvey & Carpenter, 1991];

4. the progressive hedging algorithm [Rockafellar & Wets, 1987].

For the most part, these methods do little to take advantage of the underlying network structure of the problem. An exception is the progressive hedging algorithm [Rockafellar & Wets, 1987] which works by relaxing the nonanticipativity constraint and solving an augmented Lagrangian problem. If the original problem is a linear network, the result of this decomposition is an algorithm where at each iteration it is necessary to solve a sequence of nonlinear network problems, thus retaining at least some of the original structure.

The progressive hedging algorithm has been applied to stochastic networks with random arc capacities by Kornhauser & Maslanka [1989]. Rolling horizon experiments are reported and compared to the results given in Frantzeskakis & Powell [1990] on the same dataset. Although the scenario approach did not perform as well as the SLAP algorithm of Frantzeskakis and Powell, the results were generally good. Execution times, however, were substantially higher.

Kornhauser & Chen [1990] further experimented with stochastic, capacitated multicommodity flow problems. Scenario methods have been most widely applied to stochastic dynamic networks for financial planning [Mulvey & Vladimirou, 1989; Vladimirou, 1991]. Financial problems characteristically exhibit complex correlations in the random variables which give the return from a particular investment. Scenario methods allow these correlations to be captured without requiring strong independence assumptions.

7.5. Stochastic gradient methods for complex systems

Scenario methods are effective for two-stage stochastic programs with two characteristics: a) the recourse problem $Q_1(x_0, \omega_s)$ can be formulated as an optimization problem for a particular outcome $\omega_s \in \Omega$, and b) the sample space Ω can be reasonably approximated by a small set of outcomes (scenarios). It is not uncommon for one or both of these qualifications to be violated in the context of large, complex problems. For example, it might be relatively easy to calculate the future consequences of decisions made now using a *simulation* model, but may be virtually impossible to formulate future activities using an *optimization* model. Thus, we can find a statistical estimate of $\bar{Q}_1(x_0)$, but we cannot find $\bar{Q}_1(x_0)$ exactly and, more importantly, we cannot calculate the gradient $\nabla_x \bar{Q}_1(x_0)$ exactly. The second qualification is when we are able to formulate $Q_1(x_0, \omega_1)$ as an optimization problem, but we cannot accurately approximate the sample space using a 'reasonable' number of outcomes. This problem is more common in stochastic, dynamic networks where the number of stochastic elements in each period can be quite large. Instances where $|\Omega|$ is on the order of 10^{100} are quite common, implying the sample space is effectively infinite. We can still take a small sample from this population, but even 10 or 20 scenarios can create an extremely large deterministic equivalent problem.

Stochastic approximation procedures handle these problems by successively sampling from the set of outcomes and then obtaining an *estimate* of the derivative of the objective function. Under fairly mild assumptions, convergent algorithms can be developed which are quite easy to implement. The limitation of the approach is that the rate of convergence can be quite slow.

The first stochastic approximation procedure was given in Robbins & Monroe [1951] which sought to solve a problem in regression analysis. Let $\bar{F}(x) = E_\omega F(x, \omega)$ be a monotone function of x, and assume $\bar{F}(x^*) = \theta$ for a given value of θ. The problem is to find x^*, given that we can find $F(x, \omega)$ but cannot exactly calculate $\bar{F}(x)$. Kiefer & Wolfowitz [1952] consider the related problem:

$$\min_x E_\omega F(x, \omega) \tag{128}$$

for $x \in \Re^1$ and show how the method of Robbins and Monroe could be used to find a sequence x_n that converges stochastically to x^*. Blum [1954] extends this to problems where $x \in \Re^n$ using a sequence of the form

$$x^{k+1} = x^k + \alpha_k g^k(x^k, \omega^k) \tag{129}$$

where $g^k(x^k, \omega^k)$ is a *statistical estimate* of the gradient of $\bar{F}(x)$ which satisfies

$$E_\omega g(x, \omega)^T \nabla_x \bar{F}(x) < 0 \tag{130}$$

which is to say that the expectation of the gradient must be a valid descent vector (we assume that $F(x, \omega)$ is differentiable). Blum shows that $\lim_{k\to\infty} x^k = x^*$ almost surely under the following conditions:

$$\sum_{k=1}^{\infty} \alpha_k = \infty \tag{131}$$

$$\sum_{k=1}^{\infty} \alpha_k^2 < \infty \tag{132}$$

$$|\nabla \bar{F}^2(x)| < \infty \tag{133}$$

Conditions (131) and (132) are very standard in this literature. One sequence that satisfies them is $\alpha_n = 1/n$. Condition (131) can be viewed as ensuring that the algorithm will not stall out. Condition (132), combined with certain limits on the function itself, guarantees that the *variance* of the solution x^k goes to zero as $k \to \infty$. An important characteristic of this method is that while the *expectation* of $g^k(x, \omega)$ must be a valid descent vector as required by (130) a specific *realization* $g(x, \omega)$, used in (129), may not be a descent vector. As a result, the objective function may actually increase from one iteration to the next.

Stochastic approximation methods have been studied by a number of authors. A thorough review of the field is given in Wasan [1969]. However, these methods apply only to unconstrained problems. For constrained programs (and problems where $F(x, \omega)$ is not differentiable), Ermoliev [1969] introduced the notion of stochastic quasigradient methods, drawing on the concepts behind stochastic approximation procedures and nondifferentiable optimization problems [see Shor, 1979]. One of the earliest applications of these methods to stochastic network problems is given in Powell & Sheffi [1982] and Sheffi & Powell [1982], where the approach was dubbed the method of successive averages (when $\alpha_k = 1/k$, equation (129) implies that x^k is an average of all previous gradients). An excellent review of stochastic gradient methods is given in Ermoliev [1988].

Consider the stochastic programming problem (with recourse):

$$\min_{x \in \mathcal{X}} \bar{F}(x) \tag{134}$$

where

$$\bar{F}(x) = E_\omega \left[c^T x + \min_{y \in \mathcal{Y}(x,\omega)} q^T y \right] \tag{135}$$

where $\mathcal{X}$ and $\mathcal{Y}(x, \omega)$ are compact, convex feasible regions for the first and second stage problems, where:

$$\mathcal{Y}(x, \omega) = \{ y \mid A(\omega)y = R(\omega) - B(\omega)x, \; y \geq 0 \} \tag{136}$$

for given matrices and vectors $B(\omega)$, $R(\omega)$ and $A(\omega)$. Let $g_x^k(x^k)$ be the subgradient of $\bar{F}(x)$ with respect to the first-stage decisions x. A stochastic quasi-gradient

(SQG) $g_x^k(x^k, \omega)$ is a subgradient of $\bar{F}(x)$ for a particular realization ω which satisfies:

$$E_\omega\left[g_x^k(x^k, \omega)\right] = \partial\bar{F}(x^k) + \beta^k \tag{137}$$

where $\partial\bar{F}(x^k)$ is a subgradient of $\bar{F}(x^k)$ and $\beta^k \in \mathfrak{R}^n$ is a *bias vector* which must satisfy:

$$\lim_{k\to\infty} \beta^k = 0$$

The bias vector simply generalizes the original condition (130) that the expectation of the stochastic gradient be a strict descent vector. The theory also allows $\bar{F}(x)$ to be nondifferentiable in x, requiring the use of the subgradient term in (137).

In the case of dynamic networks, it is quite easy to find $g_x^k(x^k, \omega^k)$ for a given realization ω^k. In this case, the constraint set (136) is just the flow conservation constraint for the second stage. Let x^{k-1} be the first stage flows from the previous iteration, and solve, for a given realization ω^k:

$$\min c^T y(\omega^k) \tag{138}$$

subject to

$$\sum_m y_{jm}(\omega^k) = R_j^1 + \sum_i y_{ij}^{k-1} \qquad j \in C \tag{139}$$

$$y_{ij}(\omega^k) \leq \xi_{ij}(\omega^k) \qquad\qquad i, j \in C \tag{140}$$

Let $\pi^k(\omega^k)$ be the dual of constraint (139). Note that for a particular outcome ω, we have only to solve a single stage *deterministic* network (using standard network flow algorithms). The dual $\pi_j^k(\omega^k)$ is then the standard dual for the flow conservation constraint. Our stochastic gradient is then:

$$g_x^k(x^k, \omega) = c + \pi^k(\omega^k)^T B(\omega) \tag{141}$$

where the matrix $B(\omega)$ is the node-arc incidence matrix for nodes j in the second stage and arcs in the first stage [in effect, the coefficients of x in (139)].

The stochastic gradient $g_x^k(\omega)$ is not a *feasible* descent vector. Two methods can be used to obtain a feasible descent vector. Section 7.5.1 discusses using a projection operation; Section 7.5.2 describes a stochastic linearization approach. Finally, Section 7.5.3 briefly reviews methods for using simulation techniques to estimate gradients.

7.5.1. Stochastic projection

The first method of obtaining a feasible direction is to use a simple projection operator $\Pi_{\mathcal{X}}(\ldots)$ to map points back onto the feasible region:

$$x^{k+1} = \Pi_{\mathcal{X}}\left[x^k - \alpha_k g_x^k(x^k, \omega)\right] \tag{142}$$

where α_k is the step size at iteration k. If the feasible region $\mathcal{X}$ is defined by flow conservation constraints, then the projection step is impractical. On the other hand, if we are using the flow splitting methods of Section 5.4, the projection operation is quite easy. While this method has not yet been tried, it may be quite promising.

The stepsize α_k plays an important role in smoothing the fluctuations between iterations. The only condition the stepsize sequence must satisfy is (131). Not surprisingly, the rate of convergence can be extremely poor. One step that has been found to accelerate convergence is to smooth the estimate of the gradient vector. Let

$$g_x^k(x^k, \omega) = c + \pi^k(\omega^k)^T B(\omega) \tag{143}$$

represent a sample of the gradient of iteration k. Now let

$$\bar{g}_x^{k+1}(\omega) = \beta_k \bar{g}_x^k(\omega) + (1 - \beta_k)g_x^k(x^k, \omega) \tag{144}$$

where $0 \le \beta_k \le 1$. Equation (144) performs a successive averaging of sample gradients. We then use the averaged gradient $\bar{g}_x^k(x^k, \omega)$ in (129) instead of $g_x^k(x^k, \omega)$. This approach is sometimes referred to as a mixed stochastic gradient method.

7.5.2. Stochastic linearization

As an alternative, we could solve the stochastic linearization problem (this can be thought of as a kind of 'stochastic Frank–Wolfe'), first introduced by Gupal & Bajenov [1972]. Thus we would solve:

$$\min_{y \in \mathcal{X}} \bar{g}_y^k(\omega^k)^T y \tag{145}$$

where $\bar{g}^k(\omega)$ is calculated using (144). The feasible set $\mathcal{X}$ captures the network constraints in the first period. The attraction of (145) is that it is a linear network flow problem. The second stage problem, which is captured by the vector $\pi^k(\omega^k)^T B(\omega)$, adds no complexity to solving the first stage problem. Also, obtaining the duals $\pi_k(\omega_k)$ involves solving a deterministic network. Thus, this approach is extremely amenable to solving stochastic networks. However, considerable research is needed to evaluate and improve the rate of convergence, which is known to be quite slow for most problems.

Unlike the projection operation, the stochastic linearization method requires $0 < \beta_k < 1$ in equation (144) [see Ermoliev, 1988]. Let $\hat{y}^k(\omega)$ be the optimal solution of this subproblem at iteration k. Since $\hat{x}^k(\omega)$ is feasible, we can update the current solution using:

$$x^{k+1} = x^k + \alpha_k(\hat{y}^k(\omega) - x^k) \tag{146}$$

where again, α_k is a given stepsize.

7.5.3. Simulation methods for gradient estimation

Stochastic gradient procedures have the attractive feature that they accommodate complex models which cannot be formulated as optimization problems. Assume, for example, that given a state $S_1 = A_1 x_1$ for the second stage, we can only *simulate* future activities using Monte Carlo methods, yielding a statistical estimate of the recourse function, $Q_1(S_1)$. We now need to obtain an estimate of the gradient $h_1(S_1) = \nabla_s Q_1(S_1)$. When $Q_1(S_1)$ can be formulated as an optimization problem, this estimate is obtained from the dual of a realization of the recourse function. If we are forced to simulate the recourse function, then we can obtain an estimate of $h_i = \partial Q(S)/\partial S_i$ using finite difference methods [Kiefer & Wolfowitz, 1952]

$$h_i(S, \omega) = \left(\frac{1}{\Delta}\right) [Q(S + \Delta e_i, \omega) - Q(S, \omega)] \tag{147}$$

where Δ is a stepsize, e_i is a vector of zeroes with a 1 in the ith element, and ω is a sample realization. Given a sample gradient $h(S, \omega)$, the gradient $g^k(x^k, \omega)$ is found from

$$g^k(x^k, \omega) = c + h(S^k, \omega) \cdot B(\omega) \tag{148}$$

Kiefer & Wolfowitz [1952] and Ermoliev [1969] provide the conditions required to guarantee that the sequence x^k converges to the optimum when using (147) for unconstrained and constrained problems, respectively.

Another approach for estimating gradients is to use *infinitesimal perturbation analysis* (IPA). IPA was originally introduced by Ho, Eyler & Chien [1979] and is reviewed in depth in Glasserman [1991]. It has since been studied by a number of authors, primarily in the context of queueing networks (see Chong [1991] and the references cited there). The motivation behind IPA is to obtain an estimate of the gradient $h(s)$ by using a single simulation rather than the $n + 1$ simulations (assuming $S \in \Re^n$) implicit in the finite difference equations (147). The basic concept of IPA is to perform a single simulation while keeping track of events which would trace the impact of small changes in the vector S.

7.5.4. Remarks

An attraction of stochastic gradient procedures is that they easily handle complex stochastic networks. Each iteration involves solving a single realization of a stochastic network. We can accommodate uncertainty in arc capacities, travel times, costs, and even external supplies and demands.

Stochastic gradient methods provide a rigorous foundation for solving very complex stochastic networks that can arise in practice. In addition, they are particularly well suited to networks since the linear approximations of the recourse function produce network subproblems. At this time, stochastic approximation procedures have not been applied to stochastic, dynamic networks. As a result, there is no experimental evidence supporting the overall effectiveness of the approach. Stochastic approximation procedures in general are known to be quite slow in convergence, reflecting the weak assumptions required to guarantee convergence. Thus, these methods must be viewed as an algorithm of last resort.

8. Stochastic programming models in networks and routing

The basic concepts and framework presented in Section 7 can now be applied to two problems in logistics. Section 8.1 presents several models that have been used to formulate vehicle routing problems with uncertain demands. Then, Section 8.2 shows how the dynamic fleet management problem with random demands can be formulated as a multistage dynamic network with random arc capacities. Solution methods for this special type of stochastic program are described in Section 9.

8.1. Stochastic programming for vehicle routing

Consider a formulation of the standard vehicle routing problem with deterministic demands. We might define:

$$x_{ij}^k = \begin{cases} 1 & \text{if vehicle } k \text{ goes from } i \text{ to } j, \\ 0 & \text{otherwise.} \end{cases}$$

d_i = demand at customer i.

$$(VRP) \quad \min_{x,y} F(x, y) = \sum_{i \in C} \sum_{j \in C} x_{ij}^k c_{ij} \tag{149}$$

subject to

$$\sum_{j \in C} x_{ij}^k - \sum_{l \in C} x_{jl}^k = 0 \quad \forall i \in C$$

$$\sum_{k} \sum_{j \in C} x_{ij}^k = 1 \quad \forall i \in C$$

$$\sum_{j \in C} x_{ij}^k - y_i^k = 0 \quad \forall i, k \tag{150}$$

$$\sum_{i \in C} y_i^k d_i \le V$$

Problem (VRP) is a standard formulation of the vehicle routing problem with deterministic demands. In this context, we first assume demands are known, and then determine the routing.

As described in Section 7, there are several strategies which can be used when demands are uncertain:

1. *Myopic model.* In this case, we ignore forecasted demands. The model would be run again as new demands were called in.

2. *Deterministic model.* Here, we simply forecast demands and then treat these forecasts deterministically within the model. This model would be identical to (VRP) above, except that some of the demands would represent realizations, and others would be expectations of forecasts. The model would normally be run again as each forecast became a realization.

3. *Chance constrained programming.* In this model, the vehicle capacity constraint in (150) would be replaced with a probabilistic inequality and solved as a nonlinear programming problem.

4. *Stochastic programming.* Stochastic programming is the most rigorous solution framework, but also the most complex. In this approach, we optimize the problem over the known demands, taking account the expected costs of responding in an optimal way to various demand outcomes (scenarios). Various

approximations of stochastic programs can be developed by imposing restrictions on how we allow the system to respond to specific demand outcomes.

Myopic and deterministic models, and chance constrained formulations, are all effectively deterministic models. Myopic models ignore forecasted demands; deterministic models typically use the expected demand, and chance constrained formulations, as we show below, are equivalent to solving the model with deterministic demands set at a particular percentile from the demand distribution. Stochastic programming formulations offer a much richer framework in terms of accounting for more complex behavior.

A nice review of chance constrained and stochastic programming formulations is given in Dror, Laporte & Trudeau [1989]. We first present a basic chance constrained model, and then discuss stochastic programming formulations.

8.1.1. Chance constrained models

In a chance constrained formulation [see Stewart & Golden, 1983], we replace the vehicle capacity constraint with:

$$P\left(\sum_{i \in C} y_i^k d_i(\omega) \le V \right) \ge (1 - \alpha) \tag{151}$$

Equation (151) can be replaced with an analytical expression by making suitable assumptions about the distribution of $d_k(\omega)$. For example, if the demand is normally distributed with mean μ_i and variance σ_i^2, then (151) becomes (Stewart & Golden, 1983]:

$$\sum_{i \in C} y_i^k \mu_i + \gamma \left(\sum_{i \in C} \sigma_i^2 (y_i^k)^2 \right)^{1/2} \le V \tag{152}$$

where γ is a constant chosen so that:

$$P\left[\frac{\sum_i d_i^k y_i^k - M_k}{S_k} \le \gamma \right] = 1 - \alpha$$

$$M_k = \sum_l \mu_l y_l^k \tag{153}$$

$$S_k = \sum_l \sigma_l^2 (y_l^k)^2$$

Stewart and Golden show that if (i) the demands are independent, (ii) the terms $(\sum_i d_i^k y_i^k - M_k)/S_k$ and $(d_i(\omega) - \mu_i)/\sigma_i$ have the same distribution for all i, and (iii) σ_i^2/μ_i is equal to the same constant θ for all i, then constraint (153) is equivalent to:

$$\sum_i^k \mu_i \le \bar{V} \tag{154}$$

where

$$\bar{V} = \frac{2V + \gamma^2 \theta - (\gamma^4 \theta^2 + 4V\gamma^2 \theta)^{1/2}}{2} \tag{155}$$

Thus the stochastic vehicle routing problem with chance constraints reduces to a deterministic vehicle routing problem with demands μ_i and vehicle capacity $\bar{V}$.

8.1.2. Stochastic programming with recourse

Models of stochastic, dynamic problems can be differentiated based on what do we know, when do we know it, and what can we do about it once it is known. Most tractable models are characterized by strong assumptions that allow us to do very little once demands become known. Assume that, using some type of assumption about future demands, we make an initial decision about a tour. As demands become known, we then have to decide what assumptions are we going to make regarding our response to those demands. Note that we are not talking about the possibility of rerunning the model. Rather, we are making assumptions about our response so that we will make a better decision about what to do right now (before new data arrives).

We discuss three basic recourse strategies. The first is simple recourse, where constraint violations are assessed with a penalty. In this case, no recourse action is allowed in response to random demands, and infeasible problems may result. The second is a class of strategies we call *depot return recourse* where routes are decided in advance, and are not modified as demands become known. Finally, the third assumes full reoptimization at each step in the journey.

Simple recourse. Consider the basic VRP with random demands, denoted by $d_i(\omega)$. If we substitute random demands into the vehicle capacity constraint, we obtain:

$$\sum_{i \in C} y_i^k d_i(\omega) \le V \tag{156}$$

which may of course be infeasible for a given realization $d_i(\omega)$. Now consider the addition of a recourse variable $y_i^+(\omega)$ to produce

$$\sum_{i \in C} y_i^k d_i(\omega) - y_k^+(\omega) \le V \tag{157}$$

We can now formulate the vehicle routing problem with simple recourse as:

$$\min_{x,y} F(x, y) + E_\omega \left\{ \min_{y^+(\omega)} \sum_k c_k^r y_k^+(\omega) \right\} \tag{158}$$

where c_i^r is the recourse penalty cost for violating the capacity of the kth vehicle. The minimization inside the expectation is a trivial one, since for any realization ω, $y_k^+ = \max\{0, (\sum_{i \in C} y_i^k d_i(\omega) - V)\}$. When analytically tractable assumptions are made about the distribution of $d_i(\omega)$, it is normally possible to rewrite (158) in the following form:

$$\min_{x,y} F(x, y) + \bar{Q}(x, y) \tag{159}$$

where $\bar{Q}(x, y)$ is a nonlinear, convex function of x and y (as well as being a function of the recourse penalty c^r). For most local search techniques, this problem is no harder than one with linear costs. The drawback, of course, is that the penalty for violating the vehicle capacity will vary depending on the distance of the customer to other customers and the depot.

A restricted recourse strategy. Simple recourse strategies assume that random outcomes are handled by assessing penalties. As a next step, we can consider strategies where we allow a limited response. Consider the event where the total demands on a tour happen to exceed the capacity of the vehicle. Ideally, we would like to completely reoptimize the vehicle tours. A step in this direction is to consider restricted recourse strategies, as suggested in Frantzeskakis & Powell [1989a], is to *anticipate* only a limited set of responses to random demands.

For example, assume that vehicle routes are fixed in advance (based, for example, on forecasted demands). Then, as the vehicle completes each tour, instances where total demands exceed vehicle capacity are handled by inserting breaks in the tour. Thus, if a vehicle is picking up goods and reaches capacity before reaching the end of the tour, the vehicle returns to the depot, drops off its load, and then returns to the next customer on the original tour. Laporte & Louveaux [1990] first introduced this model, and suggested a second variation where customer demands are determined before the vehicle actually starts on the tour (but after the tour has been designed). The sequence of customers is not changed, but the point at which the vehicle stops and returns to the depot is optimized to determine the best break in the tour.

Dror & Trudeau [1986] propose an even more extreme model. In the event of a break in the route, they assume that the remainder of the route consists of trips back and forth to the depot for each customer. Thus, a potential route break carriers a very high penalty.

Full VRP recourse. It is useful to consider a more complete model to illustrate the complexities of modeling dynamic demands explicitly in the context of a multistage stochastic program. First, we need to characterize demands arriving to a central facility as a stochastic process. Assume we have up to N customers and let (d_n, τ_n) be the (random) demand, and the time τ that the demand is called in. Let a specific realization of this be denoted by $(d_n(\omega), \tau_n(\omega))$ and let $N_t \in N$ be the set of customers whose demands are known prior to time t. Assume time is indexed so that $t = 0$ refers to now, $t < 0$ refers to past events and $t > 0$ index future events. Assume t is discrete, and for simplicity, assume that all travel times between customers require exactly one time period.

Now let $H_t = \{(d_n(\omega), \tau_n(\omega)) | n \in N_t\}$ be the *history* of the process up to time t. We can then define $x_t(H_t)$ as the set of routes given data known up to time t and with forecasted information about customers $n \notin N_t$. Dropping the dependence on the history (which is implicit), x_t, $t < 0$, refers to the part of the tour that has not yet been completed, x_0 represents the decision we need to make now, and x_t, $t > 0$ represents the part of the tour that is being planned but will not be implemented until later, and therefore may be changed (given proper communication with the driver).

A multistage stochastic program would take the basic form:

$$\min_{x_0}\{c^T x_0 + E_{\omega_1}\{\min\{c^T x_1 + E_{\omega_2}\{c^T x_2 + \ldots\}\}\}\} \tag{160}$$

where in each stage we would constrain the problem to satisfy basic vehicle routing constraints. A tractable multistage model has not been proposed, and remains an active area of research.

8.2. A stochastic programming formulation of dynamic fleet management

We are now ready to write out the dynamic fleet management problem as a multistage stochastic program. We begin with the basic, single commodity formulation of DVA, where the only source of uncertainty is in the forecasts of future demands. We let $\xi_t(\omega_t)$ represent a particular outcome of ξ_t. Further, we let $\hat{\xi}_0$ denote the actual market demands in period 0, which are known. The stochastic DVA with random demands can be written:

$$\max_{x_0, y_0 \in \mathcal{X}_0} r^T x_0 - c^T y_0 + E_{\omega_1} \left\{ \max_{x_1(\omega_1), y_1(\omega_1) \in \mathcal{X}_1(\omega)} r^T x_1(\omega_1) - c^T y_1(\omega_1) \right.$$

$$+ E_{\omega_2} \left\{ \max_{x_2(\omega_2), y_2(\omega_2) \in \mathcal{X}_2(\omega)} r^T x_2(\omega_2) - c^T y_2(\omega_2) \right.$$

$$\left. + E_{\omega_3} \left\{ \ldots \right\} \right\} \right\} \tag{161}$$

where $\mathcal{X}_t(\omega)$ is the set of feasible x_t, y_t defined by, for $t = 0, \ldots, P$:

$$\sum_j (x_{ijt}(\omega)) + y_{ijt}(\omega)) - \sum_k \left[x_{ki,(t-\tau_{ki}(\omega))} + y_{ki,(t-\tau_{ki}(\omega))} \right] = R_{it} \quad i \in C$$

$$x_{ijt}(\omega_t) \leq \xi_{ijt}(\omega_t) \quad i, j \in C \tag{162}$$

$$x_{ijt}(\omega_t), y_{ijt}(\omega_t) \geq 0 \quad i, j \in C$$

where

x_{ijt} = flow of loaded vehicles from i to j in time period t.

y_{ijt} = flow of empty vehicles from i to j in time period t.

$\tau_{ij}(\omega)$ = realization of a travel time from i to j for a journey initiated in time period t.

If $t = 0$, the right hand side of (162) is replaced with $\hat{\xi}_{ij0}$. The set $\mathcal{X}_t(\omega_t)$ is a function of the history of the process H_{t-1} and the current outcome ω_t.

Problem (161) formulates the stochastic DVA as a multistage dynamic network with random arc capacities. This model was first given in Powell [1988]. Random travel times are captured by using $\tau_{ki}(\omega)$, which represents an outcome of a trip terminating in period t (this is not the most general possible model, but it serves our purposes here). Random travel times are equivalent to random node-arc incidence matrices with uncertainty in the location of the (-1) in each column.

Many of the extensions to the basic DVA discussed in Section 4.5 can be easily handled in the context of this framework. This includes multiple equipment types and transportation constraints that might represent rail or ship capacities. Traffic assignment problems, where smaller shipments must be consolidated into vehicles, are also easily accommodated.

Problem (161) is of course only a formulation, and does not readily lend itself to any existing solution procedures for problems of realistic size. It does, however, provide a framework and a starting point for the development of approximations.

As a first step, we can simplify (161) by introducing a state variable S_t defined by

$$S_{it} = \text{total supply of vehicles in region } i \text{ at the beginning of stage } t.$$
$$= \sum_{k \in C} \left(x_{ki,(t-\tau_{ki})} + y_{ki,(t-\tau_{ki})} \right)$$

We assume that S_t completely summarizes the state of the system in period t. This is technically true for this model only if $\tau_{ij} = 1$, $i, j \in C$, but this problem can be handled by extending the state vector to account for vehicles in transit. Using this state vector, we can restate (161) as follows:

$$\max_{x_0, y_0 \in \mathcal{X}_0} \left[r^T x_0 - c^T y_0 + \bar{Q}_1(S_1) \right] \qquad (163)$$

where $\bar{Q}_1(S_1)$ is the expected recourse function for stage 1, defined by:

$$\bar{Q}_1(S_1) \quad = E_{\omega_1}[Q_1(S_1, \omega_1)] \qquad (164)$$
$$Q_1(S_1, \omega_1) = \max_{x_1, y_1 \in \mathcal{X}_1(\omega_1)} \left[r^T x_1 - c^T y_1 + \bar{Q}_2(S_2) \right] \qquad (165)$$

The expected recourse function $\bar{Q}_t(S_t)$ is found recursively, where $\bar{Q}_{P+1}(S_{P+1}) = 0$.

For the purposes of our presentation, we refer to formulations (161)–(162) as the *simultaneous* form of the stochastic program and (163)–(165) as the *recursive*, or *state variable*, form. While mathematically equivalent, the two formulations lead to very different lines of investigation for solving these problems.

8.3. Remarks

This section has presented stochastic models of two basic problems, vehicle routing and fleet management. At this point in time, explicit, dynamic models for vehicle routing problems are intractable, and research to date has considered only chance constrained and simple recourse models. There is not enough experimental evidence at this point to say conclusively whether these simple models are performing effectively. Furthermore, there is some anecdotal evidence that supports the use of myopic models that ignore forecasted demands, with possibly limited use (in a deterministic fashion) of forecasted data. It is quite likely, at this point, that myopic models, or dynamic models with deterministic forecasts, will provide considerable benefits, with additional incremental benefits coming from the development of models which handle stochastic demands in a more rigorous manner.

The model for dynamic fleet management is, of course, much simpler because it is linear and avoids the combinatoric properties of vehicle routing problems. As a result, we have been able to make considerably more progress in this area. These results are described in the next section.

9. Approximations for networks with random arc capacities

Section 7 reviews general results for stochastic programs that can be applied to stochastic networks. For the most part, these methods tend to do little to take advantage of the basic structure of the problem. In some cases, algorithms can be designed which require solving sequences of network subproblems. Beyond this, there are few simplifying assumptions, allowing these methods to be applied to fairly general problems. The price of this generality is algorithms that can be quite slow and cumbersome to use. For some problems, this generality is not required, and other techniques can be used to produce accurate models that can be easily solved using standard algorithms. In this section, we review specialized results for dynamic networks with random arc capacities, a framework that is well suited to modeling uncertainty in demand forecasts. In contrast with the results of Section 7, which focuses on two stage problems with recourse, we treat both two-stage and n-stage models in this section. Section 9.1 illustrates the use of several simple approximations, including simple recourse, chance constrained programming and deterministic approximations. Section 9.2 introduces the concept of null recourse, which arises in inventory-type problems. Section 9.3 reviews a method for approximating the recourse function for multistage networks using linear approximations of the recourse function. Finally, Section 9.4 introduces the use of decomposition methods for approximating the recourse functions. Two models are used to illustrate these methods. The first is the classical stochastic transportation problem. The second is the n-stage dynamic network with random arc capacities.

9.1. Simple recourse and related methods

There are several simple methods that can be used to solve stochastic networks with random arc capacities. We illustrate their use in the context of the stochastic transportation problem with random demands. Let:

$\mathcal{M}$ = set of markets, $\quad m = |\mathcal{M}|$.

$\mathcal{N}$ = set of suppliers, $\quad n = |\mathcal{N}|$.

x_{ij0} = flow from supplier i to market j in the first stage $x_0 \in \mathfrak{R}^{n \times m}$.

S_{j1} = total flow supplied to market j in the second stage $S_1 \in \mathfrak{R}^m$.

ξ_{j1} = random variable giving the market demand in the second stage.

c = vector of transportation costs, $c \in \mathfrak{R}^m$.

r = vector of market revenues, $r \in \mathfrak{R}^n$.

R_{i0} = initial inventory in market i.

We wish to solve the following demand-constrained version of the stochastic transportation problem:

$$\min c^T x_0 + \bar{Q}_1(S_1) \tag{166}$$

subject to

$$
\begin{aligned}
A_0 x_0 &= R_0 \\
B_0 x_0 - S_1 &= 0 \\
x_0, \ S_1 &\geq 0
\end{aligned}
$$

The expected recourse function captures the 'profits' to be earned from supplying amount S_{j1} to market j. The form of the recourse function depends on how we wish to model second stage activities. For a particular outcome ω, our basic model is:

$$Q_1(S_1, \omega) = \sum_{j \in M} (-r)^T \max(S_{j1}, \xi_{j1}(\omega)) \tag{167}$$

However, other models may be used which yield different solution procedures. Below, we briefly outline three simple methods.

9.1.1. Deterministic approximations

The most widely used model in practice replaces the random demands $\xi_1(\omega)$ with deterministic approximations, d_1. We may use $d_1 = E_\omega[\xi_1(\omega)]$, or choose a value d_1 that satisfies Prob $[\xi_1 \leq d_1] = \alpha$, for a given service level α. Let:

$y_{j1} =$ amount of satisfied demand.

Then we may write

$$
\begin{aligned}
S_{j1} - y_{j1} &\geq 0 \\
y_{j1} &\leq d_1 \\
y_{j1} &\geq 0
\end{aligned}
\tag{168}
$$

Now the recourse function is given simply by $\bar{Q}_1(S_1) = -r^T y_1$, and (166) reduces to a transportation problem with two demand arcs from each market node to a super sink, one with upper bound d_1 and cost $-r$ and the other serving as an overflow arc.

9.1.2. Simple recourse

If we did not use deterministic demands above, equation (168) would become

$$y_{j1} \leq \xi_{j1}(\omega) \tag{169}$$

which is not well-defined for all outcomes ω (unless we require the constraint to be defined *almost surely*, which implies that it must hold for *all* outcomes ω). As an alternative, we may introduce the following *recourse variables*:

$$y_{j1}^+(\omega) = \max[0, \xi_{j1}(\omega) - y_{j1}] \tag{170}$$

$$y_{j1}^-(\omega) = \max[0, y_{j1} - \xi_{j1}(\omega)] \tag{171}$$

$y_1^+(\omega)$ is a random variable giving the unsatisfied demand (the underage) and $y_1^-(\omega)$ gives the excess supply (the overage). We may now write the demand constraint as:

$$y_{j1} + y_{j1}^+(\omega) - y_{j1}^-(\omega) = \xi_{j1}(\omega)$$

and the flow conservation constraint as:

$$B_0 x_0 - y_1 = 0$$

Assume that excess supply costs r^- and unsatisfied demand costs r^+. Note that the satisfied demand is given by $y_{j1} - y_{j1}^-$, which produces revenue r. The conditional recourse function is now given by:

$$Q(y_1, \omega) = \sum_j -r^T(y_{j1} - y_{j1}^-(\omega)) + (r^-)^T y_1^-(\omega) + (r^+)^T y_1^+(\omega)$$

The recourse function is separable in the recourse variables $y_{j1}^+(\omega)$ and $y_{j1}^-(\omega)$, which allows the expected recourse function to be found very simply. Depending on the distribution function for $\xi_1(\omega)$, the expectations

$$\bar{y}_{j1}^+(s) = \int_\omega \max[0, \xi_{j1}(\omega) - s] \, dP(\omega)$$

$$\bar{y}_{j1}^-(s) = \int_\omega \max[0, s - \xi_{j1}(\omega)] \, dP(\omega)$$

may be found as analytical functions of the supply $s = y_{j1} = S_{j1}$, or numerically as piecewise linear functions. As a result, we may write

$$\begin{aligned}\bar{Q}(y_1) &= E_w Q_1(y_1, \omega) \\ &= -r^T y_1 + (r + r^+)^T \bar{y}_1^+(y_1) + (r^-)^T \bar{y}_1^-(y_1)\end{aligned}$$

which is separable in the vector y_1. Now the original problem is a simple nonlinear network, with a linear transportation problem augmented by nonlinear arcs from each market to the supersink. The new network is illustrated in Figure 17.

The stochastic transportation problem is a multidimensional analog of the original newsboy problem and has been studied by a number of authors [Williams, 1963; Szwarc, 1964; Cooper & Leblanc, 1977; Qi, 1985; among others]. Williams [1963] exploited this feature to develop a specialized algorithm which relaxed the *supply* constraints, and then used the result from inventory theory that the optimum amount to be supplied to each market is $(r_j^- - c_{ij})/(r_j^- - r_j)$, for $r_j^- \geq c_{ij} > r$. Williams [1963] shows that there exists a *certainty equivalent* ξ^* that gives a deterministic problem whose solution x_1^* is the same as the stochastic problem. Cooper & Leblanc [1977] demonstrate how the Frank–Wolfe algorithm can be used to solve the nonlinear network. Qi [1985] presents the forest iteration method which further exploits the structure of the problem to develop an efficient solution algorithm. Wallace [1986] presents an algorithm which groups realizations of the demand vector which share the same optimal basis. Finally, Zipkin [1982] presents results for aggregating market nodes for very large stochastic transportation problems.

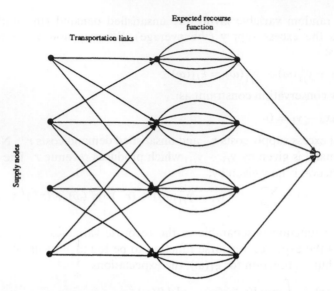

Fig. 17. Network representation of stochastic transportation problem with simple recourse.

9.1.3. Chance constrained programming

A third approach for dealing with random bounds is to use the concept of *chance constraints*, introduced by Charnes & Cooper [1959]. Here, the basic demand constraint (169) is replaced with:

$$\text{Prob}\,[y_{j1} \geq \xi_{j1}] \geq \alpha \tag{172}$$

where α is a specified *service level*. Chance constraints are useful when managers are more comfortable specifying the fraction of the market demand that should be satisfied. This is often more natural than specifying overage and underage costs.

9.1.4. Extension to multistage problems

These three solution methods can be applied to multistage stochastic networks, although their interpretation is not as natural. Consider the n-stage dynamic network with random arc capacities, given in Section 8.2. The complicating constraint in this problem is the random upper bounds on each link. In stage t, this constraint is given by:

$$x_{ijt} \leq \xi_{ijt}(\omega_t) \tag{173}$$

In a deterministic formulation, constraint (173) is replaced by a deterministic estimate d_{ijt}:

$$x_{ijt} \leq d_{ijt} \tag{174}$$

which transforms the problem into a n-period deterministic dynamic network. If $d_{ijt} = \bar{\xi}_{ijt} = E_{\omega_t}[\xi_{ijt}(\omega_t)]$, then it can be shown using Jensen's inequality that the optimal solution is a lower bound on the original n-stage recourse problem

[Frantzeskakis & Powell, 1990], but the bound can be quite loose. If the expected upper bounds are small, fractional solutions can become a problem. Frantzeskakis and Powell use the rounding procedure:

$$d_{ijt} = \lfloor \bar{\xi}_{ijt} + \gamma_{it} \rfloor$$

where $\gamma_{it}, 0 \le \gamma_{it} \le 1$, is chosen so that

$$\sum_j d_{ijt} \cong \sum_j \bar{\xi}_{ijt}$$

Alternatively, we may use simple recourse to handle the random upper bounds. As before, we introduce recourse variables $x_{ijt}^{+}(\omega_t)$ and $x_{ijt}^{-}(\omega_t)$ in a manner analogous to the stochastic transportation problem. The random arc capacity is now stated as:

$$x_{ijt} + x_{ijt}^{+}(\omega_t) - x_{ijt}^{-}(\omega_t) = \xi_{ijt}(\omega_t)$$

In the context of the dynamic vehicle allocation problem, $x_{ijt}^{+}(\omega_t)$ gives the unsatisfied demand, and $x_{ijt}^{-}(\omega_t)$ is the flow of excess vehicles, which must therefore move empty. x_{ijt} is the total flow of vehicles (loaded and empty), and $x_{ijt} - x_{ijt}^{-}(\omega_t)$ is the number of vehicles moving loaded. Flow conservation constraints are written entirely as a function of x_{ijt}. As before, if the distribution of $\xi_{ijt}(\omega_t)$ follows a simple form, the expected overage and underage $\bar{x}_t^{+}$ and $\bar{x}_t^{-}$ can be found as functions of x_{ijt}. This model was first introduced by Powell, Sheffi & Thiriez [1984] and Powell [1988]. The use of simple recourse produces an n-period nonlinear dynamic network, which can be solved using the methods described in Section 5.4.

Finally, we can enforce (173) using a chance constraint, producing a dynamic network with nonlinear arc constraints. An effective solution procedure can be developed by relaxing the chance constraint and adding it to the objective function. The difficulty with both simple recourse and chance constrained programming in the context of the DVA is the physical interpretation of the constraints. For example, simple recourse assumes that any of the x_{ijt} vehicles that cannot move loaded will move empty anyway. This is a very restrictive assumption that is likely to be very inaccurate if the expected upper bounds are small.

9.2. Null recourse

Simple recourse is a strategy where responses to random arc capacities are handled by using a parallel overflow link. As a result, while the flow on the random arc from node i to node j may be a random variable, the total flow on both the random arc and the overflow arc joining i and j is deterministic, greatly simplifying the problem. For dynamic fleet management problems, this is equivalent to assuming that if a vehicle does not move loaded between i and j, it will move empty.

A different model allocates a certain amount of flow to move over a link with a random arc capacity. If this flow is greater than the capacity of the arc for a specific realization, the excess flow is assumed to spill to the inventory link to be used in the subsequent time period. An important consequence of this policy is that future supplies of vehicles are random.

This problem can be approached by extending the basic approach developed in Section 5.4 for nonlinear dynamic networks. Although their presentation is quite different, Jordan & Turnquist [1983] effectively use this approach to solve a stochastic version of the empty freight car allocation problem. Their problem is to allocate a fleet of empty freight cars around the system to meet forecasted demands. They assume that once a car is moved empty from i to j, it remains in inventory at j until the spatially distributed, car is consumed. Thus, the problem becomes a dynamic inventory planning problem. Their model allows for uncertainty in demands and travel times. Each node has a stochastic *inventory* of cars S_{it} which is to be used to satisfy demands. Unused inventories of cars at a location are held in place until the next time period. The inventories of cars S_{it} are stochastic due to uncertainty in the travel times. For this problem, we could define the decision variables as:

θ_{ijt} = fraction of the *supply* of cars S_{it} to be moved empty from i to j at time t.

Reverting back to our original notation for the dynamic vehicle allocation problem, let y_{ijt} represent flows of empty cars where $y_{ijt} = S_{it}\theta_{ijt}$. Jordan and Turnquist assume S_{it} is normally distributed with known mean and variance, which implies y_{ijt} is also normally distributed with moments that are easily calculated. Let:

D_{it} = random variable describing the demand for cars at node i at time t.

and assume D_{it} is also normally distributed with known mean and variance. If S_{it} is the available inventory in period t, then $\max\{0, S_{it} - D_{it}\}$ is the remaining inventory to be held over to the next period. The total inventory in period $t + 1$ is then

$$S_{j,t+1} = \max\left\{0, S_{jt} - D_{jt}\right\} + \sum_i S_{i,t-\tau_{ij}}\theta_{ij,(t-\tau_{ij})}$$

Jordan and Turnquist develop approximations for the mean and variance of $S_{j,t+1}$, using normal approximations everywhere. Derivatives of the objective function with respect to the decision variables θ_{imt} are developed using backward recursions similar to that developed for deterministic flows.

Powell [1986] extends this approach to the full dynamic vehicle allocation problem with loaded and empty moves. In this model, the demands are given by

D_{ijt} = number of loads available to be moved from i to j in period t (no backlogging allowed).

The decision variables are:

α_{ijt} = fraction of the supply of available vehicles in node i at time t *allocated* to be moved loaded from i to j at time t.

β_{ijt} = fraction of the supply to be moved empty.

Let:

x_{ijt} = *actual* number of vehicles that are moved loaded from i to j in period t.

R_{it} is used as the *exogenous* supply of vehicles while S_{it} is the endogenous flow of vehicles through node i at time t. D_{ijt}, S_{it}, y_{ijt} and x_{ijt} are all random variables that are approximated using an Erlang distribution. The flow of loaded vehicles is given by:

$$x_{ijt} = \min \left\{ D_{ijt}, S_{it}\theta_{ijt} \right\} \quad i, j \in C, \ i \neq j \tag{175}$$

$$x_{iit} = \sum_{j \in C} \left[S_{it}\theta_{ijt} - x_{ijt} \right] \quad i \in C \tag{176}$$

Equation (176) expresses the assumption that if $S_{it}\theta_{ijt} > D_{ijt}$, then the excess vehicles are assumed to be held in inventory. Thus, if the flow allocated to move over the link exceeds the random upper bound, we assume that we do nothing with the excess flow (that is, we do not move capacity 'empty' as we did in simple recourse), hence the term null recourse.

It is possible to formulate this problem as a convex, nonseparable, nonlinear programming problem to maximize total expected profits over the planning horizon. Derivatives are again calculated using backward recursions, leveraging the same basic structure given in (96). The Frank–Wolfe algorithm works quite well for this problem, although it is very easy to implement other algorithms with better convergence [Bertsekas, 1979; Bertsekas, Gafni & Gallagher, 1984; Powell, Berkkam & Lustig, 1992].

The real power of this approach is its ability to exploit the dynamic structure of the problem and to handle forecasting uncertainties in a simple way. Its major limitation is that it is exceptionally difficult to develop a nonlinear cost function for strategies that are much more general than fixed flow allocations. For example, the 'move or hold' strategy implicit in the inventory equations (175)–(176) produces a fairly cumbersome set of equations. Even small generalizations can be intractable in this framework. One example is a strategy where a vehicle that cannot be moved loaded over link (i, j) is used to satisfy a demand on another link (i, k).

9.3. Successive linear approximation procedure

A different approach to solving the n-stage (161) is suggested by Powell [1987] and studied more formally by Frantzeskakis [1990] and Frantzeskakis & Powell [1990] (see also the discussion in Assad, Wasil & Lilien [1992, pp. 124–129]). This approach seeks to develop a separable, nonlinear approximation of the recourse function $Q_1(S_1)$ which would allow the first stage problem to be solved using standard optimization methods. We begin by restating the basic problem:

$$\min c^T x_0 + \bar{Q}_1(S_1) \tag{177}$$

subject to

$$A_0 x_0 = R_0$$
$$B_0 x_0 - S_1 = R_1$$
$$x_0 \leq \hat{\xi}_1$$
$$x_0 \geq 0$$

where $\bar{Q}_t(S_t)$ is defined recursively as:

$$\bar{Q}_t(S_t) \quad = E_{\omega_t}[Q_t(S_t, \omega_t)]$$
$$\bar{Q}_t(S_t, \omega_t) = \min_{x_t, S_{t+1}} c_T x_t + \bar{Q}_{t+1}(S_{t+1})$$

subject to

$$A_t x_t = R_t + S_t \tag{178}$$
$$B_t x_t - S_{t+1} = 0$$
$$x_t \le \xi_t(\omega_t) \tag{179}$$
$$x_t \ge 0$$

We now introduce the restriction that the network problem in each stage has a bipartite structure, where every link moves from a node in stage t to a node in stage $t + 1$. Thus, each stage is like a one-sided transportation problem with supply constraints (178), random arc capacities (179), and nonlinear market cost functions $\bar{Q}_{t+1}(S_{t+1})$.

The successive linear approximation procedure (SLAP) develops an approximation of $\bar{Q}_1(S_1)$ by starting at period $t = P$, and building a sequence of linear approximations. The approach is quickly illustrated in Figure 18. First, we use the structure of the last time period to calculate exactly the expected recourse function, which is a piecewise linear, convex function. Then, we replace the nonlinear recourse functions with linear approximations (these linear approximations can also be formed to create both upper and lower bounds). Finally, we add the slope of the linear approximation to the cost of each arc coming into that node. Now we

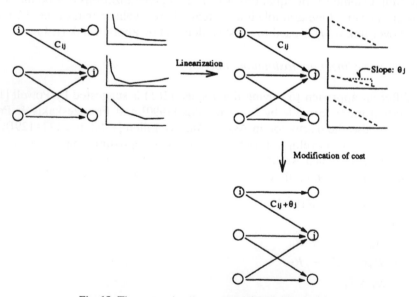

Fig. 18. The successive linear approximation procedure.

have a problem for time period $P - 1$ that looks just like time period P, and the process can be repeated.

We next demonstrate these steps in more detail. Consider the case of $t = P$, where $\bar{Q}_{P+1}(S_{P+1}) \equiv 0$. Time period P becomes a one-sided transportation problem with random arc capacities. The recourse problem for stage P is then separable in S_P allowing us to write:

$$Q_P(S_P, \omega_P) = \sum_i Q_{iP}(S_{iP}, \omega_P)$$

where

$$Q_{iP}(S_{iP}, \omega_{iP}) = \min \sum_j c_{ij} x_{ijP}$$

subject to

$$\sum_j x_{ijP} = S_{iP}$$

$$x_{ijP} \leq \xi_{ij}(\omega_P)$$

For a given realization ω_P the optimal solution is to greedily assign flow to the lowest cost links. Assume the links are ordered:

$$c_{i1} \leq c_{i2} \leq \ldots \leq c_{iL}$$

where L denotes the number of links emanating from node i. Let $s = S_{iP}$ be the total flow that must be moved over these L links (we assume that the Lth link is unbounded to ensure feasibility). Let

$$I_{ijt}(k) = \begin{cases} 1 & \text{if the } k\text{th unit of flow is assigned to link } (i, j) \text{ in period } t \\ 0 & \text{otherwise} \end{cases}$$

$$d_{ijt}(k) = \text{Prob}[I_{ijt}(k) = 1]$$

$d_{ijt}(k)$ is referred to as a *routing probability*. Let

$$U_{imt} = \sum_{j=1}^m \xi_{ijt}$$

be the cumulative capacity of the first m links. Then it is easy to verify that the event $[I_{ijt}(k) = 1]$ is equivalent to the joint event $(U_{i,j-1,t} < k \cap U_{ijt} \geq k)$. Thus

$$d_{ijt}(k) = \text{Prob}[U_{i,j-1,t} < k \cap U t_{ijt} \geq k]$$
$$= \text{Prob}[U_{i,j-1,t} < k] + \text{Prob}[U_{ijt} \geq k] -$$
$$- \text{Prob}[U_{i,j-1,t} < k \cup U_{ijt} \geq k] \tag{180}$$

The events $(U_{i,j-1,t} < k)$ and $(U_{ijt} \geq k)$ are collectively exhaustive (since $U_{i,j-1,t} \geq k$ implies $U_{ijt} \geq k$), and hence the last term of (180) is equal to 1.0:

$$d_{ijt}(k) = \text{Prob}[U_{i,j-1,t} < k] + \text{Prob}[U_{ijt} \geq k] - 1$$
$$= \text{Prob}[U_{i,j-1,t} < k] - \text{Prob}[U_{ijt} < k] \tag{181}$$

If the random variables ξ_{ijt} are independent, then the distribution of U_{ijt} is easy to compute (this is especially true if ξ_{ijt} follows a Poisson distribution). Thus the routing probabilities can be calculated relatively easily, even for large networks.

Now let

$$v_{iP}(k) = \text{expected value of the } k\text{th unit of flow}$$
$$= \sum_j d_{ijP}(k)c_{ij}$$

For integer s we can express the recourse function $Q_{iP}(S_{iP})$ as:

$$Q_{iP}(s = S_{iP}) = \sum_{k=1}^{s} v_{iP}(k)$$

The importance of this result is that we have found the expected recourse function exactly. The challenge is to exploit this result for the remaining stages. Consider next the problem encountered with stage $t = P - 1$.

Now the recourse function $\bar{Q}_{P-1}(S_{P-1})$ is no longer separable in the vector $S(N-1)$. The nonlinear recourse function $\bar{Q}_P(S_P)$ couples the routing of flow out of different nodes in stage $P - 1$. To circumvent this problem, replace $\bar{Q}_{iP}(s)$ with a linear approximation of the form:

$$\hat{Q}_{iP}(s) = a_{iP} + b_{iP}s$$

Since $\bar{Q}_{iP}(s)$ is convex, an appropriate choice of a_{iP} and b_{iP} gives a linear *support* of $\bar{Q}_{iP}(s)$, implying that $\hat{Q}_{iP}(s)$ can be constructed as a lower bound. However, the constant a_{iP} does not affect the optimal solution of stage $P - 1$. Substituting $\hat{Q}_{iP}(s)$ into the expression for $\bar{Q}_{P-1}(S_{P-1})$ gives

$$\bar{Q}_{P-1}(S_{P-1}, \omega_{P-1}) = \min \sum_i \sum_j c_{ij}x_{ij,P-1} + \sum_j [a_{jP} + b_{jP}S_{jP}]$$

$$= \min \sum_i \sum_j c_{ij}\, x_{ij,P-1} + \sum_j \left[a_{jP} + b_{jP} \sum_i x_{ij,N-1} \right]$$

$$= \min \sum_i \sum_j (c_{ij} + b_{jP})x_{ij,N-1} + \sum_j a_{jP}$$

This problem has the same structure as $Q_P(S_P, \omega_P)$ with the exception that the link costs c_{ij} are replaced by $c_{ij} + b_{jP}$. Thus, this recourse function is also separable, and can be calculated in the same manner we used to find $\bar{Q}_P(S_P)$.

This procedure can be applied recursively back to time period 1, with the last linearization occurring in time period 2. This leaves us with a piecewise linear, separable, convex approximation of $\bar{Q}_1(S_1)$. If the first stage is a network problem, then we are able to solve the original stochastic program (approximately) as a pure network with the form identical to that given in Figure 17.

The SLAP methodology easily handles P-stage stochastic transportation problems. Its novelty lies in the way it approximates the recourse function, thereby avoiding the explosion that occurs with dynamic programming or the conditioning inherent in multi-stage scenario methods. The successive linearizations, however, do introduce errors, and informal experiments have suggested that it may actually produce worse results as the number of stages is increased past a certain limit. On the other hand, it significantly outperforms the best competing heuristics (see Section 10.1).

9.4. The successive convex approximation method

The SLAP methodology linearizes the recourse function for period $t + 1$ to induce separability of the recourse function for period t. The method then takes advantage of the special structure of the optimization problem for a single node to find the expected recourse problem exactly. As a result, the methodology is highly tuned to a very specific type of stochastic network.

In this section, we use decomposition methods to develop a more accurate approximation of the expected recourse function. The approach is presented in three steps. The first result we need is the ability to find the expected recourse function over a tree with random arc capacities, given in Section 9.4.1. The method generalizes the technique used in Section 9.3 for single level trees. Next, Section 9.4.2 shows how dynamic, acyclic networks can be decomposed into stochastic trees using Lagrangian relaxation. This result allows us to approximate the recourse function for two-stage transshipment networks. Finally, Section 9.4.3 describes a backward recursion similar to SLAP which provides an approximation to multistage networks with random arc capacities.

9.4.1. Trees with random arc capacities

Powell & Cheung [1991] introduce an efficient method for finding the expected recourse function for a single stage optimization problem that has the structure of an n-level tree with random arc capacities, depicted in Figure 19. Let ξ

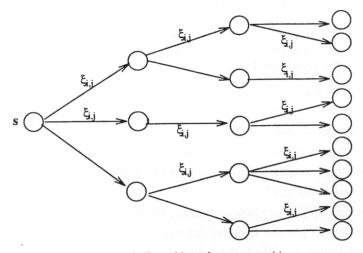

Fig. 19. Tree with random arc capacities.

represent the vector of (independent) arc capacities with a single realization $\xi(\omega)$. Let

r = index of the root node.

s = supply of flow entering the tree at node r.

$x(\omega)$ = vector of link flows conditioned on outcome ω.

$\mathcal{N}_e$ = set of ending nodes (where flow may leave the network).

$\mathcal{N}_t$ = set of pure transshipment nodes in the network ($\mathcal{N}_e \cap \mathcal{N}_t = \phi$).

The expected recourse function is given by

$$\bar{Q}(s) = E_\omega[Q(s, \omega)]$$

where the conditional recourse function is defined by :

$$Q(s, \omega) = \min_{x(\omega)} c^T x(\omega) \tag{182}$$

subject to

$$\sum_{j \in \mathcal{N}} x_{rj}(\omega) = s \tag{183}$$

$$\sum_{i \in \mathcal{N}} x_{ij}(\omega) - \sum_{k \in \mathcal{N}} x_{jk}(\omega) \begin{cases} = 0 \ \forall\, j \in N_t \\ \geq 0 \ \forall\, j \in N_e \end{cases} \tag{184}$$

$$x_{ij}(\omega) \leq \xi_{ij}(\omega) \tag{185}$$

$$x_{ij}(\omega) \geq 0 \tag{186}$$

Equation (184) allows flow to exit from any node in the set $\mathcal{N}_e$. Constraint (183) forces s units of flow to enter the network. Constraint (186) allows a random capacity on any link of the network, but we assume there exists at least one unbounded path from r to some node $j \in \mathcal{N}_e$ to ensure feasibility. There is no a priori restriction on the number of links in a path through the tree.

The solution of (182) is discussed in depth in Powell & Cheung [1991]. Here we present only the algorithm, which involves a single backward pass through the tree. To set up the algorithm, let:

n = path index.

c_n = cost of the nth path.

$l(n)$ = last node on path n.

$p(j)$ = predecessor of node j in the tree.

T^n = subtree consisting of all the links in the first n paths.

$\mathcal{N}^n$ = set of nodes in T^n.

$\hat{\xi}_{ij}^n(\omega)$ = maximum possible flow on link $(i, j) \in T_n$.

$\zeta_i^n(\omega)$ = maximum possible flow that can move through node i in the graph T_n.

$Z^n(\omega)$ = total capacity of the first n paths.

The random variables $\hat{\xi}_{ij}^n(\omega)$, $\zeta_i^n(\omega)$ and $Z^n(n)$ are defined in the algorithm below.

Expected tree recourse algorithm

Step 1. Set $n = 0$, $\hat{\xi}_{ij}^0 = 0$, $S_i = 0$.

Step 2. Set $n = n + 1$, $j = l(n)$. Here, we augment the tree T^n with the $n + 1^{st}$ path to create the subtree T^{n+1}.

Step 3. Set $i = p(j)$.

Compute the distribution of $\hat{\xi}_{ij}^n$, defined by (for $j \in \mathcal{N}^n$):

$$\hat{\xi}_{ij}^n = \begin{cases} \xi_{ij} & \text{if } j = l(n) \\ \min\{\hat{\xi}_{ij}^n, \zeta_j^n\} & \text{otherwise} \end{cases}$$

Compute the distribution of ζ_i^n by:

$$\zeta_i^n = \sum_{k \in \mathcal{N}^n} \hat{\xi}_{ik}^n$$

Step 4. Set $j = i$ and return to step 3 until $i = r$.

Step 5. Let $Z^n = \zeta_r^n$, and go to step 2 until all paths are exhausted or we reach an uncapacitated path.

Steps 2 and 3 are the main computational steps, since the distributions of $\hat{\xi}_{ij}^n$ and ζ_i^n must be calculated numerically. Note that the random variables $\hat{\xi}_{ik}^n$ are independent for $k \in \mathcal{N}$, and ζ_j^n and $\hat{\xi}_{ij}^n$ are independent. When the algorithm is completed, we assume we have computed the probability mass function of the residual path capacities Z^n, from which we can obtain the cumulative distribution. Let

$d_{in}(k) = $ probability the kth unit of flow is routed over the nth path.

Clearly, $d_{in}(k)$ is similar to the routing probabilities we introduced for the SLAP algorithm, so it should not be too surprising that we calculate them using:

$$d_{in}(k) = \text{Prob}[Z^n < k] - \text{Prob}[Z^{n-1} < k]$$

which is analogous to equation (181). The distributions of Z^{n-1} and Z^n are calculated numerically using the tree recourse algorithm. Note that even if the arc capacities are independent Poisson random variables, the random variables ξ_i^n, and hence Z^n, do not have any special form (as a result of the minimum operator in step 3).

An interesting side benefit of the tree recourse algorithm is that we are equally capable of solving a single stage, n-level tree or an n-stage stochastic program composed of single level trees. In fact, Powell & Cheung [1992] show that the n-stage formulation is actually easier to solve for large problems. To see this, consider the tth stage of an n-stage tree. For $t > 1$, this stage will consist of a *forest* of two level trees. The links in stage t form a single level network with random arc capacities. The links for stage $t + 1$ represent the piecewise linear

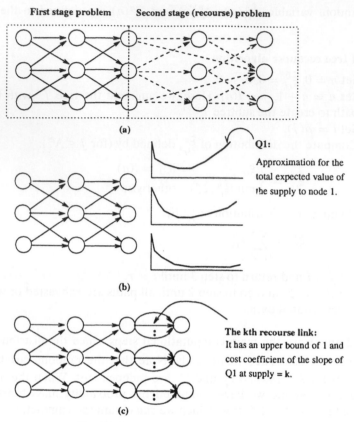

Fig. 20. Second stage stochastic network.

recourse function $\bar{Q}^{t+1}(s^{t+1})$ which is separable as a result of the tree structure (these links will each have an upper bound of 1 or ∞). We now have to apply the tree recourse algorithm to each subtree to find expected recourse functions for period t, after which we repeat the process for stage $t - 1$. At each point, we are working with relatively small trees.

9.4.2. Network recourse decomposition

Powell & Cheung [1992] demonstrate how the tree recourse concept can be used to approximate recourse functions for general (multistage) dynamic networks with random arc capacities. Consider the second stage of a two-stage stochastic dynamic network as illustrated in Figure 20. Assume flow enters the network through nodes 1,2 and 3. Treat the flow on link (i, j) that originated at node k as a separate commodity. If we denote this flow as x_{ij}^k (suppressing the stage index), then of course

$$x_{ij} = \sum_k x_{ij}^k \tag{187}$$

This problem must be solved subject to random upper bounds:

$$x_{ij} \le \xi_{ij} \tag{188}$$

Using the multicommodity formulation, (188) becomes a (random) bundle constraint, which destroys the tree structure. Assume instead that we relax (188) and add it to the objective function with penalty $\lambda_{ij}(\omega)$:

$$\min \sum_k \sum_{ij} c_{ij} x_{ij}^k(\omega) + \sum_{ij} \lambda_{ij}(\omega) \left(\sum_k x_{ij}^k(\omega) - \xi_{ij}(\omega) \right)$$

and instead enforce the constraint:

$$x_{ij}^k(\omega) \le \xi_{ij}(\omega)$$

Now the problem decomposes by commodity k, but this does not guarantee that each subproblem forms a tree, since the flows that originate at a given node k can still form a cycle (since the network itself is not a tree). Powell and Cheung [1992] introduce the concept of the extended graph which forms a tree out of a subset of the paths that originate at node k. When two paths form a cycle, the joining link is split into two links, which is effectively another level of decomposition. The conditional Lagrange multiplier $\lambda(\omega)$ can be approximated by a constant λ, which can be optimized using subgradient methods. Since we are interested only in an approximation of the expected recourse function, and not in the second stage solution itself, the multipliers λ do not have to be very precise, and it is not that important if the bundle constraints in the second stage are violated. Numerical work in Powell and Cheung [1992] shows that the Lagrangian relaxation closely approximates the exact expected recourse function (calculated using Monte Carlo simulation) after ten iterations of the subgradient optimization algorithm.

9.4.3. The SCAM algorithm for multistage networks

The network recourse decomposition method can be applied to multistage networks in the following way. Consider the three-stage network illustrated in Figure 21, with stages $t - 1, t$ and $t + 1$. We first use the network recourse decomposition method to develop an approximate, expected recourse function for stage $t + 1$, producing the functions in Figure 21b. (If the network is the simple transportation problem shown in the figure, then stage $t + 1$ is separable, and the resulting expected recourse function is exact). These convex, piecewise linear functions can be represented as a series of links (the recourse links) out of each node in stage $t + 1$, as shown in Figure 21c.

The next step is to find the expected recourse function for stage t. This is a transshipment network, where flows from different nodes in stage t must move over the same recourse links representing stage $t + 1$. We apply the network recourse decomposition method to these links, effectively unbundling the flows from different nodes within stage t, as illustrated in Figure 21d. The result is a set of trees, one for each node in stage 1, which exhibit stochastic upper bounds on the links in stage t, plus the deterministic recourse links which represent stage $t + 1$. Lagrange multipliers are introduced for each recourse link, and these

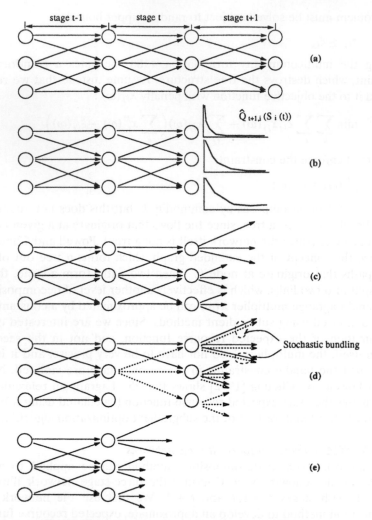

Fig. 21. The successive convex approximation methodology.

are optimized using subgradient optimization. This in turn results in a separable approximation of the recourse function for stage t, which is again modeled as a set of recourse links for each node, as shown in Figure 21e.

Thus, the SCAM methodology uses a backward recursion where for each stage, we find an approximation of the expected recourse function for a particular stage t which includes the recourse links for stage $t + 1$. In contrast with the SLAP methodology, which linearizes the recourse function for stage $t + 1$, we are able to use a convex approximation. Although the computational costs for SCAM are somewhat higher, it consistently outperforms the SLAP approach in rolling horizon experiments by one to five percent. Both of these methods, in turn, consistently outperform deterministic models.

10. Evaluating dynamic models

As this chapter has illustrated, there are often a number of ways to formulate and solve dynamic models. A rich set of approximations have evolved to handle forecasting uncertainties and model truncation in a tractable way. However, this leaves the problem of evaluating the quality of the solution that is provided. Static models, regardless of their complexity, typically enjoy the feature of a single, well-defined cost function that can be evaluated even if it is extremely difficult to find the minimum. Two heuristics can be compared by simply comparing the objective function values produced by each method. By contrast, the objective function of a dynamic model is often of little intrinsic interest, since the more relevant measure of performance is the costs produced over time through repeated applications of the model.

Dynamic models are generally applied on a rolling horizon basis, producing a stream of costs that must be evaluated over time. Two questions arise. First, how well are we optimizing the dynamic model over the chosen planning horizon? And second, how well do the recommendations of the dynamic model perform when evaluated over time? Four broad areas of research provide insight into these issues, although overall, relatively little progress has been made for evaluating the results of dynamic network models of realistic size. The first area of research addresses the first question by deriving upper and lower bounds on the expected recourse function, which helps answer the question of how well we are optimizing the dynamic model. The next three lines of investigation consider the quality of the recommendations when evaluated over time. These include the use of simulation experiments; the use of posterior bounds; and rigorous bounds for rolling horizon results. In this section, we briefly review the basic ideas behind each approach, and summarize the state of the art of research in each area.

10.1. Rolling horizon simulations

Consider a rolling horizon procedure $\mathcal{P}$, which might consist of an n-stage stochastic program, or an n-period deterministic dynamic network model. When applied in time period τ, the procedure will span time periods $t = \tau$ up to $t = \tau + P$. The procedure will operate on a *realization* of the actual data in time period τ, denoted by outcome ω_τ, and *forecasted* data for periods $\tau + 1$ up to $\tau + n$. Let the *actions* recommended by $\mathcal{P}$ in time period τ be denoted by $x_\tau(\mathcal{P}, \omega_\tau, S_\tau)$ where S_τ is the state of the system at time τ. S_τ is assumed to capture the history of the process up to time τ. A *simulation* is comprised of the sequence of outcomes and actions $\omega_1, x_1(\mathcal{P}, \omega_1, S_1), \omega_2, x_2(\mathcal{P}, \omega_2, S_2), \ldots, \omega_N, x_N(\mathcal{P}, \omega_N, S_N)$. This simulation reflects a single realization of the complete set of outcomes $\omega = \{\omega_1, \ldots, \omega_N\}$, and yields total costs:

$$C(\mathcal{P}, \omega, N) = \sum_{\tau=1}^{N} c^T x_\tau(\mathcal{P}, \omega_\tau, S_\tau)$$

Of course, $C(\mathcal{P}, \omega, N)$ is a random variable. We would like to compare two rolling horizon procedures $\mathcal{P}_1$ and $\mathcal{P}_2$ on the basis of these costs, but this requires developing an understanding of the distribution of $C(\mathcal{P}, \omega, N)$. We can reduce the variance of $C(\mathcal{P}, \omega, N)$ by increasing the simulation length or by taking a sample of n simulations, and calculating a sample average cost:

$$\bar{C}(\mathcal{P}, N) = \frac{1}{n} \sum_{i=1}^{n} C(\mathcal{P}, \omega_i, N)$$

This reduces, but does not eliminate uncertainty in the estimate of the costs.

Testing a procedure $\mathcal{P}$ in a rolling horizon experiment is the most direct method for evaluating a model. Many dynamic models are fairly robust and yield good results despite fairly coarse approximations of future events. However, they can be computationally very burdensome, and while they can provide an indication if one procedure is better than another, they provide little absolute measure of the effectiveness of either procedure (perhaps they are both poor). Finally, if a procedure yields poor results, rolling horizon simulations yield few insights with respect to how the procedure should be modified.

10.2. Bounds for stochastic programs with network recourse

In sharp contrast with the other lines of investigation, the literature on bounds for stochastic programs is extensive. The basic program considered is as follows. Consider the two-stage stochastic program:

$$\min_{x} F(x) = c^T x + \bar{Q}(x) \tag{189}$$

where $\bar{Q}(x)$ is the usual expected recourse function given by

$$\bar{Q}(x) = E_{\omega}\{Q(x, \omega)\} \tag{190}$$

where $Q(x, \omega)$ is an imbedded optimization problem. In Section 7, we considered methods for optimizing (189) which recognizes the fact that we cannot write $\bar{Q}(x)$ analytically. However, for practical problems, we cannot even calculate $\bar{Q}(x)$ for a given value of x, since the expectation in (190) is generally too difficult. As a result, given two solutions $x^{(1)}$ and $x^{(2)}$, that may be produced by two different heuristics, we are unable to say, precisely, whether $F(x^{(1)}) > F(x^{(2)})$ or $F(x^{(1)}) < F(x^{(2)})$. Note that we are not trying to optimize $F(x)$ at this point, but rather are simply trying to estimate the objective function of the dynamic model.

To address this problem, a large literature has evolved over the years on the problem of developing rigorous bounds on the expected recourse function. An excellent review of this work is given in Kall & Stoyan [1982] and Birge & Wets [1986]. The basic approach to finding a lower bound of the recourse function has been to apply Jensen's inequality which states that, for a convex function $F(\xi)$ of a random variable ξ,

$$E_{\xi}[F(\xi)] \geq F(\bar{\xi})$$

Kall & Stoyan [1982] show that Jensen's inequality can be applied to stochastic programs with recourse as long as a) the recourse function is convex in the second stage actions x_1 and the random variable ξ, for a given first stage action x_0, and b) the random variable ξ does not appear in the objective function. The appeal of Jensen's inequality is that it is so easy to apply, since it involves solving the recourse function for a single realization of the random variables. Furthermore, it is relatively straightforward to apply Jensen's inequality to multistage problems. For example, Frantzeskakis and Powell [1989b] show that when applied to an N-stage network with random arc capacities, Jensen's inequality involves solving an N-period dynamic network with upper bounds equal to the expected arc capacities.

A novel approach for bounding the recourse function for networks is proposed by Wallace [1987]. In contrast with methods that approximate the probability measure (such as the Edmundson–Madansky bound) or the recourse function (the ray approximation of Birge and Wets), Wallace's approach approximates the problem itself. Consider the second stage of a two stage stochastic network with recourse with random supplies α_i, random demands β_j, and random arc capacities ξ_{ij}. Wallace's bound takes the following form:

$$Q(\alpha, \beta, \xi) \leq P(\alpha, \beta, \xi) = \rho + \sum_i f_i(\alpha_i) + \sum_j g_j(\beta_j) + h(\xi)$$

These calculations assume a fixed first stage vector x_0. $P(\alpha, \beta, \xi)$ is an upper bound on the recourse function which is constructed to be separable in the random supplies, demands and arc capacities. ρ is a constant found using $\rho = Q(\alpha^0, \beta^0, \xi^0)$ where $(\bar{\alpha}, \bar{\beta}, \bar{\xi})$ represents a base point. We might use, for example, $(\bar{\alpha}, \bar{\beta}, \bar{\xi}) = (0, 0, 0)$ or $(E\{\alpha\}, E\{\beta\}, E\{\xi\})$. Let $\bar{x}^1$ be the flows resulting from this base point (thus, $\rho = c^T \bar{x}^1$). Now let z_α^1, z_β^1 and z_ξ^1 be random perturbations from $\bar{x}^1$ resulting from specific realizations of α, β and ξ. Assume $(\bar{\alpha}, \bar{\beta}, \bar{\xi}) = (0, 0, 0)$. Increasing realizations of α_i are captured by placing flow on successively longer (more costly) paths from node i to the root (or slack) node of the graph. Realizations of the arc capacities are handled by putting flow on a predetermined set of cycles which are ordered from least to most cost. As much flow as possible is put on the highest ranked/least cost cycle. Flow that cannot fit on the highest ranked cycle spills to lower ranked cycles. To simplify taking expectations, Wallace describes a method for splitting links that are shared by more than one cycle (since the amount of flow that can fit on one cycle is random, the amount of unused capacity that might be available for another cycle that shares the same link is also random with a potentially complex probabilistic structure). This splitting process produces an upper bound since it restricts flow on one cycle from using capacity that may have been allocated to a higher ranked cycle (but which was not fully utilized). This concept is extended in Frantzeskakis and Powell [1989a] which presents much tighter bounds for the same class of problems.

Frantzeskakis and Powell [1989b] use the SLAP concept of Section 9.3 to develop rigorous upper and lower bounds for large-scale dynamic networks with random arc capacities. Lower bounds are constructed by finding a linear support of the recourse function around some point $\hat{x}$. An upper bound is established by drawing a line between the origin and the recourse function evaluated at the maximum feasible value of x. Both of these upper and lower bounds represent linear approximations that can be used within the SLAP methodology. Experimental testing demonstrated that the SLAP upper and lower bound, as well as Jensen's bound, can be calculated very efficiently. However, the bounds are not very close. In one experiment, the upper and lower bounds were found to be as much as 50% higher and lower than the best available estimate of the recourse function (obtained using Monte Carlo methods). More promising is the network recourse decomposition method in Section 9.4.2 where subgradient optimization is used to refine the Lagrange multipliers, but this has not been tested.

10.3. Posterior bounds for rolling horizon procedures

One approach for obtaining an absolute measure of the effectiveness of a RHP is to use a device known as a posterior bound. A RHP must choose a set of actions x_τ at time τ based on the state of the system S_τ and the outcome ω_τ, but without knowledge of future events. Assume now that we know the entire set of outcomes $\omega = \{\omega_1, \ldots, \omega_N\}$ over the entire simulation. Rather than solving a sequence of N minimization problems, we could solve a single minimization problem over all N periods. Let

$$x^{(p)}(\omega, N) = \left\{ x_1^{(p)}(\omega, N), x_2^{(p)}(\omega, N), \ldots, x_N^{(p)}(\omega, N) \right\} \tag{191}$$

be the set of activities resulting from this global, posterior optimization with cost:

$$C^{(p)}(\omega, N) = \sum_{t=1}^{N} c^T x_t^{(p)}(\omega, N) \tag{192}$$

Since this posterior optimization can anticipate future events, it will produce lower overall costs, and thus:

$$C^{(p)}(\omega, N) \le C(\mathcal{P}, \omega, N) \tag{193}$$

giving us our posterior bound on the costs for a simulation. Let $C(\mathcal{P}^*, \omega, N)$ be the optimal achievable (nonanticipatory) costs found from using the best possible rolling horizon procedure $\mathcal{P}^*$. By definition

$$C^{(p)}(\omega, N) \le C(\mathcal{P}^*, \omega, N) \le C(\mathcal{P}, \omega, N) \tag{194}$$

Of course, we will never know $C(\mathcal{P}^*, \omega, N)$. However, we conjecture that for large, real-world problems, $C^{(p)}(\omega, N)$ and $C(\mathcal{P}^*, \omega, N)$ may be surprisingly close (at least within 10%). If this is the case, the gap between the posterior bound $C^{\mathcal{P}}(\omega, N)$ and the results of a given rolling horizon procedure $\mathcal{P}$ may provide an indication of the absolute effectiveness of a particular dynamic model.

10.4. Error bounds for rolling horizon procedures

It would be especially desirable to have a rigorous bound on the error produced by a finite horizon RHP over the results produced by an optimal, infinite horizon procedure. Alden [1987] and Alden & Smith [1987] derive error bounds for RHP's and demonstrate their use in the context of a particular version of the dynamic vehicle allocation problem. Let $S_i(t)$ denote the number of vehicles in city i at time t, and let $S_t = \{\ldots, S_{it}, \ldots\}$represent the vector of vehicle supplies. S_t represents the state of the system at time t. Each day, a certain number of vehicles will move loaded from one region to the next, after which the remaining vehicles may be repositioned empty to handle excess supplies of vehicles in other regions. Let π_s denote a *policy* for repositioning vehicles given that the system is in state s, and let $\mathcal{P}(\pi)$ be the one-step transition matrix using policy π.

Alden & Smith [1987] derive error bounds for a finite-stage RHP in terms of the *Doeblin coefficient* of the one-step transition matrix. If Q is a stochastic matrix with element Q_{ij}, the Doeblin coefficient β_D is defined as [Seneta, 1981]:

$$\beta_D = 1 - \sum_{j=1}^{n} \min_i Q_{ij} \tag{195}$$

If Q has at least one zero in each column, then $\beta_D = 1$. Alternatively, if Q is a stable matrix (all rows equal) then $\beta_D = 0$. β_D is a measure of the ergodicity of the transition matrix. Now define the *coupling coefficient* β_C as:

$$\beta_C = 1 - \min_\pi \sum_{j=1}^{n} \min_i Q_{ij}(\pi)$$

β_C is the largest possible Doeblin coefficient over all policies π.

Finally, let:

α = discount factor per period.

P = length of the planning horizon in the RHP.

N = number of periods being simulated.

C = maximum cost of any state transition.

$\bar{C}(\mathcal{P}^*, N) = E_\omega[C(\mathcal{P}^*, \omega, N)].$

$\bar{C}(\mathcal{P}, N) = E_\omega[C(\mathcal{P}, \omega, N)]$

The bound derived by Alden & Smith [1987] is given by:

$$\bar{C}(\mathcal{P}, N) - \bar{C}(\mathcal{P}^*, N) \leq \frac{(\alpha\beta_c)^T}{1 - \alpha\beta_c} \left[1 + \frac{\alpha(1 - \beta_c)(1 - \alpha^{N-T-1})}{1 - \alpha} \right] C \tag{196}$$

Note that as long as $\beta_c > 0$, we have a valid bound even if there is no discounting ($\alpha = 1$).

The Alden–Smith bound is significant primarily because it is the only bound for rolling horizon procedures (actually it extends an earlier result by Bean & Smith [1984]) and because it captures the effect of ergodicity on the choice of planning horizons. The concept has also been applied to the dynamic vehicle allocation problem, and in fact presents a fresh perspective on this problem. The limitation of the result is that the bound is likely to be very weak. For large problems, the coupling coefficient β_c is likely to be quite large, since it will often be the case that some states cannot be reached from at least one other state (however, steps can be taken to reduce this problem). Also, we have the classic curse of dimensionality which produces an exponentially large number of states, making the calculation of β_c difficult. These issues notwithstanding, the result is an important milestone in this line of research.

Acknowledgement

This research was supported in part by grant DDM-9102134 from the National Science Foundation, and by grant AFOSR-F49620-93-1-0098 from the Air Force Office of Scientific Research.

References

Agnihothri, U.K.S., and P. Kubat (1982), Stochastic allocation rules, *Oper. Res.* 30(3), 545–555.

Ahuja, R., T. Magnanti, and J. Orlin (1992), *Network Flows: Theory, Algorithms and Applications*, Prentice Hall, New York.

Alden, J. (1987), *Error Bounds for Rolling Horizon Procedures*, Ph.D. dissertation, Dept. of Industrial and Operations Engineering, University of Michigan.

Alden, J., and R. Smith (1992), Rolling horizon procedures in nonhomogeneous Markov decision processes. *Oper. Res.* 40(Supp. 2), 183–194.

Andreatta, G. (1987), Shortest path models in stochastic networks, in: G. Andreatta, F. Mason, and P. Serafini, eds., *Stochastics in Combinatorial Optimization*, World Scientific, Singapore.

Andreatta, G., F. Ricaldone, and L. Romeo (1985), Exploring stochastic shortest path problems, in: *ATTI Giornate di Lavoro*, Tecnoprint, Bologna.

Andreatta, G., and G. Romanin-Jacur (1987), The flow management problem, *Transp. Sci.* 21(4), 249–253.

Andreatta, G., and L. Romeo (1988), Stochastic shortest paths with recourse, *Networks* 18, 193–204.

Arabeyre, J., J. Fearnley, F. Steiger, and W. Teather (1969), The airline crew scheduling problem: A survey, *Transp. Sci.* 3, 140–163.

Aronson, J. (1980), *Forward Linear Programming*, Ph.D. dissertation, Grad. School of Industrial Administration, Carnegie–Mellon University.

Aronson, J. (1989), A survey of dynamic network flows, in: P. Hammer, ed., *Ann. Oper. Res.*, pp. 1–66. J.C. Baltzer AG.

Aronson, J. (1990), Private communication.

Aronson, J., and B. Chen (1986), A forward network simplex algorithm for solving multiperiod network flow problems, *Nav. Res. Logistics Q.* 33(3), 445–467.

Aronson, J., and B. Chen (1989), A computational study of emperical decision horizons in infinite horizon multiperiod network flow models, Working paper 89-275, Dept. of Mgmt. Sci. and Information Tech., University of Georgia.

Aronson, J., and G.L. Thompson (1984), A survey on forward methods in mathematical programming. *Large Scale Systems* 7, 1–16.

Assad, A. (1987), Multicommodity network flows: A survey, *Networks* 8(1), 37–92.

Assad, A., E. Wasil, and G. Lilien (1992), *Excellence in Management Science: A Readings Book*, Prentice Hall, New York.

Avram, J., and D. Bertsimas (1992), The minimum spanning tree constant in geometrical probability and under the independent model; an unified approach, *Ann. Appl. Probab.* 2, 113–130.

Bagchi, B., and B. Nag (1991), Dynamic vehicle scheduling: An expert systems approach, *Int. J. Phys. Distrib. Logist. Manage.* 21(2), 10–18.

Bahl, H., L. Ritzman, and J. Gupta (1987), Determining lot sizes and resource requirements: A review, *Oper. Res.* 35(3), 329–345.

Ball, M., L. Bodin, and R. Dial (1983), A matching based heuristic for scheduling mass transit crews and vehicles, *Trans. Sci.* 17, 4–31.

Ball, M., and A. Roberts (1985), A graph partitioning approach to airline crew scheduling, *Trans. Sci* 19, 107–126.

Bartholdi, J., and L. Platzman (1982), An $O(n \log n)$ planar traveling salesman heuristic based on spacefilling curves, *Oper. Res. Lett.* 1, 121–125.

Bartholdi, J., L. Platzman, R. Collins, Lee, and W. Warden (1983), A minimal technology routing system for meals on wheels, *Interfaces* 13, 1–8.

Batta, R., O. Berman, S. Chiu, R. Larson, and A. Odoni (1990), Discrete location theory, in: P. Mirchandani and R. Francis, eds., *Discrete Location Theory*, Wiley, New York.

Beale, E. (1955), On minimizing a convex function subject to linear inequalities, *J. R. Stat. Soc.* B17, 173–184.

Beale, E., G. Dantzig, and R. Watson (1986), A first order approach to a class of multi-time period stochastic programming problems, *Math. Program. Study* 27, 103–117.

Beale, E., J. Forest, and C. Taylor (1980), Multi-time period stochastic programming, in: M. Dempster, ed., *Stochastic Programming*, Academic Press.

Bean, J.C., and R.L. Smith (1985), Optimal capacity expansion over an infinite horizon, *Manage. Sci.* 31(12), 1523–1532.

Bean, J.C., and R.L. Smith (1987), Conditions for the existence of planning horizons, *Math. Oper. Res.* 9(12), 391–401.

Beardwood, J., J. Halton, and J. Hammersley (1959), The shortest path through many points, *Proc. Camb. Phil. Soc.* 55, 299–327.

Bell, W., L. Dalberto, M. Fisher, A. Greenfield, R. Jaikumar, P. Kedia, R. Mack, and P. Prutzman (1983), Improving the distribution of industrial gases with an online computerized routing and scheduling optimizer, *Interfaces* 13, 4–23.

Bellmore, M., W. Ecklof, and G. Nemhauser (1969), A decomposable transshipment algorithm for a multiperiod transportation problem, *Nav. Res. Logist. Q.* 16, 517–524.

Bellmore, M., and R. Vemuganti (1973), On multicommodity maximal dynamic flows, *Oper. Res.* 21, 10–21.

Bensoussan, A., and M. Crouhy, and J. Proth (1983), *Mathematical Theory of Production Planning*, North-Holland, Amsterdam.

Berman, O. (1981), Dynamic repositioning of indistinguishable service units on transportation networks, *Transp. Sci.* 15(2), 115–136.

Berman, O., and B. LeBlanc (1984), Location–relocation of mobile facilities on a stochastic network, *Transp. Sci.* 18(4), 315–330.

Berman, O., and D. Simchi-Levi (1986), Minimum location of a traveling salesman, *Networks* 16, 329–354.

Berman, O., and D. Simchi-Levi (1988), Finding the optimal a priori tour and location of a traveling salesman with non-homogeneous customers, *Transp. Sci.* 22, 148–154.

Berman, O., and D. Simchi-Levi (1988), A heuristic algorithm for the traveling salesman location problem on networks, *Oper. Res.* 36, 478–484.

Bertsekas, D. (1979), Algorithms for nonlinear multicommodity network flow problems, in: A. Bensoussan and J. Lions, eds., *International Symposium on Systems Optimization and Analysis*, pp. 210–224, Springer-Verlag, Boston.

Bertsekas, D. (1987), *Dynamic Programming: Deterministic and Stochastic Models*, Prentice Hall.

Bertsekas, D., E. Gafni, and R. Gallagher (1984), Second derivative algorithms for minimum delay distributed routing in networks, *IEEE Trans. Commun.* COM-32, 911–919.

Bertsekas, D., and R. Gallager (1987), *Data Networks*, Prentice Hall.

Bertsekas, D., and J. Tsitsiklis (1991), An analysis of stochastic shortest path problems, *Math. Oper. Res.* 16, 580–595.

Bertsimas, D. (1988), *Probabilistic combinatorial optimization problems*, Technical Report 194, Operations Research Center, MIT.

Bertsimas, D. (1989), On probabilistic traveling salesman facility location problems, *Transp. Sci.* 23, 184–191.

Bertsimas, D. (1992), A vehicle routing problem with stochastic demand, *Oper. Res.* 40, pp. 574–585.

Bertsimas, D., and M. Grigni (1989), Worst-case examples for the spacefilling curve heuristic for the Euclidean traveling salesman problem, *Oper. Res. Lett.* 5, 241–244.

Bertsimas, D.J., P. Jaillet, and A.R. Odoni, (1990), A priori optimization, *Oper. Res.* 38, 1019–1033.

Bertsimas, D.J., and G. van Ryzin (1991), A stochastic and dynamic vehicle routing problem in the Euclidean plane, *Oper. Res.* 39, 601–615.

Bertsimas, D.J., and G. van Ryzin (1993a), Stochastic and dynamic vehicle routing problem in the Euclidean plane with multiple capacitated vehicles, *Oper. Res.* 41, pp. 60–76.

Bertsimas, D.J., and G. van Ryzin (1993b), Stochastic and dynamic vehicle routing with general demand and interarrival time distributions, *Adv. Appl. Prob.* 25(4), 947–978.

Bes, C., and S. Sethi (1988), Concepts of forecast and decision horizons: Applications to dynamic stochastic optimization problems, *Math. Oper. Res.* 13(2), 295–310.

Bhaskaran, S., and S. Sethi (1988), Decision and forecast horizons in a stochastic environment: A survey, *Optimal Control Applic. Meth.* 8(1), 61–67.

Bhat, U. (1984), *Elements of Applied Stochastic Processes*, Wiley, New York.

Bielli, M., G. Calicchio, B. Micoletti, and S. Ricciardelli (1982), The air traffic flow control problem as an application of network theory, *Comput. Oper. Res.* 9(4), 265–278.

Birge, J., and R. Wets (1986), Designing approximation schemes for stochastic optimization problems, in particular for stochastic programs with recourse, *Math. Program. Study* 27, 131–149.

Birge, J.R., and J.K. Ho (1987), The stochastic dynamic traffic assignment problem, Technical Report 87-16, Dept. of Industrial and Operational Engineering, University of Michigan.

Blum, J. (1954), Approximation methods which converge with probability one, *Ann. Math. Stat.* 25, 382–386.

Bodin, L., B. Golden, A. Assad, and M. Ball (1983), Routing and scheduling of vehicles and crews, the state of the art, *Comput. Oper. Res.* 10, 69–211.

Bowman, E. (1956), Production scheduling by the transportation method of linear programming, *Oper. Res.* 4, 100–103.

Bradley, G. (1975), Survey of deterministic networks, *AIIE Transactions* 7(3), 222–234.

Brown, G., and G. Graves (1981), Real-time dispatch of petroleum tank trucks, *Manage. Sci.* 27, 19–32.

Brown, G., G. Graves, and D. Ronen (1987), Scheduling ocean transportation of crude oil, *Manage. Sci.* 33, 335–346.

Burness, R., and J. White (1976), The traveling salesman location problem, *Transp. Sci.* 10, 348–360.

Busacker, R., and P. Gowen, (1961), A procedure for determining a family of minimal cost network flow patterns, Oper. Res. Off., Tech. paper 15.

Carey, M. (1986), A constraint qualification for a dynamic traffic assignment model, *Transp. Sci.* 20(1), 55–58.

Carey, M. (1987), Optimal time-varying flows on congested networks, *Oper. Res.* 35(1), 58–69.

Chalmet, L., R. Francis, and P. Saunders (1982), Network models for building evacuation, *Manage. Sci.* 28, 86–105.

Charnes, A., and W. Cooper (1959), Chance constrained programming, *Manage. Sci.* 5(1), 73–79.

Chih, K. (1986), A real time dynamic optimal freight car management simulation model of the multiple railroad, multicommodity, temporal spatial network flow broblem, Ph.D. Dissertation, Dept. of Civil Engng. and Oper. Res., Princeton University.

Chih, K., M. Hornung, M. Rothenberg, and A. Kornhauser (1990), Implementation of a real time locomotive distribution system, in: T. Murthy, R. Rivier, G. List, and J. Mikolaj, eds., *Computer Applications in Railway Planning and Management*, pp. 39–50, Computational Mechanics Publications.

Choi, W., H. Hamacher, and S. Tufekci (1988), Modeling of building evacuation problems by network flows with side constraints, *Eur. J. Oper. Res.* 35, 98–110.

Chong, E. (1991), *On-line Stochastic Optimization of Queueing Systems*, Ph.D. dissertation, Dept. of Electrical Engng., Princeton University.

Christofides, N., and P. Brooker (1967), Optimal expansion of an existing network, *Math. Program.* 6, 197–211.

Colbourn, C. (1987), *The Combinatorics of Network Reliability*, Oxford University Press, New York.

Cooke, K., and E. Halsey (1966), The shortest route through a network with time-dependent internodal transit times, *J. Math. Anal. Appl.* 14, 493–498.

Cooper, L., and L. Leblanc (1977), Stochastic transportation problems and other network related convex problems, *Nav. Res. Logist. Q.* 24, 327–337.

Crainic, T., J. Ferland, and J. Rousseau (1984), A tactical planning model for rail freight transportation, *Transp. Sci.* 18, 165–184.

Crainic, T., M. Gendreau, and P. Dejax (1993), Dynamic stochastic models for the allocation of empty containers, *Oper. Res.* 41, 102–126.

Crainic, T., and J. Rousseau (1987), The column generation principle and the airline crew scheduling, *INFOR* 25(2), 136–151.

Crainic, T., and J. Roy (1988), Or tools for the tactical planning of freight transportation, *Eur. J. Oper. Res.* 33, pp. 290–297.

D'Ans, G., and D. Gazis (1976), Optimal control of oversaturated store and forward transportation networks, *Transp. Sci.* 10, 1–19.

Dantzig, G. (1955), Linear programming under uncertainty, *Manage. Sci.* 1, 197–206.

Dantzig, G., and A. Ferguson (1956), The allocation of aircrafts to routes: An example of linear programming under uncertain demand, *Manage. Sci.* 3, 45–73.

Dantzig, G., and D. Fulkerson (1954), Minimizing the number of tankers to meet a fixed schedule, *Nav. Res. Logist. Q.* 1, 217–222.

Dempster, M., M. Fisher, L. Jansen, B. Lageweg, J. Lenstra, and A.R. Kan (1981), Analytical evaluation of hierarchical planning systems, *Oper. Res.* 29, 707–716.

Derman, C. (1970), *Finite State Markovian Decision Processes*, Academic Press, New York.

Desrosiers., J. (1990), private communication.

Desrosiers, J., F. Soumis, and M. Desrochers (1984), Routing with time windows by column generation, *Networks* 14, 545–565.

Doulliez, P., and M.R. Rao (1984), Optimal network capacity planning: A shortest path scheme, *Oper. Res.* 23, 811–818.

Dreyfus, S. (1969), An appraisal of some shortest-path algorithms, *Oper. Res.* 17, 395–412.

Driscoll, J., H. Gabow, R. Shraiman, and R. Tarjan (1988), Relaxed heaps: An alternative to Fibonacci heaps with applications to parallel computation, *Commun. ACM* 31, 1343–1354.

Dror, M., M. Ball, and B. Golden (1985), A computational comparison of algorithms for the inventory routing problem, *Ann. Oper. Res.* 4, 3–23.

Dror, M., and P. Trudeau, (1986), Stochastic vehicle routing with modified savings algorithm, *Eur. J. Oper. Res.* 23, 228–235.

Dror, M., G. Laporte, and P. Trudeau (1989), Vehicle routing with stochastic demands: Properties and solution frameworks, *Transp. Sci.* 23, 166–176.

Eilon, B., C. Watson-Gandy, and N. Christofides (1971), *Distribution Management*, Hafner.

Ellis, R., and R. Rishel (1974), An application of stochastic optimal control theory to optimal rescheduling of airplanes, *IEEE Trans. Autom. Control* 139–147.

Erickson. R.E., C.L. Monma and A.F. Veinott (1987), Send-and-split method for minimum-concave-cost network flows, *Math. Oper. Res.*, 12(4), 634–664.

Erlenkotter, D. (1969), *Preinvestment Planning for Capacity Expansion: A Multi-Location Dynamic Model*, Ph.D. Dissertation, Graduate School of Business, Stanford University, Stanford, CA.

Erlenkotter, D. (1975), Capacity planning for large multi-location systems: Approximate and incomplete dynamic programming approaches, *Manage. Sci.* 22, 274–285.

Erlenkotter, D. (1981), A comparative study of approaches to dynamic location problems, *Eur. J. Oper. Res.* 6, 133–143.

Ermoliev, Y. (1969), On the stochastic quasigradient method and stochastic quasi-feyer sequences, *Kibernetika* 2.

Ermoliev, Y. (1988), Stochastic quasigradient methods, in: Y. Ermoliev and R. Wets, eds., *Numerical Methods in Stochastic Programming*, Springer-Verlag.

Ermoliev, Y., T. Krivets, and V. Petukhov (1976), Planning of shipping empty seaborne containers, *Cybernetics* 12, 664.

Farvolden, J., and W. Powell, (1991), Subgradient methods for the service network design problem, Dept. of Industrial Engng., University of Toronto.

Farvolden, J., W. Powell and I. Lustig (1993), A primal partitioning solution for the arc-chain formulation of a multicommodity network flow problem, *Oper. Res.* 41(4), 669–693.

Feeney, G. (1957), Controlling the distribution of empty cars, in: *Proc. 10th Nat. Meeting, Oper. Res. Soc. Am.*

Fisher, M.L., and M.B. Rosenwein (1989), An interactive optimization system for bulk-cargo ship scheduling, *Nav. Res. Logist.* 36, 27–42.

Flam, S., and R.J.-B. Wets (1987), Existence results and finite horizon approximates for infinite horizon optimization problems, *Econometrica* 55(5), 1187–1209.

Florez, H. (1986), *Empty-Container Repositioning and Leasing: An Optimization Model*, Ph.D. dissertation, Polytechnic Institute of New York.

Fong, C. (1974), *The Multi-Region Dynamic Capacity Expansion Problem: Exact and Heuristic Approaches*, Ph.D. Dissertation, Graduate School of Management, University of Rochester.

Fong, C., and V. Srinivasan (1981a), The multiregion dynamic capacity expansion problem. part II, *Oper. Res.* 29, 800–816.

Fong, C., and V. Srinivasan (1981b), The multiregion dynamic capacity expansion problem. part I, *Oper. Res.* 29, 787–799.

Fong, C., and V. Srinivasan (1986), The multiregion dynamic capacity expansion problem: An improved heuristic, *Manage. Sci.* 32(9), 1140–1152.

Fong, C., and V. Srinivisan (1964), Multiperiod capacity expansion and shipment planning with linear costs, *Nav. Res. Logist. Q.* 23(1), 255–260.

Ford, L., and D. Fulkerson (1956), Maximal flow through a network, *Can. J. Math.* 8, 399–404.

Ford, L., and D. Fulkerson (1958a), A suggested computation for maximal multicommodity networks flows, *Manage. Sci.* 5, 97–101.

Ford Jr., L., and D. Fulkerson (1958b), Constructing maximal dynamic flows from static flows, *Oper. Res.* 6, 419–433.

Ford, L., and D. Fulkerson (1962), *Flows in Networks*, Princeton University Press, Princeton.

Fox, K., B. Gavish, and S. Graves (1980), An *n*-constraint formulation of the (time dependent) traveling salesman problem, *Oper. Res.* 28, 1018–1021.

Frank, H. (1967), Dynamic communication networks, *IEEE Trans. Commun. Tech.* COM-15(2), 156–163.

Frank, H. (1969), Shortest paths in probabilistic graphs, *Oper. Res.* 17, 583–599.

Frank, H., and M. El-Bardai (1969), Dynamic communication networks with capacity constraints, *IEEE Trans. Commun. Tech.* COM-17(4), 432–437.

Frantzeskakis., L, (1990), *Dynamic networks with random arc capacities, with application to the stochastic dynamic vehicle allocation problem*, Ph.D. dissertation, Dept. of Civil Engng., Princeton University.

Frantzeskakis, L., and W. Powell (1989a), *An Improved Polynomial Bound on the Expected Recourse Function for Stochastic Networks*, Working paper–SOR-89-xx, Dept. of Civil Engng. and Oper. Res., Princeton University.

Frantzeskakis, L., and W. Powell (1989b), *Bounding procedures for dynamical networks with random link capacities, with application to stochastic programming*, Report sor-89-12, Princeton University.

Frantzeskakis, L., and W. Powell (1990), A successive linear approximation procedure for stochastic dynamic vehicle allocation problems, *Transp. Sci.* 24(1), 40–57.

Fredman, M., and R. Tarjan (1987), Fibonacci heaps and their uses in improved network optimization algorithms, *J. ACM* 34, 596–615.

Friesz, T., D. Bernstein, T. Smith, R. Tobin, and B. Wie (1993), A variational inequality formulation of the dynmamic network user equilibrium problem, *Oper. Res.* 41, 179–191.

Friesz, T., J. Luque, R. Tobin, and B. Wie, (1995), Dynamic network traffic assignment considered as a continuous time optimal control problem *Oper. Res.*, to appear.

Frieze, A., and G. Grimmet (1985), The shortest path problem for graphs with random arc lengths, *Discr. Appl. Math.* 10, 57–77.

Gale, D. (1959), Transient flows in networks, *Michigan Math. J.* 6, 59–63

Gallagher, R. (1978), A minimum delay routing algorithm using distributed computation, *IEEE Trans. Commun.* COM-25, 73–85.

Garey, M., and D. Johnson (1979), *Computers and Intractibility: A Guide to the Theory of NP-completeness*, Freeman, San Francisco.

Gavish, B. (1981), A decision support system for managing the transportation needs of a large corporation, *AIIE Trans.* 61–85.

Glasserman, P. (1991), *Gradient Estimation via Perturbation Analysis*, Kluwer, Norwell, Massachusetts.

Glickman, T., and H. Sherali, Large-scale network distribution of pooled empty freight cars over time, with limited substitution and equitable benefits, *Trans. Res.*, 19, 85–94.

Glover, F., G. Jones, D. Karney, D. Klingman, and J. Mote (1979), An integrated production, distribution and inventory planning system, *Interfaces* 9(5), 72–86.

Golden, B., and A. Assad, eds. (1988), *Vehicle Routing: Methods and Studies*, Studies in Management Science and Systems, Vol. 16, North-Holland, Amsterdam.

Golden, B., and J. Yee (1979), A framework for probabilistic vehicle routing, *AIIE Trans.* 11, 109–112.

Goldman, A. (1974), Optimal centre location in simple networks, *Transp. Sci.* 5, 212–221.

Gorenstein, S., S. Poley, and W. White (1971), *On the Scheduling of the Railroad Freight Operations*, Tech. Report 320-2999, IBM Philidelphia Scientific Center, IBM.

Graves, S.C., and J.B. Orlin (1985), A minimum concave-cost dynamic network flow problem with an application to lot-sizing, *Networks* 15, 59–71.

Grinold, R. (1977), Finite horizon approximations of infinite horizon linear programs, *Math. Program.* 12, 1–17.

Grinold, R. (1983a), Convex infinite horizon programs, *Math. Program.* 25, 64–82.

Grinold, R. (1983b), Model building techniques for the correction of end effects in multistage convex programs, *Oper. Res.* 31(3), 407–431.

Grinold, R., and D. Hopkins (1973), Computing optimal solutions for infinite-horizon mathematical programs with a transient stage, *Oper. Res.* 21, 179–187.

Gupal, A., and L. Bajenov (1972), Stochastic linearization, *Kibernetika*, 1.

Haghani, A. (1989), Formulation and solution of a combined train routing and makeup, and empty car distribution model, *Transp. Res.* 23B(6), 433–452.

Haimovich, M., A.R. Kan, and L. Stougie (1988), Analysis of heuristics for vehicle routing problems, in: B. Golden and A. Assad, eds., *Vehicle Routing: Methods and Studies*, North-Holland, Amsterdam.

Hakimi, S. (1964), Optimum location of switching centres and the absolute centres and medians of a graph, *Oper. Res.* 12, 450–459.

Halpern, J. (1977), Shortest path with time dependent length of edges and limited delay possibilities in nodes, *Z. Oper. Res.* 21, 117–124.

Hamacher, H., and S. Tufekci (1987), On the use of lexicographic-minimum cost flows in evacuation modeling, *Nav. Res. Logist.* 34, 487–503.

Hammer, P. (1969), Time-minimizing transportation problems, *Nav. Res. Logist. Q.* 16, 345–357.

Hammer, P. (1971), Communication on 'the bottleneck problem' and 'some remarks on the time transportation problem', *Nav. Res. Logist. Q.* 18, 487–490.

Hassin, R., and E. Zemel (1985), On shortest paths in graphs with random weights, *Math. Oper. Res.* 10, 557–564.

Hausman, W., and P. Gilmour (1967), A multiperiod truck delivery problem, *Transp. Res.* 1(4), 349–357.

Herren, H. (1973), The distribution of empty wagons by means of computer: An analytical model for the Swiss Federal Railways (SSB), *Rail Int.* 4(1), 1005–1010.

Herren, H. (1977), Computer controlled empty wagon distribution on the SSB, *Rail Int.* 8(1), 25–32.

Heyman, D., and M. Sobel (1984), *Stochastic Models in Operations Research, Volume II: Stochastic Optimization*, McGraw Hill, New York.

Higle, J.L., J. Bean, and R.L. Smith (1984), *Capacity Expansion Under Stochastic Demands*, Tech. Report 84-28, Dept. of Systems and Industrial Engng., University of Arizona.

Ho, J. (1980), A successive linear optimization approach to the dynamic traffic assignment problem, *Transp. Sci.* 14(4), 295–305.

Ho, J., and E. Loute (1983), Computational experience with advanced implementation of decomposition algorithms, *Math. Prog.* 27(3), 283–290.

Ho, J.K., and A.S. Manne (1974), Nested decomposition for dynamic models, in: E.A. Ho, ed., *Mathematical Programming 6*, North-Holland, Amsterdam.

Ho, Y.-C., M. Eyler, and T. Chien (1979), A gradient technique for general buffer storage design in a serial production line, *Int. J. Prod. Res.* 7(6), 557–580.

Hopkins, D. (1971), Infinite horizon optimality in an equipment replacement capacity expansion model, *Manage. Sci.* 18, 145–156.

Howard, R. (1971), *Dynamic Probabilistic Systems, Volume II: Semimarkov and Decision Processes*, Wiley, New York.

Hughes, R., and W. Powell (1988), Mitigating end effects in the dynamic vehicle allocation model, *Manage. Sci.* 34(7), 859–879.

Jaillet, P. (1985), *The Probabilistic Traveling Salesman Problems*, Tech. Report 185, Operations Research Center, MIT.

Jaillet, P. (1987), Stochastic routing problems, in: G. Andreatta, F. Mason, and P. Serafini, eds., *Stochastics in Combinatorial Optimization*, World Scientific, Singapore.

Jaillet, P. (1988), A priori solution of a traveling salesman problem in which a random subset of the customers are visited, *Oper. Res.* 36, 929–936.

Jaillet, P. (1991a), On some probabilistic combinatorial optimization problems defined on graphs, in: A. Odoni, L. Bianco, and G.S. "o, eds., *Flow Control of Congested Network*, Springer Verlag, Berlin.

Jaillet, P. (1991b), Probabilistic routing problems in the plane, in: H. Bradley, ed., *Operational Research 90*, Pergamon Press, London.

Jaillet, P. (1992a), Rates of convergence for quasi-additive smooth Euclidean functionals and application to combinatorial optimization problems, *Math. Oper. Res.* 17, 965–980.

Jaillet, P. (1992b), Shortest path problems with nodes failures, *Networks* 22, 589–605.

Jaillet, P. (1992c), Analysis of combinatorial optimization problems in Euclidean spaces, *Math. of Oper. Res.* 17, pp. 965–980.

Jaillet, P. (1993), Cube versus torus models for combinatorial optimization problems and the Euclidean minimum spanning tree constant, *Ann. Appl. Probab.* 3, pp. 582–592.

Jaillet, P., and A. Odoni (1988), Probabilistic vehicle routing problems, in: B. Golden and A. Assad, eds., *Vehicle Routing: Methods and Studies*, North-Holland, Amsterdam.

Jarvis, J., and H. Ratliff (1982), Some equivalent objectives for dynamic network flow problems, *Manage. Sci.* 28, 106–109.

Bookbinder, J.H., and S. Sethi (1980), The dynamic transportation problem: A survey, *Nav. Res. Logist. Q.* 27, 447–452.

Jones., K, (1992), *Multicommodity Network Flows: The Impact of Formulation on Decomposition*, Ph.D. dissertation, Dept of Civil Engng. and Oper. Res., Princeton University.

Jones, K., I. Lustig, J. Farvolden, and W. Powell (1993), Multicommodity network flows: The impact of formulation on decomposition, *Math. Program.* 62, pp. 65–117.

Jordan, W., and M. Turnquist (1983), A stochastic dynamic network model for railroad car distribution, *Transp. Sci.* 17, 123–145.

Kall, P., and D. Stoyan (1982), Solving stochastic programming problems with recourse including error bounds, *Math. Operationsforsch. Stat., Ser. Optimization*, 13(3), 431–447.

Karbowicz, C., and J. MacGregorSmith, (1984), A *k*-shortest path routing heuristic for stochastic evacuation networks, *Engng. Optimization*, 7, pp. 253–280.

Karmarkar, U. (1979), Convex/stochastic programming and multilocal inventory problems, *Nav. Res. Logist. Q.* 26(1), 1–19.

Karmarkar, U. (1981a), The multiperiod, multilocation invertory problem, *Oper. Res.* 29(2), 215–228.

Karmarkar, U. (1981b), Policy structure in multistage production/inventory problems: An application of convex analysis, *TIMS Stud. Manage. Sci.* 16, 331–352.

Karmarkar, U. (1987), The multilocation, multiperiod inventory problem: Bounds and approximations. *Manage. Sci.* 33(1), 86–94.

Karp, R. (1977), Probabilistic analysis of partitioning algorithms for the traveling salesman in the plane, *Math. Oper. Res.* 2, 209–224.

Karp, R. (1979), A patching algorithm for the non-symmetric traveling salesman problem, *SIAM J. Comput.* 8, 561–573.

Kaufman, D., and R.L. Smith (1993), Fastest paths in time-dependent networks for intelligent vehicle/highway systems application, *IVHS J.* 1, 1–11.

Kennington, J., and R. Helgason (1980), *Algorithms for Network Programming*, Wiley, New York.

Kiefer, J., and J. Wolfowitz (1952), Stochastic estimation of the maximum of a regression function, *Ann. Math. Stat.*, 23, 462–466.

Klingman, D., and J. Mote (1982), A multi-period production, distribution and inventory planning model, *Adv. Manage. Stud.* 1(2), 56–76.

Kornhauser, A., and V. Maslanka (1989), *A Scenario Aggregation Approach to the Dynamic Vehicle Allocation Problem Under Uncertainty*, Dept. of Civil Engng. and Oper. Res., Princeton University.

Kornhauser, A.L., and C. Chen (1990), *A Stochastic Multicommodity Network Optimization Model for the Dynamic Vehicle Allocation Problem*, Tech. Report, Dept. of Civil Engng. and Oper. Res., Princeton University.

Kraay, D., P. Harker, and B. Chen (1991), Optimal pacing of trains in freight railroads: Model formulation and solution, *Oper. Res.* 39(1), 82–99.

Laporte, G. (1988), Location-routing problems, in: B. Golden and A. Assad, eds., *Vehicle Routing: Methods and Studies*, North-Holland, Amsterdam.

Laporte, G., and P.J. Dejax (1989), Dynamic location-routing problems, *J. Opl. Res. Soc* 40(5), 471–482.

Laporte, G., and F. Louveaux (1990), Formulations and bounds for the stochastic capacitated vehicle routing problem with uncertain supplies, in: J. Gabzewicz, J. Richard, and L. Wolsey, eds., *Economic Decision-Making: Games, Econometrics and Optimization*, North-Holland, Amsterdam.

Laporte, G., F. Louveaux, and H. Mercure (1989), Models and exact solutions for a class of stochastic location-routing problems, *Eur. J. Oper. Res.* 39, 71–78.

Laporte, G., F. Louveaux, and H. Mercure (1992), The vehicle routing problem with stochastic travel times, *Transp. Sci.* 26(3), 161–170.

Larson, R., and A. Odoni (1981), *Urban Operations Research*, Prentice Hall, New Jersey.

Lasdon, L.S., and R. Terjung (1971), An efficient algorithm for multi-item scheduling, *Oper. Res.* 19(4), 946–969.

Lawler, E., A.R.K.J. Lenstra, and D. Shmoys, eds. (1985), *The Traveling Salesman Problem: A Guided Tour of Combinatorial Optimization*, Wiley, Chichester.

Leddon, C., and E. Wrathall (1967), Scheduling empty freight car fleets on the louisville and nashville railroad, in: *Second International Symposium on the Use of Cybernetics on the Railways, October*, pp. 1–6, Montreal, Canada.

Lin, S. (1965), Computer solutions of the traveling salesman problem, *Bell System Tech. J.* 2245–2269.

Lundin, R.A., and T.E. Morton (1975), Planning horizons for the dynamic lot size model: Zabel vs. protective procedures and computational results, *Oper. Res.* 23(4), 711–734.

Lustig, I., J. Mulvey, and T. Carpenter, (1991), Formulating stochastic programs for interior point methods, *Oper. Res.*, 39(5), 757–770.

MacGregor Smith, J., and D. Towsley, (1981), The use of queueing networks in the evaluation of egress from buildings, *Environ. Planning*, Vol. B8, pp. 125–139.

MacGregor Smith, J. (1993), Multi-objective routing in stochastic evacuation networks, in: D.-Z. Du and P. Pardalos, eds., *Network Optimization Problems*, pp. 263–281, World Scientific.

Magnanti, T., and R. Simpson (1978), *Transportation Network Analysis and Decomposition Methods*, Report No. DOT-TSC-RSPD-78-6, U.S. Dept. of Transportation.

Malandrak, C., and M.S. Daskin (1992a), Time dependent vehicle routing problems: Formulations, properties and heuristic algorithms, *Transp. Sci.* 26(3), 185–200.

Malandrak, C., and R.B. Dial (1992b), *A Dynamic Programming Heuristic Algorithm for the Time Dependent Traveling Salesman Problem*, Unpublished tech. report, Roadnet Technologies, Inc.

Malandraki, C. (1989), *Time Dependent Vehicle Routing Problems: Formulations, Solution Algorithms and Computational Experiments*, Ph.D. Dissertation, Department of Civil Engineering, Northwestern University.

Marsten, R., and F. Shepardson (1981), Exact solution of crew scheduling problems using the set partitioning model: Recent successful applications, *Networks* 11, 167–177.

Mendiratta., V, (1981), *A Dynamic Optimization Model of the Empty Car Distribution Process*, Ph.D. Dissertation, Dept. of Civil Engng., Northwestern University.

Mendiratta, V., and M. Turnquist (1982), A model for the management of empty freight cars, *Trans. Res. Rec.* 838, 50–55.

Merchant, D., and G. Nemhauser (1978a), A model and an algorithm for the dynamic traffic assignment problem, *Transp. Sci.* 12(3), 200–207.

Merchant, D., and G. Nemhauser (1978b), Optimality conditions for a dynamic traffic assignment model, *Transp. Sci.* 12(3), 183–199.

Minieka, E. (1973), Maximal, lexicographic and dynamic network flows, *Oper. Res.* 12(2), 517–527.

Minieka, E. (1974), Dynamic network flows with arc changes, *Networks* 4, 255–265.

Minoux, M. (1987), Network synthesis and dynamic network optimization, *Ann. Discr. Math.* 31, 283–324.

Mirchandani, P., and R. Francis, eds. (1990), *Discrete Location Theory*, Wiley, New York.

Misra, S. (1972), Linear programming of empty wagon disposition, *Rail Int.* 3, 151–158.

Morton, T. (1979), Infinite horizon dynamic programming models: A planning horizon formulation, *Oper. Res.* 27(4), 730–742.

Morton, T. (1981), Forward algorithms for forward thinking managers, in: R. Schultz, ed., *Applications of Management Science*.

Morton, T., and W. Wecker (1977), Discounting, ergodicity, and convergence for Markov decision processes, *Manage. Sci.* 23, 890–900.

Mulvey, J., and H. Vladimirou (1989), Solving Multistage Stochastic Networks: An Application of Scenario Aggregation, SOR-88-1, *Statistics and Operations Research Series*.

Mulvey, J., and S. Zenios (1987), *GENOS 1.0 User's Guide: A Generalized Network Optimization System*, Tech. Rep. 87-12-03, Princeton University.

Mulvey, J., and S. Zenios (1987), Real-time operational planning for the U.S. air traffic system, *Appl. Num. Math.* 3, 427–441.

Murphy, F., and A. Soyster (1986), End effects in capacity expansion models with finited horizons, *Nav. Res. Logist. Q.* 33, 373–383.

Odoni, A. (1986), The flow management problem in air traffic control, in: L.B.A.R. Odoni and G. Szego, eds., *Flow Control of Congested Networks*, pp. 269–288, Springer-Verlag, New York.

Odoni, A. (1987), Stochastic facility location problems, in: G. Andreatta, F. Mason, and P. Serafini, eds., *Stochastics in Combinatorial Optimization*, World Scientific, Singapore.

Orda, A., and R. Rom (1990), Shortest-path and minimum-delay algorithms in networks with time-dependent edge-length, *J. Assoc. Comput. Machin.* 37, 607–625.

Orlin, J. (1983), Maximum-throughput dynamic network flows, *Math. Program.* 27(2), 214–231.

Orlin, J. (1984a), Minimum convex cost dynamic network flows, *Math. Oper. Res.* 9(2), 190–207.

Orlin, J. (1984b), Some problems on dynamic/periodic graphs, *Prog. Combinat. Optim.* pp. 273–293.

Papadimitriou, C., and K. Steiglitz (1982), *Combinatorial Optimization: Algorithms and Complexity*, Prentice Hall, New Jersey.

Perakis, A., and N. Papadakis (1989), Minimal time vessel routing in a time-dependent environment, *Transp. Sci.* 23(4), 266–276.

Picard, J., and M. Queyranne (1978), The time dependent travelling salesman problem and its application to the tardiness problem in one-machine scheduling, *Oper. Res.* 26(2), 86–110.

Platzman, L., and J. Bartholdi (1989), Spacefilling curves and the planar traveling salesman problem, *J. ACM* 36, 719–737.

Powell, W. (1986), A stochastic model of the dynamic vehicle allocation problem, *Transp. Sci.* 20, 117–129.

Powell, W. (1987), An operational planning model for the dynamic vehicle allocation problem with uncertain demands, *Transp. Res.* 21B, 217–232.

Powell, W. (1988), A comparative review of alternative algorithms for the dynamic vehicle allocation problem with uncertain demands, *Vehicle Routing: Methods and Studies*, pp. 249–292.

Powell, W. (1991), Optimization models and algorithms: An emerging technology for the motor carrier industry, *IEEE Trans. Vehicular Technol.* 40, 68–80.

Powell, W., E. Berkkam, and I. Lustig (1993), On algorithms for nonlinear dynamic networks, in: D. Du and P. Pardalos, eds., *Network Optimization Problems: Algorithmis, Complexity and Applications*, pp. 203–231, World Scientific, New Jersey.

Powell, W., and R. Cheung (1994a), Stochastic programs over trees with random arc capacities, *Networks* 24, 161–175.

Powell, W., and R. Cheung (1994b), A network recourse decomposition method for dynamic networks with random arc capacities, *Networks* 24, 369–384.

Powell, W., and L. Frantzeskakis (1992), Restricted recourse strategies for stochastic, dynamic networks, *Transp. Sci.* (to appear).

Powell, W., and Y. Sheffi (1982), The convergence of equilibrium algorithms with predetermined step sizes, *Transp. Sci.* 16, 45–55.

Powell, W., and Y. Sheffi (1989), Design and implementation of an interactive optimization system for network design in the motor carrier industry, *Oper. Res.* 37(1), 12–29.

Powell, W., Y. Sheffi, K. Nickerson, K. Butterbaugh, and S. Atherton (1988), Maximizing profits for North American van lines' truckload division: A new framework for pricing and operations, *Interfaces* 18, 21–41.

Powell, W., Y. Sheffi, and S. Thiriez (1984), The dynamic vehicle allocation problem with uncertain demands, in: *Ninth Int. Symp. on Transportation and Traffic Theory*, pp. 357–374.

Psaraftis, H. (1988), Dynamic vehicle routing problems, in: B. Golden and A. Assad, eds., *Vehicle Routing: Methods and Studies*, pp. 223–248, North-Holland, Amsterdam.

Psaraftis, H., J. Orlin, D. Bienstock, and P. Thompson (1985), *Analysis and Solution Algorithms of Sealift Routing and Scheduling Problems*, Sloan Working Paper 1700-85, Sloan School of Management, MIT.

Psaraftis, H., and J. Tsitsiklis, Dynamic shortest paths with Markovian arc costs (1990), preprint.

Qi, L. (1985), Forest iteration method for stochastic transportation problem, *Math. Program. Study* 25, 142–163.

Ran, B., D. Boyce, and L. LeBlanc (1992), Toward a new class of instantaneous dynamic user-optimal traffic assignment models, *Oper. Res.* (to appear).

Ratcliffe, L., B. Vinod, and F. Sparrow (1984), Optimal prepositioning of empty freight cars, *Simulation*, 269–275.

Rhee, W., and M. Talagrand (1988), A sharp deviation inequality for the stochastic traveling salesman problem, *Ann. Probab.* 17, 1–8.

Robbins, H., and S. Monro (1951), A stochastic approximation method, *Ann. Math. Stat.* 22, 400–407.

Rockafellar, T., and R. Wets (1991), Scenarios and policy aggregation in optimization under uncertainty, *Math. Oper. Res.* 16(1), 119–147.

Russell, R., and D. Gribbin (1991), A multiphase approach to the period routing problem, *Networks* 21, 747–765.

Seneta, E. (1981), *Nonnegative Matrices and Markov Chains*, Springer-Verlag, New York.

Sethi, S., and S. Bhaskaran (1985), Conditions for the existence of decision horizons for discounted problems in a stochastic environment, *Oper. Res. Lett.* 4(2), 61–64.

Shan, Y. (1985), *A Dynamic Multicommodity Network Flow Model for Real-Time Optimal Rail Freight Car Management*, Ph.D. Dissertation, Princeton University.

Shapiro, J. (1993), Mathematical programming models and methods for production planning and scheduling, in: A.R.K.S.C. Graves and P. Zipkin, eds., *Logistics of Production and Inventory*, pp. 371–444.

Sheffi, Y., and W. Powell (1982), An algorithm for the equilibrium assignment problem with random link times, *Networks* 12, 191–207.

Shor, N. (1979), *The Methods of Nondifferentiable Op[timization and their Applications*, Naukova Dumka, Kiev.

Simchi-Levi, D., and O. Berman (1988), Heuristics and bounds for the traveling salesman location problem on the plane, *Oper. Res. Lett.* 5, 243–248.

Smith, S., and Y. Sheffi (1992), *Railroad Car Distribution: A Fast Network Model with Multiple Car Types and Delays*, CTS working paper, M.I.T.

Spaccamela, A.M., A.R. Kan, and L. Stougie (1984), Hierarchical vehicle routing problems, *Networks* 14, 571–586.

Stanley, J. (1987), A forward convex-simplex method, *Eur. J. Oper. Res.* 29, 328–335.

Steele, J. (1981), Complete convergence of short paths and karp's algorithm for the tsp, *Mathematics of Operations Research*, 6, 374–378.

Steele, J. (1981), Subadditive Euclidean functionals and nonlinear growth in geometric probability, *Ann. Probab.* 9, 365–376.

Stewart, W., and B. Golden (1983), Stochastic vehicle routing: A comprehensive approach, *Eur. J. Oper. Res.* 14(3), 371–385.

Szwarc, W. (1964), The transportation problem with stochastic demand, *Manage. Sci.* 11, 35–50.

Szwarc, W. (1966), The time transportation problem, *Zastowania Matematyki, Warsaw* 8, 231–242.

Szwarc, W. (1971a), The dynamic transportation problem, *Nav. Res. Logist. Q.* 13, 335–345.

Szwarc, W. (1971b), Some remarks on the time transportation problem, *Nav. Res. Logist. Q.* 18, 473–485.

Tapiero, C. (1971), Transportation-location-allocation problems over time, *J. Regional Sci.* 11, 377–384.

Tapiero, C. (1975), Transportation over time: A system theory approach, in: *Proceedings – XX International TIMS Meeting, Tel-Aviv, Israel, Vol 1*, pp. 239–243, Jerusalem.

Tapiero, C., and M. Soliman (1972), Multicommodities transportation schedules over time, *Networks* 2, 311–327.

Tillman, F. (1969), The multiple terminal delivery problem with probabilistic demands, *Transp. Sci.* 3, 192–204.

Trudeau, P., and M. Dror (1992), Stochastic inventory routing: Route design with stockouts and route failures, *Transp. Sci.* 26(3), 171–184.

Turnquist., M, Mov-em: A network optimization model for empty freight car distribution (1986), School of Civil and Environmental Engineering, Cornell University.

Turnquist, M.A., and B.P. Markowicz (1989), *An Interactive Microcomputer-Based Planning Model for Railroad Car Distribution*, For the CORS/ORSA/TIMS National Meeting, Cornell University.

Van Roy, T., and D. Erlenkotter (1980), *A Dual-Based Procedure for Dynamic-Facility Location*, Working paper 80-31, International Institute for Applied Systems Analysis, Laxenburg, Austria.

Van Slyke, R., and R. Wets (1969), L-shaped linear programs with applications to optimal control and stochastic programming, *SIAM J. Appl. Math.* 17, 638–663.

Veinott Jr, A. (1969), Minimum concave-cost solution of leontief substitution models of multi-facility inventory systems, *Oper. Res.* 17, 262–291.

Vemuganti, R., M. Oblak, and A. Aggarwal (1989), Network models for fleet management, *Decision Sci.* 20, 182–197.

Vladimirou, H. (1991), *Stochastic Networks: Solution Methods and Applications in Financial Planning*, Ph.D. dissertation, Dept. of Civil Engng. and Oper. Res., Princeton University.

Wagner, H., and Whithin (1958), Dynamic version of the economic lot size model, *Manage. Sci.* 5, 89–96.

Wallace, S. (1986), Decomposing the requirement space of a transportation problem, *Math. Prog. Study*, 28, 29–47.

Wallace, S. (1987), A piecewise linear upper bound on the network recourse function, *Math. Program.* 38, 133–146.

Wasan, M. (1969), Stochastic approximations, in: *Cambridge Transactions in Math. and Math. Phys.* 58, Cambridge University Press, Cambridge.

Wets, R. (1966), Programming under uncertainty: The equivalent convex program, *SIAM J. Appl. Math.* 14, 89–105.

Wets, R. (1974), Stochastic programs with fixed resources: The equivalent deterministic problem, *SIAM Rev.* 16, 309–339.

Wets, R. (1982), Stochastic programming: Solution techniques and approximation schemes, in: M.G.A. Bachen and B. Korte, eds., *Mathematical Programming: The State of the Art*, pp. 566–603, Springer-Verlag, Berlin.

Wets, R. (1988), Large scale programming techniques in stochastic programming, in: Y. Ermoliev and R. Wets, eds., *Numerical Methods in Stochastic Programming*, Springer-Verlag, Berlin.

White, W. (1972), Dynamic transshipment networks: An algorithm and its application to the distribution of empty containers, *Networks* 2(3), 211–236.

White, W., and A. Bomberault (1969), A network algorithm for empty freight car allocation, *IBM Systems J.* 8(2), 147–171.

Wilkinson, W. (1971), An algorithm for universal maximal dynamic flows in a network, *Oper. Res.* 19(7), 1602–1612.

Williams, A. (1963), A stochastic transportation problem, *Oper. Res.* 11, 759–770.

Zadeh, N. (1974), On building minimum cost communication networks over time, *Networks* 4, 19–34.

Zahorik, A., L. Thomas, and W. Trigeiro (1984), Network programming models for production scheduling in multi-stage, multi-item capacitated systems, *Manage. Sci.* 30(3), 308–325.

Zangwill, W. (1968), Minimum concave cost flows in certain networks, *Manage. Sci.* 14(7), 429–450.

Zangwill, W. (1969), A backlogging model and a multi-echelon model of a dynamic economic lot size product system – a network approach, *Manage. Sci.* 15(9), 506–527.

Zipkin, P. (1982), Transportation problems with aggregated destinations when demands are uncertain, *Nav. Res. Logist. Q.* 29, 257–270.

M.O. Ball et al., Eds., *Handbooks in OR & MS, Vol. 8*

Chapter 4

Analysis of Vehicle Routing and Inventory-Routing Problems

Awi Federgruen

Graduate School of Business, Columbia University, New York, NY 10027, U.S.A.

David Simchi-Levi

Department of Industrial Engineering & Management Sciences, Northwestern University, Evanston, IL 60208-3119, U.S.A.

1. Introduction

In many service systems such as delivery, customers pick-up, repair and maintenance services, customers have to be served by a fleet of vehicles initially located at a given set of depots. The objective is to find a set of routes which satisfy a variety of constraints and which are of minimal total length. The problem, called the vehicle routing problem (VRP), has been analyzed extensively in contemporary journals. Since the problem is NP-hard (see Garey & Johnson [1979] for a background on NP-completeness) it is unlikely that a polynomial time algorithm will be developed for its optimal solution. Consequently, a great deal of work has been devoted to the development of heuristic algorithms.

Almost all of the literature on these problems deals with *empirical analyses* of heuristic algorithms which evaluate the performance of a specific heuristic on test problems. Limited results have been obtained using an *analytical analysis* of a heuristic in attempts to characterize its theoretical behavior.

The main reason for an analytical analysis of a heuristic algorithm is to identify classes of the VRP in which a specific heuristic is highly efficient. In addition, this approach enables us to better understand models that integrate vehicle routing with other issues important to the firm such as integrating inventory control and vehicle routing.

Specifically, what we refer to as analytical analysis includes worst-case performance analysis and probabilistic analysis. While the former establishes the maximum deviation from optimality that can occur for a given heuristic algorithm, the latter establishes the average performance of the heuristic under assumptions on the distribution of the data. As was pointed out by Fisher [1980], empirical analyses and analytical analyses should be viewed as complementary approaches rather than competitive ones.

To formally define worst-case and average case analysis, we need the following notation. For a given instance I_n of the VRP with n customers, let $Z^*(I_n)$ be the cost of its optimal solution, while $Z^H(I_n)$ is the cost of a solution produced by heuristic H. The *worst-case performance ratio* of a heuristic H is given by

$$\inf\left\{r \geq 1 \,\Big|\, \frac{Z^H(I_n)}{Z^*(I_n)} \leq r, \ \text{for all } I_n\right\}.$$

That is, the worst-case performance ratio of a heuristic is the minimum upper bound on the worst-case ratio of the cost of the heuristic solution to the cost of the optimal solution.

Define now the relative error of heuristic H by

$$e^H(I_n) = \frac{Z^H(I_n) - Z^*(I_n)}{Z^*(I_n)}.$$

When the data is randomly generated, we say that H is *asymptotically optimal* if

$$\lim_{n \to \infty} e^H(I_n) = 0, \quad (\text{a.s.}).$$

That is, under some assumptions on the distribution of the data, heuristic H generates solutions whose relative error tends to zero as the number of customers tends to infinity.

Analytical analysis of vehicle routing problems has started with the seminal work of Haimovich and Rinnooy Kan [1985] on the single depot capacitated vehicle routing problem with equal demands. We review these results in Section 2. In Section 3 we introduce and analyze a model that incorporates bin-packing features into the vehicle routing problem. In Section 4 we show how such results can be used in integrating inventory control and vehicle routing costs and in Section 5 we generalize the analysis to the multi-depot case. In Section 6 we discuss some other extensions and generalizations.

We conclude this section with a brief statement on the impact of analytical analyses on the development of new algorithms for vehicle routing problems as well as its use in practice. Recently, Bramel & Simchi-Levi [1992] have used probabilistic analysis to explain the excellent performance observed in practice for the Generalized Assignment Heuristic developed by Fisher & Jaikumar [1981]. This algorithm provided the foundation for a large scale routing system developed for and implemented by Air Products and Chemicals, Inc., see Bell, Dalberto, Fisher, Greenfield, Jaikumar, Kedia, Mack & Prutzman [1983]. The insight obtained from the probabilistic analysis of the VRP was used in Bramel & Simchi-Levi [1992] to design a new and highly efficient algorithm for general routing problems. This algorithm has been used to solve large scale school bus routing problems in New York City, and the reported performance is quite impressive, see Braca, Bramel, Posner & Simchi-Levi [1993]. Finally, some of the heuristics for combined inventory control and vehicle routing described in Section 4 provided the foundation for a planning system developed for the Israeli Air Force.

2. The capacitated vehicle routing problem (CVRP) with split demands

2.0. Introduction

The single-depot capacitated vehicle routing problem (CVRP) can be stated as follows: A set of customers has to be served by a fleet of identical vehicles of limited capacity q. The vehicles are initially located at a given depot. The objective is to find a set of routes for the vehicles of minimal total length. Each route begins at the depot, visits a subset of the customers and returns to the depot without violating the capacity constraint.

The capacity constraint states that the number of customers visited by a single vehicle cannot exceed q. Without loss of generality, we assume that q is an integer. Indeed, in many vehicle routing problems the number of stops per route needs to be constrained by itself. Alternatively, if the capacity constraint is specified in terms of total volume carried by a vehicle, the constraint treated in this section implies that customers have equal demands.

The customers and the depot are represented by a set of nodes on an undirected graph $G = (N, E)$. We denote by N_c the set of customers, d_i the distance between node i and the depot; $d_{\max} \equiv \max_{i \in N_c} d_i$, is the distance to the furthest customer, and d_{ij} the distance between node i and node j. The distance matrix $\{d_{ij}\}$ is assumed to be symmetric and to satisfy the triangle inequality, i.e., $d_{ij} = d_{ji}$ for all i, j and $d_{ij} \le d_{ik} + d_{kj}$ for all i, k, j. If d_{ij} represents the length of a shortest path between i and j, the triangle inequality clearly applies. We denote the optimal solution value of the CVRP by Z^* and that of any given heuristic H by Z^H.

In what follows, the optimal traveling salesman tour plays an important role. So, for any set of nodes $S \subseteq N$, let $L^*(S)$ be the length of the optimal traveling salesman tour through the set of points S. Also, let $L^\alpha(S)$ be the length of an α-optimal traveling salesman tour through S whose length is bounded from above by $\alpha \cdot L^*(S), \alpha \ge 1$.

The graph depicted in Figure 1, which is denoted by $\mathcal{G}(t, s)$, also plays an important role in our worst-case analyses. It consists of s groups of q nodes and another $s - 1$ nodes, called *white nodes*, separating the groups. The nodes within the same group have zero interdistances and each group is connected to the depot by an arc of unit length. The white nodes are of zero distance apart and t units distance away from the depot. Each white node is connected to the two groups of nodes it separates by an arc of unit length. Note that when $0 \le t \le 2$, $\mathcal{G}(t, s)$ satisfies the triangle inequality (if an edge $i-j$ is not shown in the graph then the distance between node i and node j is defined as the length of the shortest path from i to j). Also note that whenever $0 \le t \le 2$, the tour depicted in Figure 2 is an optimal traveling salesman tour of length $2s$.

2.1. Worst-case analysis of heuristics

A simple heuristic for the CVRP, suggested by Haimovich and Rinnooy Kan [1985] and latter modified by Altinkemer & Gavish [1990], is to partition a

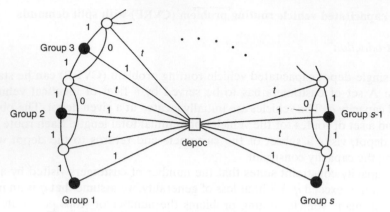

Fig. 1. Every group contains q customers with interdistance zero.

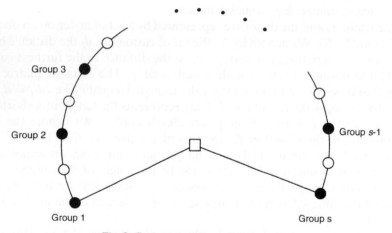

Fig. 2. Optimal traveling salesman tour.

traveling salesman tour into segments, such that each segment of customers is served by a single vehicle, i.e., each segment has no more than q points. The heuristic, called the iterated tour partitioning (ITP) heuristic, starts from a traveling salesman tour through all $n = |N_c|$ customers and the depot. Starting at the depot and following the tour in an arbitrary orientation, the customers and the depot are numbered $x^{(0)}, x^{(1)}, x^{(2)}, \ldots, x^{(n)}$ where $x^{(0)}$ is the depot. We partition the path from $x^{(1)}$ to $x^{(n)}$ into $\lceil n/q \rceil$ disjoint segments, such that each one contains no more than q customers, and connect the endpoints of each segment to the depot. The first segment contains only customer $x^{(1)}$. All the other segments contain exactly q customers, except maybe the last one. To possibly improve the generated set of tours, we repeat the above construction $q - 1$ times such that every time we move the endpoints of all but the first and last segments up by one position in the direction of the orientation. The starting (last) point of the first (last) segment is always $x^{(1)}$ ($x^{(n)}$), while its last (starting) point is moved by one position in each of the $q - 1$ constructions. We then choose the best of the q generated sets of tours.

It is easy to see that, for a given a traveling salesman tour, the running time of the ITP heuristic is $O(nq)$. The performance of this heuristic clearly depends on the quality of the initial traveling salesman tour chosen in the first step of the algorithm. Hence, when the ITP heuristic partitions an α-optimal traveling salesman tour, it is denoted by ITP(α). To establish the worst-case behavior of the algorithm, we find a lower bound on Z^*, and calculate an upper bound on the cost of the solution produced by the ITP(α) heuristic.

Lemma 2.1 [Haimovich and Rinnooy Kan, 1985].

$$Z^* \geq \max \left\{ L^*(N), \frac{2}{q} \sum_{i \in N_c} d_i \right\}.$$

Proof. Clearly, $Z^* \geq L^*(N)$ follows from the triangle inequality. To prove $Z^* \geq (2/q) \sum_{i \in N_c} d_i$, consider an optimal solution to the CVRP in which N_c is partitioned into subsets $\{N_1, N_2, \cdots, N_m\}$ where each set N_j is served by a single vehicle. Clearly,

$$Z^* = \sum_j L^*(N_j \cup \{x^{(0)}\}) \geq \sum_j 2 \max_{i \in N_j} d_i \geq \sum_j \frac{2}{|N_j|} \sum_{i \in N_j} d_i$$

$$\geq \sum_j \frac{2}{q} \sum_{i \in N_j} d_i = \frac{2}{q} \sum_{i \in N_c} d_i. \qquad \Box$$

Lemma 2.2 [Altinkemer & Gavish, 1990].

$$Z^{\text{ITP}(\alpha)} \leq \frac{2}{q} \sum_{i \in N_c} d_i + \left(1 - \frac{1}{q}\right) \alpha L^*(N).$$

Proof. We prove the Lemma by finding the cumulative length of the q solutions generated by the ITP heuristic. The ith solution consists of the following segments: $\{x^{(1)}, x^{(2)}, \cdots, x^{(i)}\}, \{x^{(i+1)}, x^{(i+2)}, \cdots, x^{(i+q)}\}, \cdots, \{x^{(i+1+\lfloor (n-i)/q \rfloor q)}, \cdots, x^{(n)}\}$. Thus, among the q solutions generated, each customer $x^{(i)}$, $2 \leq i \leq n - 1$, appears exactly once as the first point of a segment and exactly once as the last point. Therefore in the cumulative length of the q solutions the term $2d_{x^{(i)}}$ is incurred for $2 \leq i \leq n - 1$. Customer $x^{(1)}$ is the first point of a segment in each of the q solutions, and in the first one it is also the last point. Thus, the term $d_{x^{(1)}}$ appears $q + 1$ times in the cumulative length. Similarly, $x^{(n)}$ is always the last point of a segment in each of the q solutions, and once the first point. Thus, the term $d_{x^{(n)}}$ appears $q + 1$ times in the cumulative length as well. Finally, each one of the arcs $(x^{(i)}, x^{(i+1)})$ for $1 \leq i \leq n - 1$ appears in exactly $q - 1$ solutions since it is excluded from one solution. These arcs together with the $q - 1$ arcs connecting the depot to $x^{(1)}$ and $q - 1$ arcs connecting the depot to $x^{(n)}$, form $q - 1$ copies of the initial traveling salesman tour selected in the first step of the heuristic. Thus, if the initial traveling salesman tour is an α-optimal tour, the cumulative length of all q tours is

$$2 \sum_{i \in N_c} d_i + (q - 1)L^\alpha(N)$$

$$\leq 2 \sum_{i \in N_c} d_i + (q - 1)\alpha L^*(N).$$

Hence,

$$Z^{ITP(\alpha)} \leq \frac{2}{q} \sum_{i \in N_c} d_i + \left(1 - \frac{1}{q}\right)\alpha L^*(N). \quad \square$$

We have thus obtained the following result.

Theorem 2.3 [Altinkemer & Gavish, 1990].

$$\frac{Z^{ITP(\alpha)}}{Z^*} \leq 1 + \left(1 - \frac{1}{q}\right)\alpha. \qquad (2.1)$$

For example, if Christofides' [1976b] algorithm ($\alpha = 1.5$) is used to obtain the initial traveling salesman tour in polynomial time, we have

$$\frac{Z^{ITP(1.5)}}{Z^*} \leq \frac{5}{2} - \frac{3}{2q}.$$

The proof of the worst-case result for the ITP(α) heuristic suggests that if we can improve the bound in (2.1) for $\alpha = 1$, then the bound can be improved for any $\alpha > 1$. However, the following theorem, proved by Li & Simchi-Levi [1990], says that this is impossible, i.e., the bound

$$\frac{Z^{ITP(1)}}{Z^*} \leq 2 - \frac{1}{q}$$

is sharp.

Theorem 2.4. *For any integer $q \geq 1$, there exists a problem instance with $Z^{ITP(1)}/Z^* = 2 - 1/q$.*

Proof. Let us consider the graph $\mathcal{G}(0, q)$. A solution obtained by the ITP heuristic is shown in Figure 3. In this solution,

$$Z^{ITP(1)} = 2 + 2 + \underbrace{4 + 4 + \cdots + 4}_{q-2 \text{ times}} + 2 = 4q - 2.$$

Consider next the solution which has q vehicles serve the q groups of customers and the $(q + 1)^{st}$ vehicle serve the other $q - 1$ nodes. Thus,

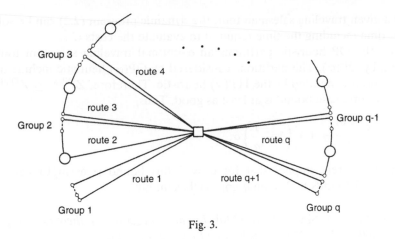

Fig. 3.

$$Z^* \leq 2q.$$

Hence,

$$\frac{Z^{\text{ITP}(1)}}{Z^*} \geq 2 - \frac{1}{q}.$$

This together with the upper bound of (2.1) completes the proof. $\square$.

Another variant of the tour partitioning heuristic is the optimal partitioning (OP) of a tour described by Beasley [1983]. The algorithm optimally partitions the traveling salesman tour into a set of feasible routes, each containing at most q customers.

Given a traveling salesman tour through the customers and the depot, the points are numbered $x^{(0)}, x^{(1)}, \ldots, x^{(n)}$ in order of appearance on the tour, where $x^{(0)}$ is the depot. Let

$$C_{jk} = \begin{cases} \text{the distance traveled by a vehicle that starts at the} \\ \text{depot, visits customers } x^{(j+1)}, x^{(j+2)}, \ldots, x^{(k)} \text{ in} \\ \text{this order, and returns to the depot,} & \text{if } k - j \leq q; \\ +\infty, & \text{otherwise.} \end{cases}$$

For $j = 0, 1, 2, \cdots, n - 1$, let

$F(j) =$ minimum total vehicle routing cost to serve customers $x^{(j+1)}, x^{(j+2)}$, $\ldots, x^{(n)}$ if each vehicle visits its assigned customers in the order they appear on the traveling salesman tour.

We are clearly interested in $F(0)$ and the associated set of routes. They can be obtained via the dynamic program:

$$F(j) = \min_{j < k \leq j+q} \{C_{jk} + F(k)\}, \qquad j = 0, 1, 2, \cdots, n - 1 \qquad (2.2)$$

For a given traveling salesman tour, the dynamic program (2.2) can be solved in $O(nq)$ time including the time required to evaluate the costs C_{jk}.

When the OP heuristic partitions an α-optimal traveling salesman tour, it is denoted by OP(α). The partitions considered by OP(α) heuristic include all q of the partitions generated by the ITP(α) heuristic. Therefore, $Z^{OP(\alpha)} \le Z^{ITP(\alpha)}$ and hence its worst-case bound is at least as good. Thus,

$$\frac{Z^{OP(\alpha)}}{Z^*} \le 1 + \left(1 - \frac{1}{q}\right)\alpha.$$

The next theorem implies that for $\alpha = 1$ this bound is asymptotically sharp, i.e., $Z^{OP(1)}/Z^*$ tends to 2 when q approaches infinity.

Theorem 2.5 [Li & Simchi-Levi, 1990]. *For any integer $q \ge 1$, there exists a problem instance with $Z^{OP(1)}/Z^*$ arbitrarily close to $2 - 2/(q+1)$.*

Proof. Consider the graph $\mathcal{G}(1, Kq + 1)$, where K is a positive integer. It is easy to check that

$$Z^{OP(1)} = 2(Kq + 1) + 2Kq.$$

This can be verified as follows. The $F(\cdot)$ function defined in (2.2) is clearly non-increasing. Also, $C_{0k} = 2$ for all $k \le q$; therefore it is optimal to assign the entire final group to a single vehicle. The remainder of the claim can be obtained similarly. On the other hand, consider the solution in which $Kq + 1$ vehicles serve the $Kq + 1$ groups of customers and another K vehicles serve the other nodes. Hence,

$$Z^* \le 2(Kq + 1) + 2K,$$

and therefore

$$\lim_{K \to \infty} \frac{Z^{OP(1)}}{Z^*} \ge 2 - \frac{2}{q+1}. \qquad \square$$

We conclude this subsection with a polynomial time ϵ-approximation scheme suggested by Haimovich and Rinnooy Kan. This heuristic produces a solution with worst-case bound of $(1 + \epsilon)Z^*$ and whose running time is polynomial in n but superpolynomial in q/ϵ. The idea is to divide the set of customers into two subsets; the first contains the k furthest customers from the depot while the second set includes all the remaining customers. The heuristic then proceeds by finding an optimal solution for the set of k customers and a near optimal solution for the second set of customers, using for instance, the ITP(α) heuristic. Haimovich and Rinnooy Kan provide a lower bound on the value of k that depends only on q/ϵ and show that the above worst-case bound is obtained by selecting any k larger than this bound. Since k is independent of n, one can find the optimal solution to the CVRP defined on the first set of customers in time independent of n.

2.2. *The asymptotic optimal solution value assuming euclidean distances*

In the remainder of this section, we assume that the customers are points in the plane and that the distance between any pair of customers is given by the Euclidean distance. The results discussed in this subsection as well as the next one are mainly based on Haimovich and Rinnooy Kan's work.

The upper bound of Lemma 2.2 has two cost components; the first component is proportional to the total 'radial' cost between the depot and the customers. The second component is proportional to the 'circular' cost; the cost of traveling between customers. This cost is related to the cost of the optimal traveling salesman tour. It is well known that, for large n the cost of the optimal traveling salesman tour grows like $\sqrt{n}$, while the total radial cost between the depot and the customers grows like n. Therefore, it is intuitive that when the number of customers is large enough the first cost component will dominate the optimal solution value. This observation is now formally proven.

Theorem 2.6. *Let x_k, $k = 1, 2, \cdots, n$ be a sequence of independent random variables having a distribution μ with compact support in $\Re^2$. Let $d(y)$ be the Euclidean distance between the depot and $y \in \Re^2$ and let*

$$E(d) = \int_{\Re^2} d(y)\,\mathrm{d}\mu(y).$$

Then, with probability one,

$$\lim_{n\to\infty} \frac{Z^*}{n} = \frac{2}{q}E(d).$$

Proof. Lemma 2.1 and the strong law of large numbers tell us that

$$\lim_{n\to\infty} \frac{Z^*}{n} \geq \frac{2}{q}E(d) \qquad (a.s.). \tag{2.3}$$

On the other hand, from Lemma 2.2,

$$\frac{Z^*}{n} \leq \frac{Z^{\text{ITP}(1)}}{n} \leq \frac{2}{nq}\sum_{i\in N_c} d_i + \left(1 - \frac{1}{q}\right)\frac{1}{n}L^*(N).$$

It is well known (see, Beardwood, Halton & Hammersley, 1959) that there exists a constant $\beta > 0$, independent of the distribution μ, such that with probability one

$$\lim_{n\to\infty} \frac{L^*(N)}{\sqrt{n}} = \beta \int_{\Re^2} f^{1/2}(x)\,\mathrm{d}x,$$

where f is the density of the absolutely continuous part of the distribution μ.

Hence,

$$\lim_{n \to \infty} \frac{Z^*}{n} \leq \frac{2}{q} E(d) \qquad \text{(a.s.)}.$$

This together with (2.3) proves the Theorem.   □

The following observation is in order. Haimovich and Rinnooy Kan prove Theorem 2.6 merely assuming $E(d) < \infty$ rather than a compact support for the distribution μ. However, the restriction to a compact support seems to be satisfactory for all practical purposes. The following is another important generalization of Theorem 2.6. Assume that a *cluster* of w_k customers (rather than a single customer) is located at point x_k, $k = 1, 2, \ldots, n$. The theorem continues to hold with

$$\lim_{n \to \infty} \frac{Z^*}{n} = \frac{2}{q} E(w) E(d),$$

where $E(w)$ is the expected cluster size, provided that the cluster size w is independent of the location x. This follows from a straightforward adaptation of Lemma 2.1 and Lemma 2.2.

2.3. Asymptotically optimal heuristics

The proof of the previous theorem (Theorem 2.6) reveals that the ITP(α) heuristic provides a solution whose cost approaches the optimal cost when n tends to infinity. Indeed, replacing $Z^{\text{ITP}(1)}$ by $Z^{\text{ITP}(\alpha)}$ in the previous proof gives the following theorem.

Theorem 2.7. *Under the conditions of Theorem 2.6 and for any fixed α, the* ITP(α) *heuristic is asymptotically optimal.*

As is pointed out by Haimovich and Rinnooy Kan, iterated tour partitioning heuristics, though asymptotically optimal, hardly exploit the special topological structure of the Euclidean plane in which the points are located. It is therefore natural to consider *region partitioning* (RP) heuristics that are more geometric in nature.

Haimovich and Rinnooy Kan thus consider three classes of regional partitioning schemes. In *rectangular region partitioning* (RRP), one starts with a rectangle containing the set of customers N_c and cuts it into smaller rectangles. In *polar region partitioning* (PRP) and *circular region partitioning* (CRP) one starts with a circle centered at the depot and partitions it by means of circular arcs and radial lines. We shall shortly discuss each one of them in detail.

In all cases the RP heuristics construct subsets of customers $N(j)$, one subset for every subregion, such that each one of them has exactly q customers except possibly one subset which has less than q customers.

Since every subset $N(j)$ has no more than q customers, each of these RP heuristics allocates one vehicle to every subregion. The vehicle then uses the following routing strategy: the first customer visited is the one closest to the depot among all the customers in $N(j)$. The rest are visited by finding an α-optimal traveling salesman tour through $N(j)$. After visiting all its customers the vehicle returns to the depot through the first (closest) customer. It is therefore natural to call each one of these heuristics $RP(\alpha)$ heuristics. In particular we have $RRP(\alpha)$, $PRP(\alpha)$ and $CRP(\alpha)$.

Lemma 2.8.

$$Z^{RP(\alpha)} \leq \frac{2}{q} \sum_{i \in N_c} d_i + 2d_{\max} + \alpha \sum_j L^*(N(j)).$$

Proof. We number the subsets $N(j)$ constructed by the $RP(\alpha)$ heuristic such that $|N(j)| = q$ for every j except maybe for $j = 1$. It follows that the total distance traveled by the vehicle that visits subset $N(j)$, for $j \geq 2$, is

$$\leq 2 \min_{i \in N(j)} d_i + \alpha L^*(N(j))$$

$$\leq \frac{2}{q} \sum_{i \in N(j)} d_i + \alpha L^*(N(j)),$$

while the total distance traveled by the vehicle that visits $N(1)$ is no more than

$$2d_{\max} + \alpha L^*(N(1)).$$

Taking the sum over all subregions we obtain the desired result. $\square$

The quality of the upper bound of Lemma 2.8 depends, of course, on the quantity $\sum_j L^*(N(j))$. This value was analyzed by Karp [1977] who showed that for any RP heuristic

$$\sum_j L^*(N(j)) \leq L(N_c) + \tfrac{3}{2} P^{RP}, \tag{2.4}$$

where P^{RP} is the sum of the perimeters of the subregions generated by the RP heuristic. For this reason we analyze the quantity P^{RP} for each of the three region partitioning heuristics.

Our analysis of the Rectangular Region Partitioning and Polar Region Partitioning is slightly different from the one performed by Haimovich and Rinnooy Kan. In their work, Haimovich and Rinnooy Kan use the partitioning schemes suggested by Karp [1977] and Spaccamela, Rinnooy Kan & Stougie [1984]. We observe that using these schemes creates some difficulties since they do not generate subregions with exactly q customers each and furthermore some customers may belong to more than one subregion. Therefore the upper bound of Lemma 2.8 may fail to hold for these schemes. For these reasons we modify their partitioning schemes and suggest the following.

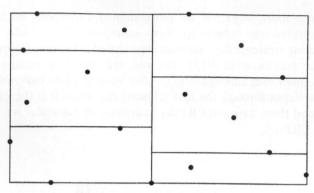

Fig. 4. $n = 17$, $q = 3$; $h = 2$, $t = 1$.

Rectangular region partitioning

The smallest rectangle with sides a and b containing the set of customers N_c is partitioned by means of horizontal and vertical lines. First, the region is subdivided by t vertical lines such that each subregion contains exactly $(h + 1)q$ points except possibly the last one. Each of these $t + 1$ subregions is then partitioned by means of h horizontal lines into $h + 1$ smaller subregions such that each contains exactly q points with one possible exception that has less than q points; see Figure 4.

Clearly h and t should satisfy

$$t = \left\lfloor \frac{n}{(h + 1)q} \right\rfloor,$$

and

$$t(h + 1)q < n \le (t + 1)(h + 1)q.$$

The unique integer that satisfies these conditions is $h = \lceil \sqrt{n/q} - 1 \rceil$. To evaluate the quantity P^{RRP}, note that the number of vertical lines added is $t \le \sqrt{n/q}$. Each of these lines is counted twice in the quantity P^{RRP}.

In the second step of the RRP we add h horizontal lines where $h \le \sqrt{n/q}$. These horizontal lines are also counted twice in P^{RRP}. It follows that

$$P^{RRP} \le 2\sqrt{\frac{n}{q}}(a + b) + 2(a + b) \le 8d_{\max}\sqrt{\frac{n}{q}} + 8d_{\max}.$$

Polar region partitioning

The circle with radius $d_{\max}$ containing the set N_c and centered at the depot is partitioned in exactly the same way as in the previous partitioning scheme, with the exception that circular arcs and radial lines replace vertical and horizontal lines. Using the same analysis as before, one shows,

$$P^{PRP} \le 6\pi d_{\max}\sqrt{\frac{n}{q}} + 2\pi d_{\max} + 2d_{\max}.$$

Circular region partitioning

The CRP scheme partitions the circle centered at the depot with radius d_{max} into h equal sectors, where h is to be determined later on. Each sector is then partitioned into subregions by means of circular arcs, such that each subregion contains exactly q customers except possibly the one closest to the depot. Thus, at most h subregions, each from one sector, have less than q customers. These subregions (with the depot on their boundary) are then repartitioned by means of radial cuts such that $h - 1$ of them have exactly q customers each, while the last one may have less than q customers.

The total length of the initial radial lines is hd_{max}. The length of an inner circular arc bounding a subregion containing a set $N(j)$ is no more than

$$\frac{2\pi}{h} \min_{i \in N(j)} d_i \le \frac{2\pi}{h} \frac{\sum_{i \in N(j)} d_i}{|N(j)|} = \frac{2\pi \sum_{i \in N(j)} d_i}{hq},$$

while the length of the outer circle is $2\pi d_{max}$. Finally, the repartitioning of the central subregions adds no more than $hd_{max}/2$. Thus,

$$P^{CRP} \le 2 \left(hd_{max} + \frac{2\pi \sum_i d_i}{hq} + \frac{hd_{max}}{2} \right) + 2\pi d_{max}.$$

Taking $h = \lceil \sqrt{(4\pi \sum_i d_i)/(3qd_{max})} \rceil$, we obtain the following upper bound on P^{CRP},

$$P^{CRP} \le 4 \sqrt{3\pi d_{max} \frac{1}{q} \sum_{i \in N_c} d_i} + (3 + 2\pi)d_{max}.$$

The reader should be aware that all of these partitioning schemes can be implemented in $O(n \log n)$ time. We now have all the necessary ingredients for an asymptotic analysis of the performance of partitioning heuristics.

Theorem 2.9. *Let x_k, $k = 1, 2, \cdots, n$ be a sequence of independent random variables having a distribution μ with compact support in $\Re^2$. Let $d(y)$ be the Euclidean distance between the depot and $y \in \Re^2$ and let*

$$E(d) = \int_{\Re^2} d(y) \, d\mu(y).$$

Then, for any fixed α, RRP(α), PRP(α) and CRP(α) are asymptotically optimal. That is, with probability one

$$\lim_{n \to \infty} \frac{Z^{RP(\alpha)}}{n} = \frac{2}{q} E(d).$$

Proof. Lemma 2.8 together with equation (2.4) provide the following upper bound on the total distance traveled by all vehicles in the solution produced by the above RP heuristics.

$$Z^{\mathrm{RP}(\alpha)} \leq \frac{2}{q} \sum_{i \in N_c} d_i + 2d_{\max} + \alpha L^*(N_c) + \frac{3}{2}\alpha P^{\mathrm{RP}(\alpha)}.$$

By the strong law of large numbers and the fact that the points are in a compact region, $(1/n) \sum_{i \in N_c} d_i$ converges almost surely to $E(d)$ while $d_{\max}/n$ converges to 0. Furthermore, $L^*(N_c)/n$ converges to 0 almost surely; see the proof of Theorem 2.6. Finally, from the analysis of each of the region partitioning heuristics and the fact that the points are in a compact region, we have that $P^{\mathrm{RP}(\alpha)}/n$ converges to zero. $\square$

2.4. Asymptotically optimal heuristics for the CVRP with general cost structures

In this section we consider the problem where the cost of a route may depend on its length (ϑ) as well as the number of points visited in the route (m), according to some general function $f(\vartheta, m)$. This general cost structure is motivated by several optimization problems in the area of *integrated inventory control* and *vehicle routing* which reduce to this class of routing problems as is explained in Section 4.

Formally, the cost of a route of length ϑ which visits m customers is given by $f(\vartheta, m)$ where $f : \Re \times \mathcal{N} \to \Re$ is a general function. We assume without loss of generality that $f(0, 0) = 0$.

In addition, we assume
 (i) $f(\vartheta, m)$ is non-decreasing in ϑ,
 (ii) $f(\vartheta, m)$ is concave in ϑ.

Let R denote the number of vehicles available at the depot where R may be fixed or variable. The jth vehicle has a capacity q_j, $j = 1, 2, \cdots, R$ and the vehicles are numbered in ascending order of their capacities, i.e., $q_1 \leq q_2 \leq \cdots \leq q_R \equiv q$. As in the previous sections, $N_c = \{x_1, x_2, \cdots, x_n\}$ denotes the set of customers; furthermore, the elements of N_c are numbered in ascending order of their radial distances from the depot, i.e., $d_1 \leq d_2 \leq, \cdots, \leq d_n$. Let Z_f^H be the cost of the solution produced by a heuristic H, under the cost function f. Similarly, let Z_f^* be the cost of the optimal solution under the cost function f.

In this section we describe a heuristic due to Anily & Federgruen [1990b] which is asymptotically optimal under mild probabilistic assumptions regarding the locations of the points, and whose complexity is at worst $O(n^2)$.

If the vehicles are assigned the sets $\{N_1, \cdots, N_R\}$, the total cost may clearly be bounded from below by

$$\sum_{j=1}^{R} f\left(\frac{2\sum_{i \in N_j} d_i}{|N_j|}, |N_j|\right) \leq \sum_{j=1}^{R} f(2 \max_{i \in N_j} d_i, |N_j|)$$

since $2\sum_{i \in N_j} d_i / |N_j| \leq 2\max_{i \in N_j} d_i \leq L^*(N_j)$ and since f is non-decreasing in its first argument. Thus, let

$$\underline{Z}_f^1 = \min \left\{ \sum_{j=1}^{R} f\left(\frac{2\sum_{i \in N_j} d_i}{|N_j|}, |N_j|\right) : |N_j| \le q_j, \ j = 1, 2, \cdots, R \right\} (2.5)$$

$$\underline{Z}_f^2 = \min \left\{ \sum_{j=1}^{R} f(2 \max_{i \in N_j} d_i, |N_j|) : |N_j| \le q_j, \ j = 1, 2, \cdots, R \right\} \quad (2.6)$$

We conclude,

Lemma 2.10. $\underline{Z}_f^1 \le \underline{Z}_f^2 \le Z_f^*$.

It follows from Chakravarty, Orlin & Rothblum [1982] and Anily & Federgruen [1991a] that $\underline{Z}_f^2$ is achieved, for any function f non-decreasing in ϑ, by a *consecutive partition* $\{N_1, N_2, \cdots, N_R\}$, i.e., the indices of the points in each set N_j are *consecutive* integers. For example $\{N_1, N_2\} = \{\{4, 5\}; \{1, 2, 3\}\}$ is a consecutive partition of $N = \{1, 2, \cdots, 5\}$.

This implies that $\underline{Z}_f^2$ may be computed via a simple shortest path calculation similar to (2.2). The complexity of the shortest path computation depends on whether R is a variable or fixed and whether the capacities q_j, $j = 1, 2, \cdots, R$, are identical or not. In the worst case the complexity is only quadratic in n, the number of customers.

Similarly, $\underline{Z}_f^1$ is achieved by a consecutive partition and can thus be computed via the same shortest path calculations provided that f is concave in its first argument. When f has additional properties significant simplifications in the shortest path calculations for $\underline{Z}_f^1$ and $\underline{Z}_f^2$ may be obtained by exploiting additional qualitative properties of the partition achieving these bounds, see Anily & Federgruen [1991a] for details.

The proposed heuristic starts with the computation of either one of the bounds $\underline{Z}_f^1$ or $\underline{Z}_f^2$ and a corresponding partition $\{N_1, \cdots, N_R\}$. For $m = 1, \cdots, q$, let $N^{(m)}$ denote the points assigned to a set of cardinality m, i.e.,

$$N^{(m)} = \{i : i \in N_j \text{ for some } j = 1, 2, \cdots, R \text{ with } |N_j| = m\},$$
$$m = 1, 2, \cdots, q.$$

The remainder of the heuristic consists of a modification of the circular region partitioning scheme described in Section 2.3. We refer to this scheme as the modified circular region partitioning scheme and apply it separately to each of the non-empty sets $N^{(m)}$, for $m = 1, 2, \cdots q$.

Modified region partitioning scheme (MCRP)

Step 1. If $N^{(m)} \ne \emptyset$, $n_m \equiv |N^{(m)}|$, $d^{(m)} \equiv \max\{d_i | i \in N^{(m)}\}$ and $v_m \equiv \lfloor n_m/(m\lfloor n_m^{1/2}\rfloor)\rfloor$. Otherwise, go to Step 4.

Step 2. Partition the circle with radius $d^{(m)}$ into $\lfloor n_m^{1/2}\rfloor$ consecutive sectors containing mv_m points each and potentially one additional sector containing

$n_m - \lfloor n_m^{1/2} \rfloor m v_m$ points. Let K_m denote the number of generated sectors. Let $S_k^{(m)}$ denote the kth generated sector, $k = 1, 2, \cdots, K_m$.

Step 3. For each $k = 1, \cdots, K_m$ partition the sector $S_k^{(m)}$ by circular cuts such that each of the subregions contains m retailers and denote by $S_{k,l}^{(m)}$ the lth subregion in the kth sector, $l = 1, \cdots, |S_k^{(m)}|/m$.

Step 4. For each of the generated subregions, determine the optimal traveling salesman tour through the depot and the m points in the subregion.

We first assess the computational complexity of the MCRP heuristic. Note that for a given value of $m = 1, 2, \cdots, q$, Steps 2 and 3 can be implemented efficiently by first ranking the points in $N^{(m)}$ according to their angle coordinate and then in each sector separately according to their radial distances. The number of operations is thus bounded by $C n_m \log n_m$ for an appropriate constant C. The total number of operations required for Steps 2 and 3 is thus bounded by $O(n \log n)$.

The total number of subregions generated is of course bounded by n, while the determination of the optimal traveling salesman tour for each subregion, takes constant time. The total amount of work in Step 4 is thus linear in n. We conclude that the computational requirements of the MCRP heuristic are, in general, dominated by the amount of work required to determine the lower bounds $\underline{Z}_f^1$ and $\underline{Z}_f^2$.

The proof of the asymptotic optimality of the MCRP heuristic starts by showing that the number of points assigned to a sector is $O(n_m^{1/2})$.

Lemma 2.11. *The number of points in each sector $S_k^{(m)}$ ($k = 1, 2, \cdots, m$) is bounded from above by $m \lfloor n_m^{1/2} \rfloor + 3$.*

Proof. v_m clearly satisfies the following inequalities:

$$v_m m \lfloor n_m^{1/2} \rfloor \le n_m < (v_m + 1) m \lfloor n_m^{1/2} \rfloor. \tag{2.7}$$

Note that the number of points in each of the sectors $S_k^{(m)}$, $k = 1, \cdots, \lfloor n_m^{1/2} \rfloor$ satisfies the following inequalities:

$$|S_k^{(m)}| = v_m m \le \frac{n_m}{\lfloor n_m^{1/2} \rfloor} \le \frac{(\lfloor n_m^{1/2} \rfloor + 1)^2}{\lfloor n_m^{1/2} \rfloor} \le \lfloor n_m^{1/2} \rfloor + 2 + \frac{1}{\lfloor n_m^{1/2} \rfloor}$$
$$\le m \lfloor n_m^{1/2} \rfloor + 3.$$

If $K_m > \lfloor n_m^{1/2} \rfloor$ we get that

$$|S_{K_m}| = n_m - \lfloor n_m^{1/2} \rfloor v_m m < m \lfloor n_m^{1/2} \rfloor$$

by (2.7). Thus, $|S_k^{(m)}| \le m \lfloor n_m^{1/2} \rfloor + 3$, $k = 1, \cdots, K_m$. $\square$

Clearly, the cost of the generated solution is given by

$$Z_f^H = \sum_{m=1}^{q} \left\{ \sum_{k=1}^{K_m} \sum_{l=1}^{|S_k^{(m)}|/m} f\left(L^* \left(S_{k,l}^{0(m)} \right), m \right) \right\}, \tag{2.8}$$

where $S_{k,l}^{0(m)} = S_{k,l}^{(m)} \cup \{\text{depot}\}$. When deriving an explicit upper bound for Z_f^H and when analyzing the asymptotic properties of the heuristic we employ for every $m = 1, 2, \cdots, q$, the upper bound (2.4) for $\sum L^*(S_{k,l}^{(m)})$ i.e.,

$$\sum_{k=1}^{K_m} \sum_{l=1}^{|S_r^{(m)}|/m} L^*(S_{k,l}^{(m)}) \le L^*(N^{(m)}) + \frac{3}{2} P^{(m)}, \qquad m = 1, 2, \cdots, q, \tag{2.9}$$

where $P^{(m)}$ is the total perimeter of the generated subregions in $N^{(m)}$, $m = 1, \cdots, q$. Explicit $O(n_m^{1/2})$ bounds for both terms to the right of (2.9) may be obtained from the following lemmas.

Lemma 2.12 [Haimovich and Rinnooy Kan, 1985]. *If a set of points N is contained in a connected polar region with area A and finite perimeter P, then*

$$L^*(N) \le (2|N|A)^{1/2} + 1.5P.$$

Lemma 2.13. $P^{(m)} \le d^{(m)} \left((4\pi + 2) \lfloor n_m^{1/2} \rfloor + 10\pi + 2 \right), \qquad m = 1, \cdots, q.$

Proof. Fix $m = 1, \cdots, q$. Let C be the circle centered at the origin with radius $d^{(m)}$. Also, let

$P_1 =$ the total length of all circle cuts in $N^{(m)}$,
$P_2 =$ the perimeter of $C = 2\pi d^{(m)}$,
$P_3 =$ the total length of all radial cuts in $N^{(m)}$.

Clearly, $P^{(m)} = 2P_1 + P_2 + 2P_3$ since each circle cut is adjacent to two subregions within a sector and each radial cut is adjacent to two sectors. The number of circle cuts performed in the kth sector is given by $|S_k^{(m)}|/m - 1$ which in view of Lemma 2.11 is bounded by $\lfloor n_m^{1/2} \rfloor + 2$. The total length of all circle cuts is thus bounded by $(\lfloor n_m^{1/2} \rfloor + 2)$ times the perimeter of C; hence, $P_1 \le 2\pi d^{(m)}(\lfloor n_m^{1/2} \rfloor + 2)$. Moreover, $P_3 \le K_m d^{(m)} \le (\lfloor n_m^{1/2} \rfloor + 1)d^{(m)}$. The lemma now follows from simple algebra. $\square$

The following theorem derives an explicit upper bound on Z_f^H. First, we need the following definitions: Let $f_m^0 = \lim_{\vartheta \downarrow 0} f(\vartheta, m)$, $m = 1, \cdots, q$, which exists since f is concave in ϑ. Also, let

$$\underline{f}(\vartheta) = \min_{1 \le m \le q} \{f(\vartheta, m)\},$$

$$\alpha_m = \left(1 - \frac{1}{m}\right)\left((2\pi)^{1/2} + 6\pi + 3\right), \qquad \beta_m = \frac{3}{2}\left(1 - \frac{1}{m}\right)(10\pi + 2).$$

Finally, let

$$\overline{Z}_f^H = \sum_{m=1}^{q}\sum_{k=1}^{K_m}\sum_{l=1}^{|S_{k,l}^{(m)}|/m} f\left(2m^{-1}\sum_{i \in S_{k,l}^{(m)}} d_i, m\right)$$

$$+ \sum_{m=1}^{q} \frac{n_m}{m}\left[f\left(\frac{m\alpha_m d^{(m)}}{\sqrt{n_m}} + \frac{\beta_m d^{(m)}}{n_m}, m\right) - f_m^0\right].$$

Theorem 2.14. $Z_f^H \le \overline{Z}_f^H$.

Proof. It follows from a similar argument to the proof of Lemma 2.2 that

$$L^*(S_{k,l}^{0(m)}) \le 2m^{-1}\sum_{i \in S_{k,l}^{(m)}} d_i + \left(1 - \frac{1}{m}\right)L^*(S_{k,l}^{(m)}).$$

Since f is non-decreasing and concave in ϑ, we have

$$f\left(L^*\left(S_{k,l}^{0(m)}\right), m\right) \le f\left(2m^{-1}\sum_{i \in S_{k,l}^{(m)}} d_i, m\right)$$

$$+ f\left(\left(1 - \frac{1}{m}\right)L^*(S_{k,l}^{(m)}), m\right) - f_m^0. \qquad (2.10)$$

Let $\Omega(m) = (1 - 1/m)\sum_{k=1}^{K_m}\sum_{l=1}^{|S_k^{(m)}|/m} L^*(S_{k,l}^{(m)})$ for $m = 1, \cdots, q$. Substituting (2.10) into (2.8) we obtain

$$Z_f^H \le \sum_{m,k,l} f\left(2m^{-1}\sum_{i \in S_{k,l}^{(m)}} d_i, m\right)$$

$$+ \sum_{m,k,l}\left[f\left(\left(1 - \frac{1}{m}\right)L^*\left(S_{k,l}^{(m)}\right), m\right) - f_m^0\right]$$

$$\le \sum_{m,k,l} f\left(2m^{-1}\sum_{i \in S_{k,l}^{(m)}} d_i, m\right)$$

$$+ \sum_{m=1}^{q} \max\left\{\sum_{k,l}\left[f(\xi_{k,l}, m) - f_m^0\right] : \sum_{k,l}\xi_{k,l} = \Omega(m)\right\}.$$

Since f is concave in its first argument, the maxima within $\{\ \}$ are achieved by equalizing all of the $\xi_{k,l}$, i.e., $\xi_{k,l} = m\Omega(m)/n_m$ for all k, l $(m = 1, \cdots, q)$. Thus,

$$Z_f^H \le \sum_{m,k,l} f\left(2m^{-1}\sum_{i \in S_{k,l}^{(m)}} d_i, m\right) + \sum_{m=1}^{q} \frac{n_m}{m}\left[f\left(\frac{m\Omega(m)}{n_m}, m\right) - f_m^0\right].$$

In view of (2.9) and Lemmas 2.12–2.13 we have:

$$
\begin{aligned}
\Omega(m) &\leq \left(1 - \frac{1}{m}\right) L^*(N^{(m)}) + 1.5 \left(1 - \frac{1}{m}\right) P^{(m)} \\
&\leq \left(1 - \frac{1}{m}\right) (2\pi n_m)^{1/2} d^{(m)} \\
&\quad + 1.5 \left(1 - \frac{1}{m}\right) d^{(m)} \left((4\pi + 2)\lfloor n_m^{1/2}\rfloor + 10\pi + 2\right) \\
&\leq d^{(m)} (\alpha_m \sqrt{n_m} + \beta_m), \qquad m = 1, \cdots, q.
\end{aligned} \tag{2.11}
$$

The Theorem thus follows from this bound and the monotonicity of f in ϑ. $\quad\square$

The bound $\overline{Z}_f^H$ is used below to prove that the MCRP heuristic is asymptotically optimal. It should be noted that alternative and potentially more accurate upper bounds may be derived which can be computed after Step 3 of the MCRP heuristic.

For this purpose, let $\delta_m = \partial^+ f(\vartheta, m)/\partial\vartheta$, the right side partial derivative of f with respect to ϑ, for $m = 1, \cdots, q$. Since f is concave, $\partial^+ f(\vartheta, m)/\partial\vartheta$, exists for all $\vartheta > 0$ and is non-increasing. Define

$$
\overline{Z}_f^2 = \sum_{m=1}^q \sum_{k=1}^{K_m} \sum_{l=1}^{|S_{k,l}^{(m)}|/m} f\left(2m^{-1} \sum_{i \in S_{k,l}^{(m)}} d_i, m\right) + \sum_{m=1}^q d^{(m)} \delta_m (\alpha_m \sqrt{n_m} + \beta_m).
$$

Theorem 2.15. $Z_f^H \leq \overline{Z}_f^2$.

Proof. It follows from (2.8), the monotonicity of f in ϑ and the definition of δ_m that

$$
\begin{aligned}
Z_f^H &\leq \sum_{m,k,l} f\left(2m^{-1} \sum_{i \in S_{k,l}^{(m)}} d_i, m\right) + \sum_{m=1}^q \delta_m \left(1 - \frac{1}{m}\right) \sum_{k,l} L^*(S_{k,l}^{(m)}) \\
&= \sum_{m,k,l} f\left(2m^{-1} \sum_{i \in S_{k,l}^{(m)}} d_i, m\right) + \sum_{m=1}^q \delta_m \Omega(m),
\end{aligned}
$$

where $\Omega(m)$ is defined in the proof of Theorem 2.14. Use (2.11) to complete the proof. $\quad\square$

The following theorem shows that the MCRP heuristic is asymptotically optimal.

Theorem 2.16. *Let $\{x_1, x_2 \cdots\}$ be a sequence of random points whose radial distances are i.i.d. Let $\eta = E f(2d)$.*

(a) $\underline{\lim}_{n\to\infty} \overline{Z}_f^2/n \geq \underline{\lim}_{n\to\infty} \underline{Z}_f^1/n \geq \eta/q$ *(a.s.).*

(b) If, in addition, $\eta > 0$ and the radial distances are uniformly bounded by ρ, the MCRP heuristic is asymptotically optimal and the lower bounds $\underline{Z}_f^1$ and $\underline{Z}_f^2$ and the upper bounds Z_f^H and $\overline{Z}_f^H$ are asymptotically accurate.

Proof. (a) Note that $f(\cdot, m)$ is continuous for all $m = 1, \cdots, q$, except possibly for $\vartheta = 0$. Thus $f(\vartheta)$ is continuous for $\vartheta > 0$ (since it is concave) so that f is integrable and η exists. For any given set of customers N_c, let $\{N_1, N_2, \cdots N_R\}$ be a partition achieving $\underline{Z}_f^1$. Thus, since f is concave in ϑ,

$$\underline{Z}_f^1 = \sum_{l=1}^{R} f\left(\frac{2\sum_{i \in N_j} d_i}{|N_j|}, |N_j|\right) \geq \sum_{l=1}^{R} \frac{1}{|N_l|} \sum_{i \in N_l} f(2d_i, |N_l|)$$

$$\geq \frac{1}{q} \sum_{l=1}^{R} \sum_{i \in N_l} \underline{f}(2d_i) = \frac{1}{q} \sum_{i=1}^{n} \underline{f}(2d_i).$$

Part (a) now follows from the law of large numbers.

(b) In view of part (a) and Theorem (2.14) it suffices to show that

$$\lim_{n \to \infty} \frac{1}{n} \left[\overline{Z}_f^H - \underline{Z}_f^1\right] = 0 \quad \text{(a.s.)}.$$

Fix a realization of the sequence $\{x_1, x_2, \cdots\}$. We write $\overline{Z}_f^H = W_1 + W_2$, where

$$W_1 = \sum_{m,k,l} f\left(2m^{-1} \sum_{i \in S_{k,l}^{(m)}} d_i, m\right)$$

and

$$W_2 = \sum_{m=1}^{q} \frac{n_m}{m} \left[f\left(\frac{m\alpha_m d^{(m)}}{\sqrt{n_m}} + \frac{\beta_m d^{(m)}}{n_m}, m\right) - f_m^0\right].$$

We first show that $\lim_{n \to \infty} W_2/n = 0$ (a.s.). Since f is non-decreasing in ϑ, and the expressions in $[\ \ ]$ are non-negative, assume to the contrary that for some $m = 1, \cdots, q$, $\overline{\lim}_{n \to \infty} a(n) = \gamma > 0$, where

$$a(n) = \frac{|N^{(m)}(n)|}{nm} \left[f\left(\frac{m\alpha_m d^{(m)}(n)}{|N^{(m)}(n)|^{1/2}} + \frac{\beta_m d^{(m)}(n)}{|N^{(m)}(n)|}, m\right) - f_m^0\right].$$

Let $\{n_k\}_{k=1}^{\infty}$ be a sequence of integers such that $\lim_{k \to \infty} |N^{(m)}(n_k)|$ exists and $\lim_{k \to \infty} a(n_k) = \gamma$. If $\lim_{k \to \infty} |N^{(m)}(n_k)| = \infty$, we have in view of $|N^{(m)}|/n \leq 1$, $d^{(m)} \leq \rho$, the definition of f_m^0 and the monotonicity and concavity of f in ϑ,

$$0 < \gamma \leq \frac{1}{m} \lim_{k \to \infty} \left[f\left(\frac{m\alpha_m \rho}{|N^{(m)}(n_k)|^{1/2}} + \frac{\beta_m \rho}{|N^{(m)}(n_k)|}, m\right) - f_m^0\right] = 0.$$

Thus, $\lim_{k \to \infty} |N^{(m)}(n_k)| < \infty$. But then, since f is monotone in ϑ, we have

$$0 < \gamma \leq \lim_{k \to \infty} \frac{|N^{(m)}(n_k)|}{mn_k} \left[f(m\alpha_m \rho + \beta_m \rho, m) - f_m^0\right] = 0,$$

which leads to a contradiction.

It remains to be shown that

$$\lim_{n\to\infty}[W_1 - \underline{Z}_f^1] = 0 \qquad \text{(a.s.).} \tag{2.12}$$

We assume that the MCRP heuristic is implemented on the basis of a partition $\{N_1, \cdots, N_R\}$ achieving $\underline{Z}_f^1$. (A similar proof applies if it is implemented on the basis of a partition achieving $\underline{Z}_f^2$.) Let

$$W_1^{(m)} = \sum_{k=1}^{K_m} \sum_{l=1}^{|S_k^{(m)}|/m} f\left(\frac{2\sum_{i\in S_{k,l}^{(m)}} d_i}{m}, m\right), \qquad m = 1, \cdots, q,$$

and

$$\underline{V}^{1(m)} = \sum_{l:|N_l|=m} f\left(\frac{2\sum_{i\in N_l} d_i}{m}, m\right) \qquad m = 1, \cdots, q,$$

To prove (2.12) it suffices to show that

$$\lim_{n\to\infty} \frac{1}{n}[W_1^{(m)} - \underline{V}^{1(m)}] = 0 \qquad \text{for all } m = 1, \cdots, q.$$

Thus, fix m $(1 \le m \le q)$. In the following we drop the superscripts (m) whenever possible.

Consider the circle which is centered at the origin and has a radius $d^{(m)}$. Partition this circle into $\lceil n_m^{1/4}\rceil$ so-called 'major circle rings' as follows. Define $\Delta = d^{(m)}/\lceil n_m^{1/4}\rceil$. Let C_1 be the closed circle, centered at the origin with radius Δ. For $j = 1, \cdots, \lceil n_m^{1/4}\rceil$ let C_j denote the half-open circle ring bordered by the two circles centered at the origin, with radius $(j-1)\Delta$ and $j\Delta$ and with only the outer circle included in the ring; see Figure 5.

Also let $C_{j,k}$ be the intersection of C_j with the kth sector $(j = 1, \cdots, \lceil n_m^{1/4}\rceil, k = 1\cdots, K_m)$.

The proof is established by the determination of an upper bound $U \ge W_1^{(m)}$ and a lower bound $L \le \underline{V}^{1(m)}$ which, unlike $W_1^{(m)}$ and $\underline{V}^{1(m)}$, are easily compared with each other. Define $E_j = \{l : \text{the lowest indexed point in } N_l \text{ belongs to } C_j\}$.

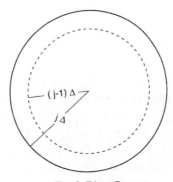

Fig. 5. Ring C_j.

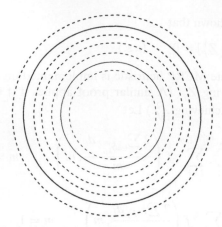

Fig. 6. The ring between the bold-faced circles represents a circle ring C_j. The rings between the dotted circles represent the consecutive sets of cardinality m for which the intersection with C_j is nonempty.

Note that the collection $\{E_j, j = 1, \cdots \lceil n_m^{1/4} \rceil\}$ partitions $\{l : N_l \subset N^{(m)}\}$. Also let $v_j = |\{i : i \in N^{(m)}$ and i belongs to $C_j\}|$ $(j = 1, \cdots, \lceil n_m^{1/4} \rceil)$. Since the partition $\{N_1, \cdots, N_R\}$ is consecutive, all the points in C_j belong to some set N_l with $l \in E_j$ with the possible exception of the $(m - 1)$ closest points in C_j; see Figure 6. Hence

$$m|E_j| \geq v_j - m + 1, \qquad j = 1, \cdots, \lceil n_m^{1/4} \rceil. \tag{2.13}$$

Moreover, since f is monotone in ϑ and in view of (2.13)

$$
\begin{aligned}
\underline{V}^{1(m)} &= \sum_{j=1}^{\lceil n_m^{1/4} \rceil} \sum_{l \in E_j} f\left(\frac{2 \sum_{i \in N_l} d_i}{m}, m\right) \\
&\geq \sum_{j=1}^{\lceil n_m^{1/4} \rceil} |E_j| f(2(j-1)\Delta, m) \\
&\geq \sum_{j=1}^{\lceil n_m^{1/4} \rceil} \left(\frac{v_j}{m} - 1\right) f(2(j-1)\Delta, m).
\end{aligned}
$$

Similarly, define for all $k = 1, \cdots, K_m$ and $j = 1, \cdots, \lceil n_m^{1/4} \rceil$ the sets $D_{j,k} = \{l :$ the highest indexed point in $S_{k,l}$ belongs to $C_{j,k}\}$ and $D_j = \bigcup_{k=1}^{K_m} D_{j,k}$. Note that the collection $\{D_{j,k} : j = 1, \cdots, \lceil n_m^{1/4} \rceil\}$ partitions the set of subregions generated in the kth sector, $k = 1, \cdots, K_m$. Also, let $v_{j,k} = |\{i : i \in N^{(m)}$ and i belongs to $C_{j,k}\}|$ $j = 1, \cdots, \lceil n_m^{1/4} \rceil$ and $k = 1, \cdots, K_m$.

Since within a given sector the sets $\{S_{k,l}\}$ are all consecutive and since they all have cardinality m, only the points in $C_{j,k}$ and possibly the $(m - 1)$ highest indexed

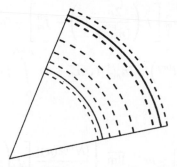

Fig. 7. The area between the bold-faced lines represents $C_{j,k}$: the intersection of the circle ring C_j with the kth sector. The areas between the dotted arcs represent the subregions of cardinality m in the kth sector ($S_{k,1}^{(m)}$) for which the intersection with $C_{j,k}$ is nonempty.

points in $C_{j-1,k}$ may belong to a set $S_{k,l}$ with $l \in D_{j,k}$; see Figure 7. Thus,

$$\nu_{j,k} + (m-1) \geq m|D_{j,k}| \qquad j = 1, \cdots, \lceil n_m^{1/4} \rceil, k = 1, \cdots, K_m,$$

and hence, in view of $K_m \leq \sqrt{n_m} + 1$,

$$\nu_j + (m-1)(\sqrt{n_m} + 1) \geq \sum_{k=1}^{K_m} (\nu_{j,k} + (m-1)) \geq m|D_j|. \qquad (2.14)$$

Moreover, since f is monotone in ϑ and in view of (2.14)

$$
\begin{aligned}
W_1^{(m)} &= \sum_{j=1}^{\lceil n_m^{1/4} \rceil} \sum_{k=1}^{K_m} \sum_{l \in D_{j,k}} f\left(2 \sum_{i \in S_{k,l}} d_i / m, m \right) \\
&\leq \sum_{j=1}^{\lceil n_m^{1/4} \rceil} |D_j| f(2j\Delta, m) \\
&\leq \sum_{j=1}^{\lceil n_m^{1/4} \rceil} m^{-1} \nu_j f(2j\Delta, m) + (\sqrt{n_m} + 1) \sum_{j=1}^{\lceil n_m^{1/4} \rceil} f(2j\Delta, m) \\
&\leq \sum_{j=1}^{\lceil n_m^{1/4} \rceil} m^{-1} \nu_j f(2j\Delta, m) + \Delta^{-1}(\sqrt{n_m} + 1) \int_{\Delta}^{\rho+\Delta} f(2\vartheta, m)\, d\vartheta.
\end{aligned}
$$

We thus obtain, since f is concave in ϑ

$$
\begin{aligned}
W_1^{(m)} - \underline{V}^{1(m)} &\leq m^{-1} \sum_{j=1}^{\lceil n_m^{1/4} \rceil} \nu_j \big[f(2(j-1)\Delta + 2\Delta, m) - f(2(j-1)\Delta, m) \big] \\
&\quad + \sum_{j=1}^{\lceil n_m^{1/4} \rceil} f(2(j-1)\Delta, m) + \Delta^{-1}(\sqrt{n_m} + 1) \int_{\Delta}^{\rho+\Delta} f(2\vartheta, m)\, d\vartheta
\end{aligned}
$$

$$\leq m^{-1} n_m \left[f\left(\frac{2\rho}{\lceil n_m^{1/4} \rceil}, m \right) - f_m^0 \right] + \lceil n_m^{1/4} \rceil f(2\rho, m)$$

$$+ \left(d^{(m)} \right)^{-1} \lceil n_m^{1/4} \rceil (\sqrt{n_m} + 1) \int_\Delta^{\rho + \Delta} f(2\vartheta, m)\, d\vartheta.$$

Thus,

$$\overline{\lim_{n \to \infty}} \frac{1}{n} \left[W_1^{(m)} - \underline{V}^{1(m)} \right] \leq m^{-1} \overline{\lim_{n \to \infty}} \left\{ \frac{|N^{(m)}(n)|}{n} \left[f\left(\frac{2\rho}{\lceil n_m^{1/4} \rceil}, m \right) - f_m^0 \right] \right\} = 0,$$

where the equality is verified in complete analogy to the above proof for $\lim_{n \to \infty} n^{-1} W_2$. $\square$

3. The CVRP with unequal demands

3.0. Introduction

In this section we consider the capacitated vehicle routing problem with unequal demands (UCVRP). In this version, each customer i has a demand w_i and the capacity constraint states that the total amount delivered by a single vehicle cannot exceed q.

If the demand of a customer can be split over several vehicles, the problem is reduced to the equal-weight case by treating a customer with demand w as w customers with unit size demand all at the same location. This is easily verified for the worst case analyses. It also holds for probabilistic analyses, assuming demands and customer locations are independent; see our remark following Theorem 2.6. Therefore, we consider the case where the demand of a customer may not be divided among vehicles.

As with other routing problems, two possible objectives may be considered or combinations thereof: minimizing the number of vehicles used (MV) or minimizing the total distance traveled (MD). The first objective is trivially solved in the equal weight case. In the unequal weight case the problem is known as the bin-packing problem. For an excellent survey on probabilistic analysis of packing algorithms, see the recent book by Coffman & Lueker [1991] and on worst-case analysis, see Coffman, Garey & Johnson [1984].

It is interesting to compare the optimal solutions for both objectives in terms of both performance measures. Thus, let Z^V and Z^D denote the total distance traveled in an optimal solution for the MV-problem and MD-problem respectively. Similarly, let K^V and K^D denote the number of vehicles used in the optimal solutions for the MV and MD problems. In case several optimal solutions exist for the MD problem, we choose the one which, among all optimal solutions, uses the minimum number of vehicles. The following relationship is easily verified:

$$K^V \le K^D < 2K^V.$$

The first inequality is immediate. To prove the second one, assume the set of routes $\{N_1, N_2, \cdots, N_{K^D}\}$ is exploded by the optimal solution of the MD problem. Clearly, $\sum_{i \in N_j} w_i + \sum_{i \in N_{j+1}} w_i > q$ for every $j = 1, 2, \cdots, K^D$ and where $K^D + 1 \equiv 1$. This is true, since otherwise a pair of routes (N_j, N_{j+1}) can be merged into a single route without increasing the total distance traveled while reducing the number of vehicles, contradicting the definition of K^D. Adding all the K^D inequalities, we obtain:

$$2q K^V \ge 2 \sum_{i \in N} w_i > q K^D.$$

Thus, $K^D < 2K^V$.

On the other hand, the following example shows that the ratio Z^V/Z^D can grow linearly with the number of customers. Consider the following problem with $2m$ customers, m of which are located at a single point, one unit of distance from the depot while customers $m + 1, m + 2, \cdots, 2m$ are located at the depot. Let $w_i = q2^{-i}$ and $w_{m+i} = q - w_i$ for $i = 1, 2, \cdots, m$. Clearly, $Z^D = 2$ since the first m customers can be assigned to a single vehicle. The number of vehicles is minimized by pairing customer i and $m + i$ for all $i = 1, 2, \cdots, m$. Thus $Z^V = 2m$ and $Z^V/Z^D = m$.

We thus confine ourselves in the next subsections to the MD-objective. To simplify notation and for the sake of consistency we use Z_u^* to denote the optimal solution value of the UCVRP when the objective is to minimize total distance traveled.

3.1. Worst-case analysis of heuristics

In the worst-case analysis presented here, we assume that the numbers $w_1, w_2, \cdots, w_n$ and q are rationals, hence, without loss of generality, q and w_i are assumed to be integers. Furthermore, we may assume that q is even; otherwise one can double q as well as each w_i, $i = 1, 2, \cdots, n$, without affecting the problem. The following two-phase heuristic was suggested by Altinkemer and Gavish [1987]. In the first phase, we relax the requirement that the demand of a customer cannot be split. Each customer i is replaced by w_i unit demand points that are of zero distance apart. We then use the ITP(α) heuristic while setting the vehicle capacity at $q/2$. In the second phase, we convert the solution obtained in phase I to a feasible solution to the original problem without increasing the total cost. This heuristic is called the unequal-weight iterated tour partitioning (UITP(α)) heuristic.

We now describe the second phase procedure. Our notation follows the one suggested by Haimovich, Rinnooy Kan & Stougie [1988]. Let $m = \sum_{i \in N_c} w_i$ be the number of demand points. Recall that in the first phase an arbitrary orientation of the tour is chosen. The customers are then numbered $x^{(0)}, x^{(1)}, x^{(2)}, \ldots, x^{(n)}$

in order of their appearance on the tour, where $x^{(0)}$ is the depot. The ITP(α) heuristic partitions the path from $x^{(1)}$ to $x^{(n)}$ into $\lceil 2m/q \rceil$ disjoint segments such that each one contains no more than $q/2$ demand points and connect the endpoints of each segment to the depot. The segments are indexed by j, $j = 1, 2, \cdots, \lceil 2m/q \rceil$, such that the first customer of the jth segment is $x^{(b_j)}$ and the last customer is $x^{(e_j)}$. Hence, the jth segment, denoted by S_j, includes customers $\{x^{(b_j)}, \cdots, x^{(e_j)}\}$. Obviously, if $x^{(e_j)} = x^{(b_{j+1})}$ for some j then the demand of customer $x^{(e_j)}$ is split between the jth and $(j + 1)$st segments; therefore these are not feasible routes. On the other hand, if $x^{(e_j)} \neq x^{(b_{j+1})}$ for all j, then the set of routes is feasible.

We now transform the solution obtained in the first phase into a feasible solution without increasing the total distance traveled. We use the following procedure:

Procedure A:

Step 1. Set $S_j' = \emptyset$, for $j = 1, 2, \cdots, \lceil 2m/q \rceil$
Step 2. For $j = 1$ to $\lceil 2m/q \rceil - 1$ do begin
    If $x^{(e_j)} = x^{(b_{j+1})}$ then
       if $\sum_{i=b_j}^{b_{j+1}} w_{x^{(i)}} \leq q$ then let $S_j' = \{x^{(b_j)}, \cdots, x^{(e_j)}\}$ and $x^{(b_{j+1})} = x^{(b_{j+1}+1)}$
            else, let $S_j' = \{x^{(b_j)}, \cdots, x^{(e_j-1)}\}$ and $x^{(b_{j+1})} = x^{(e_j)}$
       else, let $S_j' = \{x^{(b_j)}, \cdots, x^{(e_j)}\}$
   end.

We first observe that the procedure succeeds in generating feasible sets S_j' ($j = 1, 2, \cdots, \lceil 2m/q \rceil$). Note that the jst set can be enlarged only in the $(j - 1)$st and jth iterations (if at all). Moreover if it is enlarged in the jth iteration, it is clearly done feasibly in view of the test $\sum_{i=b_j}^{b_{j+1}} w_{x^{(i)}} \leq q$. On the other hand, if S_j is enlarged in the $(j - 1)$st iteration, at most $q/2$ demand points are added thus ensuring feasibility. This can be verified as follows: Assume to the contrary that in the $(j - 1)$st iteration more than $q/2$ demand points are transferred from S_{j-1}' to S_j so that in the $(j - 1)$st iteration $x^{(e_{j-1})} = x^{(b_j)}$. Since the original set S_{j-1} contains at most $q/2$ demand points we must have shifted demand points in the $(j - 2)$th iteration from S_{j-2} to S_{j-1} (and in particular $x^{(b_{j-1})} = x^{(e_{j-2})}$) part of which are now being transferred to S_j. This implies that $x^{(*)} \equiv x^{(e_{j-2})} = x^{(b_{j-1})} = x^{(e_{j-1})} = x^{(b_j)}$, where $e_{j-2}, b_{j-1}, e_{j-1}$ and b_j refer to the original sets S_{j-2}, S_{j-1} and S_j. In other words at the beginning of the $(j - 1)$st iteration the set S_{j-1}' contains a single customer $x^{(*)}$. But then, shifting $x^{(b_j)} = x^{(*)}$ backwards to S_{j-1}' is feasible, contradicting the fact that more than $q/2$ demand points need to be shifted forward from S_{j-1}' to S_j'.

Theorem 3.1 [Altinkemer & Gavish, 1987].

$$\frac{Z^{\text{UITP}(\alpha)}}{Z_u^*} \leq 2 + \left(1 - \frac{2}{q}\right)\alpha.$$

Proof. Using the bound of Lemma 2.2 we obtain the following upper bound on the length of the tours generated in phase I of the UITP(α) heuristic,

$$\frac{4}{q} \sum_{i \in N_c} d_i w_i + \left(1 - \frac{2}{q}\right) \alpha L^*(N). \tag{3.1}$$

In the second phase of the algorithm, the tour obtained in the first-phase is converted into a feasible solution with total length no more than (3.1). To verify this, we need only to analyze those segments whose endpoints are modified by the procedure.

Suppose that S_j and S'_j differ in their starting point; then S'_j must start with $x^{(b_j+1)}$. This implies that arc $(x^{(b_j)}, x^{(b_j+1)})$ which is part of the phase I solution, does not appear in the jth route. The triangle inequality ensures that the sum of the length of arcs $(x^{(0)}, x^{(b_j)})$ and $(x^{(b_j)}, x^{(b_j+1)})$ is no smaller than the length of arc $(x^{(0)}, x^{(b_j+1)})$. A similar argument can be applied if S_j and S'_j differ in their terminating point. Consequently, for every segment j, for $j = 1, 2, \cdots, \lceil 2m/q \rceil$, the length of the jth route according to the new partition is no larger than the length of the jth route according to the old partition. Hence,

$$Z^{\text{UITP}(\alpha)} \leq \frac{4}{q} \sum_{i \in N_c} d_i w_i + \left(1 - \frac{2}{q}\right) \alpha L^*(N).$$

Clearly, $Z_u^* \geq Z^*$, and therefore using the lower bound on Z^* developed in Lemma 2.1 completes the proof. $\square$

· *Remark.* The worst-case bound of Theorem 3.1 would be improved if the capacity in Phase I was set at a level larger than $q/2$. Unfortunately, such a change fails to generate feasible tours at the end of the Phase II procedure.

The UITP heuristic was divided into two phases to prove the above worst-case result. However, if the optimal partitioning heuristic is used in the unequal weight model, the actual implementation is a one step process. This version of the heuristic, called the unequal-weight optimal partitioning (UOP) heuristic, also has $Z^{\text{UOP}(\alpha)}/Z^* \leq 2 + (1 - 2/q)\alpha$. The following theorem, proved by Li & Simchi-Levi [1990], implies that when $\alpha = 1$, this bound is asymptotically tight as q approaches infinity.

Theorem 3.2. *For any integer $q \geq 1$, there exists a problem instance with $Z^{\text{UOP}(1)}/Z_u^*$ (and therefore $Z^{\text{UITP}(1)}/Z_u^*$) arbitrarily close to $3 - 6/(q + 2)$.*

Proof. We modify the graph $\mathcal{G}(2, Kq + 1)$, where K is a positive integer, as follows. Every group now, instead of containing q customers, contains only one customer with demand q. The other Kq customers have unit demand. The optimal traveling salesman tour is again as shown in Figure 2, and the solution obtained by the

UOP(1) heuristic is to have $2Kq + 1$ vehicles, each one of them serving only one customer. Thus

$$Z^{\text{UOP}(1)} = 2(Kq + 1) + 4Kq.$$

The optimal solution to this problem has $Kq + 1$ vehicles serve those customers with demand q, and K other vehicles serve the unit demand customers. Hence,

$$Z_u^* = 2(Kq + 1) + 4K.$$

Therefore,

$$\lim_{K \to \infty} \frac{Z^{\text{UOP}(1)}}{Z_u^*} = \lim_{K \to \infty} \frac{2(Kq + 1) + 4Kq}{2(Kq + 1) + 4K} = 3 - \frac{6}{q + 2}. \qquad \square$$

3.2. The asymptotic optimal solution value

Our probabilistic analysis will use results from the bin-packing problem, a problem that has been analyzed extensively in the literature. An instance of the bin-packing problem is composed of the bin capacity (equal to 1) and a set of items each with a prespecified size no larger than 1. The problem is to find the smallest number of bins in which these items can be packed, subject to the constraint that the total size of items assigned to a single bin does not exceed 1.

In the probabilistic analysis of the capacitated vehicle routing problem with unsplit demands we assume, without loss of generality, and in accordance to the convention in Coffman, Lueker and Rinnooy Kan [1988], that the vehicles' capacity q equals 1, and the demand of each customer is no more than 1. Thus, vehicles and demands in the capacitated vehicle routing problem correspond to bins and item sizes (respectively) in the bin-packing problem. Hence, for every routing instance there is a unique corresponding bin-packing instance.

Without loss of generality, we center the plane at the depot and denote by $\mathfrak{R}_0^2 = \mathfrak{R}^2 \setminus \{\text{depot}\}$. Assume the demands $w_1, w_2, \cdots, w_n$ are independent and identically distributed with pdf Φ defined on $[0, 1]$. In this section we find the asymptotic optimal solution value for *any* distribution of the demands Φ. This is done by showing that an asymptotically optimal algorithm for the bin-packing problem, with item sizes distributed like Φ, can be used to solve the capacitated vehicle routing problem with unsplit demands.

Given the demands $w_1, w_2, \cdots, w_n$, let b^* be the number of bins used in the optimal solution for the corresponding bin-packing problem. As demonstrated in Rhee & Talagrand [1987], there exists a constant $\gamma > 0$ such that

$$\lim_{n \to \infty} \frac{b^*}{n} = \gamma \qquad \text{(a.s.)}. \tag{3.2}$$

The following theorem was proved by Simchi-Levi & Bramel [1990a].

Theorem 3.3. *Let* x_k, $k = 1, 2, \cdots, n$ *be a sequence of independent random variables having a distribution* μ *with compact support in* $A \subset \Re_0^2$. *Let* $d(y)$ *be the Euclidean distance between the depot and point* $y \in \Re^2$ *and let*

$$E(d) = \int_{\Re^2} d(y) \, d\mu(y).$$

Let the demands w_k, $k = 1, 2, \cdots, n$ *be a sequence of i.i.d. random variables having a distribution* Φ *with support on* $[0, 1]$ *and assume that the demands and the locations of the customers are independent of each other. Then, almost surely,*

$$\lim_{n \to \infty} \frac{1}{n} Z_u^* = 2\gamma E(d),$$

where γ *is defined in (3.2).*

Thus, Theorem 3.3 fully characterizes the asymptotic optimal solution value of the capacitated vehicle routing problem, for any distribution function Φ. The restriction that μ has support in $\Re_0^2$ is without loss of generality, since any customer located at the depot can be served with zero cost.

An interesting observation concerns the case where the distribution of the demands allows perfect packing, that is, when the wasted space in the bins tends to become a small fraction of the number of bins used. Formally, Φ is said to allow perfect packing if $\lim_{n \to \infty} (b^*/n) = E(w)$ (a.s.). Karmarkar [1982] proved that a non-increasing probability density function (with some mild regularity conditions) allows for perfect packing. Rhee [1988] completely characterizes the class of distribution functions Φ which allow for perfect packing. Clearly, in this case $\gamma = E(w)$. Thus, Theorem 3.3 indicates, surprisingly, that allowing the demands to be split or not does not change the asymptotic objective function value. That is, the capacitated vehicle routing problem with unsplit demands and the capacitated vehicle routing problem when demands can be split are *asymptotically equivalent*, when Φ allows for perfect packing.

To prove Theorem 3.3, we start by presenting in Section 3.2.1 a lower bound on the optimal objective function. In Section 3.2.2, we present a heuristic algorithm for the capacitated vehicle routing problem with unsplit demands based on a simple region partitioning scheme. We show that the cost of the solution produced by the heuristic converges to our lower bound for any distribution Φ, thus proving the main theorem of the section.

3.2.1. A lower bound

In this section, we introduce a lower bound on the optimal objective function value Z_u^*. Define $d_{\sup} = \sup_{y \in A} \{d(y)\}$ and for a given fixed positive integer r, partition the circle with radius $d_{\sup}$ centered at the depot, into r circle rings of equal width. Thus, let $\underline{d}_j = (j-1)(d_{sup}/r)$ for $j = 1, 2, \cdots, r, r+1$, and construct the following $2r$ sets of customers:

$$S_j = \left\{ x_k \in N_c \mid \underline{d}_j < d_k \leq \underline{d}_{j+1} \right\} \qquad \text{for } j = 1, \cdots, r,$$

and

$$F_j = \bigcup_{i=j}^{r} S_i \qquad \text{for } j = 1, \cdots, r.$$

Note that $F_r \subseteq F_{r-1} \subseteq \cdots \subseteq F_1 = N_c$ since $d_k > 0$ for all $y_k \in N_c$.

In the lemma below, we show that $|F_r|$ grows to infinity, almost surely, as n grows to infinity. This implies that $|F_j|$ also grows to infinity, almost surely, for $j = 1, 2, \cdots, r$, since $|F_{j+1}| \leq |F_j|$, for $j = 1, 2, \cdots, r-1$.

Lemma 3.4.

$$\lim_{n \to \infty} \frac{|F_r|}{n} = p \text{ (a.s.)} \quad \text{for some real constant } p > 0.$$

Proof. Let $D = \{y \in A \mid \underline{d}_r = (r-1)(d_{sup}/r) < d(y)\}$. Note that if a customer is located in D, then the customer is in F_r. Define $p = \int_D d\mu(y)$, that is p is the probability that a given customer falls in the region D. Since A is the support of μ and $d_{sup} = \sup_{y \in A}\{d(y)\}$, $p > 0$. So, $|F_r|$ is binomial (n, p). $\square$

For any set of customers $T \subseteq N_c$, let $b^*(T)$ be the minimum number of vehicles needed to serve the customers in T, i.e., $b^*(T)$ is the optimal solution to the bin-packing problem defined by item sizes equal to the demands of the customers in T. We can now present a family of lower bounds on Z_u^* that hold for different values of $r > 1$.

Lemma 3.5.

$$Z_u^* \geq 2\frac{d_{sup}}{r} \sum_{j=2}^{r} b^*(F_j) \qquad \text{for any } r > 1.$$

Proof. Given an optimal solution to the capacitated vehicle routing problem with unsplit demands, let K_r^* be the number of vehicles in the optimal solution that serve at least one customer from S_r, and for $j = 1, 2, \cdots, r-1$, let K_j^* be the number of vehicles in the optimal solution that serve at least one customer in the set S_j, but do not serve any customers in F_{j+1}. Also, let V_j^* be the number of vehicles in the optimal solution that serve at least one customer in F_j. By these definitions, $V_j^* = \sum_{i=j}^{r} K_i^*$, for $j = 1, 2, \cdots, r$, hence $K_j^* = V_j^* - V_{j+1}^*$ for $j = 1, 2, \cdots, r-1$ and $K_r^* = V_r^*$.

Note that $V_j^* \geq b^*(F_j)$, for $j = 1, 2, \cdots, r$, since V_j^* represents the number of vehicles used in a feasible packing of the demands of customers in F_j, while $b^*(F_j)$ represents the number of bins used in an optimal packing.

By the definition of K_j^* and $\underline{d}_j$, $Z_u^* > 2\sum_{j=1}^{r} \underline{d}_j K_j^*$, and therefore

$$Z_u^* > 2\underline{d}_r V_r^* + \sum_{j=1}^{r-1} 2\underline{d}_j \left\{ V_j^* - V_{j+1}^* \right\}$$

$$= 2\underline{d}_1 V_1^* + \sum_{j=2}^{r} 2(\underline{d}_j - \underline{d}_{j-1}) V_j^*$$

$$= 2\sum_{j=2}^{r} (\underline{d}_j - \underline{d}_{j-1}) V_j^* \qquad \text{(since } \underline{d}_1 = 0)$$

$$\geq 2\sum_{j=2}^{r} (\underline{d}_j - \underline{d}_{j-1}) b^*(F_j) \qquad \text{(since } V_j^* \geq b^*(F_j))$$

$$= 2\sum_{j=2}^{r} \frac{d_{\sup}}{r} b^*(F_j) \qquad \text{(since } \underline{d}_j - \underline{d}_{j-1} = \frac{d_{\sup}}{r}). \qquad \square$$

Note that Lemma 3.5 provides a deterministic lower bound, that is, no probabilistic assumptions are involved. Lemma 3.4 and Lemma 3.5 are both required to provide a lower bound on $(1/n)Z_u^*$ that holds almost surely.

Lemma 3.6. *Under the conditions of Theorem 3.3, we have*

$$\lim_{n\to\infty} \frac{1}{n} Z_u^* \geq 2\gamma E(d) \qquad (a.s.).$$

Proof. Lemma 3.5 implies that

$$\lim_{n\to\infty} \frac{1}{n} Z_u^* \geq 2\frac{d_{\sup}}{r} \lim_{n\to\infty} \sum_{j=2}^{r} \frac{b^*(F_j)}{n} = 2\frac{d_{\sup}}{r} \sum_{j=2}^{r} \lim_{n\to\infty} \frac{b^*(F_j)}{|F_j|} \lim_{n\to\infty} \frac{|F_j|}{n}.$$

From Lemma 3.4, $|F_j|$ grows to infinity almost surely as n grows to infinity, for $j = 1, 2, \cdots, r$. Moreover, since demands and locations are independent of each other, the demands in F_j, $j = 1, 2, \cdots, r$ are distributed like Φ. Therefore,

$$\lim_{n\to\infty} \frac{b^*(F_j)}{|F_j|} = \lim_{|F_j|\to\infty} \frac{b^*(F_j)}{|F_j|} = \gamma \qquad (a.s.).$$

Hence, almost surely

$$\lim_{n\to\infty} \frac{1}{n} Z_u^* \geq 2\frac{d_{\sup}}{r} \sum_{j=2}^{r} \gamma \lim_{n\to\infty} \frac{|F_j|}{n} = 2\frac{d_{\sup}}{r} \gamma \lim_{n\to\infty} \frac{1}{n} \sum_{j=2}^{r} |F_j|.$$

Since,

$$F_j = \bigcup_{i=j}^{r} S_i \qquad \text{for } j = 1, 2, \cdots, r,$$

we have $|F_j| = \sum_{i=j}^{r} |S_i|$, hence, almost surely,

$$\lim_{n\to\infty} \frac{1}{n} Z_u^* \geq 2 \frac{d_{\text{sup}}}{r} \gamma \lim_{n\to\infty} \frac{1}{n} \sum_{j=2}^{r} \sum_{i=j}^{r} |S_i| = 2 \frac{d_{\text{sup}}}{r} \gamma \lim_{n\to\infty} \frac{1}{n} \sum_{j=2}^{r} (j-1)|S_j|.$$

By the definition of $\underline{d}_j$,

$$\lim_{n\to\infty} \frac{1}{n} Z_u^* \geq 2\gamma \lim_{n\to\infty} \frac{1}{n} \sum_{j=2}^{r} \underline{d}_j |S_j| = 2\gamma \lim_{n\to\infty} \frac{1}{n} \sum_{j=1}^{r} \underline{d}_j |S_j|,$$

since $\underline{d}_1 = 0$ and $|S_1| \leq n$. By the definition of $\underline{d}_j$ and S_j, $\underline{d}_j \geq d_k - d_{\text{sup}}/r$, for all $x_k \in S_j$. Then almost surely,

$$\begin{aligned}
\lim_{n\to\infty} \frac{1}{n} Z_u^* &\geq 2\gamma \lim_{n\to\infty} \frac{1}{n} \sum_{x_k \in N_c} \left(d_k - \frac{d_{\text{sup}}}{r}\right) \\
&= 2\gamma \lim_{n\to\infty} \frac{1}{n} \sum_{x_k \in N_c} d_k - 2\gamma \frac{d_{\text{sup}}}{r} \\
&= 2\gamma E(d) - 2\gamma \frac{d_{\text{sup}}}{r}.
\end{aligned}$$

This lower bound holds for an arbitrarily large integer r; hence,

$$\lim_{n\to\infty} \frac{1}{n} Z_u^* \geq 2\gamma E(d) \qquad \text{(a.s.)}. \qquad \square$$

In the next section we show that this lower bound is tight by presenting an upper bound that asymptotically approaches the same value.

3.2.2. An upper bound

We prove Theorem 3.3 by analyzing the cost of the following three step heuristic which provides an upper bound on Z_u^*. In the first step, we partition the area A into subregions. Then, for each of these subregions, we find the optimal packing of the customers' demands in the subregion, into bins of unit size. Finally, for each subregion, we allocate one vehicle to serve the customers in each bin.

The region partitioning scheme. For any *fixed* number $h > 0$, let $G(h)$ be an infinite grid of squares with side $h/\sqrt{2}$ and edges parallel to the system coordinates. Recall that A is the compact support of the distribution function μ, and let $A_1, A_2, \cdots, A_{t(h)}$ be the intersection of the squares of $G(h)$ with the compact support A for which $p \equiv \int_{A_i} d\mu(y) > 0$. Note $t(h) < \infty$ since A is compact and $t(h)$ is independent of n.

Let $N(i)$ be the set of customers located in subregion A_i, and define $n(i) = |N(i)|$. For every $i = 1, 2, \cdots, t(h)$, let $b^*(i)$ be the minimum number of bins needed to pack the demands of customers in $N(i)$. Finally, for each subregion $A_i, i = 1, 2, \cdots, t(h)$, let $n_j(i)$ be the number of customers in the jth bin of this optimal packing, for each $j = 1, 2, \cdots, b^*(i)$.

We now proceed to find an upper bound on the value of our heuristic. Recall that for each bin produced by the heuristic, we send a single vehicle to serve all the customers in the bin. First, the vehicle visits the customer closest to the depot in the subregion to which the bin belongs, then serves all the customers in the bin in any order, and the vehicle returns to the depot through the closest customer again. Let $\underline{d}(i)$ be the distance from the depot to the closest customer in $N(i)$, i.e., in subregion A_i. If $N(i)$ is empty, let $\underline{d}(i) = 0$. Note that since each subregion A_i is a subset of a square of side $h/\sqrt{2}$, the distance between any two customers in A_i is no more than h. Consequently, using the method just described, the distance traveled by the vehicle that serves all the customers in the jth bin of subregion A_i is no more than

$$2\underline{d}(i) + h(n_j(i) + 1).$$

Therefore,

$$Z^* \leq \sum_{i=1}^{t(h)} \sum_{j=1}^{b^*(i)} \{2\underline{d}(i) + h(n_j(i) + 1)\} \leq 2\sum_{i=1}^{t(h)} b^*(i)\underline{d}(i) + 2nh. \tag{3.3}$$

This inequality will be coupled with the following lemma to find an almost sure upper bound on the cost of this heuristic.

Lemma 3.7. *Under the conditions of the Theorem 3.3, we have*

$$\overline{\lim_{n\to\infty}} \frac{1}{n} \sum_{i=1}^{t(h)} b^*(i)\underline{d}(i) \leq \gamma E(d) \quad (a.s.).$$

Proof. Recall that $p_i = \int_{A_i} d\mu(y)$ is the probability that a given customer x_k falls in subregion A_i. Since $p_i > 0$, then by the strong law of large numbers, $\lim_{n\to\infty}(n(i)/n) = p_i$ almost surely and therefore $n(i)$ grows to infinity almost surely, as n grows to infinity. Thus, we have

$$\lim_{n\to\infty} \frac{b^*(i)}{n(i)} = \lim_{n(i)\to\infty} \frac{b^*(i)}{n(i)} = \gamma \quad (a.s.).$$

Hence,

$$\overline{\lim_{n\to\infty}} \frac{1}{n} \sum_{i=1}^{t(h)} b^*(i)\underline{d}(i) = \overline{\lim_{n\to\infty}} \frac{1}{n} \sum_{i=1}^{t(h)} \frac{b^*(i)}{n(i)} n(i)\underline{d}(i)$$

$$\leq \overline{\lim_{n\to\infty}} \frac{1}{n} \sum_{i=1}^{t(h)} \frac{b^*(i)}{n(i)} \sum_{x_k \in N(i)} d_k$$

$$(\text{ since } \underline{d}(i) \leq d_k, \forall x_k \in N(i))$$

$$= \sum_{i=1}^{t(h)} \overline{\lim_{n\to\infty}} \frac{b^*(i)}{n(i)} \overline{\lim_{n\to\infty}} \frac{1}{n} \sum_{x_k \in N(i)} d_k$$

$$= \gamma \overline{\lim_{n\to\infty}} \frac{1}{n} \sum_{x_k \in N_c} d_k.$$

Using the strong law of large numbers, we have

$$\overline{\lim_{n \to \infty}} \frac{1}{n} \sum_{i=1}^{t(h)} b^*(i)\underline{d}(i) \leq \gamma E(d) \qquad \text{(a.s.)},$$

which completes the proof of this lemma. $\square$

Remark. A simple modification of the proof of Lemma 3.7 shows that the inequality that appears in the statement of the Lemma can be replaced by an equality.

We can now finish the proof of the Theorem 3.3. From equation (3.3) we have,

$$\frac{1}{n} Z_u^* \leq \frac{2}{n} \sum_{i=1}^{t(h)} b^*(i)\underline{d}(i) + 2h.$$

Taking limits and using Lemma 3.7, we obtain

$$\overline{\lim_{n \to \infty}} \frac{1}{n} Z^* \leq 2\gamma E(d) + 2h \qquad \text{(a.s.)}.$$

Since this inequality holds for arbitrarily small h, we have

$$\overline{\lim_{n \to \infty}} \frac{1}{n} Z^* \leq 2\gamma E(d) \qquad \text{(a.s.)}.$$

This upper bound combined with the lower bound of Lemma 3.6 proves the main theorem.

3.3. Probabilistic analysis of heuristics

The upper bound used in the proof of Theorem 3.3 is non-constructive in the sense that it does not lend itself to a polynomial time algorithm. This is mainly due to the lack of a polynomial time algorithm for solving the bin-packing problem to optimality.

Thus, it is important to find asymptotically optimal heuristics for the CVRP with unsplit demands. A great deal of work has been devoted to the development of heuristics for the CVRP; see, e.g., Christofides [1985] or Fisher [1992]. Of special interest to us is the class of heuristics called (by Christofides, 1976a and 1985) two-phase methods. These heuristics are of one of two types: (i) cluster first-route second, or (ii) route first-cluster second. In the first category, customers are clustered into groups and assigned to vehicles (phase I) and then efficient routes are designed for each cluster (phase II). In the second category, one constructs a traveling salesman tour through all the customers (phase I) and then partitions the tour into segments (phase II). One vehicle is assigned to each segment and visits the customers according to their appearance on the traveling salesman tour.

Bienstock, Bramel & Simchi-Levi [1993] analyze the average performance of heuristics that belong to the latter class. To present their result, we need a precise definition of the class of route first-cluster second method. Define this class as all those heuristics that first order the customers according to their locations and then partition this ordering to produce feasible clusters. These clusters consist of sets of customers that are consecutive in the initial order. Customers are then routed within their cluster depending on the specific heuristic.

Observe that this definition of the class of route first-cluster second heuristics is more general than classical definitions. It is also clear that the UITP(α) and the UOP(α) heuristics described in Section 3.1 belong to this class of heuristics. The sweep algorithm suggested by Gillett & Miller [1974] can also be viewed as a route first-cluster second type of heuristic. In this algorithm, an arbitrary customer is selected as a starting customer. The other customers are ordered according to the angle between them, the depot and the starting customer. Customers are then assigned to vehicles following this initial ordering and efficient routes are designed for each vehicle.

Bienstock, Bramel & Simchi-Levi [1993] show that the performance of any heuristic in this class is strongly related to the performance of a non-efficient bin-packing heuristic called Next-Fit (NF). Thus, heuristics in this class can never be asymptotically optimal for the CVRP with unsplit demands. The Next-Fit bin-packing heuristic can be described in the following manner. Given a list of n items where the size of item i is w_i, start with item 1 and place it in bin 1. Suppose we are packing item j, let bin i be the highest indexed non-empty bin. If item j fits in bin i, then place it there, else place it in a new bin indexed $i + 1$. Thus, the NF heuristic assigns items to bins according to the order they appear in without using any knowledge of subsequent items in the list.

In their seminal work on the use of martingale inequalities for NP-Complete problems, Rhee & Talagrand [1987] show that for any distribution of the item sizes, there exists a constant $\gamma^{NF} > 0$ such that $\lim_{n \to \infty} (b^{NF}/n) = \gamma^{NF}$ almost surely, where b^{NF} is the number of bins produced by the NF packing and γ^{NF} depends only on the distribution of the item sizes. This constant is used in the following Theorem which is presented without a proof.

Theorem 3.8 [Bienstock, Bramel & Simchi-Levi, 1993].

(i) Let H be a generic route first-cluster second heuristic, that is, H is a heuristic that starts by ordering the customers in some manner depending only on their relative locations and not on their demands. Then, under the assumptions of Theorem 3.3 we have

$$\lim_{n \to \infty} \frac{1}{n} Z^H \geq 2\gamma^{NF} E(d) \qquad (a.s.).$$

(ii) The UOP(α) heuristic is the best possible heuristic in this class, that is, for any fixed $\alpha \geq 1$ we have

$$\lim_{n \to \infty} \frac{1}{n} Z^{UOP(\alpha)} = 2\gamma^{NF} E(d) \qquad (a.s.).$$

In view of Theorems 3.3 and 3.8 it is interesting to compare γ^{NF} to γ since the asymptotic error for any heuristic H in the class of route first-cluster second satisfies

$$\frac{\lim\limits_{n\to\infty} Z^H}{Z_u^*} \geq \frac{\lim\limits_{n\to\infty} Z^{\mathrm{UOP}(\alpha)}}{Z_u^*} = \frac{\gamma^{\mathrm{NF}}}{\gamma}.$$

This ratio was characterized by Karmarkar [1982] for the case where the item sizes are uniformly distributed on an interval $[0, a]$ for $0 < a \leq 1$. For instance, for a satisfying $1/2 < a \leq 1$, we have

$$\frac{\gamma^{\mathrm{NF}}}{\gamma} = \frac{2}{a}\left\{\frac{1}{12a^3}(15a^3 - 9a^2 + 3a - 1) + \sqrt{2}\left(\frac{1-a}{2a}\right)\tanh\left(\frac{1-a}{\sqrt{2a}}\right)\right\},$$

so that when the item sizes are uniform $[0, 1]$ the above ratio is 4/3 which implies that UOP(α) converges to a value which is 33.3 % more than the optimal cost; a very disappointing performance for the best heuristic currently available in terms of worst-case behavior.

To our knowledge, no polynomial time algorithm which is asymptotically optimal is known for the general CVRP with unsplit demands and a general Φ distribution. We now describe such a heuristic for the case where Φ is uniform on the interval $[0, 1]$. This heuristic, called *optimal matching of pairs* (OMP), considers only feasible solutions in which each vehicle visits no more than two customers. Among all such feasible solutions, the heuristic finds the one with minimum cost. This can be done by formulating the following integer linear program.

For every $x_k, x_l \in N_c$, let

$$c_{kl} = \begin{cases} d_k + d_{kl} + d_l, & \text{if } k \neq l \text{ and } w_k + w_l \leq 1; \\ 2d_k, & \text{if } k = l; \\ \infty, & \text{otherwise.} \end{cases}$$

The integer program that we solve is

Problem P

$$\text{Min } \sum_{k \leq l} c_{kl} X_{kl}$$

subject to

$$\sum_{l \geq k} X_{kl} + \sum_{l < k} X_{lk} = 1 \qquad \forall k = 1, 2, \cdots, n \tag{3.4}$$

$$X_{kl} = 0, 1 \qquad \forall k \leq l. \tag{3.5}$$

For $k < l$, X_{kl} equals one if a vehicle delivers items to customers x_k and x_l and zero otherwise. Constraint (3.4) ensures that each customer is visited.

It is not hard to see that Problem P can be solved in polynomial time since it is no more than the classical weighted matching problem defined on the following graph $\overline{G} = (\overline{N}, \overline{E})$, where each customer x_k is represented by two nodes v_k and v'_k, for $k = 1, 2, \cdots, n$. The set of edges of $\overline{G}$ is defined as follows:

$$
\begin{aligned}
\overline{E} = \;& \{(v_k, v'_k) \mid x_k \in N_c\} \\
& \cup \{(v_k, v_l) \mid x_k \in N_c, \ x_l \in N_c, \ k \neq l, \ w_k + w_l \leq 1\} \\
& \cup \{(v'_k, v'_l) \mid x_k \in N_c, \ x_l \in N_c, \ k \neq l, \ w_k + w_l \leq 1\}.
\end{aligned}
$$

Thus, $\overline{G}$ has $2n$ vertices. The length of edge (v_k, v_l), for $k \neq l$, is c_{kl}, of edge (v_k, v'_k) is c_{kk} and of edge (v'_k, v'_l) is zero, for all k and l.

Note that any given feasible solution to Problem P can be transformed into a feasible solution to the matching problem on $\overline{G}$ with the same cost: For any feasible solution to Problem P, choose edge (v_k, v'_k) if customer k is served by a vehicle that does not serve any other customer and choose edges (v_k, v_l) and (v'_k, v'_l) if customers x_k and x_l are visited together. Similarly, any feasible solution to the matching problem can be transformed into a feasible solution to Problem P with the same cost. Hence, the two problems are equivalent.

The optimal matching on $\overline{G}$ can be found in $O(n^3)$ using Lawler's [1976] algorithm.

Theorem 3.9 [Simchi-Levi & Bramel, 1990b]. *Let x_k, $k = 1, 2, \cdots, n$ be a sequence of independent random variables having a distribution μ with compact support in $\Re^2$. Let $d(y)$ be the Euclidean distance between the depot and point $y \in \Re^2$ and let*

$$
E(d) = \int_{\Re^2} d(y) \, d\mu(y).
$$

Let the demands w_k, $k = 1, 2, \cdots, n$ be a sequence of independent random variables having a uniform distribution on $[0, 1]$ and assume that the demands and the location of the customers are independent of each other. Then, the OMP heuristic is asymptotically optimal. That is, with probability one,

$$
\lim_{n \to \infty} \frac{Z^*}{n} = \lim_{n \to \infty} \frac{Z^{\mathrm{OMP}}}{n} = E(d).
$$

To prove that the OMP heuristic is asymptotically optimal, we approximate its performance by that of the *sliced region partitioning heuristic with parameters h and r* (SRP(h, r)). For any *fixed* positive integer r, the set N_c is partitioned into the following $2r$ disjoint subsets, some of which may be empty:

$$
N_j = \left\{ x_k \in N \ \Big| \ 0.5\left(1 - \frac{j+1}{r}\right) < w_k \leq 0.5\left(1 - \frac{j}{r}\right) \right\}
$$
$$
j = 1, 2, \cdots, r - 1,
$$

and

$$
N^j = \left\{ x_k \in N \ \Big| \ 0.5\left(1 + \frac{j-1}{r}\right) < w_k \leq 0.5\left(1 + \frac{j}{r}\right) \right\}
$$
$$
j = 1, 2, \cdots, r - 1.
$$

Also

$$N_0 = \left\{ x_k \in N \;\middle|\; 0.5\left(1 - \frac{1}{r}\right) < w_k \le 0.5 \right\},$$

and

$$N^r = \left\{ x_k \in N \;\middle|\; 0.5\left(1 + \frac{r-1}{r}\right) < w_k \right\}.$$

The number of customers in each N_j (respectively, N^j) is denoted by n_j (respectively, n^j) for all possible values of j.

Note that for any $j = 1, 2, \cdots, r - 1$, one vehicle can deliver the demand of a customer from N_j together with the demand of exactly one customer from N^j. The SRP(h, r) heuristic generates pairs of customers, one customer from N_j and one from N^j, for every $j = 1, 2, \cdots, r - 1$, using the same region partitioning scheme used in the proof of Theorem 3.3 (Section 3.2.2). The customers in $N_0 \cup N^r$ are served separately; a single vehicle is assigned to each of these customers.

For every subregion A_i, $i = 1, 2, \cdots, t(h)$, generated by the grid $G(h)$ (see Section 3.2.2) and for every $j = 1, 2 \cdots, r - 1$, let $N_j(i)$ (respectively, $N^j(i)$) be the subset of points in N_j (respectively, N^j) that fall in subregion A_i. Also, let $n_j(i) = |N_j(i)|$ and $n^j(i) = |N^j(i)|$.

In each subregion A_i, $i = 1, 2, \cdots, t(h)$, and for any $j = 1, 2, \cdots, r - 1$, we arbitrarily match one customer from $N_j(i)$ with exactly one customer from $N^j(i)$; one vehicle serves each such pair. If $n_j(i) = n^j(i)$, then all customers in $N_j(i) \cup N^j(i)$ are matched and therefore visited in pairs. If, however, $n_j(i) \ne n^j(i)$, then we can match exactly $\min\{n_j(i), n^j(i)\}$ pairs of customers. The remaining $|n_j(i) - n^j(i)|$ customers in $N_j(i) \cup N^j(i)$ that have not yet been matched, are served each by one vehicle. Thus the total number of vehicles used in subregion A_i is

$$n_0(i) + n^r(i) + \sum_{j=1}^{r-1} \max\{n_j(i), n^j(i)\}.$$

The heuristic clearly generates a feasible solution for the UCVRP. Moreover, this solution is feasible for Problem P, as each vehicle visits at most two customers. Thus

$$Z^{\mathrm{OMP}} \le Z^{\mathrm{SRP}(h,r)} \qquad \text{for any } r \text{ and } h. \tag{3.5}$$

We now proceed by finding an upper bound on $Z^{\mathrm{SRP}(h,r)}$. The same type of analysis as in Section 3.2.2 shows that the total distance traveled by all vehicles is no more than

$$2 \sum_{i=1}^{t(h)} \underline{d}(i) \left\{ n_0(i) + n^r(i) + \sum_{j=1}^{r-1} \max\{n_j(i), n^j(i)\} \right\} + 2nh.$$

We now show that

$$\lim_{n(i)\to\infty} \frac{1}{n(i)}\left\{n_0(i) + n^r(i) + \sum_{j=1}^{r-1}\max\{n_j(i), n^j(i)\}\right\} = \frac{r-1}{2r} + \frac{1}{r} \quad \text{(a.s.).}$$

$$(3.6)$$

The remainder of the proof is identical to the proof of the upper bound of Theorem 3.3, letting r tend to infinity. But (3.6) holds since

$$\lim_{n(i)\to\infty} \frac{n_j(i)}{n(i)} = \lim_{n(i)\to\infty} \frac{n^j(i)}{n(i)} = \frac{1}{2r} \quad \text{(a.s.)} \qquad \text{for all } j = 1, 2, \cdots, r.$$

3.4. Rate of convergence to the asymptotic value

While the results in the two previous sections completely characterize the asymptotic optimal solution value of the CVRP with unsplit demands, they do not say anything about the rate of convergence to the asymptotic solution value. See Psaraftis [1984] for an informal discussion of this issue.

To get some intuition on the rate of convergence, it is interesting to determine the expected difference between the optimal solution for a given number of customers, n, and the asymptotic solution value (i.e., $2\gamma E[d]$). This can be done for the special case, called the *Uniform Model*, in which the uniform distribution applies to the customers locations as well as their demands.

In this case, Bramel, Coffman, Shor & Simchi-Levi [1992] and independently Rhee [1991] proved the following strong result.

Theorem 3.10. *Let* y_k $k = 1, 2, \cdots, n$ *be a sequence of independent random variables uniformly distributed in the unit square* $[0, 1]^2$. *Let the demands* w_k $k = 1, 2, \cdots, n$ *be drawn independently from a uniform distribution on* $[0, 1]$. *Then,*

$$E[Z_n^*] = nE[d] + \Theta(n^{2/3}).$$

The proof of Theorem 3.10 relies heavily on the theory of 3-dimensional stochastic matching which is outside the scope of our survey. We refer the reader to Coffman & Lueker [1991, chapter 3] for an excellent review of matching problems.

Rhee [1991] has also found an upper bound on the rate of convergence to the asymptotic solution value, for general distributions of the customers' locations and their demands. Using a new matching theorem developed together with M. Talagrand she proved,

Theorem 3.11. *Under the assumptions of Theorem 3.3, we have*

$$2n\gamma E[d] \le E[Z_n^*] \le 2n\gamma E[d] + O\big((n\log n)^{2/3}\big).$$

4. Inventory-routing models

4.0. Introduction

The models of the previous sections all assume that the frequency, timing and size of customer deliveries are predetermined. There are however many distribution systems in which the vehicle schedules and the timing and size of the customer deliveries are (or should be) simultaneously determined. This is clearly the case in *internal* distribution systems in which the 'depot' and the 'customers' represent (part of) consecutive layers in the distribution network of a single company.

In addition, the need to integrate inventory control and routing decisions arises in many *external* distribution processes in which deliveries need to be made to *external* customers. An example is the gas industry where the gas producers install tanks at their customers' locations and assume the responsibility for maintaining an adequate inventory level by determining the replenishment frequency and delivery sizes of all customers. Suppliers of supermarkets and department stores, to give another example, often acquire shelf space and are given the responsibility for replenishing the stock. They often adopt the complete inventory management function of their retailer customers. By billing a retailer only at the time it makes a sale to a consumer, the capital costs associated with the retailer's inventories are borne by the supplier. The supplier is given the responsibility to replenish the retailer's inventory at its discretion while guaranteeing a given fill rate or being charged for any lost sales or backlogs.

This arrangement alleviates the industrial retailer of its costly inventory investments and the intricacies of inventory planning; the supplier has the advantage of being able to determine when and in what quantities to deliver to its retailer customers. Moreover, when demands are subject to a considerable degree of uncertainty, the system as a whole derives additional benefits from this arrangement because the supplier can meet a given service level with an aggregate safety stock significantly smaller than the sum of the safety stocks required by the individual retailers.

There are many potential models integrating inventory control and vehicle routing problems. To date, the literature has confined itself to the following two variants: the first is a single-period model with customers having stochastic demands; here deliveries serve to replenish inventories to levels that appropriately balance inventory carrying and shortage costs but thereby vehicle routing costs are incurred as well. The second variant is an infinite horizon model with customers each having demands at a (customer specific) constant and *deterministic* rate. Here one needs to determine (infinite horizon) replenishment *policies* for all customers as well as efficient vehicle routes.

Since demands are deterministic it is easily verified that it is optimal to replenish the depot and all customers only when their respective inventories are down to zero. Such policies are referred to as Zero-Inventory Ordering Policies. Thus, under deterministic demands, the problem reduces to determining a sequence of replenishment epochs for all locations and to combine the corresponding deliveries into efficient routes.

The analysis of multi-period models with more general, time-varying and/or stochastic demands, remains an open challenge.

4.1. Single-period inventory-routing models

Federgruen & Zipkin [1984] consider the following single-period model. At the beginning of the period, the initial inventory (perhaps supply remaining from the previous period) for each location is reported to the central depot. This information is used to determine the allocation of the available product among the locations. At the same time, the assignment of customers to vehicles and their routes are determined. After the deliveries are made, the demands occur and inventory-carrying and shortage costs are incurred at each location proportional to the end-of-the-period inventory level. Observe, that in this model it is possible not to visit some of the locations. Following Federgruen and Zipkin, we use the following notation:

K = number of vehicles.

$F_i(\cdot)$ = cumulative distribution function of the one period demand at location i which is assumed to be strictly increasing.

h_i^+ = inventory carrying cost per unit at location i.

h_i^- = inventory shortage cost per unit at location i.

β_i = initial inventory at location i.

W = total amount of product available at the central depot.

b = vehicle capacity.[1]

We use the same notation for the set of locations (customers) and their distances from the depot as those employed in Section 2. Unlike in the VRP, w_i, the amount delivered to customer i, is now a variable. Let $y_{ik} = 1$ if delivery point i is assigned to route k, and $y_{ik} = 0$ otherwise. Let $x_{ijk} = 1$ if vehicle k travels directly from location i to location k, and $x_{ijk} = 0$ otherwise.

The stock cost function $s_i(\cdot)$ is given by

$$s_i(w_i) = \int_{\beta_i + w_i}^{\infty} h_i^-(u - \beta_i - w_i) \, dF_i(u) + \int_0^{\beta_i + w_i} h_i^+(\beta_i + w_i - u) \, dF_i(u)$$

for all i, while its derivative is denoted by $s_i'(\cdot)$. It is easy to see that $s_i(\cdot)$ is strictly convex and continuously differentiable.

The single-depot single-period inventory-routing problem can now be formulated in the following way:

---

[1] The use of b as the vehicle capacity is a departure from our standard notation and is only done in Section 4. We do so since some of the models analyzed in this section have, in addition, constraints on the number of customers in each tour, for which we use the notation q.

Problem P_1

$$\text{minimize} \quad \sum_{i,j,k} c_{ij} x_{ijk} + \sum_i s_i(w_i)$$

$$\text{subject to} \quad \sum_i w_i y_{ik} \leq b \qquad \forall k = 0, 1, \cdots, K \qquad (4.1)$$

$$\sum_i w_i \leq W \qquad (4.2)$$

$$w_i \geq 0 \qquad (4.3)$$

$$\sum_{k=1}^{K} y_{0k} = K \qquad (4.4)$$

$$\sum_{k=0}^{K} y_{ik} = 1 \qquad i = 1, \cdots, n \qquad (4.5)$$

$$0 \leq y_{ik} \leq 1 \text{ and integer} \qquad (4.6)$$

$$\sum_i x_{ijk} = y_{jk} \qquad \forall j, k \qquad (4.7)$$

$$\sum_j x_{ijk} = y_{ik} \qquad \forall i, k \qquad (4.8)$$

$$\sum_{ij \in S} x_{ijk} \leq |S| - 1 \qquad \forall S \subseteq \{1, \cdots, n\} \qquad (4.9)$$

$$0 \leq x_{ijk} \leq 1 \text{ and integer} \qquad (4.10)$$

Constraint (4.1) ensures that the total load assigned to a vehicle is no more than the vehicle capacity. Constraint (4.2) guarantees that the amount shipped is no more than W, the quantity available at the depot, while constraint (4.3) guarantees that this amount is non-negative. The remaining constraints ensure that feasible tours are constructed.

Observe that when the variables y_{ik} are fixed, the problem decomposes into $(K + 1)$ simpler subproblems: an inventory allocation problem and K traveling salesman problems. Let $Y_k = \{i : y_{ik} = 1\}$, $\forall k$, i.e., Y_k is the set of customers assigned to the kth route. The inventory allocation problem can be written as follows:

Problem P_2

$$\text{minimize} \quad \sum_i s_i(w_i)$$

$$\text{subject to} \quad \sum_{i \in Y_k} w_i \leq b \qquad k = 0, 1, \cdots, K \qquad (4.11)$$

$$\sum_i w_i \leq W \qquad (4.12)$$

$$w_i \geq 0 \qquad i = 1, \cdots, n \qquad (4.13)$$

The kth traveling salesman problem consists of determining the traveling salesman tour through the depot and the set Y_k.

Projecting the problem onto new variables $W_k = \sum_{i \in Y_k} w_i$, the total loads on the vehicles, we define

Problem P_3

$$S_k(W_k) = \text{minimize} \quad \sum_{i \in Y_k} s_i(w_i)$$

$$\text{subject to} \quad \sum_{i \in Y_k} w_i \leq W_k \tag{4.14}$$

$$w_i \geq 0 \qquad i \in Y_k \tag{4.15}$$

Problem P_2 is equivalent, in an obvious sense, to the following model:

Problem P_4

$$\text{minimize} \quad \sum_{k=1}^{K} S_k(W_k)$$

$$\text{subject to} \quad \sum_{k=1}^{K} W_k \leq W \tag{4.16}$$

$$0 \leq W_k \leq b \qquad k = 1, \cdots, K \tag{4.17}$$

Problem P_3 consists of minimizing a separable (strictly) convex and continuously differentiable function subject to a single budget constraint for which many efficient methods can be used. Moreover, each $S_k(\cdot)$ function is strictly convex and continuously differentiable. Similar methods can be used to solve Problem P_4 itself although additional complications arise from the upper bound on the W variables and the fact that the $S_k(\cdot)$ functions are defined implicitly.

Based on these observations Federgruen & Zipkin [1983] present an efficient method to solve Problem P_2 whose complexity reduces to $O(n \log n)$ elementary operations and at most n determinations of the root of a non-increasing function. The method yields a (unique) optimal solution $\overline{w}$ and the (not necessarily unique) dual variables, denoted by $\overline{\vartheta}_k$, $k = 1, 2, \cdots, K$, and $\overline{\rho}$ associated with constraints (4.11) and (4.12) respectively. In addition, the method is well suited for recovery of optimality when y is changed, especially when it is changed slightly.

Following a number of successful approaches for the TSP as well as for the VRP, Federgruen & Zipkin [1984] suggest an interchange heuristic for the single-period inventory-routing problem. For this purpose, we need to examine the change in the optimal solution of Problem P_2 when some elements of Y_{k_1} and Y_{k_2} are traded, for some k_1 and k_2. Specifically, we want an estimate of the change in total inventory cost, i.e., the solution to Problem P_2, when we switch a set of locations $J_1 \subseteq Y_{k_1}$ with a set of locations $J_2 \subseteq Y_{k_2}$.

For $i \in J_1$, let $\overline{w}'_i$ be the solution to $s'_i(w) = \overline{\rho} + \overline{\vartheta}_{k_2}$ if the solution is positive and zero otherwise. Similarly, let $\overline{w}'_i$ solve $s'_i(w) = \overline{\rho} + \overline{\vartheta}_{k_1}$ or $\overline{w}'_i = 0$ for $i \in J_2$.

It is easy to verify that the change in the cost of Problem P_2 from the prior optimal solution to the new solution of Problem P_2 (with $J_1 \subseteq Y_{k_2}$ and $J_2 \subseteq Y_{k_1}$) is bounded from below by the quantity:

$$\Delta P_2 = \sum_{i \in J_1 \cup J_2} [s_i(\overline{w}_i') - s_i(\overline{w}_i)]$$

$$+ (\overline{\rho} + \overline{\vartheta}_{k_1})\left(\sum_{i \in J_1} \overline{w}_i - \sum_{i \in J_2} \overline{w}_i'\right) + (\overline{\rho} + \overline{\vartheta}_{k_2})\left(\sum_{i \in J_2} \overline{w}_i - \sum_{i \in J_1} \overline{w}_i'\right).$$

We now show how interchange methods such as the r-opt method of Croes [1958] and Lin & Kernhigan [1973] for the TSP, and Christofides & Eilon [1969] for the VRP, can be adapted to accommodate inventory allocation considerations. For the sake of notational simplicity we concentrate on the 2-opt method. In 2-opt, two existing links (x_i, y_i) and (x_j, y_j) are exchanged for two new links (x_i, y_j) and (x_j, y_i). Depending on whether or not the two old links belong to different tours, the partition of the customers between the different vehicles changes or remains the same. In the latter case, only the sequence in which the vehicle visits its customers might change.

When it does change, Problem P_2 changes so we must also assess the change in inventory cost. While we could resolve Problem P_2 for each potential switch, a more attractive approach is to use the lower bound approximation ΔP_2 and to resolve Problem P_2 only when considering implementing a switch.

Federgruen & Zipkin [1984] suggest in addition an exact algorithm using a generalized Benders' decomposition approach. This method may be viewed as an adaptation of the method of Fisher and Jaikumar (see Fisher 1992) for the standard vehicle routing problem. It relies on the fact that the TSP can be viewed as a linear program, whose feasible set is defined implicitly as the convex hull of all feasible solutions to the TSP (the so-called traveling salesman polytope). From that standpoint, Problem P_1 is a (non-linear) mixed integer program, where the integer variables are the y_{ik}'s only. Since the inventory allocation problem is non-linear, one must use generalized Benders' decomposition, see Geoffrion [1972].

A different approach for a similar model is due to Chien, Balakrishnan & Wong [1989], who suggest a Lagrangian relaxation based heuristic. In their model however, they assume that the maximum customer demand is known before routes are designed, and inventory costs are specified by a linear function of the number of items delivered to a customer. The approach they take is to relax the problem so that it decomposes into an inventory allocation problem and a customer assignment/vehicle utilization subproblem. Because of the assumed linear structure of the inventory costs, the allocation problem reduces to a simple continuous knapsack problem which can be solved efficiently by a greedy procedure. Solving these two problems provides a lower bound on the total cost which can be improved by updating the Lagrangian multipliers using a subgradient search method. A heuristic procedure based on the Lagrangian solution is also developed.

The single-period inventory-routing problem, while important in its own right, arises as a subproblem in multi-period inventory-routing problems. Dror & Ball

[1987] and Chien, Balakrishnan & Wong [1989] suggest decomposing the multi-period problem into a series of single-period problems using a cost adjustment in each single-period model to reflect the effect of decisions made in one time period on later time periods. For further discussion of the multi-period inventory-routing problem the reader is referred to Golden, Assad & Dahl [1984], Assad, Golden, Dahl & Dror [1982], Dror, Ball & Golden [1986], Dror & Ball [1987] and Chandra & Fisher [1990].

4.2. Infinite horizon inventory-routing models

Consider now a distribution system with a single depot and retailers with external demands for a single item occurring at a constant (but retailer-dependent) deterministic rate. The depot places orders with an outside supplier. Goods are distributed from the depot to the retailers by a fleet of identical vehicles, each having a limited capacity.

We distinguish between settings where central inventories may be kept at the depot and those without central stock. In systems *without* central stock, the depot serves as a coordinator of the replenishment process or alternatively as a trans-shipment point which all vehicle routes depart from and return to. In such systems, one has to determine replenishment policies for all retailers as well as matching vehicle routes. In systems *with* central stock, these problems are compounded by that of determining a replenishment strategy for the warehouse 'optimally' integrated with that of each retailer and synchronized with the transportation schedule.

Inventory carrying costs are incurred at a constant rate per unit of time and per unit stock. This rate is identical for all retailers but, when applicable, may be different at the warehouse. Transportation costs include a fixed cost per route for each vehicle and variable costs proportional to the total Euclidean distance of all routes. In systems with central stock, a fixed-plus-linear order cost is incurred for each replenishment of the warehouse inventory. The objective is to minimize the system-wide long-run inventory, transportation and order costs.

Optimal policies can be very complex; note e.g. that even when an extremely simplistic inventory strategy is chosen, specifying replenishments to all facilities at times $0, T, 2T, \cdots$ for some $T > 0$, a capacitated vehicle routing problem needs to be solved. The complexity of optimal policies makes them difficult, if not impossible, to implement even if they could be computed efficiently. We thus restrict ourselves to a class of replenishment strategies Ψ with the following properties: a replenishment strategy specifies a collection of regions (subsets of outlets) covering all outlets. If a retailer belongs to several regions, a specific fraction of its sales/operations is assigned to each of these regions. Each time one of the outlets in a given region receives a delivery, this delivery is made by a vehicle who visits (in an efficient sequence) all outlets in the region and none outside the region. We use the terms regions and routes interchangeably.

We note that a large amount of flexibility is preserved within the class Ψ by allowing retailers to be assigned to several regions, i.e., by allowing regions to overlap. On the other hand, under a strategy in Ψ, all regions are controlled independently of each other. Thus, if an outlet belongs to two regions, it is treated as two separate sub-outlets each responsible for a specific fraction of the sales; it is therefore possible that a delivery is made to one sub-outlet at an epoch at which the other sub-outlet continues to have stock. However, the proposed heuristics generate regions in which only a *few* retailers are split in this way.

Also, note that under strategies in Ψ, outlets assigned to *different* regions, are never served in a common route even though in an *optimal* strategy any given outlet may be served in *varying* rather than *constant* combinations of other outlets. This is illustrated by the example in Hall [1991]. For further discussion regarding the merits of this restrictive approach and a review of other joint replenishment problems for which a similar restriction has been employed, see Anily & Federgruen [1991b].

In addition to the vehicle capacity constraint, it is often necessary or desirable to consider (within the above class of replenishment strategies):
  – An upper bound on the frequency with which deliveries may be made in each of the subregion.
  – An upper bound on the total sales volume in each of the regions.

Following the notation in Anily & Federgruen [1990a, 1992] let:

μ_j = demand rate of retailer j ($j = 1, 2, \cdots, n$). The demand rates are assumed to be integer multiples of some common quantity $\mu > 0$, i.e., $\mu_j = k_j \mu$ where k_j is an integer between 1 and K for some $K \geq 1$ ($j = 1, 2, \cdots, n$).

h^+ = inventory holding cost rate at the retailers.

c = fixed cost per route driven. We assume, without loss of generality, that the variable transportation cost per mile is normalized to be one.

b = capacity of the vehicle. In an uncapacitated model, we let $b = +\infty$.

f^* = upper bound on the frequency with which a given route may be driven.

q = maximum number of demand points per region. We define a *demand* point as a point in the plane facing a demand rate of μ.

Each retailer j, $j = 1, 2, \cdots, n$ may thus be viewed as consisting of k_j demand points, all facing identical demand rates of μ units per unit of time and all located at the same geographic point as retailer j. In the class Ψ, we assume that if retailer j is split among several regions, its k_j demand points are assigned to these regions. Thus, let $\{x_1, \cdots, x_N\}$ denote the collection of demand points numbered in ascending order of their radial distance from the depot; note $N \leq Kn$. Let L denote the number of routes (regions) to be employed. L may be variable or fixed; in the latter case, the maximum number of demand points per region (q) may be region-dependent, i.e., $q = q_l$. Below we consider the case where all $q_l = q$; see Anily & Federgruen [1990a] for a treatment of the more general case.

4.2.1. Systems without central stock

For a given strategy in Ψ, consider a specific region with m demand points and route length ϑ. It is easily verified that the region is optimally replenished at constant intervals of length (say) T, so that a constant quantity (μT) is delivered to each demand point at each delivery. The choice of T is constrained by the vehicle capacity $(m\mu T \le b)$ and the frequency constraint $(T^{-1} \le f^*)$. Note that a feasible range for T exists if and only if $m \le (f^*b)/\mu$. The long-run average holding cost for the region is given by

$$\frac{(\vartheta + c)}{T} + \frac{1}{2}h^+m\mu T. \tag{4.18}$$

The optimal value of T is obtained by minimizing (4.18) subject to the bounds on T. The resulting optimal long-run average cost can be computed in closed-form and is given by

$$f(\vartheta, m) = \begin{cases} \dfrac{h^+\mu m}{(2f^*)} + f^*(\vartheta + c), & \text{if } \vartheta + c \le \dfrac{\mu m h^+}{(2f^{*2})}; \\[2ex] (2h^+\mu m(\vartheta + c))^{1/2}, & \text{if } \dfrac{\mu m h^+}{(2f^{*2})} \le \vartheta + c \le \dfrac{b^2 h^+}{(2\mu m)}; \\[2ex] h^+\dfrac{b}{2} + \dfrac{\mu m}{b}(\vartheta + c), & \text{otherwise.} \end{cases}$$

The problem of finding an optimal strategy in Ψ thus reduces to solving the routing problem with respect to demand points $\{x_1, x_2, \cdots, x_N\}$, route cost function $f(\vartheta, m)$ and route capacity $q^* = \min\{q, \lfloor f^*b/\mu \rfloor\}$. The function is concave and non-decreasing in ϑ, as is easily verified. This has the following implications:

(i) The bounds $\underline{Z}_f^1$ and $\underline{Z}_f^2$ developed in Section 2.4 provide lower bounds for Z^*, the minimum cost among all strategies in Ψ.

(ii) The modified region partitioning scheme (MCRP) may be applied to generate the regions, and associated constant replenishment intervals can be determined for each region, as described above. The complexity of this procedure amounts to $O(N \log N) = O(Kn \log n)$ elementary operations and $O(n)$ evaluations of a square root.

(iii) The lower bound on the cost of the MCRP heuristic is almost surely asymptotically optimal within the class Ψ if the locations of the retailers have i.i.d. random distances from the depot.

(iv) Additional properties of the function $f(\vartheta, m)$ referred to in Section 2.4 allow for $\underline{Z}_f^1$ to be computed in closed form. In fact, if L is variable, one obtains that *every* region in the partition for which this lower bound is achieved consists of q^* consecutive demand points with the possible exception of *one* region containing the $N - \lfloor N/q^* \rfloor q^*$ demand points that are closest to the depot. If a specific number of regions L is required, we have two integers $0 \le k \le k'$ such that in the partition achieving $\underline{Z}_f^1$, a specific region is established for each of the k closest demand points, $\{x_{k+1}, \cdots, x_{k'}\}$ is the $(k + 1)^{\text{st}}$ region and all remaining regions consist of q^* consecutive demand points.

(v) Extensive numerical results verify that the gap between the lower bound on the cost of any policy in Ψ and the cost of the proposed heuristic is small even when the number of retailers is of moderate size. For example, for problems with 500 *demand* points uniformly distributed in a given square, the average optimality gap is only 7% and for problems with 100 demand points the average gap is only 14%.

4.2.2. Systems with central stock

Additional cost components need to be considered when stock may be kept at the warehouse. Let

h_0 = the inventory holding cost rate per unit stored at the warehouse.

K_0 = the fixed cost per order placed by the warehouse.

h = $h^+ - h_0$ is the retailer *echelon holding cost rate*. We assume that $h > 0$. Since the holding cost rate tends to increase with the (cumulative) value added, this assumption is almost always satisfied.

The replenishment strategy of any given region can no longer be determined independently of that of other regions, when (limited) inventories are kept by the warehouse. The reason is that each region's replenishment strategy needs to be carefully synchronized with that of the warehouse. In fact, for a *given* choice of regions, the remaining problem reduces to that of determining an optimal replenishment strategy in a one-warehouse multiple-retailer system in which each region acts as a single (super) retailer.

The structure of a fully optimal inventory strategy for this system can be very complex in spite of the relatively simple demand processes and cost structure. No efficient algorithm is known to determine such a strategy, see Roundy [1985] and Muckstadt & Roundy [1992]. On the other hand, Roundy [1985] has shown that at least in the absence of constraints on regional replenishment frequencies (see below), a simple, so-called *power-of-two* policy exists whose cost is guaranteed to be within 6% of optimality. Under a power-of-two policy, the warehouse (region l) replenishes its inventory every T_0 (T_l) time units when its inventory reaches zero ($l = 1, 2, \cdots, L$); also ($T_0, T_1, \cdots, T_L$) are power-of-two multiples of a base planning period T^B.

Assume therefore that the warehouse and every region is replenished at constant intervals. As explained above, T_l, the replenishment interval for region l with m_l demand points is constrained by the bounds $1/f^* \le T_l \le b/\mu m_l$. For the sake of notational simplicity, we choose the units of demand so that $\mu = 2$. We also make the following parameter restrictions: (i) $bf^*/2$ is either integer or $+\infty$; (ii) if $q < bf^*/2 < \infty$, bf^* is a power-of-two multiple of q.

Anily & Federgruen [1992] have shown that a set of regions and an associated power-of-two inventory policy can be constructed whose cost is almost surely guaranteed to be within 6% of Z^*, the minimum cost (among all strategies in Ψ), for all n sufficiently large and assuming the retailers are located at i.i.d. radial distances from the depot. Similarly under the same assumptions a lower bound $\underline{Z}^1 \le Z^*$ can be constructed which almost surely is guaranteed to come within 6% of being accurate for all n sufficiently large. Moreover, the complexity of determining the heuristic solution and the lower bound is $O(Kn \log n)$.

The analysis proceeds as follows: For a given partition of the collection of demand points into sales regions $\Pi = \{X_1, \cdots, X_L\}$ and a given power of two policy $T = \{T_0, T_1, \cdots, T_L\}$, we denote the corresponding *average cost* by $Z_\Pi(T) = K_0/T_0 + \sum_{l=1}^{L} D_{T_0}(T_l, \theta_l, m_l)$. Here $D_{T_0}(T_l, \theta_l, m_l)$ represents the average cost per unit of time of replenishing region l which depends on T_0, T_l, $m_l = |X_l|$ and θ_l, the length of the traveling salesman tour through X_l and the depot. D_{T_0} includes the transportation costs which are incurred for X_l as well as the carrying costs for the inventories at the region's demand points and part of the warehouse inventory which is destined to be shipped to X_l. It is easily verified, see Roundy [1985], that

$$D_{T_0}(T_l, \theta_l, m_l) = (\theta_l + c)/T_l + m_l(hT_l + h_0 \max\{T_0, T_l\}). \tag{4.19}$$

Lemma 4.1 below states that $\inf\{Z_\Pi(T) : T_0 > 0 \text{ and } 1/f^* \le T_l \le b/2m_l, l = 1, 2, \cdots, L\}$ provides a lower bound on the minimum long-run average costs over all feasible policies that employ the regions in Π, i.e., not just the power-of-two policies.

Lemma 4.1. *For any given partition* $\Pi = \{X_1, \cdots, X_L\}$ $\underline{Z}_\Pi \equiv \inf\{Z_\Pi(T) : T_0 > 0 \text{ and } 1/f^* \le T_l \le b/2m_l, l = 1, 2, \cdots, L\}$ *is a lower bound for the minimum long-run average costs over all feasible policies employing the regions in* Π.

Proof. See Lemma 1 in Anily & Federgruen [1992] which extends the lower bound theorem in Roundy [1985]. $\square$

Thus, a lower bound on Z^* is obtained by solving

$$(P) \min_\Pi \underline{Z}_\Pi = \inf_{T_0 > 0} \left\{ \frac{K_0}{T_0} + \min\left[\sum_{l=1}^{L} f_{T_0}(\theta_l, m_l) : \Pi = \{X_1, \cdots, X_L\} \right] \right\} \tag{4.20}$$

where

$$f_{T_0}(\theta_l, m_l) \equiv \min\left[D_{T_0}(T_l, \theta_l, m_l) : \frac{1}{f} \le T_l \le \frac{b}{2m_l} \right].$$

The inner minimization in (4.20) represents, for any given warehouse replenishment interval T_0, a routing problem with general route cost function $f_{T_0}(\theta, m)$.
   The function f_{T_0} can easily be obtained in closed form: Let

$$\tau'(\theta, m) = \left(\frac{\theta + c}{m(h' + h_0)} \right)^{1/2};$$

$$\tau(\theta, m) = \left(\frac{\theta + c}{mh'} \right)^{1/2}.$$

If $T_0 < 1/f^*$ then

$$f_{T_0}(\theta, m) = \begin{cases} \dfrac{2m(\theta + c)}{b} + \dfrac{1}{2}b(h' + h_0), & \text{if } \dfrac{b}{2m} \leq \tau'; \\[2ex] 2[(\theta + c)m(h' + h_0)]^{1/2}, & \text{if } \dfrac{1}{f^*} \leq \tau' \leq \dfrac{b}{2m}; \\[2ex] (\theta + c)f^* + m(h + h'), & \text{if } \tau' < \dfrac{1}{f^*}; \end{cases} \quad (4.21)$$

if $1/f^* < T_0 < b/2m$

$$f_{T_0}(\theta, m) = \begin{cases} \dfrac{2m(\theta + c)}{b} + \dfrac{1}{2}b(h' + h_0), & \text{if } T_0 \leq \dfrac{b}{2m} \leq \tau'; \\[2ex] 2[(\theta + c)m(h' + h_0)]^{1/2}, & \text{if } T_o < \tau' \leq \dfrac{b}{2m}; \\[2ex] \dfrac{(\theta + c)}{T_o} + m(h' + h_0)T_0, & \text{if } \tau' \leq T_0 \leq \tau; \\[2ex] 2[(\theta + c)mh']^{1/2} + mh_0 T_0, & \text{if } \dfrac{1}{f^*} \leq \tau < T_0; \\[2ex] (\theta + c)f^* + \dfrac{mh'}{f^*} + mhT_0, & \text{if } \tau < \dfrac{1}{f^*} \leq T_0; \end{cases} \quad (4.22)$$

and if $1/f^* \leq b/2m < T_0$ then

$$f_{T_0}(\theta, m) = \begin{cases} \dfrac{2m(\theta + c)}{b} + \dfrac{1}{2}bh' + mh_0 T_0, & \text{if } \dfrac{b}{2m} < \tau; \\[2ex] 2[(\theta + c)mh']^{1/2} + mh_0 T_0, & \text{if } \dfrac{1}{f^*} \leq \tau < \dfrac{b}{2m}; \\[2ex] (\theta + c)f^* + \dfrac{mh'}{f^*} + mh_0 T_0, & \text{if } \tau < \dfrac{1}{f^*}. \end{cases} \quad (4.23)$$

The functions $f_{T_0}(\theta, m)$ are non-decreasing and concave in θ for every $T_0 > 0$ as is easily verified by inspecting (4.21)–(4.23). This implies that for every choice of $T_0 > 0$, the routing problem with general cost function $f_{T_0}(\cdot, \cdot)$ can be solved by the method in Section 2.4, resulting in a lower bound and heuristic solutions which are almost surely asymptotically accurate and optimal, respectively under the mild probabilistic assumptions discussed there.

Returning to the problem of determining Z^*, since $f_{T_0}(\theta, m)$ is non-decreasing in θ, we obtain a further lower bound for Z^* by replacing in (4.20) each of the route lengths $\{\theta_1, \theta_2, \cdots, \theta_L\}$ by twice the average distance between the depot and the points in the region.

Lemma 4.2.

$$\underline{Z}^1 \equiv \inf_{T_0 > 0} \left\{ \dfrac{K_0}{T_0} + \min\left[\sum_{l=1}^{L} f_{T_0}\left(\dfrac{2\sum_{i \in X_l} d_i}{m_l}, m_l \right) : \right.\right.$$

$$\left.\left. \Pi = \{X_1, \cdots, X_L\} \right] \right\} \leq Z^* \quad (4.24)$$

Since $f_{T_0}(\theta, m)$ is concave in θ, it follows that the partition achieving the inner minimum in (4.24) is easily obtained. Moreover, due to additional properties of the $f_{T_0}(\cdot, \cdot)$ functions referred to in Section 2.4, the partition $\Pi^* = \{X_1^*, \cdots, X_L^*\}$ described in the previous subsection (4.2.1) can be shown to achieve the minimum in (4.24) for all $T_0 > 0$. Thus,

$$\underline{Z}^1 = \inf_{T_0 > 0} \left\{ \frac{K_0}{T_0} + \sum_{l=1}^{L} f_{T_0} \left(\frac{2 \sum_{i \in X_l^*} d_i}{|X_l^*|}, |X_l^*| \right) \right\}. \tag{4.25}$$

It remains to be shown how the minimum over $T_0 > 0$ is to be computed. Note, in view of the shape of the function $f_{T_0}(\cdot, \cdot)$ that $\underline{Z}^1(T_0)$, the function within $\{\ \}$ in (4.25), is of the form

$$\sum_{l=1}^{L} \left[\frac{\alpha_l(T_0)}{T_0} + \beta_l(T_0) + \gamma_l(T_0) T_0 \right],$$

where $\alpha_l(T_0)$, $\beta_l(T_0)$ and $\gamma_l(T_0)$ are *piecewise constant* for all $l = 1, \cdots, L$. Moreover, $\underline{Z}^1(T_0)$ is convex and continuously differentiable in T_0 everywhere except possibly in at most three points. Thus, $\underline{Z}^1$ is achieved in one of those three points or the unique point T_0^* where $d\underline{Z}^1(T_0)/dT_0 = 0$, i.e., $T_0^* = [\alpha(T_0^*)/\gamma(T_0^*)]^{1/2}$. Also, the piecewise constant functions $\alpha_l(T_0)$, $\beta_l(T_0)$ and $\gamma_l(T_0)$ change values at most $(2L + 3)$ times. These observations suggest a simple $O(N \log N)$ algorithm for the minimization of $\underline{Z}^1(T_0)$, see appendix A in Roundy [1985]. Queyranne [1987] proposes an alternative, *linear* time, procedure based on a linear time median finding procedure.

We are now ready to describe a complete solution procedure for the inventory-routing problem with central stock.

Algorithm A

Step 1. Determine the partition Π^*.

Step 2. Compute the value T_0^* which achieves the minimum in (4.25) and hence $\underline{Z}^1$.

Step 3. Construct a set of regions Π^H with the modified circular region partitioning scheme (MCRP), as described in Sections 2.4 and 4.2.1. Let m_l^H, θ_l^H denote the number of points in the lth region and the length of the traveling salesman tour through these points and the depot, respectively.

Step 4. Determine

$$\underline{Z}^H \equiv \inf_{T > 0} Z_{\Pi^H}(T) = \inf_{T_0 > 0} \left\{ \frac{K_0}{T_0} + \sum_{l=1}^{L^H} f_{T_0}(\theta_l^H, m_l^H) \right\} \tag{4.26}$$

and the corresponding vector of replenishment intervals $T^H = \{T_0^H, T_1^H, \cdots, T_{L^H}^H\}$ as follows: T_0^H is the unique value of T_0 which achieves the minimum in (4.26) and

T_l^H is the unique value of T_l achieving

$$f_{T_0^H}(\theta_l^H, m_l^H) = \inf_{1/f^* \le T_l \le b/2m_l^H} D_{T_0^H}(T_r, \theta_l^H, m_l^H).$$

Step 5. Set the base planning period $T^B = 1/f^*$ or $T^B = b/2q$ if $f^* = +\infty$ and chose T^B arbitrarily if $f^* = b = +\infty$. Round each of the components of the vector T^H to one of the two closest power-of-two multiples of T^B. Let $\overline{T}^H$ denote the resulting power-of-two vector of intervals. Implement the power-of-two policy with respect to the set of regions Π^H.

Theorem 4.3. *Assume all retailers are located at i.i.d. distances from the depot. Algorithm A generates a lower bound $\underline{Z}^1$, a set of regions Π^H and a power-of-two policy $\overline{T}^H$ with associated cost $Z^H \equiv Z_{\Pi^H}(\overline{T}^H)$ such that $\underline{Z}^1 \le Z^* \le Z^H$ and $\lim_{n\to\infty} Z^H/\underline{Z}^1 \le 1.06$ (a.s.).*

Proof. Let $\hat{Z} \equiv K_0/T_0^* + \sum_{l=1}^{L^H} f_{T_0^*}(\theta_l^H, m_l^H)$. We have

$$\underline{Z}^1 \le Z_{\Pi^H}(T^H) \le \hat{Z} \tag{4.27}$$

and

$$Z_{\Pi^H}(T^H) \le Z^H. \tag{4.28}$$

The first inequality in (4.27) follows from $\underline{Z}^1 \le \inf_{\Pi} \inf_{T} Z_{\Pi}(T) \le Z_{\Pi^H}(T^H)$; The second inequality follows from the definition of $\hat{Z}$ and the fact that T_0^H achieves the minimum in (4.26). Equation (4.28) follows from the definition of T^H. Consider now the routing problem with general route cost function $f_{T_0^*}(\theta, m)$. It follows from Theorem 2.16 that $\lim_{n\to\infty} \hat{Z}/\underline{Z}^1 = 1$ a.s. It then suffices to show that $\overline{\lim}_{n\to\infty} Z^H/Z_{\Pi^H} \le 1.06$, i.e., that the impact of the rounding procedure in Step 2 is asymptotically restricted to at most a 6% increase in system wide cost. This inequality, which is established in Lemma 2 in Anily & Federgruen [1992], is an extension of the same bound for the one warehouse multiple retailer inventory systems without interval restrictions, see Roundy [1985]. □

As in the case for systems without central stock, the algorithm's complexity is $O(Kn \log n)$. Indeed, Steps 1 and 3 are part of the procedure of Section 4.2.1; as explained above, Step 2 requires $O(L \log L) = O(Kn \log n)$ work and the same is true for Step 4. The rounding procedure in Step 2 clearly has linear time complexity.

As in the case of systems without central stock, extensive numerical results indicate that the gap between the lower bound and the cost of the proposed strategy is small even when the number of retailers is of moderate size. The observed (bounds for the) optimality gap are almost always smaller than those obtained for the corresponding systems without central stock even though the theoretical asymptotic bounds are worse. For example, with 100 identical retailers and $q^* = 4$, the average observed (bound for the) optimality gap is 9.5% and for problems with 500 identical retailers and $q^* = 4$, the average observed gap is 6.3%.

4.3. The effectiveness of direct shipping strategies

In view of the worst-case results for the CVRP described in Sections 2 and 3 one wonders whether similar results can be obtained for some, or all, of the inventory-routing models described above. Here we provide a partial answer for the infinite horizon inventory-routing problem without central stock. For this model, Gallego & Simchi-Levi [1990] characterize the effectiveness of so-called *direct shipping* strategies.

Consider the model of Section 4.2.1 with $f^* = +\infty$, $q = +\infty$ and $c = 0$. (The results described below hold in fact for any $c \geq 0$.) In addition to the cost components considered there, assume that a fixed (retailer specific) set-up cost K_i is incurred every time retailer i places an order.

A lower bound on the long-run average cost over all inventory-routing strategies is obtained by combining lower bounds on the long-run average ordering and holding costs and a lower bound on the long-run average transportation cost.

Lemma 4.4. *A lower bound, $\underline{Z}$, on the long-run average cost over all inventory routing strategies is given by*

$$\underline{Z} = \sum_{i \in N_c} \left\{ \sqrt{2\mu_i K_i h_i^+} + \frac{2 d_i \mu_i}{b} \right\}. \tag{4.29}$$

Proof. Let $\underline{Z}$ denote the lower bound obtained by minimizing separately
   (i) the ordering and holding costs, and
   (ii) the total vehicle routing costs required to allow all retailers to meet their demands, and then adding these two values.

The minimum for (i) is given by the average costs of n independent EOQ models, i.e., by $\sum_{i \in N_c} \sqrt{2\mu_i K_i h_i^+}$. To find a lower bound for (ii) we use a similar analysis to the proof of Lemma 2.1. Consider the distance traveled by vehicles of capacity b serving a set N_c of geographically dispersed retailers located at distances d_i from the depot and facing demands w_i. A lower bound on the total distance traveled is given by

$$\frac{2}{b} \sum_{i \in N_c} d_i w_i. \tag{4.30}$$

Let us now consider the distance traveled up to time T. The cumulative demand at retailer i up to time T is $\mu_i T$, and since no shortages or backlogging is allowed, the minimal amount shipped to retailer i up to time T is $\mu_i T$ for all $i \in N_c$. Therefore the minimal distance traveled up to time T is obtained by substituting $\mu_i T$ for w_i in (4.30). Consequently, a lower bound on the distance traveled per unit time is given by $(2/b) \sum_{i \in N_c} d_i \mu_i$. Adding this expression to the lower bound on the long-run inventory ordering and holding cost we obtain (4.29). $\square$

We now analyze the cost of supplying all retailers independently. We call this the class of direct shipping strategies. An important subclass, called fully loaded direct shipping strategies, consists of all shipments being made by fully loaded vehicles. We obtain an upper bound on the optimal cost in this class of policies. This bound, together with the lower bound of Lemma 4.4, characterizes (an upper bound on) the worst-case performance of direct shipping.

Let Q_i be the lot size for retailer i, $i \in N_c$. The cost per unit of time for retailer $i \in N_c$ is given by

$$z_i(Q_i) = \frac{K_i \mu_i}{Q_i} + \frac{h_i^+ Q_i}{2} + 2d_i \left\lceil \frac{Q_i}{b} \right\rceil \frac{\mu_i}{Q_i}.$$

Let $Z^d = \sum_{i \in N_c} z_i(Q_i)$ be the total cost per unit of time corresponding to the order quantities $\{Q_1, Q_2, \cdots, Q_n\}$. We find an upper bound on Z^d by restricting the choice of lot sizes to fully loaded vehicles, i.e., the order quantities are restricted to the set $F = \{mb : m = 1, 2, \cdots\}$.

Clearly Z^d is separable, so it is enough to find an upper bound on z over F, where z is identical to z_i omitting the index i. Let $f(Q) = K\mu/Q + h^+ Q/2$ and note that, in F, the functions f and z differ only by the constant $2d\mu/b$. Thus, Q^f, the minimizer of f over F, is also the minimizer of z over the same set. Finally, let $Q^e = \sqrt{2K\mu/h^+}$, $\eta = \max\{b/Q^e, \sqrt{2}\}$ and $e(\eta) = 2/(\eta + 1/\eta)$.

Lemma 4.5.

$$z(Q^f) < \frac{f(Q^e) + 2d\mu/b}{e(\eta)}.$$

Proof. It is easily verified that $Q^f = mb$ minimizes $f(Q)$ over F if and only if

$$\sqrt{(m-1)/m} \leq \frac{Q^e}{Q^f} \leq \sqrt{(m+1)/m}. \tag{4.31}$$

Assume now that $Q^e \geq b/\sqrt{2}$ then by (4.31) $1/\sqrt{2} \leq Q^e/Q^f \leq \sqrt{2}$. Since f is convex and $f(Q^e \sqrt{2}) = f(Q^e/\sqrt{2})$ we obtain

$$f(Q^f) \leq \frac{f(Q^e)}{e(\eta)}. \tag{4.32}$$

If $Q^e < b/\sqrt{2}$ then $Q^f = b$. Hence

$$f(Q^f) = \frac{f(Q^e)}{e(\eta)} \tag{4.33}$$

Combining (4.32), (4.33), $e(\eta) < 1$ and the definition of z we obtain

$$z(Q^f) = f(Q^f) + \frac{2d\mu}{b} < \frac{f(Q^e) + 2d\mu/b}{e(\eta)}. \qquad \square$$

We are now ready to characterize the worst-case performance of direct shipping. For this purpose, let $\eta_i = \max\{b/Q_i^e, \sqrt{2}\}$, $\eta = \max_{i \in N_c}\{\eta_i\}$ and $Z^f = \sum_{i \in N_c} z(Q_i^f)$. It is easy to see that the lower bound $\underline{Z}$ obtained in Lemma 4.4 together with the upper bound of Lemma 4.5 yields

Theorem 4.6. *For any instance, $Z^f/\underline{Z} \leq 1/e(\eta)$.*

Due to Theorem 4.6, the worst-case ratio of the cost of direct shipping to a lower bound on the optimal cost is no more than 1.061 whenever the economic lot sizes exceed 71% of the vehicle capacity, i.e., whenever $Q_i^e \geq b/\sqrt{2}$ for all $i \in N_c$. The worst-case ratio increases as the economic lot sizes decrease. For instance, if the minimum lot size is 50% (respectively 33%) of the vehicle capacity then the worst-case ratio is 1.25 (respectively 1.68).

5. The multi-depot CVRP

5.0. Introduction

The multi-depot capacitated vehicle routing problem (MCVRP) is a generalization of the CVRP in which the vehicles are initially located at a given set of depots. As in the CVRP the customers and the depots are presented as a set of nodes on an undirected graph $G = (N, E)$. We denote by N_c and N_d the set of customers and the set of depots, respectively, and d_i denotes the distance between node i and its closest depot.

In our formulation of the MCVRP, a vehicle can depart from any depot. There are two variants of the problem: a vehicle can be required to depart from and return to the same depot (this is called the fixed destination MCVRP) or allowed to return to some other depot (this is called the non-fixed destination MCVRP). The non-fixed destination model occurs, e.g., as a subproblem in a multi-period vehicle routing model in which a vehicle leaves a depot, visits several customers and enters a depot (not necessarily its origin depot) and then repeats this for several periods. The only restriction is that at the end of the overall cycle it returns to its home depot. A description of such a large scale system is given in Gavish [1981].

We denote the optimal solution values of the non-fixed destination MCVRP by Z_n^* and for the fixed destination MCVRP by Z_f^*. As in the previous sections, for any heuristic H, Z^H denotes the value of the solution obtained by the heuristic.

5.1. Worst-case analysis of heuristics

The worst-case analysis for the multi-depot capacitated vehicle routing problem is based on the work of Li & Simchi-Levi [1990]. They distinguish between the equal-weight case and the unequal weight case. Similar to the single depot model, and without loss of generality, we assume in the equal-weight case that $w_i = 1$ for every $i \in N_c$.

5.1.1. The non-fixed destination equal-weight MCVRP

The ITP(α) heuristic can be generalized to solve the non-fixed destination equal weight MCVRP. The heuristic, called ITP$_n(\alpha)$, starts by obtaining a new undirected complete graph $\overline{G} = (\overline{N}, \overline{E})$ on which only one depot p exists and $\overline{N} = N_c \cup \{p\}$. Every edge $i-j \in \overline{E}$ such that $i, j \in N_c$ has a length d_{ij}, the shortest distance from i to j obtained from G. For every $i \in N_c$, let the length of edge $i-p$ in $\overline{E}$ be d_i, the distance between i and its nearest depot. In other words, we combine all the depots in N_d to form a new single depot p. Notice that distances in $\overline{G}$ do not necessarily satisfy the triangle inequality. Hence, our next step is to compute for every $i, j \in \overline{N}$ the value $\overline{d}_{ij}$ which is the length of the shortest path from i to j in $\overline{G}$.

Clearly, $\overline{d}_{ip} = d_i$ for all $i \in \overline{N} \setminus \{p\} = N_c$. However, $\overline{d}_{ij}$ may be different from d_{ij} for some $i, j \in N_c$, that is, $\overline{d}_{ij} = \min\{d_{ij}, d_i + d_j\}$. If $\overline{d}_{ij} < d_{ij}$ $(i, j \in N_c)$, we say that edge $i-j$ is a *dummy* edge. If $i-j$ is a dummy edge then there exist two different depots in N_d, one closest to i and the other closest to j such that $d_i + d_j < d_{ij}$.

We now present the ITP$_n(\alpha)$ heuristic.

The ITP$_n(\alpha)$ heuristic

Step 1. Obtain the graph $\overline{G} = (\overline{N}, \overline{E})$ and its associated shortest distance matrix $\{\overline{d}_{ij}\}$.

Step 2. Use the ITP(α) heuristic to solve the single depot CVRP on $\overline{G}$ using the shortest distance matrix $\{\overline{d}_{ij}\}$.

Step 3. Let $\overline{T}^{(k)}$ denote the route driven by the kth vehicle ($k = 1, 2, \ldots, \lceil n/q \rceil$) in $\overline{G}$. We transform this solution to a solution in the original network G as follows: For $k = 1, 2, \ldots, \lceil n/q \rceil$,
- delete the pair of edges connecting the first and the last customers in $\overline{T}^{(k)}$ with the depot p;
- delete all dummy edges in $\overline{T}^{(k)}$ (if any); and
- connect the endpoints of each segment produced in this way to its closest depot in G.

Theorem 5.1.

$$\frac{Z^{\text{ITP}_n(\alpha)}}{Z_n^*} \leq 1 + \left(1 - \frac{1}{q}\right)\alpha.$$

Proof. Notice that the value of the solution obtained by the ITP$_n(\alpha)$ heuristic is equal to the value of the solution obtained by the ITP(α) heuristic for the CVRP on $\overline{G}$, and an optimal solution to the non-fixed destination MCVRP on G can be transformed into a feasible solution to the CVRP on $\overline{G}$ (by combining the depots) without changing the objective value. Hence,

$$Z^{\text{ITP}_n(\alpha)} = Z^{\text{ITP}(\alpha)}(\overline{G}) \tag{5.1}$$

and
$$Z_n^* \geq Z^*(\overline{G}), \tag{5.2}$$

where $Z^{\mathrm{ITP}(\alpha)}(\overline{G})$ is the value of the solution obtained by the ITP(α) heuristic on $\overline{G}$ and $Z^*(\overline{G})$ is the value of the optimal solution to the CVRP on $\overline{G}$. Therefore,

$$\frac{Z^{\mathrm{ITP}_n(\alpha)}}{Z_n^*} \leq \frac{Z^{\mathrm{ITP}(\alpha)}(\overline{G})}{Z^*(\overline{G})} \leq 1 + \left(1 - \frac{1}{q}\right)\alpha. \qquad \square$$

The OP heuristic for the CVRP can be generalized in a similar way to solve the non-fixed destination MCVRP by replacing ITP(α) in Step 2 of the algorithm by OP(α). This generalized OP heuristic is denoted by $\mathrm{OP}_n(\alpha)$. Since the $\mathrm{OP}_n(\alpha)$ heuristic always produces solutions no worse than the $\mathrm{ITP}_n(\alpha)$ heuristic, its worst-case bound is at most $1 + (1 - 1/q)\alpha$. Using the examples given in Theorems 2.4 and 2.5, it is easy to see that this bound is sharp for the $\mathrm{ITP}_n(1)$ heuristic and asymptotically sharp for the $\mathrm{OP}_n(1)$ heuristic.

5.1.2. The fixed destination equal-weight MCVRP

We can now modify the solution produced by the $\mathrm{ITP}_n(\alpha)$ heuristic to solve the fixed destination equal-weight MCVRP. Let m be the number of tours produced by the $\mathrm{ITP}_n(\alpha)$ heuristic, and let $T^{(k)} = p_1 - v_1 - v_2 - \cdots - v_r - p_2$ be the kth tour, where $p_1, p_2 \in N_d$ are the depots and $v_i \in N_c$ ($i = 1, 2, \ldots, r$) are the customers in the tour. We modify $T^{(k)}$ ($k = 1, 2, \ldots, m$) to form a feasible tour for the fixed destination MCVRP as follows:

If $p_1 = p_2$ then this tour is feasible, otherwise replace this tour

by $\begin{cases} p_1 - v_1 - v_2 - \cdots - v_r - p_1, & \text{if } d_{p_1 v_1} + d_{p_1 v_r} \leq d_{p_2 v_1} + d_{p_2 v_r}; \\ p_2 - v_1 - v_2 - \cdots - v_r - p_2, & \text{if } d_{p_1 v_1} + d_{p_1 v_r} > d_{p_2 v_1} + d_{p_2 v_r}. \end{cases}$

This modified heuristic is called the $\mathrm{ITP}_f(\alpha)$ heuristic. We now prove the following result.

Theorem 5.2.
$$\frac{Z^{\mathrm{ITP}_f(\alpha)}}{Z_f^*} \leq 1 + \left(2 - \frac{1}{q}\right)\alpha.$$

Proof. Consider the increase in the length of tour $T^{(k)}$ obtained by the transformation of $T^{(k)}$ to a feasible solution for the fixed destination MCVRP. We have

length of the modified kth tour
$\leq$ (length of segment $v_1 - \cdots - v_r$) $+ \frac{1}{2}(d_{p_1 v_1} + d_{p_1 v_r} + d_{p_2 v_1} + d_{p_2 v_r})$
$\leq$ (length of $v_1 - \cdots - v_r$) $+ \frac{1}{2}(d_{p_1 v_1} + d_{p_2 v_r})$
$\quad + \frac{1}{2}(d_{p_1 v_1} + \text{length of } v_1 - \cdots - v_r) + \frac{1}{2}(d_{p_2 v_r} + \text{length of } v_1 - \cdots - v_r)$
$= 2 \cdot (\text{length of } v_1 - \cdots - v_r) + d_{p_1 v_1} + d_{p_2 v_r}.$

Hence, the length of the modified kth tour is at most the length of tour $T^{(k)}$ plus the length of the segment $v_1 - v_2 - \cdots - v_r$. Denote by $L^\alpha(\overline{G})$ the length of the α-optimal tour in $\overline{G}$ used by the $ITP_n(\alpha)$ heuristic to obtain $T^{(k)}$, $k = 1, 2, \ldots, m$ and by $L^*(\overline{G})$ the length of the optimal traveling salesman tour in $\overline{G}$. Thus,

$$Z^{ITP_f(\alpha)} \leq Z^{ITP_n(\alpha)} + L^\alpha(\overline{G}), \tag{5.3}$$

Since $L^\alpha(\overline{G}) \leq \alpha \cdot L^*(\overline{G})$ and by Theorem 5.1,

$$Z^{ITP_f(\alpha)} \leq Z_n^* \left[1 + \left(1 - \frac{1}{q}\right)\alpha\right] + \alpha L^*(\overline{G}).$$

Clearly, $Z_n^* \leq Z_f^*$ and $L^*(\overline{G}) \leq Z^*(\overline{G}) \leq Z_n^* \leq Z_f^*$ where the second inequality is justified by (5.2). Hence,

$$Z^{ITP_f(\alpha)} \leq Z_f^*[1 + (2 - \frac{1}{q})\alpha].$$

This completes the proof. $\square$

The $OP_n(\alpha)$ heuristic can be modified in a similar way to solve the fixed destination MCVRP. The heuristic, denoted by $OP_f(\alpha)$, always gives a solution no worse than the one obtained by the $ITP_f(\alpha)$. Thus

$$Z^{OP_f(\alpha)} \leq Z_f^* \left[1 + \left(2 - \frac{1}{q}\right)\alpha\right].$$

Can this bound be improved? The following theorem says that the bound is asymptotically sharp for the $OP_f(1)$ heuristic (and thus for the $ITP_f(1)$ heuristic also) as q tends to infinity.

Theorem 5.3. *There exists an example in which $Z^{OP_f(1)}/Z_f^* \to 3$ as $q \to \infty$.*

Proof. Consider the following example with rq depots and rq^2 customers, where $q = 2^r$ and $r \geq 3$ is an integer. The customers are located at the nodes of an $rq \times 2r$ grid graph. Every horizontal/vertical edge in the grid is of unit length. There is one customer located at every node in the 1st and $(2r)$th columns, and there are 2^{j-2} customers (with zero distance apart) located at each node in the jth and $(2r + 1 - j)$th columns ($j = 2, 3, \ldots, r$). The depots are connected to two opposite sides of the boundary of the grid with unit length edges as shown in Figure 8 in which we duplicate every depot to simplify the presentation. Every arc $i-j$ that does not appear in Figure 8 has a length given by the shortest distance from i to j.

Consider the following feasible solution in which rq vehicles serve all the customers, where the ith vehicle departs from depot i, visits all q customers in row i and returns to depot i. Hence,

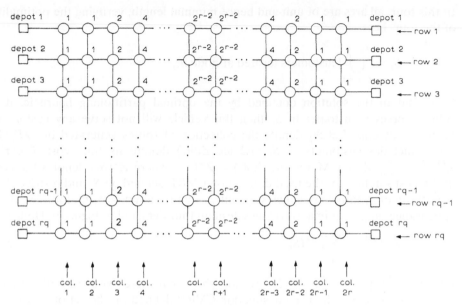

Fig. 8. The label at a node represents the number of customers located at that node.

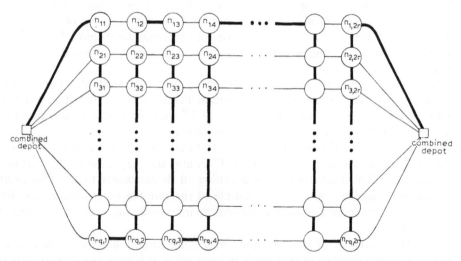

Fig. 9. The darkened lines represent the optimal traveling salesman tour.

$$Z^* \leq rq(2r+1).$$

Let n_{ij} denote the node located at row i and column j. After combining the depots, we obtain an optimal traveling salesman tour as shown in Figure 9 in which once again we duplicate the combined depot to simplify the presentation.

In this tour, all arcs are of unit and hence minimal length, verifying the optimality of the tour. Let

$$S_j = \bigcup_{i=q+1}^{rq-q} \{\text{customers located at node } n_{ij}\}.$$

Note that in the solution obtained by the optimal partitioning heuristic, if a vehicle serves a customer in S_j then this vehicle will not serve any customers in other columns. Let $\mathfrak{R}_j$ denote the collection of routes generated by $OP_f(1)$ which includes customers in S_j and let $Z(\mathfrak{R}_j)$ denote its total cost. Clearly, $Z^{OP_f(1)} \geq \sum_j Z(\mathfrak{R}_j)$. Moreover, $Z(\mathfrak{R}_j) \geq Z^*(S_j)$, where $Z^*(S_j)$ denotes the cost of the best solution for the fixed destination CVRP defined on S_j and restricted to a partition of S_j into consecutive segments, i.e., a partition of S_j that follows their appearance on the optimal traveling salesman tour depicted in Figure 9. Hence,

$$Z^{OP_f(1)} \geq \sum_j Z^*(S_j). \tag{5.4}$$

For any fixed j, $j = 1, 2, \cdots, 2r$, we now show that there exists an optimal consecutive solution for the fixed destination CVRP defined on the set of customers S_j for which every tour is fully loaded up to the vehicle capacity. Assume to the contrary that an optimal consecutive solution chooses a set of routes some of which are less than full capacity. Let T^0 denote the string of nodes visited by the first route R^0, with initial load less than q; T^{-1} the set of nodes of the previous route R^{-1} (if any) and T^1 the set of nodes of the subsequent route R^1. Note that T^1 must exist, since otherwise the total number of customers visited in S_j would not be a multiple of q.

Since all routes up to R^{-1} are of full capacity, and q is a multiple of the number of customers in each route, R^{-1} must contain all customers of its last visited node. Thus R^0 covers all customers in all nodes of T^0 with the possible exception of its last node. We thus distinguish between two cases:

(i) Only part of the customers of the last node in T^0 are visited together: In this case we have that the last node of T^0 is also the first node of T^1; this node can now be eliminated from T^1 by assigning all its customers to R^0. This clearly maintains feasibility, since q is a multiple of the number of customers (as there are 2^{j-2} customers in S_j) and it eliminates an inter-node-arc without affecting any other arc.

(ii) All customers of the last node in T^0 are visited together: Since q is a multiple of the number of customers in each node, it is now possible to shift the customers in the first node of T^1 to R^0. Let n_{ij} be this node. Clearly, the depot that belongs to the original route R^1 must be either depot i or a depot with a larger index, i.e., $i + 1, i + 2$, and so on. Suppose first that R^1 is served by depot i. Since n_{ij} is transferred to R^0, depot $i+1$ can serve R^1 with at most the same radial cost thus saving at least one unit distance of not moving from n_{ij} to $n_{(i+1)j}$. This unit cost plus the distance from the depot serving R^0 to its last original customer is no less than the distance from this depot to node n_{ij}; hence the total cost is either reduced or unchanged. Suppose now that R^1 is served by a depot with index $> i$, then shifting n_{ij} to R^0 reduces the radial cost from the depot serving R^1 to its first node $(n_{(i+1)j})$ by one unit. This unit together with the distance from the depot serving R^0 to its last original customer in R^0 is no less than the distance from this depot to n_{ij}.

By repeating the above argument we generate a collection of fully loaded tours which are consecutive and have minimum total cost $Z^*(S_j)$. This collection of tours uses m_j vehicles all of total length l_j, where

$$m_j = \begin{cases} r - 2, & \text{if } j = 1 \text{ or } 2r, \\ (r - 2) \cdot 2^{j-2}, & \text{if } j = 2, 3, \ldots, r, \\ (r - 2) \cdot 2^{2r-1-j}, & \text{if } j = r + 1, r + 2, \ldots, 2r - 1, \end{cases}$$

and

$$l_j = \begin{cases} 2q, & \text{if } j = 1 \text{ or } 2r, \\ 2\left(\dfrac{q}{2^{j-2}} + j - 1\right), & \text{if } j = 2, 3, \ldots, r, \\ 2\left(\dfrac{q}{2^{2r-1-j}} + 2r - j\right), & \text{if } j = r + 1, r + 2, \ldots, 2r - 1. \end{cases}$$

This is true, since nodes $n_{ij}, n_{(i+1)j}, \cdots, n_{i'j}$ is optimally served from one depot with index k, $i \le k \le i'$ in a route which covers the rectangle depot $k \to$ depot $i \to n_{ij} \to n_{i'j} \to$ depot $i' \to$ depot k.

Therefore, by (5.4)

$$Z^{OP_f(1)} \ge \sum_{j=1}^{2r} m_j l_j$$

$$> \sum_{j=2}^{2r-1} m_j l_j$$

$$= 2 \sum_{j=2}^{r} m_j l_j \qquad \text{(by symmetry)}$$

$$= 2 \sum_{j=2}^{r} (r - 2) 2^{j-2} \cdot 2\left(\frac{q}{2^{j-2}} + j - 1\right)$$

$$= 4(r - 2) \sum_{j=2}^{r} [q + (j - 1)2^{j-2}]$$

$$= 4(r - 2)[q(r - 1) + (r2^{r-1} - 2^r + 1)]$$

$$= 4(r - 2)[(\frac{3r}{2} - 2)q + 1].$$

Thus, by Theorem 5.2

$$3 - \frac{2}{q} \ge \frac{Z^{OP_f(1)}}{Z^*} > \frac{4(r - 2)[(3r/2 - 2)q + 1]}{rq(2r + 1)},$$

and

$$3 \geq \lim_{r \to \infty} \frac{Z^{\mathrm{OP}_f(1)}}{Z^*} \geq \lim_{r \to \infty} \frac{4(r-2)[(3r/2-2)q+1]}{rq(2r+1)} = 3,$$

which implies

$$\lim_{r \to \infty} \frac{Z^{\mathrm{OP}_f(1)}}{Z^*} = 3. \quad \square$$

5.1.3. The unequal-weight case

Let us now consider the multi-depot capacitated vehicle routing problem with unequal demands (UMCVRP). We assume that splitting the demand of a customer over several vehicles is not allowed.

The non-fixed destination UMCVRP can be solved by first combining the depots and calculating the new distance matrix $\{\bar{d}_{ij}\}$. We then, again, apply one of the two heuristics in Subsection 5.1.1 for the non-fixed destination equal-weight MCVRP, i.e., $\mathrm{ITP}_n(\alpha)$ and $\mathrm{OP}_n(\alpha)$, where in Step 2, $\mathrm{ITP}(\alpha)$ or $\mathrm{OP}(\alpha)$ need to be replaced by $\mathrm{UITP}(\alpha)$ or the $\mathrm{UOP}(\alpha)$ to maintain feasibility. We denote these versions by $\mathrm{UITP}_n(\alpha)$ and $\mathrm{UOP}_n(\alpha)$, respectively. It is easy to check by a minor adaptation of the proof of Theorem 5.1 that

$$\frac{Z^{\mathrm{UITP}_n(\alpha)}}{Z_n^*} \leq 2 + \left(1 - \frac{2}{q}\right)\alpha$$

and

$$\frac{Z^{\mathrm{UOP}_n(\alpha)}}{Z_n^*} \leq 2 + \left(1 - \frac{2}{q}\right)\alpha.$$

Again, the example from Theorem 3.2 can be used to show that both bounds are asymptotically sharp when $\alpha = 1$.

The fixed destination UMCVRP can be solved by modifying the tours obtained by the $\mathrm{UITP}_n(\alpha)$ or $\mathrm{UOP}_n(\alpha)$ similarly to what we have done in Section 5.1.2. We denote these versions of the heuristics by $\mathrm{UITP}_f(\alpha)$ and $\mathrm{UOP}_f(\alpha)$. By following the same argument as in the derivation of the worst-case bound for the $\mathrm{ITP}_f(\alpha)$ and $\mathrm{OP}_f(\alpha)$ in Section 5.1.2, we obtain

$$\frac{Z^{\mathrm{UITP}_f(\alpha)}}{Z_f^*} \leq 2 + \left(2 - \frac{2}{q}\right)\alpha$$

and

$$\frac{Z^{\mathrm{UOP}_f(\alpha)}}{Z_f^*} \leq 2 + \left(2 - \frac{2}{q}\right)\alpha.$$

If $\alpha = 1$ then

$$\frac{Z^{\mathrm{UOP}_f(1)}}{Z_f^*} \leq 4 - \frac{2}{q}$$

which is bounded by the constant 4 for all values of q. The following theorem says that this constant cannot be less than $3\frac{29}{42} \approx 3.69$.

Theorem 5.4. *There exists an example with* $Z^{\text{UOP}_f(1)}/Z_f^* \to 3\frac{29}{42}$ *as* $q \to \infty$.

Proof. Consider the following example with rq depots and $rq(2r+1)$ customers, where $q = 84 \cdot 2^r$ and $r \geq 3$ is an integer. There is one customer located at every node of an $rq \times (2r+1)$ grid graph. Every horizontal/vertical edge is of unit length. The customers in the rth, $(r+1)$st and $(r+2)$nd columns have demands $q/3+1$, $q/2+2$ and $q/7+1$, respectively, and the customers in the jth and $(2r+2-j)$th columns have demand 2^j $(j = 1, 2, \ldots, r-1)$. The depots are connected to two opposite sides of the boundary of the grid with unit length edges as in the proof of Theorem 5.3. The graph is depicted in Figure 10 in which we duplicate every depot to simplify the presentation.

Consider the following feasible solution for which there are rq vehicles such that the ith vehicle departs from depot i, visits all $2r+1$ customers (with a total demand q) in row i and returns to depot i. Hence,

$$Z_f^* \geq rq(2r+2).$$

Let n_{ij} denote the node located at row i and column j. After combining the depots, we obtain an optimal traveling salesman tour as shown in Figure 11 in which the combined depot is duplicated to simplify presentation. (We note that the network is identical to the one used in the proof of Theorem 5.3 and therefore has the same optimal traveling salesman tour.) Let

$$S_j = \{\text{customer at node } n_{ij} \mid q+1 \leq i \leq rq - q\}.$$

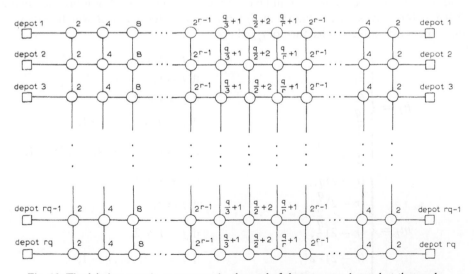

Fig. 10. The label at a node represents the demand of the customer located at that node.

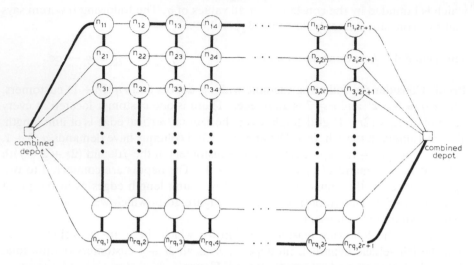

Fig. 11. The darkened lines represent the optimal traveling salesman tour.

As in the proof of Theorem 5.3 we have $Z^{\mathrm{UOP}_f(1)} \geq \sum_j Z^*(S_j)$ where $Z^*(S_j)$ denotes the cost of the best solution for the fixed destination CVRP defined on S_j and restricted to a partition of S_j into consecutive segments, i.e., a partition of S_j that follows their appearance on the optimal traveling salesman tour depicted in Figure 11. Since for all $j \neq r, r + 1$ and $r + 2$ the vehicle capacity is a multiple of the common demand shared by all customers in S_j, one verifies as in the proof of Theorem 5.3 that a solution with cost $Z^*(S_j)$ consists of a number of full-capacity routes visiting consecutive segments. A similar argument verifies that for S_r, S_{r+1} and S_{r+2} each vehicle visits 2, 1 and 6 customers, respectively. A solution achieving $Z^*(S_j)$ thus serves all customers in S_j with m_j vehicles, each of which serves h_j customers and travels a distance of l_j, where

$$
h_j = \begin{cases}
1, & \text{if } j = r + 1, \\
2, & \text{if } j = r, \\
6, & \text{if } j = r + 2, \\
\dfrac{q}{2^j}, & \text{if } j = 1, 2, \ldots, r - 1, \\
\dfrac{q}{2^{2r+2-j}}, & \text{if } j = r + 3, r + 4, \ldots, 2r + 1,
\end{cases}
$$

$$
m_j = \begin{cases}
(r - 2)q, & \text{if } j = r + 1, \\
(r - 2)\dfrac{q}{2}, & \text{if } j = r, \\
(r - 2)\dfrac{q}{6}, & \text{if } j = r + 2, \\
(r - 2) \cdot 2^j, & \text{if } j = 1, 2, \ldots, r - 1, \\
(r - 2) \cdot 2^{2r+2-j}, & \text{if } j = r + 3, r + 4, \ldots, 2r + 1,
\end{cases}
$$

and

$$
l_j = \begin{cases}
2r + 2, & \text{if } j = r + 1, \\
2r + 2, & \text{if } j = r, \\
2r + 10, & \text{if } j = r + 2, \\
2\left(j - 1 + \dfrac{q}{2^j}\right), & \text{if } j = 1, 2, \ldots, r - 1, \\
2\left(2r + 1 - j + \dfrac{q}{2^{2r+2-j}}\right), & \text{if } j = r + 3, r + 4, \ldots, 2r + 1.
\end{cases}
$$

Therefore,

$$
Z^{\mathrm{UOP}_f(1)} \geq \sum_{j=1}^{2r+1} m_j l_j
$$

$$
= 2\sum_{j=1}^{r-1} m_j l_j + m_r l_r + m_{r+1} l_{r+1} + m_{r+2} l_{r+2} \quad \text{(by symmetry)}
$$

$$
= \left[2\sum_{j=1}^{r-1} 2^j \cdot 2\left(j - 1 + \frac{q}{2^j}\right) \right.
$$
$$
\left. + \frac{q}{2}(2r + 2) + q(2r + 2) + \frac{q}{6}(2r + 10) \right](r - 2)
$$

$$
= \left[16\sum_{j=1}^{r-1}(j - 1)2^{j-2} + 4(r - 1)q \right.
$$
$$
\left. + \frac{q}{2}(2r + 2) + q(2r + 2) + \frac{q}{6}(2r + 10) \right](r - 2)
$$

$$
= \left\{ 16\left[(r - 1)2^{r-2} - 2^{r-1} + 1\right] + 4(r - 1)q \right.
$$
$$
\left. + \frac{q}{2}(2r + 2) + q(2r + 2) + \frac{q}{6}(2r + 10) \right\}(r - 2)
$$

$$
= \left[16(r - 3)2^{r-2} + 16 + \left(\frac{22}{3}r + \frac{2}{3}\right)q \right](r - 2)
$$

$$
= \left[\frac{q}{21}(r - 3) + \left(\frac{22}{3}r + \frac{2}{3}\right)q + 16 \right](r - 2)
$$

$$
= \left[\left(\frac{155}{21}r + \frac{11}{21}\right)q + 16 \right](r - 2).
$$

Thus,

$$
\frac{Z^{\mathrm{UOP}_f(1)}}{Z_f^*} \geq \frac{\left[\left(\frac{155}{21}r + \frac{11}{21}\right)q + 16\right](r - 2)}{rq(2r + 2)} \rightarrow \frac{155}{42} \quad \text{as } r \rightarrow \infty. \qquad \square
$$

5.2. *Asymptotic analysis of heuristics*

In this section we briefly describe asymptotic results related to the multi-depot capacitated vehicle routing problem. We start with the most general result on the asymptotic optimal solution value, whose proof is identical to the proof of Theorem 3.3.

Theorem 5.5. *Let x_k, $k = 1, 2, \cdots, n$ be a sequence of independent random variables having a distribution μ with compact support in $\Re^2$. Let $d(y)$ be the Euclidean distance between point $y \in \Re^2$ and its closest depot and let*

$$E(d) = \int_{\Re^2} d(y)\, \mathrm{d}\mu(y).$$

Let the demands w_k, $k = 1, 2, \cdots, n$ be a sequence of independent random variables having a distribution Φ with support on $[0, 1]$, and let $\lim_{n \to \infty} b^*/n = \gamma$ almost surely where b^* denotes the minimum number of bins required to pack the demands of all customers. Then, almost surely,

$$\lim_{n \to \infty} \frac{1}{n} Z_u^* = 2\gamma E(d).$$

Next, we turn our attention to finding asymptotically optimal heuristics for the equal weight multi-depot case. It is not hard to show that the $\mathrm{ITP}_n(\alpha)$ and the $\mathrm{ITP}_f(\alpha)$ are asymptotically optimal for the non-fixed destination model and the fixed destination model, respectively. The proof is based on the following two lemmas:

Lemma 5.6. $Z^{\mathrm{ITP_f}(\alpha)} \geq Z^{\mathrm{ITP_n}(\alpha)} \geq (2/q) \sum_{i \in N_c} d_i$.

Proof. Follows from a simple variation of the proof of Lemma 2.1. □

Lemma 5.7. $Z^{\mathrm{ITP_n}(\alpha)} \leq Z^{\mathrm{ITP_f}(\alpha)} \leq (2/q) \sum_{i \in N_c} d_i + \alpha(2 - 1/q)L^*(N_c) + O(1)$.

Proof. Recall the definition of the graph $\overline{G}$, its shortest distance matrix $\{\overline{d}_{ij}\}$ as well as $Z^{\mathrm{ITP}(\alpha)}(\overline{G})$ and $L^*(\overline{G})$. For the latter two definitions, see the proof of Theorem 5.1 and Theorem 5.2, respectively. From equations (5.1) and (5.3) we have

$$Z^{\mathrm{ITP_f}(\alpha)} \leq Z^{\mathrm{ITP}(\alpha)}(\overline{G}) + \alpha L^*(\overline{G})$$

$$\leq \frac{2}{q} \sum_{i \in N_c} d_i + \alpha \left(1 - \frac{1}{q}\right) L^*(\overline{G}) + \alpha L^*(\overline{G}).$$

To bound $L^*(\overline{G})$, let tour $T = v_1 - v_2 - \cdots - v_n - v_1$ be the optimal traveling salesman tour through all customers in the set N_c using the original shortest distance matrix $\{d_{ij}\}$ and whose length is $L^*(N_c)$. We transform this tour into a feasible solution to the traveling salesman problem on $\overline{G}$ in the following way: Connect customers v_1 and v_2 to the single depot p in $\overline{G}$ to form a traveling salesman tour $T' = p - v_2 - v_3 - \cdots - v_n - v_1 - p$. The length of this tour, using the shortest distance matrix $\{\bar{d}_{ij}\}$ is denoted by $\overline{L}(T')$ and satisfies

$$L^*(\overline{G}) \leq \overline{L}(T') \leq L^*(N_c) + d_{v_1} + d_{v_2}.$$

The left hand side inequality is justified by the optimality of $L^*(\overline{G})$ on $\overline{G}$ while the right hand side is justified by the inequality $\bar{d}_{ij} \leq d_{ij}$ for every i and j. Clearly $d_{v_1} + d_{v_2} = O(1)$. Hence,

$$L^*(\overline{G}) \leq L^*(N_c) + O(1).$$

This completes the proof of the Lemma. □

As in the single depot case, it is easy to see that for n large enough, the quantity $(2/q) \sum_{i \in N_c} d_i$ dominates the upper bound (Lemma 5.7). This observation completes the proof that the $\text{ITP}_n(\alpha)$ and the $\text{ITP}_f(\alpha)$ are asymptotically optimal for the non-fixed destination model and the fixed destination model, respectively.

5.3. Asymptotic optimal design for distribution systems

The previous results characterize the theoretical behavior of a number of heuristics for the MCVRP as well as the optimal solution. These results enable us to develop analytical models to assist the design and control of distribution systems. This approach is suggested in Simchi-Levi [1992].

As an example, consider a company that delivers consumer goods to a number of stores located in an area of size A. The company has decided to open a number of warehouses in the region and has carried out a market survey to estimate the number of potential customers, denoted by n, and the probability distribution of their demands. At this preliminary stage of the analysis the company assumes that all potential customers have the same demand distribution with $\overline{w}$ being the expected customer's demand.

Based on the information available from the survey, the company wants to determine the number and locations of service stations; how to allocate customers to depots and what the routing strategies should be so as to minimize total system-wide costs. This cost measure includes a component associated with the average distance traveled by all vehicles plus a fixed set-up cost, denoted by c, for each established depot.

The insight obtained from the analysis of the CVRP can be used to propose a three-stage hierarchical approach in which decisions about the number of centers and their locations (first stage), customers allocations (second stage) and routing strategies (third stage) are combined to reduce total system-wide costs.

As we have seen, the total radial cost between the depot and the customers dominates the objective function. This cost is, of course, related to the cost of the M-median problem. For a survey on the M-median problem, see Mirchandani & Francis [1990]. Thus, the M-median problem provides an insight to our model. For instance, when the demand of a customer can be split over several vehicles, the total distance traveled in a distribution system with M centers optimally located in the area is asymptotically

$$\frac{2\overline{w}n}{q}\beta\sqrt{\frac{A}{M}}, \tag{5.5}$$

where $\beta = 0.377196\ldots$ and q is the vehicles' capacity. Furthermore, this asymptotic value is achieved by placing centers in a regular hexagonal pattern throughout the area and each service center serves all the customers inside its hexagon.

It follows that, for large enough n, the best number of service stations is the minimizer of the following function (recall that c is a fixed set-up cost for establishing a depot)

$$TC(M) = \frac{2\overline{w}n}{q}\beta\sqrt{\frac{A}{M}} + cM.$$

Let $\delta = (\beta(\overline{w}n/qc)\sqrt{A})^{2/3}$, then the best number of stations is $\lceil\delta\rceil$ or $\lfloor\delta\rfloor$ whichever yields the best $TC(M)$. Furthermore, the analysis also shows where to locate the centers and how to allocate customers to each center, thus providing answers to the strategic and tactical problems.

What should be the routing strategy used on a daily basis? Note that in this model, we assume that every working day the servers have exact information on the customers that need service and their actual demands. Hence, every working day, each center faces an instance of the single depot capacitated vehicle routing problem, for which efficient heuristics exist. For example, by using the ITP(α) heuristic on a daily basis the minimal total cost (5.5) can actually be achieved when the number of customers n is large enough.

We can now summarize the hierarchical design: choose the number of stations as the one that minimizes equation (5.5), locate the facilities at the center of hexagonal patterns (strategic decision), each customer will be served by its closest service station (tactical decision) and use the ITP(α) heuristic on a daily basis to find efficient routes for servers (operational control).

6. Generalizations and extensions

6.0. Introduction

In the previous sections we analyze models related to the vehicle routing problem with a capacity constraint on the total load delivered by a vehicle. In practice, other constraints may also exist. For example, in some cases there may be an additional constraint that bounds the total distance traveled by a vehicle. This constraint is related to the number of working hours allowed for each driver, which is sometimes imposed by union regulations. In other circumstances time window constraints must be taken into account, that is, constraints that impose lower and upper bounds on the arrival time of vehicles to the customers. In this section we show that some of the previous analysis can be carried over to models with these types of constraints although the results are not quite as strong as in the CVRP.

6.1. The distance constrained vehicle routing problem

The distance constrained vehicle routing problem is the vehicle routing problem with only an upper bound λ on the distance traveled by a vehicle, i.e., no capacity constraint is assumed. Two possible objective functions are important in practice [Assad, 1988]: minimize the total distance traveled or minimize the number of vehicles used. Exact algorithms for the former are described in Laporte, Desrochers & Nobert [1984] and Laporte, Nobert and Taillefer [1988].

Our analysis of this routing problem is based on the paper by Li, Simchi-Levi & Desrochers [1992], who show that optimal solutions for the two objective functions are closely related. This is done by introducing simple lower bounds on the total distance traveled in each case. These bounds are used to prove that when minimizing the total distance traveled, the number of vehicles in the optimal solution is less than twice the minimal possible number of vehicles used over all feasible solutions. Similarly, when minimizing the number of vehicles, the total distance traveled in the optimal solution is less than twice the minimal possible total distance.

Li, Simchi-Levi and Desrochers also study the worst-case performance of some heuristics for the model. First, they show that when the objective is to minimize the number of vehicles, no polynomial time heuristic exists in which the worst-case ratio is less than two, unless $\mathcal{P} =$ NP. Nevertheless, they suggest a polynomial time heuristic algorithm whose worst-case bound is independent of the number of customers. Clearly, a feasible solution exists iff $d_i \leq \lambda/2$ for all $i \in N_c$. The heuristic in Li, Simchi-Levi & Desrochers [1992] provides a good worst-case result when the radial distance from the depot to its farthest customer is much less than half of the distance constraint. The heuristic is based on the tour partitioning algorithm introduced in Section 2. When some customers are far away from the depot, the tour partitioning heuristic can behave arbitrarily bad, i.e., there exists an instance of the problem in which the worst-case ratio tends to infinity.

To formally prove these results we need the following notation. As in the CVRP, the customers and the depot are presented as the nodes of the undirected graph $G = (N, E)$. Each customer has to be visited by a vehicle. The jth vehicle starts from the depot and returns to the depot after visiting a subset $N_j \subseteq N \setminus \{p\} = N_c$. The distance constraint λ says that the total distance traveled by each vehicle should be no more than λ units of distance (i.e., $L(N_j) \leq \lambda$ for any j).

We consider two possible objective functions: (i) minimize the total distance traveled (MD), and (ii) minimize the number of vehicles used (MV). Let K be the number of vehicles used by a specific solution and Z be the total length traveled. To distinguish between optimal solutions of the two versions of the problem, we use the indices V and D. The pair K^V, Z^V denotes K and Z under the MV objective function, while K^D and Z^D have similar meaning under MD.

Clearly, when $\lambda \geq L(N)$, the optimal traveling salesman tour solves the problem with the MD and MV objectives. We therefore assume, without loss of generality, that $\lambda < L(N)$.

6.1.2. On the relationship between optimal solutions for MD and MV

Given an optimal solution for MD, we check whether the number of vehicles used can be reduced without increasing the distance traveled. For any two vehicles i and j $(i \neq j)$, we check whether $L(N_i) + L(N_j) \leq \lambda$. If so, we reduce K^D by one; subsets N_i and N_j are to be served by a single vehicle that travels a total distance of $L(N_i) + L(N_j)$. This reduction does not change Z^D. From now on, we assume that any optimal solution to MD has a number of vehicles K^D that cannot be reduced by this simple procedure.

We continue by introducing bounds on Z^V and Z^D that are used later to establish the relationship between the two versions of the problem.

Theorem 6.1. $\frac{1}{2}K^\pi\lambda < Z^\pi \leq K^\pi\lambda$,   for $\pi = V, D$.

Proof. The second inequality is trivial. To prove the first one, note that by definition of K^V and the above procedure used for MD

$$L(N_j) + L(N_{j+1}) > \lambda, \qquad j = 1, 2, \ldots, K^\pi$$

(where $K^\pi + 1 \equiv 1$). Otherwise, we could reduce K^π by one vehicle. Taking the sum over j, we get

$$2 \sum_{j=1}^{K^\pi} L(N_j) > \lambda \cdot K^\pi.$$

Hence, $Z^\pi > \frac{1}{2}\lambda K^\pi$. $\square$

Theorem 6.1 provides the basis for the following result on the relationship between the two objective functions.

Corollary 6.2.
   (I) $K^D < 2K^V$ (or $K^D \leq 2K^V - 1$).
   (II) $Z^V < 2Z^D$.

Proof. We first prove (I). From Theorem 6.1,

$$\frac{1}{2}K^D\lambda < Z^D \leq Z^V \leq K^V\lambda.$$

Hence, $K^D < 2K^V$. To prove (II), use Theorem 6.1 again to get

$$Z^V \leq K^V\lambda \leq K^D\lambda < 2Z^D.   \square$$

Li, Simchi-Levi and Desrochers provide examples to show that these bounds are the best possible.

6.1.3. Worst-case heuristic analysis

For a given heuristic algorithm H, let K^H, Z^H be the number of vehicles and the total distance traveled, respectively. Theorem 6.1 provides an important tool for analyzing heuristic algorithms. It shows that if one of the two problems has a heuristic with a constant worst-case optimality gap, so does the other one. For example, if there exists a heuristic algorithm which satisfies $K^H/K^V \leq \eta > 1$ for any feasible instance, then this heuristic also satisfies $Z^H/Z^D < 2\eta$.

The next theorem rules out the possibility of such an efficient heuristic with $\eta < 2$.

Theorem 6.3. *If $\mathcal{P} \neq$ NP, then any polynomial time heuristic for MV has a worst-case bound of at least two.*

Proof. By contradiction, assume that there exists a polynomial time heuristic algorithm with $K^H/K^V < 2$. We use this heuristic to solve the traveling salesman decision problem defined as follows: Given a set of cities and the shortest distance between any two of them, does there exist a traveling salesman tour of total length no more than B? It is well known that this decision problem belongs to the class of NP-complete problems [see Garey & Johnson, 1979]. Now choose one of the points as the depot and let $\lambda = B$.

Observe that if the traveling salesman decision problem has a tour of length $\leq B$ then the polynomial time heuristic algorithm for MV must generate a feasible solution with $K^H < 2K^V = 2$, or $K^H = 1$. On the other hand, $K^H = 1$ implies that the answer to the traveling salesman problem is 'yes'. $\square$

We now turn our attention to finding a feasible solution by means of a partitioning heuristic. The heuristic, called the *greedy tour partitioning* (GTP) heuristic, starts with the optimal traveling salesman tour through all the customers and the depot. In an arbitrary orientation of this tour, the nodes are numbered $(x^{(0)}, x^{(1)}, \ldots, x^{(n)})$ in order of appearance, where $n = |N_c|$, $x^{(0)} \equiv p$ is the depot, and $x^{(1)}, \ldots, x^{(n)}$ are the customers. We break the tour into K^{GTP} segments and connect the end-points of each segment to the depot. This is done in the following way. Each vehicle j, $1 \leq j < K^{\mathrm{GTP}}$, starts by traveling from the depot to the first customer $x^{(l)}$ with minimum l such that $x^{(l)}$ has not been visited by the previous $j - 1$ vehicles and then visits the *maximum* number of customers according to the traveling salesman tour without violating the distance constraint λ upon returning to the depot.

We now present an upper bound on K^{GTP}.

Lemma 6.4.

$$K^{\mathrm{GTP}} \leq \min\left\{n, \left\lceil \frac{L(N) - 2d_{\max}}{\lambda - 2d_{\max}} \right\rceil\right\}.$$

Proof. The above heuristic breaks the traveling salesman tour into K^{GTP} segments. Let the first segment $S_1 = \{x^{(0)}, x^{(1)}, \ldots, x^{(l(1))}\}$ include customers $x^{(1)}, \ldots, x^{(l(1))}$

and the depot, the second segment be $S_2 = \{x^{(l(1)+1)}, \ldots, x^{(l(2))}\}$, etc., and the last segment be $S_{K^H} = \{x^{(l(k^H-1)+1)}, \ldots, x^{(n)}, x^{(0)}\}$. Each subset of customers S_j is visited by a vehicle in their order of appearance in S_j so that $x^{(l(j))}$ is the last customer visited by the jth vehicle, $1 \leq j \leq K^{GTP}$. Let $T(S_j)$ be the length of the jth segment, $1 \leq j \leq K^{GTP}$. Note that $T(S_1)$ $(T(S_{K^{GTP}}))$ includes the distance between the depot and the first (last) customer in S_1 $(S_{K^{GTP}})$. All other $T(S_j)$, $j = 2, \ldots, K^{GTP} - 1$, do not include any distance from the depot to the customers. Since $x^{(l(1))}$ is the last customer visited by the first vehicle,

$$T(S_1) + d_{x^{(l(1))}x^{(l(1)+1)}} + d_{x^{(l(1)+1)}} > \lambda.$$

Also, for any $1 < j < K^{GTP}$, we have

$$d_{x^{(l(j-1)+1)}} + T(S_j) + d_{x^{(l(j))}x^{(l(j)+1)}} + d_{x^{(l(j)+1)}} > \lambda.$$

For the last vehicle, the triangle inequality implies

$$T(S_{K^{GTP}}) - d_{x^{(l(K^{GTP}-1)+1)}} \geq 0.$$

Taking the sum over $j = 1, 2, \ldots, K^{GTP}$, we obtain

$$\sum_{j=1}^{K^{GTP}} T(S_j) + \sum_{j=1}^{K^{GTP}-1} d_{x^{(l(j))}x^{(l(j)+1)}} + 2 \sum_{j=1}^{K^{GTP}-2} d_{x^{(l(j)+1)}} > (K^{GTP} - 1)\lambda$$

or

$$L(N) + 2 \sum_{j=1}^{K^{GTP}-2} d_{x^{(l(j)+1)}} > (K^{GTP} - 1)\lambda$$

which implies

$$L(N) + 2d_m(K^{GTP} - 2) > (K^{GTP} - 1)\lambda$$

or

$$K^{GTP} < \frac{L(N) - 2d_{\max}}{\lambda - 2d_{\max}} + 1.$$

Hence, by the integrality of K^{GTP},

$$K^{GTP} \leq \left\lceil \frac{L(N) - 2d_{\max}}{\lambda - 2d_{\max}} \right\rceil. \qquad \square$$

The worst-case performance of the heuristic can now be found. Again, Z^{GTP} is the total distance traveled by all vehicles using the greedy tour partitioning heuristic.

Theorem 6.5.
  (I) $K^{GTP}/K^V < 1 + \beta/(\beta - 2)$.
  (II) $Z^{GTP}/Z^D < 1 + \beta/(\beta - 2)$,
where $\beta = \lambda/d_{\max}$.

Proof. By the triangle inequality, $K^V \lambda \geq L(N)$. Hence,

$$\frac{K^{\text{GTP}}}{K^V} < \left[\frac{L(N) - 2d_{\max}}{\lambda - 2d_{\max}} + 1 \right] \cdot \frac{\lambda}{L(N)}$$

$$\leq 1 + \frac{\lambda}{\lambda - 2d_{\max}}.$$

To prove (II), notice that $L(N) \leq Z^D$ and $Z^{\text{GTP}} \leq K^{\text{GTP}} \lambda$, and therefore the worst-case bound is obtained. □

Notice that the greedy tour partitioning heuristic is based on the availability of an optimal traveling salesman tour. Since finding such a tour is NP-hard, we replace it by an α-optimal traveling salesman tour.

Let $K^{\text{GTP}(\alpha)}$, $Z^{\text{GTP}(\alpha)}$ be K and Z when the greedy partitioning heuristic is based on an α-optimal tour. Hence, we obtain (similarly to Lemma 6.4) that

$$K^{\text{GTP}(\alpha)} \leq \min \left\{ n, \left\lceil \frac{\alpha L(N) - 2d_{\max}}{\lambda - 2d_{\max}} \right\rceil \right\},$$

and the worst-case bounds of Theorem 6.5 become

$$\frac{K^{\text{GTP}(\alpha)}}{K^V} < 1 + \frac{\alpha \beta}{\beta - 2}$$

and

$$\frac{Z^{\text{GTP}(\alpha)}}{Z^D} < 1 + \frac{\alpha \beta}{\beta - 2}.$$

This type of heuristic provides a satisfactory bound when $\lambda \gg 2d_{\max}$ (i.e., $\beta \gg 2$). In this case, the expected number of vehicles used is relatively small. However, when $\lambda \approx 2d_{\max}$, the bound is useless. So it is interesting to find out how bad a tour partitioning heuristic can be in actual practice. Indeed, Li, Simchi-Levi & Desrochers [1992] demonstrate that the greedy tour partitioning algorithm can be arbitrarily bad for both MV and MD. This holds if the heuristic starts with either an optimal traveling salesman tour or with Christofides' algorithm, nearest insertion algorithm or the heuristic based on minimum spanning tree to obtain an initial traveling salesman tour partitioned by the greedy heuristic.

6.2. The vehicle routing problem with time window constraints

Analysis of heuristics for vehicle routing problems with time window constraints was initiated by Solomon [1986] who showed that a variety of heuristics tend to produce solutions that can be arbitrarily bad. Recently, Bramel, Li & Simchi-Levi [1994] analyzed the vehicle routing problem with time window constraints. In this model, a set of customers require service by vehicles initially located at a central depot. Each customer provides a time period when they can be serviced. The service may be repair work or unloading/loading the vehicle. The objective is to find tours for the vehicles, such that each customer receives service in its specified time period, and the total distance traveled by the vehicles is as small as possible. In this paper, they present a polynomial time algorithm for a stylized version of the problem, and show that it is asymptotically optimal when customers are independently and identically distributed in a given region.

In this model, the operating day has W working time units and is divided into P equally long time periods, each of length T time units, such that $PT = W$. Each customer picks independently one time window from this set of P possible time windows.

Associated with customer y_k is a service time s_k which is the length of the service required for that customer and during which the vehicle serving this customer does not travel. The service time must start *after* the beginning of his time window and must end *before* the end of his time window. Obviously, feasibility requires $s_k \leq T$, $\forall k = 1, 2, \cdots, n$. The service time is assumed to be uniformly distributed between 0 and T/p for some integer $p \geq 1$. Without any loss of generality, we can assume that the vehicles travel at unit speed, so that distances can represent times. They further assume that a vehicle may leave or return to the depot at any time, and that a vehicle incurs no cost for being idle. Hence, a feasible solution always exists, by assigning one vehicle to each customer. In addition, they assume that the customers are located in a compact connected region with bounded perimeter. They develop a Region Partitioning heuristic based on solving maximum cardinality matching problems for a number of subregions and show that the heuristic is asymptotically optimal.

6.3. Dynamic vehicle routing problems

The vehicle routing problems that we have analyzed so far can all be viewed as static, deterministic problems. In many practical applications, however, their is a significant dynamic component to the problem. For instance, a distributor that delivers a product from a central depot to customers based on orders that arrive in real time, and a utility firm responsible for maintaining geographically dispersed facilities and operating in response to a facility failure can be modeled as dynamic vehicle routing problems.

Bertsimas and Van Ryzin [1991, 1993] analyze the following version of the dynamic vehicle routing problem. Demands for service arrive according to a renewal process with intensity λ to a connected, bounded Euclidean service region A. Demand locations are assumed to be a sequence of independent and identically distributed random variables according to a given continuous density function $f(\cdot)$ defined over A. Demands are served by m vehicles each of which has a capacity q, i.e., the number of demand points that a vehicle can visit before it has to go back to the depot is no more than q.

When a vehicle arrives at a customer, it spends s units of time serving that customer. The service times are assumed to be independent and identically distributed random variables with finite first and second moments. The objective is to find a policy to service the demands over an infinite horizon so as to minimize expected system time of all demands.

They first analyze the case where the region is served by a single uncapacitated vehicle and the demand locations are uniformly distributed. They find a policy which is optimal in light traffic and several policies that have a system time within a constant factor of the optimum in heavy traffic. These results are then extended to the capacitated model with m vehicles.

Acknowledgement

Research supported in part by ONR Contract N00014-90-J-1649, N00014-95-1-0232, NSF Contracts DDM-8922712 and DDM-9322828.

References

Altinkemer, K., and B. Gavish (1987). Heuristics for unequal weight delivery problems with a fixed error guarantee. *Oper. Res. Letters* 6, 149–158.

Altinkemer, K., and B. Gavish (1990). Heuristics for delivery problems with constant error guarantees. *Transp. Sci.* 24, 294–297.

Anily, S., and A. Federgruen (1990a). One warehouse multiple retailer systems with vehicle routing costs. *Manage. Sci.* 36, 92–114.

Anily, S., and A. Federgruen (1990b). A class of Euclidean routing problems with general route cost functions. *Math. Oper. Res.* 15, 268–285.

Anily, S., and A. Federgruen (1991a). Structured partitioning problems. *Oper. Res.* 39, 130–149.

Anily, S., and A. Federgruen (1991b). Rejoinder to: Comments on one warehouse multiple retailer systems with vehicle routing costs. *Manage. Sci.* 37, 1497–1499.

Anily, S., and A. Federgruen (1992). Two-echelon distribution systems with vehicle routing costs and central inventories, to appear.

Assad, A.A. (1988). Modeling and implementation issues in vehicle routing, in: Golden, B.L. and A.A. Assad (eds.). *Vehicle Routing: Methods and Studies,* Elsevier Science Publishers, B.V., pp. 7–45.

Assad, A., B. Golden, R. Dahl and M. Dror (1982) Design of an inventory/routing system for a large propane distribution firm, in: C. Gooding (ed.), *Proc. 1982 Southeast Tims Conference, Myrtle Beach,* pp. 315–320.

Beardwood, J., J.L. Halton and J.M. Hammersley (1959). The shortest path through many points. *Proc. Cam. Phil. Soc.* 55, 299–327.

Beasley, J. (1983). Route first–cluster second methods for vehicle routing. *Omega* 11, 403–408.

Bell, W.J., Dalberto, L.M., Fisher, M.L., Greenfield, A.J., Jaikumar, J., Kedia, P., Mack, R.G. and P.J. Prutzman (1983). Improving the distribution of industrial gases with on-line computerized routing and scheduling optimizer. *Interfaces* 13, 4–23.

Bertsimas J.D., and G. van Ryzin (1991). A stochastic and dynamic vehicle routing problem in the Euclidean plane. *Oper. Res.* 39, 601–615.

Bertsimas J.D., and G. van Ryzin (1993). Stochastic and dynamic vehicle routing in the Euclidean plane with multiple capacitated vehicles. *Oper. Res.* 41, pp. 60–76.

Bienstock, D., Bramel, J. and D. Simchi-Levi (1993). A probabilistic analysis of tour partitioning heuristics for the capacitated vehicle routing problem with unsplit demands. *Math. Oper. Res.* 18, pp. 786–802.

Braca J., Bramel, J., Posner B. and D. Simchi-Levi (1993). A computerized approach to the manhattan school bus routing project. Submitted to *IIE Trans.*

Bramel, J., E.G. Coffman, Jr., P.W. Shor and D. Simchi-Levi (1992). Probabilistic analysis of the capacitated vehicle routing problem with unsplit demands. *Oper. Res.* 40 pp. 1095–1106.

Bramel, J., C.L. Li and D. Simchi-Levi (1994). Probabilistic analysis of heuristics for the vehicle routing problem with time windows. *Am. J. Math. Manage. Sci.* 13, pp. 267–322.

Bramel, J., and D. Simchi-Levi (1992). A location based heuristic for general routing problems. Working Paper, Columbia University.

Chakravarty, A.K., J.B. Orlin and V.G. Rothblum (1982). A partitioning problem with additive objective with an application to optimal inventory groupings for joint replenishment. *Oper. Res.* 30, 1018–1020.

Chandra P., and M.L. Fisher (1990). Coordination of production and distribution planning. Working Paper, 90-11-06, The Wharton School, University of Pennsylvania.

Chien, T.W., A. Balakrishnan and R.T. Wong (1989). An integrated inventory allocation and vehicle routing problem. *Transp. Sci.* 23, 67–76.

Christofides, N. (1976a). The vehicle routing problem. *R.A.I.R.O. Rech. Oper.* 10, 55–76.

Christofides, N. (1976b). Worst-case analysis of a new heuristic for the travelling salesman problem. Report 388 Graduate School of Industrial Administration, Carnegie-Mellon University.

Christofides, N. (1985). Vehicle routing, in: E.L. Lawler, J.K. Lenstra, A.H.G. Rinnooy Kan and D.B. Shmoys (eds.), *The Traveling Salesman Problem*, John Wiley & Sons Ltd., pp. 431–448.

Christofides, N., and S. Eilon (1969). An algorithm for the vehicle dispatching problems. *Oper. Res. Q.* 20 pp. 309–318.

Coffman, Jr., E.G., M.R. Garey and D.S. Johnson (1984). Approximation algorithms for bin-packing — an update survey, in: G. Ausiello, M. Lucertini and P. Serafini (eds.), *Algorithm Design for Computer System Design,* Springer- Verlag, pp. 49–106.

Coffman, E.G., Jr., G.S. Lueker and A.H.G. Rinnooy Kan (1988). Asymptotic methods in the probabilistic analysis of sequencing and packing heuristics. *Manage. Sci.* 34, 266–290.

Coffman, E.G., Jr., and G.S. Lueker (1991). *Probabilistic Analysis of Packing and Partitioning Algorithms,* John Wiley & Sons Ltd., New York, N.Y.

Croes, A. (1958). A method for solving traveling salesman problems. *Oper. Res.* 5, 791–812.

Dror, M., M. Ball and B. Golden (1986). A computational comparison of algorithms for the inventory routing problem. *Ann. Oper. Res.* 4, 3–23.

Dror, M., and M. Ball (1987). Inventory/routing: reduction from an annual to a short-period problem. *Nav. Res. Logist. Q.* 34, 891–905.

Federgruen, A., and P. Zipkin (1983). Solution techniques for some allocation problems. *Math. Program.* 25, 13–24.

Federgruen, A., and P. Zipkin (1984). A combined vehicle routing and inventory allocation problem. *Oper. Res.* 32, 1019–1032.

Fisher, M. (1980). Worst-case analysis of algorithms. *Manage. Sci.* 26, 1–17.

Fisher, M. (1992). Vehicle routing, in: M.O. Ball, T.L. Magnanti, C.L. Monma and G.L. Nemhauser (eds.), *Handbooks on Operations Research and Management Science,* Vol. 8, Network Routing, Elsevier Science, B.V., Amsterdam, Chapter 1, this volume.

Fisher, M., and R. Jaikumar (1981). A generalized assignment heuristic for vehicle routing. *Networks* 11, 109–124.

Gallego, G., and D. Simchi-Levi (1990). On the effectiveness of direct shipping strategy for the one warehouse multi-retailer R-systems. *Manage. Sci.* 36, 240–243.

Garey, M.R., and D.S. Johnson (1979). *Computers and Intractability: A Guide to the Theory of NP-Completeness,* Freeman, San Francisco, Calif.

Gavish, B. (1981). A decision support system to manage the transportation needs of a large corporation. *AIIE Trans.* 13(1), 61–85.

Geoffrion, A.M. (1972). Generalized Benders decomposition. *J. Optimization Theory Appl.* 10, 237–260.

Gillett, B.E., and L.R. Miller (1974). A heuristic algorithm for the vehicle dispatch problem. *Oper. Res.* 22, 340–349.

Golden, B., Assad, A. and R. Dahl (1984). Analysis of a large scale vehicle routing problem with inventory component. *Large Scale System* 7, 181–190.

Haimovich, M., and A.H.G. Rinnooy Kan (1985). Bounds and heuristics for capacitated routing problems. *Math. Oper. Res.* 10, 527–542.

Haimovich, M., A.H.G. Rinnooy Kan and L. Stougie (1988). Analysis of heuristics for vehicle routing problems, in: B.L. Golden and A.A. Assad (eds.), *Vehicle Routing: Methods and Studies,* Elsevier Science, B.V., Amsterdam, pp. 47–61.

Hall, R.W. (1991). Comments on: One-warehouse multiple retailer systems with vehicle routing costs. *Manage. Sci.* 37, 1496–1497.

Karmarkar, N. (1982). Probabilistic analysis of some bin-packing algorithms. *Proc. 23rd Annu. Symp. on Foundations of Computer Science,* pp. 107–111.

Karp, R.M. (1977). Probabilistic analysis of partitioning algorithms for the traveling salesman problem. *Math. Oper. Res.* 2(3), 209–224.

Laporte, G., M. Desrochers and Y. Nobert (1984). Two exact algorithms for the distance-constrained vehicle routing problem. *Networks* 14, 161–172.

Laporte, G., Y. Nobert and S. Taillefer (1988). Solving a family of multi-depot vehicle routing and location routing problems. *Transp. Sci.* 22, 161–172.

Lawler, E.L. (1976). *Combinatorial Optimization: Networks and Matroids.* Helt, Rinehart and Winston, New York.

Li, C.L., and D. Simchi-Levi (1990). Analysis of heuristics for the multidepot capacitated vehicle routing problems. *ORSA J. Comput.* 2, 64–73.

Li, C.L., D. Simchi-Levi and M. Desrochers (1992). On the distance constrained vehicle routing problem. *Oper. Res.* 40, 790–800.

Lin, S., and B. Kernighan (1973). an effective heuristic algorithm for the traveling salesman problem. *Oper. Res.* 21, 498–513.

Mirchandani, P.B., and R.L. Francis (1990). *Discrete Location Theory.* John Wiley and Sons, Inc, New York.

Muckstadt, J.M., and R.O. Roundy (1992). Analysis of multistage production systems, in: S.C. Graves, A.H.G. Rinnooy Kan and P.H. Zipkin (eds.), *Handbooks of Operations Research and Management Science,* Vol. 4, Logistics of Products and Inventory, Elsevier Science, B.V., Amsterdam, Chapter 2.

Psaraftis, H.N. (1984). On the practical importance of asymptotic optimality in certain heuristic algorithms. *Networks* 14, 587–596.

Queyranne, M. (1987). Finding 94%-effective policies in linear time for some production inventory systems. University of British Columbia, Working Paper.

Rhee, W.T. (1988). Optimal bin packing with items of random sizes. *Math. Oper. Res.* 13, 140–151.

Rhee, W.T. (1991). Probabilistic analysis of a capacitated vehicle routing problem. Working Paper, The Ohio State University.

Rhee, W.T., and M. Talagrand (1987). Martingale inequalities and NP-complete problems. *Math. Oper. Res.* 12, 177–181.

Roundy, R. (1985). 98% effective integer ratio lot sizing for one warehouse multi-retailer systems. *Manage. Sci.* 31, 1416–1430.

Simchi-Levi, D., and J. Bramel (1990a). On the optimal solution value of the capacitated vehicle routing problem with unsplit demands. Working Paper, Columbia University.

Simchi-Levi, D., and J. Bramel (1990b). Probabilistic analysis of heuristics for the capacitated vehicle routing problem with unsplit demands. Working Paper, Columbia University.

Simchi-Levi, D. (1992). Hierarchical planning for probabilistic distribution systems. *Manage. Sci.* 38, 198–211.

Solomon M.M. (1986). On the worst-case performance of some heuristics for the vehicle routing and scheduling problem with time window constraints. *Networks* 14(4), 571–586.

Spaccamela, A.M., A.H.G. Rinnooy Kan and L. Stougie (1984). Hierarchical vehicle routing problems. *Networks* 14(4), 571–586.

M.O. Ball et al., Eds., *Handbooks in OR & MS, Vol. 8*

Chapter 5

Arc Routing Methods and Applications

Arjang A. Assad and Bruce L. Golden

College of Business and Management, University of Maryland, College Park, MD 20742, U.S.A.

1. Introduction

The term 'arc routing' refers to routing problems where the key service activity is to cover arcs of a transportation network. In contrast to node routing, where the key service activity occurs at the nodes (customer sites) and arcs are of interest only as elements of paths that connect the nodes, arc routing focuses on the traversal of edges. Practical examples include the routing of street sweepers, snow-plowing, salt gritting (covering streets with salt grit), inspection of streets for maintenance, postal delivery, and meter reading. In all examples, each street segment must be covered in its entirety. In meter reading, postal delivery, and the delivery of telephone books, the density of the customer locations along a street segment is sufficiently high to consider the street segment as a whole as the service entity. We stress that the distinction between node routing and arc routing is a matter of modeling. One must base such modeling decisions on a careful consideration of the nature of the service or delivery activities. In general, arc routing models must be considered serious contenders when the density of service points along links of a network is high, or when the characteristics of the traversal of edges are of direct interest. The goal of this chapter is to review the literature in arc routing with special emphasis on applications.

Plan of the paper. This chapter is divided into two main parts. The first part, comprising Sections 2–6, covers the basic methodology of arc routing by discussing generic arc routing models and their solution techniques. The second part, which spans Sections 7–10, focuses on specific areas of application where arc routing approaches have proved to be especially appropriate. Given the focus of this chapter on applications, this part addresses the issues, constraints and complications of the various operational settings in some detail. In doing so, we hope that we can convey some of the challenges of translating well-known routing models into effective routing systems that can gain approval in day-to-day operations.

Section 2 of this chapter covers the well-known postman problems that can be regarded as the classical problems of arc routing. The basic objective of these problems is to cover *all* arcs of a given network with an Euler cycle of minimum cost. Section 3 focuses on the *rural postman* problem where only a given *subset* of the arcs of the network requires coverage. Sections 4 and 5 consider the *capacitated* arc routing problem, where each vehicle tour (cycle) must observe an explicit capacity constraint. In Section 4, we have opted to cover lower bounding techniques for this problem in some detail to provide an example of a topic within arc routing where there has been a sustained and connected body of research over the years. Section 6 is devoted to variations on classical postman problems arising from the use of different cost structures or priority schemes. Sections 7 and 8 describe two major applications areas of arc routing — sanitation services and postal delivery operations. The other two areas of meter reading and snow control are grouped together in Section 9. While all of these applications arise in transportation, arc routing has also proven useful in manufacturing operations. Section 10 describes two manufacturing problems that can be modeled as rural postman problems and also mentions other routing applications. In the remainder of the first section, we compare arc and node routing problems, sketch the early history of arc routing, and introduce the terminology and notation of the subsequent sections.

1.1. Comparison of arc and node routing

The general arc routing problem involves servicing a set R of required edges (or arcs) in an underlying network $G = (V, A \cup E)$ with nodes V, directed arcs A and undirected edges E. The graph G may be composed entirely of directed arcs ($E = \phi$), of undirected edges ($A = \phi$), or contain a mixture of both. Moreover, G can be a multigraph, so that more than a single copy of an edge (or arc) may exist between the same pair of nodes in V. Edge costs play an important role in all arc routing problems: the cost of servicing and the cost of traversing an edge (without servicing it) are specified as input. As long as the cost of servicing an edge is a constant that is not affected by the routing decisions, the sum of these costs over edges in R can be viewed as a sunk cost. Typically, the objective of an arc routing problem is to minimize the sum of the costs of traversing non-required edges. In general, one seeks a set of cycles that cover all edges (or arcs) in R at minimum cost.

Table 1 summarizes some important characteristics of well-known arc routing problems discussed in the first half of this chapter. These problems differ in the underlying graph $G = (V, A \cup E)$ and the required edges R. G_R denotes the graph induced by edges or arcs in R and $n = |V|$. The first three problems are distinguished by the requirement that all edges or arcs require service and are commonly referred to as Chinese postman problems for historical reasons outlined in the next section. The last column of the table refers to the section of this work devoted to the problem in question. Similar tables and classifications are available in Lenstra and Rinnooy Kan [1981] and Su [1992].

One can consider the node routing analogues of the problems in Table 1. One way to make this translation is to take the required service elements in R to be nodes rather than edges on the same underlying network. In that

Table 1
Well-known arc routing problems

Program	A	E	Required edges R	Complex-ity	Comments	See
Undirected postman problem (UPP)	ϕ	$*$	$R = E$	P	Solved by matching algorithm in $O(n^3)$ or less	2.1
Directed postman problem (DPP)	$*$	ϕ	$R = A$	P	Solved by flow problem in $O(n^3)$	2.2
Mixed postman problem (MPP)	$*$	$*$	$R = A \cup E$, $A \neq \phi, E \neq \phi$	NPC	Remains $\mathcal{NP}$-complete even if all edge costs are equal; solvable if G is an even connected graph	2.3
Rural postman problem (RPP)	ϕ	$*$	$R \subset E$	NPC	Solvable if G_R is connected	3
Directed rural postman problem (DRPP)	$*$	ϕ	$R \subset A$	NPC	Solvable if G_R is connected	3
Stacker crane problem (SCP)	$*$	$*$	$R = A$	NPC	Remains $\mathcal{NP}$-complete even if all edge costs are equal	3.1
Windy postman problem (WPP)	ϕ	$*$	$R = E$	NPC	Edge costs depend on direction of traversal; solvable if G is connected and even	6.1
Capacitated arc routing problem (CARP)	ϕ	$*$	$R \subset E$	NPC	Explicit capacity constraint on each cycle with respect to edge demands (as in VRP)	4.5
Capacitated postman problem (CAPP)	ϕ	$*$	$R = E$	NPC	Same as the CARP but all edges must be serviced	5

Legend: $*$ = arbitrary set, P = polynomial, NPC = $\mathcal{NP}$-complete, n = number of nodes.

case, the first six problems of the table would all correspond to symmetric or asymmetric traveling salesman problems (TSPs). Clearly, much of the richness of arc routing problems disappears in this translation. In fact, in node routing problems, the underlying network matters only to the extent of providing shortest paths between pairs of nodes that require service. Moreover, one can always restrict the graph to the required nodes. The correspondence between node and arc routing problems becomes more appropriate when explicit constraints (such as capacity or precedence constraints) are added. Thus, just as the vehicle routing problem (VRP) is a capacitated version of the TSP, CARP can be viewed as a capacitated version of the RPP.

A node routing problem can be transformed into an arc routing problem simply by replacing each node that requires service with an edge. Conversely, an arc routing problem can be transformed into an equivalent node routing problem as described by Pearn, Assad & Golden [1987]. This transformation converts the CARP with m_r required arcs into a VRP with $3m_r + 1$ nodes. In practice, of course, this increase in size may not be acceptable and it may be preferable to attack the arc routing version directly. As mentioned before, the choice of an arc routing or node routing model for a given application must consider the distinct nature of the operations as well as the solution techniques.

Most would agree that node routing problems have been studied with much greater intensity than their arc routing counterparts. Numerous surveys of node routing problems have appeared in the last decade: General overviews of the area include Christofides [1985], Bott & Ballou [1986], Golden & Assad [1986, 1988], Assad [1988], Christofides & Mingozzi [1989], and Laporte [1992a, b]. More specialized reviews have focused on specific applications areas [Golden & Wasil, 1987; Sutcliffe & Board, 1991; Ronen, 1993], or classification schemes [Desrochers, Lenstra & Savelsbergh, 1990]. In contrast, comprehensive overviews of arc routing problems are quite rare: Bodin et al. [1983] review some of the early arc routing applications, while Benavent, Campos, Corberan & Mota [1990] summarize arc routing developments in a paper written in Spanish. One must acknowledge the encyclopedic work of Fleischner [1990, 1991] on Eulerian graphs. This author discusses postman problems and some arc routing applications, but he emphasizes graph-theoretic results of a more abstract nature. Apart from these, we know of no other published review of arc routing problems. The recent comprehensive review of postman problems by Eiselt, Gendreau & Laporte [1995a, b] came to our attention as we were completing the present manuscript. These reports concentrate on the methodology for the most part and provide a good coverage of the recent work performed at the University of Montreal on arc routing problems.

1.2. The early history of arc routing

It is customary to trace the origin of graph theory to the classic paper Euler published in 1736, regarding the seven bridges over the River Pregel in Königsberg. "Concerning these bridges", wrote Euler, "it was asked whether anyone could arrange a route in such a way that he would cross each bridge once and only once". Since the key activity is that of traversing certain arcs of a transportation network, this paper may also be viewed as the first arc routing application!

Euler solved the bridges problem by creating a graph where the edges corresponded to the bridges and the nodes represented land areas separated by the river. He then showed that the question was equivalent to finding a path that contains each edge of the graph exactly once — what is now called an Euler path. Euler's main result was that an Euler path in a connected graph has at most two nodes of odd degree. Thus, for an Euler cycle to exist in a graph all nodes must have even degree. Biggs, Lloyd & Wilson [1976] reproduce Euler's original paper, accompanied by an article of Hierholzer's, published in 1873, that provided the first formal proof of the existence of an Euler cycle when all nodes have even degree.

If a graph has nodes of odd degree, an Euler cycle does not exist unless some edges are traversed more than once. The *postman problem* is to find an Euler cycle of minimum cost that traverses each edge of the graph at least once. This problem was first introduced by Meigu Guan [1962a], or Mei-Ko Kwan in the old transliteration. In the honor of this Chinese mathematician, the problem is often called the *Chinese postman problem*. Guan [1962a] himself provided the applied setting of mail delivery by writing: "A mailman has to cover his assigned segment before returning to the post office. The problem is to find the shortest walking distance for the mailman". Guan realized that the problem involves adding a set of edges to the graph to ensure even degree at every node of the augmented graph. The key theorem in his pioneering paper characterized the optimal tour by two conditions: that no edges be duplicated more than once and that the cost of added edges on every cycle not exceed half the total cost for the cycle. Since one can remove redundant edges and modify the cycles that violate the latter condition to reduce the total cost of a solution, Guan suggested that these improvement steps be used successively to solve the problem. Realizing that this procedure is not efficient, Guan [1962b] was able to limit the class of cycles that needed to be checked but even this modification does not produce a polynomial algorithm for the postman problem. Fleischner [1990] provides a formal account of Guan's characterization and related result.

In a different strand of research, Hedetniemi [1968] considered the cycle of smallest possible length that covers all edges of a given connected undirected graph G. If this minimal length (that is, the smallest number of edges) is denoted by $\sigma_C(G)$ and $m = |E|$, he proved the relation $m \leq \sigma_C(G) \leq 2m$ for every connected graph G. Moreover, if G has exactly two nodes of odd degree, Hedetniemi showed that $\sigma_C(G) = m + d$, where d is the shortest path length between the nodes of odd degree, and the shortest path is computed with all edge costs equal to 1. This paper provides more general relations and computes $\sigma_C(G)$ for graphs with special structure. Goodman & Hedetniemi [1973, 1974] continued this line of investigation to study Eulerian covering cycles (which cover all edges of G) and cited the connection between their work and that of Guan. In their work, the length (number of edges) in the smallest Euler cycle was specifically identified as $m + M(G)$, where $M(G)$ is the smallest possible total number of edges in all collections of paths connecting pairs of odd-degree nodes in G. However, these authors provided no efficient mechanism for finding $M(G)$. Of course, in retrospect, it is easy to see that $M(G)$ coincides with the value of the perfect matching on all nodes of odd degree in G when all edge costs equal one.

It was Edmonds [1965] who realized that Guan's problem of adding a set of edges of minimum cost to obtain an even graph can be efficiently solved as a matching problem. The full treatment of the problem for undirected, directed, and mixed even graphs appeared in the celebrated paper of Edmonds & Johnson [1973]. Christofides [1973] also drew attention to the use of matching in postman problems. Orloff [1974] unified various arc routing and traditional node routing problems by defining the *general routing problem* (GRP). In this problem, one seeks a minimum-cost set of tours to cover a set of required nodes *and* a set of required edges. Orloff also discussed the *rural postman problem* — a special case of the GRP where there are no required nodes (except the depot). Apart from Orloff's original procedure, other algorithmic approaches to the GRP were proposed by Orloff & Caprera [1976] and Male, Liebman & Orloff [1977]. The complexity of the GRP and various arc routing special cases were analyzed by Lenstra and Rinnooy Kan [1976].

Early applications of arc routing methods involved public services. In 1970, Stricker analyzed municipal waste collection and snow plowing problems for the city of Cambridge, Massachusetts. Stricker [1970] noted such practical complications of real problems as mixed graphs, vehicle capacities, street priorities and multiple vehicles. Stricker used a partitioning approach to the trash collection problem by dividing the service area into subareas, each with an even number of odd-degree nodes. A standard postman problem was then solved for each subarea. Further details of this approach appear in Marks & Stricker [1971]. In the early seventies, several other researchers considered variants of the postman approach for the waste collection problem: Male [1973] considered the m-postman problem, Wathne [1972] investigated routing for waste collection in collaboration with J. Liebman. In a different research project, Shuster & Schur [1974] developed heuristics to handle the added restrictions of waste collection problems (such as U-turns). Another early study of routing was performed by Liebling [1970, 1973] who used the directed postman problem in his study of street cleaning for the city of Zurich. For this study, information on 8000 street segments was collected and analyzed.

Arc routing applications may be said to have come of age with the work of Beltrami & Bodin [1974]. This study, which is discussed in detail in Section 7.1, was the culmination of work initiated in 1970 on the waste collection problem for New York City [see Bodin, 1990]. The careful attention to practical requirements and constraints, the comprehensive solution technique, and the geometrical appeal of the techniques have caused this work to be cited as a classic paper on modeling and implementation.

1.3. Routing and geographical information systems

Routing systems ultimately rely on accurate information about the underlying transportation network and the characteristics of the demand associated with the routing application. In vehicle routing, for example, not only is basic topological connectivity information vital to the generation of paths, but the nature of the demand may depend on such information as the number of households along street segments or specific demographic classifications over the area of interest.

Early implementations of routing systems attempted to by-pass the obstacles of obtaining detailed geographical data. For example, true (network) distances were based on Euclidean distances computed from (x, y) coordinates of locations of interest. A popular method was to use regression to fit the actual travel distances with approximations based on linear combinations of Euclidean or other norms. A summary of this line of research appears in Assad [1988]; more recent examples appear in Stokx & Tilanus [1991] and Brimberg & Love [1992]. In general, this

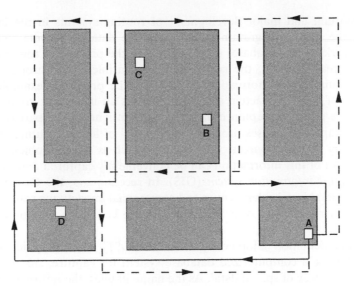

Fig. 1. Two paths through points A–D. Solid path requires no U-turns.

methodology is more appropriate for the longer haul, such as intercity distance computations. For intra-city and urban street networks a much finer level of geographical detail is often required. The following example highlights the need for this refined information.

Consider the points A, B, C and D in Figure 1 and suppose that we seek to construct a vehicle tour visiting all four points through the street network. Assume that all streets are two-way and that the sequence A–B–C–D–A minimizes the total travel distance of either the Euclidean or the Manhattan (ℓ_1) metric. To transform this sequence into a useable (or 'driveable') route, we must take the direction of travel and side of street into account. For example, the path from A to B must approach B from above so that the vehicle would be on the same side of the street as B when servicing it. Otherwise, an illegal U-turn is required for the vehicle to switch sides. Proceeding from A to B, B to C and C to D in this fashion produces the dashed path in Figure 1. However, the dashed path is clearly inferior to the solid path that visits the points in the order A–D–C–B–A, thereby making the original sequence sub-optimal. Of course, it is possible to find the optimal sequence by solving the traveling salesman problem over the four points using a modified distance matrix that *does* consider turn penalties and side-of-street effects. The main point of the example is to show the need for more detailed geographical information, a simple example of which is the association of a demand location with a side of a street segment. The preceding example is adapted from Bodin, Fagan & Levy [1992b] who argue for the importance of routing with true street network information. Similar examples appear in Lapalme, Rousseau, Chapleau, Cormier, Cossette & Roy [1992] and Sections 7.2 and 8.3 of this chapter.

An important and major development of routing systems in the last decade has been the closer integration with geographic databases containing detailed street networks. Such databases are important in the practice of routing decisions for several reasons. First, the construction of feasible paths through the street network requires such information as shown in the preceding example. Second, the display of routes on actual street networks is an important visualization aid that provides greater acceptance, flexibility and insight on the part of the users. Third, geographic databases provide the potential of linking routing information with a variety of other sources of information such as demographics, zoning and districting. All of this information can be conveniently captured and manipulated by a *geographic information system* (GIS). In recent years, the GIS industry has grown considerably in both revenues and commercial impact: one source estimates 1991 industry revenues of over $3.5 billion in the U.S. (The Economist, 3/21/92, p. 69).

Using the definition adopted by Huxhold [1991], a geographic information system (GIS) is 'a computerized database system for capture, storage, retrieval, analysis and display of spatial data'. As the name implies, the information in a GIS is organized and related spatially. Huxhold's broad definition of a GIS comprises several classes of tools as follows:

(a) automated mapping technology for flexible manipulation of map information,

(b) database management tools for managing attribute data,

(c) land records information with specific geographic references and locations (street segments, census tracts, land parcels, buildings),

(d) topological data structures capturing the spatial relations among points, lines and polygons, and

(e) spatial analysis capabilities for retrieving and displaying map data.

Item (c) above is exemplified by a *base map* that contains the coordinate system, cartographic data and location identifiers used as a common reference by various users of a GIS. Additional attributes are then defined and attached to the map based on the needs of each group of users. Geocoding (attaching geographical identifiers to points), nearest-neighbor capabilities, the processing of points within a polygon, or the generation of a shortest path through the street network and its display are examples of (e).

From the perspective of routing, the key topological structure in (d) above is the structure used to capture the street network at the block face level. Street segments are viewed as *center lines* between intersection points, and blocks are defined by a boundary composed of lines. Attached to each street segment (arc) in these files are the address ranges on the left and right sides of the street as it is traversed in the direction of increasing addresses. Also provided are the blocks situated to the left and to the right of this segment. In this way, not only is connectivity data captured in full, but it is also possible to associate each street address with an exact location on the appropriate side of the street.

The creation of DIME files by the U.S. Bureau of the Census in the 1970s provided a major impetus for the availability of digital street maps. While the original DIME files covered only 2% of the land area of the U.S., the TIGER files are slated to cover the entire nation with a total of over 50 million line segments [Huxhold, 1991, p. 167]. One estimate of the development costs of the TIGER files is $182 million [Antennucci, Brown, Croswell, Kevany & Archer, 1991, p. 39]. To take another indicator of the level of effort required, Clysdale [1992] states that the data gathering effort for the GIS used by Canada Post yields 700 square kilometers urban area per person-year. It is also important to note that the resulting street networks at the block-face level of detail can be large. Lapalme, Rousseau, Chapleau, Cormier, Cossette & Roy [1992] mention that for Toronto, a city with over 20,000 streets and 3 million people, the number of arcs in the network is approximately 96,000. For Montréal, the number of streets is about 11,200 and the number of arcs is 65,700.

We believe that a GIS is the natural platform on which to operate a routing system. It captures detailed network information that is crucial to the estimation of appropriate route characteristics (load, travel distance and so on) as well as the acceptability of the final routes. A GIS can significantly facilitate the interaction of the user with the routing algorithms. Bodin & Levy [1994] particularly stress the advantage of visualizing the output of a routing system, for which the mapping and display features of a GIS can be very helpful. The other major benefit derives from the sharing of geographic data across a variety of business applications. For example, a GIS makes it easy for an ad-mail campaign to be linked to the delivery of special advertising materials to individual households. In fact, one of the attractions of a GIS is that it can provide the base information for multiple applications. Huxhold [1991] describes the case of the City of Milwaukee, where the GIS supported such applications as waste collection, zoning, housing regulations and building inspections. Mannering & Kilareski [1986] give an example of the wealth of attributes a GIS can incorporate in the context of roadway maintenance. Finally, Lee [1989] argues for the use of GIS in transportation services and describes its use within a dynamic vehicle routing system.

General references that describe the power of GISs and the breadth of the applications are plentiful: examples include Huxhold [1991], Antennucci, Brown, Croswell, Kevany & Archer [1991] and Castle [1993]. The last source contains an overview by Landis [1993] and sections on individual industry applications such as transportation [see Badillo, 1993]. The area of geopostal systems, or GIS for postal operations, has been particularly active in the last decade. Clysdale [1992] and Landry [1992] describe the development of geopostal applications within Canada Post Corporation, while Harrison and Deegan [1992] review the interplay of GIS and routing in the Irish Postal Service (An Post).

1.4. Notation

This section briefly introduces the notation used in the remainder of this chapter. A graph $G = (V, E \cup A)$ is specified by a node set $V = 1, \ldots, n$, directed arcs A, and undirected edges E. G is called *directed* if E is empty, *undirected* if

A is empty, and *mixed* if both E and A are non-empty. In a directed graph, edge (i, j) is directed from node i to node j. The undirected edge between nodes i and j is denoted by $e(i, j)$ without regard to the order of i and j. We assume that a set of edge costs $c_{ij} \geq 0$ is available for all edges $e(i, j) \in E$ and $(i, j) \in A$. For any pair of nodes in G, the cost of the shortest path from i to j with respect to the edge costs c_{ij} is denoted by $SP(i, j)$. Note that $SP(i, j)$ may not coincide with c_{ij}.

Given a node i of a mixed graph $G = (V, E \cup A)$, let $\delta^+(i) = \{i \in V \mid (i, j) \in A\}$ and $\delta^-(i) = \{j \in V \mid (j, i) \in A\}$ and define $d^+(i) = |\delta^+(i)|$ and $d^-(i) = |\delta^-(i)|$. $d^-(i)$ and $d^+(i)$ are called the *in-degree* and *out-degree* of node i, respectively. Similarly let $\delta(i) = \{j \in V \mid e(i, j) \in E\}$ and $d(i) = |\delta(i)|$ so that $d(i)$ denotes the number of undirected edges that meet node i, that is, the degree of node i in (V, E). Thus, the total number of edges meeting node i in $G = (V, E \cup A)$ is $d(i) + d^-(i) + d^+(i)$. G is called *even* if this total degree is even for all nodes. Let $b(i) = d^-(i) - d^+(i)$ for $i \in V$. G is called *balanced* (symmetric) if $b(i) = 0$ for all i.

In many arc routing problems, multiple edges between the same pair of nodes arise naturally. It is therefore convenient to allow the structure G to be a multi-graph. We do not introduce explicit notation for this and rely on the context to make the distinction. Naturally, in a multigraph, the in-degree and out-degree of node i count all arcs directed into or out of i with the appropriate multiplicities.

A *tour*, or a *cycle*, in a graph $G = (V, E \cup A)$ is a sequence of nodes and edges $(i_1, e_1, i_2, e_2, \ldots, e_p, i_{p+1})$, such that $i_1 = i_{p+1}$ and e_k is either directed from i_k to i_{k+1} or is an undirected edge joining these two nodes. This definition allows for multiple incidences of the same edge in a tour. A tour is called an *Euler tour* if it contains each edge of G exactly once and a *postman tour* if each edge occurs at least once in the tour. Following Fleischner [1983], we call a tour (cycle) that covers all edges and arcs of G a *covering tour*. Given a subset $R \subset E \cup A$ of arcs and edges, a covering tour for R is a tour that covers all arcs and edges of R at least once. By suppressing the requirement that $i_1 = i_{p+1}$, we can extend the preceding definitions for Euler or postman *paths*. A graph G is called *Eulerian* (or *unicursal*) if it possesses an Euler tour. Note that this definition depends upon the original edge set specified for G. As we shall see, one may transform a given graph into an Eulerian one by augmenting its edge set. If G_0 is any graph with an even number of nodes $MP(G_0)$ denotes the optimal cost of the minimum perfect matching on G_0.

Given a graph $G = (V, E \cup A)$ and a required set of edges and arcs denoted by $R \subset E \cup A$, we denote by $G_R = (V_R, R)$ the subgraph of G induced by R. The various degrees of node i in G_R are denoted by $d^+(i, R)$, $d^-(i, R)$ and $d(i, R)$. Thus, $d(i, R)$ denotes the number of edges incident to node i in G_R.

In analyzing heuristics, we let V_H stand for the value produced by the heuristic and denote the optimum by $v*$. Finally, for any graph $G = (V, E \cup A)$, we reserve n and m throughout to denote the number of nodes and links of G. Thus, $n = |V|$ and $m = |E| + |A|$.

2. The classical postman problems

2.1. The undirected postman problem (UPP)

Consider an undirected graph $G = (V, E)$ and suppose that an edge cost $c_e \geq 0$ is provided for traversing each edge e of G. The *undirected postman problem* (UPP) is to find the postman tour of minimum cost in a connected graph G. Given any postman tour T, let x_e = (number of times edge e occurs in T) − 1. Since any postman tour covers every edge e at least once, $x_e \geq 0$. The postman problem reduces to minimizing the objective $\sum_{e \in T} c_e x_e$ over all postman tours. Clearly, if G possesses an Euler tour, this problem is solved by setting $x_e = 0$ for all e. Otherwise, some edges must be covered more than once, creating an unproductive or *non-required* traversal that is often called *deadheading*. Edmonds & Johnson [1973] show that the optimal choice of x_e can be determined by solving a matching problem.

More specifically, let V_0 = the set of all nodes of odd degree in G (nodes incident to an odd number of edges in E). Let (a_{ie}) be the node–edge incidence matrix for G, so that $a_{ie} = 1$ if edge e meets node i, and 0 otherwise.

Consider the integer program

$$\text{Min} \sum_{e \in E} c_e x_e$$

subject to

$$\sum_e a_{ie} x_e - 2w_i = \begin{cases} 1 & \text{if } i \in V_0 \\ 0 & \text{otherwise,} \end{cases}$$

$$x_e, \ w_i \geq 0 \text{ and integer.}$$

The main constraint of this problem simply ensures that once the additional copies of all edges e for which $x_e > 0$ are added to the original graph, each node i will have even degree, that is

$$\sum_{e \in E} a_{ie}(1 + x_e) = 0 \,(\text{mod } 2)$$

for all i. Edmonds and Johnson observe that, after solving some shortest path problems, this program can be solved as a matching problem as described below. Edmonds and Johnson also describe a direct approach that is based on matching concepts.

The undirected postman problem algorithm

INPUT: Undirected graph $G = (V, E)$; edge costs $c_e = c_{ij} = c_{ji}$ for all $e(i, j) \in E$.

Step 1. Identify the set of all nodes of odd degree V_0 and compute shortest paths between all pairs of nodes in V_0. Let $SP(i, j)$ = cost of the shortest path from i to j for $i, j \in V_0$.

Step 2. Construct a new graph G_o by forming the complete graph with node set V_o. Attach the edge cost $SP(i, j)$ to the edge (i, j) in this graph and record the edges of E lying on this shortest path.

Step 3. Find an optimal matching in G_o, that is, a minimum cost subset M of edges of G_o such that every node $v \in V_o$ is incident to exactly one edge in M.

Step 4. For each edge (i, j) in the optimal matching M^*, set $x_e = 1$ for all edges that lie on the shortest path from i to j obtained in Step 1.

Step 5. Form the augmented graph $\tilde{G}$ by adding all edges with $x_e = 1$ to E. $\tilde{G}$ has even degree at every node.

Step 6. Trace an Euler tour through $\tilde{G}$. This corresponds to the optimal postman tour in G.

Note that Step 3 of this procedure solves a matching problem in which the matching edges correspond to shortest paths. One can show that no edge e will occur in the shortest paths corresponding to two different matching edges of the optimal matching M^*. Therefore, Step 4 is well-defined. Step 6 requires an algorithm for tracing an Euler tour through an undirected even graph (where all nodes have even degree). Edmonds and Johnson provide two different algorithms for this purpose. We do not discuss these procedures here but refer the reader to the comprehensive account in Chapter 10 of Fleischner [1990] which is devoted to algorithms for identifying Eulerian trails in Eulerian graphs.

2.2. The directed postman problem (DPP)

In traversing a postman tour, one must be able to leave every node that the tour visits. In a directed graph, this means that the number of edges leading into a given node must equal the number of edges directed out of that node. It is well known that this condition is necessary and sufficient for the existence of an Euler tour. Since the original graph may not satisfy this condition, additional copies of certain edges must be added to bring this about. We call this operation *balancing* (or symmetrizing) and outline an optimal procedure to perform it.

Let $G = (V, A)$ be a directed graph. Compute $b(i) = d^-(i) - d^+(i)$ for all nodes i in V. Any node with non-zero $b(i)$ has an imbalance. To solve the postman problem, we need to add additional copies of edges of G to balance the graph. Let x_{ij} be the number of additional copies of edge (i, j) to be added to G. The optimal choice of the edges to add can be determined by solving the following minimum cost flow problem.

$$\text{Min} \sum_{(i,j) \in A} c_{ij} x_{ij} \tag{2.1}$$

subject to

$$\sum_{j \in \delta^+(i)} x_{ij} - \sum_{j \in \delta^-(i)} x_{ji} = b(i) \quad \text{for all } i \tag{2.2}$$

$$x_{ij} \geq 0 \text{ and integer.} \tag{2.3}$$

This formulation appears in Edmonds & Johnson [1973]. Since the optimal flow can be decomposed into a collection of paths from nodes with positive supply ($b(i) > 0$) to those with net demand ($b(i) < 0$), this observation leads to the following procedure for solving the DPP. The use of the transportation problem in Step 3 below is due to Beltrami & Bodin [1974].

The DPP algorithm

INPUT: Directed $G = (V, A)$; edge costs c_{ij} for all $(i, j) \in A$.

Step 1. Define the subsets I and J of V consisting of nodes with positive or negative net degree:

$$I = \{i \in V \mid b(i) > 0\} \text{ and } J = \{j \in V \mid b(j) < 0\}.$$

Using the edge costs c_{ij}, compute the shortest paths from i to j for all pairs $i \in I, j \in J$. Record the cost $SP(i, j)$ of each part.

Step 2. Solve the transportation problem with the set I as supply nodes and the nodes in J as demand nodes. Each $i \in I$ has supply $b(i)$ while $j \in J$ has a demand equal to $|b(j)|$. The edge cost $SP(i, j)$ is attached to arc $(i, j) \in I \times J$. Let y_{ij} be the optimal flow on the arc (i, j) in the solution to the transportation problem.

Step 3. For any arc (p, q), set $x_{pq} = \sum_{(i,j) \in H_{pq}} y_{ij}$ where H_{pq} is the set of all pairs (i, j) for which (p, q) lies on the shortest path from i to j.

Step 4. Augment the original graph G with x_{pq} copies of edge $(p, q) \in A$ to obtain the balanced augmented graph $\tilde{G}$.

Step 5. Trace an Euler tour through $\tilde{G}$. This is the optimal directed postman tour.

The preceding algorithm for the DPP has a complexity of $O(mn^2)$ on a graph with n nodes and m arcs. The complexity of the algorithm of Edmonds & Johnson [1973] is $O(n^3)$. Lin & Zhao [1988] provide a flow algorithm for the DPP that maintains dual feasibility, complementary slackness and the flow balance relations and iteratively achieves primal feasibility. They show that this procedure requires $(m - n + 1)O(n^2)$ computations and argue that this may be preferable to the above two procedures if $m - n$ is small.

2.3. The mixed postman problem (MPP)

Consider a mixed graph $G = (V, E \cup A)$ that contains both directed and undirected edges. A necessary condition for this graph to be Eulerian (or unicursal) is that G be connected and even. (Recall that G is even if the total number of edges meeting each node, regardless of direction, is even.) Ford & Fulkerson [1962] provide the following necessary and sufficient conditions for the existence of an Euler tour. A mixed graph $G = (V, E)$ has an Euler tour if and only if

(a) G is connected and even, and

(b) for any subset S of the nodes,

$$N_d(S, \bar{S}) - N_d(\bar{S}, S) \le N_u(S, \bar{S})$$

where $N_d(S, \bar{S})$ is the number of directed edges originating in a node of S and ending at a node in $\bar{S}$, while $N_u(S, \bar{S})$ is the number of undirected edges with exactly one endpoint in each of the sets S and $\bar{S}$. Note that by interchanging the role of S and $\bar{S}$, we can rewrite condition (b) as

$$|N_d(S, \bar{S}) - N_d(\bar{S}, S)| \le N_u(S, \bar{S}),$$

and for the specific choice of $S = \{i\}$ for a node $i \in V$, the latter condition states that the imbalance between $d^-(i)$ and $d^+(i)$ cannot exceed the number of undirected edges that meet node i. In fact, Ford and Fulkerson show that if (b) holds, one can use a network flow algorithm to orient some of the undirected edges so as to balance the resulting graph G'.

If G is even and balanced, the optimal postman tour coincides with the Euler tour, which is known to exist. If G is even, Edmonds & Johnson [1973] give an optimal algorithm for balancing the graph by orienting some undirected edges and adding certain directed or undirected edges to the original graph. To state this algorithm, it is convenient to consider the two directed pairs of arcs corresponding to each (undirected) edge e in E. Consider a mixed graph $G = (V, A \cup E)$. Assign an arbitrary direction to all $e \in E$ and call the set of all resulting directed arcs A_f. We call this the forward orientation for ease of reference. The set of arcs directed in the opposite direction for all $e \in E$ is denoted by A_r, and is said to have the reverse orientation. Thus, each edge e results in two directed edges $a_f(e)$ and $a_r(e)$ in A_f and A_r, respectively. We can now state the Mixed Postman Algorithm for even graphs following Minieka [1978; see also Evans and Minieka, 1992].

The mixed postman algorithm for even graphs

INPUT: Even connected graph $G = (V, A \cup E)$; edge costs c_e for all $e \in a \cup E$.

Step 1. Arbitrarily assign a direction to each undirected edge $e \in E$ to obtain the directed graph $G_0 = (V, A \cup A_f)$. Calculate the net degree $b(i) = d^-(i) - d^+(i)$ in G_0. If all $b(i) = 0$, set $G_2 = G_0$ and proceed to Step 5.

Step 2. Construct a graph $G_1 = (V, A_1)$ with edge set $A_1 = A \cup A_f \cup A_r \cup F$ as follows: A_f and A_r contain directed arcs with opposite orientations for each $e \in E$ with cost c_e and no capacity. F contains a directed arc (j, i) with zero cost and a capacity of 2 for every (i, j) in A_f. Attach a net supply of $b(i)$ to each node i of G_1.

Step 3. Solve a minimum cost flow problem on G_1 to satisfy all node demands. Let x_{ij} be the optimal flow on arc $(i, j) \in A_1$.

Step 4. Create a graph $G_2 = (V, A_2)$ with the following arcs. For each $(i, j) \in A \cup A_f \cup A_r$, place $x_{ij} + 1$ copies of the arc (i, j) in A_2. For each arc (j, i) in F, place a copy of (i, j) in A_2 if $x_{ji} = 0$ and add a copy of (j, i) if $x_{ji} = 2$. G_2 is a directed even graph that is balanced.

Step 5. Identify an Euler tour in G_2. This corresponds to the optimal postman tour for the original graph.

The key balancing operation is performed optimally by the minimum cost flow algorithm in Step 3. One can show that the non-zero optimal flows are even numbers and that G_2 is an even graph if G is even. On the other hand, if the minimum cost flow problem has no feasible solution, then G has no postman tour.

We remarked that the preceding algorithm requires the original graph G to be even. If G is not even, one can apply the matching algorithm to duplicate certain edges of G to achieve even degree at all nodes, minimizing the cost of the added edges. One can then proceed to use the preceding algorithm. However, this two-stage algorithm may not be optimal even though each stage is performed optimally. In the words of Edmonds & Johnson [1973], "the optimum way to duplicate edges so as to make even degrees may not be the best way to do so if we later have to duplicate more edges to symmetrize the graph". In fact, Papadimitriou [1976] proves that the mixed postman problem is *NP*-complete. Thus, the mixed problem is inherently harder than its fully directed or fully undirected counterparts. We discuss this result further at the close of this section.

Even though the two-stage algorithm mentioned above is not optimal, it can serve as the basis of an effective heuristic. In fact, Edmonds & Johnson [1973] suggest such a procedure where a minimum cost matching pairs up nodes of odd degree (ignoring arc directions) to obtain an even graph. A network flow algorithm is then solved to balance the node degrees by duplicating certain arcs and orienting some of the edges. Frederickson [1979] shows that this flow problem may fail to preserve even degrees, but that its output can always be modified to obtain a solution of equal cost that forms a symmetric even graph. This yields a three-step procedure called MIXED1 for the MPP comprised of matching, flow problem, and adjustment steps. A second algorithm, called MIXED2, results if the balancing is performed first using the flow problem and the matching problem is solved next to achieve evenness. Frederickson proves that MIXED1 and MIXED2 have complexity $O[\text{Max}\{|V|^3, |A|(\max\{|A|, |E|\})^2\}]$, and a guaranteed error bound of 2 (so that $v_H/v* \leq 2$). However, if the two procedures are run in tandem and the best solution is selected, the performance bound improves to $5/3$. A detailed exposition of these results is given by Frederickson [1979] and a summary presentation may be found in Brucker [1981].

It is possible to formulate the mixed postman problem (MPP) as an integer program. Minieka [1979] outlines how the MPP can be transformed into an integer network flow problem with gains but does not explore the algorithmic implications of this transformation. Kappauf & Koehler [1979] present an integer programming formulation of the MPP and prove that the extreme points of the polyhedron corresponding to the linear programming relaxation of this formulation are half-integral (comprised of component values of 0, 1/2, or some positive integer). The same result has been obtained more directly by Ralphs [1993]. We outline the integer programming formulation given by Ralphs.

Let $G = (V, A \cup E)$ be a mixed graph that is strongly connected. We can consider the MPP as the problem of finding a minimum cost Eulerian augmentation of G. Let y_a be the number of additional-copies of each arc $a \in A$ included in the augmentation. For any edge $e \in E$, one must choose the orientation with which e appears in the augmentation. As defined above, let A_f and A_r denote the sets of directed counterparts to edge e with opposite orientations. Let

u_e^f and y_e^f denote the first and additional copies of the oriented copy $a_f(e)$ of edge e ($a_f(e) \in A_f$). Let u_e^r and y_e^r denote similar variables for A_r. The integer programming formulation of Ralphs for the MPP is:

$$(P): \quad \text{Min} \sum_{a \in A \cup A_f \cup A_r} c_a y_a + \sum_{a \in A} c_a + \sum_{a \in A_f} c_e u_e^f + \sum_{a \in A_r} c_e u_e^r \tag{2.4}$$

subject to

$$u_e^f + u_e^r \geq 1 \qquad \text{for all } e \in E \tag{2.5}$$

$$x_a = 1 + y_a \qquad \text{for all } a \in A \tag{2.6.1}$$

$$x_a = u_e^f + y_e^f \qquad \text{for all } a = a_f(e) \in A_f \tag{2.6.2}$$

$$x_a = u_e^r + y_e^r \qquad \text{for all } a = a_r(e) \in A_r \tag{2.6.3}$$

$$\sum_{a \in O(i)} x_a - \sum_{a \in I(i)} x_a = 0 \qquad \text{for all } i \in V \tag{2.7}$$

$$y_a, \ y_e^r, \ y_e^f \geq 0 \text{ and integer} \qquad \text{for all } e \tag{2.8}$$

$$u_e^f, \ u_e^r \in \{0, 1\} \qquad \text{for all } e. \tag{2.9}$$

The objective function in (2.4) is the total cost of the augmentation. The constant term $\sum_{a \in A} c_a$ appears since each arc in A must be retained in the augmentation at least once. Constraint (2.5) states that each edge $e \in E$ must also appear in the augmentation at least once, but there is a choice as to its orientation. The variable x_a simply counts the total number of times an arc appears with a specific direction. The balance equations appear in (2.7): $I(i)$ and $O(i)$ denote the arcs directed into or out of node i. These constraints ensure that the augmentation is symmetric.

Let (P) denote the integer program in (2.4)–(2.9) and denote by $(\bar{P})$ its linear programming relaxation. It is not difficult to see that (2.5) holds as an equality for any basic solution to $(\bar{P})$. This allows us to treat the last two sums of (2.4) as a constant as well (they sum to $\sum_{e \in E} c_e$). Thus, only the first sum and the variables y_a survive in the objective function for $(\bar{P})$. By redefining $x_a \leftarrow x_a - 1$ for $a \in A$, we can re-write $(\bar{P})$ as:

$$(\bar{P}): \quad \text{Min} \sum_{a \in A \cup A_f \cup A_r} c_a y_a \tag{2.10}$$

$$x_a = y_a \qquad \text{for all } a \in A \tag{2.11.1}$$

$$x_a = u_e^f + y_e^f \qquad \text{for all } a \in A_f \tag{2.11.2}$$

$$x_a = u_e^r + y_e^r \qquad \text{for all } a \in A_r \tag{2.11.3}$$

$$\sum_{a \in O(i)} x_a - \sum_{a \in I(i)} x_a = b(i) \qquad \text{for all } i \tag{2.12}$$

$$y_a, \ y_e^r, \ y_e^f \geq 0 \tag{2.13}$$

$$0 \leq u_e^f, \ u_e^r \leq 1 \qquad \text{for all } e \tag{2.14}$$

$$u_e^f + u_e^r = 1 \qquad \text{for all } e, \tag{2.15}$$

where it is important to note the change in the right-hand side of (2.12) as compared to (2.7).

The main advantage of this reformulation is that (2.10)–(2.14) is a network flow problem; $(\bar{P})$ reduces to a flow problem if the complicating constraint (2.15) is relaxed. Let $(\bar{P}_0)$ denote the flow problem defined by (2.10)–(2.14). Then $(\bar{P}_0)$ clearly has an integer optimal solution. Moreover, since (2.15) can only fail to be satisfied if $u_e^f = u_e^r$ in this solution, one can set $u_e^f = u_e^r = 1/2$ for each edge e whenever u_e^f and u_e^r turn out to be equal in the solution to $(\bar{P}_0)$. Since (2.15) is satisfied after this re-definition, we see that any integral solution to $(\bar{P}_0)$ can be easily transformed into a solution *of equal cost* for $(\bar{P})$ that is *half-integral*. This produces a half-integral optimum for $(\bar{P})$ for an arbitrary cost function in (2.10), establishing that every extreme point of $(\bar{P})$ is half-integral.

Christofides, Benavent, Campos, Corberan & Mota [1983] use Lagrangean relaxation to solve the MPP exactly. They use an integer programming formulation similar to (2.4)–(2.9) and relax either the symmetry constraints (2.7) or the edge traversal constraints (2.5). Lagrangean bounds from these two relaxations are incorporated into a branch-and-bound scheme that solves MPPs with up to 50 nodes, 70 arcs and 40 edges.

We have already cited the result of Papadimitriou [1976] that proves MPP to be *NP*-complete. His proof reduces the 3-satisfiability problem to the mixed postman problem and shows that the latter problem remains *NP*-complete even if the graph is planar, has no node of degree greater than 3 (counting both directed and undirected incident edges), and all edge costs are equal. In view of this result, it is interesting that the problem of finding the minimal Eulerian graph that contains a given mixed graph is tractable. Given a mixed graph $G = (V, E)$, this problem is to find a set of directed edges D of minimum cardinality such that $\tilde{G} = (V, E \cup D)$ is Eulerian. Papadimitriou [1976] shows that this problem is essentially equivalent to a maximum flow problem with unit edge capacities and provides a polynomial algorithm based on this observation. The output of this algorithm also determines an orientation for each undirected edge of G. Clearly, this algorithm can also be used to test the original graph for unicursality, thereby providing a more efficient alternative to the procedure in Ford & Fulkerson [1962].

In postman problems, one seeks an Eulerian *supergraph* of the original graph. It is also meaningful to pose the opposite notion of finding a *spanning Eulerian subgraph* of a given undirected graph $G = (V, E)$. Richey, Parker & Rardin [1985] show that deciding whether an arbitrary graph G has a spanning Eulerian subgraph is *NP*-complete. However, if G is series-parallel, it is possible to find the spanning Eulerian subgraph with the maximum number of edges in polynomial time. Richey & Parker [1991] consider the maximum weight Eulerian subgraph in a directed graph with arbitrary arc weights. Once again, the authors prove that while this problem is *NP*-complete in general, it can be solved in cubic time for series-parallel graphs. Finally, Guan [1984] poses the problem of finding the set of edge-disjoint cycles with maximum total weight in an undirected connected graph G with given edge weights. Guan calls this problem the maximum-weight *cycle packing* problem and shows that it is equivalent to the undirected postman problem.

Finally, we should cite the work of Korach & Penn [1992] that provides a duality relation for a certain generalization of the solutions to the undirected postman problem (UPP). Also, Saruwatari & Matsui [1994] describe an algorithm for finding the K best solutions of the UPP.

3. The rural postman problem

3.1. Introduction and formulation

In many routing applications, only a subset of all available arcs requires service, whereas classical postman problems assume that all arcs and edges of the graph must be traversed. Let $G = (V, A \cup E)$ be a mixed graph and let R denote the subset of *required* arcs or edges in $G(R \subset A \cup E)$. Furthermore, assume that the cost c_{ij} of traversing each arc (i, j) [or edge $e(i, j)$] is specified. The *rural postman problem* (RPP) is to find a minimum-cost cycle in G that traverses all edges in R. This problem was introduced by Orloff [1974]. We call the arcs (or edges) in R required and refer to the other arcs or edges of the RPP solution as *non-required* edges. In transportation applications, the traversal of non-required edges is often called deadheading. Correspondingly, we sometimes call non-required arcs deadhead arcs. We call the subgraph $G_R = (V_R, R)$ induced by the arcs and edges of R the *required subgraph*.

Note that connectivity alone adds a new source of complication that is absent from the classical postman problems: the graph G in the classical problems is always assumed to be connected to ensure the existence of a postman tour. When G_R is disconnected, the rural postman tour must link the connected components of G_R with minimum cost, thereby assuming the features of a traveling salesman problem. The early work on the RPP already mentions that the difficulty of this problem increases with the number of connected components in G_R [Orloff, 1976]. More formally, Frederickson [1979] points out that there exists a recursive algorithm for the RPP, exponential only in the number of components $c(R)$ of G_R. Furthermore, Lenstra and Rinnooy Kan [1976] proved that RPP is *NP*-complete by reducing the Hamiltonian circuit problem — a well known member of the class of *NP*-complete problems — to the directed RPP as follows. Let $G_0 = (N_0, A_0)$ be an instance of the Hamiltonian circuit problem with $|N_0| = n$. Form the node set N_1 by placing a copy of i_0 (called i_1) in N_1 for all $i_0 \in N_0$. Add two arcs (i_0, i_1) and (i_1, i_0) of zero cost between each pair of twin nodes and attach a cost of 1 to all arcs in A_0. Consider the RPP on the graph $G = (N_0 \cup N_1, A_0 \cup Q \cup R)$ where $R = \{(i_0, i_1); i = 1, \ldots, n\}$ is the set of required arcs and Q contains the reverse counterpart edges to those in R. Then RPP has a solution of cost n if and only if G_0 has a Hamiltonian circuit.

An early version of the RPP investigated by Frederickson, Hecht & Kim [1978] is the *stacker crane problem* (SCP). Given a mixed graph $G = (V, A \cup E)$, together with arc and edge costs c_{ij}, the problem is to find a rural postman tour that covers all arcs in A. The required arcs $R = A$ in this problem correspond to movements that a crane or a vehicle must perform. Frederickson, Hecht and Kim prove that this problem is *NP*-complete by transforming the traveling salesman problem into an instance of the SCP. They proceed to describe two heuristic algorithms for this problem that run in $O[\max(n^3, m^3)]$ and provides a worst-case performance guarantee of $v_H \leq (9/5)v^*$. These authors also study a multiple-vehicle version of the SCP called the k-SCP where k tours must be constructed to minimize the length of the longest tour (a min–max problem). The authors provide an approximation algorithm for the k-SCP with the same complexity as above and a guaranteed performance bound of $v_H \leq (14/5 - 1/k)v^*$.

We now return to the directed version of the RPP to provide a problem formulation and two intuitively appealing heuristics for solving the problem. Consider the directed graph $G = (V, A)$ with arc cost c_{ij} and required arcs $R \subset A$. A feasible solution to RPP is a cycle consisting of all arcs in R and a set of non-required (or deadhead) arcs A_d such that the multigraph $G_s = (V, R \cup A_d)$ is (a) balanced [that is $d^-(i) = d^+(i)$], for all nodes (i), and (b) connected. Selecting the set A_d such that both conditions are satisfied at minimum cost is complicated by the interaction between the two conditions.

Ball & Magazine [1988] provide a natural integer programming formulation of the *directed rural postman problem* (DRPP) by using the decision variable x_{ij} that counts the number of times edge (i, j) is traversed in the non-required mode. This formulation requires the notion of a *postman cut* to ensure connectivity of the solution $R \cup A_d$. Let $K_1, \ldots, K_{c(R)}$ denote the node sets of the connected components of G_R. Define the set of all postman cuts K as follows:

$$KP(G) = \{K \subset V \mid \text{each } K_i \text{ is completely contained in } K \text{ or } \bar{K};$$
$$\text{and } V_R \cap K \neq \phi, \; V_R \cap \bar{K} \neq \phi\}.$$

Thus, any cut K in $KP(G)$ contains from one to $c(R) - 1$ component nodes sets K_i completely so that K and $\bar{K}$ do not *both* intersect any K_i. Finally, let $b(i) = d^-(i, R) - d^+(i, R)$ be the imbalance between the in-degree and out-degree of node i in the required subgraph $G_R = (V, R)$. Define x_{ij} as the flow on the non-required arc (i, j) to obtain the following formulation:

(DRPP):

$$\text{Min} \sum_{(i,j) \in A} c_{ij} x_{ij} \tag{3.1.1}$$

subject to

$$\sum_{j \in \delta^+(i)} x_{ij} - \sum_{j \in \delta^-(i)} x_{ji} = b(i) \quad \text{for all } i \in V \tag{3.1.2}$$

$$\sum_{\substack{i \in K, \, j \in V-K \\ (i,j) \in A}} x_{ij} \geq 1 \quad \text{for all cut sets } K \in KP(G) \tag{3.1.3}$$

$$x_{ij} \geq 0 \tag{3.1.4}$$

$$x_{ij} \text{ integer.} \tag{3.1.5}$$

Constraint (3.1.2) ensures that the solution is balanced by adding arcs into and out of unbalanced nodes as required. Constraint (3.1.3) requires that the graph formed by selecting all edges with positive values of x_{ij} *and* the required arc set R

be connected. Stated otherwise, the cuts in (3.1.3) ensure that the graph formed by shrinking all components K_i to single nodes remains connected via non-required arcs with $x_{ij} > 0$. In principle, there is an exponential number of constraints in (3.1.3). However, the problem is particularly easy to solve if (3.1.3) is relaxed since (3.1.1)–(3.1.2) and (3.1.4)–(3.1.5) define a network flow problem. This observation is the basis of an intuitively appealing heuristic for the DRPP described in the following section.

Corberan & Sanchis [1990] provide an integer programming formulation for the undirected version of the RPP, denoted here by URPP. Consider the graph $G = (V, E)$ with the required edges $R \subset E$. Let the decision variable $x_{ij}(i < j)$ associated with each edge $e(i, j)$ of E count the number of times this edge is traversed in non-required mode (traversal without servicing). For convenience, let $x_{ji} = x_{ij}$ if $i < j$ and recall that $d(i, R)$ denotes the degree of node i in G_R. The URPP formulation follows:

(URPP):

$$\text{Min} \sum_{(i,j),\, i<j} c_{ij} x_{ij} \tag{3.2.1}$$

$$\sum_{j \in \delta(i)} x_{ij} = 2w_i - d(i, R) \quad \text{for all } i \in V \tag{3.2.2}$$

$$\sum_{\substack{i \in K,\, j \in \bar{K} \\ e(i,j) \in E}} x_{ij} \geq 2 \quad \text{for all cut sets } K \in KP(G) \tag{3.2.3}$$

$$x_{ij} \geq 0 \text{ and integer} \tag{3.2.4}$$

$$w_i \geq 0 \text{ and integer.} \tag{3.2.5}$$

Constraint (3.2.2) ensures that the multigraph with edge set $R \cup A_d$ is even, where A_d is comprised of x_{ij} copies of edge $e(i, j)$. Another way to state (3.2.2) is that its left-hand side should have the same parity as d(i,R). Constraint (3.2.3) forces connectivity using the family of postman cuts $KP(G)$ defined earlier in this section. Note that the right-hand side of (3.2.3) is now 2, as compared to 1 in (3.1.3).

3.2. The balance-and-connect procedure and related algorithms

We have mentioned that the key constraints of the DRPP require balanced node degrees [constraint (3.1.2)] and connectivity (3.1.3). It is natural to propose sequential heuristics that carry out balancing and connecting in tandem. We review two such heuristics and then describe a dual ascent procedure that considers the interaction between balancing and connecting. We begin with a simple algorithm studied by Ball & Magazine [1988].

The balance-and-connect (B&C) heuristic for RPP

INPUT: Directed graph $G = (V, A)$; required arcs $R \subset A$; edge costs c_{ij} for all $(i, j) \in A$.

Step 1. *Balancing step*: Compute $b(i) = d^-(i, R) - d^+(i, R)$ for all nodes i in the required subgraph $G_R = (V, R)$. Solve the network flow problem (3.1.1), (3.1.2) and (3.1.4) to select the edge set A_B so that $G_B = (V, R \cup A_B)$ is balanced.

Step 2. Find the connected components of G_B denoted by $K_1, \ldots, K_c$. If $c = 1$, output $R \cup A_B$ and stop. Otherwise, form the complete undirected graph $G_K = (N_K, E_K)$ on the node set $N_K = \{1, \ldots, c\}$ with edge costs $\tilde{c}_{pq} =$ MIN $_{i \in K_p, j \in K_q} \{SP(i, j) + SP(j, i)\}$ where $SP(i, j)$ is the length of the shortest path from node i to node j in G.

Step 3. *Connecting step*: Find the minimum cost spanning tree in G_K. Let A_c be the set of all directed arcs of G that lie on the shortest paths corresponding to the edges of G_K on the minimum spanning tree. Let $\tilde{G} = (V, R \cup A_B \cup A_c)$.

Step 4. Identify an Euler tour on $\tilde{G}$ and output it as the RPP solution.

Clearly, this algorithm uses a sequential approach that first constructs a balanced network G_B and then proceeds to connect the components of G_B if required. Note that the connection costs represent the 'round-trip' costs of traveling from one component to another and back.

While the B&C algorithm can be applied to any RPP, it may fail to identify good solutions because its sequential approach cannot address the interaction between the balancing and connecting steps. Figure 2 provides a simple illustration. The balancing step for this graph selects the arcs $A_B = \{(2, 3), (5, 4)\}$ so that $R \cup A_B$ has two connected components. The two-way connection of these components requires the additional arcs $A_c = \{(3, 4), (4, 3)\}$. The total cost of non-required arcs in the B&C algorithm is therefore 12 while the optimal solution only adds arcs (2,4) and (5,3) for a total cost of 5. In this simple example, carrying out the connecting step is advantageous as it also takes care of the balance requirement. However, as we describe in Section 10.2, Ball & Magazine [1988] prove that the B&C algorithm is optimal for a special class of problems arising in the manufacturing of printed circuit boards.

Given the interaction between the balancing and connecting steps, it is natural to compare B&C with a connect-and-balance (C&B) algorithm that reverses the sequence of these operations. The general steps of this procedure follow:

The connect-and-balance (C&B) heuristic for RPP

INPUT: Directed graph $G = (V, A)$; required arcs R; edge costs c_{ij} for all $(i, j) \in A$.

Step 1. Find the connected components of $G_R = (V, R)$. If G_R is connected, set $A_C = \emptyset$ and go to Step 3.

Step 2. *Connecting step*: Select a set of arcs A_C such that $R \cup A_C$ is connected.

Step 3. *Balancing step*: Solve a network flow problem to select arcs A_B whose addition to $R \cup A_C$ results in the balanced network $\tilde{G} = (V, R \cup A_C \cup A_B)$ at minimum cost.

Step 4. Identify an Euler tour on $\tilde{G}$.

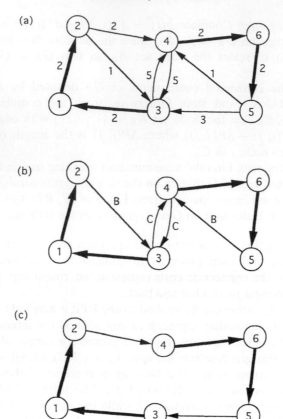

Fig. 2. (a) Graph G with required arcs R in solid lines; (b) balance-and-connect solution; (c) connect-and-balance solution.

Recently Su [1992], in joint work with Michael Ball, carried out a detailed comparison of the B&C and C&B algorithms. In this investigation, Su considered different implementations of the two classes of algorithms. For instance, the connecting step in C&B can be implemented by using an undirected minimum spanning tree algorithm, or a procedure for solving the asymmetric traveling salesman problem. We now briefly review some of the refinements considered by Su.

The connecting step of the C&B algorithm must use paths consisting of deadhead arcs from the underlying graph G to connect the distinct components of G_R. Denote the $c(R)$ components of G_R by $K_1, \ldots, K_{c(R)}$. For each component $p = 1, \ldots, c(R)$, introduce a pair of supersource and supersink nodes (s_p, t_p) and additional directed arcs (s_p, v), (v, t_p) joining these to every node v of component p. Let $\tilde{c}_{pq} = \tilde{c}_{qp} = SP(s_p, t_q) + SP(s_q, t_p)$ where $SP(v, w)$ is the shortest path cost of going from node v to node w in G. One can find the minimum spanning tree in the graph G_K where each node p corresponds to component K_p as in the B&C procedure using the edge costs $\tilde{c}_{pq}$. Alternatively, the connections in G_K

can be made by solving an asymmetric TSP and deleting the arc of maximum cost, thereby arriving at a directed Hamiltonian path for G_K. In this procedure, $\tilde{c}_{pq}$, the cost of connecting nodes (components) p and q in G_K, is defined as $SP(s_p, t_q)$ so that $\tilde{c}_{pq} \neq \tilde{c}_{qp}$.

The use of the supersource and supersink nodes offers additional flexibility in the connecting step. First, in contrast to Step 2 of B&C, the connection cost for the two components K_p and K_q does not have to use shortest paths between the same pair of nodes i and j in G. For example, the two components of R in Figure 2 can be joined by arcs (2,4) and (5,3) with $\tilde{c}_{12} = SP(s_1, t_2) + SP(s_2, t_1) = c_{24} + c_{53} = 2 + 3 = 5$. Second, this approach can easily accommodate node penalties that reflect balancing considerations by encouraging the use of node i as a start node if $b(i) > 0$ and as an end node if $b(i) < 0$. Thus, in Figure 2, it is desirable in the connecting step to encourage the use of arcs that lead out of nodes 2 and 5 or lead into nodes 3 and 4. This can be accomplished by placing penalties on arcs (s_p, i) and (i, t_p). For example, for a supply node i (a node with $b(i) > 0$), one can attach a cost of zero to the arc (s_p, i) to encourage the use of deadhead arcs that originate in i, and give a large cost, such as $\text{Max}_{(i,j) \in A} c_{ij}$, to the arc (i, t_p) to discourage deadheading into node i. In this way, node penalties enter the shortest path calculations that determine the edge costs of G_K. Ideally, these penalties should steer the connecting step towards selecting those edges that also improve the node balances, thereby helping the subsequent balancing step.

Su [1992] experimented with four versions of the C&B procedure: either the MST or the ATSP problem can be solved in the connecting step, and each can be implemented with or without node penalties. The ATSP was solved with a simple nearest neighbor technique. Generally, Su found that the MST version performed better, but that the use of node penalties did not provide a consistent advantage. Overall, the procedure exhibiting the best results in these tests is a modified B&C algorithm with a preprocessing scaling step and an improvement post-processing step applied after the Euler tour is generated. The effect of the preprocessing scaling step is to reduce the costs of deadhead arcs so as to encourage the balancing step to achieve greater connectivity, so that the graph G_B passed on to the connecting step has fewer connected components. The improvement step involves replacing deadhead path segments of the Euler tour produced in Step 4 of the B&C algorithm with shortest paths in the underlying network. With these refinements, the resulting B&C approach can improve upon the best of the C&B solutions by 5 to 10%.

The scaling step just mentioned is closely related to a dual ascent algorithm that Ball and Su devised to obtain lower bounds for the RPP. Their procedure is based on the dual of the RPP formulation (3.1.1)–(3.1.4) where the integrality constraints are relaxed. If u_i and w_k denote the dual variables associated with constraints (3.1.2) and (3.1.3), the dual problem is

(DRPP):

$$\text{Max} \sum_i b(i) u_i + \sum_K w_K \qquad (3.3.1)$$

subject to

$$u_i - u_j + \sum_{K \in K(i,j)} w_K \le c_{ij} \quad \text{for all } (i, j) \in A \tag{3.3.2}$$

$$w_k \ge 0 \quad\quad\quad\quad\quad\quad\quad\quad \text{for all cuts } K. \tag{3.3.3}$$

In this problem, u_i's are unrestricted in sign and $K(i, j)$ in (3.3.2) denotes the set of all cuts that include arc (i, j). One heuristic for solving this dual is to modify only one of the two sets of variables $\{u_i\}$ and $\{w_K\}$ at a time, holding the other set fixed. For example, if w_K's all vanish, (3.3.1)–(3.3.3) coincide with the dual of the minimum cost flow problem used in balancing so that the first sum in (3.3.1) can be easily optimized. On the other hand, for a fixed set of values of $\{u_i\}$, (3.3.2) may be written as

$$\sum_{K \in K(i,j)} w_K \le \bar{c}_{ij}, \tag{3.4}$$

where $\bar{c}_{ij} = c_{ij} - u_i + u_j$. The dual ascent on the cut variables attempts to increase the sum $\sum_K w_K$ in (3.3.1) by increasing a cut variable w_K to its maximum value of Min$_{(i,j) \in K} \bar{c}_{ij}$. This maximum value is attained when (3.4) becomes binding for at least one arc of the cut. This value is then subtracted from the reduced costs of all arcs (i, j) in the cut and the ascent procedure identifies a new cut for which this can be repeated. The details of the cut processing rules and their role in the scaling step of B&C are available in Su [1992].

Su also reports the results of testing the RPP algorithms and lower bounding procedures on real, Euclidean, and random networks with up to 250 nodes, where 10% to 50% of the arcs are required. These tests contain some interesting results on the factors that affect the performance of heuristics for the RPP. First, the number of components in the required network affects the difficulty of the problem as observed before. This number decreases as the percentage of required arcs increases. For instance, in a real network tested by Su, the number of components goes from 36 to 6 as the percentage of required arcs is increased from 10% to 50% (the underlying network has 211 nodes and 576 arcs). The B&C heuristic performs well when it succeeds in reducing the number of components in the balancing step, so that the connecting step has fewer components to connect. When this does not occur, the B&C and C&B procedures may have comparable performances.

3.3. Other exact algorithms for the RPP

One of the earliest exact approaches to the RPP was proposed by Christofides, Campos, Corberan & Mota [1981] who used branch-and-bound with bounds produced by Lagrangean relaxation. Their formulation of the URPP is similar to (3.2.1)–(3.2.5) but uses a right-hand side of 1 in (3.2.3). They relax the even-degree constraints (3.2.2). However, to obtain useful bounds, they bound the variables of interest as follows. It is clear that $x_{ij} \le 1$ if $e(i, j) \in R$ and $x_{ij} \le 2$, otherwise. These inequalities can be used to show that $\lceil 1/2 d(i, R) \rceil \le d(i)$, where $d(i)$ is the degree of node i in $G = (V, E)$. In this way, these variables can be set at their lower or upper bounds in the Lagrangean subproblem. For the variables x_{ij} corresponding to the non-required edges, the subproblem is a minimum spanning tree problem on the graph whose nodes correspond to the connected components K_i. Moreover, since the 2-connectivity of the solution in $(x_{ij})_{e(i,j)\in E}$ is no longer guaranteed in the relaxed problem, constraints of the form (3.2.3) that are violated can also be penalized in a Lagrangean fashion. The branching strategy based on achieving connectivity first and then satisfying as many degree requirements as possible turns out to be superior. The authors report computational results on randomly generated problems with 9–84 nodes and 13 to 184 edges, one-third to one-half of which are required. The number of connected components of G_R ranges from 2 to 8.

Christofides, Campos, Corberan & Mota [1986] solve the DRPP with a Lagrangean relaxation procedure imbedded within a branch-and-bound search. Their procedure relaxes the degree balancing constraints (3.1.2). However, they repeat the connectivity constraints (3.1.3) by placing each connected component K_h first in K and then in $V - K$ for each postman cut $K \in KP(G)$. Such repetition creates redundant connectivity constraints in the presence of (3.1.2) but *not* when these constraints are relaxed. Accordingly, the Lagrangean relaxation relaxes one of these two sets of constraints. The Lagrangean subproblem with respect to each component K_h is a shortest spanning arborescence over a condensed graph. The authors also use a connect-and-balance heuristic to obtain good feasible solutions (for upper bounds). Of course, the connect step of this heuristic also solves a spanning arborescence problem. The computational results are on test problems with n ranging from 13 to 80, m from 24 to 180, and $|R|$ from 7 to 74. The number of components of G_R ranges from 2 to 8, but equals 4, 5 or 6 in most test problems.

Corberan & Sanchis [1994] study the polyhedron $P(G)$ associated with the edge variables x_{ij} of the URPP in (3.2.1)–(3.2.5). They provide conditions for constraints (3.2.3) and (3.2.4) to be facet inducing. In particular, $x_{ij} \geq 0$ defines a facet if $e(i, j)$ is not a cut-edge of G, and (3.2.4) defines a facet if the subgraphs induced by K and $\bar{K}$ are connected. Both conditions are also sufficient. The authors show that the latter conditions also ensure that the following constraint is facet-inducing:

$$\sum_{i\in K, j\in\bar{K}} x_{ij} \geq 1 \quad \text{for } K \subset V \quad \text{with } |R \cap (K, \bar{K})| \text{ odd.} \tag{3.5}$$

This constraint considers cut sets $(K, \bar{K})$ that contain an odd number of edges from R. The sum in (3.5) is over all edges $e(i, j)$ in the cut set $(K, \bar{K})$ and hence includes required edges as well. Finally, they developed a new class of constraints, called K–C constraints, that are not implied by those mentioned above.

In their algorithm for the URPP, Corberan and Sanchis solve the linear program with a subset of the constraints in (3.2.2) and (3.2.3) with $K = K_i$, and identify violated facet-defining constraints (3.2.3), (3.5) and K–C constraints. This procedure was applied to the test problems of Christofides, Campos, Corberan & Mota [1981] where the largest problem had $n = 84$, $m = 180$, $|R| = 74$

and 8 connected components. They also tested a graph representing the street network of the city of Albaida (Valencia, Spain) with 113 nodes, 171 edges and 11 components. Generally, the algorithm was able to solve all problems to optimality.

In joint work with M. Ball, Gün [1993] has recently studied the polyhedral structure of the DRPP. This work contains conditions for the non-negativity and connectivity constraints (postman cuts) to be facet-inducing. The conditions for the postman cuts turn out to be complicated and are not described here.

Gün also studies special cases of the DRPP where the non-required arcs form a special sparse network. In the Tree special case, non-required arcs occur in anti-parallel counterpart pairs between nodes in such a way as to give a tree when collapsed into undirected edges. There is no condition on the required edges. Gün provides a linear programming formulation for this problem, proves that it has integral extreme points and shows that the B&C algorithm of Section 3.2 can be used to solve the linear program optimally. Similar results are provided for the cycle special case (where the underlying non-required graph forms a cycle), and for the case of the Manhattan (ℓ_1) metric.

4. The capacitated arc routing problem

4.1. Formulation

The classical and rural postman problems discussed so far do not involve capacity restrictions and may be likened to TSP or m-TSP problems where the total demand or duration associated with a tour is not limited. The *capacitated arc routing problem* (CARP) is the arc-routing analogue of the classical vehicle routing problem in that it places an explicit capacity restriction on the tour associated with each postman (vehicle). We introduce this problem by describing an integer programming formulation following Golden & Wong [1981].

Let $G = (V, E)$ be an undirected graph with edge costs $c_{ij} = c_{ji}$. Associate a demand $q_{ij} \geq 0$ with each edge $e(i, j)$ of E. Edges with positive demand are called required arcs and $R = \{e(i, j) \mid q_{ij} > 0\}$. K identical vehicles (postmen) of capacity Q are available. We assume that Q exceeds q_{ij} for all (i, j). We denote the nodes of G by $\{1, \ldots, n\}$ and reserve node 1 for the depot. $N(i)$ denotes the nodes adjacent to i in G.

Introduce two sets of variables: x_{ij}^k is 1 if vehicle k traverses $e(i, j)$ from i to j and 0 otherwise; $\ell_{ij}^k = 1$ if vehicle k services $e(i, j)$ while traversing it from i to j. The formulation is:

(CARP):

$$\text{Min} \sum_{k=1}^{K} \sum_{(i,j)} c_{ij} x_{ij}^k \tag{4.1.1}$$

subject to

$$\sum_{j \in N(i)} (x_{ji}^k - x_{ij}^k) = 0 \qquad \text{for all } i \text{ in } V, \ k = 1, \ldots, K \tag{4.1.2}$$

$$x_{ij}^k \geq \ell_{ij}^k \qquad\qquad\qquad\qquad \text{for all } (i, j), \text{ and all } k \qquad (4.1.3)$$

$$\sum_{k=1}^{K} (\ell_{ij}^k + \ell_{ji}^k) = 1 \qquad\qquad\qquad \text{if } e(i, j) \in R \qquad (4.1.4)$$

$$\sum_i \sum_j \ell_{ij}^k q_{ij} \leq Q \qquad\qquad\qquad \text{for all } k \qquad (4.1.5)$$

$$\left.\begin{array}{l} \displaystyle\sum_{(i,j)\in(S,S)} x_{ij}^k - n^2 y_s^k \leq |S| - 1 \\[2ex] \displaystyle\sum_{(i,j)\in(S,\bar{S})} x_{ij}^k + u_s^k \geq 1 \\[2ex] u_s^k + y_s^k \leq 1, \ u_s^k, \ y_s^k \in \{0, 1\} \end{array}\right\} \quad \begin{array}{l}\text{for every non-empty} \\ \text{subset } S \text{ of } \{2, \ldots, n\} \\ \text{and all } k\end{array} \qquad (4.1.6)$$

$$x_{ij}^k, \ \ell_{ij}^k \in \{0, 1\} \qquad\qquad\qquad \text{for all } (i, j) \text{ and } k. \qquad (4.1.7)$$

Constraint (4.1.2) ensures route continuity for each vehicle. Constraint (4.1.4) forces all required arcs to be serviced; (4.1.3) states that serviced edges must be traversed. The explicit capacity constraint (4.1.5) applies to each vehicle. The group of constraints in (4.1.6) eliminates disconnected subtours but allows tours that include two more or more closed cycles. Of course, the exponential number of constraints in (4.1.6) makes their direct use impractical. Another formulation of the CARP on an undirected network has been proposed by Belenguer & Benavent [1992].

One advantage of the CARP formulation is its generality. If $\sum_{(i,j)} q_{ij} \leq Q$, so that one vehicle has sufficient capacity to handle all of the demand, CARP becomes the classical undirected postman problem if $R = E$ and the RPP if R is a proper subset of E. When vehicle capacity is limited and $q_{ij} > 0$ for all $e(i, j)$, we obtain the capacitated version of the postman problem we call the *capacitated postman problem* (CAPP), also known as the capacitated Chinese postman problem (CCPP). Golden and Wong also point out that the classical UPP can also be viewed as a special case of CARP by splitting each node and placing the node demand on the undirected edge joining the two copies. They also prove that the 0.5-approximate CAPP is *NP*-hard by reducing the partition problem to this problem. This establishes the difficulty of finding solutions to the CAPP with a guaranteed error bound of $v(H) \leq 1.5v^*$ where $v(H)$ and v^* are the objective function values associated with the heuristic and optimal solutions. We note that bounds for the vehicle routing problem on a tree derived by Labbé, Laporte & Mercure [1991] also apply to the CARP on a tree structure.

More recent results on heuristics with guaranteed error bounds for the CARP are provided by Jansen [1992, 1993]. Jansen [1993] extends the analysis of two well-known VRP heuristics to the General Routing Problem, which includes CARP as a special case. He proves a bound of $v(H) \leq (2 - 1/Q)v^*$, where H stands for the IOTP heuristic of Haimovich and Rinnooy Kan [1985], when G_R is connected and Eulerian, all edges in the CARP have equal demand (of unity) and the vehicle capacity is Q. For unequal demands, the bound of $(2 - 1/Q)$ deteriorates to $3 - 4/\beta(Q)$ where $\beta(Q) = 2Q$ if Q is even, and $Q + 1$ if Q is odd.

Given these results, it is worthwhile to investigate lower bounding procedures for both CAPP and CARP. This is the topic of the remainder of this section.

4.2. Matching-based lower bounds for CARP

In the undirected postman problem with no capacity constraints, we can use the matching algorithm to identify a minimum cost set of non-required edges whose addition turns the original graph into an Eulerian graph. The non-required edges form connecting paths between pairs of odd-degree nodes. In the CARP, it is necessary to add other non-required edges since several vehicle tours are based at the depot. This is easy to see in the extreme case where the underlying graph has no nodes of odd degree but at least K vehicles are needed to meet demand. If $d(1)$, the degree of the depot node, is smaller than $2K$, then it must be raised to $2K$ (by adding non-required edges) because K cycles must be based at the depot node. This section discusses a sequence of lower bounds that rely upon the matching algorithm to bound the cost of the non-required edges added on account of both even degree and capacity constraints. The following notation is used in this section.

Let $G = (V, E)$ be an undirected graph with edge costs c_{ij}. A demand q_{ij} is associated with each edge in E and the vehicle capacity is Q. Edges with nonzero demand constitute the required arcs R. Let $G_R = (V_R, R)$ be the subgraph induced by these edges and define

$$M_0 = \left\lceil \frac{\sum\limits_{(i,j)} q_{ij}}{Q} \right\rceil, \qquad J = 2M_0 - d(1, R)$$

where

$d(i, R)$ = the degree of node i in G_R (number of edges in R that meet node i),

$V_0(R)$ = the set of nodes of odd degree in $G_R = \{i \in V \mid d(i, R) \text{ is odd}\}$,

$c_0(1, R)$ = Min $\{SP(1, i) \mid i \text{ meets } R\}$.

For any graph G_0 with an even number of nodes, $MP(G_0)$ denotes the cost of the minimum-cost *perfect matching* on G_0. For any subset of edges $F \subset E$, $c(F) = \sum_{e(i,j) \in F} c_{ij}$. In particular, $c(R)$ denotes the cost of traversing the required edges of G and constitutes the fixed portion of the CARP objective. For ease of exposition, we assume that $d(1, R)$ is even and that $J \geq 0$ throughout this section.

Golden & Wong [1981] proposed the first matching-based lower bound technique for the CARP based on the observation that the degree of the depot node must be raised by at least J units since any solution to the CARP must have at least $2M_0$ edges (required or non-required) incident to the depot. The matching problem is solved on the graph G_0 defined as follows. The node set of G_0 is $V_0(R) \cup A \cup B$ where the sets of additional nodes $A = \{a_1, \ldots, a_J\}$ and $B = \{b_1, \ldots, b_J\}$ correspond to additional copies of the depot. Its edge set consists of the following: (1) edges (i, j) for i in $V_0(R)$ and j in $V_0(R) \cup A$ with edge cost $d(i, j) = SP(i, j)$ for j in $V_0(R)$ and $d(i, a) = SP(i, 1)$ for $a \in A$; (2) edges (a_k, b_k) for all $k = 1, \ldots, J$ with $d(a, b) = c_0(1, R) = \text{Min } \{SP(1, i) \mid i \in V_R\}$; (3) edges of zero cost between any pair of nodes from B.

Theorem 4.1 [Golden & Wong, 1981]. *Let $MP(G_0)$ be the cost of the minimum-cost perfect matching on G_0 with edge costs $d(i, j)$ defined above. Then $MLB = MP(G_0) + c(R)$ is a valid lower bound for the CARP called the matching lower bound (MLB).*

Example 4.1. Consider the graph G in Figure 3. Assume that all nine edges of G require service with $q_{26} = q_{34} = q$ and $q_{ij} = 1$ for all other edges. The vehicle capacity is $Q = 6$. For $q = 5$, $M_0 = 3$ as $\sum q_{ij} = 17$. As $d(1, R) = 2$, $J = 4$ so four copies of the depot appear in the set A. The matching network G_0 is shown in Figure 4, where the solid edges form the optimal solution. Note that only

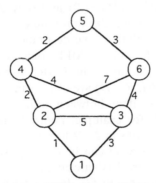

Fig. 3. Original CARP network G with edge costs.

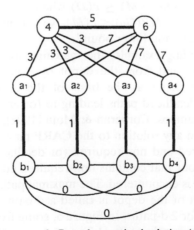

Fig. 4. The MLB matching network G_0 and an optimal solution (given by the solid edges).

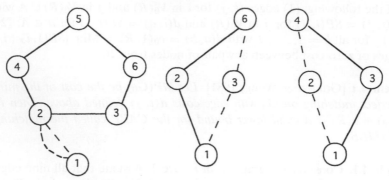

Fig. 5. An optimal solution to G with three cycles (artificial edges are dashed).

nodes 4 and 6 of the original graph appear in G_0. Clearly, $MP(G_0) = 9$ so that MLB $= 9 + 30 = 39$. An optimal solution to the problem is shown in Figure 5, with a cost of $12 + 30 = 42$. Note, also, that if $q = 2$ so that the total demand decreases to 11, then two vehicles suffice ($M_0 = 2$) and only $J = 2$ copies of the depot appear in G_0. In this case, adding the non-required edges $(5,6)$, $(1,2)$ and $(1,2)$ to G provides an Euler cycle that can be decomposed into two capacity-feasible tours. For $q = 2$, therefore, the value of MLB coincides with the optimum (37).

In general, the optimal solution to the CARP is composed of a set of edges $R \cup D$. Clearly, $d(i, r)$ and $d(i, R \cup D)$ are of the same parity. One can form paths out of contiguous edges of D alone between nodes in $V_0(R)$ or a node in $V_0(R)$ and the depot. In Figure 5, for example, nodes 4 and 6 are connected to two copies of the depot via such paths in D. When all nodes of $V_0(R)$ are matched in this way, any remaining edges in D can be decomposed into cycles based at the depot (the cycle 1–2–1 in Figure 5 is an example). Each cycle of this kind can be represented by two edges (a_j, b_j) in G_0. In this fashion, the edges in D define edges of no greater cost to be selected in the matching problem. Stated differently, the construction ensures that $c(M) \le c(D)$ where M is the matching found by following the procedure. As $MP(G_0) \le c(M)$, we conclude that $MP(G_0) \le c(D)$. This shows that MLB is a valid lower bound.

Clearly, in the preceding example, the matching solution underestimated the cost of deadheading to and from the depot by using only the minimum-cost edge $(1,2)$ leaving the depot as the basis of the estimate. This motivates a closer examination of deadhead paths leading to (or emanating from) the depot. Following Benavent, Campos, Corberan & Mota [1992], we introduce the notion of *d-paths*. The edges in any solution to the CARP may be represented as $R \cup D$ — the union of required and non-required (or deadheading) edges. Starting at the depot, follow any tour that contains non-required edges incident to the depot until the first edge in R is encountered. The maximal path consisting of contiguous non-required edges out of the depot is called a d-path. In general, each vehicle tour can generate 0, 1, or 2 d-paths. In Figure 5, going from left to right, the three cycles generate the four d-paths $(1,2)$, $(1,2)$, $(1,3,6)$, and $(1,2,4)$.

If node i is the terminal vertex of a d-path, then the 'next' edge on the vehicle tour (incident to i) is a required edge by the maximality of a d-path. This results in the following useful observation.

Lemma 4.2. *There are at least J d-paths in any solution to the CARP. If NP(i) denotes the number of d-paths that terminate at node i, the relation $NP(i) \leq d(i, R)$ holds for all $i > 1$.*

Several observations on the MLB now follow: Let M^* be the smallest number of vehicles required in any optimal solution to CARP. The matching construction requires a lower bound for M^*, which may be set at $M_0 = \lceil \sum q_{ij}/Q \rceil$ in the absence of a better (tighter) estimate. A better choice is to solve the bin packing problem for items of size $\{q_{ij} : (i, j) \in R\}$ and bin capacity Q. However, as the bin packing problem is NP-complete, it is generally impractical to solve it for bounding purposes alone. Assad, Pearn & Golden [1987] show that the value of the MLB bound increases with M, the estimate provided for the number of required vehicles. The deviation between M^* and M may be called the 'packing error'. A second source of error — the 'routing error' — has already been discussed. The cost $SP(1, i)$ of a d-path originated at node i is seriously underestimated by the edge cost $c_0(1, R)$ attached to edges in G_0. Assad, Pearn & Golden [1987] show that the packing and routing errors can each result in arbitrarily bad worst-case performance of the MLB. For the case of CAPP where all edge demands satisfy $0 < q_{ij} \leq \alpha Q$ for some fixed $\alpha \in (0, 1)$, these authors provide the bound

$$\frac{v^*}{\text{MLB}} \leq \left\lceil \frac{m}{\lfloor 1/\alpha \rfloor} \right\rceil$$

where v^* denotes the optimum and m is the number of edges in G. Clearly, for small α and large m, this bound equals αm.

It is easy to see that MLB deteriorates in the presence of edges with large demands (relative to Q): In the extreme case where $q_{ij} > Q/2$ for all edges $(i, j) \in R$, so that each vehicle serves exactly one edge, deadhead paths must join the end points of each edge to the depot: $d(i, R)$ such paths run between node i and the depot, each following the shortest path at a cost of $SP(1, i)$. Thus, the total cost $c(D)$ of non-required edges equals $\sum_{i=2}^{n} SP(1, i)d(i, R)$. For the graph G in Figure 3, this sum equals 56 while MLB $= 21$ even if it is supplied with the correct number of vehicles ($M^* = 9$). We now describe a simple bounding technique that is based on this observation.

The *node scanning lower bound* (NSLB) was first suggested by Pearn [1984] [see Assad, Pearn & Golden, 1987, and Pearn, 1988]: Arrange the nodes of G_R in increasing order of their shortest path cost to the depot and re-number if necessary to obtain $SP(1, 2) \leq SP(1, 3) \leq \ldots \leq SP(1, n)$. Let $L = \text{MIN}\{\ell \mid \sum_{i=2}^{\ell} d(i, R) \geq J\}$ so that L denotes the smallest number of nodes, taken in sequence of $SP(1, i)$ required to accumulate a sum of node degrees greater than $J - 1$. Set

$$\text{NSLB} = c(R) + \sum_{i=2}^{L-1} SP(1, i)d(i, R) + SP(1, L)\Big(\sum_{i=2}^{L} d(i, R) - J\Big).$$

The preceding expression shows that NSLB simply computes the total cost of the J 'shortest' d-paths, considering that node i can contribute at most $d(i, R)$ such paths. Assad, Pearn & Golden [1987] observe that this bound is optimal if each vehicle serves a single edge and show that $\text{NSLB} \geq \text{MLB} - MP(G_0)$ where G_0 is the graph induced by all nodes in $V_0(R)$. Clearly, if no nodes of odd-degree are present, $\text{NSLB} \geq \text{MLB}$, so that a tighter bound than MLB obtains. However, experiments performed by Pearn [1984] indicate that the NSLB rarely outperforms the MLB unless the graph G has few nodes of odd-degree and contains many nodes 'far' from the depot. However NSLB is very easy to compute. A new lower bounding procedure proposed by Pearn [1988], which we call the *matching path lower bound* (MPLB), combines the key ideas of NSLB and MLB to obtain superior results.

The matching path lower bound

INPUT: Undirected graph $G = (V, E)$; a subset R of required edges; edge costs $c_{ij} = c_{ji}$ for all $e(i, j)$ in E.

Step 1. Set $J = 2M_0 - d(1, R)$ and let n_0 be the cardinality of $V_0(R)$.

For $p = 0, 2, \ldots, n_0$ and *even*, perform Steps 2–4.

Step 2. Construct the graph G_p: Let $A_p = \{a_1, \ldots, a_p\}$ denote p copies of the depot. The G_p is the complete graph on node set $V_0(R) \cup A_p$ with symmetric edge costs

$$d(i, j) = \begin{cases} SP(i, j) & \text{if } i, j \text{ are in } V_0(R), \\ SP(i, 1) & \text{if } i \in V_0(R) \text{ and } j = a_k \\ \infty & \text{otherwise.} \end{cases}$$

Step 3. Solve the matching problem to obtain $MP(G_p)$. Denote by M_p the set of edges in E corresponding to this optimal matching.

Step 4. Compute the NSLB for the augmented graph $\tilde{G}(p) = (V, E \cup M_p)$ with $\tilde{R}(p) = R \cup M_p$ and $J(p) = 2M_0 - d(\tilde{R}(p), 1)$. Denote the resulting value by $\text{NSLB}(p)$. Let $LB(p) = c(R) + MP(G_p) + \text{NSLB}(p)$.

Step 5. Output the bound $\text{MPLB} = \text{Min}\{LB(p)\}$, the minimum being taken over all even values of p between 0 and n_0.

As with MLB, the MPLB procedure above is based on raising the degree of the depot from $d(1, R)$ to $2M_0$. However, the MPLB separates out the p artificial paths that connect (or match) the depot to nodes in $V_0(R)$ and explicitly searches over p. For a given p, Step 3 of the procedure finds the best way to match p copies of the depot with nodes in $V_0(R)$ while Step 4 selects $J(p)$ additional d-paths to the depot. Since the precise value of p in the optimal solution is unknown, Step 5 reports the minimum of the bounds obtained for various admissible p values. The properties of the MPLB are stated in the following result.

Theorem 4.3 [Pearn, 1988]. *The MPLB provides a valid lower bound for CARP that satisfies MLB $\leq$ MPLB. Moreover, if $V_0(R)$ is empty, so that there are no nodes of odd degree in G_R, MPLB reduces to the NSLB.*

To see that MPLB dominates MLB, note that p copies of the depot are matched to nodes in $V_0(R)$ in Step 3 of MPLB. This corresponds to p matching edges $(i, a) \in V_0(R) \times A$ in the matching graph G_0 of MLB. Next, NSLB(p) generates $J(p)$ artificial paths into the depot that can be associated with $J(p)$ edges (a_k, b_k) in G_0. Since these edges are given a cost of $c_0(1, R)$ in MLB, this association does not increase the costs. In this way, Steps 3 and 4 of MPLB define a feasible solution to the matching problem in G_0 with cost no greater than $MP(G_p) + J(p) * c_0(1, R) \leq MP(G_p) + \text{NSLB}(p)$. Now, observe that $MP(G_0)$ cannot exceed the left side of this inequality, so that $c(R) + MP(G_0) \leq c(R) + MP(G_p) + \text{NSLB}(p) = LB(p)$ holds for all even p. This shows MLB $\leq$ MPLB.

4.3. Refined lower bounds for the CARP

Recent work on lower bounds for the CARP has considerably refined the original ideas behind the matching lower bound. In this section, we review the comprehensive work of Benavent, Campos, Corberan & Mota [1992] which integrates several strands of earlier research on lower bounds.

Win [1988] proposed several refinements of the MLB, one of which extends the procedure of Golden & Wong [1981] to arbitrary cut sets $(U, \bar{U})$ separating the depot from other nodes. Let U be a subset of the node set V that includes the depot $(1 \in U \cup V)$. In the general case, the subgraph of G induced by $V - U$ has one or several connected components. Consider one such component c with node set V_c. Let r_c be the number of required edges in the cut set $(V_c, \bar{V}_c)$ and denote by Q_c the total load of all edges with at least one endpoint in V_c. Clearly, $p_c = \lceil Q_c/Q \rceil$ vehicles are required to serve this demand so that $J_c = 2p_c - r_c$ non-required edges must cross the boundary defined by the cut set. In this way, just as the MLB considers the required increase in the degree of the depot node resulting from non-required edges, Win performs the MLB analysis for each component V_c. In fact, the matching problem is solved on the node set $[V_0(R) \cap V_c] \cup A_c \cup B_c$ where the sets A_c and B_c each contain J_c artificial nodes. It is important to note that for $U = \{1\}$, this procedure reproduces the MLB. Win's algorithm initializes U with the depot node but proceeds to successively expand U. The bound is re-computed for each choice of U and the maximum value is retained as the final lower bound value.

Benavent, Campos, Corberan & Mota [1992] develop four new lower bounds for the CARP that subsume the bounds of Win [1988] and Pearn [1988]. We describe the first of these lower bounds (*LB*1 in their paper) and briefly mention the others.

Recall that the MPLB bound developed by Pearn [1988] computes a lower bound on the cost of non-required edges in the optimal solution by adding the two terms $MP(G_p)$ and NSLB($_p$), these estimates do not interact except through p

— the number of edges incident to the depot. By duplicating nodes of G_R explicitly in their matching graph, Benavent and coworkers take this interaction into account. For this reason, we call their bounds *node-duplicated matching path bounds* and denote them by NDMPB1, NDMPB2 and NDMPB3.

The node-duplicated matching path bound 1 (NDMPB1)

INPUT: Same as for MPLB.

Step 1. Order the nodes in G_R in increasing order of $SP(1, i)$ so that $SP(1, 2) \leq SP(1, 3) \leq \dots$.

Step 2. Let $J = 2M_0 - d(1, R)$ and compute $L = \text{Min}\{\ell \mid \sum_{i=2}^{\ell} d(i, R) \geq J\}$. Let $V_L = \{2, \dots, L\}$.

Step 3. Construct the matching graph $G_1 = (V_1, E_1)$ as the complete graph on node set $V_1 = A \cup W \cup S_1$ where $A = \{a_1, \dots, a_J\}$ contains J copies of the depot; W contains $d(i, R)$ copies of each node i in V_L, and $S_1 = S - V_L$. The edge cost $d(i, j) = SP(i, j)$ for all i and j in V_1 unless both belong to A; $d(a_i, a_j) = \infty$ for all i and j.

Step 4. Find the optimal matching on G_1 and its cost $MP(G_1)$.

OUTPUT: NDMPB1 $= c(R) + MP(G_1)$.

Theorem 4.4 [Benavent, Campos, Corberan & Mota, 1992]. *The value NDMPB1 obtained by the preceding algorithm is a valid lower bound for the CARP. Moreover, MPLB $\leq$ NDMPB1 so that this bound dominates the matching path lower bound.*

The proofs of the two statements in this theorem are rather long and include an enumeration of several cases. To establish the validity of the bound, the d-paths in an optimal solution are transformed to provide a matching of no higher cost in the graph G_1 defined in Step 3 of the bounding procedure. The second assertion of the theorem is proved by showing that the relation $LB(p) \leq$ NDMPB1 must hold for some even integer $p \leq |V_0(R)|$, where $LB(p)$ is the quantity computed in Step 4 of the MPLB procedure. The following example illustrates the bound NDMPB1.

Example 4.2. Consider the graph G in Figure 6 due to Benavent, Campos, Corberan & Mota [1992]. All edges of this graph are required and labeled with (c_{ij}, q_{ji}). Note that $c(R) = 14$. Let the vehicle capacity $Q = 5$. As $\sum q_{ij} = 18$, $M_0 = 4$ and $J = 8 - 2 = 6$. The nodes of G are already numbered in ascending order of $SP(1, i)$. Because $d(2) + d(3) = 6$, $L = 2$. Therefore, $V_L = \{2, 3\}$ and each of the nodes 2 and 3 is duplicated three times in W. The set S_1 contains the remaining nodes of odd degree: 4 and 5. An optimal solution to the matching problem on G_1 appears in Figure 7 with a cost of 10. Thus, NDMPB1 $= 14 + 10 = 24$. It is easy to check that MLB $= 14 + 9 = 23$. Performing the computations for MPLB on this graph provides values of 14 plus 13, 9 or 11 for $LB(0)$, $LB(2)$, $LB(4)$, respectively. Thus MPLB $= \text{Min}_p\{LB(p)\} = \text{Min}\{27, 23, 25\} = 23$, a value that is inferior to NDMPB1. Finally, we note that an optimal solution to this problem involves four tours as shown in Figure 6 and costs $\{14 + 18 = 32\}$.

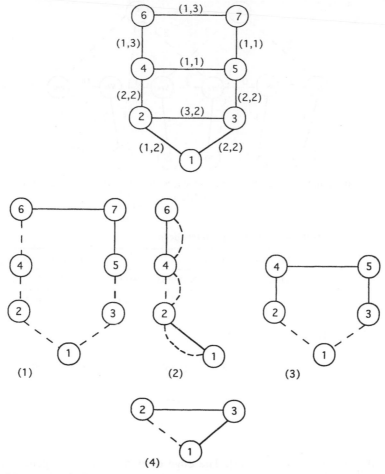

Fig. 6. A CARP network G with (c_{ij}, q_{ij}) shown on each edge, and an optimal solution composed of four cycles.

Having developed the first bound NDMPB1, Benavent, Campos, Corberan & Mota [1992] refine it further by using Win's idea of successive cut sets. This results in their second bound NDMPB2 (or LB2 in their notation) that dominates both NDMPB1 and Win's bound. Their third bound, NDMPB3, also dominated NDMPB1 but is valid only when the number of vehicles equals M_0 (the minimum possible). Correspondingly, their computational results on comparing the bounds (outlined below) comprises instances of the CARP where a feasible solution with M_0 vehicles exists.

Benavent and coworkers test all three bounds, as well as the bounds developed by Pearn [1988] and Win [1988] on networks where the number of nodes (n) ranges from 24 to 50, the number of edges (m) from 39 to 97 and all edges are

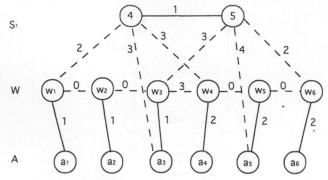

Fig. 7. Matching graph for NDMLB1 (not all edges are shown). Solid edges form an optimal solution.

Table 2
Comparison of NDMPB lower bounds on two test beds

Bounding technique	Test problems A (24)				Test problems B (10)			
	MPLB	NDMPB1	NDMPB2	NDMPB3	MPLB	NDMPB1	NDMPB2	NDMPB3
Average % under UB	6.4	6.3	6.1	5.8	26.1	22.5	19.7	20.5
No. of times best LB	11	12	14	22	0	2	5	7
Average % above MPLB	0	1.74	9.03	14.85	0	5.13	9.55	8.22
Average % above NDMPB1	–	0	0.3	0.55	–	0	4.06	2.87

required (have positive demand). The edge costs are randomly generated in the range 1 to 15. Table 2 summarizes the results for two test beds A and B. Test bed A contains 24 problems with different vehicle capacities to obtain a range of 2 to 5 for M_0. Clearly, NDMPB3 performs best in this set (best among the four lower bounds listed in the table). However, this bounding technique is seven times more time consuming than NDMPB2 to compute on the average. Interesting, NDMPB2 not only outperforms MPLB by approximately 9%, it is also about four times faster.

In test bed B, the capacity of the vehicles is further reduced so that M_0 ranges from 8 to 10 and each vehicle serves only a few edges. As expected, the performance of all lower bounds deteriorates significantly and NDMPB1 produces a larger improvement over MPLB. Over all, the authors conclude that NDMPB2 has the best performance when both accuracy and computational burden are considered. These results suggest that the performance of the lower bounds may be quite sensitive to the problem data, a remark that is also suggested by the computational results of Section 5.

Finally, we wish to mention the lower bound Saruwatari, Hirabayashi & Nishida [1992a] have developed and called the *node duplication lower bound* (NDLB). As the name suggests, each node $i(i \neq 1)$ of the network G_R is duplicated $d(i, R)$ times to obtain a graph $G_1 = (V_1, E_1)$ where V_1 is the set of all duplicated nodes together with J copies of the depot. The edge set E_1 of the graph consists of demand and non-demand edges. For each edge $e(i, j)$ of the original CARP network, a single demand edge joins a single copy of node i with a copy of node j and is given an infinite cost. Other copies of i and j are joined by edges of zero demand with cost $SP(i, j)$. The value of the lower bound is NDLB $= c(R) + MP(G_1)$.

It is easy to see the similarity of this procedure to the NDMPB1 bound. Essentially, the latter is more parsimonious in its duplication of nodes. In fact, Benavent, Campos, Corberan & Mota [1992] note that, in its basic form, NDLB coincides with NDMPB1. Saruwatari, Hirabayashi and Nishida independently prove the relation MPLB $\leq$ NDLB and also give a modified version of NDLB that uses a 'prohibiting rule'. The effect of this rule is to prohibit matching edges that join two demand edges for which the sum of the demands exceeds Q. In the computational results of these authors, this additional constraint improves the lower bound considerably. For example, on 10 test problems where $|R|$ equalled 20, the authors found the ratio $100($NDLB$/$MPLB $- 1)$ to average 18% for the basic NDLB but approximately 50% when the prohibiting rule is applied. The ratio of the modified bound to the basic NDLB averages 1.27 for the same problems. On the other hand, for the test cases in Benavent, Campos, Corberan & Mota [1992], the demands were not large enough for the prohibiting rule to apply.

5. Algorithms for capacitated arc routing problems

In the early applications of arc routing, researchers handled vehicle capacity or route length constraints on a problem-specific basis. A number of practice-oriented reports describe ad hoc methods for obtaining capacity-feasible tours in the context of specific applications (see Male [1973], Beltrami & Bodin [1974], Male & Liebman [1978] or Stern & Dror [1979] for examples). The work of Christofides [1973], which introduced the construct–strike heuristic for the CARP, initiated the design and evaluation of heuristics for the *generic* capacitated problem. In this section, we describe the heuristics introduced for the general CARP, summarize the computational experience reported for these procedures, and briefly mention the work on developing exact algorithms for the CARP.

5.1. Heuristic algorithms for the CARP

This section is devoted to heuristics developed for the CARP and the capacitated postman problem (CPP). To facilitate the description of the heuristics, we use the following notation throughout this section.

$G = (V, E)$ = undirected graph with node 1 reserved for the depot,

(c_{ij}, q_{ji}) = cost and demand of edge $e(i, j) \in E$,

Q = the vehicle capacity,

R = the set of edges that require service ($q_{ij} > 0$),

RE = the set of remaining (uncovered) edges that still require service at any intermediate point of a procedure.

RE is always initialized to R and is updated as edges receive service.

The path-scanning algorithm [Golden, DeArmon & Baker, 1983]

Step 1. Form a cycle C by sequentially adding the edge in RE chosen according to the edge selection criterion (see below). Stop when the vehicle capacity is exhausted and return to the depot via the shortest path.

Step 2. Remove C from RE and STOP if RE is empty. Otherwise, return to Step 1.

The main appeal of this heuristic is its simple structure. For the CPP, Golden and coworkers proposed five different edge selection criteria: If i is the current last node of the partial cycle C, extend C with the next edge (i, j) that fits within capacity and minimizes (or maximizes) $c_{ij}/$(remaining demand); minimizes (or maximizes) $SP(j, 1)$; maximizes $SP(j, 1)$ if the vehicle is less than half-full and minimizes this quantity otherwise. The patch-scanning procedure generates a complete solution with each of the five criteria and selects the one with the lowest cost. As an alternative, Pearn [1989] proposes that the edge selection criterion be chosen randomly at each step following a given probability distribution. In this way, p_i, the expected fraction of edges selected by criterion i is an input parameter for $i = 1, \ldots, 5$. Instead of five complete solutions, Pearn generates k solutions and picks the best. We call this procedure the *random path-scanning algorithm*.

The augment–merge algorithm [Golden & Wong, 1981; Golden, DeArmon & Baker, 1983]

Step 1. Construct a separate cycle to serve each edge in R.

Step 2. *Augment*: Starting with the longest cycle, extend service to the edges on the cycle as long as vehicle capacity permits. Discard the cycles that served these edges.

Step 3. *Merge*: In descending order of the savings produced, merge pairs of cycles subject to the capacity constraint. Continue until no further positive savings remain.

The augment–merge algorithm is clearly analogous to the Clarke–Wright heuristic for vehicle routing. Golden and coworkers discuss a number of different implementations of this basic procedure, varying the number of solutions generated and the form of the savings term. Pearn's modification of this procedure is called the *augment–insert algorithm*. This version, outlined below, attempts to capture some of the advantages of the Parallel Insert Algorithm, and is designed to perform well on relatively sparse graphs with large edge demands (relative to Q).

The augment–insert algorithm [Pearn, 1991]

Step 1. *Initialize*: Set $RE = R$ and construct a list L of edges $e(i, j)$ in RE in descending order of $SP(1, i) + SP(j, 1)$. Find the least-cost cycle $C(i, j)$ that covers edge $e(i, j)$.

Step 2. *Augment*: Select the first edge $e(i, j)$ on L that resides in RE and has an associated cycle $C(i, j)$. If no such edge is found, go to Step 3. Augment $C(i, j)$ by assigning edges e' in RE following L until the vehicle capacity is exhausted. Complete the cycle by joining its endpoints to the depot via shortest paths. Remove the serviced edges from RE. Repeat this step as long as RE is not empty.

Step 3. *Insert*: Select the cycle $C(i, j)$ corresponding to the first edge $e(i, j)$ on L that remains in RE. Following L, and as long as the insertion cost does not exceed CLIM, successively insert edges $e(k, \ell)$ in RE into $C(i, j)$ until the vehicle capacity is exhausted or the depot is encountered. Complete the cycle and update RE as in Step 2. Repeat this step until RE is empty.

Pearn [1991] proposes two versions of this algorithm: the above is Version I. Version II differs from the above by using the least-demand cycle $CQ(i, j)$ in lieu of $C(i, j)$ throughout. CLIM is an input parameter that controls the length of the cycles. A large value of CLIM creates longer cycles with substantial detours. Note that the priority list L favors assigning far-away edges [those with large $SP(1, i) + SP(1, j)$].

The next procedure is motivated by the matching algorithm for the classical postman problem: successive cycles are identified for removal and struck from the list of required edges. Periodically, non-required edges are added to the network to ensure the existence of new cycles until all required edges are covered by cycles.

The construct–strike algorithm [Christofides, 1973]

Step 1. *Construct*: By sequentially adding edges, form a capacity-feasible cycle C such that $RE - C$ remains connected.

Strike: Remove the cycle C and set $RE \leftarrow RE - C$. Repeat this step until RE is empty or no more cycles are found.

Step 2. Remove any non-required edges added by Step 3 in a previous iteration.

Step 3. Solve the matching problem on nodes of odd degree and two copies of the depot. Add the non-required paths identified by the matching solution to obtain an Eulerian graph with positive even degree for the depot. Return to Step 1.

While the original paper of Christofides [1973] did not specify an edge selection rule for Step 1, any of the rules discussed for the path scanning heuristic can be used. Pearn [1989] selects the edge $e(i, j)$ to maximize the total demand along the path from j back to the depot with the smallest total demand (thus, j solves a max–min problem). Pearn suggests that the restriction of maintaining connectivity in Step 1 is not necessary and gives examples where the solution value generated by construct–strike exceeds the optimum by 50% or more. His *modified construct–strike* procedure uses the edge selection rule just described in Step 1 and continues to remove cycles until RE becomes disconnected. When this occurs, a connect-and-balance heuristic (see Section 3) is applied to the subgraph induced by RE to produce an Eulerian graph. Step 1 is repeated if feasible cycles exist in this graph. Otherwise, an arbitrary cycle of demand edges is formed until the vehicle capacity is exhausted. This cycle is removed and the preceding steps are repeated until RE becomes empty.

The parallel insert algorithm [Chapleau, Ferland, Lapalme & Rousseau, 1984]

   Step 0. *Initialize*: Create one route containing the furthest edge from the depot.
   Step 1. *Insert I*: Select the furthest edge e from RE. Insert e into the route with the least insertion cost among the existing routes with sufficient capacity to absorb e. If no route with sufficient capacity exists, create a new route as long as the number of routes remains no longer than NRL. Stop if RE becomes empty at any point.
   Step 2. *Insert II*: Select the route with the lowest load. For this route, find the edge e in RE with the minimum insertion cost subject to the capacity constraint. Declare a route 'closed' if no further edges can be placed on the route without violating capacity. Continue as long as RE is non-empty and not all routes are closed.
   Step 3. *Exchange*: Consider exchanges that reduce the total length of routes. Return to Step 1 to perform new insertions if feasible.
   Step 4. *Relaxation*: Relax the capacity and route length constraints by a fixed percentage. Return to Step 3.
   Step 5. STOP if the current solution is no better than the incumbent. Otherwise, save the current solution as the incumbent. Eliminate the route with the lowest load and move the edges it covered back to RE. Return to Step 3.

   Clearly, the parallel insert algorithm is inspired by the insertion heuristics used for the traveling salesman and other node routing problems (see Frederickson, Hecht & Kim [1978] in particular). Steps 1 and 2 use two different insertion strategies, labeled I and II, that differ in the order of determining the edge to be inserted and the candidate route for insertion. Starting with a single route formed in Step 0, the algorithm forms new routes in Step 1 but the total number of routes in this step cannot exceed NRL — a pre-specified lower bound on the number of routes. Step 2 assigns additional edges to fill out these routes, always favoring the least-filled route to achieve balance of the routes's loads. Chapleau, Ferland, Lapalme & Rousseau [1984] state that based on their computational experience, Step 1 allocates 60–70% of the total demand during the first pass, and that this percentage increases to 85–90% during Step 2. The remaining edges are assigned in the other steps or during repeat visits to Steps 1 and 2. The authors give additional details on the interaction of Steps 3–5 with the insertion steps (1 and 2).
   By relaxing the capacity and the limit on route length, Step 4 allows overfilling of certain routes. Subsequent exchanges attempt to recover feasibility. Step 5 drops

the route with the lowest utilization and can be expected to result in a reduction of about 5% in the distance traveled. Additional constraints may be checked in the insertion steps: Chapleau et al. check *both* vehicle capacity and route length limits. Moreover, a control parameter limits insertions to those between the last edge on the route and the final destination once the route length goes beyond a pre-specified threshold. This rule eliminates undue zigzagging since only short detours are allowed in the final segment of the vehicle's approach towards its destination (when the threshold is crossed).

The next heuristic, due to Benavent, Campos, Corberan & Mota [1990], uses a 'route-first–cluster-second' strategy by assigning clusters of edges to the available vehicles. Each cluster is capacity feasible: the total demand of its edges is no larger than Q. An interesting feature of the procedure is that while the clusters are first obtained heuristically, the final assignment of edge demands to individual vehicles is determined by solving the generalized assignment problem (GAP) as in the algorithm of Fisher & Jaikumar [1981] for the standard vehicle (node) routing problem. We call this heuristic the *cycle assignment algorithm*.

The cycle assignment algorithm [Benavent, Campos, Corberan & Mota, 1990]

INPUT: K = the number of available vehicles

Step 1. Locate K geographically dispersed nodes of V_R to serve as seed points $i_1, \ldots, i_k$ for the K vehicles. Declare all vehicles *open*.

Step 2. Use the connect-and-balance procedure to augment G_R into an Eulerian graph $\tilde{G}$. Let $\tilde{G}_k$ be the subgraph of nodes and edges assigned to vehicle k and initialize $\tilde{G}_k = \{i_k\}$.

Step 3. *Assign cycles*: Select the open vehicle with the largest remaining capacity. Using a shortest path algorithm, find the minimum load cycle C containing any vertex of $\tilde{G}_k$. If no such cycle exists or if its load exceeds the available capacity, declare the vehicle *closed*. Otherwise, add the edges of C to $\tilde{G}_k$ and remove them from $\tilde{G}$. Reset $RE \leftarrow RE - C$. Repeat this step while open vehicles remain and RE is non-empty.

Step 4. *Improve clusters*: Perform exchanges of edges and paths between any $\tilde{G}_k$ and $\tilde{G}$ that reduce the total demand of the unassigned required edges in $\tilde{G}$.

Step 5. GAP: Solve the *generalized assignment problem* (GAP) to determine the assignment of each edge e of R to exactly one of the K vehicles. Let R_k be the set of edges assigned to vehicle k in the GAP solution.

Step 6. *Route generation*: For each k, augment R_k to R_k^1 by adding an artificial edge joining two copies of the depot. If R_k^1 is connected, solve the undirected postman problem on R_k^1 to obtain the route for vehicle k. Otherwise, use the connect-and-balance heuristic to solve the rural postman problem on G with $R = R_k^1$.

We conclude this description with some brief comments. In Step 1, the criterion for widely dispersing seed points over the graph is to maximize the product of the distances among the seedpoints and the depot. In particular, if $i_0 = 1$ denotes the depot and seeds $i_1, \ldots, i_k$ have already been selected, seed i_{k+1} is chosen to maximize $\prod_{h=0}^{k} SP(i, i_h)$ over all nodes i that are incident to some edge in R.

While Steps 3 and 4 assign edges to the vehicle clusters, some required edges may remain unassigned after Step 4. In Step 5, however, all required edges are assigned. The GAP decision variables y_{ek} are defined as 1 if edge e is assigned to vehicle k and 0 otherwise. It is important to note that the information about the clustering in Steps 3 and 4 passes to Step 5 through the costs c_{ek} attached to the variables y_{ek} in the GAP objective: The cost c_{ek} equals 0 for all edges already assigned to cluster k at the end of Step 4. For unassigned edges, c_{ek} is set equal to the insertion cost of e into the cycle containing the depot and seed i_k. Clearly, the costs set to zero provide an incentive for the GAP to retain the clusters formed in Steps 3 and 4. Finally, since the cycles assigned to cluster k in Step 3 do not necessarily go through the depot, Step 6 creates an artificial edge that forces the vehicle tour to go through the depot.

5.2. Computational experience with CARP heuristics

In this section, we review the available computational experience with CARP heuristics, using the acronyms shown below.

CS	= construct–strike	MCS	= modified CS
PS	= path-scanning	RPS	= random path-scanning
AM	= augment–merge	AI1, AI2	= augment–insert I, II
CA	= cycle assignment	PI	= parallel insert

Before describing the computational experience, it is worthwhile to consider the computational complexity of these heuristics. Let n and m respectively denote the number of nodes and edges in the CARP network. Heuristics PS, RPS, AM, AI1 and AI2 do not use a matching procedure; their complexity is driven by the all-pairs shortest path computations. As noted by Pearn [1991], a routine implementation of these heuristics has $O(n^3)$ complexity. On the other hand, the main burden of procedures such as CS and MCS lies in solving the matching problem: CS has complexity $O(mn^3)$ and for MCS, the complexity increases to $O(mn^4)$.

The systematic testing of CARP heuristics starts with the work of Christofides [1973] who tested the CS heuristic on five test problems with $(n, m) = (n, 2.5n)$, where the number of nodes n ranged from 10 to 50. In these problems, all edge demands q_{ij} were positive and drawn from the uniform distribution on [10,30] and Q equalled 100. Christofides found that the CS solutions were 3.6 to 9.0% above the lower bound provided by the uncapacitated postman solution (UPP). In some preliminary tests, Golden & Wong [1981] ran the AM heuristic on CPP instances where (n, m) ranged from (11,19) to (13,26). On six of the seven test problems, the AM solution was 13 to 18% over the matching lower bound (MLB) value.

Table 3 combines and summarizes the computational work reported by Golden, DeArmon & Baker [1983] and Pearn [1989, 1991]. Four sets of test problems appear in this table. The density of the underlying graph is an important char-

Table 3

Characteristics of test problems and performance of heuristics listed
as percentage over the lower bound

		GDB	P89A	P89B	P91
Test beds					
Number of problems		23	20	15	30
Number of nodes (n)		7–27	11–17	11–17	13–27
Number of edges (m)		11–55	55–136	45–116	23–51
Density (%)		13–100	100	70–90	15–30
Algorithms (% over LB)					
Construct–strike	CS	17.91	2.29	3.85	71.95
Path-scanning	PS	11.03	3.13	3.59	57.97
Augment–merge	AM	9.16	–	–	56.78
Random path scan	RPS	8.05	2.76	3.16	51.30
Modified CS	MCS	7.59	0.04	0.70	57.32
Augment–insert I	AI1	12.39	–	–	44.17
Augment–insert II	AI2	13.95	–	–	44.97

acteristic of the test problems. In Table 3, this density is measured by the ratio
$m/[n(n-1)/2]$, expressed as a percentage. This measure compares the number of
edges of each test problem with the corresponding number in a complete network
on the same number of nodes. The performance measure tabulated is the average
percentage by which the heuristic solution value v_H exceeds the lower bound *LB*.
The value of the lower bound is the larger of the two bounds MLB and NSLB
described in the preceding section.

The first column of Table 3 focuses on the test bed GDB. This set consists of
23 of the 25 problems tested by Golden, DeArmon & Baker [1983], a complete
description of which is available in DeArmon [1981]. In their tests, Golden and
coworkers found that the heuristics performed better on complete networks with
arc demands that remain small compared to Q. The procedures CS and PS had
roughly comparable running times, while AM took 6 to 80 times longer. It should
be noted that each application of the AM chose the best of the solutions obtained
by 24 runs with different parameter settings and rules. Running AM, PS, and CS
on their 25 test problems, Golden and coworkers found that these procedures find
the best value (among the three) 16, 12 and 5 times respectively.

Pearn [1989] proposed modifications to PS and CS that improved their performance on dense (nearly complete) networks with relatively small arc demands and
constructed two problem sets — P89A and P89B — with these characteristics. The
very small average percentages reported for MCS in the corresponding columns
of Table 3 reflect the superior performance of MCS: it obtained the best solution
(among the four algorithms tested) in all of the 35 problems in P89A and P89B.
In fact, the MCS solution equalled the lower bound in 19 of the 20 problems
of P89A. Note that the lower bounds are extremely tight for this group of dense
problems. Pearn [1989] states that the running time of MCS is, on the average,

3 times that of the original CS. However, this factor is quite variable and can be as large as 25 times on some problems. In summary, Table 3 shows that RPS and MCS do improve upon their counterparts (PS and CS) and that this improvement is generally greater for MCS.

Test bed P91 contains sparse problems with large ratios of q_{ij}/Q. Pearn [1991] constructed this set of 30 problems to test AI1- and AI2-heuristics he had developed specifically for sparse problems with large demands on which the earlier heuristics performed poorly. As Table 3 reveals, AI1 and AI2 outperform the other heuristics. Pearn reports the solutions for different values of the parameter CLIM. However, even for a single value of CLIM (5), AI1 and AI2 together find the best solution (among all seven heuristics) in 25 of the 30 problems. Moreover, AI1 and AI2 are 10 to 20 times faster than AM, CS or MCS. Pearn [1991] also remarks that the running time is very sensitive to the vehicle capacity Q and that this effect is more pronounced when arc demands are large relative to Q. As Q decreases, each vehicle can serve only a few required arcs and many cycles have to be generated.

The computational results of Table 3 show the marked difference between dense and sparse CARP graphs. The MCS heuristic and the lower bound both perform very well on dense problems but this performance deteriorates substantially on sparse graphs with large arc demands.

Benavent, Campos, Corberan & Mota [1990] have tested the cycle assignment (CA) algorithm they designed on test problems where (n, m) ranges from (24,39) to (50,97) — corresponding to densities of 7.5 to 14%. In these test problems, all edges have positive demands generated between 1 to 15 and the vehicle capacity is generally between 120 and 250. On a test bed of 24 problems, the CA heuristic produced solutions that were 6.4% above the lower bound on the average, although this gap exceeded 8% in 7 of the 24 problems.

An independent computational test of CARP heuristics has been carried out by Coutinho-Rodrigues, Rodrigues & Climaco [1991; see also Rodrigues, 1990]. Several heuristics, including AM, RPS, MCS and PI, are tested on the 23 problems of GDB. The overall trends appear to reproduce the results of Pearn [1989], but there are discrepancies in the solution values of specific problems. Surprisingly, a variation of the PS approach (in which the next edge is selected randomly) gives the best results (best solution value in 19 of the 23 problems). This computational testing formed the initial step in the implementation of a solid waste collection (sanitation) routing procedure. Data was collected for 5 cities with (n, m) ranging from (97,147) to (705,1022) and the authors report average savings of 49% in the total distance traveled.

Other computational experience with CARP heuristics in the context of specific applications appears in the work of Roy & Rousseau [1989], Chapleau, Ferland, Lapalme & Rousseau [1984] and Dror, Leung & Mullaseril [1993]. The latter use various CARP heuristics to generate capacitated cycles as an intermediate step of their routing problem.

Another variant of the CARP, called the *fleet size and mix problem for capacitated arc routing*, is addressed by Ulusoy [1985]. In this problem, fixed costs are attached to various types of vehicles that can be used in the fleet and the objective is to minimize the sum of fixed vehicle and variable routing costs. As such, the problem is the directed arc routing counterpart to the node routing problem solved by Golden, Assad, Levy & Gheysens [1984]. Ulusoy's approach is a route-first–cluster-second procedure that partitions a covering Euler tour through *all* required edges into feasible vehicle tours by using a shortest path problem in a manner very similar to the decomposition approach of Golden, Assad, Levy & Gheysens [1984]. To allow the algorithm wider choice, Ulusoy runs the decomposition procedure on several Euler tours and selects the best solution. He reports running times on 10 test problems with (n, m) ranging from (10,18) to (30,128). However, no solution values or measures of solution quality are reported in this study.

5.3. Exact algorithms for the CARP

Researchers have only recently turned to the design of *exact* algorithms for the CARP. In this section, we sketch two different approaches for the optimal solution of the CARP. While space considerations preclude a detailed description of the algorithms, we review the computational results reported for these algorithms.

The first body of research involves two branch-and-bound algorithms that capitalize on the *node duplication lower bound* (NDLB) of Saruwatari, Hirabayashi & Nishida [1992a], which we briefly described in Section 4.3. The *tour construction algorithm* (TCA) is described in detail by Hirabayashi, Saruwatari & Nishida [1992] and a brief account of the *subtour elimination algorithm* (SEA) appears in Saruwatari, Hirabayashi & Nishida [1992b]. In both approaches, a node of the branch-and-bound search tree corresponds to a partial solution where certain edges are fixed into, or out of the candidate solution. The NDLB procedure is applied to the graph with these restrictions to obtain a lower bound on the cost of any completion of the partial solution. In TCA, the branching rule selects, from all the free edges in the NDLB matching solution, the edge whose exclusion provides the highest lower bound value. This edge is then fixed in or excluded in the two descendants of the current node. In the SEA branch-and-bound tree, however, each node generates multiple descendants (or subproblems) as follows. A property of the matching graph constructed for NDLB is that any matching solution produces alternating paths $(e_1, e_2, \ldots, e_{2p+1})$, where all edges with odd indices are not required. If this path violates the vehicle capacity, $p + 1$ descendant nodes are formed in which the rth node excludes the edge e_{2r+1} from the solution but fixes the 'preceding' non-required edges $\{e_1, \ldots, e_{2r-1}\}$ into the partial solution ($0 \leq r \leq p$).

TCA and SEA share certain routines. For example, the estimated number of required vehicles is re-computed as successive nodes of the search tree further constrain the CARP solution. This is important since NDLB requires the number of vehicles as input and the initial estimate $\lceil \sum q_{ij}/Q \rceil$ of the fleet size may be far from tight. There are other routines to reduce the size of the matching problem and to prohibit the assignment of two edges to the same cycle if their combined demands exceed Q.

Hirabayashi, Saruwatari & Nishida [1992] observe that the computational bur-
den of CARP depends on $|R|$ — the number of required edges. They report
computational results for the TCA on graphs with randomly generated arc costs
and demands where $|R|$ ranges from 15 to 50. The authors observe that the
number of nodes in the search tree remains relatively flat as $|R|$ increases beyond
25 and that the run times do not grow very rapidly: the average run time for
problems with $|R| = 50$ was five times that of $|R| = 25$. Saruwatari, Hirabayashi
& Nishida [1992a] provide computational experience for both the SEA and the
lower bounding procedure NDLB on test problems with $|R| = 10, 15$ and 20 (ten
of each). The number of nodes in the search tree exceeded 3500 in two of the
ten problems with $|R| = 20$. Across the twenty problems with $|R| = 15$ or 20, the
percentage gap above the lower bound — $100[(OPT/LB - 1)]$ — averages 12.6%
but this value is highly variable: it is no larger than 8% in 11 of the 20 cases but
exceeds 15% in 5 cases.

The other exact approach to the CARP reported in the literature is due to
Belenguer & Benavent [1992]. This work uses an integer programming formulation for the undirected version of CARP with K vehicles that is similar to
the one in Section 4.1, with additional surrogate constraints for evenness. They
develop constraints to describe the facets of the polyhedron associated with
the CARP, particularly in the case of unit edge demands. The case of $q_{ij} = 1$
avoids the demand-related facets that can be derived from the generalized assignment problem formulation. Their preliminary computational experience with the
branch-and-cut algorithm based on this formulation involves two test problems
with $(n, m, m_r) = (16, 26, 20)$ and $(24,34,34)$. Based on the research cited in this
section, we conclude that the area of exact solutions for the CARP is still largely
unexplored.

6. Variants of postman problems

The preceding two sections were devoted to arc routing models with capacity constraints. In this section, we focus on three variants of the uncapacitated
postman problem that introduce additional considerations. The first model, called
the windy postman problem, allows each edge to be traversed in either direction, but with different costs. The second problem introduces a precedence
constraint on the order in which the edges of a network are serviced. In the
third model, the benefit resulting from servicing a subset of the edges must be
maximized.

6.1. The windy postman problem

Minieka [1979] introduced a class of postman problems characterized by
asymmetric costs of traversing undirected edges in a network. In traversing an
undirected edge, a postman may encounter different costs depending upon the
direction of the traversal (with or against the wind). More precisely, given an

undirected graph $G = (V, E)$ and a pair of costs c_{ij} and c_{ji} for each edge e between nodes i and j, the *windy postman problem* (WPP) is to find a postman tour of minimum cost where the cost of edge e in the tour is recorded as c_{ij} if traversed from i to j, and c_{ji} otherwise.

Guan [1984] established that this problem is *NP*-hard by reducing the mixed postman problem to the WPP. The construction is simple and involves adding a counterpart arc (j, i) with a large arc cost M for every directed edge of the mixed network. The optimal WPP solution then solves the mixed problem by avoiding all the added counterpart arcs. Given the complexity of the WPP, it is natural to ask if there are polynomially solvable special cases. Guan identified one solvable class of the WPP that relies on the following condition: The cost of any cycle in G does not change if the direction of traversing the cycle is reversed. In the presence of this condition, the WPP is simply solved by solving the undirected postman problem for G with edge cost $1/2(c_{ij} + c_{ji})$ for the undirected edge between i and j

The other polynomially solvable case of the WPP occurs when the underlying network is Eulerian (connected and even). Win [1989] provided the following algorithm that uses ideas similar to the Edmonds and Johnson postman procedure for mixed even graphs.

The windy postman algorithm for connected even graphs

Given: $G = (V, E)$ an undirected, connected even graph; a pair of costs (c_{ij}, c_{ji}) for each edge $e(i, j)$ in E.

Step 1. Convert G into a directed graph $G_0 = (V, A_0)$ by orienting each edge along the direction that yields the smaller of the edge costs c_{ij} and c_{ji}. Let $b(i) = d^-(i) - d^+(i)$ for all nodes i in G_0.

Step 2. Construct a directed graph $G_1 = (V, A_1)$ with edge set $A_1 = A_0 \cup A_r \cup F$ where A_r and F each have a copy of the arc (j, i), for each $(i, j) \in A_0$, with costs c_{ji} and $1/2(c_{ji} - c_{ij})$ respectively.

Step 3. Assign infinite capacity to all edges in $A_0 \cup A_r$ and a capacity of 2 to each edge in F. Find a minimum cost flow in G_1 with supply (or demand) $b(i)$ attached to each node $i \in V$. Let y^* be the resulting optimal flows.

Step 4. Construct $G_2 = (V, A_2)$ by placing $y^*_{ij} + 1$ copies of (i, j) in A_2 if $y^*_{ji} = 0$ for $(j, i) \in F$, and $y^*_{ji} + 1$ copies of (j, i) if $y^*_{ji} = 2$. G_2 is Eulerian.

Step 5. Identify an Euler circuit in G_2. This is the optimal windy postman tour.

The similarity of this procedure to the mixed postman algorithm for even graphs is apparent. In proving the optimality of the WPP algorithm, Win [1989] shows that the flows y^* of Step 3 are even and that y^*_{ji} is either 0 or 2 for any edge (j, i) in F. Moreover arcs (i, j) and (j, i) do not simultaneously carry flow in the optimal solution. Given these facts, the rules described in Step 4 cover all possible cases.

Win also provides an interesting description of the WPP polyhedron. Let the integer variable x_{ij} count the number of times edge (i, j) is traversed from i and j in the windy postman tour. Then each tour corresponds to a vector $(x_{ij}, x_{ji})_{e(i,j) \in E}$

and the WPP formulation is

$$\text{Min} \sum_{e(i,j)\in E} (c_{ij}x_{ij} + c_{ji}x_{ji}) \tag{6.1.1}$$

subject to

$$x_{ij} + x_{ji} \geq 1 \qquad \text{for all } e(i,j) \in E \tag{6.1.2}$$

$$\sum_{j\in N(i)} (x_{ij} - x_{ji}) = 0 \quad \text{for all } i \in V \tag{6.1.3}$$

$$x_{ij}, x_{ji} \geq 0 \tag{6.1.4}$$

$$x_{ij}, x_{ji} \text{ integer.} \tag{6.1.5}$$

Here, $N(i)$ is the set of nodes j joined to node i via some edge e in E. Let P_1 be the polyhedron defined by (6.1.1)–(6.1.4). Win shows that P_1 coincides with the polyhedron defined by the convex hull of all incidence vectors $(x_{ij}, x_{ji})_{e(i,j)\in E}$ corresponding to windy postman tours in G. For a general graph G, the vertices of P_1 are vectors whose components equal 0, 1/2 or a positive integer. If G is Eulerian, then every vertex of P_1 is integral while for a non-Eulerian graph G, P_1 possesses at least one fractional vertex. Therefore, the following three conditions are equivalent: (i) G is Eulerian, (ii) P_1 is an integer polyhedron, and (iii) P_1 coincides with the polyhedron of WPP tours of G.

The preceding algorithm can also be used to obtain a heuristic solution procedure for the WPP on a general connected, undirected graph G: Apply the matching-based algorithm for the undirected postman problem on G with edge costs $c_e = 1/2(c_{ij} + c_{ji})$ for all $e(i,j)$ in E. This procedure identifies the edges to be added to G to obtain an Eulerian graph $\tilde{G}$. Next, solve the WPP on $\tilde{G}$ using the WPP for Eulerian graphs. While the second step is exact, the procedure is approximate due to 'averaging' of arc costs $c_{ij} + c_{ji}$ in the balancing step. However, Win proves that this approximate procedure has a worst-case bound of 2. Grötschel and Win [1992] have conducted a comprehensive computational study of the WPP using a cutting plane algorithm. This algorithm uses a new class of constraints these authors call the *odd cut inequalities*: Let W be any subset of V and let $\delta(W)$ denote the set of edges in the cut $(W, \bar{W}) = \{e(i,j) \mid i \in W, j \in \bar{W}\}$. For any odd cut $\delta(W)$, the following inequality defines a facet of the WPP if the subgraphs of G induced by W and $\bar{W}$ are connected:

$$\sum_{e(i,j)\in\delta(W)} (x_{ij} + x_{ji}) \geq |\delta(W)| + 1 \text{ for } W \subset V \text{ with odd } |\delta(W)|. \tag{6.1.6}$$

Let $WP(G)$ denote the WPP polyhedron defined by (6.1.2)–(6.1.5). As before, P_1 is the polyhedron associated with (6.1.2)–(6.1.4) and P_2 is the polyhedron defined by (6.1.2)–(6.1.4) plus (6.1.6). Then $WP(G) \subseteq P_2 \subseteq P_1$. Grötschel and Win solve the WPP by solving the linear program that minimizes (6.1.1) over P_2. They use test problems where n ranges from 52 to 264 and m from 78 to 489. The algorithm produces optimal results in 31 of the 36 test problems, thereby showing that P_2 provides a good relaxation of $WP(G)$. We note that these test problems also included instances of the mixed postman problem solved as a WPP.

Another close relative of the Eulerian WPP is the *minimum cost Eulerian orientation problem* (MCEOP): Given an undirected Eulerian graph $G = (V, E)$ and a pair of edge costs (c_{ij}, c_{ji}) for each $e(i, j) \in E$, orient all edges of G so that the resulting directed graph $G = (V, A)$ has an Eulerian tour of minimum cost. Stated otherwise, one seeks a windy postman tour of minimum cost subject to the additional condition that each edge be traversed exactly once. Guan & Pulleyblank [1985] showed that this problem can be solved as a minimum cost circulation problem. They also investigated the relation between various possible Eulerian orientations of a graph G by constructing a graph where the nodes correspond to orientations of G and adjacency of two nodes is defined by the transformation of one orientation into another through the reversal of the orientations of arcs lying on a simple directed cycle.

Win [1989] shows that the WPP algorithm for Eulerian graphs can be easily modified to solve the MCEOP as well. Moreover, the incidence vectors $(x_{ij}, x_{ji})_{e(i,j)\in E}$ corresponding to possible Eulerian orientations of G are 0–1 vectors that define a polyhedron described by constraints (6.1.2)–(6.1.4) with equality used in (6.1.2).

6.2. Postman problems with arc priorities

In classical postman problems, the required edges of the network can be traversed and serviced in any order. In some arc routing applications, however, a subset of the edges may have a higher priority and must be serviced before the other edges. The prominent example of this is snow plowing where the main streets must be cleared before the secondary roadways (see Section 9). In Section 10.1, we describe a manufacturing problem where arc traversal corresponds to cutting a sheet with a torch flame [Manber & Israni, 1984]. In this problem, the stability of the sheet requires that 'inner' edges of the network be traversed (cut) first.

One of the earliest algorithms for tour construction in the presence of priorities appears in Lemieux & Campagna [1984]. These authors considered a snow plowing problem in which a subset of the streets is given higher priority (priority = 1) to indicate preference for servicing these street segments first. Because each street segment must be plowed on both sides, the resulting travel network is a directed network with counterpart arcs in opposite directions between intersection nodes. Since this directed network is already Eulerian, deadheading is not an issue. Instead, the chief objective of Lemieux and Campagna is to identify an Euler cycle that visits all edges of priority 1 first. Their heuristic algorithm for constructing this cycle has two versions. In one version, the heuristic attempts to strictly observe the priorities, even if this means that a segment just covered in one direction is immediately traversed in the opposite direction (implying a U-turn). The second version relaxes this: priorities are observed unless an immediate return is involved. In this way, the algorithm can heuristically trade off the importance of observing priorities against the undesirability of U-turns. Both versions of the algorithm may violate the priority structure. For example, a violation occurs if selecting the higher priority arc out of a node causes connectivity problems in the construction of the Euler tour. If the arcs are deleted when traversed by the Euler cycle, this infeasibility corresponds to making the remaining subgraph disconnected. In any case, there is no guarantee that the algorithm will identify a sequence of traversals that obeys priorities in full without introducing any deadheading.

In contrast to the heuristic approach of Lemieux and Campagna, the postman problem with precedence relations on arcs is systematically analyzed by Dror, Stern & Trudeau [1987]. They consider a partition of the edges of the graph $G = (V, E)$ into k priority classes $E_1, \ldots, E_k$ and impose the constraint that any edge $e \in E_i$ must be traversed for the first time before any of the edges in $E_{i+1}, \ldots, E_k$. The objective is to find an optimal postman tour that traverses all edges of E subject to the precedence relations defined by the E_i's. The existence of such a tour in the case of an undirected graph G is established by the following.

Theorem 6.1 [Dror, Stern & Trudeau, 1987]. *Given an undirected graph $G = (V, E)$ and precedence classes $E_i (1 \le i \le k)$, let F_i be the subgraph of G induced by the edges in $\bigcup_{h=1}^{i} E_h$. Then a feasible postman tour exists on G if and only if F_i is connected for all i. If G is a directed graph, a sufficient condition for the existence of a feasible tour is that F_i be strongly connected for all i.*

Dror, Stern and Trudeau provide an exact algorithm for finding the optimal tour in an undirected network when all the edge sets E_i are connected. The motivation for this procedure can be described by means of the example in Figure 8 where three precedence classes E_1, E_2 and E_3 are specified. Any feasible tour must start at node 1 (the depot in this example) and proceed to service the edges $e(1, 2)$, $e(2, 3)$ and $e(2, 4)$ that comprise E_1. To move on to the required edges E_2, the tour must end up at one of the two 'entry nodes' for E_2 — node 3 or 4 — from which the first edge in E_2 is serviced. In servicing E_2, the tour can use the edges of both E_1 and E_2 for deadheading. We can, therefore, consider this stage as a rural postman problem on F_2 with E_2 as the required edges. By fixing the entry node for each precedence class, the overall problem can be decomposed into subproblems corresponding to the individual precedence classes.

To state the preceding approach more formally, let $G_i = (V_i, E_i)$ be the subgraph induced by the edges in $E_i (V_i$ is defined as the set of nodes incident to E_i). Let $Y_i = (\bigcup_{h=1}^{i-1} V_h) \cap V_i$ be the set of possible entry nodes for G_i. In Figure 8, $Y_1 = \{1\}$, $Y_2 = \{3, 4\}$, $Y_3 = \{1, 2\}$. Let $P[E_i, u, v]$ denote the subproblem of finding the Euler path of minimum cost that starts at node u, covers all edges in E_i and ends at node v, using only edges in F_i. Let N_i be the number of nodes in F_i. Dror, Stern and Trudeau show that $P[E_i, u, v]$ can be solved in $O(N_i^3)$ steps when both E_i and F_i are connected. In this case, one only needs to solve a matching problem to pair the nodes of odd degree in the subgraph induced by $E_i \cup \{e(u, v)\}$. Thus, if $n = |V|$, the subproblem $P[E_i, u, v]$ requires no more than $O(n^3)$ steps for each i. With these observations, we can state the algorithm for the postman problem with priorities. Note that the algorithm selects the entry and exit nodes u and v for each precedence class considering the cost $C_i(u, v)$. The

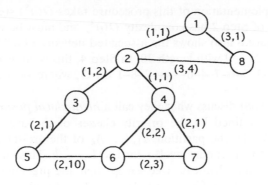

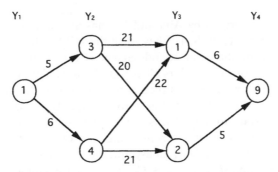

Fig. 8. A network with three precedence classes (top) and its associated graph G_p. The edge labels give the priority class followed by the edge cost (top).

network G_p of Step 1 is composed of 'bipartite' layers corresponding to various (u, v) pairs for each E_i.

Algorithm for postman problem with precedence classes

Given: Undirected graph $G = (V, E)$ with k precedence classes $E_1, \ldots, E_k$; each E_i is connected; depot node specified as node 1; $n = |V|$.

Step 1. Define the directed network $G_p = (V_p, A_p)$ where $V_p = \bigcup_{i=1}^{k+1} Y_i$, with $Y_1 = \{1\}$, $Y_i = (\bigcup_{h=1}^{i-1} V_h) \cap V_i$ for $1 < i < k$, and $Y_{k+1} = \{n + 1\}$, where $n + 1$ is a second copy of the depot node. Define the arc-set of G_p as $A_p = \{(u, v) \mid u \in Y_i, v \in Y_{i+1}$ for $i = 1, \ldots, k\}$.

Step 2. For each $(u, v) \in A$ solve the subproblem $P[E_i, u, v]$ and let $C_i(u, v)$ be the cost of the resulting tour from u to v. Assign arc (u, v) the cost $C_i(u, v)$.

Step 3. Using the arc costs $C_i(u, v)$ on G_p, find the shortest path from 1 to $n + 1$, which can be written as $v_1 = 1, v_2, \ldots, v_{k+1} = n + 1$, where $(v_i, v_{i+1}) \in A_p$ and $v_i \in Y_i$ for all i.

Step 4. Construct the optimal tour on G by concatenating the paths in G corresponding to each $C_i(v_i, v_{i+1})$ in the optimal path of Step 3.

A routine implementation of this procedure takes $O(n^5)$ steps as the subproblem $P[E_i, u, v]$ of Step 2 has complexity $O(n^3)$ and must be executed for $O(n^2)$ pairs (u, v). Figure 8 also shows the associated network G_p. The shortest path in G_p is $1 \rightarrow 3 \rightarrow 2 \rightarrow 9$ with length 30. In Step 4, this is converted to the optimal tour 1–2–4--2–3–5–6–7–4–6--4--2–8–1 on G, where -- denotes a deadhead edge in the tour.

Alfa & Liu [1988] discuss what they call a *hierarchical postman problem* where the hierarchy is defined by the priority classes of the arcs: Given a directed network $G = (V, A)$, the partition $A_1, \ldots, A_k$ of the arc-set A specifies all arcs of priority class (or hierarchy level) i for $i = 1, \ldots, k$. This structure is similar to what Dror, Stern and Trudeau used except that the precedence constraint Alfa and Liu employ is $ST_i < ST_{i+1}$ and $CT_i < CT_{i+1}$ for all i, where ST_i and CT_i are the start and completion times for servicing the entire class A_i. This is a weaker precedence relation than the strict precedence used by Dror, Stern and Trudeau.

The algorithm described by Alfa and Liu has a simple interpretation. Starting with the highest priority class A_1, they solve the RPP with required arcs $A_R = A_1$ on $G = (V, A)$ using the connect-and-balance (C&B) heuristic of Section 3. This results in an Euler tour that covers all arcs of A_1, as well as some arcs of the other priority classes. This procedure is then repeated with all the untreated arcs of the next priority class A_2, and continues through A_k. Alfa and Liu do not provide computational results or any discussion of the interaction among the various classes of arcs. We note that the connecting step of the C&B heuristic connects the components of G_R by solving an asymmetric traveling salesman problem using the costs $\tilde{c}_{pq} = \text{Min}\{SP(i, j) : i \in C_p, j \in C_q\}$ for connecting components p and q. This step can use arcs from other priority classes (including these of lower priority). In contrast, Dror, Stern & Trudeau [1987] limited the deadhead arcs to arcs already serviced so that a lower-priority arc was never traversed out of turn.

Finally, we point out that the Postman problem with general precedence relations, where the edge classes E_i follow a general partial ordering, is proved to be *NP*-complete in Dror, Stern & Trudeau [1987].

6.3. The maximum benefit postman problem

The previous section considered the postman problem in the presence of a priority structure on the order in which arcs of the network are traversed. A different scheme of introducing priorities uses costs to differentiate the utility of visiting a given arc of the network. In the *maximum benefit postman problem* (MBPP) introduced by Malandraki & Daskin [1993], a benefit is realized each time an arc is traversed and the objective is to find the tour of maximum *net* benefit. In contrast to the classical postman problems of Section 2, not every arc may be traversed in an optimal MBPP tour, reflecting the option of leaving out arcs of high net cost or arcs that are costly to access. Furthermore, the MBPP allows multiple traversals of the same arc while providing service. This corresponds to such applications as snow plowing, in which primary streets may receive service more than once. The formulation of MBPP, which is discussed next, makes the cost structure more explicit.

Let $G = (V, A)$ be a directed graph. Let c_{ij}^s and c_{ij}^d denote the costs of traversing arc (i, j) while providing service and in deadhead mode. We assume that $c_{ij}^s > c_{ij}^d$. For each arc (i, j), the input to MBPP specifies the *marginal benefit* of traversing the arc for the pth time. Denote this by b_{ij}^p for $p = 1, 2, \ldots$. It is natural to assume that repeated traversals of the same arc exhibit diminishing returns, so that b_{ij}^p is decreasing in p for $p \geq 1$. The *net cost* of traversing arc (i, j) for the pth time is equal to $c_{ij}^p = c_{ij}^s - b_{ij}^p$ for $p \geq 1$ and, for $p = 0$, we define $c_{ij}^0 = c_{ij}^d$. The decision variables in MBPP are

$$x_{ij}^p = \begin{cases} 1 & \text{if arc (i,j) is traversed for the } p\text{th time } (p \geq 1) \\ 0 & \text{otherwise} \end{cases}$$

$x_{ij}^0 = $ number of times link (i, j) is traversed in deadheading mode.

We can now represent MBPP as a network flow problem with additional constraints for subtour elimination by considering the preceding variables as flow variables. For each (i, j) in A, create $\bar{p}(i, j)$ parallel arcs between nodes i and j with costs c_{ij}^p for $p = 0, 1, \ldots, \bar{p}(i, j)$ where $\bar{p}(i, j) = \text{Max} \{p \mid c_{ij}^p < c_{ij}^d\}$. Let x_{ij}^p be the flow on the arc with cost c_{ij}^p and attach an upper bound of 1 to this arc for $p \geq 1$. The deadhead arcs have no upper bound and can be covered more than once. To ensure that the MBPP tour goes through the depot, split the depot node D into two nodes D and D' and connect these copies with an arc of zero cost with a lower bound of 1. The MBPP formulation is as follows:

(MBPP):

$$\text{Min} \sum_{p=0}^{P} \sum_{(i,j)\in A} c_{ij}^p x_{ij}^p \tag{6.2.1}$$

subject to

$$\sum_{p=0}^{P} \left(\sum_i x_{ij}^p - \sum_k x_{jk}^p \right) = 0 \quad \text{for all } j \tag{6.2.2}$$

$$0 \leq x_{ij}^p \leq 1 \quad \text{for all } (i, j), \ p \geq 1 \tag{6.2.3}$$

$$0 \leq x_{ij}^0 \tag{6.2.4}$$

$$1 \leq x_{DD'}^0 \tag{6.2.5}$$

$$x_{ij}^p \text{ integer} \quad \text{for all } (i, j) \text{ and all } p \tag{6.2.6}$$

$$(x_{ij}^p) \in S. \tag{6.2.7}$$

In this formulation, Equations (6.2.2) require flow conservation at each node j. The last set of constraints schematically represents the subtour elimination constraints that rule out subcycles disconnected from the depot. P is the largest value of $\bar{p}(i, j)$ encountered. $x_{DD'}^0$ denotes the flow on the arc joining D and D', for which $c_{DD'}^0 = 0$.

Malandraki and Daskin define a subtour as any connected subgraph of G that is disconnected from the depot node. Let C be the node set of a subtour ($C \subset V \sim \{D\}$) and define the binary variable w_c as follows:

$$w_C = \begin{cases} 1 & \text{if any node in } C \text{ is selected in the optimal solution} \\ 0 & \text{otherwise.} \end{cases}$$

Define the additional constraints

$$\sum_{(i,j)\in(C,C)} \sum_{p=0}^{P} x_{ij}^p \leq Bw_C \tag{6.3.1}$$

$$\sum_{(i,j)\in(C,\bar{C})} x_{ij}^1 \geq w_C \tag{6.3.2}$$

where B is a suitably large number. Note that if $w_C = 1$, (6.3.1) is not binding but (6.3.2) forces some arc of positive flow out of the subtour and since the arc corresponding to x_{ij}^1 has the most advantageous net cost (c_{ij}^1), this is the only variable included in (6.3.2). Thus (6.3.1) and (6.3.2) together form subtour elimination constraints which can be included in the MBPP formulation in lieu of (6.2.7).

To solve the MBPP, Malandraki and Daskin use the minimum cost flow formulation (6.2.1)–(6.2.6) within a branch-and-bound procedure that invokes constraints (6.3.1)–(6.3.2) when a subtour is encountered. More specifically, let C_0 be a subtour detected at some node of the branch-and-bound tree. Select a cut $(C, \bar{C})$ with $C_0 \subset C$ such that $(C, \bar{C})$ has the smallest number q of free arcs in the extended network $\bar{G}$. Create $q + 1$ descendant nodes in the search tree. In the first q descendants, each of the q free arcs is included in the solution so that (6.3.2) becomes binding with $w_C = 1$. Stated otherwise, each of the q free arcs leads out of the subtour in one of the descendent nodes.

The last descendant node corresponds to the choice $w_C = 0$ with (6.3.1) binding. In this case, none of the nodes in C_0 can be visited by the subsequent solutions. The authors report computational results for this branch-and-bound algorithm on a network with 25 nodes and 95–139 links after expansion. They claim that the minimum cost flow relaxation (6.2.1)–(6.2.6) is unlikely to produce subtours if the network represents a realistic road network where primary streets have positive net benefits.

7. Arc routing for sanitation services

7.1. Background and early work

Sanitation and municipal waste management is an important and costly component of public services. The increasing costs of waste collection and disposal, and the growing importance of recycling make sanitation a natural candidate for productivity improvement studies. While this section focuses on the routing component of waste collection, it is useful to give an indication of the magnitude of the problem, following a series of informative articles on New York City's sanitation operations by Riccio [1984], Riccio & Litke [1986], and Riccio, Miller & Litke [1986].

In 1986, New York City's Department of Sanitation had a budget of $500 million and 12,000 employees. Close to 7000 of these employees were sanitation workers. The city collected about 12,000 tons of household refuse daily and handled the disposal of another 10,000 tons collected by private carters. The department is also responsible for cleaning the streets, snow removal, and enforcing the sanitation code. The scope of the street cleaning activity alone is noteworthy: Each day, 4 to 5 million pieces of litter are deposited on the city's 6000 miles of streets. A fleet of over 350 mechanical sweepers is dispatched to clean the streets on a daily basis.

Street sweeping provides one of the earliest practical instances of an arc routing problem. A mechanical broom (or sweeper vehicle) traverses the streets of the city. When the broom is engaged, the vehicle is sweeping a street segment that requires service. At other times, the vehicle is deadheading (traveling but not sweeping). Apart from street sweeping, the traversal of entire street segments also arises in the collection of residential refuse. Both street sweeping and residential refuse collection are modeled as arc routing problems. However, node routing problems also arise in sanitation applications. In bulk collection, for instance, the vehicle visits specific sites where refuse has accumulated in large metal containers with a volume of 6–12 cubic yards — (the equivalent of 40–80 normal garbage cans according to Beltrami & Bodin [1974]). In this setting, the vehicle only needs to visit the nodes representing the container sites.

One of the earliest and most comprehensive treatments of sanitation routing to appear in print is due to Beltrami & Bodin [1974] and reports their work with the New York City Department of Sanitation. The detailed street sweeping algorithm that developed out of this work is described later in this section. For the bulky pickup problem, Beltrami and Bodin used a modified Clarke–Wright heuristic to design truck routes to collect refuse from 1000 container sites. Two complications in this node routing problem deserve mention as they also occur in other sanitation problems.

(1) *Multiple dumps.* Each truck must visit a dump site (or landfill) to unload when it is filled to capacity. Since the pick up loads are not known with certainty, the scheduling of this trip is based on estimated loads. Moreover, when several dump sites are available, the routing algorithm has to select the most appropriate site for unloading.

(2) *Service frequencies.* Nodes in the network receive service with different frequencies: some require daily service, while others may be visited twice or only once a week. This gives rise to a *period-routing* problem where pickup points must be assigned to days of the week in such a way as to ensure both efficient routes and minimum fleet size. (For more general discussions of period-routing, see Ball [1988], chapter 5 of Chao [1993], and Chapter 4 of this volume.)

Another application described by Beltrami and Bodin for New York City is the routing of barges and tugboats to carry refuse from eight marine disposal locations around the city to the landfill on Staten Island. A similar and more recent application which focuses on the development of a planning tool for sizing the fleet of barges and tugs is described by Larson, Minkoff & Gregory [1988]. Going beyond routing problems, Bodin [1990] describes the manpower scheduling procedure he developed in 1970 to assign work scheduling to the 10,000 sanitation workers of New York City.

Apart from the work of Bodin and his co-workers, a number of other studies addressed sanitation problems in the early 1970s. Clark [1973] gives an overview of the early work in this area and Coyle [1973] reviews several planning systems for refuse collection. Wyskida & Gupta [1972] describe a project conducted by faculty and students of the University of Alabama for the city of Huntsville. At the time of the study, the city provided refuse collection services to 33,000 customers twice a week. The team developed a procedure for constructing routes that considered vehicle capacity and trips to the landfill and also observed the eight-hour limit on route duration. The project resulted in a 12% reduction in the number of routes. Turner & Hougland [1975] describe a node routing refuse collection problem for the town of Blacksburg, Virginia. The problem size in this application is small: two trucks have to visit 100 pickup points and make three visits to the disposal site (one truck has two trips and the other only one). A procedure to solve the m-TSP was developed by the authors to construct the routes. Moreover, the location of the transfer site (used for disposal) was optimized by solving a median problem.

Clark & Gillean [1975] report major cost reductions in the cost of solid waste collection and disposal in Cleveland, Ohio: The city was able to reduce its sanitation work force from 1640 workers to 850 and cut its annual budget from $14.8 million in 1970 to $8.8 million in 1972. Most of this reduction was due to a careful analysis of collection practices based on extensive data analysis and reductions in the sizes of collection vehicle crews. A simulation model was used to generate the number of routes and vehicles needed each day by considering collection loads, trips to the disposal sites, queueing delays for unloading, and route duration restrictions.

As part of the Cleveland study, a management information system was developed to collect daily information on collection routes based on daily trip tickets. Clark & Lee [1976] describe quantitative planning models that rely upon this information for short-term and long-range analysis. In particular, using regression models, they relate the total collection time to such factors as the density of the refuse collected, the average weight of solid waste generated by a household, and the number of household units per linear mile.

Several early studies in arc routing arose from sanitation applications. Liebling [1973] analyzes street sweeping operations for the City of Zurich. This application uses the directed postman problem to construct vehicle tours. Shuster & Shur [1974] looked at tour construction procedures that specifically dealt with the reduction or elimination of U-turns and left turns. Such turns are undesirable for sanitation vehicles and can be reduced using procedures discussed later in this section. Liebman, Male & Wathne [1975] reviewed the use of arc routing techniques for sanitation problems, focusing primarily on postman problems and algorithms. Male & Liebman [1978] describe a tour construction algorithm for refuse collection that builds up vehicle routes by combining a number of small cycles. In this approach, the cycles are constructed first to cover non-overlapping regions corresponding to residential blocks. In this way, they define a 'checker-board' pattern over the entire service area. To combine the small cycles into vehicle tours the authors solve a capacitated spanning tree problem: Each cycle is represented by a single node in the auxiliary network, with the depot appended as an extra node. Cycle nodes are connected via edges of zero cost if they have a node in common in the street network. The depot node is connected to each cycle node with an edge cost equal to the shortest round trip cost between the depot and a node belonging to the cycle. Once the auxiliary network is constructed, the procedure searches for m subtrees leaving the depot that form a spanning tree for the auxiliary network when combined. Each of the m subtrees corresponds to a single vehicle tour that leaves the depot, services the streets represented by the cycle nodes in the subtree, and returns to the depot. A key consideration in forming the subtrees is that the total load must observe the capacity constraints. The subtree problem is solved following heuristics initially developed by Male [1973]. Male & Liebman [1978] tested this procedure on networks ranging from 21 nodes and 37 edges to those with 83 nodes and 153 edges. The number of districts (m) ranged from 4 to 6. However, their computational results do not include criteria for judging the quality of the solutions. They also used the algorithm for one of the 24 collection areas into which the city of Knoxville, Tennessee was partitioned. This subdivision was represented by a street network with 110 nodes and 330 edges, making it the largest problem reported in this work.

7.2. Algorithms for sanitation routing

We begin our discussion of sanitation algorithms with a detailed description of an algorithm for street sweeping due to Bodin & Kursh [1978]. This description will serve as a basis for covering subsequent extensions and refinements to the original algorithm.

It is useful to establish the following terminology. Let $G = (V, A)$ be a directed network representing the underlying street network for the area of service. Typically G contains detailed information at the block face level so that V corresponds to all street intersections. We call G the *underlying network* or *travel network*. In general, only a subset R of arcs in A requires service. The graph $G_R = (V_R, R)$ induced by these edges is the *required network* or *service network* (V_R is the set of nodes incident upon edges in R). The routing problem is to find a set of vehicle tours based at the depot that covers all required arcs in R with minimum deadheading. In this notation, one may consider the routing problem as a multi-vehicle version of the rural postman problem (RPP) discussed in Section 3. The need for multiple vehicles arises from limited vehicle capacity or restrictions on route duration.

In arc routing applications, a pivotal modeling question is the appropriate representation of the travel and service networks. For instance, the choice of the edges as directed or undirected clearly affects the solution technique used. For the street sweeping problem, Bodin & Kursh [1978] argue that "because a street sweeper can only cover one side of a street at a time and must proceed in the direction of traffic" the network G_R is directed.

The definition of the required network G_R is also problem-dependent. In New York City, parking regulations constitute a key set of constraints for the routing problem: the city has approximately 70 different parking codes that ban parking on a given side of a street during certain days of the week, or certain hours of the day. Naturally, a street can only be swept when parking is banned since, to be effective, sweepers need to come close to the curb. Given the parking regulations for each street segment requiring service, the street sweeping problem is decomposed by time period. The day is broken into a collection of non-overlapping time intervals I and a required set of arcs $R(I)$ is defined for each time interval. For example, in the 8:00–9:00 AM time interval, $R(I)$ is extracted from the set of all street segments for which parking is banned over 8:00–9:00 AM or any time interval including 8:00–9:00 AM. In this way, the same segment may be part of the required network for some I and part of the underlying network during other time intervals.

We now describe the details of the street sweeping algorithm due to Bodin & Kursh [1978, 1979], which developed out of the work Beltrami & Bodin [1974] initially carried out for New York City. In this description, we devote specific attention to practical operational issues, as these are shared by other arc routing applications described in the subsequent sections. The reader interested in a pedagogically-motivated description of the algorithm with a detailed numerical example may refer to Tucker & Bodin [1983].

An algorithm for street sweeping [Bodin & Kursh, 1978]

INPUT: Underlying network $G = (V, A)$, the set of required arcs R; parking regulations and load for each arc in R; cost and penalty data.

Initialize: Set I = time interval index = 1.

Step 1. Extract the set of required arcs to be swept in time interval I by checking the parking regulations. Call this set $R(I)$.

Step 2. Balance the network $G_R = (V_R, R(I))$ by solving a transportation problem over the nodes in G_R. The solution gives the minimum-cost set of deadhead arcs $D(I)$ so that $\tilde{G}_R = (V_R, R(I) \cup D(I))$ is balanced.

Step 3. Pair the in-bound and out-bound arcs at each node by solving an assignment problem. The pairing defines a covering cycle $C(\tilde{G}_R)$ in $\tilde{G}_R$ (the cycle need not be simple or connected).

Step 4. Within each component of $C(\tilde{G}_R)$, check for subcycles. If any exist, remove them by successively merging subcycles that share a common node. This yields an Euler tour in each component.

Step 5. Combine the tours in the different components into a single Euler tour $T(I)$ for interval I.

Step 6. Break up the tour $T(I)$ into feasible vehicle routes in interval I.

Step 7. Set $I \leftarrow I + 1$ and repeat Steps 1–6 until all time intervals are processed.

Step 8. Form daily vehicle schedules by patching together vehicle routes in successive time intervals.

The preceding steps outline the *route-first–cluster-second* approach of Bodin & Kursh [1978, 1979]. As the name implies, the procedure performs routing first by constructing a covering Euler cycle that covers all the required arcs in $R(I)$ and then breaks it down into feasible routes for individual vehicles. The routing Steps 2–5 essentially solve the directed rural postman problem for the required arcs $R(I)$, ignoring vehicle capacity issues. The clustering stage appears in Step 6 when this single-vehicle tour is transformed into a multi-vehicle solution. We now comment on certain steps of the algorithm in greater detail.

Step 2 involves solving the transportation problem defined by calculating the degree imbalance $b(i) = d^-(i) - d^+(i)$ for each node i, as shown in Section 2.2. Given the set of deadhead arcs that balance the network G_R at minimum cost, one can trace an Euler cycle in each component of $\tilde{G}_R$ using one of several cycle-tracing algorithms, such as the one given in Edmonds & Johnson [1973]. However, Bodin and Kursh found that the resulting Euler tour may contain too many U-turns or left-hand turns. To control this, at a given node they attach penalties to undesirable pairings of inbound and outbound arcs as follows:

Movements	Points
U-turns	8
Deadhead to sweep	5
Sweep to deadhead	5
Left-hand turn	4
Right-hand turn	1
Straight-ahead (no turn)	0

To illustrate the use of these penalties, consider node 1 in Figure 9. This node is in balance as it has two arcs (A and E) entering and two arcs (B and F) leaving it. The pairing A–B, E–F involves two right-hand turns for a total penalty of 2 while the pairing A–F, B–E involves a single U-turn for 8 penalty units. If B and E were deadhead arcs, the pairings A–B and E–F would each incur another 5 units of penalty as the sweeper needs to switch between deadhead and sweep operations in traversing node 1.

In general, the pairing of arcs a_{ki} and $a_{i\ell}$ into and out of node i receives a penalty of $d_{k\ell}$ as defined above. The optimal pairing at each node is found by solving an assignment problem that pairs all in-bound and out-bound arcs in such a way as to minimize $\sum_k \sum_\ell d_{k\ell}$.

It is clear that the pairing solution can be myopic since it operates one node at a time. In particular, as pointed out by Bodin & Kursh [1979], it may create sub-cycles that Step 4 would need to remove. For instance, in Figure 9, the assignment problem may select the pairings A–B, E–F, B–C, and H–E to avoid U-turns. This results in the two subcycles ABCD and EFGH. In Step 4, these two subcycles are merged to obtain the larger cycle ABEFGHCD, which contains one U-turn at node 2. This example shows that some of the benefits of the pairings may be lost when the subsequent consideration of obtaining an Euler cycle is addressed. If $\tilde{G}_R$ is disconnected, Step 4 may terminate with disjoint cycles in several distinct components. To combine these into a single covering

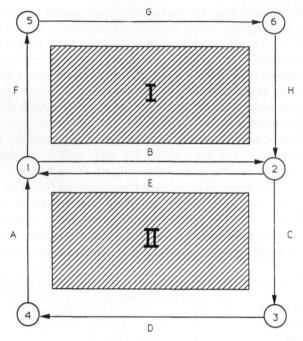

Fig. 9. Routing around two blocks I and II.

Euler tour, the cycles in distinct components must be connected. Bodin and Kursh connect components p and q by finding the shortest 'round-trip' path of cost $\bar{c}_{pq} = \mathrm{Min}_{(i,j)}\{SP(i, j) + SP(j, i)\}$, with the minimum taken over all pairs of nodes i and j in components p and q respectively. We remark that this is exactly the procedure used in the balance-and-connect heuristic for the RPP described in Section 3.

In Step 6, the covering Euler tour is broken into feasible routes for the individual sweeper vehicles. A tour duration constraint may apply at this stage. For example, if interval I is one hour long (8–9 AM, say), each vehicle tour must run no more than 60 minutes. Another consideration is the distribution of deadheading along the covering tour. Naturally, it is useful to 'cut up' the covering tour in such a way as to eliminate the deadhead stretches. This is accomplished by positioning the deadhead stretches at the initial and terminal portions of each vehicle tour, so that they can be dropped. A similar deadheading consideration arises in joining the routes of the same vehicle in successive time intervals I and $I + 1$ (Step 8).

As we describe, the overall strategy of the route-first–cluster-second algorithm is to first solve a single-vehicle rural postman problem to obtain an Euler cycle that is subsequently broken down into routes feasible for the m street-sweeping vehicles. Bodin & Kursh [1978, 1979] also describe a *cluster-first–route-second* procedure that decomposes the problem into m single-vehicle problems by dividing the area of service into m regions a priori. One then solves m smaller RPP's, one for each region. The eight-step procedure described above can be easily modified to describe the cluster-first approach as well: Steps 2–5 are run once for each set $R_k(I)$ of required arcs, where k runs from 1 through m (the number of vehicles). $R_k(I)$ is chosen to represent the amount of service for a single vehicle. Once all required arcs are processed in this way, Step 6 is no longer required.

In their work, Bodin and Kursh found that the route-first approach results in less deadheading. For example, in one region of New York City, the route-first procedure generated routes with 30% less deadheading and used only 7 routes as compared to 8 routes required by the cluster-first procedure. This behavior is not surprising as the objective of the RPP is to minimize deadheading and the route-first approach solves the RPP 'globally' over the entire network G_R. A potential disadvantage of the routes produced by the route-first procedure is the extent of overlapping among routes. By partitioning the network into separate service areas, one for each vehicle, the cluster-first approach reduces overlapping considerably. Therefore, in sanitation problems where clean boundaries between different services areas are desirable, possibly for such administrative reasons as clearly demarcating areas of responsibility, the cluster-first approach is advantageous.

In a short paper, McBride [1982] proposes certain steps for controlling U-turns and left turns within the basic procedure outlined above. A key observation of this work is that such turns can be eliminated by introducing forced deadheading, over and above what deadheading the balancing step identifies. The example in Figure 10 illustrates the basic approach. Consider the street network in Figure 10a and assume that both sides of Street A and the right-hand side of Street C require service. Let node k denote the intersection of Streets A and C. Clearly, this node is balanced so that the balancing step does not have to introduce deadhead arcs incident to this node. The assignment problem in Step 3 of the Bodin and Kursh algorithm will select the pairing A1–C2, C1–A2 which involves two left turns, avoiding the U-turn implied by the A1–A2 pairing. Introducing forced deadheading in the form of the two arcs C3 and C4 eliminates these left turns because, once these arcs are added, the assignment problem selects the pairing C1–C2, C4–A2, A1–C3 as shown in Figure 10b. This corresponds to two right turns and one straight movement. McBride points out that these forced deadheads must be introduced prior to the balancing step. In effect, they augment the set of required arcs. He also gives heuristics rules for combining cycles in Step 4 to avoid bad turns.

While the Bodin and Kursh algorithm was developed for street sweeping originally, it can be modified for other refuse collection applications. Bodin, Fagan, Welebny & Greenberg [1989] describe a computerized sanitation system for the town of Oyster Bay, located in Nassau County, New York, that relies upon a modified version of this algorithm. This town spans an area of 114 square miles and has a population of approximately 327,000 people. Curbside refuse collection services are provided twice weekly to some 64,000 residential and commercial stops. On each day, approximately 625 tons of solid waste are collected.

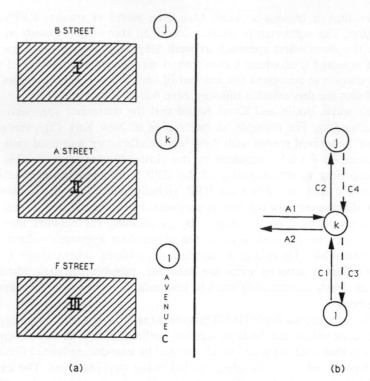

Fig. 10. (a) Avenue C requires service only on the side adjacent to blocks I, II and III. (b) Forcing deadheading on arcs C3 and C4 eliminates left turns.

In Oyster Bay, the crews worked 4 days a week under a *task completion agreement*, whereupon a crew completed its designated route regardless of its duration. The incorporation of new housing divisions into existing routes and negotiations between the town and the sanitation union combined to result in unbalanced routes: routes varied widely in duration, number of stops per route, and collection volumes (loads). A key motivation of the routing study was to determine a new set of routes that were balanced, efficient, and compact.

The required street segments in Oyster Bay fall into two groups. The *delivery network* consists of required segments that the vehicle can traverse in either direction, collecting refuse on both sides of the street during this traversal. This type of movement is termed *meandering* by Bodin and coworkers. These street segments are represented by undirected edges R_u. The second group of street segments can be traversed only in the direction of traffic with collection on only one side of the street during the traversal. Each of the street segments, which form the *special delivery network*, is represented by two directed arcs (i, j) and (j, i). The set of these directed arcs is denoted by R_d. In this fashion, the required network G_R becomes a mixed graph with $R = R_u \cup R_d$.

The routing system requires the following basic information for each street segment:
- the length of the street segment,
- the number of parcels on the segment,
- the class of each parcel (identified as residential, commercial, industrial, entertainment, and so forth),
- the total refuse to be collected on the segment,
- the total collection time and deadheading time on the segment.

The refuse load is estimated by the system based on the number of collection stops for the segment and estimates of the load per collection point, based on past data. The information listed above can be conveniently placed into a geographical information system (GIS) where each street segment has a collection of attributes.

In basic outline, the solution procedure follows the route-first–cluster-second approach described in Bodin & Kursh [1978]. Starting from an undirected postman problem solution for the undirected edges R_u alone, Bodin and coworkers construct a covering tour through the augmented network $\tilde{G}(R_u)$ composed of the edges R_u and the deadhead edges D_u added by the minimum cost matching solution. In tracing an Euler tour through $R_u \cup D_u$, turn penalties are considered in pairing edges entering and leaving each node. Next, edges in the special delivery network adjacent to a portion of the Euler cycle are appended to the covering tour by simply 'walking' each edge in both directions. In this way, the covering Euler tour ultimately covers all required arcs $R_u \cup R_d$. Once this tour is constructed, the total number of stops, weight collected, and elapsed time are calculated so that the covering tour can be broken down into a collection of feasible routes.

Bodin, Fagan, Welebny & Greenberg [1989] report that the system generated routes that required 37 vehicles, saving 3 vehicles over the past practice. They estimate annual cost savings of approximately $200,000 for the town of Oyster Bay.

Eglese & Murdock [1991] describe another interesting street sweeping application carried out for Lancashire County in the United Kingdom. In this county, there were very few one-way streets, so that all roads could be treated as two-way streets. This results in a balanced service network for which a postman tour with no deadheading exists. However, the problem had several complications as follows:

(1) *Sweeper capacity.* Because of limited sweeper capacity, each sweeper must be emptied several times on each day. Once full, the sweeper must travel to one of several available 'tipping sites', where it can be emptied. A similar issue arises with trips to landfills for refuse collection vehicles when capacity limitations apply.

(2) *Service frequencies.* The roads of the county require sweeping with different frequencies: some roads are swept 12 times a year while others are swept only once.

(3) *Service constraint.* Both sides of the street should be swept on the same day.

As compared to the New York City application in Bodin & Kursh [1978], this application has a simpler routing component. First, the county is already divided into districts, each of which is served by a single vehicle out of a local depot. Second, as the required network G_R consists entirely of two-way streets, it is automatically Eulerian (if connected). This means that in each component of G_R a solution with no deadheading is available. Unfortunately, this uncapacitated postman solution is not usable due to vehicle capacity constraints. Even for the single-vehicle district, the capacity constraint transforms the problem into a capacitated arc routing problem (CARP) and adds deadheading for the trips to the tipping (dump) sites. Eglese & Murdock [1991] describe a simple routing algorithm that capitalizes on the presence of two-way streets, and takes constraints (1) and (3) into account through simple decision rules.

To describe the procedure, let us call a street unswept if it has not been swept at all, and half-swept if it has been swept on only one side. The basic selection rule is the following: At node i, select the next unswept road segment (i, j). If no such segment exists, then return on the segment last selected, sweeping its other (as yet unswept) side. Before selecting (i, j), a simple test establishes if the remaining time in the day is sufficient to sweep (i, j), complete the sweeping of all half-swept streets, make the trip to the tipping site, and return to the local depot. In this way, the procedure embarks on a new road only if it has enough time to complete sweeping all selected roads on *both* sides. This simple rule ensures that (3) is observed. At the end of the day, when the vehicle has to return to the depot, the tipping site selected is the one that minimizes $SP(i, s) + SP(s, 1)$ over all sites s, where i is the terminal node of the last road swept and 1 is the local depot. The next day's new route starts as close as possible to node i.

Eglese and Murdock point out that the routine application of the preceding rule may lead to inefficient routes. For instance, the route may bypass some required segments, thereby creating isolated required edges that can only be served by incurring extra deadheading. They propose priority rules for selecting the next segment to serve. While these rules help avoid the undesirable routing behavior, the resulting procedure is not immune to occasional pathological cases and manual overrides may be required.

Eglese and Murdock test their algorithm on one district of Lancashire County, the network for which has 624 nodes, 807 edges, and 4 tipping sites. Using the computerized procedure with some user overrides, a single vehicle can complete all required sweeping, at the specified frequencies, in 62 working days in their solution. They state that the vehicle spent roughly half of its annual workdays serving the same area in the County's past practice. A substantial reduction in the cost of street sweeping is, therefore, possible.

Muralidharan & Pandit [1991] describe another interesting application of arc routing to the collection of recyclable products. According to the authors, about 80% of the 300 billion pounds of municipal solid waste generated each year in the United States is taken to landfills and not re-used. It is common for vehicles to collect such recyclables as plastics from voluntary drop-off or buy-back centers. However, the recovery rate can be significantly increased if a curb-side collection program is implemented. From the vantage point of vehicle routing, this means that the vehicle services arcs as well as nodes of the street network. The authors, therefore, formulate the collection of recyclable materials as a general routing problem where the entities that require service comprise both nodes and arcs.

Muralidharan & Pandit [1991] use a tour partitioning heuristic to solve the capacitated general routing problem. First, an Euler tour covering all required nodes and street segments is constructed. This covering tour is then broken into individual vehicle tours that observe both capacity and route length (duration) constraints. A more detailed outline of the algorithm follows.

The underlying street network is represented as a mixed graph $G = (V, A \cup E)$ with both directed arcs (A) and undirected edges (E). A subset V_r of the nodes require service. These nodes represent such pick-up locations as convenience stores of fast-food restaurants where large volumes of recyclables accumulate. Let A_r and E_r be the subsets of A and E corresponding to required street segments. The required network is $G_R = (V_R \cup V_r, A_r \cup E_r)$. Each edge (i, j) in G_R has a load q_{ij} and a pick-up time of t_{ij}. Each node i in V_r has a load of q_i and time t_i required for pick-up. The capacity of vehicle k is Q_k. Each vehicle has a limit T on the total route length. The algorithm consists of the following steps.

(1) Connect step. — Solve a minimum spanning tree problem to connect the disconnected components of G_R. Denote the resulting graph by G_1.

(2) Create a strongly connected graph. — Add arcs to G_1 to ensure that a path exists between every pair of nodes in G_1. Call the resulting strongly connected graph G_2.

(3) Create an even graph. — Solve the minimum cost matching problem on vertices of odd degree in G_2. Add the arcs selected by the matching solution to obtain the even graph G_3.

(4) Solve the mixed postman problem on the even graph G_3 to obtain an Euler tour in G_3.

(5) Partition the Euler tour into feasible tours (with respect to capacity and duration) for the individual vehicles.

Both Steps (1) and (2) focus on creating a network of desired connectivity. Step (1) is performed in the usual way by shrinking connected components into nodes and constructing a minimum spanning tree ignoring the direction of arcs on shortest paths joining the components. In Step (2), one must find a minimum cost augmentation of the graph G_1, that yields a strongly connected graph. Since this problem is NP-complete, the authors solve it heuristically. First, the strongly connected components of G_1 are identified and condensed into nodes of an auxiliary graph $F = (N, B)$. For any pair of nodes p and q in N, B contains two arcs, one from p to q and the other in the opposite direction (these arcs are found by solving shortest paths in G). F is thus a complete directed graph. Following the additive bounding procedure of Fischetti & Toth [1989], the authors solve an arborescence problem and an anti-arborescence problem with node 1 (the depot) as the root node. The arcs selected by these two arborescence solutions provide paths from node 1 to every other node and from every node into node 1.

Steps (3) and (4) of the procedure involve creating an even balanced graph so that an Euler tour can be constructed. Thus the overall procedure may be viewed as a connect-and-balance heuristic. Step (4) uses the algorithm of Edmonds & Johnson [1973] for the MPP described in Section 2.3. Given an even connected graph, this algorithm orients the undirected edges in such a way as to obtain a balanced graph. In Step (5), the Euler tour is followed until the vehicle capacity or route length constraints force the termination of a route and the return of the vehicle to the depot. The tour for the next vehicle starts with the next arc of the Euler tour.

Muralidharan and Pandit have tested the preceding algorithm on problems with 50 nodes and 100 arcs. In their test problems, they varied both the mix of directed and undirected arcs and the percentage of arcs that require service as these may be expected to affect the performance of the procedure. The authors report the solution obtained by the capacitated general routing procedure, as well as the matching lower bound (MLB) for the CARP. However, the computational results indicate that the ratio of the heuristic solution value to MLB is usually larger than 1.75 and frequently over 2.0. Therefore, MLB does not appear to be a strong bound for these test problems.

8. Postal delivery routing problems

8.1. Introduction

Few organizations are engaged in delivery operations on as massive a scale as the United States Postal Service (USPS). The USPS delivers over 166 billion pieces of mail each year to over 100 million delivery locations, thereby accounting for about 40 percent of the volume of mail worldwide. There are over 275,000 individual carriers in the USPS who are engaged in delivery activities on a daily basis [see Wright, 1992].

The USPS has consistently pursued productivity improvements through automation and the use of new technology. Many of the technological changes have occurred in the processing and sorting of mail including automated sorting based on optical character readers. A recent overview of some of these developments is available in Cebry, DeSilva & DiLisio [1992]. The USPS has also depended on geographical databases and information systems to support its field operations. In this section, we focus on routing problems of the USPS and its Canadian counterpart — Canada Post Corporation (CPC).

In the context of the USPS, the daily workload of a carrier may be divided into three portions:

(i) initial collection and sorting at the postal facility for the day's delivery,

(ii) driving from the postal facility to the delivery area or from one delivery area to another, and

(iii) the actual delivery of the mail by vehicles or on foot.

Of these, the third activity accounts for most of the time: about 5 hours per day. Driving between delivery areas, or to and from the post office, typically consumes only a small fraction of an hour. More efficient sorting operations implemented by the USPS will likely result in a shift of approximately one hour from activity (i) to (iii), thereby making more time available for deliveries on the road. Given the large percentage of the carrier's time consumed by the delivery portion, it is clear that the optimization of delivery routes can result in very significant savings.

To give an idea of the delivery route structure, it is useful to start with the *park and loop problem* (PALP). In this problem, the postal vehicle is used to drive from one delivery area to another. In each area, the carrier stations the vehicle at a *parking location* and delivers mail to the surrounding area by walking along cycles or loops based at the parking location. These loops are called *walking loops* (or cycles) and are approximately one hour in duration, reflecting the amount of mail the carrier can comfortably carry in the mailbag. At the end of each loop, the carrier extracts the mail for the next loop out of the vehicle. Once all loops are traversed, the carrier *drives* the vehicle to a new parking location where this process is repeated. The *route* of the carrier corresponds to a daily work schedule and may involve two or three parking locations. This problem and an algorithm for devising efficient routes are described in Section 8.2.

In addition to the park-and-loop problem, Bodin, Fagan & Levy [1992a] describe other delivery routing problems that arise in the USPS. These problems differ in the mode of delivery used by the carrier and are summarized in Table 4. In the *curbline* problem, all the deliveries are made in the driving mode, while in the *relay box* problem, the carrier extracts mail from relay boxes that play the role of parking locations. It is also possible for the carrier to combine the park-and-loop and curbline operations by delivering mail both on foot and while driving. Since the deliveries occur along the entire length of street segments, all of these must be regarded as arc routing problems. Bodin, Fagan & Levy [1992a] and Walsh, Bodin, Levy & Fagan [1994] describe a system called GEOMOD, developed for the USPS, that solves these arc routing problems over street networks.

In contrast to the above, the *collection box* problem of the USPS is a node routing problem. In this problem, vehicle routes must be designed to visit collection boxes with specified time windows. A key consideration here is that the vehicles should transport mail back to the postal facility early in the day to ensure an even load for sorting and handling mail. Consequently, multiple pickups at the collection boxes and drop-offs at the postal facility are scheduled in the course

Table 4
Characteristics of arc routing problems for mail delivery

Routing problem	Park-and-loop	Curbline	Relay box
Mode of mail delivery	Walking	Driving	Walking
Mail replenishment locations	Postal facility parking locations	Postal facility	Relay box locations
Structure of walking cycles	Loops based at each parking location		Loops based at relay boxes
Vehicle route structure	Loops based at postal facility with stops at parking locations	Loops based at postal facility servicing arcs	–

of a single day. We note that node routing algorithms have been used by other researchers to optimize postal delivery routes. One example of this approach is the work performed for An Post of the Republic of Ireland as described by Deegan & Harrison [1990] and Harrison & Deegan [1992]. These authors discuss the interplay of geographical information and the delivery planning system and state that average savings of 15% may be expected from the route optimization.

8.2. Solving the park-and-loop problem (PALP)

In this section, we describe an algorithm for the PALP and provide greater detail about the operational characteristics of this problem. This problem is characterized by the condition that all mail is delivered by the carrier on foot. This condition leads to an interesting structure for the underlying travel network $G = (V, E)$ for the PALP.

First, we note that $G = (V, E)$ must be considered *undirected* since the carrier can walk any street segment in either direction. This remains true even if the street network has directed arcs for vehicle movements. The set R of required street segments consists chiefly of edge pairs between the same pair of nodes since the carrier must traverse each street segment twice, delivering mail on one side of the street at a time. Thus, given nodes i and j, two copies of undirected edges $e_1(i, j)$ and $e_2(i, j)$ are placed in R. When only one side of the street requires service (the other side borders a park, for instance), only one undirected edge $e(i, j)$ appears in R. Levy & Bodin [1988, 1989] call $e_1(i, j)$ and $e_2(i, j)$ *counterpart edges*. Their algorithm for PALP capitalizes on the fact that such counterpart edges comprise most of R.

Following Levy & Bodin [1988], PALP may be stated as follows: Find a set of delivery routes and associated parking locations and walking loops that meet the following objectives and constraints.

Objectives:
(a) the cycles use minimum deadheading,
(b) the work schedules for the carriers are balanced,
(c) the number of parking locations is minimized,
(d) the number of returns to the vehicle at each parking location is minimized.
Constraints:
(e) the duration of each walking loop cannot exceed a given upper bound TWL,
(f) the cycles cover all edges in R,
(g) the two edges in a counterpart pair must be serviced within T_s time units of one another.

In reality, PALP is a multicriterion problem in which objectives (a)–(d) appear in the order of importance. The balance objective in (b) is of great practical importance to the USPS. To be acceptable, the routes assigned to carriers should be approximately equal in duration. Bodin and Levy impose a lower bound and an upper on the durations of all work schedules in order to achieve balance. Criterion (c) aims to keep the driving time and the unproductive time associated with parking a vehicle at a minimum during the day. Criterion (d) may be motivated as follows. If in covering a cycle that takes 60 minutes to walk, the carrier expects to pass by the vehicle after about 20 minutes (say), he or she is likely to load only the mail needed for the first 20 minutes and to extract the rest from the vehicle during the interim visit. Such multiple re-loads are deemed unproductive. Consequently, there is an advantage to having hour-long *simple* cycles based at the parking location.

Constraint (e) reflects the limit on the load the carrier can carry in one loop. The typical value of 60 minutes for TWL uses a one-hour duration as a surrogate for this load weight or volume limit. To observe (d), one can also place a (soft) lower bound on the duration of the loop. Constraint (g) reflects a customer service consideration: serving one side of the street but delaying the service to the other side until 3 hours later is likely to generate complaints. Typically T_s is about one hour. Note that if TWL equals one hour as well, (g) is automatically satisfied if counterpart edges are *always* placed in the *same* loop.

As defined above, PALP is closely related to two well-known classes of routing problems. In view of constraint (e), the problem can be viewed as a multiple-depot capacitated arc routing problem, where the depots correspond to the parking locations. However, as the parking locations are not fixed in advance and must be determined, PALP is also a *location-routing problem*. The latter problem has been studied in the node routing context (see the review by Laporte [1988]) and PALP constitutes an arc routing version.

An arc partitioning algorithm for the PALP

In the original work on PALP, Levy [1987] used two arc partitioning algorithms that are formally described by Levy & Bodin [1989] and Bodin & Levy [1991]. The basic objective of these algorithms is to identify clusters of service edges over which Euler tours for walking cycles are formed. Let $G = (V, E)$ be the underlying street network and let $G_R = (V_R, R)$ denote the service network. Because R consists mainly of counterpart edges, G_R is a multigraph. Let D_1 consist of counterparts to those required edges that occur singly in R (those with no required counterparts). Set $R_1 = R \cup D_1$ and form the subgraph $G_1 = (V_R, R_1)$. G_1 consists entirely of counterpart edges and we assume that it is connected. Attach a service time $t_s(i, j)$ to every edge $e(i, j)$ and a deadhead time $t_d(i, j)$ to each edge in D_1.

A *partition* of the edge set R_1 is a mutually exhaustive and exclusive collection of subsets $P_1, \ldots, P_n$ of R_1. We call each P_k a *partition cluster*, or simply a *cluster* (Levy and Bodin just use the term partition). We limit the choice of partitions to those that always place both members of a counterpart pair in the same cluster, thereby allowing no splitting of counterparts edges into different P_k's. The *workload* for cluster P_k is defined as

$$W_k = w(P_k) = \sum_{e(i,j)\in P_k\cap R} t_s(i, j) + \sum_{e(i,j)\in P_k\cap D_1} t_d(i, j). \tag{8.1}$$

It is desirable for the workloads of all partition clusters to satisfy pre-specified lower and upper bounds:

$$T_L \leq W_k \leq T_U \quad \text{for all } k. \tag{8.2}$$

Instead of imposing (8.2) directly, we invoke it as a soft constraint by allowing violations at a penalty of

$$\text{PEN}(k) = a(W_k - T_U)^+ + b(T_L - W_k)^+,\tag{8.3}$$

where $(x)^+$ denotes the positive part of x or $\text{Max}(x, 0)$. The parameters b and a are weights attached to unit violations of the desired lower and upper bounds. A partition is called *balanced* if (8.2) holds. Clearly, this is equivalent to $\text{PEN}(k) = 0$ for all k.

The *edge partitioning algorithm* (EPA) of Levy & Bodin [1989] is designed to find a balanced partition of R_1 and an associated set of Euler cycles, one for each partition cluster. The main steps of the EPA are shown in Figure 11. Motivated by the PALP, EPA grows the clusters out of seed nodes that correspond to the parking locations. Each seed point acts as the base node of one or several clusters, but the total number of clusters is n. We now describe each main step of EPA in greater detail.

1. *Seed point selection.* The objective of this step is to select a well-separated set of seed nodes dispersed over G_1. The seeds are selected by following a priority list L for eligible nodes i of G_1. L is arranged in descending order of the attractiveness measure

$$U(i) = \alpha_1 \left(\frac{d(i, R_1)}{2}\right) + \alpha_2 \sum_j t_s(i, j)$$

computed for each node i, where α_1 and α_2 are positive weighting factors with $\alpha_1 + \alpha_2 = 1$. Note that the first factor in $U(i)$ favors nodes with large degree since

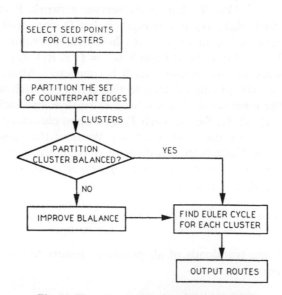

Fig. 11. The edge partitioning algorithm (EPA).

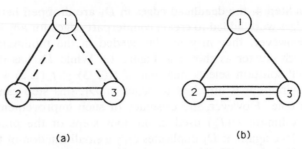

(a)   (b)

Fig. 12. (a) Counterpart edges in G_1; (b) postman solution for required edges only with less deadheading (all dashed edges represent deadheading).

the procedure will attempt to grow a cluster out of each counterpart edge pair emanating out of node i. The second factor measures the total service time for all edges incident to node i. While the two measures are clearly correlated, Levy and Bodin tested different settings of α_1 and α_2 on postal data and found that values of α_2 close to 1 give the best results.

The selection of seed points proceeds from list L until $(1/2) \sum_{i \in S} d(i, R_1)$ exceeds n, where S is the set of seed points selected already. To be selected, however, any seed node must satisfy two conditions for given parameters K_S and K_B.

(i) the node must be separated from any node in S by at least K_S edge pairs, and

(ii) the node must be at least K_B edge pairs away from any border node of G_1.

Condition (i) ensures that the seed points are widely dispersed over the region. A *border node* of G_1 is a node adjacent to exactly one other node in G_1 (these are pendant nodes in the graph obtained by coalescing counterpart edges). Condition (ii) therefore avoids placing seed nodes in the vicinity of border nodes as this may hamper the expansion of clusters grown from that seed. Clearly, both conditions are easily checked by solving shortest path problems with unit edge costs.

2. *Partitioning.* This step creates partition clusters by allocating each edge pair of R_1 to a unique cluster. Initially, each edge pair incident to the seed point serves as the root of a different cluster. At each subsequent stage, the cluster with the smallest workload $w(P_k)$ is chosen for expansion. Among all edge pairs adjacent to this cluster, the pair with the largest summed service time is added to the cluster. Partition clusters, therefore, expand concurrently until all edge pairs in R_1 are assigned.

3. *Balancing.* If the partition obtained in Step 2 is not well-balanced, improve the balance by performing edge pair moves from one partition to another. At each stage, choose the cluster with the largest penalty PEN (k) for improvement. Evaluate candidate edge pair moves and carry out the swap that decreases $PS = \sum_k \text{PEN}(k)$ most. STOP, if $PS = 0$. (The edge pair exchanges are described below.)

4. *Finding Euler cycles.* For each cluster P_k obtained in Step 3, solve the undirected postman problem (UPP) over all required edges $P_k \cap R$.

Note that in Step 4, the deadhead edges in D_1 are dropped before solving the UPP. These edges were added to create counterpart edges for any edge appearing in R singly. However, they may not be needed in the minimum deadheading solution for each cluster as shown in Figure 12. While D_1 has three deadhead edges, the UPP solution selects only one if $t_d(2, 3) \leq t_d(1, 2) + t_d(1, 3)$. Since no service requirement is attached to edges in D_1, this step can only improve the deadhead time. However, the existence of such improvements implies that the workload estimate $w(P_k)$ used in previous steps of the procedure is only approximate. Once again, if D_1 duplicates only a small fraction of the edges in R, this inaccuracy does not affect the quality of the solutions.

We now briefly describe the edge pair exchanges of Step 3. As implemented by Levy & Bodin [1989], all exchanges satisfy two conditions: counterpart edges always move together (from one cluster to another), and no exchange may cause a cluster to become disconnected. Subject to these conditions, Levy and Bodin use a data structure that traces an Euler cycle through each cluster to guide the subset of all possible exchanges to consider. We use Figure 13 to illustrate the main idea for a small portion of the graph G_1. In this figure, three clusters are identified by the labels A, B, and C on each edge pair. Nodes 1 and 9 are the seed points for these partition clusters. The arrows indicate an Euler walk on each cluster based at each seed point. Clearly, since all edges appear in counterpart pairs at this stage, such a walk always exists as long as each cluster remains connected.

The simplest type of an exchange is a *leaf swap* where a single edge pair is moved. In Figure 13, the pair (11,12) can be moved from cluster B to cluster C. This can be iterated: If this exchange is implemented, then moving (12,3) from B to C constitutes another leaf swap. If, on the other hand, the pair (3,12) is considered first, cluster B will become disconnected. In this case the arrows are followed until this pair is encountered again and all pairs encountered en route are moved together (that is, (3,12) and (12,11) move from B to C simultaneously). This move is called a *branch swap*. In a *cycle swap*, only the first pair in a possible group of edge pairs is moved. An example is the edge pair (8,7) considered for a move to cluster C. While cluster A remains connected if this move occurs, the sense of the walk is violated since the pairs (7,5), (5,4), and (4,6) are traced before returning to (7,8). However, since node 4 along this path is also incident to the pair (2,4), which also lies in A, one can 're-constitute' a connected walk: after (7,8) moves to C, edge pairs (4,5) and (5,7) are traced after (2,4) and a 'U-turn' occurs at node 7. Levy & Bodin describe the detailed rules governing the order in which they perform leaf, branch, and cycle swaps.

Edge partitioning applied to the postal problem

Given the preceding statement of the EPA, its application to the park-and-loop problem (PALP) can proceed in two directions. In a straightforward way, one can use EPA to find all parking locations and the associated one-hour walking loops in one pass. In this case, EPA balances the durations of the walking cycles, keeping them all under TWL. These cycles can then be combined to form a number of five-hour daily routes satisfying (8.2). Levy & Bodin [1988] call this

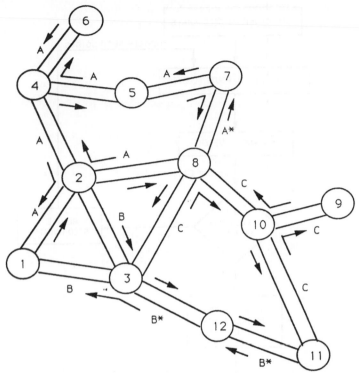

Fig. 13. Branch exchanges in G_1. Labels A, B, and C indicate the partition cluster to which each edge pair belongs. The arrows define a walk through each cluster. Nodes 1 and 9 are seed points. Asterisks denote candidates considered for exchange: the pair (7,8) from A to C, for example.

the *conventional algorithm* and state that it reflects the past manual approach of USPS to route design. The aggregation of the walking loops into daily routes can be accomplished by solving a vehicle routing problem: Each walking cycle is represented by a node placed at the parking location of the loop. The service time for this node equals the time required to walk the loop. Using the vehicle depot as the base, the vehicle node routing problem finds routes to serve all the 'loop nodes' subject to a route duration constraint of 5 hours. It is easy to incorporate driving time between distinct parking locations into this approach as well.

Levy & Bodin [1988] also used a more involved procedure called the *composite algorithm* to solve the EPA. As shown in Figure 14, the basic idea of this two-stage procedure is to obtain five-hour routes first, and then to divide each route into a number of walking cycles. Clearly, the second stage can be solved by applying the EPA to the required edges of each five-hour cluster. However, an iterative seed point selection scheme is used to ensure that the clusters are balanced. Bodin & Levy [1991] show how the partitioning into five-hour clusters can also be carried out through a simple modification of the partitioning step of the EPA: Let N be the desired number of five-hour clusters and assume that N seed points have been iden-

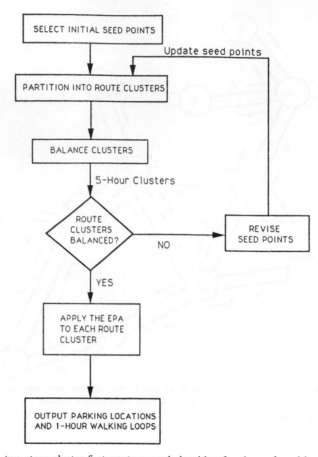

Fig. 14. The two-stage cluster-first–route second algorithm for the park-and-loop problem.

tified over the entire service area. Connect a 'supernode' to each seed point by a pair of counterpart edges with zero service times. Treat each such pair as the root edge pair of a different partition cluster. The EPA then treats the supernode as a single (artificial) parking location out of which N partition clusters are grown. Once these clusters are balanced, the supernode and the seed points are discarded and each cluster serves as the input network G_1 for the EPA in the second stage of the composite algorithm. Note that since only the required workload of each route cluster is of interest in the first stage, $t_d(i, j)$ is set equal to zero for all edges $e(i, j)$ in D_1.

8.3. Canadian postal problems

In postal operations, it is not unusual to see different delivery schemes as one moves from one country to the next. The park and loop problem (PALP) described in the last section reflects U.S. postal practices and specifically addresses the case where the carrier uses a combination of driving and walking. Roy & Rousseau [1989] describe the corresponding routing problem for Canada Post Corporation (CPC) which they call the *Canadian postman problem* (CPP).

The workday of a letter carrier at the CPC is divided into a morning portion and an afternoon portion. The carrier starts at the postal station, returns to it for the lunch break and ends the day by returning to the station. The duration of the morning route is approximately five hours while only three hours are available for the afternoon route. Transportation times from and to the station are included in these durations as are morning and afternoon breaks (17.2 and 15 minutes, respectively). The carrier may use public transport for the commute to the station and, therefore, does not need to drive a postal vehicle (in contrast to the PALP). Moreover, while in the PALP the capacity constraints on the walking loops reflected load considerations, at CPC the carrier can pick up the mail to be delivered from relay boxes placed along the route. Therefore, the main capacity consideration in the Canadian postman problem applies to route duration.

The preceding description suggests that the CPC routing problem can be formulated as a capacitated arc routing problem (CARP) with route duration limits playing the role of capacities. Since morning and afternoon routes have different durations, this formulation must allow a route-dependent capacity $Q(k)$ for route k. Let $t_s(i, j)$ and $t_d(i, j)$ be the service and deadhead (traversal without service) times for edge $e(i, j)$ as in Section 8.2. Let K be the number of routes required. Typically K equals twice the number of carriers. The CARP formulation of Section 4.1 can be modified to describe the CPP as follows. First, the objective (4.1.1) is replaced by

$$\text{MIN} \sum_{k=1}^{K} \sum_{i,j} \left\{ t_d(i, j)(x_{ij}^k - \ell_{ij}^k) + t_s(i, j)\ell_{ij}^k \right\}. \tag{8.4}$$

The CPP problem is to minimize (8.4) subject to the constraints (4.1.2)–(4.1.7) except that (4.1.5) is replaced by the route duration constraint

$$\sum_{i,j} \left\{ t_d(i, j)(x_{ij}^k - \ell_{ij}^k) + t_s(i, j)\ell_{ij}^k \right\} \leq Q(k) \quad \text{for all } k. \tag{8.5}$$

Note that in both (8.4) and (8.5), the deadhead travel time is incurred only if $x_{ij}^k > \ell_{ij}^k$, that is $x_{ij}^k = 1$ and $\ell_{ij}^k = 0$.

Based on the preceding CARP formulation, Roy & Rousseau [1989] state the CPP as a location-routing problem by allowing each route to select its starting node i. In this way, the model selects multiple bases out of which routes are formed. If route k uses node i as the base, the deadhead time $t_0(i) = SP(1, i) + SP(i, 1)$ is incurred in going from the true depot (node 1) to node i and back. The following assumptions are made in the location-routing model:

(1) the service network is connected,
(2) the required edges occur in counterpart pairs,
(3) both edges of a counterpart pair appear in the same route,
(4) each route starts and ends at the same node.

With these assumptions, the route structure involves a round trip between the depot and node i in deadhead mode, and a service portion where all edges on the route are serviced so that $x_{ij}^k = \ell_{ij}^k$. Since the x and ℓ variables coincide for all edges traversed and serviced from the base at node i, one can eliminate the x variables from the CARP formulation. To see this, introduce the new variables

$$s_i^k = \begin{cases} 1 & \text{if route } k \text{ starts (and ends) at node } i, \\ 0 & \text{otherwise.} \end{cases}$$

Suppose that route k starts at node i^*. Then the first sum in (8.5) simply represents the deadheading in the route between the depot and i^* and therefore equals $t_0(i^*)$. The second sum in (8.5) represents the time for servicing all edges in the cycle based at i^*. While this sum varies from route to route, the second summation in (8.4) is a constant as it equals the service time of all edges. Leaving out this constant term, Roy and Rousseau obtain the following location-routing formulation.

$$\text{MIN} \quad \sum_{k=1}^{K} \sum_{i=1}^{n} t_0(i) s_i^k \tag{8.6.1}$$

subject to

$$\sum_{j \in \delta(i)} \ell_{ji}^k - \sum_{j \in \delta(i)} \ell_{ij}^k = 0 \qquad \text{for all } i \text{ and } k \tag{8.6.2}$$

$$\sum_{k=1}^{K} (\ell_{ij}^k + \ell_{ji}^k) = 1 \qquad \text{for all required edges } e(i, j) \tag{8.6.3}$$

$$\sum_{i,j} t_s(i, j) \ell_{ij}^k + \sum_{i=1}^{n} t_0(i) s_i^k \leq Q(k) \qquad \text{for all } k \tag{8.6.4}$$

$$\sum_{i=1}^{n} s_i^k = 1 \qquad \text{for all } k \tag{8.6.5}$$

$$s_i^k \leq \sum_{j \in N(i)} \ell_{ij}^k \qquad \text{for all } i \text{ and } k \tag{8.6.6}$$

$$s_i^k, \ \ell_{ij}^k \in \{0, 1\} \qquad \text{for all } i, j, k \tag{8.6.7}$$

plus subtour breaking constraints.

Roy & Rousseau [1989] describe a heuristic to solve the CPP which can be summarized as follows.

Heuristic for the Canadian postman problem

Step 1. *Generate routes.*

(a) Create basic units by aggregating edges with the same postal code and their counterpart edges.

(b) Aggregate basic units into cycles. Merge or augment cycles to create routes. Fix cycles into morning or afternoon routes. Decompose cycles that are not fixed into basic units.

(c) Insert any unselected basic units into the routes while observing the duration limits. If the insertion fails to insert all units, relax the duration constraint progressively until all units are incorporated into routes.

Step 2. *Sequence routes*. Construct an Euler tour through each cycle that minimizes the penalties for road crossings and street changes. To obtain the sequence, solve the TSP node-routing representation of the edges with penalties attached to TSP edges.

Step 3. *Generate workdays*. Solve the assignment algorithm to pair morning and afternoon routes to create $L = K/2$ workdays, where $L =$ the number of available letter carriers.

The present summary leaves out some of the interesting features of the individual steps of the algorithm which Roy and Rousseau describe in detail. A key idea of these authors is to state the heuristic steps as general operations on objects (cycles, routes, etc.). The criterion for selecting the objects is treated as a parameter so that it can be changed for greater flexibility. For example, in seeking two cycles $C1$ and $C2$ to merge in Step 1b, one may select $C1$ to have the maximum duration and $C2$ the minimum duration when constructing a morning route and reverse the selection criteria for afternoon routes. This augmentation of cycles is stated generally so that two cycles $C1$ and $C2$ may be merged directly or $C1$ may be merged with a portion E of $C2$ that is adjacent to $C1$, as long as $C2$ is not disconnected by the extraction of E. Similarly, Step 1c is structured so that it can execute both insertions and generalized exchanges in a recursive fashion. In these manipulations, the groupings created by Step 1a move as a unit.

In Step 2, the transformation described by Pearn, Assad & Golden [1987] is used to represent the construction of the Euler tour as a TSP. The relevant traversal penalties can then be represented as costs C_{pq} for joining nodes p and q of the TSP. Roy and Rousseau use a simple nearest neighbor heuristic to solve this TSP. In Step 3, the assignment cost of pairing morning route $AM(i)$ with the afternoon route $PM(j)$ is defined as

$$C_{ij} = [(T_{ij} - 480)^+ + (465 - T_{ij})^+]^{3/2} + [(d_{ij} - 1500)^+]^2,$$

where

$T_{ij} =$ sum of the durations of routes $AM(i)$ and $PM(j)$,

$d_{ij} =$ the distance between the starting nodes of the two routes,

and $(x)^+ = \text{Max}\{x, 0\}$ simply denotes the positive part of x. The second term attaches a penalty to inter-route distances above 1500 feet. The first term penalizes workday lengths that go beyond 480 minutes or fall short of 465 minutes. The inclusion of these penalties encourages the construction of a balanced set of workdays.

Roy & Rousseau [1989] provide computational results on applying the algorithm to two postal areas. The first area has a total service time of about 172 hours and is represented by a network with 353 nodes and 1116 arcs. The service time of the second area is 170.63 hours, with 683 nodes and 1972 arcs in its network. The algorithm reduced the total transportation time by 4.3% for the first area and maintained the same number of carriers (25). In the second area, a 48% reduction in travel time allowed the number of carriers to decrease by one (21 to 20). Roy and Rousseau conclude that important savings opportunities exist for the CPC in this area.

Mailbox collection routes

At Canada Post Corporation (CPC) trucks based at a postal station carry out two key operations on a daily basis: deliveries to relay boxes and mailbox collections. The latter operation occurs at different times of the day and with different frequencies. In a city like Montréal (of about 3 million people), there are 3084 mailboxes on the streets. Mail from these boxes is collected from one to four times daily and delivered to 71 postal stations. According to Laporte, Chapleau, Landry & Mercure [1989], one to three vehicle routes may be based at a postal station, each visiting 20–50 mailboxes. However, the routes and clusters may change over different collection periods.

The collection problem is clearly a node routing problem: the nodes correspond to the mailboxes to be visited by each vehicle. Once the clustering problem of assigning a subset of mailboxes to each vehicle is solved, a minimum-length tour through the nodes of this subset must be found for each vehicle. Laporte and coworkers concentrate on this second routing step. Their contribution is to explicitly address the dependence of the shortest path distance matrix on the itinerary of the vehicle. Since this dependence raises interesting issues regarding the use of geographical databases on the block-face level, we provide a brief description of the problem.

Consider the shortest path between two mailboxes A and B in Figure 15, both of which are located at street intersections. This distance depends on the direction of travel of the vehicle upon its arrival at A. The simplest case is path P1 where both A and B are visited by following a single street. On the other hand, if the vehicle approaches A along path P2, mailbox A is served along the edge AG and the vehicle must go around block III to complete the traversal from A to B. Alternatively, if left turns are allowed, path P3 may be followed to go around block II and serve B along the edge DB. Path P4 shows how the direction of approach to A can be changed by executing a U-turn. If street crossing is allowed, it is possible to avoid the U-turn and only perform a left turn, but this involves stopping at the other side of the street (opposite to A) and is considered undesirable. The path P5 corresponding to this does not appear in the figure.

To compute the shortest distances corresponding to these paths, one can replace A and B with A1, A2, B1, and B2 as shown in Figure 15. This results in attaching a copy of A to each of the streets defining the intersection, A1 to the segment AG and A2 to AH; and similarly for B. The distance of the paths in Figure 15 can then be computed as $SP(A2, B2)$ for P1, $SP(A1, B2)$ for P2, $SP(A1, B1)$ for P3 and so on. Clearly, turn penalties can be incorporated into these distance computations.

The node duplication for mailboxes at intersections simplifies the distance computations but complicates the routing (sequencing) problem: Since only one

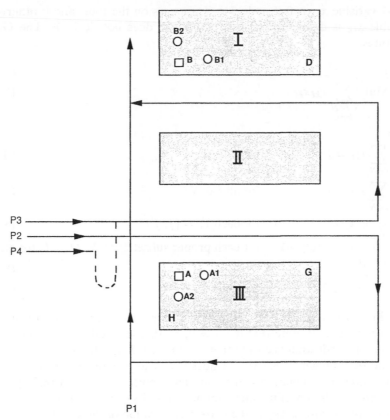

Fig. 15. Shortest path between mailboxes A and B located at street intersections. Paths P1–P4 show four ways of traveling from A to B while servicing both.

of the two copies requires service, a generalized TSP must be solved for each vehicle, where only one of the two nodes corresponding to the same mailbox appears in the tour. Laporte, Chapleau, Landry & Mercure [1989] formulate this problem as follows.

Let the underlying street network be represented by the directed graph $G = (V, A)$. Let A_o be the set of arcs $a_i = (k, \ell)$ in G that contain one or more mailboxes at some *intermediate* point along the street segment joining k and ℓ. Let $N(A_o)$ be the set of all nodes i created to represent the arcs a_i in A_o. Next, let J be the set of all mailboxes located at street intersections (nodes) of G. For each $j \in J$, let $T_j = \{j, j'\}$ contain two copies of node j, and let $T = \bigcup_{j \in J} T_j$. Finally let d be an additional dummy node.

The generalized traveling salesman problem (GTSP) on the node set $N = N(A_o) \cup T \cup \{d\}$ seeks a minimum cost subtour that visits all nodes in $N(A_o) \cup \{d\}$, as well as exactly one node from each subset T_j. For k and ℓ in N with $k \neq \ell$, the binary variable $x_{k\ell}$ is 1 if the arc (k, ℓ) is the tour and 0 otherwise. For j in

T, the 0–1 variable x_{jj} is 0 if node $j \in J$ appears on the tour, and 0 otherwise. The variable x_{kk} is defined to be zero if node k does not lie in T. The GTSP formulation is:

(GTSP):

$$\text{Min} \sum_{\substack{k,\ell \in N \\ k \neq \ell}} c_{k\ell} x_{k\ell} \tag{8.7.1}$$

subject to

$$\sum_{\ell \in N} x_{k\ell} = 1 \qquad \text{for all } k \text{ in } N \tag{8.7.2}$$

$$\sum_{k \in N} x_{k\ell} = 1 \qquad \text{for all } \ell \in N \tag{8.7.3}$$

$$x_{jj} + x_{j'j'} = 1 \qquad \text{for each } T_j = \{j, j'\} \tag{8.7.4}$$

$$\sum_{\substack{k,\ell \in Q \\ k \neq \ell}} x_{k\ell} \leq |Q| - 1 \qquad \text{for each proper subset } Q \subset N, \, Q \neq T_j \text{ for all } j \tag{8.7.5}$$

$$x_{k\ell} = 0 \text{ or } 1 \qquad \text{for all } k, \ell. \tag{8.7.6}$$

Constraints (8.7.2) and (8.7.3) are modified versions of the usual assignment constraints for the TSP. Note that if $x_{jj} = 0$, so that the node $j \in T$ appears in the solution, these constraints ensure that j is preceded and followed by two other nodes in the solution while (8.7.4) precludes a visit to j'. Constraint (8.7.5) is the well-known subtour elimination constraint. The choice $Q = T_j$ in (8.7.5) is not required as (8.7.4) already eliminates subtours on pairs of nodes in T_j.

The choice of arc costs is key to the correct application of the GTSP to the mailbox problem. Let h and i be nodes in $N(A_o)$ representing arcs $a_h = (e, f)$ and $a_i = (k, \ell)$ of A_o. Then $c_{hi} = SP(f, k)$ and $c_{ih} = SP(\ell, e)$ with shortest path distances computed in G. If $j \in J$, then $c_{ij} = c_{ij'} = SP(\ell, j)$ and $c_{ji} = c_{j'i} = SP(j, k)$. Finally, since Laporte, Chapleau, Landry & Mercure [1989] originally start with a Hamiltonian path problem between two specified nodes in $N(A_o) \cup T$, node d is used to convert the path problem to a GTSP tour problem by setting c_{ds} and c_{td} equal to 0 for the endpoints s and t of the path, and infinity for all other choices.

In the CPC application, Laporte, Chapleau, Landry & Mercure [1989] perform the initial clustering (for geographically compact clusters of stops) using an arbitrarily chosen copy of each node in J. Once the clusters are defined, however, each sequence of stops is examined to account for the effect of selecting the 'best' copy (j or j') of each $j \in J$. Specifically, given a sequence $(i_1, \ldots, i_n)$ of stops, the authors consider all subsequences of the form $(i_s, \ldots, i_{s+d+1})$ where the first and last stops are in $N(A_o)$ but all intermediate stops $\{i_u : s + 1 \leq u \leq s + d\}$ lie in T. A complete search determines whether replacing any $i_u = j$ (or j') by its alternate copy j' (or j) reduces the total cost. As the number of intermediate stops d rarely exceeds 5 in practice, the total number of evaluations required (2^d) is not prohibitive. Given this initial solution, Laporte and coworkers used the GTSP as a postoptimizer for each cluster. However, the computational burden of solving the GTSP limited its application to clusters of up to 30 nodes, corresponding to 20 original mailbox locations (ten of which are duplicated so that $|J| = 10$). As the CPC routes often involve 45–50 stops, a route may have to be broken into 2 or 3 segments before the GTSP is applied. Laporte and coworkers used 'natural breaking points' that arise from long driving stretches for this purpose. In their computational work, the GTSP was able to reduce the distance by 3 to 14.5%, but produced no reduction in 3 out of the 10 clusters. On the whole, the authors concluded that the GTSP provided a useful tool for these postal problems.

Locating postal relay boxes

Earlier in this section, we mentioned that CPC carriers replenish their mailbags from relay boxes located along their routes. The need for relay boxes is driven by the maximum weight W of mail the carrier can carry (W is approximately 35 lbs). Bouliane & Laporte [1992] use a set covering model to solve the problem of minimizing the number of relay box locations subject to this weight constraint.

In this formulation, the street segments (or arcs) $a_i (i \in I)$ correspond to the rows and the nodes $n_j (j \in J)$ of the street network form the columns. The input data allows one to specify the matrix $A = (a_{ij})$ where $a_{ij} = 1$ if arc a_i covers node j and 0 otherwise. The set covering problem becomes

(GTSP):

$$\text{Min} \sum_{j \in J} y_j \tag{8.8.1}$$

subject to

$$\sum_{j \in J} a_{ij} y_j \geq 1 \quad \text{for all } i \in I \tag{8.8.2}$$

$$y_j \in \{0, 1\}. \tag{8.8.3}$$

Since $y_j = 1$ means that at least one relay box is located at node n_j, the objective (8.8.1) simply minimizes the number of locations for relay boxes. It remains to show how the matrix A is specified. Bouliane & Laporte solve the relay box problem assuming that the carrier routes are already fixed. Thus each arc a_i is served by a unique carrier $k(i)$. Consider a node n_j and all the routes that pass through this node. Consider a carrier k and any arc a_i served by k at some point of its route subsequent to n_j. We say that n_j covers a_i and set $a_{ij} = 1$ if and only if the total load delivered by carrier k on the segment between node n_j and the endpoint of a_i is no larger than W. In other words, if a relay box is located at n_j, carrier k can feasibly deliver to all arcs on route k from n_j through a_i. This shows that one can determine the set of all arcs covered by n_j simply by tracing through the routes of all carriers passing through n_j and accumulating the delivery loads of each carrier.

Bouliane & Laporte [1992] applied this algorithm to a postal territory in Montréal with $|J| = 138$ nodes and $|I| = 384$ arcs, in which CPC used 33 relay boxes. The set covering algorithm identified a feasible solution with 13 relay boxes and established a lower bound of 12 on the objective. While this marked reduction is noteworthy, it is important to realize that the covering problem solves only one part of a more complicated problem. For example, the authors point out that when the replenishment costs of the relay boxes are considered, it is possible to reduce costs by locating more boxes! Clearly, there is a complex interaction among the structure of carriers' routes, replenishment routes, and the cost of locating relay boxes. The objective of minimizing the number of relay sites is an approximate surrogate for the more complex problem.

9. Routing for meter reading and snow control

9.1. Meter reading

Electric power and water companies routinely send meter readers to record the consumption of utilities at households. The design of efficient routes for meter readers is an arc routing problem since almost all residential and commercial units along street segments need to be visited. Generally, these routes are designed on a periodic basis, so that each neighborhood is re-visited every w weeks, where w is the cycle time specified by the company.

The earliest full-scale study of routing for meter readers is the analysis that Stern & Dror [1979] carried out for the city of Beersheva in Israel. At the time of this study, the city was divided into eight neighborhoods and the limit T on each walking tour was 5 hours. In their paper, Stern and Dror present a typical neighborhood with 42 nodes and 62 edges. Using their arc routing heuristic, Stern and Dror were able to reduce the number of tours from 24 to 15, thereby producing time savings of approximately 40% for the neighborhood.

The algorithm designed by Stern and Dror is a 'route-first–cluster-second' approach: first a covering Euler tour that traverses all required arcs is identified; next, it is partitioned into a number of walking paths observing the 5-hour limit on the duration of each path. Since we have described the tour partitioning approach in Section 7.2, we do not reproduce the details of the Stern–Dror procedure here. Rather, we point out two key characteristics of their application:

(1) The underlying network for the city is taken to be undirected since a reader can cover any street segment in either direction. Narrow streets can be covered with a single traversal in 'meander' mode. Not all street segments are required, but G_R is connected. U-turns are allowed only at intersections.

(2) Each reader is driven by car from the central office to the first address on her route but returns by public transportation (bus). The routes are therefore walking *paths* (rather than cycles). Moreover, since service times of some edges are comparable to T, walking paths may begin or end at intermediate points of an edge. In their heuristic, Stern and Dror cut the covering Euler tour 'within' an edge if necessary. Clearly, if such cut points occur at an intermediate point of a deadhead edge, the entire edge can be eliminated by moving the end of the path back to the terminal node of the last serviced edge. All other factors being equal, one may try to partition the Euler tour in such a way as to take advantage of this.

Finally, we point out that, in this application, the ratio of the service time to the deadhead time — $t_s(i, j)/t_d(i, j)$ — ranged from 10 to 40. For example, a short street segment that can be traversed in 3 minutes without servicing meters can take as long as one hour to 'read' due to the presence of large apartment complexes. These large ratios show that reductions in deadhead time may not be of great concern; the key objective is to minimize the number of tours required.

A more recent application of routing to meter reading is the study by Wunderlich, Collette, Levy & Bodin [1992] conducted for the Southern California Gas Company (SOCAL). Comparing this study with that of Stern and Dror highlights the advances in computerized routing over a period of a dozen or so years. The scales of the two studies are also different: SOCAL spends over $15 million a year on meter reading. Each of SOCAL's 10,042 routes is read once a month, an ongoing operation that requires over 475 meter readers.

Prior to the study, SOCAL had compiled a PC-based database of meter-reading standards that revised service times downwards based on negotiations between the company and the meter-reading union. This reduction in the amount of time allocated to each street segment and growth in certain service areas necessitated the generation of a new set of routes. Historically, the company had been slow to restructure routes and had a four-year backlog of routes requiring revision. The main objective of Wunderlich, Collette, Levy & Bodin [1992] was to assess the potential benefits of using a computerized routing system based on optimization algorithms.

For the purposes of their study, Wunderlich and coworkers focused on a sample region that represented 2.5% of the total area served by SOCAL. The region extended from Santa Monica to Hollywood and included a variety of terrains (hills, grades, etc.). SOCAL covered this region with 242.5 routes that required walking, driving, and mixtures of the two. The sample region comprised 6900 street segments (centerlines). The routing system produced routes over this network and compared them with SOCAL's existing routes. The main routing algorithm is an enhanced version of the Arc Partitioning Algorithm (APA) designed by Bodin & Levy [1991] described in Section 8.2. This procedure divides the service region into balanced routes, each with a workload W_k bracketed by given bounds as in (8.1) and (8.2). The APA requires the number of routes (or partitions) as input. In their work for SOCAL, Wunderlich and coworkers estimated this number, denoted by n_r, as follows:

Let

$$\text{WTOT} = \text{total workload for the partition network}$$
$$= \sum_{e(i,j)\in R} t_s(i, j) + \sum_{e(i,j)\in D_1} t_d(i, j)$$

and

$$\text{UTS} = 8 \text{ hours} - \text{office time} - \text{average depot travel time}.$$

Compute n_r as

$$n_r = \left\lceil \frac{\text{WTOT}}{\text{UTS}} \right\rceil \tag{9.1}$$

and set

$$\text{WAUX (auxiliary workload)} = \text{WTOT} - \left\lfloor \frac{\text{WTOT}}{\text{UTS}} \right\rfloor \cdot \text{UTS}. \tag{9.2}$$

As defined above, UTS stands for the useable time for service; it represents the total time the meter reader can spend covering the service territory once the initial preparation time and the travel time on deadhead legs from and to the depot are subtracted out. With these definitions, it is clear that at least n_r routes are needed.

To obtain routes that fully utilize the meter readers' time, Wunderlich, Collette, Levy & Bodin [1992] use the notion of an *auxiliary or remnant route* that absorbs the remaining work once the other routes are filled to UTS. More specifically, consider the case where the ratio WTOT/UTS in (9.1) is non-integer. The APA is directed to construct $n_r - 1$ routes of equal size once the auxiliary route is allocated a workload of approximately WAUX, as defined by (9.2). The algorithm grows the auxiliary route first, out of a seed point specified by the user. In practice, the user may decide to place the auxiliary route in an area of service that anticipates high growth (so that the workload for the route may grow to fill an entire workday). Alternatively, the route may be located on the boundary of one region, so that it may be merged with an auxiliary route from an adjacent region. The user locates the seedpoint to guide the formation of the auxiliary route; the algorithm constructs the remaining routes to achieve balance.

The results of the APA for the SOCAL sample region are shown in Table 5. In the APA, the route duration bounds — T_L and T_U in (8.2) — were set at 7.9 and 8.0 hours respectively (this includes the office time and depot travel time). With this choice of bounds, APA produced routes that ranged from 7.77 to 8.11 hours in duration, as compared to SOCAL's range of 7.03 to 8.56. APA reduced the number of whole routes from 235 to 230 and decreased the number of auxiliary routes to 6 (from 7.5). Clearly, APA was able to create full routes close to the 8-hour limit with less overtime and less deadheading. Approximately 60 minutes of deadheading were used to connect the components of the partition network.

To estimate the savings for the entire region served by SOCAL, Wunderlich et al. extended the percentage savings of the sample region to the entire area. For example, using a daily cost of $159.98 for a meter reader, and recalling that each route is read once a month, the monthly savings associated with the 6.5 fewer routes in the sample region is $1039.87. As the entire service area is $10,042/242.5 = 41.41$ times larger, one can project the annual savings of route reduction as $12 \times 41.41 \times \$1039.87$ or $516,736 for SOCAL. This number appears in the last column of Table 5, (B). Wunderlich and coworkers point out that the $252,103 in savings obtained in a similar way by comparing the overtime in SOCAL's routes to the automated routes is an underestimate and must be revised upwards to $357,450. The reason for this correction is that SOCAL's routes for the sample region contain less overtime (per route) than the average overtime per route over the entire region. Indeed, 3827 of SOCAL's routes required overtime, and the average overtime was 24 minutes as compared to 10.9 minutes in the sample region [Table 5, (A)]. The company savings were expected to exceed $874,000 after this correction. To this, one must add the savings associated with

Table 5

Summary of results for SOCAL's meter-reading operation [adapted from Wunderlich, Collette, Levy & Bodin, 1992]

(A) Characteristics of sample region		
	Sample region	Entire service area
Number of routes	242.5	10,042
Number of sections	10	over 800
Number of centerlines	6900	over 0.5 million
Average overtime per overtime route (min)	10.9	24

(B) APA results and associated savings				
	SOCAL routes	APA routes	% Savings	Annual saving for entire area
Number of routes	242.5	236	2.7	$516,736
Average route length (hr)	7.82	7.95		
Overtime (min)	293.4	110.4	183	$252,103
Deadhead times (min)				
– walking	693.4	63.1	630	–
– driving	545.5	133.2	412	–

replacing the time-consuming manual route alignment methods that were slow and suffered from a large backlog.

This summary does not fully reflect the implementation challenges of this study for SOCAL. The use of the system across SOCAL's entire service region involves a major data collection and management effort. To mention one challenge pointed out by Wunderlich and coworkers, the SOCAL street data were not compatible with the input requirements for the routing system. The algorithms require information on centerlines (or arcs) — a term that refers to both sides of the street segment between two intersections (nodes). SOCAL, on the other hand, traditionally dealt with street segments that comprised several centerlines so that their basic unit of service was an aggregation of street segments. This required a manual breaking up of SOCAL's segments into centerlines, a procedure that gave rise to several complications as follows:

(a) Not all centerlines within a SOCAL segment required service; some had no meters. In the sample region, the time associated with centerlines that could be eliminated was nearly 467 minutes.

(b) The service time associated with each SOCAL segment had to be divided among the centerlines. Due to the uneven distribution of meters and differences in type of meters, simple pro-rating was inappropriate; a new count of meters on each centerline was conducted.

(c) The connectivity of the service region was affected by the disaggregation. For example, removing a centerline from the set of required arcs may disconnect a network previously connected at all times.

9.2. Snow and ice control

A well-known application of arc routing involves snow plowing and the treatment of roads for snow and ice control. For certain regions of the U.S., the cost of this activity is very high. According to Haslam & Wright [1991], for example, the Indiana Department of Highways budgeted over $15 million during the 1987–1988 winter season to maintain approximately 11,400 miles of roadway (corresponding to 29,000 lane-miles) throughout the state. This operation required 1500 trained personnel and over a thousand maintenance vehicles. According to Campbell & Langevin [1993], Montréal is a global leader in snow removal operations. This city, which covers 187 square kilometers, has 2000 kilometers of streets and 3200 kilometers of sidewalks from which snow must be cleared. There are ten major snow storms each winter with an average snowfall of 223 cm. (s.d. 63 cm.) per storm. Snow removal requires the services of 60 crews who operate up to 1500 vehicles for spreading, plowing and hauling snow. As Haslam and Wright point out, the uncertainties of snow episodes, the required rapidity of response, the simultaneous deployment of a large fleet of vehicles and the interaction with ongoing traffic and parked vehicles make highway snow and ice control the most difficult public-sector routing problem.

Snow removal has been analyzed from a systems analysis perspective in a number of studies. The well-known article of Savas [1973] describes the analysis carried out for the mayor of New York City in the wake of the highly disruptive 15-inch snowfall on February 9, 1969 in the city. The study analyzed the capability of the city to clear and salt streets following a pre-established priority classification. In the recommended scheme, the highest priority was attached to a primary network that stretched over 1600 linear miles (out of a total of 5839 miles for the city) and accounted for 33% of the total plow miles. The actual workload resulting from a snowfall depends on a large number of factors that fall under the broad categories of climate, snowfall characteristics, road network, and traffic [see Minsk, 1979]. In their simulation model of snow removal, Tucker & Clohan [1979] use these factors to estimate the fleet requirements and the total plow time required for each truck. Their model accepts the truck routes as input and simulates the plowing activity over street segments. These authors also present a validation study of the simulation using the town of Newington, Connecticut. We point out that there is a substantial literature on 'engineering' issues in snow control — a representative sample of which appears in the handbook edited by Gray & Male [1981]. In particular, Minsk [1981] reviews the various types of snow removal vehicles. Facility location studies have been conducted to find the best configuration of snow disposal sites. Capacity constraints play an important role in the selection of such sites. In one of the earlier studies, Leclerc [1985] used a linear programming formulation for the site allocation problem. More recent investigations are cited in the review by Campbell & Langevin [1993].

From the perspective of routing, snow and ice control present the following practical complications.

(1) *Priorities*. Priorities reflect the different service requirements of individual streets and roads. According to Haslam & Wright [1991], for example, the state roads of Indiana are categorized into three classes based on historical *average daily traffic* (ADT) on the roads. Class I roads include major traffic arteries, interstates, and associated roads with ADT > 5000. Roads in this class receive continuous plowing and spreading of chemicals to keep the road surface bare. Generally, this translates to service every two hours during the snow event. On Class II roads, which have ADT values ranging from 1000 to 5000, only the center portion of the road is maintained, typically requiring service every 3 hours. Class III roads are generally served every four hours with the objective of maintaining passable routes. As Haslam and Wright point out, this classification is for planning purposes: the actual service provided is determined in response to the specific snowfall conditions while the operation is in progress. Other priority schemes include the identification of an emergency subnetwork (of the street network) that is fully maintained to ensure basic transportation needs.

(2) *Class continuity*. It is sometimes desirable that each route service streets with the same priority classification. Thus, if a lower-class street is included in a Class I route, its service level may be upgraded. For example, one may wish to assign a Class II road with a steep slope to a Class I route.

(3) *Tandem service*. The service provided to an individual street may involve traversal of two vehicles in tandem.

(4) *Multiple lanes*. Most arterial roads have multiple lanes that require separate passes. For this reason, the total workload is measured in lane-miles.

(5) *Service times*. The service time of a street depends on a variety of factors including 3 and 4 above. Tucker & Clohan [1979] provide some expressions for estimating the plowing time. The accumulation rate plays a major role in determining the service time. Since the traversal of a road is subject to a minimum plowing depth constraint, the accumulation rate also determines how soon the truck can repeat a route that has already been serviced.

(6) *Service-dependent travel times*. An interesting feature of snow control and removal is that the deadheading time on an arc depends on whether the arcs on the path have been treated. Clearly, it is faster and safer for trucks to treat a street segment prior to its use for deadheading. This dependence also occurs for service times if one considers that some of the snow plowed is cast onto the next pass lane on the right and must be treated on the next pass [see Tucker & Clohan, 1979].

(7) *Turn constraints*. The impact of undesirable turns is generally greater in routing snow plows as compared to sanitation operations. Certain locations may be pre-specified as turn-around locations while turns may be simply prohibited in some regions.

(8) *Presence of traffic*. Generally, plowing and spreading operations are carried out most efficiently within certain optimal ranges of speeds. When public traffic

or obstructive parking interfere with the movement of the snow control trucks, efficiency deteriorates and serious safety concerns may arise.

A single routing algorithm cannot hope to incorporate all of the preceding complications. However, most of the literature on routing for this application has attempted to capture some constraints in the context of algorithms that construct postman tours. In one of the earliest papers in the area, Marks & Stricker [1971] note the shortcoming of the postman problem as an appropriate model of snow plowing, and emphasize the need for observing priorities. Lemieux & Campagna [1984] were also motivated by the snow plowing problem to devise their procedure for observing priorities as discussed in Section 6.

Haslam & Wright [1991] consider the snow plowing problem as an m-RPP with priorities for which they construct route generation and improvement procedures. The improvement procedures are standard exchange and elimination heuristics, which are applied to an existing set of routes. The generation procedure, however, considers class priorities and invokes the class continuity constraint, treating the latter as a secondary objective (the problem would decompose by road class if class contiguity is strictly enforced). The heuristic generation procedure starts by computing a lower bound on the number of required routes by dividing the lane-miles in each class of roads by the maximum distance a truck may travel when servicing that class. The user then provides seeds with designated classes out of which snow routes are grown. Given a seed point and its class, the procedure constructs a shortest path from the seed to the depot and another shortest path in the reverse direction. Arcs on this path that are of the same class as the seed are added to the route in both directions. Next, the route is expanded by adding contiguous arcs that have the same class. Finally, in the last stage of the procedure, the class constraint is relaxed and routes are expanded by adding arcs with lower classification (so that these are upgraded). In all phases, the expanding route is checked for feasibility against the maximum travel distance constraints.

A key shortcoming of the preceding procedure is that the routes it produces may fail to cover all arcs of the network. Haslam and Wright state the need for a better procedure that can relax the distance constraints. They also acknowledge that the procedure performs no optimization other than producing shortest paths. Given the complexities of growing compact regions out of seed nodes in the partitioning algorithm of Section 8.2, the required modifications to the procedure of Haslam and Wright may be far from trivial. These authors also applied the improvement and elimination procedures to the road network for the Fowler subdistrict, a test site for the Indiana Department of Highways (INDOT). The network for this test area has 99 nodes and 362 arcs, and a total of 903.5 lane-miles. Moreover, the set of required arcs forms a nearly balanced graph.

In further work for INDOT, Wang & Wright [1992] describe an interactive route design system called CASPER. This system uses a heuristic to generate an initial set of routes, and then applies a local improvement procedure based on tabu search. The details of these procedures appear in Wang [1992]. Apart from its algorithmic component, the system relies upon user-directed control of route features and decisions regarding the extent of upgrading. The authors provide the results obtained with the system for service areas from two districts in Indiana. The system can reduce deadheading and eliminate routes without reducing the level of service. In another study, Evans [1990] describes a software system for the routing of snow and ice control vehicles that was tested in Butler County, Ohio. The road network for this county has 185 road segments and a total of 284 lane miles and was originally served by 13 trucks. The system reduced the fleet by four vehicles and cut the total time by 40%.

Apart from snow plowing, some authors have considered snow and ice removal through salt spreading or the application of other abrasives to clear the road surfaces. In this problem, the routing procedure must consider the vehicle capacity for salt and the location of salt piles if these do not exclusively reside at the depot. Cook & Alprin [1976] considered the salt spreading problem in the city of Tulsa. At the time of their study, Tulsa had no plowing equipment and depended solely on spreading for snow and ice control. Since most streets on emergency routes are major arteries, it takes two passes to spread salt on the street. The spreader truck can therefore turn around after spreading salt over half the street to cover the other half unless the street is one-way. This operation avoids deadheading at the expense of requiring U-turns. Cook and Alprin argue that the key objective is to balance the workload rather than to minimize deadheading. They define a street segment as the distance along which a street can be salted on both sides "at the application rate with one truckload of salt". With this definition, only one street segment is serviced in each spreader tour and the deadhead time from and to the salt pile facility becomes important. Note, however, that this notion of a street segment is sensitive to the weather conditions so that new segments are necessary if the storm is severe. Cook and Alprin propose a simple algorithm to balance the total workloads assigned to the different spreader trucks. The procedure, which they call the *closest street heuristic*, assigns the untreated street segment that is closest to the salt pile facility to the truck. This rule is used each time the truck returns to the salt pile facility. According to the authors, one advantage of this procedure is that each time a truck sets out from the depot, it deadheads over streets that are already salted — this is desirable in view of constraint (6) mentioned earlier.

Cook and Alprin evaluated the performance of the closest street heuristic within a simulation model of the salt spreading operations for Tulsa, Oklahoma. The simulation model simulates a snow emergency in which trucks are equipped with spreaders, loaded with salt and dispatched to the street segment chosen by the heuristic. Once the salting is performed, the truck goes to the salt pile facility for re-loading, whereupon the service cycle is repeated. The simulation also incorporates waiting times incurred by trucks when queueing for the safety check (after spreaders are mounted) or for re-loading at the salt pile. The heuristic was able to reduce total spreading time by approximately three hours (a reduction of 36%) as compared to Tulsa's pre-assigned routes. The authors also used the simulation model to evaluate the benefits of adding two new salt pile facilities and decided against opening them. Finally, we note that in the absence of priorities, the salt spreading routing problem is very similar to the sanitation problem, with the salt pile facility playing the same role as the landfill. Therefore, the techniques of Section 7 apply to the problem solved by Cook and Alprin and extend to the case where routes contain multiple serviced arcs.

Eglese & Li [1992] describe a more recent application of routing for salt spreading (called winter gritting in their work). Their goal is to treat roads within a certain number of hours from 'call out'. This response time incorporates priorities: important roads must be cleared within two hours of call out while other roads may wait up to 4 hours for treatment. In evaluating the efficiency of their routing procedure, Eglese and Li computed the ratio of salting distance (treated miles) to the total distance covered. This ratio varied from 56% to 75% in the 31 county divisions they studied. An upper bound on this ratio can be obtained by evaluating it for the single uncapacitated postman solution. Even this upper bound rarely exceeded 75% in their test cases, a fact that the authors explain by noting the large incidence of 'T-junctions' — nodes at which three roads meet — in the rural areas they were considering. Such nodes of degree 3 create deadheading that is reflected in the low efficiency ratios. In contrast, city networks are often more nearly balanced and yield higher ratios.

As mentioned before, two key constraints drive the salt spreading problem: (a) the time by which roads in each class are treated and (b) the capacity of the vehicles. The time constraint applies to the completion time of the gritting operation on each route and does not consider the return trip to the depot. The vehicle capacity constraint translates into a maximum gritting distance. As Eglese [1994] and Li & Eglese [1994b, c] point out, one can view the salt gritting problem as a distance and time-constrained CARP. Eglese [1994] describes an algorithm for this problem that uses the procedure developed by Male & Liebman [1978] for refuse collection as its initial step. This procedure groups edges into minicycles that are subsequently aggregated into vehicle routes based on the time and capacity constraints (see Section 7.1). Eglese employs simulated annealing to improve the initial solution by considering movements of nodes on the adjacency tree defined on the minicycles. The objective uses the number of routes as well as penalties associated with violating the time and distance limits. Li & Eglese [1994b] compare the performance of simulated annealing and tabu search on salt gritting problems for service areas with 77, 140 and 254 nodes and 111, 203 and 380 road segments, respectively. Li & Eglese [1994a] describe an algorithm based on the farthest insertion procedure with additional decision rules to determine if the vehicle should head back towards the depot. In tests on the three areas mentioned above, they report that the interactive procedure using input from the user outperforms the automated procedure, reducing the number of vehicles from 12 to 10 in one case, and from 20 to 17 in another.

Li & Eglese [1994c] have also defined the *time constrained arc routing problem* (TCARP) to capture the time considerations of the salt gritting problem. In this problem, the time limit is applied to each route up to the end of the last required movement (gritted segment). Li & Eglese derive a lower bound for TCARP based on an analysis of d-paths similar to that of the NDMPB1 bound of Section 4.3.

In a recent mathematical modeling competition, teams from various colleges and universities were presented with the problem of designing snow routes for a district spanning about 50 square miles in Wicomico County, Maryland. The county roads were two-lane and formed a road network with 140 nodes and 424 directed arcs. Two snow plow trucks operating out of a single garage were available. A map of the area was provided, but all road distances and travel times had to be estimated by the teams. The teams were able to construct good solutions to the problem by combining analysis with intelligence that took advantage of the problem-specific features of the network. From a pedagogical point of view, the range of assumptions and approaches adopted by the different groups is of interest. We now present a brief summary of these approaches.

Atkins, Dierckman & O'Bryant [1990] first partitioned the service network into two approximately equal subnetworks and then constructed an Euler circuit for each subnetwork. Their procedure for constructing the circuit involves tracing a walk on a spanning tree for the region, into which the non-tree edges are incorporated in a simple fashion. Chernak, Kustiner & Phillips [1990] considered a priority scheme where the primary run of the trucks services only the main roads, relegating the other roads to a second run. They also considered the width of the plow blade explicitly and investigated its impact on the routes. Robinson, Ogawa & Frickenstein [1990] used the route-first–cluster-second approach of Section 7.2, paying specific attention to U-turns. Finally, Hartman, Hogenson & Miller [1990] simultaneously built two Euler tours emanating from two seed points that served as entry points into the service region. Their procedure tries to grow the tours in a balanced way so as to end up with two balanced (more or less equal distance) tours for the entire region. While this procedure performed well for the network under investigation, there is no guarantee that the desired balance will be achieved in general. In summary, this collection of papers shows that small arc routing problems can be solved quite effectively with simple algorithmic techniques. Naturally, for larger problems, more systematic procedures are required.

10. Applications in manufacturing and other areas

It is natural to think of arc routing problems in the context of transportation or distribution applications. In this section, we describe applications that arise in a manufacturing context. An interesting feature of these applications is the different network structure that may result. For instance, in road networks, any required arc is also available for deadheading. In manufacturing applications, however, the set of available non-required arcs may be markedly different from (or smaller than) the set of required arcs. The network itself may have no physical counterpart: the network of the circuit testing problem described later in this section corresponds to the state transition graph for a finite state device. In general, the contrast between required and non-required arcs is sharper outside the sphere of transportation problems. In the last part of this section, we return to transportation applications in which a possible arc routing formulation completes with other formulations or approaches to the problem.

10.1. Pierce point minimization for cutting operations

Manber & Israni [1984] describe an interesting application of arc routing in manufacturing cutting processes where required parts with specified shapes are cut out of a rectangular sheet of stock. The stock material can be glass, sheet metal, textile or lumber and the cutting apparatus may be a plasma arc, laser beam, or flame torch (the authors focus on the torch). In the well-known cutting-stock problem, the layout of parts on the sheet, called *part nesting*, is obtained by considering trim losses. Manber & Israni take the layout as fixed input and turn their attention to the cost of the cutting operation. They argue that the number of pierce points drives an important component of this cost and they seek to minimize this number. A *pierce point* occurs when the flame torch has to blow through the stock sheet anew, at a point that is not contiguous to the most recent cutting path of the torch. Such new 'starts' take time and money and must be minimized. Generally, the required parts are defined by polygonal boundaries as in Figure 16. To minimize the number of pierce points, one must cut along the edges of all required parts using paths that are as 'contiguous' as possible. The key decision is to select non-required paths that connect disjoint cutting paths, thereby creating greater contiguity.

To make this more concrete, consider the required parts shown in Figure 16. Eight parts with various geometrical shapes are identified. First, we simplify the network associated with this part nesting by aggregating edges that are always covered contiguously in any Euler cycle. This corresponds to replacing edges by a single edge by eliminating intermediate nodes of degree 2. For instance, along the path 7–14–15–10, nodes 14 and 15 (which have degree 2) can be removed to form the single edge (7,10). The reduced graph appears in Figure 17 and has multiple edges between the node pairs (3,4) and (11,12). In general, the aggregation results in a multi-graph.

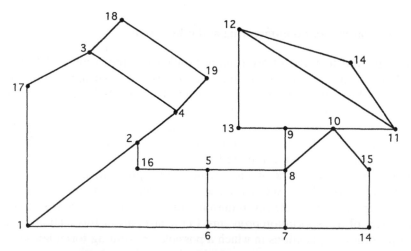

Fig. 16. Part nesting showing eight required parts.

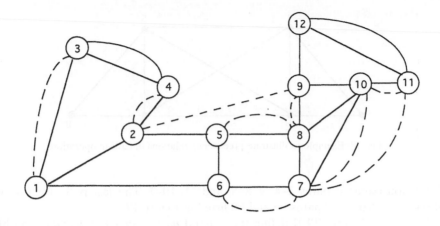

PATHS

P1	1-3-4-3
P2	4-2-5-6-1-2
P3	6-7-8-10-7
P4	5-8-9-12-11-12
P5	9-10-11

Fig. 17. The required network G_R (in bold) and five paths that cover it. Available deadhead edges are dashed.

Since the graph G_R in Figure 17 has 10 nodes of odd degree, all of its edges can be covered by 5 paths as listed in the same figure. These paths also identify the five pierce points. For example, starting with the path 1–3–4–3, the torch flame cuts through to node 3 and subsequently moves on to start the next path 4–2–5–6–1–2. Nodes 1 and 4 are the pierce points associated with the start nodes of the two paths (since no cutting occurs in going from node 3 to node 4). Naturally, the number of pierce points is reduced if the torch continues to cut along the non-required path that connects the end of one path with the start of the next.

Non-required movements in the flame torch problem correspond to cutting along non-required edges. Manber and Israni call these *trim margin edges* and assume that they are specified as input. To identify allowed trim edges, one must consider the possible re-use of the sheet remnants once the required parts are extracted. In Figure 16, for example, the edge (2,9) can be a trim edge since the sheet region bounded by the path 4–2–16–5–8–9 is unfit for re-use. This means that one may cut through this region without creating additional trim loss. Similarly, trim margin edges close to the boundary of the sheet — such as edges (6,7), (7,10) and (10,11) — do not impact re-use adversely. In general, one expects the set of available non-required edges to form a very sparse network. In Figure 17, the dashed edges represent trim margin edges provided for G_R. In the example, we can reduce the number of contiguous paths from 5 to 2 by using the

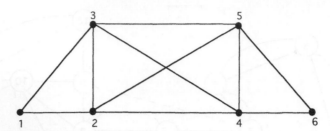

Fig. 18. Example to illustrate precedence relations in cutting operations.

three non-required paths 2–4, 9–8–5, and 11–10–7. The last path, for instance, allows us to 'connect' paths P3 and P5 listed in Figure 17.

Manber & Israni [1984] define the general *pierce point minimization* problem (PPM) as follows. Given an undirected graph $G = (V, R \cup A_d)$, where R denotes the set of required edges, and A_d the available non-required (trim margin) edges, find the minimum number k of paths $P_1, \ldots, P_k$ that satisfy the following:
  (i) the paths are edge disjoint, and
  (ii) together, the paths cover all edges in R.
Note that this problem can be viewed as a path version of the rural postman problem (RPP) of Section 3.

To find the non-required paths that reduce the number of pierce points, one has to connect odd vertices of $G_R = (V_R, R)$. Let V_o denote the vertices of odd degree in G_R and let $G_d = (V, A_d)$ be the graph of the non-required edges alone. Manber and Israni show that if G_d has c connected components and h_i denotes the number of nodes from V_o that lie in component i, then one can find $\sum_{i=1}^{c} \lfloor h_i/2 \rfloor$ non-required paths that connect nodes in V_o. Their algorithm for finding such paths uses a spanning tree in each component of G_d to construct the paths.

The authors also consider the problem of sequencing the cutting operation subject to the manufacturing constraint that the outer perimeter of a piece be cut only after all the edges within it are cut. This precedence relation anchors the pierce to avoid heat distortion. In Figure 18, for instance, edges (2,3), (3,4), (2,5) and (4,5) must be cut before the edges on the outside perimeter of the nested piece, as accomplished by following the cutting path 1–2–3–4–5–2–4–6–5–3–1. Manber and Israni show that a suitable path obeying the traversal precedence constraints always exists for Eulerian graphs. Intuitively, the path 'peels' the complex piece from within, moving progressively towards the outside boundary.

10.2. Optimal insertion sequence in printed circuit board assembly

Ball & Magazine [1988] describe an interesting application of the rural postman problem in the context of manufacturing printed circuit boards (PCBs). The insertion step of this manufacturing process consists of a collection of pick-and-

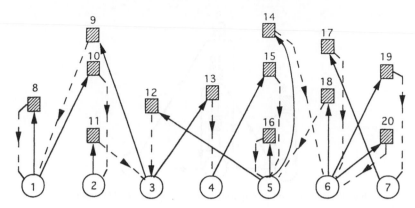

Fig. 19. Chip insertion with feeder nodes 1–7 and chip location nodes 8–20. The path 1–8–1–10–12–11–3–13–4–15–5–16–14–6–19–7–17–6–20–6–18–5–12–3–9–1 is a feasible Euler tour (deadhead arcs are dashed).

place operations which may be performed in any sequence. The basic required task for the automated placement machine is to move a required component from its feeder location to a specified chip location for insertion. The objective of the problem is to obtain the best sequence of insertions in the sense of minimizing the total insertion time (for all tasks). It is assumed that the travel time of the placement device (or *head*) is proportional to the distance traveled (that is, constant speed of travel).

The chip insertion problem can be represented as a directed network. The nodes correspond to feeder and chip locations and the required movements form arcs from the feeder nodes to the chip nodes as shown in Figure 19. To create an Euler cycle that covers all required arcs, one must add non-required (or deadhead) edges from the chip locations *back to* the feeder nodes. Figure 19 shows one such cycle, where all deadhead arcs are dashed.

Given the chip insertion network $G = (V, A)$ and a subset of required arcs $R \subset A$, the rural postman problem (RPP), as defined in Section 3, is to select a set D of non-required edges in A so that $\tilde{G} = (V, R \cup D)$ is Eulerian and the sum of edge costs in D is minimized. Note that non-required edges may be covered more than once so that $\tilde{G}$ is a multi-graph. A key contribution of Ball & Magazine [1988] is to observe the significant simplification that results from using the Manhattan (or ℓ_1) metric defined by $d(x, y) = \sum_{i=1}^{n} |x_i - y_i|$ for vectors x, y in R^n. The key advantage of this metric is that any movement of the head can be decomposed into the sum of a vertical movement followed by a horizontal one.

Consider, for example, the path 3–9–1 in Figure 19 composed of the required arc (3,9) and the non-required arc (9,1). Decomposing the latter arc into the vertical move (9,2) and horizontal move (2,1) does not alter the RPP objective. Moreover, since a movement from node 9 back to *any* feeder node *must* incur the vertical movement, one can regard the compulsory move represented by (9,2) as

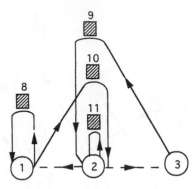

Fig. 20. Redefinition of required arcs (1,8), (1,10), (2,11) and (3,9).

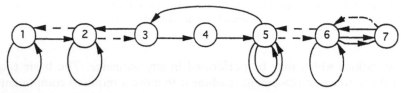

Fig. 21. Required network (in bold) and solution to the RPP.

an extension of the required arcs as shown in Figure 20. In general, any required arc (i, j) is extended to the path (i, j, k) where k is the feeder node directly below chip location j. This path is represented by a single required arc (i, k) between feeder nodes alone.

The preceding transformation represents $G_R = (V_F, R)$ as a network on the set V_F of feeder nodes given in Figure 21. To solve the RPP for G_R, the non-required arcs must also be specified. Non-required arcs are added in both directions between every pair of consecutive feeder nodes to form the graph Ball and Magazine call a *linear network*. Once again, the Manhattan metric allows one to restrict the use of non-required arcs to these 'short' non-required arcs since any 'long' non-required move (i, k) can be decomposed into a sequence of consecutive short non-required edges $(i, i+1), (i+1, i+2), \ldots, (k-1, k)$. Let A_F be the set of all non-required arcs in the linear network plus the required arcs in R. The chip insertion problem now reduces to the RPP on $\tilde{G}_F = (V_F, A_F)$.

In Section 3, we described the balance-and-connect (B&C) algorithm for the general RPP. The main result for the performance of this algorithm on the chip insertion problem is as follows.

Theorem 10.1 [Ball & Magazine, 1988]. *The B&C heuristic finds the optimal solution to the chip insertion RPP when the Manhattan (ℓ_1) metric is used. For arbitrary metrics, the value v_H of the B&C heuristic obeys $v_H \leq v^* + c_0$ where v^* is the RPP optimum and c_0 is the cost of the cycle that starts at the left-most feeder node, proceeds to the right-most node, and returns to its starting point.*

To motivate this result, it is useful to see the B&C heuristic applied to the required network G_R of Figure 21. Let $b(i)$ be the imbalance (in-degree minus out-degree) of node i considering required arcs alone and denote by $B_k = \sum_{i=1}^{k} b(i)$ the running sum of the imbalances accumulated from left to right. Clearly, $(b(1), \ldots, b(7)) = (1, 2, -1, 0, 0, -1, +1)$ and $(B_1, \ldots, B_7) = (-1, 1, 0, 0, 0, -1, 0)$. If x_{ij} denotes the flow on edge (i, j), a solution to the balancing step is $x_{21} = x_{23} = x_{76} = 1$ since these values satisfy the balancing constraints (3.1.2) of Section 3 at minimum cost. However, once the set of non-required arcs $A_B = \{(2, 1), (2, 3), (7, 6)\}$ is added to G_R, the resulting network has two connected components: $K = \{1, \ldots, 5\}$ is a cut-set that violates (3.1.3). To connect these components, the connecting step of B&C adds the non-required arcs (5,6) and (6,5) to obtain the final solution of Figure 21. This solution on the linear network can in turn be expanded into the Euler path for the original problem as shown in Figure 19.

In the general case, summing the balancing constraints (3.1.2) over $i = 1, \ldots, k$ gives the relation

$$x_{k,k+1} - x_{k+1,k} = B_k$$

for all k. This relation implies that any feasible solution to (3.1.2), (3.1.4) satisfies

$$(x_{i,i+1}, x_{i+1,i}) = \begin{cases} (B_i + r_i, \; r_i) & \text{if } B_i \geq 0, \\ (s_i, \; s_i - B_i) & \text{if } B_i \leq 0, \end{cases} \tag{10.1}$$

for some set of values $r_i, s_i \geq 0$. The particular choice of zero for all r_i, s_i yields $(x_{i,i+1}, x_{i+1,i}) = ((B_i)^+, (B_i)^-)$, where $(B_i)^{\pm}$ are the positive and negative parts of B_i. When the arc costs are all non-negative, this solution solves the balancing step of B&C optimally for the linear network. However, if the set of arcs A_B selected by the balance step (those with $x_{ij} > 0$) fails to produce a connected network $G_B = (V_F, R \cup A_B)$, as in the example just described, additional non-required edges are needed to establish connectivity. Ball and Magazine show that the addition of these arcs corresponds to raising the values of some r_i and s_i to a positive level and prove that the minimum spanning tree solution of the connect step selects precisely those arcs for which r_i and s_i can be increased with the least cost. This argument establishes the optimality of the B&C heuristic for this special RPP. As mentioned in Section 3.3, Gün [1993] provides an alternate demonstration of this result by exhibiting a linear programming formulation for the entire problem that is optimally solved by the B&C heuristic.

10.3. Test generation for VLSI sequential circuits

Very large scale integrated (VLSI) circuits must be tested routinely to ensure reliability. However, because of the large number of parts and connections in such circuits, testing can be costly and efficient methods for performing the tests are highly desirable. Recently, Adams & Hochbaum [1993] have shown that, under certain conditions, efficient test generation can be formulated as an arc routing problem that they call the *tour covering problem* (TCP): Consider a directed graph $G = (V, A)$ and a collection of path sets P_j for $j = 1, \ldots, J$, each of which consists of a set of paths in G. The TCP seeks a cycle of minimum length that starts at a prespecified node in V and traverses at least one path from each of the path sets P_j. Conceptually, one may think of the TCP as a two-stage problem: once the representative paths from the sets P_j are selected, the selected paths define a set of required arcs in G. The RPP can then be solved in G to identify a cycle that covers all the required arcs at minimum cost. As Adams and Hochbaum put it, the TCP combines the features of the set covering and rural postman problems. We now proceed to describe how the TCP arises as a combinatorial problem in test generation.

A sequential circuit is a collection of combinatorial gates and flip-flops. One can associate a finite-state machine with the behavior of the circuit. The *state* of this machine corresponds to a bit vector with components that represent the memory elements (flip-flops) of the circuit. The behavior is fully described by the *state transition graph* (STG), the nodes of which correspond to all possible (reachable) states of the circuit. An arc (k, ℓ) in STG bearing the label <in, out> indicates that a transition to state ℓ occurs and the output 'out' is produced when the circuit is given the input 'in' in state k. Given the state space S and the input and output alphabets I and O, the mapping $\delta: S \times I \to S$ and $\lambda: S \times I \to O$ specify the *next state* $\delta(k, \text{in})$ and output $\lambda(k, \text{in})$. The circuit is assumed to have a *reset state* that corresponds to the logic values of the states when the circuit is first switched on. This state can be reached from any node of the STG simply by turning the circuit off and then on again. Since the nodes included in the STG are those reachable from the reset state, STG is a strongly connected digraph.

The notion of a valid test for the circuit is based on the fault model that applies. Faults in a circuit can alter the output $\lambda(k, \text{in})$ or the next state $\delta(k, \text{in})$. A *stuck-at* fault occurs when one of the logic values of the circuit is stuck at a value of 0 or 1 and does not change. Adams and Hochbaum assume that the circuit contains a *single stuck-at fault*. To see if this fault is present, the circuit must be taken to a state whose output is altered by this fault, that is, the output in this state in the presence of the fault is different from the fault-free result. An *excitation vector* for a fault f is any assignment of values to the current state values and input variables that produces faulty output or a faulty next-state when a fault is present. An *excitation state* is the state in which this faulty behavior occurs. Since the states of the circuit are not directly observable or controllable, faulty behavior can only be observed through the output of the circuit. This means that a faulty next-state transition must be propagated forward, by applying additional inputs, until the effect of the fault is manifested in a faulty output. The sequence of inputs that takes the circuit from a reset state to an excitation state is called a *justification sequence*. A *differentiating sequence* of inputs applied to the excitation states propagates the effects of the fault to the output. Together, these sequences form a test for the fault. Additional details on the testing of circuits is available in Ghosh, Devadas & Newton [1991] or Cho, Hachtel & Somenzi [1991].

The circuit testing problem can be interpreted as an arc traversal problem on the graph $G = (V, A)$ of the STG. Aho, Dahbura, Lee & Uyar [1988], for instance, use the rural postman problem as a procedure for covering the required arcs corresponding to differentiating sequences. Building on this and other work, Adams and Hochbaum formulate the testing problem as a tour covering problem (TCP) on G: If there are J potential defects in the circuit, P_j is the collection of paths in G that checks for defect j $(j = 1, \ldots, J)$. Since traversing any path in P_j reveals defect j (or its absence), at least one path from each class P_j must be selected. The path sets P_j correspond to the set of differentiating sequences originating in the excitation state for defect j. In practice, the cardinality of P_j is large since many different paths can test for the same defect. The cost structure chosen by Adams and Hochbaum is a simple one. Each arc has a cost of one reflecting the fact that an input vector must be supplied to the circuit. Reset arcs that take the circuit back to the reset state may be more costly, but a cost of unity is attached to these arcs as well. In this way, the path cost is simply equal to the number of arcs in the path. Finally, since the arcs in a path must be covered contiguously, each path is represented by a *superarc* whose initial and terminal vertices coincide with those of the path. The TCP on G is the problem of finding a cycle of minimum cost that starts at the reset node and traverses at least one superarc from each collection P_j of superarcs.

Adams & Hochbaum [1993] propose and compare three different heuristics for solving the TCP. Here, we only outline their first heuristic which relies on the classical directed postman (DPP) algorithm to generate subtours. Initially, the DPP is solved over the network comprised of the union of all superarcs. Of the subtours produced by the DPP solution, the one with the largest ratio of the number of defects covered to the length of the subtour is selected and the path sets corresponding at all defects j covered by this subtour are removed. The remaining path sets define a smaller network over which the DPP is solved in the next iteration. At each iteration, the criterion for selecting the subtour considers the per-arc number of *new* defects covered by the subtour. The procedure continues until all defects are covered.

In their preliminary computational tests, Adams and Hochbaum used a test bed of problems where each circuit possesses just one excitation state for each fault. For such problems, only one application of the DPP is required since all generated subtours are selected. The circuits in the test bed had 300–1500 faults. For a given number of input vectors, the DPP-based heuristic is able to detect 90% of the faults, on the average. This can be compared to 63% for the *random method*, a naive testing procedure that uses random sequences of input vectors without considering interactions. A key lesson of the test results is the advantage of starting a new test 'on the tail of' a completed test *without* returning to the reset node — an advantage that is not available to the random method. Adams and Hochbaum also tested a greedy procedure for the TCP that outperformed the DPP heuristic, suggesting possible combinations of the two methods that they are currently exploring. Other extensions include devising tests that check for the presence of more than one fault at a time.

10.4. Other applications

In their recent work, Dror, Leung & Mullaseril [1993] identify an interesting application of the capacitated postman problem to the curious setting of livestock feed distribution. In a feedyard, animals are brought in as calves and fed during their growth cycle with specially formulated feed that depends on the age of the cattle. A feed mixing plant produces feed on the premises, one feed type at a time. The distribution problem is to carry the various types of feed from the plant to the feeding pens.

One may visualize pens as arrays of parallel strips on a feed lot. A truck feeding the pen must traverse its length while discharging feed from the left. Therefore, the distribution problem involves covering segments along the pens in a given direction. Dror and his coworkers model this problem as a capacitated postman problem as follows. Consider a directed graph $G = (V, A)$ where the arcs correspond to road segments along each pen and additional service roads. (As usual, the node set consists of all road segment intersections.) On a given day, only a subset of all the pens requires feed delivery based on the cattle needs. This subset defines the set of required arcs $R \subset A$. Moreover, there are five different feed types in the application described by Dror and coworkers and each pen receives only one type of feed. Thus, R can be partitioned by feed type t as $R = \bigcup_t R_t$. Since the trucks have finite capacity, one can view the distribution of each feed type t as a capacitated arc routing problem (CARP) — a model discussed in Section 4.

Dror, Leung and Mullaseril use and modify several existing heuristics for CARP to generate capacity-feasible postman tours to cover each R_t. These tours are fed into an assignment module that assigns tours to trucks, creating daily truck schedules. Ultimately, the truck routing problem must be integrated with the production scheduling problem that decides the start and finish times for mixing each type of feed. The livestock feed problem is, thus, a rich combined production-distribution problem the distribution component of which involves arc routing.

Full-load movement of trucks or trailers is another routing application that can be modeled as an arc routing problem. The demand in this problem corresponds to trips to be covered for known origin-destination pairs. A full load is transported from the origin to the destination. Ball, Golden, Assad & Bodin [1983] pointed out that this problem can be viewed as a directed rural postman problem (DRPP), where the required arcs correspond to the trips and the objective is to minimize the empty movements of the vehicles. They used this approach alongside two other procedures to solve a problem with 400–800 trips for a chemical firm.

More recently, Fisher, Tang & Zheng [1994] solve a similar pickup and delivery problem for the Shanghai Truck Transportation Corporation. The problem is noteworthy for its large size: 6000–7000 orders are transported daily using 3400 trucks. A detailed description of the operations appears in Fisher, Huang & Tang [1986]. We note that several vehicles may be associated with the same trip so that the required network forms a multigraph. Fisher, Tang and Zheng use a network flow algorithm to solve the flow balance problem. Since the optimal solution to this problem may include too many cycles, or cycles that do not go through the depot, they describe procedures for splitting big cycles and merging small ones. The largest problem based on real data in their test set has 481 trips and uses 53 vehicles. The algorithm provides feasible solutions within 1% of the lower bound on the total distance traveled (including the required movements).

11. Conclusions

Our objective in this chapter was to provide an overview of arc routing problems and applications. The broad scope of the material covered in this chapter attests to the importance of this class of problems, particularly when one considers the various urban and commercial operations that can benefit from the techniques of arc routing.

The future developments of arc routing cannot be considered in isolation from other vehicle routing problems. On the methodological level, new advances are often proposed and developed in the context of node routing. For example, the traveling salesman problem has been the prototypical example of the use of polyhedral techniques in branch-and-bound and cutting plane algorithms. These techniques are now being tested on arc routing problems and show promise [see Grötschel & Win, 1992; Corberan & Sanchis, 1994]. Thus, one may expect to see a continued migration of novel combinatorial optimization techniques from stylized routing problems into the area of arc routing.

Another development shared by node and arc routing is the development of geographic information systems (GIS). We argued in Section 1 that these systems provide the natural environment for routing algorithms and systems. In addition to providing detailed geographical data, a GIS presents an exciting opportunity for adopting a new philosophy of algorithmic design that can capitalize on and enrich the spatial analysis capabilities of a GIS. To our knowledge, little explorative work on this topic has been conducted in the traditional academic setting.

An important issue that deserves greater attention in our opinion is the interplay between aggregate and detailed planning in routing. Many of the applications covered in Sections 7–9 of this chapter use heuristics based on a partitioning step that divides the overall area of service into smaller zones to which a detailed arc routing procedure may be applied [see, for example, Bodin & Levy, 1991]. The characteristics of the partitioning technique, however, may be crucial to how well the routing algorithms perform within each partitioned zone. Over time, we hope that experience with practical problems and more systematic research will reveal greater insights into such interactions. It may be that the type of aggregate analysis represented by the work of Robuste, Daganzo & Souleyrette [1990] for node routing will shed light on this question. In fact, to take the simpler example of the rural postman problem described in Section 3, the interaction between the balancing and connecting steps raises a wealth of methodological and computational issues.

In conclusion, we believe that arc routing will continue to grow as a field of both research and application. Over the past three decades, operations researchers and transportation analysts have successfully catalogued the many complications and constraints of arc routing applications. The greater power and sophistication of computer systems, geographical databases and optimization techniques can now be merged to address the full scope of such practical problems.

Acknowledgments

We are very grateful to our colleague Michael Ball for his helpful comments on the manuscript and his patience in handling the various versions of this paper. We also thank Florence Kemerer for her care in typing and processing this paper.

References

Adams, J.B., and D.S. Hochbaum (1993). A new and fast approach to a very large scale integrated sequential circuit test generation. Unpublished manuscript.

Aho, A.V., A.T. Dahbura, D. Lee and M.U. Uyar (1988). An optimization technique for protocal conformance test generation based on UIO sequences and rural Chinese postman tours. *Proc. IFIP WG 6.1 8th Int. Symp. on Protocol Specification, Testing and Verification*, Atlantic City, N.J., pp. 75–86.

Alfa, A.S., and D.Q. Liu (1988). Postman routing problem in a hierarchical network. *Eng. Optim.* 14, 127–138.

Antennucci, J.C., K. Brown, P.L. Croswell, M.J. Kevany and H. Archer (1991). *Geographic Information Systems: A Guide to the Technology*, Van Nostrand Reinhold, New York, N.Y.

Assad, A.A. (1988). Modeling and implementation issues in vehicle routing, in: B. Golden and A. Assad (eds.). *Vehicle Routing: Methods and Studies*, North-Holland, New York, N.Y., pp. 7–45.

Assad, A.A., W.L. Pearn and B.L. Golden (1987). The capacitated Chinese postman problem: Lower bounds and solvable cases. *Am. J. Math. Manage. Sci.* 7(1–2), 63–88.

Atkins, J.E., J.S. Dierckman and K. O'Bryant (1990). A real snow job. *UMAP J.* 11(3), 231–239.

Badillo, A.S. (1993). Transportation and navigation, in: G.H. Castle, III (ed.). *Profiting From a Geographic Information System*, GIS World, Fort Collins, Co., pp. 161–176.

Ball, M.O. (1988). Allocation/routing: Models and algorithms, in: B. Golden and A. Assad (eds.), *Vehicle Routing: Methods and Studies*, North-Holland, New York, N.Y., pp. 199–221.

Ball, M.O., and M.J. Magazine (1988). Sequencing of insertions in printed circuit board assembly. *Oper. Res.* 36(2), 192–201.

Ball, M.O., B.L. Golden, A.A. Assad and L.D. Bodin (1983). Planning for truck fleet size in the presence of a common carrier option. *Decision Sci.* 14, 103–120.

Ballou, R.H. (1990). A continued comparison of several popular algorithms for vehicle routing and scheduling. *J. Business Logist.* 11(1), 111–126.

Bastian, C., and A.H.G. Rinnooy Kan (1992). The stochastic vehicle routing problem revisited. *Eur. J. Oper. Res.* 56(3), 407–412.

Belenguer, J.M., and E. Benavent (1992). Polyhedral results on the capacitated arc routing problem. Unpublished manuscript.

Beltrami, E., and L. Bodin (1974). Networks and vehicle routing for municipal waste collection. *Networks* 4(1), 65–94.

Benavent, E., V. Campos, A. Corberan and E. Mota (1983). Problemas de rutas por arcos. *Qüestiió* 7(3), 479–490.

Benavent, E., V. Campos, A. Corberan and E. Mota (1985). Analisis de heurísticos para el problema del cartero rural. *Trab. Estad. Invest. Oper.* 36(2), 27–38.

Benavent, E., V. Campos, A. Corberan and E. Mota (1990). The capacitated arc routing problem — A heuristic algorithm. *Qüestiió* 14(1–3), 107–122.

Benavent, E., V. Campos, A. Corberan and E. Mota (1992). The capacitated arc routing problem: Lower bounds. *Networks* 22(4), 669–690.

Biggs, N.L., E.K. Lloyd and R.J. Wilson (1976). *Graph Theory 1736–1936*, Clarendon Press, Oxford.

Bodin, L.D. (1990). Twenty years of routing and scheduling. *Oper. Res.* 38(4), 571–579.

Bodin, L.D., and S. Kursh (1978). A computer-assisted system for the routing and scheduling of street sweepers. *Oper. Res.* 26(4), 525–537.

Bodin, L.D., and S.J. Kursh (1979). A detailed description of a computer system for the routing and scheduling of street sweepers. *Comput. Oper. Res.* 6, 181–198.

Bodin, L.D., and L.S. Levy (1991). The Arc Partitioning Problem. *Eur. J. Oper. Res.* 53(3), 393–401.

Bodin, L.D., and L.S. Levy (1994). Visualization in vehicle routing and scheduling problems. *ORSA J. Comput.* 6(3), 261–269.

Bodin, L.D., B. Golden, A. Assad and M. Ball (1983). Routing and scheduling of vehicles and crews: the state of the art. *Comput. Oper. Res.* 10(2), 63–211.

Bodin, L.D., G. Fagan and L. Levy (1992a). The GEOMOD system. *Proc. USPS Advanced Technology Conference*, Vol. 1, United States Postal Service, pp. 413–418.

Bodin, L.D., G. Fagan and L. Levy (1992b). Vehicle routing and scheduling problems over street networks. *Proc. USPS Advanced Technology Conference*, Vol. 2, United States Postal Service, pp. 625–641.

Bodin, L.D., G. Fagan, R. Welebny and J. Greenberg (1989). The design of a computerized sanitation vehicle routing and scheduling system for the town of Oyster Bay, New York. *Comput. Oper. Res.* 16(1), 45–54.

Bott, K., and R. Ballou (1986). Research perspectives in vehicle routing and scheduling. *Transp. Res.* 20A, 239–243.

Bouliane, J., and G. Laporte (1992). Locating postal relay boxes using a set covering algorithm. *Am. J. Math. Manage. Sci.* 12(1), 65–74.

Brimberg, J., and R.F. Love (1992). A New Distance Function for Modeling Travel Distances in a Transportation Network. *Transp. Sci.* 26(2), 129–137.

Brucker, P. (1981). The Chinese postman problem for mixed graphs. *Proc. Int. Workshop, Lecture Notes in Computer Science* 100, 354–366.

Campbell, J.F., and A. Langevin (1993). Operations management for urban snow removal and disposal. GERAD Working Paper No. G-93-34, H.E.C., Montréal.

Castle, G.H. III, ed. (1993). *Profiting From a Geographic Information System*, GIS World, Fort Collins, Colorado.

Cebry, M.E., A.H. DeSilva and F.J. DiLisio (1992). Management science in automating postal operations: Facility and equipment planning in the United States Postal Service. *Interfaces* 22(1), 110–130.

Chao, I.M. (1993). Algorithms and solutions to multi-level vehicle routing problems. Ph.D. Dissertation, Applied Mathematics Program, University of Maryland at College Park.

Chapleau, L., J.A. Ferland, G. Lapalme and J.-M. Rousseau (1984). A parallel insert method for the capacitated arc routing problem. *Oper. Res. Lett.* 3(2), 95–99.

Chernak, R., L.E. Kustiner and L. Phillips (1990). The snowplow problem. *UMAP J.* 11(3), 241–250.

Cho, H., G.D. Hatchel and F. Somenzi (1991). Fast sequential ATPG based on implicit state enumeration. *Int. Test Conf. 1991*, IEEE Paper 3.2, pp. 67–74.

Christofides, N. (1973). The optimum traversal of a graph. *Omega* 1(6), 719–732.

Christofides, N. (1985). Vehicle routing, in: E.L. Lawler et al. (eds.). *The Traveling Salesman Problem*, John Wiley & Sons, New York, N.Y., pp. 431–448.

Christofides, N., and A. Mingozzi (1989). Vehicle routing: Practical and algorithmic aspects, in: C.F.H. van Rijn (ed.). *Logistics: Where Ends Have to Meet*, Pergamon Press, Oxford, pp. 30–48.

Christofides, N., V. Campos, A. Corberan and E. Mota (1981). An algorithm for the rural post-man problem. Unpublished Report IC.OR.81.5, Department of Management Science, Imperial College, London (revised 1982).

Christofides, N., E. Benavent, V. Campos, A. Corberan and E. Mota (1983). The mixed Chinese postman problem, in: P. Thoft-Christensen (ed.). *Systems Modeling and Optimization*, Lecture Notes in Control and Information Systems, Vol. 59, Springer-Verlag, Berlin, pp. 641–649.

Christofides, N., V. Campos, A. Corberan and E. Mota (1986). An algorithm for the rural postman problem on a directed graph. *Math. Program. Study* 26, 155–166.

Clark, R.M. (1973). Solid waste: Management and models, in: R.A. Deininger (ed.). *Models for Environmental Pollution Control*, Ann Arbor, Michigan, pp. 269–305.

Clark, R.M., and J.I. Gillean (1975). Analysis of solid waste management operations in Cleveland, Ohio: A case study. *Interfaces* 6(1), 32–42.

Clark, R.M., and J.C. Lee, Jr. (1976). Systems planning for solid waste collection. *Comput. Oper. Res.* 3, 157–173.

Clysdale, D. (1992). The natural affinity of GIS and postal administrations: A Canada post perspective. *Proc. USPS Advanced Technology Conf.*, Vol. 3, United States Postal Service, pp. 1421–1435.

Cook, T.M., and B.S. Alprin (1976). Snow and ice removal in an urban environment. *Manage. Sci.* 23(3), 227–234.

Corberan, A., and J.M. Sanchis (1994). A polyhedral approach to the rural postman problem. *Eur. J. Oper. Res.* 79(1), 95–114.

Coutinho-Rodrigues, J., N.V. Rodrigues and J.N. Climaco (1991). Urban routing of sanitation vehicles: A successful case study. Unpublished manuscript, University of Coimbra, Coimbra.

Coyle, R.G. (1973). Computer-based design of refuse collection systems, in: R.A. Deininger (ed.), *Models for Environmental Pollution Control*, Ann Arbor, Michigan, pp. 307–325.

DeArmon, J. (1981). A comparison of heuristics for the capacitated Chinese postman problem. Unpublished Masters Thesis, University of Maryland at College Park.

Deegan, A.J., and H.C. Harrison (1991). The optimization of the collections and deliveries of the Irish Postal Service, in: H.E. Bradley (ed.), *Operational Research '90*, Pergamon Press, New York, N.Y., pp. 277–290.

Desrochers, M., and T.W. Verhoog (1991). A new heuristic for the fleet size and mix vehicle routing problem. *Comput. Oper. Res.* 18(3), 263–274.

Desrochers, M., J.K. Lenstra and M.W.P. Savelsbergh (1990). A classification scheme for vehicle routing and scheduling problems. *Eur. J. Oper. Res.* 46(3), 322–332.

Desrosiers, J., G. Laporte, M. Sauvé, F. Soumis and S. Taillefer (1988). Vehicle routing with full loads. *Comput. Oper. Res.* 15(3), 219–226.

Dror, M., H. Stern, and P. Trudeau (1987). Postman tour on a graph with precedence relation on arcs. *Networks* 17(3), 283–294.

Dror, M., J. Leung, and P.A. Mullaseril (1993). Large livestock feed production and distribution. Paper presented at the 36th National ORSA/TIMS Meeting, Phoenix, Arizona.

Edmonds, J. (1965). The Chinese postman's problem. *Oper. Res.* 13, Supplement 1, B-73.

Edmonds, J., and E.L. Johnson (1973). Matching, Euler Tours and the Chinese postman. *Math. Program.* 5, 88–124.

Eglese, R.W. (1994). Routing winter gritting vehicles. *Discrete Appl. Math.* 48(3), 231–244.

Eglese, R.W., and L.Y.O. Li (1992). Efficient routing for winter gritting. *J. Oper. Res. Soc.* 43, 1031–1034.

Eglese, R.W., and H. Murdock (1991). Routing road sweepers in a rural area. *J. Oper. Res. Soc.* 42(4), 281–288.

Eiselt, H.A., M. Gendreau and G. Laporte (1995a). Arc routing problems. Part I: The Chinese postman problem. *Oper. Res.* 43(2), 231–242.

Eiselt, H.A., M. Gendreau and G. Laporte (1995b). Arc routing problems. Part II: The rural postman problem, forthcoming in *Oper. Res.*

Eswaran, K.P., and R.E. Tarjan (1976). Augmentation problems. *SIAM J. Comput.* 5(4), 653–665.

Evans, J.R. (1990). Design and implementation of a vehicle routing and planning system for snow and ice control. *Proc. 1990 Annual Meeting*, Vol. 2, Decision Sciences Institute, San Diego, Calif., pp. 1832–1834.

Evans, J.R., and E. Minieka (1992). *Optimization Algorithms for Networks and Graphs*, Marcel Dekker, New York, N.Y.

Ferland, J., and G. Gudnette (1990). Decision support system for the school districting problem. *Oper. Res.* 38(1), 15–21.

Fischetti, M., and P. Toth (1989). An additive bounding procedure for combinatorial optimization problems. *Oper. Res.* 37, 319–328.

Fisher, M.L., and R. Jaikumar (1981). A generalized assignment heuristic for vehicle routing. *Networks* 11(2), 109–124.

Fisher, M.L., J. Huang and B.-X. Tang (1986). Scheduling bulk pick-up delivery vehicles in Shanghai. *Interfaces* 16(2), 18–23.

Fisher, M.L., B. Tang and Z. Zhen (1994). A network-flow based heuristic for bulk pickup and delivery routing. *Transp. Sci.* 29(1), 45–55.

Fleischner, H. (1983). Eulerian Graphs, in: L.W. Bieneke and R.J. Wilson (eds.), *Selected Topics in Graph Theory*, Academic Press, New York, N.Y., pp. 17–53.

Fleischner, H. (1990). *Eulerian Graphs and Related Topics*, Part 1, Volume 1. *Annals of Discrete Mathematics* 45, North-Holland, Amsterdam,.

Fleischner, H. (1991). *Eulerian Graphs and Related Topics*, Part 1, Volume 2. *Annals of Discrete Mathematics* 50, North-Holland, Amsterdam,.

Ford, L.R. Jr., and D.R. Fulkerson (1962). *Flows in Networks*, Princeton University Press, Princeton, N.J.

Frederickson, G.N. (1979). Approximation algorithms for some postman problems. *J. ACM* 26(3), 538–554.

Frederickson, G.N., M.S. Hecht and C.E. Kim (1978). Approximation algorithms for some routing problems. *SIAM J. Comput.* 7(2), 178–193.

Gastou, G., and E.L. Johnson (1986). Binary group and Chinese postman polyhedra. *Math. Program.* 34, 1–33.

Ghosh, A., S. Devadas and A.R. Newton (1991). Test generation and verification for highly sequential circuits. *IEEE Trans. Comput. Design*, 10(5), 652–667.

Golden, B.L., and A. Assad (1986). Perspectives on vehicle routing: Exciting new developments. *Oper. Res.* 34(5), 803–810.

Golden, B.L., and A.A. Assad, eds. (1988). *Vehicle Routing: Methods and Studies*, North Holland, New York, N.Y.

Golden, B.L., and E.A. Wasil (1987). Computerized vehicle routing in the soft drink industry. *Oper. Res.* 35(1), 6–17.

Golden, B.L., and R.T. Wong (1981). Capacitated arc routing problems. *Networks* 11(3), 305–315.

Golden, B.L., J.S. DeArmon and E.K. Baker (1983). Computational experiments with algorithms for a class of routing problems. *Comput. Oper. Res.* 10(1), 47–59.

Golden, B.L., A. Assad, L. Levy and F. Gheysens (1984). The fleet size and mix vehicle routing problem *Comput. Oper. Res.* 11(1), 49–66.

Goodman, S.E., and S.T. Hedetniemi (1973). Eulerian walks in graphs. *SIAM J. Comput.* 2(1), 16–27.

Goodman, S.E., and S.T. Hedetniemi (1974). On Hamiltonian walks in graphs. *SIAM J. Comput.* 3(3), 214–221.

Gray, D.M., and D.H. Male, eds. (1981). *Handbook of Snow*, Pergamon Press, Toronto, Ont.

Grötschel, M. and Z. Win (1992). A cutting plane algorithm for the windy postman problem. *Math. Program.* 55, 339–358.

Guan, M. (Kwan, M-K.) (1962a). Graphic programming using odd or even points. *Chinese Math.* 1(3), 273–277.

Guan, M. (1962b). Improvement on graphic programming. *Chinese Math.* 1(3), 278–287.

Guan, M. (1984). On the windy postman problem. *Discr. Appl. Math.* 9, 41–46.

Guan, M. (1984). The maximum weighted cycle-packing problem and its relation to the Chinese postman problem, in: J. Bondy and A. Murty (eds.). *Progress in Graph Theory*, Academic Press, Toronto, Ont., pp. 323–326.

Guan, M., and W. Pulleyblank (1985). Eulerian orientations and circulations. *SIAM J. Algebraic Discr. Math.* 6(4), 657–664.

Gün, H. (1993). Polyhedral structure and efficient algorithms for certain classes of the directed rural postman problem. Unpublished Ph.D. Dissertation, Applied Mathematics Program, University of Maryland at College Park, Md.

Haimovich, M., and A.H.G. Rinnooy Kan (1985). Bounds and heuristics for capacited routing problems. *Math. Oper. Res.* 10(4), 527–542.

Harrison, H.C., and A.J. Deegan (1992). A geographical database approach to the optimization of the Irish postal networks. *Proc. USPS Advanced Technology Conference*, Vol. 2, United States Postal Service, pp. 981–990.

Hartman, C., K. Hogenson and J.L. Miller (1990). Plower power. *UMAP J.* 11(3), 261–272.

Haslam, E., and J.R. Wright (1991). Application of routing technologies to rural snow and ice control. *Transp. Res. Rec.* 1304, 202–211.

Hedetniemi, S. (1968). On minimum walks in graphs. *Nav. Res. Logist. Q.* 15, 453–458.

Hirabayashi, R., Y. Saruwatari and N. Nishida (1992). Tour construction algorithm for the capacitated arc routing problems. *Asia Pacific J. Oper. Res.* 9(2), 155–175.

Huxhold, W.E. (1991). *An Introduction to Urban Geographic Information Systems*, Oxford University Press, New York, N.Y.

Itai, A., and M. Rodeh (1978). Covering a graph by circuits. *Lecture Notes Comput. Sci.* 62, 289–299.

Itai, A., R.J. Lipton, C.H. Papadimitriou and M. Rodeh (1981). Covering graphs by simple circuits. *SIAM J. Comput.* 10(4), 746–750.

Jansen, K. (1992). An approximation algorithm for the general routing problem. *Inf. Process. Lett.* 41, 333–339.

Jansen, K. (1993). Bounds for the general capacitated routing problem. *Networks* 23(3), 165–173.

Kappauf, C.H., and G.J. Koehler (1979). The mixed postman problem. *Discr. Appl. Math.* 1, 89–103.

Korach, E., and M. Penn (1992). Tight integral duality gap in the Chinese postman problem. *Math. Program.* 55, 183–191.

Kwan, M-K. (1962). see Guan, Meigu.

Labbé, M., G. Laporte and H. Mercure (1991). Capacitated vehicle routing on trees. *Oper. Res.* 39(4), 616–622.

Landis, J.D. (1993). GIS capabilities, uses, and organizational issues, in: G.H. Castle, III (ed.). *Profiting From a Geographic Information System*, GIS World, Fort Collins, Co., pp. 23–53.

Landry, P.-E. (1992). Geopostal applications at Canada Post Corporation (CPC). *Proc. USPS Advanced Technology Conference*, Vol. 2, United States Postal Service, pp. 969–980.

Lapalme, G., J.-M. Rousseau, S. Chapleau, M. Cormier, P. Cossette and S. Roy (1992). A geographic information system for transportation application. *Commun. ACM* 35(1), 80–88.

Laporte, G. (1988). Location-routing problems, in: B. Golden and A. Assad (eds.). *Vehicle Routing: Methods and Studies*, North-Holland, New York, N.Y., pp. 163–197.

Laporte, G. (1992a). The traveling salesman problem: An overview of exact and approximate algorithms. *Eur. J. Oper. Res.* 59(2), 231–247.

Laporte, G. (1992b). The vehicle routing problem: An overview of exact and approximate algorithms, *Eur. J. Oper. Res.* 59(3), 345–358.

Laporte, G., S. Chapleau, P.-E. Landry and H. Mercure (1989). An algorithm for the design of mailbox collection routes in urban areas. *Transp. Res.* 23B(4), 271–280.

Larson, R., A. Minkoff and P. Gregory (1988). Fleet sizing and dispatching for the Marine Division of the New York City Department of sanitation, in: B. Golden and A. Assad (eds.). *Vehicle Routing: Methods and Studies*, North-Holland, New York, N.Y., pp. 395–423.

Leclerc, G. (1985). Least-cost allocation of snow times to elimination sites: Formulation and post-optimal analysis. *Civil Eng. Systems* 2, 217–222.

Lee, J. (1989). Dynamic delivery of transportation services with GIS. *URISA J.*

Lemieux, P.F., and L. Campagna (1984). The snow ploughing problem solved by a graph theory algorithm. *Civil Eng. Systems* 1, 337–341.

Lenstra, J.K., and A.H.G. Rinnooy Kan (1976). On general routing problems. *Networks* 6(3), 273–280.

Lenstra, J.K., and A.H.G. Rinnooy Kan (1981). Complexity of vehicle routing and scheduling problems. *Networks* 11(2), 221–227.

Levy, L., and L. Bodin (1988). Scheduling the postal carriers for the United States Postal Service: An application of arc partitioning and routing, in: B. Golden and A. Assad (eds.) *Vehicle Routing: Methods and Studies*, North-Holland, New York, N.Y., pp. 359–394.

Levy, L., and L. Bodin (1989). The arc oriented location routing problem. *INFOR* 27(1), 74–93.

Li, L.Y.O., and R.W. Eglese (1994a). An interactive algorithm for vehicle routing for winter gritting. Unpublished manuscript.

Li, L.Y.O., and R.W. Eglese (1994b). A comparison of simulated annealing and Tabu search for a constrained arc routing problem. Unpublished manuscript.

Li, L.Y.O., and R.W. Eglese (1994c). A lower bound for the time constrained arc routing problem. Unpublished manuscript.

Liebling, T.M. (1970). *Graphentheorie in Planungs-und-Tourenproblemen*, Lecture Notes in Operations Research Mathematics Systems, Vol. 21, Springer-Verlag, New York, N.Y.

Liebling, T.M. (1973). Routing problems for street cleaning and snow removal, in: R.A. Deininger (ed.). *Models for Environmental Pollution Control*, Ann Arbor, Michigan, pp. 363–375.

Liebman, J.C., J.W. Male and M. Wathne (1975). Minimum cost residential refuse vehicle routes. *J. Environ. Div. ASCE* 101(EE3), 399–412.

Lin, Y., and Y. Zhao (1988). A new algorithm for the directed Chinese postman problem. *Comput. Oper. Res.* 15(6), 577–584.

Lucena, A. (1990). Time-dependent traveling salesman problem — The delivery man case. *Networks* 20(6), 753–763.

Malandraki, C., and M.S. Daskin (1993). The maximum benefit Chinese postman problem and the maximum benefit traveling salesman problem. *Eur. J. Oper. Res.* 65, 218–234.

Male, J.W. (1973). A heuristic solution to the m-postmen's problem. Unpublished Ph.D. Dissertation, The Johns Hopkins University, Baltimore, Md.

Male, J.W., and J.C. Liebman (1978). Districting and routing for solid waste collection. *J. Environ. Eng. Div.* 104(EE1), 1–14.

Male, J.W., J.C. Liebman and C.S. Orloff (1977). An improvement of Orloff's general routing problem. *Networks* 7, 89–92.

Manber, U., and S. Israni (1984). Pierce point minimization and optimal torch path determination in flame cutting. *J. Manuf. Systems* 3(1), 81–89.

Mannering, F.L., and W.P. Kilareski (1986). The common structure of geo-based data for roadway information systems. *ITE J.* July, 43–49.

Marks, D.H., and R. Stricker (1971). Routing for public service vehicles. *J. Urban Planning Dev. Div. ASCE* 92(UP2), 165–178.

McBride, R. (1982). Controlling left and U-turns in the routing of refuse collection vehicles. *Comput. Oper. Res.* 9(2), 145–152.

Minieka, E. (1978). *Optimization Algorithms for Networks and Graphs*, Marcel Dekker, New York, N.Y.

Minieka, E. (1979). The Chinese postman problem for mixed networks. *Manage. Sci.* 25(7), 643–648.

Minsk, L.D. (1979). A systems study of snow removal, in: *Special Report 185: Snow Removal and Ice Control Research*, TRB, National Research Council, Washington, D.C., pp. 220–225.

Minsk, L.D. (1981). Snow removal equipment, in: D. Gray and D. Male (eds.). *Handbook of Snow*, Pergamon Press, Toronto, Ont., pp. 648–670.

Mittenthal, J., and C.E. Noon (1992). An insert/delete heuristic for the traveling salesman subset-tour problem with one additional constraint. *J. Oper. Res. Soc.* 43(5), 277-283.

Muralidharan, B., and R. Pandit (1991). Routing and scheduling of trucks to minimize the cost of collection of post-consumer recyclable products. Working Paper, Department of Industrial & Manufacturing Systems Engineering, Iowa State University.

Nemhauser, G.L., and L.A. Wolsey (1988). *Integer and Combinatorial Optimization*, John Wiley & Sons, New York, N.Y.

Noon, C.E., and J.C. Bean (1991). A Lagrangian based approach for the asymmetric generalized traveling salesman problem. *Oper. Res.* 39(4), 623–632.

Orloff, C.S. (1974). A fundamental problem in vehicle routing. *Networks* 4(1), 35–64.

Orloff, C.S. (1976). On general routing problems: Comments. *Networks* 6(3), 281–284.

Orloff, C.S., and D. Caprera (1976). Reduction and solution of large scale vehicle routing problems. *Transp. Sci.* 10(4), 361–373.

Papadimitriou, C.H. (1976). On the complexity of edge traversing. *J. ACM* 23(3), 544–554.

Park, Y.B., and C.P. Koelling (1989). An interactive computerized algorithm for multicriteria vehicle routing problems. *Comput. Ind. Eng.* 16(4), 477–490.

Pearn, W.L. (1984). The capacitated Chinese postman problem. Unpublished Ph.D. Dissertation, University of Maryland at College Park, Md.

Pearn, W.L. (1988). New lower bounds for the capacitated arc routing problem. *Networks* 18(3), 181–191.

Pearn, W.L. (1989). Approximate solutions for the capacitated arc routing problem. *Comput. Oper. Res.* 16(6), 589–600.

Pearn, W.L. (1991). Augment–insert algorithms for the capacitated arc routing problem. *Comput. Oper. Res.* 18(2), 189–198.

Pearn, W.L., A. Assad and B.L. Golden (1987). Transforming arc routing into node routing problems. *Comput. Oper. Res.* 14(4), 285–288.

Potvin, J.-Y., G. Lapalme and J.-M. Rousseau (1990). Integration of AI and OR techniques for computer-aided algorithmic design in the vehicle routing domain. *J. Oper. Res. Soc.* 41(6), 517–525.

Raghavendra, A., T.S. Krishnakumar, R. Muralidhar and D. Sarvanan (1992). A practical heuristic for a large scale vehicle routing problem. *Eur. J. Oper. Res.* 57(1), 32–38.

Ralphs, T.K. (1993). On the mixed Chinese postman problem. *Oper. Res. Lett.* 14, 123–127.

Riccio, L.J. (1984). Management science in New York's Department of Sanitation. *Interfaces* 14(2), 1–13.

Riccio, L.J., and A. Litke (1986). Making a clean sweep: Simulating the effects of illegally parked cars on New York City's mechanical street-cleaning efforts. *Oper. Res.* 34(5), 661–666.

Riccio, L.J., J. Miller and A. Litke (1986). Polishing the Big Apple: How management science has helped make New York streets cleaner. *Interfaces* 16(1), 83–88.

Richey, M.B., and R.G. Parker (1991). A cubic algorithm for the directed Eulerian subgraph problem. *Eur. J. Oper. Res.* 50(3), 345–352.

Richey, M.B., R.G. Parker and R.L. Rardin (1985). On finding spanning Eulerian subgraphs. *Nav. Res. Logist. Q.* 32(3), 443–455.

Robinson, J.D., L.S. Ogawa and S.G. Frickenstein (1990). The two-snowplow routing problem. *UMAP J.* 11(3), 251–259.

Robuste, F., C.F. Daganzo and R.R. Souleyrette, II (1990). Implementing vehicle routing models. *Transp. Res.* 24B(4), 263–286.

Rodrigues, N.V. (1990). Arc routing problems — A study of particular cases. Unpublished M.S. Thesis, University of Coimbra, Coimbra.

Roy, S., and J.M. Rousseau (1989). The capacitated Canadian postman problem. *INFOR* 27(1), 58–73.

Russell, R.A., and P.E. Challinor (1988). Effective methods for petroleum tank truck dispatching. *Comput. Oper. Res.* 15(4), 323–331.

Saruwatari, Y., and T. Matsui (1994). A note on K best solutions to the Chinese postman problem, forthcoming.

Saruwatari, Y., R. Hirabayashi and N. Nishida (1992a). Node duplication lower bounds for the capacitated arc routing problem. *J. Oper. Res. Soc. Japan* 35(2), 119–133.

Saruwatari, Y., R. Hirabayashi and N. Nishida (1992b). Subtour elimination algorithm for the capacitated arc routing problem, in: W. Bueler et al. (eds.). *DGOR Operations Research: Proceedings of Operations Research 1990*, 19th Annual Meeting, Springer-Verlag.

Savas, E.S. (1973). The political properties of crystalline H_2O: Planning for snow emergencies in New York. *Manage. Sci.* 20(2), 137–145.

Shuster, K.A., and D.A. Schur (1974). Heuristic routing for solid waste collection vehicles. U.S. Environmental Protection Agency Publication SW-113, U.S. Government Printing Office, Washington, D.C.

Soumis, F., M. Sauvé and Beau, L.L. (1991). The simultaneous origin–destination assignment and vehicle routing problem. *Transp. Sci.* 25(3), 188–200.

Stern, H.I., and M. Dror (1979). Routing electric meter readers. *Comput. Oper. Res.* 6, 209–223.

Stokx, C.F.M., and C.B. Tilanus (1991). Deriving route lengths from radial distances: Empirical evidence. *Eur. J. Oper. Res.* 50(1), 22–26.

Stricker, R. (1970). Public sector vehicle routing: The Chinese postman's problem. Unpublished Masters Thesis, Massachusetts Institute of Technology, Cambridge, Mass.

Su, S.I. (I.) (1992). The general postman problem — Models and algorithms. Unpublished Ph.D. Dissertation, College of Business & Management, University of Maryland, Md.

Sutcliffe, C., and J. Board (1991). The ex-ante benefits of solving vehicle routing problems. *J. Oper. Res. Soc.* 42(2), 135–143.

Thiriez, H. (1991). Modeling of an interactive scheduling system in a complex environment. *Eur. J. Oper. Res.* 50(1), 37–47.

Tucker, A.C., and L.D. Bodin (1983). A model for municipal street sweeping operations, in: W.F. Lucas, F.S. Roberts and R.M. Thrall (eds.). *Modules in Applied Mathematics*, Vol. 3, Discrete and System Models, pp. 76–111.

Tucker, W.B., and G.M. Clohan (1979). Computer simulation of urban snow removal. in: *Snow Removal and Ice Control Research*, Special Report No. 185, Transportation Research Board, National Research Council, Washington, D.C., pp. 293–302.

Turner, W., and E. Hougland (1975). The optimal routing of solid waste collection. *AIIE Trans.* 7, 427–431.

Ulusoy, G. (1985). The fleet size and mix problem for capacitated arc routing. *Eur. J. Oper. Res.* 22(3), 329–337.

Walsh, B., L. Bodin, L. Levy and G. Fagan (1994). The United States Postal Service's Geomod system. *Proc. USPS Advanced Technology Conference*, United States Postal Service, forthcoming.

Wang, J-Y. (1992). Computer-aided system for planning efficient routes. Unpublished Ph.D. Dissertation, Purdue University, West Lafayette, Indiana.

Wang, J-Y., and J.R. Wright (1991). Automating the design of service routes. Unpublished manuscript.

Wathne, M. (1972). Optimal routing of solid waste collection vehicles. Unpublished Ph.D. Dissertation, The Johns Hopkins University, Baltimore, Md.

Win, Z. (1988). Contributions to routing problems. Unpublished Ph.D. dissertation, Universität Augsburg.

Win, Z. (1989). On the windy postman problem on Eulerian graphs. *Math. Program.* 44(1), 97–112.

Wright, J.W., ed. (1992). *The Universal Almanac 1993*, Andrews and McMeel, Kansas City, Mo.

Wunderlich, J., M. Collette, L. Levy and L. Bodin (1992). Scheduling meter readers for southern California Gas Company. *Interfaces* 22(3), 22–30.

Wyskida, R., and J. Gupta (1972). IE's improve city's solid waste collection. *Ind. Eng.* 46, 12–15.

M.O. Ball et al., Eds., *Handbooks in OR & MS, Vol. 8*

Chapter 6

Network Equilibrium Models and Algorithms

Michael Florian

Départment IRO and Centre de Recherche sur les Transports, Université de Montréal, Case
Postale 6128-Succursale A, Montréal, H3C 3J7 Canada

Donald Hearn

Center for Applied Optimization, ISE Department, Weil Hall, University of Florida, Gainesville,
FL 32611, U.S.A.

1. Introduction

The quantitative analysis of certain transport phenomena over physical networks gives rise to network models that represent the physical infrastructure and aim to compute the flows of one or more commodities on the links of the network. Such phenomena arise in electrical networks, water pipe networks, urban traffic networks, air and rail intercity networks and interregional or international freight networks, among others. The issues of interest that generate the need to formulate, analyze and solve network models in these various contexts are addressed by stating principles which characterize the way in which the commodities transported use routes from origins to destinations on the corresponding network. Such models may be descriptive, attempting to reproduce behavior patterns, or have their origins in physical laws, and are generally referred to as network equilibrium models. Alternatively, they may be normative, attempting to prescribe how to best use the network infrastructure.

The practical contexts which give rise to such models are varied, however the network equilibrium model captures a wide variety of problem classes that involve the transportation of goods, persons, vehicles, water, electricity and data. Whenever congestion phenomena are present, the cost functions associated with the links of the network model are nonlinear; in most applications they are convex or monotone. In many cases, the resulting models may be formulated as nonlinear cost network optimization models. Other versions of such problems may be formulated as variational inequality or nonlinear complementarity problems with an embedded network structure.

This chapter introduces the formulation, analysis and solution of the network equilibrium model (NEM). In the sequel, the terminology used will be that associated with the movement of vehicles over a network of streets, such as in an urban traffic network, since this context is a very familiar one. Nevertheless, the use of the model for problems arising in different physical contexts will be outlined, once the model formulation, analysis of its properties, and solution algorithms have been presented. The material in this chapter is organized in the following way. First, a general version of the network equilibrium problem is formulated and its mathematical structure is analyzed, in order to characterize the existence and uniqueness of solutions. Then, solution algorithms for versions of the network equilibrium problem which may be formulated as nonlinear cost network optimization problems are presented. This is followed by solution algorithms for network equilibrium problems which are formulated as variational inequality problems with an embedded network structure. Then, a relatively short presentation of stochastic network equilibrium models is given. The final section outlines the applicability of the network equilibrium problem in other physical contexts.

1.1. Model formulation

Traffic equilibrium models are descriptive models which aim to predict the link flows and travel times that result from the way in which travellers, usually drivers of vehicles, choose routes from their origins to their destinations on a transportation network. The behavioral assumption, known as Wardrop's user-optimal principle, postulates that the travel times on all the routes actually used are equal to or less than those which would be experienced by a vehicle on any unused route. The traffic flows that satisfy this principle are usually referred to as 'user-optimized' flows, since each user chooses the route that he (or she) perceives to be the best. In order to formulate the model, the following notation is used.

Consider a transportation network which permits the flows of one type of traffic on its links. The nodes $i, i \in N$, represent origins, destinations, and intersections; the arcs $a, a \in A$, represent the transportation links. The origin to destination (O–D) demands give rise to link flows $v_a, a \in A$ and the cost of travelling on a link is given by a user cost function $s_a(v)$, where v is the vector $(v_a)_{a \in A}$ of link flows over the entire network.

The cost functions may model the time delay for travel on an arc, or more general costs such as tolls or fuel consumption, and are assumed to be nonnegative.

The demand between origin-destination pair p, t_p, may use directed paths k. The path flow h_k satisfies conservation of flow and nonnegativity:

$$\sum_{k \in K_p} h_k = t_p \qquad \forall p \in P \tag{1}$$

$$h_k \geq 0 \qquad \forall k \in K, \tag{2}$$

where P is the set of origin-destination pairs, K_p is the set of paths for O–D pair p, and K is the set of all paths, that is, $K = \bigcup_{p \in P} K_p$. The corresponding link flows are given by

$$v_a = \sum_{p \in P} \sum_{k \in K_p} \delta_{ak} h_k \qquad \forall a \in A \tag{3}$$

where

$$\delta_{ak} = \begin{cases} 1 & \text{if link } a \text{ belongs to path } k \\ 0 & \text{otherwise.} \end{cases}$$

Let Δ denote the $|A| \times |K|$ arc-path incidence matrix (δ_{ak}); hence $v = \Delta h$, where h is the vector $(h_k)_{k \in K}$ of path flows for all O–D pairs.

The cost $s_k (= s_k(h))$ of each path k is the sum of the user costs on the arcs that belong to k.

$$s_k = \sum_{a \in A} \delta_{ak} s_a(v) = \sum_{a \in A} \delta_{ak} s_a(\Delta h) \qquad \forall k \in K_p, \ p \in P. \tag{4}$$

Let $u_p (= u_p(h))$ be the cost of the least cost path for O–D pair p:

$$u_p = \min_{k \in K_p} s_k \qquad \forall p \in P. \tag{5}$$

For each $p \in P$, the (travel) demand t_p is given by a function $t_p(u)$ where u is the vector of least cost travel values, $(u_p)_{p \in P}$, for all the O–D pairs of the network:

$$t_p = t_p(u) \qquad \forall p \in P. \tag{6}$$

Also, let Γ be the $|K| \times |P|$ path O–D pair incidence matrix $(\gamma_{kp})_{k \in K, p \in P}$ where

$$\gamma_{kp} = \begin{cases} 1 & \text{if path } k \text{ joins O–D pair } p \\ 0 & \text{otherwise.} \end{cases}$$

A network equilibrium that satisfies Wardrop's user optimal principle for path flows $h^* = (h_k^*)$ and O–D travel costs $u^* = (u_p^*)$ is achieved when

$$\text{if } h_k^* > 0 \text{ then } s_k(h^*) = u_p^* \qquad \forall k \in K_p, \ p \in P \tag{7}$$

$$u_p^* \le s_k(h^*) \qquad \forall k \in K_p, \ p \in P \tag{8}$$

$$h^* \ge 0, \ u^* \ge 0 \tag{9}$$

and (1) and (2) are satisfied.

These conditions are reformulated as the *network equilibrium model* (NEM): Determine h^* and u^* (hence v^* and t^*) such that the following conditions are satisfied:

$$(s_k(h^*) - u_p^*)h_k^* = 0 \qquad \forall k \in K_p, \ p \in P \tag{10}$$

$$s_k(h^*) - u_p^* \ge 0 \qquad \forall k \in K_p, \ p \in P \tag{11}$$

$$\sum_{k \in K_p} h_k^* - t_p(u^*) = 0 \qquad \forall p \in P \tag{12}$$

$$h^* \ge 0, \ u^* \ge 0. \tag{13}$$

The equivalence follows since (10) is a restatement of (7).

By contrast, 'system-optimized' flows are characterized by Wardrop's normative principle, which states that the average travel time (or, equivalently, the total travel time) is minimized, that is, system-optimized flows are solutions to:

$$\min \sum_{a \in A} v_a s_a(v) \tag{14}$$

subject to   (1), (2) and (3).

1.2. Examples of the NEM

Consider the network given in Figure 1. It consists of four nodes and five arcs and it has two O–D pairs: (1,3) which is given index 1 and (4,3) which is given index 2, each with a demand of 2. The arc flow variables are indicated on the links of the network, and the arc costs are $s_1(v_1) = 2$, $s_2(v_2) = 15$, $s_3(v_3) = 4$, $s_4(v_4) = 16$ and $s_5(v_5) = v_5 + v_5^2$. K_1 contains two paths: (1,3) and (1,2,3) and K_2 contains two paths as well: (4,3), (4,2,3). The path flows h^* and O–D travel times u^* which satisfy the network equilibrium conditions are:

O–D pair 1: $h^*_{(1,3)} = 0$; $s_{(1,3)} = 15$

$\qquad\qquad\quad h^*_{(1,2,3)} = 2$; $s_{(1,2,3)} = 14$ $u_1^* = 14$

O–D pair 2: $h^*_{(4,3)} = 1$; $s_{(4,3)} = 16$

$\qquad\qquad\quad h^*_{(4,2,3)} = 1$; $s_{(4,2,3)} = 16$ $u_2^* = 16$

with the corresponding arc flows v^* are:

$$v_1^* = 2; \quad v_2^* = 0; \quad v_3^* = 1; \quad v_4^* = 1; \quad v_5^* = 3.$$

It is easy to verify that (10)–(13) are satisfied for this solution and it is relevant to note that path (1,3) for O–D pair 1 has no flow since its cost is greater than the path (1,2,3) which is assigned all of the flow.

In general, the 'user optimal' flows are different than the 'system optimal' flows. For instance, the flows that minimize the total cost for this example are

$$v_1 = 1.77; \quad v_2 = .23; \quad v_3 = 0; \quad v_4 = 2; \quad v_5 = 1.77.$$

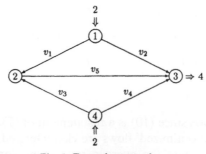

Fig. 1. Example network.

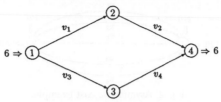

Fig. 2. Braess network.

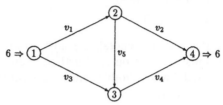

Fig. 3. Braess network with additional link.

As can be easily seen, the differences are significant. Another, more striking, example of that difference is the so-called Braess network, which is shown in Figure 2. The arc costs are $s_1(v_1) = 10v_1$, $s_2(v_2) = v_2 + 50$, $s_3(v_3) = v_3 + 50$ and $s_4(v_4) = 10v_4$.

There is only one O–D pair, (1,4), and the demand is 6. The equilibrium flows are 3 units on the path (1,2,4) and 3 units on the path (1,3,4), each of which has cost 83. If link 5 is now added to the network (see Figure 3) with cost function $v_5 + 10$, the equilibrium flows are 2 units on each of the paths of the network. Each path cost is 92. Thus, the addition of an arc increases the travel cost to each user.

This paradox, known as Braess' paradox, is understood by examining the system optimal solution which is 3 units each along the paths (1,2,4) and (1,3,4) even when arc 5 is present. Thus for the NEM there is no reason to anticipate that the travel costs may be reduced when a link is added to the network, since these costs are not minimized by the user optimal flows.

The occurrence of Braess' paradox is an illustration of the principle upon which the NEM is based and its consequences regarding the resulting flows. While this example may appear to be artificially constructed it may occur in networks that arise in practical applications.

As a final example consider Figure 4 where the network has two nodes and five arcs.

There are two O–D pairs, (1,2) and (2,1) with demands 210 and 120, respectively. Here each path contains just one arc. The arc cost functions are:

$$s_1(v_1, v_2) = 10v_1 + 5v_2 + 1000$$
$$s_2(v_1, v_2) = 20v_2 + 2v_1 + 1000$$
$$s_3(v_3, v_4) = 15v_3 + 5v_4 + 950$$

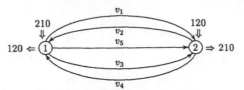

Fig. 4. Asymmetric cost example.

$$s_4(v_3, v_4) = 25v_4 + v_3 + 1300$$
$$s_5(v_5) = 20v_5 + 3000.$$

This example illustrates that, in general, the cost on arc a may depend on v, the full vector of arc flows, not just on v_a. Further, these dependencies may be asymmetric as in arcs 1 and 2 where both s_1 and s_2 are functions of v_1 and v_2 but the relationship is not symmetric.

The network equilibrium flows for this example are:

$$v_1^* = 120; \quad v_2^* = 70; \quad v_3^* = 90; \quad v_4^* = 50; \quad v_5^* = 0,$$

and the equilibrium travel times are 2550 for O–D pair (1,2) and 2640 for O–D pair (2,1).

Asymmetric cost models are important in accurately reflecting urban traffic conditions where traffic flow on a link may influence flows on other, nearby, links.

2. Equivalent problems — existence and uniqueness of solutions

Alternative problem statements for the NEM facilitate the establishment of existence and uniqueness conditions regarding the solutions and the development of algorithms. In this section, the NEM is formulated as a variational inequality problem, a nonlinear complementarity problem, and as a fixed point problem. The equivalences are stated and proven as three propositions, and these lead to two further propositions which establish the existence and uniqueness of solutions for the NEM.

2.1. Variational inequality formulation

The *variational inequality* (VI) problem seeks a vector $x^* \geq 0$ such that:

$$F(x^*)(x - x^*) \geq 0 \qquad \forall x \geq 0 \tag{15}$$

where $x = (h, u)$ is a column vector, F is a vector-valued function on $\mathbb{R}^n$ (and a row vector), $n = |K| + |P|$, and $F(x)$ has the first $|K|$ components of the form $s_k(h) - u_p$, $k \in K_p$, $p \in P$, and the last $|P|$ components of the form $\sum_{k \in K_p} h_k - t_p(u)$, $p \in P$.

Proposition 1. *The NEM and the VI formulations are equivalent, provided* $s_k(h) > 0$ *and* $t_p(u) \geq 0$, *for all* $k \in K_p$, $p \in P$, $h \in \mathbb{R}_+^{|K|}$, $u \in \mathbb{R}_+^{|P|}$.

Proof. First, suppose that $x^* = (h^*, u^*)$, $h^* = (h_k^*)$, $u^* = (u_p^*)$, solves the NEM. Then, from the definition of $F(x)$ and equations (10) and (12):

$$F(x^*)x^* = \left(s_k(h^*) - u_p^*, \sum_{k \in K_p} h_k^* - t_p\right)(h^*, u^*)$$

$$= \sum_{p \in P} \sum_{k \in K_p} (s_k(h^*) - u_p^*)h_k^* + \sum_{p \in P} \left(\sum_{k \in K_p} h_k^* - t_p(u^*)\right)u_p^*$$

$$= 0.$$

Also, $F(x^*)x = \sum_{p \in P} \sum_{k \in K_p} (s_k(h^*) - u_p^*)h_k \geq 0$, since $h_k \geq 0$ and $s_k(h^*) - u_p^* \geq 0$, $\forall k \in K_p$, $p \in P$, by (11). Thus, $F(x^*)(x - x^*) = F(x^*)x - F(x^*)x^* \geq 0$, and x^* solves the VI problem.

Now, suppose x^* solves the VI problem, and let $\hat{x} = x^* + e_i$, where e_i is the ith unit vector in $\mathbb{R}^n$. Then $\hat{x} \geq 0$ and $0 \leq F(x^*)(\hat{x} - x^*) = F(x^*)e_i =$ the ith component of $F(x^*)$. Since $i \in [1, n]$ is arbitrary, every component of $F(x^*) = F((h^*, u^*))$ is nonnegative, that is, for all $k \in K_p$, $p \in P$: $s_k(h^*) - u_p^* \geq 0$ and $\sum_{k \in K_p} h_k^* - t_p(u^*) \geq 0$.

For $x_i^* = x^* e_i > 0$, let $\hat{x} = x^* - \alpha e_i$ with $x_i^* - \alpha > 0$, and $\alpha > 0$. Then: $0 \leq F(x^*)(\hat{x} - x^*) = (-\alpha)F(x^*)e_i$ implying that $F(x^*)e_i \leq 0$. But from the above result, it must be that $F(x^*)e_i = 0$. Further, since $i \in [1, n]$ may be arbitrarily chosen, then $s_k(h^*) - u_p^* = 0$ or $\sum_{k \in K_p} h_k^* - t_p(u^*) = 0$, according to the choice of i. Therefore, for all $k \in K_p$, $p \in P$,

$$h_k^* > 0 \Rightarrow s_k(h^*) - u_p^* = 0, \quad \text{and}$$

$$u_p^* > 0 \Rightarrow \sum_{k \in K_p} h_k^* - t_p(u^*) = 0.$$

Hence, $x^* = (h^*, u^*)$ solving the VI problem implies:

$$(s_k(h^*) - u_p^*)h_k^* = 0, \qquad \forall k \in K_p, \; p \in P \tag{16}$$

$$s_k(h^*) - u_p^* \geq 0, \qquad \forall k \in K_p, \; p \in P \tag{17}$$

$$\left(\sum_{k \in K_p} h_k^* - t_p(u^*)\right)u_p^* = 0, \qquad \forall k \in K_p, \; p \in P \tag{18}$$

$$\sum_{k \in K_p} h_k^* - t_p(u^*) \geq 0, \qquad \forall p \in P \tag{19}$$

$$h^* \geq 0, \; u^* \geq 0. \tag{20}$$

Now, if $x^* = (h^*, u^*)$ does not solve the NEM, comparing conditions (16)–(20) and (10)–(13), shows that only condition (12) could be violated. That is, for some $\tilde{p}$ in P, $\sum_{k \in K_{\tilde{p}}} h_k^* > t_{\tilde{p}}(u^*)$ which implies $u_{\tilde{p}}^* = 0$, by (18). By assumption $t_{\tilde{p}}(u^*)$ is nonnegative. So, $\sum_{k \in K_{\tilde{p}}} h_k^* > 0$ and thus $h_{\tilde{k}}^* > 0$, for some $\tilde{k}$ in $K_{\tilde{p}}$ and $s_{\tilde{k}}(h^*) = u_{\tilde{p}}^*$ by (16). Therefore, $s_{\tilde{k}}(h^*) = 0$, contrary to the hypothesis, and the proof is complete. □

2.2. Nonlinear complementarity formulation

Note that conditions (16)–(20) may be stated as: find an x^* that solves:

$$F_i(x)x_i = 0 \tag{21}$$
$$F_i(x) \geq 0 \tag{22}$$
$$x_i \geq 0 \tag{23}$$

for all $i = 1, 2, \ldots, n = |K| + |P|$, where x and the function F are as defined above. This is the *nonlinear complementarity* (NC) formulation of the NEM. From the proof of the preceding proposition, it is seen that if x^* solves the VI problem, then x^* solves the NC problem. Conversely, suppose x^* solves the NC problem. Then $F(x^*)x^* = 0$. Also, since $F(x^*) \geq 0$, then $F(x^*)x \geq 0$, for any $x \geq 0$. Thus, $F(x^*)(x - x^*) = F(x^*)x - F(x^*)x^* \geq 0$. This proves:

Proposition 2. x^* *solves the NC problem if and only if* x^* *solves the VI problem.*

2.3. Fixed point formulation

A general fixed-point problem seeks a solution to the equation $x = H(x)$ for some vector-valued function H. As previously noted, in the NEM, the demand functions are dependent on the minimum path costs between the O–D pairs, which in turn are demand dependent. Consider that for a fixed demand vector $\hat{t}$, an equilibrium solution may be found given by $\hat{h}(\hat{t})$ and $\hat{u}(\hat{t})$. However, $\hat{u}(\hat{t})$ may determine a different demand vector, that is, $t(\hat{u}(\hat{t}))$ may not equal $\hat{t}$. Equilibrium occurs precisely when $t(\hat{u}(\hat{t})) = \hat{t}$.

These observations motivate a third formulation of the NEM, as a *fixed point (FP)* problem.

Define the function $H : \mathbb{R}_+^n \rightarrow \mathbb{R}_+^n$ $n = |K| + |P|$, by $H(h, u) = \mathrm{Proj}_{\mathbb{R}_+^n}[(h, u) - F(h, u)]$, where $h \in \mathbb{R}_+^{|K|}, u \in \mathbb{R}_+^{|P|}$, F is the function defined above, and $\mathrm{Proj}_X(y) = z$ means that z is the closest point to y in the set X, or the projection of y onto X.

The FP problem then is to find an x^* that solves:

$$x = H(x) = \mathrm{Proj}_{\mathbb{R}_+^n}(x - F(x)). \tag{24}$$

Note that:

$$x^* = H(x^*) \Longleftrightarrow x^* \text{ solves } \min_{\mathbb{R}_+^n} (x - (x^* - F(x^*)))(x - (x^* - F(x^*)))$$
$$\Longleftrightarrow (x^* - (x^* - F(x^*)))(x - x^*) \geq 0, \quad \forall x \geq 0$$
$$\Longleftrightarrow F(x^*)(x - x^*) \geq 0.$$

This establishes:

Proposition 3. x^* *solves the FP problem if and only if* x^* *solves the VI problem.*

The relationships established by these three propositions are summarized by the following diagram:

NEM $\overset{s_k > 0,\, t_p \geq 0}{\Longleftrightarrow}$ VI problem $\Longleftrightarrow$ FP problem

$\updownarrow$

NC problem

2.4. Existence of a solution to the NEM

The Brouwer fixed point theorem guarantees a solution to a fixed point problem $x = H(x)$ provided H is a continuous function whose domain is closed, convex and bounded. Therefore, to ensure the existence of a solution to the NEM, it is assumed that: $s_k(h)$ is continuous for all k in K, and $t_p(u)$ is continuous and bounded for all p in P. Continuity of the functions s_k and t_p for all paths k and all O–D pairs p implies that the map $H(h, u) = \text{Proj}_{\mathbb{R}^n_+}[(h, u) - F(h, u)]$, where $F(h, u) = [s_k(h) - u_p, \sum_{k \in K_p} h_k - t_p(u)]$ is continuous. Moreover, it is natural in the context of the equilibrium problem that the demand functions $t_p(u)$ be bounded and this permits bounding the domain without affecting the equilibrium solution. Specifically, let $\alpha_1 > \max_p \max_{u \geq 0}(t_p(u))$, and $\alpha_2 > \max_k \max_{0 \leq h \leq \alpha_1}(s_k(h))$, where α_2 exists since s_k is continuous on a closed, bounded set. Then, rather than projection onto the space given by $(h, u) \geq 0$, consider the projection onto a large cube C defined by:

$$0 \leq h \leq e\alpha_1$$
$$0 \leq u \leq \hat{e}\alpha_2$$

where e and $\hat{e}$ are appropriately-dimensioned vectors of ones. The new projection map $\hat{H}(h, u) = \text{Proj}_C[(h, u) - F(h, u)]$ has a fixed point $(\hat{h}, \hat{u})$, and by choice of α_1 and α_2, $\hat{h} < \alpha_1$ and $\hat{u} < \alpha_2$. Thus $(\hat{h}, \hat{u})$ will be a fixed point for the original problem. Hence:

Proposition 4. *If s_k is positive and continuous, and t_p is nonnegative, continuous and bounded from above for each $k \in K$, and each $p \in P$, the NEM has a solution.*

The following example illustrates that, when the link cost functions are not continuous, flows that satisfy the equilibrium conditions do not exist. The network consists of one O–D pair and two links as shown in Figure 5. The demand from 1 to 2 is 2 and the arc cost functions are

$$s_1(v_1) = v_1$$
$$s_2(v_2) = \begin{cases} v_2, & v_2 < 1; \\ v_2 + 1, & v_2 \geq 1. \end{cases}$$

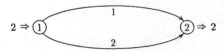

Fig. 5. Network with discontinuous costs.

When $v_1 = v_2 = 1$, the cost of link 2 is greater than the cost of link 1, but for $v_1 = 1 + \epsilon$ and $v_2 = 1 - \epsilon$, the cost of link 1 is greater than the cost of link 2.

2.5. Uniqueness of the NEM solution

A vector function $f : X \subseteq \mathbb{R}^n \to \mathbb{R}^n$ is *monotone* on X if

$$(f(x_1) - f(x_2))(x_1 - x_2) \geq 0$$

for any x_1 and x_2 in X. It is *strictly monotone* if for any distinct x_1 and x_2 in X,

$$(f(x_1) - f(x_2))(x_1 - x_2) > 0.$$

Expanding the inner product in the VI formulation (15) of the NEM gives:

$$\sum_{p \in P} \sum_{k \in K_p} \left[s_k(h^*) - u_p^* \right](h_k - h_k^*) + \sum_{p \in P} \left[\sum_{k \in K_p} h_k^* - t_p(u^*) \right](u_p - u_p^*) \geq 0.$$

Expanding the left-hand-side and using (4) gives

$$\sum_{p \in P} \sum_{k \in K_p} \left[\sum_{a \in A} \delta_{ak} s_a(v^*) - u_p^* \right](h_k - h_k^*) +$$

$$+ \sum_{p \in P} \left[\sum_{k \in K_p} h_k^* u_p - \sum_{k \in K_p} h_k^* u_p^* - t_p(u^*)(u_p - u_p^*) \right] \geq 0$$

or

$$\sum_{p \in P} \left[\sum_{k \in K_p} \sum_{a \in A} \delta_{ak} s_a(v^*)(h_k - h_k^*) - \right.$$

$$\left. - \sum_{k \in K_p} u_p^* h_k + \sum_{k \in K_p} h_k^* u_p - t_p(u^*)(u_p - u_p^*) \right] \geq 0.$$

Now assume that $(\hat{h}, \hat{u})$ and (h^*, u^*) are two solutions to the NEM. Then, from above:

$$\sum_{p \in P} \left[\sum_{k \in K_p} \sum_{a \in A} \delta_{ak} s_a(\hat{v})(h_k^* - \hat{h}_k) - \right.$$

$$\left. - \sum_{k \in K_p} \hat{u}_p h_k^* + \sum_{k \in K_p} \hat{h}_k u_p^* - t_p(\hat{u})(u_p^* - \hat{u}_p) \right] \geq 0.$$

Also:

$$\sum_{p \in P} \left[\sum_{k \in K_p} \sum_{a \in A} \delta_{ak} s_a(v^*)(\hat{h}_k - h_k^*) - \right.$$

$$\left. - \sum_{k \in K_p} u_p^* \hat{h}_k + \sum_{k \in K_p} h_k^* \hat{u}_p - t_p(u^*)(\hat{u}_p - u_p^*) \right] \geq 0.$$

Adding these last two inequalities and changing the order of summation gives:

$$\sum_{a \in A} \left(s_a(\hat{v}) - s_a(v^*) \right)(\hat{v}_a - v_a^*) + \sum_{p \in P} \left(t_p(u^*) - t_p(\hat{u}) \right)(\hat{u}_p - u_p^*) \leq 0. \quad (25)$$

Now assume that $s(v)$ and $-t(u)$ are monotone. Then both terms in (25) are nonnegative, implying that:

$$(s(\hat{v}) - s(v^*))(\hat{v} - v^*) = 0$$
$$(t(\hat{u}) - t(u^*))(\hat{u} - u^*) = 0.$$

Assuming further that $s(v)$ and $-t(u)$ are strictly monotone, it is clear from these two equations that $\hat{v} = v^*$ and that $\hat{u} = u^*$. This uniqueness condition results in:

Proposition 5. *If s and $-t$ are strictly monotone, then the NEM has a unique solution in the arc flows v^* and in the minimum path costs u^*.*

3. Some optimization reformulations

In the preceding section, the NEM was formulated as a variational inequality problem. Using that formulation, we now present optimization formulations which will be used in the descriptions of algorithms for solving the NEM.

To simplify the presentation, it will be assumed throughout that the conditions for the existence and uniqueness of a NEM solution, as given in Propositions 4 and 5, are met.

3.1. The VI formulation with arc flows and inverse demand functions

To facilitate the conversion of the VI problem into equivalent optimization problems, arc flows and inverse demand functions are explicitly introduced into the formulation.

Let (h^*, u^*) be an equilibrium solution. Then, it is immediate from the NEM conditions (10), (11), and (13), that:

$$(s_k(h^*) - u_p^*)(h_k - h_k^*) \geq 0,$$
$$\forall h = (h_k), \ u = (u_p) \geq 0, \ k \in K_p, \ p \in P. \quad (26)$$

Now, suppose (26) holds. If h_k^* equals 0, then $s_k(h^*) - u_p^* \geq 0$, since $h_k \geq 0$. On the other hand, if h_k^* is positive, then for any $0 \leq \hat{h}_k < h_k^*$, $(s_k(h^*) - u_p^*)(\hat{h}_k - h_k^*) \geq 0$ implies that $s_k(h^*) - u_p^*$ is zero. In either case, (10) and (11) hold.

Summing (26) over all $k \in K_p$, $p \in P$, gives:

$$\sum_{p \in P} \sum_{k \in K_p} s_k(h^*)(h_k - h_k^*) \geq \sum_{p \in P} \sum_{k \in K_p} u_p^*(h_k - h_k^*). \quad (27)$$

The right-hand-side of (27) equals:

$$\sum_{p \in P} u_p^* \left[\sum_{k \in K_p} h_k - \sum_{k \in K_p} h_k^* \right] = \sum_{p \in P} u_p^* [t_p(u) - t_p(u^*)].$$

Also, using the fact that $s_k(h) = \sum_{a \in A} \delta_{ak} s_a(v)$, and $v_a = \sum_{k \in K} \delta_{ak} h_k$ and by rearranging the order of summation, the left-hand-side of (27) equals:

$$\sum_{k \in K} \sum_{a \in A} \delta_{ak} s_a(v^*)(h_k - h_k^*) = \sum_{a \in A} s_a(v^*) \left[\sum_{k \in K} \delta_{ak} h_k - \sum_{k \in K} \delta_{ak} h_k^* \right]$$

$$= \sum_{a \in A} s_a(v^*)(v_a - v_a^*).$$

Thus, (27) now becomes:

$$\sum_{a \in A} s_a(v^*)(v_a - v_a^*) \geq \sum_{p \in P} u_p^*(t_p(u) - t_p(u^*)). \tag{28}$$

Strict monotonicity of $-t$ implies that $t = t(u)$ is invertible. Let w denote its inverse, that is, $w(t) = u$; or, componentwise, $w_p(t) = u_p$. Then (28) may be written:

$$\sum_{a \in A} s_a(v^*)(v_a - v_a^*) \geq \sum_{p \in P} w_p(t^*)(t_p(u) - t_p(u^*))$$

where $t^* = t(u^*)$.

In vector notation this is

$$s(v^*)(v - v^*) - w(t^*)(t - t^*) \geq 0 \tag{29}$$

and, in the case of fixed demand, is just

$$s(v^*)(v - v^*) \geq 0. \tag{30}$$

It is now possible to include arc flows and inverse demand functions in the VI problem:

Proposition 6. (v^*, t^*) *solves the NEM problem if and only if* (v^*, t^*) *solves:*

$$\sum_{a \in A} s_a(v^*)(v_a - v_a^*) - \sum_{p \in P} w_p(t^*)(t_p - t_p^*) \geq 0 \tag{31}$$

subject to

$$\sum_{k \in K_p} h_k = t_p, \qquad \forall\, p \in P \tag{32}$$

$$h_k \geq 0, \qquad \forall\, k \in K \tag{33}$$

$$v_a = \sum_{k \in K} \delta_{ak} h_k, \qquad \forall\, a \in A. \tag{34}$$

Proof. The necessity follows from the preceding discussion. To establish the converse, let an (h^*, u^*) correspond to the solution (v^*, t^*). Construct a flow pattern $h = (h_k)$, which differs from h^* in (exactly) one path, say $\hat{k} \in K_{\tilde{p}}$, so that:

$$h_{\hat{k}} = h_{\hat{k}}^* + \delta,$$

for any $\delta \geq -h_{\hat{k}}^*$. Then

$$\sum_{k \in K_{\tilde{p}}} h_k = t_{\tilde{p}}(u^*) + \delta.$$

and applying (31) yields

$$0 \leq s_{\hat{k}}(h^*)\delta - w_{\tilde{p}}(t^*)\delta = (s_{\hat{k}}(h^*) - u_{\tilde{p}}^*)\delta. \tag{35}$$

If $h_{\hat{k}}^* = 0$, then $\delta \in [0, \infty)$, and (35) implies

$$(s_{\hat{k}}(h^*) - u_{\tilde{p}}^*) \geq 0.$$

If $h_{\hat{k}}^* > 0$, then $\delta \in [-h_{\hat{k}}^*, \infty)$, and (35) implies

$$s_{\hat{k}}(h^*) - u_{\tilde{p}}^* = 0$$

since δ may take on negative and positive values in this interval. In either case, we have

$$(s_{\hat{k}}(h^*) - u_{\tilde{p}}^*)h_{\hat{k}}^* = 0$$

and

$$s_{\hat{k}}(h^*) - u_{\tilde{p}}^* \geq 0.$$

Since such a $\tilde{p} \in P$ and $\hat{k} \in K_{\tilde{p}}$ may be arbitrarily chosen, (10) and (11) hold and the proof is complete. $\square$

3.2. The symmetric NEM

Let $\nabla s(v)$ and $\nabla w(t)$ denote, respectively, the Jacobians of $s(v)$ and $w(t)$. And suppose that $\nabla s(v)$ and $\nabla w(t)$ are symmetric, that is:

$$\frac{\partial s_a(v)}{\partial v_{\hat{a}}} = \frac{\partial s_{\hat{a}}(v)}{\partial v_a}, \quad \text{and} \quad \frac{\partial w_p(t)}{\partial t_{\hat{p}}} = \frac{\partial w_{\hat{p}}(t)}{\partial t_p},$$

for all a, $\hat{a}$ in A, and all p, $\hat{p}$ in P.

Then, by Green's Lemma, the vectors $s(v)$ and $w(t)$ can be viewed as the gradients of the line integrals $\oint_0^v s(x)\,dx$ and $\oint_0^t w(x)\,dx$, respectively.

Monotonicity of s implies that $\nabla s(v)$ is positive semi-definite. Similarly, monotonicity of $-t$ implies that $\nabla w(t)$ is negative semi-definite. These, in turn, imply that

$$\oint_0^v s(x)\,dx - \oint_0^t w(x)\,dx$$

is a convex function in (v, t).

Consider the following result from optimization theory:

If $f : \mathbb{R}^n \to \mathbb{R}$ is a convex, differentiable function, and X is a nonempty, convex set in $\mathbb{R}^n$, then $x^* \in X$ is an optimal solution to the problem of minimizing $f(x)$ subject to $x \in X$ if and only if

$$\nabla f(x^*)(x - x^*) \geq 0, \qquad \forall x \in X,$$

where $\nabla f(x)$ denotes the gradient of f at x.

It follows from this that the vector (v^*, t^*) (and corresponding (h^*, u^*)) solves the convex optimization problem:

$$\min \oint_0^v s(x)\,dx - \oint_0^t w(x)\,dx \tag{36}$$

subject to (32), (33) and (34)

if and only if it satisfies the variational inequality problem (31)–(34), and hence, solves the NEM.

As an example of the NEM with equivalent convex optimization problem formulations, consider the case when the cost function on a particular arc depends only on the flow through that arc, that is:

$$s_a(v) = s_a(v_a) \qquad \forall\, a \in A.$$

Suppose, moreover, that the demand functions are such that:

$$t_p(u) = t_p(u_p) \qquad \forall\, p \in P.$$

Such functions are called *separable*.

In this case, the strict monotonicity assumptions on s and $-t$ translate into $s_a(v_a)$ being strictly increasing and $t_p(u_p)$ being strictly decreasing. Further, the Jacobians of s and w are diagonal matrices and hence, symmetric.

Condition (31) then takes the form:

$$\sum_{a \in A} s_a(v_a^*)(v_a - v_a^*) - \sum_{p \in P} w_p(t_p^*)(t_p - t_p^*) \geq 0. \tag{37}$$

And the equivalent convex optimization problem simply becomes

$$\min\ S(v) - W(t) \tag{38}$$

subject to   (32), (33) and (34),

where $S(v) = \sum_{a \in A} \int_0^{v_a} s_a(x)\,dx$ and $W(t) = \sum_{p \in P} \int_0^{t_p} w_p(y)\,dy$. In the case of fixed demand, the problem takes the classic form:

$$\min \; S(v) = \sum_{a \in A} \int_0^{v_a} s_a(x) \, dx \qquad\qquad (39)$$

$$\text{subject to} \quad (32), (33) \text{ and } (34)$$

with $t_p = \bar{t}_p$, that is constant, in (32).

These two convex minimization problems have been the focus of much attention in the development of numerical algorithms for solving the symmetric NEM and Section 4 contains a discussion of a selection of these methods.

3.3. The asymmetric NEM with fixed demand

The success of optimization algorithms in solving practical applications of the symmetric NEM has motivated researchers to seek optimization formulations for the more general NEM where the symmetry assumptions are not made. This final section will give a summary of some of the results obtained thus far for the fixed demand case.

Consider again finding an arc-flow vector v^* that satisfies (30), (32), (33), and (34), that is, a v^* in Θ so that:

$$s(v^*)(v - v^*) \geq 0, \qquad \forall\, v \in \Theta \qquad\qquad (40)$$

where:

$$\Theta = \Big\{ v = (v_a) : \exists h = (h_k) \geq 0$$
$$\text{with } v_a = \sum_{k \in K} \delta_{ak} h_k, \; \sum_{k \in K_p} h_k = \bar{t}_p, \; \forall p \in P, \; a \in A \Big\}$$

and $\bar{t}_p$ denotes the fixed demand for O–D pair p.

For any $\tilde{v}$ in Θ, define $G(\tilde{v})$ by:

$$G(\tilde{v}) = \max_{v \in \Theta} s(\tilde{v})(\tilde{v} - v)$$

Note that (40) implies that $s(v^*)(v^* - v) \leq 0$, $\forall v \in \Theta$. Therefore, unless $\tilde{v}$ solves (40), $G(\tilde{v}) > 0$, and, $G(\tilde{v}) = 0$ if and only if $\tilde{v} = v^*$ solves (40). Therefore, solving (40) is equivalent to finding a v^* which solves:

$$\min_{\tilde{v} \in \Theta} G(\tilde{v}). \qquad\qquad (41)$$

This may also be written as:

$$\min_{\tilde{v} \in \Theta} \{ \max_{v \in \Theta} \{ s(\tilde{v})(\tilde{v} - v) \} \} \qquad\qquad (42)$$

where for each $\tilde{v}$, the inner maximization problem is a linear program.

In the context of the NEM, $G(\tilde{v})$ has an important interpretation which makes (41) intuitively clear. To see this, note that

$$G(\tilde{v}) = s(\tilde{v})\tilde{v} + \max_{v \in \Theta} -s(\tilde{v})v \tag{43}$$

$$= s(\tilde{v})\tilde{v} - \min_{v \in \Theta} s(\tilde{v})v \tag{44}$$

$$= \sum_{a \in A} s_a(\tilde{v}_a)\tilde{v}_a - \sum_{p \in P} \tilde{u}_p \bar{l}_p \tag{45}$$

where $\tilde{u}_p = u_p(\tilde{v})$ is, by definition (5), the least cost path, given $\tilde{v}$, for O–D pair p. The first term in (45) is the 'system optimal' objective per Wardrop's normative principle [cf. (14)], and the second term is the total system cost if all demand is routed along minimum paths *with respect to* $\tilde{v}$. Clearly these two terms are equal if and only if $\tilde{v} = v^*$, the user optimal flow vector.

The function G is known as the *gap function*. When $\tilde{v} \neq v^*$, $G(\tilde{v})$ provides a measure of how far the arc flow vector $\tilde{v}$ is from the user optimal flow vector, v^*. As will be seen later, the gap function is often employed as a stopping criteria for algorithms.

Conditions have been derived under which G is a convex function, and, therefore, (41) is a convex network problem:

Proposition 7. *Let $s(v)v$ be convex and assume that each component of $s(v)$ is concave in v for all $v \in \Theta$. Then G is convex on Θ.*

An important case in which the cost function satisfies these conditions is when s is *affine*, i.e., when $s(v) = Cv + b$, where C is a matrix and b is a vector.

The extent to which G is differentiable is also important for algorithms. The results are given by:

Proposition 8. *If G is convex, then the subdifferential of G at $\tilde{v}$ is equal to the convex hull of $\{s(\tilde{v}) + \nabla s(\tilde{v})(\tilde{v} - v) : v \in \Phi(\tilde{v})\}$, where $\Phi(\tilde{v})$ is the set of solutions to the problem of maximizing $s(\tilde{v})(\tilde{v} - v)$ over all $v \in \Theta$. Thus G is differentiable at $\tilde{v}$ if and only if $\Phi(\tilde{v})$ is a singleton.*

In general, the conditions of these two propositions will not hold and (41) is not a differentiable convex programming problem. However, the gap function has proven useful in the development of several algorithms for the asymmetric NEM, which are described in Section 5.

A second function which has been useful for certain algorithms is based on the variational inequality result that s being monotone implies that v^* satisfies (40) if and only if:

$$s(v)(v - v^*) \geq 0, \qquad \forall v \in \Theta. \tag{46}$$

For any $\tilde{v}$ in Θ, define:

$$g(\tilde{v}) = \min_{v \in \Theta}\{s(v)(v - \tilde{v})\},$$

which is known as the *dual* gap function. From (46), note that $g(\tilde{v}) < 0$ unless $\tilde{v}$ solves (40) and that $g(\tilde{v}) = 0$ if and only if $\tilde{v} = v^*$ solves (40). Thus, solving (40) is equivalent to finding a v^* which solves the maximization problem:

$$\max_{\tilde{v} \in \Theta} g(\tilde{v}), \tag{47}$$

which may be written as:

$$\max_{\tilde{v} \in \Theta} \{ \min_{v \in \Theta} s(v)(v - \tilde{v}) \}. \tag{48}$$

Notice that for every $\tilde{v}$ in Θ, the inner minimization problem in (48) is, in general, nonlinear, in contrast to the linear inner maximization problem in (42). However, since $g(v^*)$ is the pointwise minimum of functions linear in $\tilde{v}$, g is concave in Θ. This gives

Proposition 9. *Problem (47) is a concave programming problem equivalent to the NEM.*

The function g has differentiability properties as summarized by:

Proposition 10. *The subdifferential of g at $\tilde{v}$ is equal to the convex hull of $\{-s(v) : v \in \phi(\tilde{v})\}$, where $\phi(\tilde{v})$ is the set of solutions to the problem of minimizing $s(v)(v - \tilde{v})$ over all v in Θ. Thus the function g is differentiable at $\tilde{v}$ if and only if $\phi(\tilde{v})$ is a singleton.*

The development of additional gap functions which will allow converting the NEM to an optimization problem continues to be an objective of research. Below two of these will be discussed briefly; see the notes for additional references.

First, for large-scale *column generation* methods, there is a family of such functions that requires the set Θ to be the convex hull of known extreme points, say, $\Theta_1, \Theta_2, \ldots, \Theta_m$.

To introduce these, first define, for any real number β, $\Psi(\beta) = \max(0, \beta)$, and for any integer q, $q \geq 1$, denote $(\Psi(\beta))^q$ by $\Psi^q(\beta)$. Now, define the following family of functions:

$$G_q(v) = \sum_{i=1}^m \Psi^q(s(v)(v - \Theta_i)), \qquad v \in \Theta.$$

Notice that $G_q(v) > 0$ unless v solves (40), and when $v = v^*$, $G_q(v^*) = 0$. Hence, solving (40) is equivalent to finding v^* that solves:

$$\min_{v \in \Theta} G_q(v) \tag{49}$$

This problem has the nice feature that the objective is differentiable on Θ for $2 \leq q < \infty$ since Ψ^q is a $q - 1$ differentiable function. There are also certain additional conditions, not given here, for which the functions G_q are convex.

The final optimization problem equivalent to the NEM that is presented in this section has not yet been employed in large-scale algorithms, but it has much appeal due to the properties stated in the propositions below.

First, for some positive definite matrix Q, consider the quadratic program where $\tilde{v} \in \Theta$ is fixed:

$$\min s(\tilde{v})(v - \tilde{v}) + \tfrac{1}{2}(v - \tilde{v})Q(v - \tilde{v}) \tag{50}$$

$$\text{subject to} \quad v \in \Theta.$$

Let $G_Q(\tilde{v})$ be the negative of the optimal value of (50). Then it can be established that

Proposition 11. $G_Q(\tilde{v}) \geq 0$ for all $\tilde{v}$ in Θ, with equality if and only if $\tilde{v}$ solves (40).

It follows then that solving (40) is equivalent to solving:

$$\min_{\tilde{v} \in \Theta} G_Q(\tilde{v}). \tag{51}$$

Differentiability properties of G_Q are given by:

Proposition 12. If s is continuously differentiable, then G_Q is likewise continuously differentiable, with gradient given by:

$$\nabla G_Q(\tilde{v}) = s(\tilde{v}) - (\nabla s(\tilde{v}) - Q)(H(\tilde{v}) - \tilde{v})$$

where $H(\tilde{v})$ is the projection in the Q-norm of $(\tilde{v} - Q^{-1}s(\tilde{v}))$ on the set Θ.

Since s is assumed strictly monotone, the Jacobian $\nabla s(v)$ is positive definite for all v in Θ. This positive definiteness of $\nabla s(v)$ can be used to prove the following:

Proposition 13. If $\tilde{v}$ satisfies the first order optimality conditions for (51), that is, if

$$\nabla G_Q(\tilde{v})(v - \tilde{v}) \geq 0, \qquad \forall v \in \Theta,$$

then $\tilde{v} = v^*$ is a global optimal solution of (51), and hence, solves (40).

This final result is especially important since it is not known that $G_Q(\tilde{v})$ is a convex function, and most available algorithms will only find first order stationary points when applied to non-convex functions.

4. Algorithms for the symmetric NEM

As was shown in the previous section, if the Jacobians of s and w are symmetric, a NEM which has a unique solution may be formulated as an equivalent convex cost differentiable optimization problem. Since the feasible region of this problem is determined by (32)–(34), which are conservation of flow and nonnegativity constraints, and the only interaction between the arc (path) flows for an origin or an origin-destination pair occurs in the objective function, this facilitates the construction of a wide variety of algorithms for the solution of this problem, each based on a particular decomposition of the problem which exploits its special structure.

It is possible to classify the algorithms for the symmetric NEM according to the way in which the problem is decomposed, which may be by O–D pair p, by origin r or by using simplicial decomposition based on the extreme points of the feasible region. For decomposition by O–D pair p, there are algorithms which employ the path flows as explicit variables, while decomposition by origin r employs the arc flows v_a^r as explicit variables and simplicial decomposition methods use arc flows v_a as explicit variables.

In this section, the algorithmic approaches that have been proposed for the solution in the space of path flows, which we call path equilibration algorithms, are presented first. These are followed by algorithms which employ a decomposition by origin and then by algorithms based on the special structure of the extreme points in the space of arc flows. The section is concluded by the presentation of dual ascent algorithms which also use the arc flows v_a and the appropriate dual variables as explicit variables.

Convergence properties of these methods are briefly noted, but not proven; the references may be consulted for details as well as for empirical computational results. It may be said that the symmetric NEM is a well studied problem, in the sense that many efficient algorithms for its solution are available and problems with literally thousands of arcs have been successfully solved.

4.1. Path equilibration algorithms

In this algorithmic approach, the NEM is decomposed by O–D pair p and a sequence of problems for each O–D pair is solved in the space of path flows. This general approach, which is equivalent to a Gauss–Seidel decomposition (or relaxation) is also referred to as a 'cyclic decomposition', since in one 'cycle' of the algorithm a single O–D NEM is solved for each O–D pair, by keeping the flows for all other O–D pairs fixed. The algorithm terminates when no improvement in the solution occurs for all the problems solved in a cycle.

The cyclic (or Gauss–Seidel) decomposition solves the following subproblem for each O–D pair p, where for simplicity of the exposition, the fixed demand NEM is considered:

$$\text{Min} \quad \sum_{a \in A} \int_0^{v_a^p + \bar{v}_a} s_a(x)\,dx \tag{52}$$

subject to

$$\sum_{k \in K_p} h_k = \bar{t}_p \tag{53}$$

$$h_k \geq 0, \qquad \forall k \in K_p \tag{54}$$

where

$$\bar{v}_a = \sum_{p' \neq p} \sum_{k \in K_{p'}} \delta_{ak} h_k \tag{55}$$

and

$$v_a^p = \sum_{k \in K_p} \delta_{ak} h_k, \qquad \forall a \in A. \tag{56}$$

This solution strategy may be stated as follows:

Cyclic decomposition by O–D pair

Step 0. Given an initial solution set $l = 0$, $l' = 0$.

Step 1. If $l' = |P|$, the total number of O–D pairs, stop; otherwise set $l = l \bmod |P| + 1$ and continue.

Step 2. If the current solution is optimal for the lth subproblem (52)–(56), set $l' = l' + 1$ and return to step 1; otherwise, solve the lth subproblem, update the current flows, set $l' = 0$, and return to step 1.

The convergence of this decomposition strategy is ensured since the objective function is strictly convex and it may be shown to have sufficient decrease after the solution of each subproblem. A proof of convergence is not given here since this is a well known result in the theory of mathematical programming.

Path equilibration algorithms used to solve (52)–(56) operate in the space of path flows and obtain a solution where all used paths are of equal cost. The simplest such algorithm finds the shortest path and longest path, and transfers flow between these paths in order to equalize their cost. Since the number of paths grows exponentially with the network size ($|N|$, $|A|$), path equilibration algorithms are usually implemented by using a restriction strategy, where the paths used are generated as they become required. Let $K_p^+ = \{k \in K_p \mid h_k > 0\}$ be the set of paths that have positive flow. This algorithm may be stated as follows:

Path equilibration algorithm

Step 0. Find an initial solution v_a^p; $s_a = s_a(v_a^p + \bar{v}_a)$ and the initial K_p^+.

Step 1. Compute the costs of the currently used paths s_k, $k \in K_p^+$. Find k_1 such that $s_{k_1} = \min_{k \in K_p^+} s_k$ and k_2 such that $s_{k_2} = \max_{k \in K_p^+} s_k$. If $(s_{k_2} - s_{k_1}) \leq \epsilon$, go to step 4; otherwise define the direction $d_{k_1} = (h_{k_2} - h_{k_1})$ for path flow k_1 and $d_{k_2} = (h_{k_1} - h_{k_2})$ for path flow k_2.

Step 2. Find the step size λ which redistributes the flow $h_{k_1} + h_{k_2}$ between the paths k_1 and k_2 in such a way that their costs become equal, that is, solve

$$\text{Min} \quad \sum_{a \in A} \int_0^{v_a^p + \lambda y_a + \bar{v}_a} s_a(x) \, dx \tag{57}$$

$$0 \leq \lambda \leq \left(\frac{-h_{k_2}}{d_{k_2}} \right) \tag{58}$$

where

$$y_a = \delta_{ak_1} d_{k_1} + \delta_{ak_2} d_{k_2} \tag{59}$$

Step 3. $h_k = h_k + \lambda d_k$, $k = k_1, k_2$; $v_a^p = v_a^p + \lambda y_a$; $s_a = s_a(v_a^p + \bar{v}_a)$.

Step 4. Compute the shortest path $\tilde{k}$ with cost $\tilde{s}_k = \min_{k \in K_p} s_k$; if $\tilde{s}_k < \min_{k \in K_p^+} s_k$, then the path $\tilde{k}$ is added to the set of kept paths, $K_p^+ = K_p^+ \cup \tilde{k}$ and return to step 1; otherwise stop.

It can be shown that when the arc cost functions are linear, the computation of the step size λ may be obtained analytically.

The algorithm presented above is just one of many possible path equilibration schemes. For instance, the adaptation of the reduced gradient or the projected gradient algorithms for the subproblem (52)–(56) results in an equilibration step on all the paths k, $k \in K_p^+$. Also, each subproblem need not be solved to ϵ optimality; only one or two equilibration steps may be performed for each O–D pair.

This algorithm, which is simple in the sense that it attempts to equalize the cost of the used paths directly, is probably the most intuitive one as well. Its application is suitable for the NEM when the number of paths used for an O–D pair is relatively small, such as in an airline or railway network, and when the number of O–D pairs is relatively small as well. For large scale networks, the storage requirements of the paths and their flows may become prohibitively large.

The path equilibration algorithm and its variants which are mentioned above, converge, under certain conditions, to the set of equilibrium path flows.

4.2. *Linear approximation methods*

One of the simplest convergent algorithms for minimizing a convex function subject to linear constraints is the linear approximation method. Its adaptation for the solution of the NEM yields an algorithm which requires only the computation of shortest paths and a one-dimensional minimization of a convex function.

The details of the adaptation of the method to solve both the fixed and variable demand NEM are presented next, followed by the presentation of some of its variants, both for fixed and variable demand problems.

Recall from (39) that the separable fixed demand NEM is

$$\text{Min } S(v) = \sum_{a \in A} \int_0^{v_a} s_a(x)\, dx$$

subject to

$$\sum_{k \in K_p} h_k = \bar{t}_p \qquad \forall p \in P$$

$$h_k \geq 0 \qquad \forall k \in K$$

and

$$v_a = \sum_{p \in P} \sum_{k \in K_p} \delta_{ak} h_k \qquad \forall a \in A.$$

Starting from an initial feasible solution, the linear approximation method generates a feasible descent direction by solving a subproblem which is obtained by linearizing the objective function. Then, an improved solution is found on the line segment between the current solution and the solution of the subproblem. For the NEM with fixed demands, the linearized approximation of the objective function at an intermediate iteration l, when the current solution is $v^{(l)}$, is

$$S(v^{(l)}) + \nabla S(v^{(l)})(y - v^{(l)}). \tag{60}$$

Since $S(v^{(l)})$ and $\nabla S(v^{(l)})v^{(l)}$ are constants, the linearized subproblem to be solved reduces to

$$\text{Min} \sum_{p \in P} \sum_{k \in K_p} \sum_{a \in A} s_a(v_a^{(l)}) \delta_{ak} h_k \tag{61}$$

subject to

$$\sum_{k \in K_p} h_k = \bar{t}_p \quad \forall p \in P \tag{62}$$

$$h_k \geq 0 \quad \forall k \in K. \tag{63}$$

By changing the order of summation in (61) and by using (4) the objective becomes

$$\text{Min} \sum_{p \in P} \sum_{k \in K_p} s_k^{(l)} h_k. \tag{64}$$

As the terms of the objective function (64) may be separated now by O–D pair p, the solution of the linearized subproblem may be obtained by computing shortest paths for each O–D pair p and allocating the demand $\bar{t}_p$ to the links of these paths. Such a demand allocation or assignment is referred to as an 'all-or-nothing' assignment. This yields the arc flow vector

$$y_a^{(l)} = \sum_{k \in K} \delta_{ak} h_k^{(l)} \quad \forall a \in A \tag{65}$$

and the direction of descent is

$$d_a^{(l)} = (y_a^{(l)} - v_a^{(l)}) \quad \forall a \in A. \tag{66}$$

An iteration of the linear approximation algorithm is completed by finding the solution of

$$\min_{0 \leq \lambda \leq 1} S(v^{(l)} + \lambda d^{(l)}) \tag{67}$$

or, equivalently, by annulling its derivative, that is, finding λ, $0 \leq \lambda \leq 1$, for which

$$\sum_{a \in A} s_a(v_a^{(l)} + \lambda d_a^{(l)}) d_a^{(l)} = 0 \tag{68}$$

unless the minimum of (67) is attained for $\lambda = 0$ or $\lambda = 1$.

The adaptation of the linear approximation method results then in the following algorithm:

Linear approximation method

Step 0. Find an initial solution $v^{(1)}$; $s^{(1)} = s(v^{(1)})$; $l = 1$.

Step 1. Perform an 'all-or-nothing' assignment based on the current arc costs $s(v^{(l)})$ to obtain the arc flow vector $y^{(l)}$. Let $d^{(l)} = (y^{(l)} - v^{(l)})$.

Step 2. Verify if a predetermined stopping criterion is satisfied. If it is, stop; otherwise continue to

Step 3. Find the optimal step size $\lambda^{(l)}$ by solving (68).

Step 4. Update arc flows $v^{(l+1)} = v^{(l)} + \lambda^{(l)} d^{(l)}$ and arc costs $s^{(l+1)} = s(v^{(l+1)})$; set $l = l + 1$, and return to step 1.

The algorithm has several advantages for the solution of the fixed demand symmetric NEM. The paths that are used to compute the descent direction at a given iteration are generated as required and need not be kept in successive iterations. Thus the storage requirements are modest and do not increase with the number of iterations. At each iteration only the flows $v^{(l)}$ and the link costs $s^{(l)}$ need to be kept, in addition to the network link data.

Since $S(v)$ is a convex function and $\nabla S(v) = s(v)$,

$$S(v^*) \geq S(v^{(i)}) + s(v^{(i)})(y^{(i)} - v^{(i)}), \qquad i = 1, 2, \ldots, l. \tag{69}$$

The right hand side of (69) provides a lower bound on $S(v^*)$ at each iteration. The best lower bound (BLB) obtained up to a current iteration l is

$$\text{BLB} = \max_{i=1,2,\ldots,l} S(v^{(i)}) + s(v^{(i)})(y^{(i)} - v^{(i)}). \tag{70}$$

Hence a natural stopping criterion, usually denoted as the relative gap (RGAP) is

$$\text{RGAP} = \frac{S(v^{(l)}) - \text{BLB}}{S(v^{(l)})} \cdot 100 \tag{71}$$

due to its relation with the gap function introduced in Section 3, since $S(v^{(l)}) -$ BLB is the best current estimate of the minimum value of the gap function. Thus the computations would be terminated if $\text{RGAP} \leq \epsilon$, a predetermined parameter.

Other stopping criteria that may be used are a maximum number of iterations or the gap function itself, that is

$$\frac{s(v^{(l)})v^{(l)} - s(v^{(l)})y^{(l)}}{\sum_p \bar{t}_p}, \tag{72}$$

which has the physical interpretation of the difference between average trip costs on currently used paths and the average trip costs on current shortest paths. Since the gap function is not decreasing monotonically with the number of iterations, this stopping criterion must be used with this property in mind.

It is worthwhile to note that this algorithm has the intuitive interpretation of the travelers adjusting their route choice from congested routes to less congested routes. This explains its close relation to many heuristic algorithms that had been suggested and used for solving this problem. On the other hand, the principal drawback of the linear approximation method is that it often exhibits slow convergence in the vicinity of the optimal solution due to the fact that its asymptotic rate of convergence is arithmetic. This fact has motivated the development of variants of this algorithm which attempt to improve its convergence rate. One of these variants is presented later in this section.

The variable demand NEM (38) may be solved by a *partial* linear approximation method which proceeds by linearizing only some of the variables of the objective function. For this case, it is natural to linearize only the arc cost functions. The resulting subproblem at iteration l is

$$\text{Min} \sum_{p \in P} \sum_{k \in K_p} \sum_{a \in A} s_a(v_a^{(l)}) \delta_{ak} y_k - \sum_{p \in P} w_p(t_p^{(l)}) z_p \tag{73}$$

subject to

$$\sum_{k \in K_p} y_k - z_p = 0 \qquad \forall p \in P \tag{74}$$

$$y_k \geq 0 \quad \forall k \in K, \qquad z_p \geq 0 \quad \forall p \in P. \tag{75}$$

This subproblem is solved by determining $u_p^{(l)}$, $p \in P$ to be the costs of the shortest paths based on the current link costs $s(v^{(l)})$ and then simplifying (73) by using (74) and (75) to solve:

$$\text{Min} \sum_{p \in P} (u_p^{(l)} - w_p(t_p^{(l)})) z_p \tag{76}$$

subject to

$$z_p \geq 0 \qquad \forall p \in P. \tag{77}$$

By applying the optimality conditions to (76) and (77), the $z_p^{(l)}$ are determined analytically as follows:

$$z_p^{(l)} = \begin{cases} t_p(u_p^{(l)}) & \text{if } t_p(u_p^{(l)}) \geq 0 \\ 0 & \text{otherwise.} \end{cases} \tag{78}$$

The demands $z_p^{(l)}$ are then assigned to the shortest paths in order to obtain $y_a^{(l)}$, $a \in A$, and the direction of descent $d^{(l)} = \{(y^{(l)} - v^{(l)}); (z^{(l)} - t^{(l)})\}$.

The convergence criteria for the partial linear approximation method may be based on the best lower bound or the maximum number of iterations. In particular, the lower bound obtained from the partial linear approximation for the variable demand NEM is

$$\text{BLB} = \max_l S(v^{(l)}) + s(v^{(l)})(y^{(l)} - v^{(l)}) - W(t^{(l)}) - w(t^{(l)})(z^{(l)} - t^{(l)}) \tag{79}$$

where $y^{(l)}$ is the arc flow vector, as before.

Although the solutions obtained with the linear approximation method are often acceptable for the solution of the fixed (or variable) demand symmetric NEM for large scale problems, the slow convergence of the method in the neighborhood of the optimal solution has stimulated interest in constructing variants that would improve its rate of convergence while maintaining its simplicity and advantages. One of these is the parallel tangent (PARTAN) method, which was originally developed for the unconstrained minimization of a differentiable function, but which may be adapted for constrained minimization problems. The motivation for exploring this variant is that for unconstrained minimization problems the PARTAN algorithm is equivalent to the conjugate gradient algorithm and in the framework of a linearly constrained optimization problem, it may provide a richer variety of directions with relatively little overhead. The PARTAN variant alternates an ordinary iteration of the linear approximation algorithm with a direction generated by using every other solution, $v^{(l-1)}$ and $v^{(l+1)}$. Thus, the new solution is generated by finding $\alpha^{(l)}$, $\alpha^{(l)} \leq \alpha_{max}^{(l)}$ which minimizes the objective function for the solution $v^{(l-1)} + \alpha(v^{(l+1)} - v^{(l-1)})$, where $\alpha_{max}^{(l)}$ is the largest step size that maintains nonnegativity of the path flows. The efficient computation of $\alpha_{max}^{(l)}$ is crucial for this variant, since the direction is not generated towards an extreme point of the feasible region. The algorithm may be stated as follows:

Linear approximation with PARTAN

Step 0. Find a feasible solution $v^{(1)}$; $s^{(1)} = s(v^{(1)})$; $y^{(0)} = v^{(1)}$; $l = 1$.

Step 1. Perform an 'all-or-nothing' assignment based on the current arc costs to obtain the flow vector $y^{(l)}$. Let $d^{(l)} = (y^{(l)} - v^{(l)})$.

Step 2. Verify if a predetermined stopping criterion is satisfied. If it is, stop; otherwise continue to

Step 3. Find the optimal step size $\lambda^{(l)}$ by solving (68).

Step 4. Update arc flows $\tilde{v}^{(l)} = v^{(l)} + \lambda^{(l)} d^{(l)}$.

Step 5. If $l = 1$, then $v^{(l+1)} = \tilde{v}^{(l)}$; $s^{(l+1)} = s(v^{(l+1)})$; $l = l + 1$ and return to step 1; otherwise, the PARTAN direction is $d_{PAR}^{(l)} = (\tilde{v}^{(l)} - v^{(l-1)})$.

Step 6. Find optimal PARTAN step size $\alpha^{(l)}$ as the solution of

$$\min \ S(v^{(l-1)} + \alpha d_{PAR}^{(l)})$$
$$\text{s.t.} \ \ 0 \leq \alpha \leq \alpha_{max}^{(l)}.$$

Step 7. Update arc flows $v^{(l+1)} = v^{(l-1)} + \alpha^{(l)} d_{PAR}^{(l)}$ and arc costs $s^{(l+1)} = s(v^{(l+1)})$; $l = l + 1$ and return to step 1.

The maximum step size $\alpha_{max}^{(l)}$ is given by the formula

$$\alpha_{max}^{(l)} = \frac{1}{1 - \bar{\lambda}^{(l)} \bar{\lambda}^{(l-1)} \bar{\alpha}^{(l-1)}}, \tag{80}$$

where $\bar{\lambda}^{(l)} = 1 - \lambda^{(l)}$ and $\bar{\alpha}^{(l-1)} = 1 - \alpha^{(l-1)}$. Let $co[\Theta]^{(l)}$ denote the convex hull of the extremal flows $y^{(0)}, y^{(1)}, \ldots, y^{(l)}$, where it is assumed that $y^{(0)} (= v^{(1)})$ is chosen to be an extremal flow. Thus $y^{(0)} \in co[\Theta]^{(0)}$ and $v^{(2)} \in co[\Theta]^{(1)} =$

$co\{y^{(0)}, y^{(1)}\}$. If at iteration l, $v^{(l)} \in co[\Theta]^{(l)}$ then $\alpha_{\max}^{(l)}$ is the largest step size to ensure the feasibility of $v^{(l+1)}$ in $co[\Theta]^{(l)}$. It is not necessarily the largest step that may be taken within the convex hull of all extremal flows, because the generation of all extremal flows is prohibitive due to their very large cardinality. It can be shown that if $\alpha_{\max}^{(l)}$ is attained at an iteration, this is equivalent to removing an extreme point from $[\Theta]^{(l)}$.

The global convergence of the algorithm is ensured, since at each iteration $\tilde{v}^{(l)}$ is generated by a linear approximation step and each time that a PARTAN step is taken the objective function does not increase.

The additional storage requirements of the PARTAN variant are the previous solution vectors $v^{(l)}$, $v^{(l-1)}$ and step sizes $\lambda^{(l)}$, $\lambda^{(l-1)}$, $\alpha^{(l-1)}$. These may be considered to be relatively modest.

4.3. Restricted simplicial decomposition

This algorithm is an extension of the original simplicial decomposition method which applies to nonlinear programs with pseudo-convex, continuously differentiable objective functions and linear constraints. Here the extended version is applied to the separable cost, fixed-demand NEM.

Again, consider the problem:

$$\min S(v) = \sum_{a \in A} \int_0^{v_a} s_a(x) \, dx \tag{81}$$

subject to

$$\sum_{k \in K_p} h_k = \bar{t}_p, \qquad\qquad \forall p \in P \tag{82}$$

$$h_k \geq 0, \qquad\qquad \forall a \in A \tag{83}$$

$$v_a = \sum_{k \in K} \delta_{ak} h_k. \tag{84}$$

It is assumed that the feasible region is bounded. Thus, there are a finite number of extreme points of the feasible region Θ and every element of Θ can be written as a convex combination of these extreme points. Let Θ_y be a set of retained extreme points of Θ and let Θ_v be a set which is either empty or contains one of the iterates. Additionally, let $q \geq 1$ be an integer parameter to control the number of extreme points at any iteration. Then the algorithm is:

Restricted simplicial decomposition

Step 0. Let $v^{(0)}$ be a feasible point. Set $\Theta_y^{(0)} = \emptyset$, $\Theta_v^{(0)} = \{v^{(0)}\}$, and $l = 0$.
Step 1. Solve the subproblem:

$$\min \sum_{a \in A} s_a(v_a^{(l)}) y_a$$

subject to

$$\sum_{k \in K_p} h_k = \bar{t}_p, \qquad \forall \, p \in P$$

$$h_k \geq 0, \qquad\qquad \forall \, k \in K$$

and

$$y_a = \sum_{k \in K} \delta_{ak} h_k.$$

This can be solved as in Step 1 of the linear approximation algorithm of Section 4.2, which involves finding the shortest path for each O–D pair p.

Denote the solution by $y^{(l)}$. If $s(v^{(l)})(y^{(l)} - v^{(l)}) \geq 0$, then $v^{(l)}$ is a solution, so terminate. Otherwise, if $|\Theta_y^{(l)}| < q$, set: $\Theta_y^{(l+1)} = \Theta_y^{(l)} \cup \{y^{(l)}\}$, and $\Theta_v^{(l+1)} = \Theta_v^{(l)}$. If $|\Theta_y^{(l)}| = q$, replace the element of $\Theta_y^{(l)}$ that has the minimal weight in the expression of $v^{(l)}$ in the convex combination of elements of $\Theta^{(l)}$ with $y^{(l)}$ to obtain $\Theta_y^{(l+1)}$ and let $\Theta_v^{(l+1)} = \{v^{(l)}\}$. Set $\Theta^{(l+1)} = \Theta_y^{(l+1)} \cup \Theta_v^{(l+1)}$. Go to Step 2.

Step 2 (Master problem). Let $v^{(l+1)} = \text{argmin}\{S(v) : v \in co(\Theta^{(l+1)})\}$. Express $v^{(l+1)}$ as the convex combination $v^{(l+1)} = \sum_{i=1}^m \lambda_i z_i$, where $m = |\Theta^{(l+1)}|$ and $z_i \in \Theta^{(l+1)}$. Drop all elements z_i with $\lambda_i = 0$ from $\Theta_y^{(l+1)}$ and $\Theta_v^{(l+1)}$. Set $l = l + 1$ and go to Step 1.

The use of the word 'restricted' is appropriate, since the user chooses the number q of extreme points to be retained, whereas in the original simplicial decomposition version, the number of extreme points retained may grow to be quite large. The efficiency of the algorithm lies to a large extent on the solution of the master problem in Step 2.

As presented above, it appears that at every iteration, the nonlinear master is solved exactly. However, to achieve convergence, all that is needed is that there be a sufficient decrease in the value of the objective for successive iterations. One method to ensure this is to quadratically approximate the objective function. Let $\hat{S}(v \mid z)$ quadratically approximate $S(v)$ at z, that is:

$$\hat{S}(v \mid z) = S(z) + \nabla S(z)(v - z) + \tfrac{1}{2}(v - z)Q(v - z).$$

Then the solution to $\min\{\hat{S}(v \mid z) : v \in \Theta\}$ is the projection of the vector $[z - Q^{-1}\nabla S(z)]$ onto the set Θ with respect to the Q norm. Under appropriate assumptions on Q, there exists an almost closed form expression for the solution of the quadratic approximation of S. Therefore, Step 2 of the restricted simplicial decomposition algorithm can be replaced by letting:

$$v^{(l+1)} = \text{argmin}\{\hat{S}(v \mid v^{(l)}) : v \in co(\Theta^{(l+1)})\}.$$

4.4. Dual ascent

The following algorithm addresses the dual of the separable-cost, fixed-demand NEM with arc flow capacity constraints. It is a form of the dual cutting plane algorithm with a line search step between iterates.

First the fixed demand NEM is stated in terms of the flows from origins $r \in R$ as the following equivalent optimization problem:

$$\min \sum_a \int_0^{v_a} s_a(x)\,dx \tag{85}$$

subject to

$$\Omega v^r = \bar{t}^r, \qquad \forall \text{ origin nodes } r \in R \tag{86}$$

$$v^r \geq 0 \tag{87}$$

$$\sum_{r \in R} v_a^r = v_a, \qquad \forall a \in A \tag{88}$$

$$0 \leq v_a \leq b_a \qquad \forall a \in A \tag{89}$$

where Ω is the node-arc incidence matrix, and b_a is a bound on the flow on arc a.

Since $s_a(v)$ is nonnegative and increasing, so is $\int_0^{v_a} s_a(x)\,dx$.

Thus, (88) can be written as: $\sum v_a^r \leq v_a$, $a \in A$, where the summation is taken over all origin nodes r, since the minimization (85) will force this to be an equality.

The constraints (88) couple flow on a particular arc with flow from all origin nodes. Therefore, these constraints are relaxed in the dual problem. The associated multiplier vector is denoted as $u \in \mathbb{R}^{|A|}$. And the dual problem is:

$$\max_{u \geq 0} L(u)$$

where:

$$L(u) = \inf \left\{ \sum_{a \in A} \int_0^{v_a} s_a(x)\,dx + u\left(\sum_{r \in R} v^r - v\right) \right\}$$

where the infimum is taken over all v^r such that (86), (87) and (89) hold. Notice that

$$\sum_{a \in A} \int_0^{v_a} s_a(x)\,dx + u\left(\sum_{r \in R} v^r - v\right)$$

equals

$$\sum_{a \in A} \left[\int_0^{v_a} s_a(x)\,dx - v_a u_a \right] + \sum_{r \in R} \sum_{a \in A} u_a v_a^r$$

Therefore, $L(u)$ is separable in v_a and v^r, and

$$L(u) = \sum_{a \in A} \min_{0 \leq v_a \leq b_a} \left[\int_0^{v_a} s_a(x)\,dx - v_a u_a \right] +$$

$$+ \sum_{r \in R} \min \left\{ uv^r : \Omega v^r = \bar{t}^r, \ v^r \geq 0 \right\}.$$

The first minimization is a one variable problem, and the second is obtained by evaluating as many minimum path tree problems as there are origin nodes. Therefore, there are available methods for evaluating $L(u)$.

For notational convenience, write (85)–(89) as:

$$\min_v \; S(v)$$

subject to

$$f(v) \leq 0$$

$$v \in \Theta$$

where $v = (v^1, \ldots, v^r, \ldots, v^{|R|})$, $|R|$ denotes the number of origin nodes,

$$\Theta = \{v^r : \Omega v^r = \bar{t}^r, \; v_a^r \geq 0, \; 0 \leq v_a \leq b_a\},$$

and $f(v) \leq 0$ corresponds to: $\sum_{r \in R} v_a^r \leq v_a$. The algorithm is as follows:

Dual ascent algorithm

Step 0. Let $v^{(0)} \in \Theta$ such that $f(v^{(0)}) < 0$. Set $u^{(0)} = 0$ and $l = 1$.
Step 1. Solve the master (linear) problem:

$$\max_{x,q} \; x$$

subject to

$$x \leq S(v^{(i)}) + q \, f(v^{(i)}), \quad i = 1, \ldots, l - 1$$

$$q \geq 0$$

to obtain: $(x^{(l)}, q^{(l)})$.
Step 2. Set $d^{(l)} = q^{(l)} - u^{(l-1)}$. If $l = 1$, set $\lambda^{(l)} = 1$. Otherwise, define:

$$\lambda_{\max} = \arg \max \{L(u^{(l-1)} + \lambda d^{(l)}) : 0 \leq \lambda \leq \lambda_{up}\}$$

where :

$$\lambda_{up} = \max\{\lambda : u^{(l-1)} + \lambda d^{(l)} \geq 0\}.$$

If $\lambda_{\max} \leq 1$, then choose $\lambda^{(l)} \in (\lambda_{\max}, 1]$, else choose $\lambda^{(l)} \in [1, \lambda_{\max}]$. Let $u^{(l)} = u^{(l-1)} + \lambda^{(l)} d^{(l)}$.
Step 3. Solve the subproblem:

$$L(u^{(l)}) = \min \{S(v) + u^{(l)} f(v) : v \in \Theta\}.$$

Let $v^{(l)}$ be an optimal solution. If $v^{(l)}$ is not unique choose $v^{(l)}$ so that:

$$S(v^{(l)}) + q^{(l)} f(v^{(l)}) \leq S(v^{(l)}) + u^{(l)} f(v^{(l)}) = L(u^{(l)}). \tag{90}$$

Step 4. If $x^{(l)} = L(u^{(l)})$, terminate. Otherwise, set $l = l + 1$ and go to Step 1.

This algorithm uses the direction between iterates together with an inexact line search to achieve an improved version of the dual cutting plane algorithm. Choosing $v^{(l)}$ as in Step 3 guarantees convergence, and it can be shown that the choice is always possible.

5. Solving the asymmetric NEM

As before, it will be assumed that the conditions for the existence and unique-ness of a solution to the NEM, as stated in Propositions 4 and 5, are satisfied. In the following, for the simplicity of the exposition, we present algorithms for the fixed demand NEM. The variable demand NEM (37) may be solved by including the (inverse) demand term in the variational inequality formulation.

5.1. Projection algorithms

This algorithm is presented for the fixed demand NEM (40), that is, for the problem of finding v^*, $v^* \in \Theta$ which satisfies

$$s(v^*)(v - v^*) \geq 0 \qquad \forall\, v \in \Theta \tag{91}$$

where the feasible region Θ is defined as previously.

Let Q be a symmetric positive definite matrix and let $\bar{v}$, $\bar{v} \in \Theta$ be a feasible flow. A new cost function $\hat{s}(v)$ is defined over Θ as follows:

$$\hat{s}(v) = Qv + c \tag{92}$$

where

$$c = \rho s(\bar{v}) - Q\bar{v} \tag{93}$$

and ρ, $\rho > 0$ is a constant. The choice of appropriate values of ρ is discussed below. Since $\hat{s}(v)$ is linear and has a symmetric Jacobian the solution $\hat{v}$ of

$$\hat{s}(\hat{v})(v - \hat{v}) \geq 0 \qquad \forall\, v \in \Theta \tag{94}$$

may be obtained as the solution of a symmetric NEM by using one of the algorithms of Section 4. Inequality (94) is equivalent to

$$(Q\hat{v} + \rho s(\bar{v}) - Q\bar{v})(v - \hat{v}) \geq 0. \tag{95}$$

Let M_ρ be the map on Θ which takes $\bar{v}$ into $\hat{v}$, that is, $\hat{v}$ solves (95). Since for $\bar{v} \in \Theta$ the solution of (95) is unique, M_ρ is well defined. This implies that every fixed point of M_ρ is a solution of (91). To see this let $\bar{v}$ be such a fixed point, that is, $M_\rho(\bar{v}) = \bar{v}$. Then setting $\hat{v} = \bar{v}$ in (95) gives $\rho s(\bar{v})(v - \bar{v}) \geq 0 \ \forall\, v \in \Theta$ and dividing by $\rho > 0$ shows that, indeed, $\bar{v}$ is a solution of (91).

The resulting algorithm is a projection method since $\hat{v}$ is precisely the projection of the point $\bar{v} - \rho Q^{-1} s(\bar{v})$ onto the set Θ, where the projection is defined with respect to the Q norm, that is, $\hat{v} = \mathrm{Proj}_{Q,\Theta}(\bar{v} - \rho Q^{-1} s(\bar{v}))$. The algorithm may be stated as follows:

Projection algorithm

Step 0. Find an initial solution $v^{(1)}$. Let $\hat{s}(v^{(1)}) = Qv^{(1)} + c$ and set $l = 1$.

Step 1. Determine $v^{(l+1)}$ as $\mathrm{Proj}_{Q,\Theta}(v^{(l)} - \rho Q^{-1} s(v^{(l)}))$ or equivalently solve

$$\min\ (v - v^{(l)})\hat{s}(v^{(l)}) + \frac{1}{2\rho}(v - v^{(l)})Q(v - v^{(l)})$$

subject to

$$v \in \Theta.$$

Step 2. If $\| v^{(l+1)} - v^{(l)} \| \le \epsilon$, stop. Otherwise, let $l = l + 1$; $\hat{s}(v^{(l+1)}) = Qv^{(l+1)} + c$, and go to Step 1.

The convergence of the algorithm is ensured if s is required to be continuously differentiable and strongly monotone, that is,

$$(s(v_1) - s(v_2))(v_1 - v_2) \ge \alpha |v_1 - v_2|^2 \qquad \forall \ v_1, \ v_2 \in \Theta \tag{96}$$

where $\alpha > 0$ is the modulus of strong monotonicity, provided that ρ is sufficiently small. This technical condition may be difficult to verify in practice because ρ depends on both α and the maximum eigenvalue (over Θ) of the matrix $\nabla s(v) Q^{-1} \nabla s(v)$.

The asymmetric NEM may also be solved by combining a Gauss–Seidel 'cyclic decomposition' by O–D pair with a projection algorithm used to solve the resulting subproblem for each O–D pair in the space of path flows. Consider the transformation $v = \Delta h$ where $h \in \Theta$. Equation (91) may be rewritten as

$$s(\Delta h^*)(\Delta h - \Delta h^*) \ge 0$$

or

$$\Delta^T s(\Delta h^*)(h - h^*) \ge 0. \tag{97}$$

The algorithm based on the projection

$$h^{(l+1)} = \text{Proj}_{Q,\Theta} \left(h^{(l)} - \rho Q^{-1} \Delta^T s(\Delta^T h^{(l)}) \right)$$

may be shown to converge if s is strongly monotone and Θ is polyhedral. Q is as usual a symmetric positive definite matrix. This approach differs from the projection algorithm in the space of arc flows since $\Delta^T s \Delta$ is not necessarily strongly monotone. Let $\bar{s}(h) = \Delta^T s(\Delta h)$. If Q is chosen to be a block diagonal matrix, with each block Q_p corresponding to an O–D pair p, the resulting algorithm is relatively simple. The projection iteration can be implemented by solving the convex optimization problem

$$\min \ (h - h^{(l)}) \bar{s}(h^{(l)}) + \frac{1}{2\rho}(h - h^{(l)}) Q(h - h^{(l)}) \tag{98}$$

subject to

$$h \in \Theta.$$

Since both the constraint set and the matrix Q may be decomposed by O–D pair, (98) may be solved as a sequence of smaller quadratic programs, one for each O–D pair p. If each Q_p is diagonal, these problems have the form

$$\min \ \sum_{k \in K_p^+} (h_k - h_k^{(l)}) \bar{s}_k(h^{(l)}) + \frac{q^k}{2\rho}(h_k - h_k^{(l)})^2 \tag{99}$$

subject to

$$\sum_{k \in K_p^+} h_k = \bar{t}_p \tag{100}$$

where q^k are the diagonal elements of Q_p for $k \in K_p^+$, and K_p^+ is, as defined earlier, the set of paths which have a positive flow. Thus, the algorithm is applied with a restriction strategy, where the paths used are generated as required. The algorithm may be stated as follows:

Projection algorithm — path flows

Step 0. For each $p \in P$, find an initial set of paths K_p^+ such that $h_k^{(1)} > 0$ if $k \in K_p^+$ and $h_k = 0$ if $k \notin K_p^+$, $l = 1$.

Step 1. For each $p \in P$ compute a shortest path $\tilde{k}_p$ based on arc costs $s(\Delta h^{(l)})$. $K_p^+ = K_p^+ \cup \tilde{k}_p$, $\forall p \in P$.

Step 2. Solve, for each $p \in P$, the problem (99)–(100). If $\bar{h}_k^{(l)}$ is the solution of this problem set

$$h_k^{(l+1)} = \begin{cases} \bar{h}_k^{(l)} & \text{for } k \in K_p^+ \\ 0 & \text{otherwise.} \end{cases}$$

Step 3. Apply a suitable convergence test. If optimal it is reached, stop. Otherwise, continue to

Step 4. $s^{(l+1)} = s(\Delta h^{(l+1)})$, $l = l + 1$ and return to step 1.

Although this algorithm is applied in the path space, it converges to the solution of the asymmetric NEM. The proof is based on the general property that if $M : \mathbb{R}^m \to \mathbb{R}^m$ is Lipschitz continuous and strongly monotone and the set X is polyhedral, then

$$x^{(l+1)} = \text{Proj}_X\left(x^{(l)} - \alpha \bar{M}(x^{(l)})\right)$$

is convergent, provided that α is sufficiently small, $\bar{M} = B^T M B$ and $B : \mathbb{R}^n \to \mathbb{R}^m$ is a linear mapping.

5.2. Relaxation methods

A large class of algorithms for the asymmetric NEM, which are referred to as relaxation methods, results when the cost mapping is iteratively relaxed by fixing the interactions between blocks of variables and hence removing, at each iteration, the asymmetric part of the cost mapping. These algorithms include the nonlinear Jacobi method and the nonlinear Gauss–Seidel method. They are sometimes called diagonalization methods, since the resulting Jacobians of the relaxed cost mappings are diagonal.

In order to present a relaxation algorithm it is convenient to introduce a smooth function $\hat{s}(v, \tilde{v}) : \Theta \times \Theta \to \mathbb{R}^n$ with the properties that $\hat{s}(v, v) = s(v)$ and $\nabla_v \hat{s}(v, \tilde{v})$ is positive definite and symmetric. Hence, if $v^{(l+1)} = v^{(l)}$, then $v^{(l)}$

solves the asymmetric NEM and is the unique solution of the variational inequality problem

$$\hat{s}(v^{(l+1)}, v^{(l)})(v - v^{(l+1)}) \geq 0, \qquad v \in \Theta, \tag{101}$$

which is equivalent to the strictly convex differentiable optimization problem

$$v^{(l+1)} = \operatorname*{argmin}_{v \in \Theta} \sum_{a \in A} \int_0^{v_a} \hat{s}_a(x, v^{(l)}) \, \mathrm{d}x. \tag{102}$$

Various algorithms result from various choices made for the function $\hat{s}(v, \tilde{v})$. The nonlinear Jacobi method is obtained for

$$\hat{s}_a(v, v^{(l)}) = s_a\big(v_1^{(l)}, \ldots, v_a, \ldots, v_{|A|}^{(l)}\big) \tag{103}$$

and the nonlinear Gauss–Seidel method results for

$$\hat{s}_a(v, v^{(l)}) = s_a(v_1^{(l+1)}, \ldots, v_{a-1}^{(l+1)}, v_a, v_{a+1}^{(l)}, \ldots, v_{|A|}^{(l)}) \tag{104}$$

An algorithm defined by (102) is globally convergent if

$$\|\nabla_v s^{-\frac{1}{2}}(v^1, \tilde{v}^1) \cdot \nabla_{\tilde{v}} s(v^2, \tilde{v}^2) \cdot \nabla_v s^{-\frac{1}{2}}(v^3, \tilde{v}^3)\|_2 < 1,$$
$$\forall v^i, \tilde{v}^i \in \Theta, \ i = 1, 2, 3. \tag{105}$$

This sufficient condition for convergence of the relaxation methods is difficult to verify and rather restrictive. The intuitive interpretation of the convergence criterion is that the Jacobian of the cost function should have weak asymmetries.

In summary, one way to state a class of relaxation algorithms is the following:

Relaxation algorithm
   Step 1. Find a feasible solution $v^{(1)}, l = 1$.
   Step 2. Determine $v^{(l+1)}$ as the solution of (102).
   Step 3. If $\| v^{(l+1)} - v^{(l)} \| \leq \epsilon$, stop. Otherwise, continue to
   Step 4. $l = l + 1$ and go to Step 1.

5.3. *The dual cutting plane method*

It was seen in Section 3.3 that solving the NEM is equivalent to maximizing the dual gap function g. Below a cutting plane method is given for the problem with an asymmetric cost function and fixed demands under the assumption that the set Θ of feasible arc flows is a compact polyhedron.

Consider

$$\max g(\tilde{v}) \quad \text{subject to} : \tilde{v} \in \Theta \tag{106}$$

where $g(\tilde{v})$ is defined by:

$$g(\tilde{v}) = \min_{v \in \Theta}\{s(v)(v - \tilde{v})\}.$$

Note that (106) may be written:

$$\max \ x$$

subject to

$$x \le s(v)(v - \tilde{v}), \qquad \forall v \in \Theta$$
$$\tilde{v} \in \Theta.$$

The cutting plane strategy to solve this formulation is as follows:

Dual cutting plane algorithm

Step 0. Let $v^{(1)} \in \Theta$. Set $l = 1$.

Step 1. Solve the subproblem (SP_l):

$$\max \ x$$

subject to

$$x \le s(v^{(i)})(v^{(i)} - \tilde{v}), \qquad i = 1, 2, \cdots, l$$
$$\tilde{v} \in \Theta$$

and let $(x^{(l)}, \tilde{v}^{(l)})$ denote the solution.

Step 2. If $x^{(l)} \le 0$, terminate. Otherwise, find $v^{(l+1)}$ which satisfies

$$s(v^{(l+1)})(v - v^{(l+1)}) \ge 0 \qquad \forall v \in [v^{(l)}, \tilde{v}^{(l)}],$$

increment l and go to Step 1.

Both steps of this method may be accomplished relatively easily. The subproblem of Step 1 is a linear program which may be solved by a variation of the simplex method. Step 2 is a *one-dimensional* variational inequality problem and, as such, is solvable as an optimization problem.

The algorithm may terminate or iterate indefinitely. In the first case, note that v^* is the (unique) NEM solution if and only if for any v in Θ,

$$s(v^*)(v^* - v) \le 0. \tag{107}$$

Also, the strict monotonicity of s implies that

$$s(v^*)(v^* - v) > s(v)(v^* - v), \qquad \forall v \in \Theta, \ v \ne v^*;$$

so that

$$s(v)(v - v^*) > 0, \qquad \qquad \forall v \in \Theta, \ v \ne v^*. \tag{108}$$

Hence, if the algorithm terminates after l iterations, then $x^{(i)} > 0$, $i = 1, \ldots, l - 1$, and, therefore, by (107), $v^* \ne v^{(i)}$, $i = 1, \ldots, l - 1$. Since v^* is feasible to (SP_l) and $x^{(l)} = 0$ it follows from (108) that $\tilde{v}^{(l)} = v^*$.

In the other case, suppose the algorithm generates the infinite sequences $\{x^{(l)}\}$ and $\{v^{(l)}\}$. The sequence $\{x^{(l)}\}$ is on a compact set, and by appeal to general cutting plane results, it can be shown that the limit of any convergent subsequence of $\{x^{(l)}\}$ is 0. Since v^* is feasible to (SP_l), for any l, then $0 < \min_{1 \le i \le l}[s(v^{(i)})(v^{(i)} - v^*)] \le x^{(l)}$. Hence, there is a subsequence of $\{v^{(l)}\}$ which converges to v^*.

5.4. Simplicial decomposition

This algorithm addresses the fixed demand NEM. As previously shown, the problem is to find $v^* \in \Theta$ such that:

$$s(v^*)(v - v^*) \geq 0, \quad \forall v \in \Theta \tag{109}$$

where:

$$\Theta = \left\{ v : \sum_{k \in K_p} h_k = \bar{t}_p, \ h_k \geq 0, \ v_a = \sum_{k \in K} \delta_{ak} h_k \right\}.$$

The algorithm is in the spirit of the restricted simplicial decomposition method of Section 4.3, but with two important differences. First, the number of generated columns is not restricted, and hence may become large, and second, the gap function $G(\tilde{v})$ is used to monitor the progress of the algorithm.

Simplicial decomposition

Step 0. Let $\Theta^{(1)}$ be a set initialized to contain one extreme point of Θ. Let $\delta > 0$ be chosen. Set $\bar{G}^{(1)} = \infty$ and $l = 1$. Let $\{\epsilon_l\}$ be a strictly monotonically decreasing positive sequence converging to zero.

Step 1. Find $v^{(l)} \in co(\Theta^{(l)})$ such that $s(v^{(l)})(v - v^{(l)}) \geq -\epsilon_l$, for every $v \in co(\Theta^{(l)})$. Let $D^{(l)}$ denote the elements of $\Theta^{(l)}$ with zero weight in the expression of $v^{(l)}$ as a convex combination of the extreme points of $\Theta^{(l)}$.

Step 2. Solve:

$$\min s(v^{(l)})y$$

subject to

$$y \in \Theta.$$

Denote the solution of the linear program by $y^{(l)}$. If $G(v^{(l)}) = s(v^{(l)})(v^{(l)} - y^{(l)}) = 0$, stop. Otherwise:(i) if $G(v^{(l)}) \geq \bar{G}^{(l)} - \delta$, let $\Theta^{(l+1)} = \Theta^{(l)} \cup \{y^{(l)}\}$; (ii) if $G(v^{(l)}) < \bar{G}^{(l)} - \delta$, let $\Theta^{(l+1)} = (\Theta^{(l)} - D^{(l)}) \cup \{y^{(l)}\}$.
Let $\bar{G}^{(l+1)} = \min\{\bar{G}^{(l)}, G(v^{(l)})\}$. Set $l = l + 1$. Go to Step 1.

Note that Step 1 involves a 'relaxed' variational inequality, implying that it only need be approximately solved. Any of the methods available can be used, but the projection method is often employed because the domain is a simplex. Step 2 is just the familiar calculation of shortest paths as in the linear approximation method of Section 4.2. For $\tilde{v} = v^{(l)}$, the gap function $G(\tilde{v})$ defined in Section 3.3 is

$$G(v^{(l)}) = \max_{y \in \Theta} s(v^{(l)})(v^{(l)} - y) = s(v^{(l)})(v^{(l)} - y^{(l)})$$

since $y^{(l)}$ solves $\min_{y \in \Theta} s(v^{(l)})y$. Thus, $G(v^{(l)}) = 0$ if and only if $v^{(l)}$ solves (109). Since $\{G(v^{(l)})\}$ is not monotonic, $\bar{G}^{(l)}$ is introduced to obtain a monotonically decreasing sequence and this makes it possible to prove convergence of the method.

5.5. A gap descent Newton method

This section gives a globally convergent Newton method for solving monotone variational inequalities based on the direct minimization of the associated gap function.

Again, the problem is to find v^* in the set of feasible arc flow vectors Θ such that:

$$s(v^*)(v - v^*) \geq 0, \qquad \forall v \in \Theta.$$

Equivalently, there is the problem of minimizing $G(\tilde{v})$ over the set Θ, where G is the gap function given by $G(\tilde{v}) = \max_{v \in \Theta} s(\tilde{v})(\tilde{v} - v)$. Moreover, $\tilde{v}$ is the equilibrium solution if and only if $G(\tilde{v}) = 0$.

It is assumed that s is continuously differentiable. Essentially, the method generates a sequence such that each iterate is the solution to a variational inequality problem involving a linear approximation of s at the previous iterate.

The algorithm is as follows:

Gap descent Newton method

Step 0. Let $v^{(1)} \in \Theta$. Set $l = 1$.

Step 1. Solve for $\hat{v}^{(l)}$ so that:

$$[s(v^{(l)}) + \nabla s(v^{(l)})(\hat{v}^{(l)} - v^{(l)})](v - \hat{v}^{(l)}) \geq 0, \qquad \forall v \in \Theta$$

Step 2. If $G(\hat{v}^{(l)}) \leq 0.5\, G(v^{(l)})$, set $\lambda^{(l)}$ to 1. Otherwise, let $\lambda^{(l)} \in \operatorname{argmin}_{0 \leq \lambda \leq 1} G[(1 - \lambda)v^{(l)} + \lambda \hat{v}^{(l)}]$.

Step 3. Set $v^{(l+1)} = (1 - \lambda^{(l)})v^{(l)} + \lambda^{(l)}\hat{v}^{(l)}$. Increment l by 1 and go to Step 1.

To see that the $\hat{v}^{(l)} - v^{(l)}$ is a strict descent direction for the gap function at iteration l, let $G'(v, d)$ denote the directional derivative of G at v in the direction d. Then, by a standard result in minimax theory:

$$G'(v, d) = \max_{z \in \Upsilon(v)} d\big[s(v) + \nabla s(v)(v - z)\big]$$

where $\Upsilon(v) = \{z : G(v) = s(v)(v - z)\}$. So, for $v = v^{(l)}$ and $d^{(l)} = \hat{v}^{(l)} - v^{(l)}$,

$$G'(v^{(l)}, d^{(l)}) = \max_{z \in \Upsilon(v^{(l)})} (\hat{v}^{(l)} - v^{(l)})\big[s(v^{(l)}) + \nabla s(v^{(l)})(v^{(l)} - z)\big]$$

which may be written

$$\begin{aligned}
&\big[s(v^{(l)}) + \nabla s(v^{(l)})(\hat{v}^{(l)} - v^{(l)})\big](\hat{v}^{(l)} - v^{(l)}) \\
&\quad - \big[\nabla s(v^{(l)})(\hat{v}^{(l)} - v^{(l)})\big](\hat{v}^{(l)} - v^{(l)}) + \\
&\quad + \max_{z \in \Upsilon(v^{(l)})} (v^{(l)} - z)\nabla s(v^{(l)})(\hat{v}^{(l)} - v^{(l)}).
\end{aligned} \tag{110}$$

The second term of (110) is nonpositive since s is monotone, and the third term is

$$\max_{z \in \Upsilon(v^{(l)})} \big\{\big[s(v^{(l)}) + \nabla s(v^{(l)})(\hat{v}^{(l)} - v^{(l)})\big](v^{(l)} - z) + s(v^{(l)})(z - v^{(l)})\big\}. \tag{111}$$

But $G(v^{(l)}) > 0$. Therefore, the second term in (111) is strictly negative. Hence, $G'(v^{(l)}, d^{(l)})$ is strictly less than

$$[s(v^{(l)}) + \nabla s(v^{(l)})(\hat{v}^{(l)} - v^{(l)})](\hat{v}^{(l)} - v^{(l)}) +$$
$$+ \max_{z \in \Upsilon(v^{(l)})} [s(v^{(l)}) + \nabla s(v^{(l)})(\hat{v}^{(l)} - v^{(l)})](v^{(l)} - z)$$

which equals

$$\max_{z \in \Upsilon(v^{(l)})} [s(v^{(l)}) + \nabla s(v^{(l)})(\hat{v}^{(l)} - v^{(l)})](\hat{v}^{(l)} - z). \tag{112}$$

Since $\hat{v}^{(l)}$ solves the subproblem in Step 2, (112) is less than or equal to zero. Hence, indeed, $\hat{v}^{(l)} - v^{(l)}$ is a strict descent direction for the gap function and from this result global convergence of the algorithm can be established.

5.6. Another gap descent method

The final method to be given in this section addresses the fixed demand NEM which, as was discussed in Section 3.3, can be solved by finding a *stationary* point of the problem:

$$\min_{\tilde{v} \in \Theta} G_Q(\tilde{v}).$$

Here, Q is a fixed positive definite matrix and $G_Q(\tilde{v})$ is the negative of the optimal value of the quadratic program:

$$\min s(\tilde{v})(v - \tilde{v}) + \tfrac{1}{2}(v - \tilde{v})Q(v - \tilde{v}) \text{ subject to}: v \in \Theta.$$

A method for determining a stationary point is given by

G_Q descent method

Step 0. Start from any $v^{(1)} \in \Theta$. Set $l = 1$.
Step 1. Determine $H(v^{(l)})$, the projection in the Q-norm of $(v^{(l)} - Q^{-1}s(v^{(l)}))$ on the set Θ, and let $d^{(l)} = H(v^{(l)}) - v^{(l)}$.
Step 2. Let $\lambda^{(l)} \in \text{argmin}_{0 \leq \lambda \leq 1} s(v^{(l)} + \lambda d^{(l)})$.
Step 3. Set $v^{(l+1)} = v^{(l)} + \lambda^{(l)} d^{(l)}$. Increment l by 1 and go to Step 1.

The convergence is proven by establishing that the choice of $d^{(l)}$ given in step 1 is, in fact, a descent direction for G_Q at the current iterate $v^{(l)}$.

In effect, this algorithm is somewhat conceptual in nature because the calculations necessary in Steps 1 and 2 may be very difficult. There are, however, several more practical variants and extensions of this method under development. See the reference notes for further details.

6. Stochastic network equilibrium

Stochastic network equilibrium models (SNEM) are based on the assumption that travelers make systematic errors in their perception of the length of paths.

In the NEM it is implicitly taken for granted that the travelers have perfect knowledge of link travel times and, hence, of path times. The probability density functions which are taken to represent the systematic perception errors result in variants of the SNEM. It may be argued that, when the network is not subject to a high level of congestion, such models may be better suited for representing path choice. The equilibrium definition for SNEM simply states that no user can reduce his (or her) *perceived* travel time by unilaterally changing routes.

6.1. Formulation of the SNEM

In order to formulate stochastic network equilibrium models it is necessary to introduce pr_k, the probability that an individual randomly chosen from the population t_p will choose path k, $k \in K_p$, where

$$pr_k \overset{\Delta}{=} pr_k(Z_p), \qquad k \in K_p, \forall p \tag{113}$$

where $Z_p = (z_1, \ldots, z_k, \ldots)$ is the vector of *perceived* travel times of all paths k for an origin-destination pair p. The perceived travel time on link a is assumed to be given by a probability density function

$$z_a \sim D(s_a, \theta s_a) \tag{114}$$

where s_a is the actual travel time on link a and θ is a constant. Thus, the path choice probability may be expressed as

$$pr_k = Pr(z_k = \min_{l \in K_p}\{z_l\}), \qquad k \in K_p, \forall p. \tag{115}$$

Various stochastic network equilibrium models result from different assumptions made on which probability distribution governs the systematic errors in the perceived travel times. The model formulation is completed by requiring conservation of flows and nonnegativity of the path flows.

If $D(s_a, \theta s_a)$ is assumed to be the normal distribution, then the vector of perceived path travel times (Z_p) is multivariate normally distributed. The weak law of large numbers implies that *on the average* the path flows h_k satisfy

$$pr_k = \frac{h_k}{t_p}, \qquad k \in K_p, \forall p. \tag{116}$$

It is possible to show that this SNEM is equivalent to solving the problem

$$\underset{v}{\text{Min}} \sum_{p \in P} t_p \cdot E\left(\min_{k \in K_p}\{z_k \mid s^p\}\right) + \sum_{a \in A} v_a s_a(v_a) - \sum_{a \in A} \int_0^{v_a} s_a(x)\,\mathrm{d}x \tag{117}$$

subject only to nonnegativity constraints,

where the first term contains, for each p, the expected value of the minimum of the perceived path costs, given the actual path cost vector s^p. It can be shown that the resulting link flows are unique and are feasible.

The solution algorithm for this problem employs a simulation in order to obtain a descent direction for the objective function (117). In order to evaluate the objective function exactly, an exhaustive path enumeration for all O–D pairs of the network would be necessary. This is clearly prohibitive from a computational viewpoint. An algorithm implements the basic step

$$v^{l+1} = v^l + \alpha_l \cdot d^l, \tag{118}$$

where α_l, the step size at iteration l satisfies the conditions

$$\sum_{l=1}^{\infty} \alpha_l = \infty, \qquad \sum_{l=1}^{\infty} \alpha_l^2 < \infty \tag{119}$$

and d^l, the direction of descent, is determined by simulation. This simulation is performed by sampling all links for a travel time realization, computing shortest paths and performing an 'all-or-nothing' assignment on these paths; this procedure is repeated several times and then a flow vector $\hat{y}_a$ is obtained by averaging the link flows over the number of repetitions of this procedure. The number of times that the procedure is repeated determines the variance of the estimator $\hat{y}_a$. The direction of descent d_a^l is then computed as $(\hat{y}_a - v_a)$.

It may be shown that when the step sizes α_l satisfy (119) the basic algorithm (118) converges to the unique solution of this SNEM. Since $\alpha_l = 1/l$ satisfies (118), this choice is often made, and the method is referred to as 'the method of successive averages.' The algorithm lacks a natural stopping criterion and the reduction of the value of the objective function is not monotone.

An interesting special case of the SNEM occurs when the path probabilities are given by a logit function:

$$pr_k = \frac{\exp(-\theta s_k)}{\displaystyle\sum_{l \in K_p} \exp(-\theta s_l)}, \qquad k \in K_p, \ p \in P. \tag{120}$$

In this particular case it is easy to show that the equivalent optimization problem is

$$\underset{h}{\text{Min}} \ \sum_{p \in P} \sum_{k \in K_p} h_k \ln h_k + \theta \sum_{a \in A} \int_0^{v_a} s_a(x) \, dx \tag{121}$$

subject to the usual conservation of flow and nonnegativity constraints.

This problem may also be solved by the method of successive averages, without requiring simulation in order to obtain a direction of descent. At each iteration l, the direction of descent is obtained by computing shortest paths based on current link costs and performing an 'all-or-nothing' assignment on these paths. As it is prohibitive to compute values of the objective function (121), the method lacks a suitable stopping criterion for this problem as well.

It can be shown, both theoretically and empirically, that the flow pattern that results from SNEM tends towards the flow pattern obtained with the NEM as the network becomes congested. This may be recognized intuitively by inspecting the terms of the objective functions (117) and (121).

7. Other network equilibrium models

The focus of this chapter has been the theory and algorithms for network equilibrium models that have become prominent in transportation science and operations research. In this section we will briefly introduce network equilibrium models from three other disciplines: economics, electrical engineering and civil engineering. It will be seen that the formulations are closely related to those introduced earlier and that the models can be solved by algorithms which have been described in Sections 3 and 4. The purpose here is only to demonstrate these relationships between models; the references should be consulted for in-depth treatments in each case.

Separate notation will be introduced for each model in conformance with conventions in the different disciplines.

7.1. A spatial price equilibrium model

Consider two spatially separated markets where the same commodity is traded in each. If the markets operate independently, then, by economic theory, the commodity price in each will reach an equilibrium in accordance with the laws of supply and demand. The equilibrium prices π_1^* and π_2^* for the two markets occur at the intersection of the local supply and demand curves, denoted, respectively, by S and D, as shown in Figure 6a.

If these two markets are merged into a single system (Figure 6b), a flow of the commodity between the markets can occur if the cost of transporting the commodity from the less expensive market to the more expensive market is not excessive. The result will be a new equilibrium price system where both market prices and the amount of the commodity transported between markets are to be determined. To express this mathematically, let t_{12}^* be the quantity transported from Market 1 to Market 2 at a per unit cost c_{12} and define t_{21}^* and c_{21} similarly. Then the prices and quantities obey the following conditions:

$$\pi_1^* + c_{12} = \pi_2^* \ \text{ if } t_{12}^* > 0$$
$$\pi_1^* + c_{12} \geq \pi_2^* \ \text{ if } t_{12}^* = 0$$
$$\pi_2^* + c_{21} = \pi_1^* \ \text{ if } t_{21}^* > 0$$
$$\pi_2^* + c_{21} \geq \pi_1^* \ \text{ if } t_{21}^* = 0.$$

This example illustrates the basic idea of the general spatial price equilibrium model for a single commodity. The full model requires the following notation:

X = the set of supply centers, the ith center is denoted by x_i, $i = 1, 2, \ldots, m$,

Y = the set of demand centers, the jth center is denoted by y_j, $j = 1, 2, \ldots, n$,

s_i = the level of supply at supply center x_i,

s = $(s_i)_{i=1}^m$, a vector of supply levels,

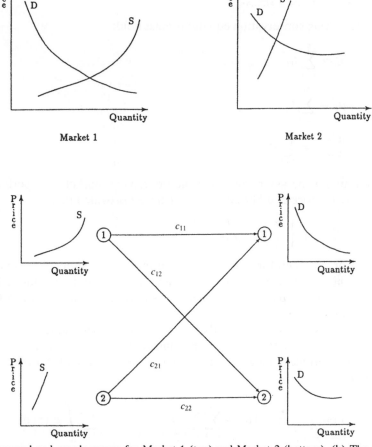

Fig. 6. (a) Demand and supply curves for Market 1 (top) and Market 2 (bottom). (b) The spatial price equilibrium problem.

π_i = $\pi_i(s)$, a continuous, positive, increasing function which gives the supply price at x_i,

$\pi(s)$ = $(\pi_i)_{i=1}^m$,

d_j = the level of demand at demand center y_j,

d = $(d_j)_{j=1}^n$,

ρ_j = $\rho_j(d)$, a continuous, positive, decreasing function which gives the demand price at y_j,

$\rho(d)$ = $(\rho_j)_{j=1}^n$,

t_{ij} = the (nonnegative) amount of the commodity shipped from x_i to y_j,

t = $(t_{ij})_{i=1,\dots,m;\ j=1,\dots,n}$,

$c_{ij} = c_{ij}(t)$, a continuous, positive, increasing function which gives the transportation cost for shipping the commodity from x_i to y_j,

$c(t) = (c_{ij})_{i=1,\ldots,m;\ j=1,\ldots,n}$.

The following conservation equations must hold:

$$s_i = \sum_{j=1}^{n} t_{ij} \tag{122}$$

$$d_j = \sum_{i=1}^{m} t_{ij} \tag{123}$$

$$\sum_{i=1}^{m} s_i = \sum_{j=1}^{n} d_j. \tag{124}$$

Then, under the assumption that the underlying market is in perfect competition, a spatial price equilibrium will be achieved provided that:

$$\pi_i^* + c_{ij}^* \begin{cases} = \rho_j^* & \text{if } t_{ij}^* > 0 \\ \geq \rho_j^* & \text{if } t_{ij}^* = 0. \end{cases} \tag{125}$$

The problem, therefore, consists of determining the supply levels at the supply centers, the demand levels at the demand centers, and the shipped quantities between every i, j pair such that the respective prices and transportation costs satisfy (125).

The simplest model assumes that π, ρ, and c are separable: $\pi_i = \pi_i(s_i)$, $\rho_j = \rho_j(d_j)$, and $c_{ij} = c_{ij}(t_{ij})$. Since these functions are monotone, the spatial price equilibrium problem may be formulated as an equivalent convex optimization problem:

$$\min \sum_{i=1}^{m} \int_0^{s_i} \pi_i(x)\,dx + \sum_{i=1}^{m}\sum_{j=1}^{n} \int_0^{t_{ij}} c_{ij}(y)\,dy - \sum_{j=1}^{n} \int_0^{d_j} \rho_j(z)\,dz$$

subject to: (122), (123), (124) and $t_{ij} \geq 0 \ \forall \ i, j$.

This optimization problem has constraints with a network structure and its derivation is similar to the separable traffic equilibrium model developed in Section 2.2. Even when the functions are not separable, the spatial price equilibrium problem may be converted to a traffic network equilibrium problem as we now illustrate.

Given X, Y, π, ρ, c, and a triple (s, t, d) satisfying (122)–(124), an equivalent traffic equilibrium network is constructed with $1 + m + n$ nodes labeled $o, x_1, x_2, \ldots, x_m, y_1, y_2, \ldots, y_n$. There are $m + mn$ arcs, constructed by connecting o to each x_i and each x_i to every y_j. Denote these arcs by (o, x_i) and (x_i, y_j), respectively. There are n O–D pairs $p_j = o/y_j$. Note that any path joining p_j passes through exactly one x_i. (See Figure 7 for the case $m = 2, n = 3$.) Letting the demand for O–D pair p_j be d_j, a feasible flow pattern consists of path flows h_{ij} corresponding to the commodity flows t_{ij}, that is, the flow on the path $o \rightarrow x_i \rightarrow y_j$ is t_{ij}.

The path flows induce the following arc flows:

$$v_{(o,x_i)} = \sum_{j=1}^{n} h_{ij} = \sum_{j=1}^{n} t_{ij} = s_i$$

$$v_{(x_i,y_j)} = h_{ij} = t_{ij}.$$

Now, define the cost associated with arc (o, x_i) to be $\pi_i(s)$, and define the cost associated with arc (x_i, y_j) to be $c_{ij}(t)$. Thus, a unit of the commodity 'traveling' from o to y_j via x_i incurs a 'travel cost' $u_{p_j} = \pi_i(s) + c_{ij}(t)$. Wardrop's principle for traffic equilibrium states that at equilibrium,

$$\pi_i(s^*) + c_{ij}(t^*) = u_{p_j}^* \quad \text{if } t_{ij}^* > 0$$
$$\pi_i(s^*) + c_{ij}(t^*) \geq u_{p_j}^* \quad \text{if } t_{ij}^* = 0$$

where $u_{p_j}^*$ corresponds to $\rho_j(d^*)$ in the spatial price equilibrium problem.

Thus, any given spatial price equilibrium problem is isomorphic to a traffic network equilibrium problem and the results and algorithms appearing in the previous sections are applicable. In particular, the spatial price equilibrium conditions are equivalent to a variational inequality problem. It may be verified that an $x^* = (s^*, d^*, t^*)$ satisfying equations (122)–(124) will satisfy the equilibrium condition (125) if and only if it satisfies the variational inequality:

$$\phi(x^*) \cdot (x - x^*) \geq 0$$

for all feasible x, where $\phi(x) = (\pi(s), -\rho(d), c(t))$. Furthermore, this result may be generalized to the multi-commodity case.

We conclude this section with a numerical example corresponding to Figure 7. Given are two supply centers x_1, x_2 and three demand centers y_1, y_2, y_3 and the following information:

$\pi_1(s) = 3s_1$	$c_{11}(t) = t_{11} + 1$	$d_1 = 3$
$\pi_2(s) = s_2$	$c_{12}(t) = t_{12} + 2$	$d_2 = 5$
	$c_{13}(t) = 2t_{13} + 1$	$d_3 = 8$
	$c_{21}(t) = 4t_{21} + 3$	
	$c_{22}(t) = 2t_{22} + 1$	
	$c_{23}(t) = 3t_{23} + 2.$	

The solution of this problem is:

$$h_1^* = 2.7385 = h_{11}^* = t_{11}^*$$
$$h_2^* = 2.2615 = h_{21}^* = t_{21}^*$$
$$h_3^* = 0 \qquad\quad = h_{12}^* = t_{12}^*$$
$$h_4^* = 3 \qquad\quad = h_{22}^* = t_{22}^*$$
$$h_5^* = 3.3385 = h_{13}^* = t_{13}^*$$
$$h_6^* = 4.6615 = h_{23}^* = t_{23}^*.$$

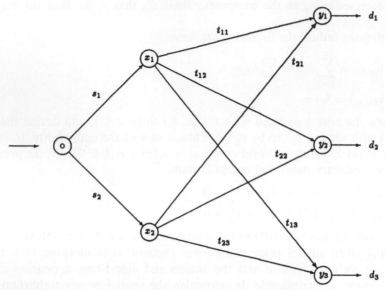

Fig. 7. A spatial price equilibrium network.

Therefore, the supply s_1^* at center x_1 equals $t_{11}^* + t_{12}^* + t_{13}^* = 6.077$ and the supply s_2^* at center x_2 equals $t_{21}^* + t_{22}^* + t_{23}^* = 9.923$. The spatial equilibrium prices are $\pi_1(6.077) = 18.231$ and $\pi_2(9.923) = 9.923$ and the equilibrium demand prices are:

$$\begin{aligned}
\rho_1^* &= u_1^* = 21.9695 \\
\rho_2^* &= u_2^* = 16.9231 \\
\rho_3^* &= u_3^* = 25.9077
\end{aligned}$$

The equilibrium transportation costs are:

$$\begin{aligned}
c_{11}(t_{11}^*) &= 3.7385 \\
c_{12}(t_{12}^*) &= 2 \\
c_{13}(t_{13}^*) &= 7.677 \\
c_{21}(t_{21}^*) &= 12.046 \\
c_{22}(t_{22}^*) &= 7 \\
c_{23}(t_{23}^*) &= 15.9845
\end{aligned}$$

It can be verified that the flow conservation and equilibrium conditions (122)–(125) are satisfied.

7.2. Electrical networks

It has long been recognized that the equilibrium current flows in the branches of an electrical circuit obey Kirchoff's laws:

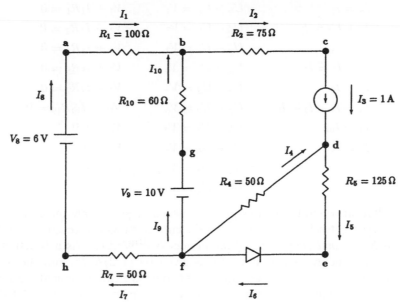

Fig. 8. An electrical network.

- current flow at nodes is conserved: flow in = flow out, and
- the potential difference between adjacent nodes is equal to the voltage on the branch connecting the nodes.

Figure 8 provides an example of a circuit with 10 branches where the current flows are labeled $I_1, I_2, \ldots, I_{10}$. Six of the branches have resistors with resistances in ohms designated by R_1, R_2, R_4, R_5, R_7 and R_{10}. Two have voltage sources V_8 and V_9. There is a diode in branch 6, and branch 3 has a current source which supplies a constant current of one ampere.

The direction of flow in branch 6 is controlled by the diode. In the other branches the directions are arbitrarily assigned — these are reversed whenever the numerical solution shows a current flow to be negative.

An equilibrium model of a particular circuit such as in Figure 8 can be constructed from Kirchoff's laws and from other relations which reflect the electrical properties of the devices on the branches. For example, Ohm's law states that the voltage in a resistor branch is the product of current and resistance, and in a diode the branch voltage is nonnegative and positive only if the current is zero. Letting

I_k = current (in amperes) in branch k

U_i = potential (in volts) at node i

V_k = voltage (in volts) on branch k

R_k = resistance (in ohms) on branch k

the model for the circuit given is to find the I_k, V_k and U_i which satisfy:

$$I_8 = I_1 \qquad\qquad U_b - U_a = V_1 \qquad\qquad V_1 + I_1 R_1 = 0$$

$$I_1 + I_{10} = I_2 \qquad\qquad U_c - U_b = V_2 \qquad\qquad V_2 + I_2 R_2 = 0$$

$$I_2 = I_3 \qquad\qquad U_d - U_c = V_3 \qquad\qquad V_4 + I_4 R_4 = 0$$

$$I_3 + I_4 = I_5 \qquad\qquad U_d - U_f = V_4 \qquad\qquad V_5 + I_5 R_5 = 0$$

$$I_5 = I_6 \qquad\qquad U_e - U_d = V_5 \qquad\qquad V_7 + I_7 R_7 = 0$$

$$I_6 = I_4 + I_7 + I_9 \qquad U_f - U_e = V_6 \qquad\qquad V_{10} + I_{10} R_{10} = 0$$

$$I_9 = I_{10} \qquad\qquad U_h - U_f = V_7 \qquad\qquad V_6 I_6 = 0$$

$$I_7 = I_8 \qquad\qquad U_a - U_h = V_8 \qquad\qquad V_6 \geq 0$$

$$U_g - U_f = V_9 \qquad\qquad I_6 \geq 0$$

$$U_b - U_g = V_{10}.$$

An alternative model for this problem which requires only the I_k as variables can be constructed by recognizing that the current flows will reach an equilibrium *for which the total energy loss of the devices is minimized.* As a function of the current, the energy loss in a resistor is given by $(1/2) R_k I_k^2$, and the gain in a voltage source is $V_k I_k$. Diodes and current sources neither dissipate nor provide energy to the circuit. Therefore, the problem of Figure 8 can be modeled as the following network optimization problem:

$$\min \ 0.5(100I_1^2 + 75I_2^2 + 50I_4^2 + 125I_5^2 + 50I_7^2 + 60I_{10}^2) - (6I_8 + 10I_9)$$

subject to

$$I_1 - I_8 = 0$$

$$I_1 - I_2 + I_{10} = 0$$

$$I_2 = 1$$

$$I_4 - I_5 = -1$$

$$I_5 - I_6 = 0$$

$$-I_4 + I_6 - I_7 - I_9 = 0$$

$$I_9 - I_{10} = 0$$

$$I_7 - I_8 = 0$$

$$I_6 \geq 0.$$

It can be demonstrated that the optimality conditions for this problem, with the multipliers corresponding to the node potentials, are exactly the conditions of the first model above. To illustrate, the solution of this problem gives the current flows:

$$I_1^* = 0.2667 \qquad\qquad I_6^* = 0.2857$$

$$I_2^* = 1 \qquad\qquad\quad I_7^* = 0.2667$$

$$I_3^* = 1 \qquad\qquad\quad I_8^* = 0.2667$$

$$I_4^* = -0.7143 \qquad\quad I_9^* = 0.7333$$

$$I_5^* = 0.2857 \qquad\quad I_{10}^* = 0.7333.$$

Moreover, the multipliers α_i, $i = 1, 2, \ldots 9$, corresponding to the nine constraints in the above minimization problem are: $\alpha_1^* = 101.6667$, $\alpha_2^* = 75$, $\alpha_3^* = 0$, $\alpha_4^* = 144.7143$, $\alpha_5^* = 109$, $\alpha_6^* = 109$, $\alpha_7^* = 119$, $\alpha_8^* = 95.667$, and $\alpha_9^* = 0$.

Note that:

$$-R_1 I_1^* \quad = -(100)(0.2667) \quad = -26.67 = \alpha_2^* - \alpha_1^*$$
$$-R_2 I_2^* \quad = -(75)(1) \quad = -75 \quad = \alpha_3^* - \alpha_2^*$$
$$-R_4 I_4^* \quad = -(50)(-0.7143) \quad = 35.71 = \alpha_4^* - \alpha_6^*$$
$$-R_5 I_5^* \quad = -(125)(0.2857) \quad = -35.71 = \alpha_5^* - \alpha_4^*$$
$$-R_7 I_7^* \quad = -(50)(0.2667) \quad = -13.33 = \alpha_8^* - \alpha_6^*$$
$$-R_{10} I_{10}^* \quad = -(60)(0.7333) \quad = -44 \quad = \alpha_2^* - \alpha_7^*$$

$$V_8 = \quad 6 = \alpha_1^* - \alpha_8^*$$
$$V_9 = 10 = \alpha_7^* - \alpha_6^*.$$

Thus, setting $U_a = \alpha_1^*$, $U_b = \alpha_2^*$, $U_c = \alpha_3^*$, ..., $U_h = \alpha_8^*$, the voltage-current relationships on the branches with the resistors and voltage sources are satisfied. Moreover, since the current, I_6^*, on the branch with the diode is nonzero, the voltage, V_6, must be zero. Indeed, $U_f - U_e = \alpha_6^* - \alpha_5^* = 0 = \alpha_9^*$. Thus, the multiplier associated with the nonnegativity condition $I_6 \geq 0$ in fact gives the voltage on that branch.

7.3. Water pipe networks

A water pipe network model consists of a set of edges representing pipes, pumps, valves and other such conduits, and a set of nodes representing reservoirs and pipe intersections. The water in the pipe network system possesses energy in the form of hydraulic pressure, or, as it is often called, hydraulic 'head'. A primary energy source is the potential energy of water in a reservoir due to its elevation. Energy is added to the system by pumps and it is lost in turbines and because of friction in the pipes.

Similar to the principles governing electrical circuits, the laws which determine the steady state flow of a system assert that the energy loss between two nodes of the network is equal to the pressure difference between the nodes. A model of equilibrium flows can be constructed from the laws, using known relations regarding energy loss or gain in the arcs and, of course, conservation of flow at the nodes.

For the model notation define an edge set E and a node set N. Elements of E represent pipes, pumps and other conduits while elements of N represent reservoirs, pipe junctions, withdrawal points, etc. Further, let

(i, j) = edge from node i to node j

Q_{ij} = flow rate of water discharged over edge (i, j)

H_n = hydraulic head at node n in N

R = the set of reservoir nodes, a subset of N.

If edge (i, j) represents a pipe, then the head loss due to friction obeys the relation:

$$\Phi_{ij}(Q_{ij}) = Z_{ij} Q_{ij}^{1.85} = H_i - H_j$$

where Z_{ij} is a constant which depends on the size, condition and type of the pipe. If edge (i, j) represents a pump then:

$$\Phi_{ij}(Q_{ij}) = a Q_{ij}^2 + b Q_{ij} + c = H_j - H_i$$

where the coefficients a, b, and c are constants depending on the type of pump.

The direction of flow Q_{ij} is assumed known and positive, which may require that some edges be replaced by two directed edges. Reservoir nodes have constant heads denoted by H_n^* for all n in R. Nodes representing inputs or withdrawals have a water requirement denoted by r_n which is positive or negative, respectively. Simple junction nodes have $r_n = 0$. The usual formulation requires that the sum of the requirements over all nodes be zero. To accomplish this, a *ground* node g is introduced with two oppositely directed edges connecting it to every reservoir node. Thus, the edge set E is augmented by the set $E' = \{(n, g), (g, n) : n \in R\}$, creating a circulation problem. Hence the requirement at g is the negative sum of the requirements at all other nodes.

Thus, the network equilibrium equations are:

$$\sum_{(n,j) \in E \cup E'} Q_{nj} - \sum_{(i,n) \in E \cup E'} Q_{in} = r_n \qquad \forall n \in N \cup \{g\} \tag{126}$$

$$H_i - H_j = \Phi_{ij}(Q_{ij}) \qquad\qquad \forall (i, j) \in E \tag{127}$$

$$H_n = H_n^* \qquad\qquad\qquad\qquad \forall n \in R. \tag{128}$$

Analogous to the minimum energy loss model for electrical networks is the minimum 'content' model for water networks which has as variables only the flows Q_{ij}:

$$\min \sum_{(i,j) \in E} \int_0^{Q_{ij}} \Phi_{ij}(t)\, dt - \sum_{(g,n) \in E'} \int_0^{Q_{gn}} H_n^*\, dt + \sum_{(n,g) \in E'} \int_0^{Q_{ng}} H_n^*\, dt$$

subject to   (126) and $Q_{ij} \geq 0 \quad \forall n \in N \cup \{g\}$.

Since the Φ_{ij}'s are continuous and monotonically increasing, this is a convex optimization problem with network constraints. Its optimality conditions, with the multipliers as the nodal heads, are (126), (127), and (128).

The water pipe network in Figure 9 shown has three reservoirs denoted as nodes 1, 6, and 7. Arcs (1,2), (7,4) and (6,5) represent three pumps. Node 8 is the ground node introduced to satisfy (126). The Z_{ij}'s necessary to determine the arc functions are given below:

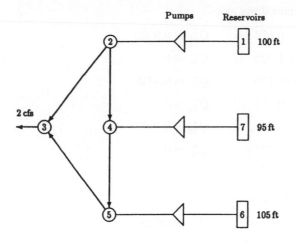

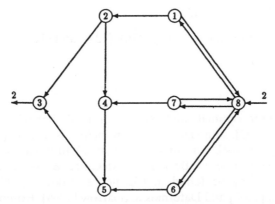

Fig. 9. A water pipe network.

Pipe (arc (i, j))	Z_{ij}
(2,3)	56.19527
(2,4)	28.04517
(4,5)	6.920492
(5,3)	13.82288
(1,2)	6.920492
(7,4)	28.04517
(6,5)	28.04517

All the pumps in this example have the following coefficients: $a = 10.328$, $b = -2.823$, and $c = -22.289$.

The equilibrium flows are:

$$Q_{23}^* = 0.674 \qquad Q_{81}^* = 0.8503$$
$$Q_{53}^* = 1.326 \qquad Q_{18}^* = 0$$
$$Q_{24}^* = 0.176 \qquad Q_{87}^* = 0.4219$$
$$Q_{45}^* = 0.598 \qquad Q_{78}^* = 0$$
$$Q_{12}^* = 0.8503 \qquad Q_{86}^* = 0.7277$$
$$Q_{74}^* = 0.4219 \qquad Q_{68}^* = 0$$
$$Q_{65}^* = 0.7277,$$

and the heads H_i's are:

$$H_1^* = 100 \qquad H_5^* = 108.29$$
$$H_2^* = 112.096 \qquad H_6^* = 105$$
$$H_3^* = 85 \qquad H_7^* = 95$$
$$H_4^* = 110.97 \qquad H_8^* = 0.$$

It can be verified that the above satisfy (126), (127), and (128).

8. Reference notes

Section 1. The 'user optimal' and 'system optimal' behavioral principles are due to Wardrop [1952], although the first principle is also implicit in concepts exposed by Knight [1924]. The first mathematical formulation of the network equilibrium model is due to Beckmann, McGuire & Winsten [1956]. Other early formulations may be found in Charnes & Cooper [1958, 1961], Jewell [1967], Beckmann [1967] and Dafermos & Sparrow [1969]. Rosenthal [1973] gives an equilibrium formulation in integer variables. Smith [1984] gives alternative behavioral interpretations of the 'user optimal principle.' The differences between the flows that satisfy the user optimal and system optimal principles was first illustrated by Braess [1968]. See also Murchland [1970]. Other networks that exhibit the Braess paradox were presented by Fisk [1979] and Stewart [1980]. The occurrence of the paradox in a network arising from an application was reported by Fisk & Pallotino [1981]. Frank [1981, 1989], Steinberg & Zangwill [1983] and Steinberg & Stone [1988] as well as Dafermos & Nagurney [1984c] engage in analytical study of specific networks which exhibit the Braess paradox. Our example with asymmetric cost functions is from Dafermos [1980].

Section 2. Dafermos [1980] identified the traffic assignment model of Smith [1979b] as a variational inequality problem and her work helped to inspire the rapid development of variational inequality theory, applications, and solution methods during the 1980's. See the comprehensive survey by Harker & Pang

[1990] for a review of these developments and the works of Patricksson [1991, 1994], which chronicle it for traffic equilibrium models.

Our presentation of the complementarity and fixed point formulations and the existence results follows the development for traffic equilibria by Aashtiani & Magnanti [1981] and Magnanti [1984], who credit Asmuth [1978] and Smith [1979b] with the basic uniqueness results. Existence results are also due to Braess & Koch [1979] and Harker [1986]. The variational inequality formulation of network equilibrium models is also presented by Fang [1982].

Section 3. The VI formulation in the space of arc flows was first identified by Smith [1979b], for the fixed demand NEM, and then extended by Dafermos [1982a] for the variable demand NEM. Our presentation follows that development. Fisk & Boyce [1983] and Dafermos & Nagurney [1984d] present, by using different proofs, VI formulations of the variable demand NEM in both the space of arc flows and path flows.

The equivalent convex cost differentiable optimization problem that corresponds to the symmetric NEM was first derived by Beckmann, McGuire & Winsten [1956]. Dafermos [1971] showed the importance of symmetry in converting a two-way traffic model to an optimization formulation.

The gap function G was developed by Hearn [1978, 1979, 1982] in the context of convex programs and the traffic assignment problem. Zuhovickiĭ, Poljak & Primak [1969] and Auslender [1976] develop both the gap function and the dual gap function g in the context of variational inequalities. The family of G_q gap functions was introduced by Smith [1983a, b] who employs them in simplicial decomposition algorithms (Section 5.4) which optimize over the convex hull of known extreme points. Hearn & Nguyen [1982] discuss the duality and related saddlepoint problems. Hearn, Lawphongpanich & Nguyen [1984] provide an analysis of the various gap functions and show that G is a limiting case of the G_q family. Fukushima [1992] has provided the first differentiable gap function, G_Q, which depends on a projection operation in the Q-norm. This approach has been generalized to a new optimization formulation by Wu, Florian & Marcotte [1991] which is solved with the linearized Jacobi method. See also Larsson & Patricksson [1991] who study the generalized gap function introduced by Auchmuty [1989].

Section 4. A path equilibration algorithm for quadratic arc cost functions (linear user cost functions) was first proposed by Dafermos & Sparrow [1969]. Gibert [1968a, b] developed a similar method for nonlinear cost functions. Leventhal, Nemhauser & Trotter [1973] combined a path equilibration approach with a restriction and column generation strategy. Nguyen [1974] showed that the adaptation of the convex simplex method in the space of arc flows results in a path equilibration algorithm identical to that of Dafermos & Sparrow [1969]. Soumis [1978] solves the path equilibration problem by a projected gradient algorithm. Bertsekas [1976, 1979, 1980] develops the Goldstein–Levitan–Polyak gradient projection algorithm for path equilibration and considers the problem in the context of communication networks. Florian & Nguyen [1974] combine Benders decomposition with a path equilibration algorithm, based on an adaptation of the reduced gradient method, for the variable demand NEM. Another path equilibration scheme for the same problem is developed by Nagurney [1988].

The adaptation of the linear approximation method of Frank & Wolfe [1956] and its variants is by far the most common approach for solving fixed and variable demand symmetric NEM's. Bruynooghe, Gibert & Sakarovitch [1969] were the first to propose the method, however the subsequent work of LeBlanc, Morlok & Pierskalla [1974, 1975] and Nguyen [1976] made the method popular. The validation studies of Florian & Nguyen [1976a] and Dow & Van Vliet [1979] led to the use of the method in the practice of transportation planning. In the context of 'packet switching' networks, the same algorithm was also proposed by Fratta, Gerla & Kleinrock [1973]. See also Ruiter [1973, 1974] and Gartner [1977]. Computer codes are widely available to solve the fixed demand version of this problem.

The partial linear approximation method for solving the variable demand NEM is inspired by the work of Evans [1976] for solving the combined distribution-assignment problem. The computational results of LeBlanc & Farhangian [1981] demonstrate that, empirically, this approach is superior to a straightforward application of the linear approximation method.

The sublinear rate of convergence of the linear approximation method [see Canon & Cullum, 1968; Wolfe, 1970] generated interest in developing variants that preserve the simplicity of the method while improving the rate of convergence. Wolfe [1970] suggested the 'away' step variant, which was applied by Florian [1977a] and further refined by Guélat & Marcotte [1986] for the fixed demand NEM.

The PARTAN variant of the linear approximation method, motivated by the work of Shah, Buehler & Kempthorne [1964], for minimizing an unconstrained convex function was first suggested by LeBlanc, Helgason & Boyce [1985]. This adaptation was further refined by Florian, Guélat, Spiess [1987] and Arezki & Van Vliet [1990]. Other variants on the linear approximation method were suggested by Fukushima [1984a], Weintraub, Ortiz & Gonzáles [1985] and Lupi [1986]. All of the latter are supported by empirical, but not theoretical, results of improved rates of convergence.

A class of algorithms not covered in this section is a cyclic decomposition approach where the problem is decomposed by origin r and the arc flows v_a^r are employed as explicit variables. In effect, each subproblem is a one commodity convex cost network optimization problem without arc capacities. Such algorithms were developed by Nguyen [1974], based on an adaptation of the convex simplex method, Ferland [1974], Weintraub [1974], Weintraub & Gonzáles [1980], and Serra & Weintraub [1981], all based on algorithms which locate negative cost cycles and Petersen [1975] who developed a primal-dual algorithm for this approach. It is relevant to comment that all algorithms that were devised to solve the one commodity convex cost network flow problem, even with capacity constraints, may be used within this cyclic decomposition approach. Worthy of note are the algorithms developed by Weintraub [1974], Klessig [1974], Helgason & Kennington [1978], Dembo & Klincewicz [1981], Klincewicz [1983], Dembo [1987], and

Dembo & Tulowitzki [1988]. The last three papers contain algorithms based on Newton and truncated Newton methods.

Simplicial decomposition methods have their roots in the methods proposed by Holloway [1974] and Cantor & Gerla [1974] and they were developed for nonlinear programming by von Hohenbalken [1975, 1977]. Restricted simplicial decomposition is due to Hearn, Lawphongpanich & Ventura [1985, 1987] and Lawphongpanich & Hearn [1986] who use the projected Newton method (Bertsekas, 1982) to solve the master problem. Dantzig, Maier & Landsdowne [1976] explore other decomposition approaches for the solution of the NEM.

The dual ascent algorithm is from Hearn & Lawphongpanich [1990] whose dual formulation is the same as Goffin [1987]. Fukushima [1984b] also considers a dual approach.

Section 5. The projection algorithm in the space of arc flows is due to Dafermos [1980] and that in the space of path flows to Bertsekas & Gafni [1982]. See also Fisk & Nguyen [1982]. Another projection method is that developed by Fukushima [1986]. Relaxation algorithms include the nonlinear Jacobi method and nonlinear Gauss–Seidel methods for solving VI problems. The nonlinear Jacobi method and conditions for its convergence were studied by Ahn [1979], Ahn & Hogan [1982], Dafermos [1982b], Florian & Spiess [1982] and Dupuis & Darveau [1986], among others. Aashtiani & Magnanti [1982] develop a linearization and decomposition approach. Mahmassani & Mouskos [1988] provide computational results based on an application of the nonlinear Jacobi method, and Harker [1988] described accelerated diagonalization and projection methods..

Nonlinear Gauss–Seidel methods were studied by Dafermos [1982b] and Pang [1985]. Dafermos [1983] provided the sufficient condition of Section 5.2 for the convergence of the nonlinear Jacobi and the nonlinear Gauss–Seidel method.

The dual cutting plane method of Section 5.3 is from Nguyen & Dupuis [1984], who show it to be an improved extension of the method by Zuhovickiĭ, Poljak and Primak [1969]. Marcotte [1985] also utilizes a cutting plane approach in developing a gap descent method. The simplicial decomposition method in Section 5.4 was originally developed by Lawphongpanich & Hearn [1984]. Independently Pang & Yu [1984] gave a similar scheme in which the subproblem is solved by a one-step linear approximation method. See also Smith [1983b].

Marcotte & Dussault [1987] first recognized that Newton's method for monotone variational inequalities could be made globally convergent by adding a line search of the gap function; their method is the one given in Section 5.5. See also Marcotte [1985] and Marcotte & Dussault [1985]. Marcotte & Guélat [1988] report computational results with this method for the asymmetric NEM.

The method of Section 5.6 is from Fukushima [1992]. Fukushima [1992], Wu, Florian & Marcotte [1991] and Larsson & Patriksson [1991] provide new, but as yet untried, descent algorithms based on their new optimization formulations.

Section 6. Burrell [1969] and Dial [1971] were the first to propose stochastic route choice models for networks with constant arc costs. Burrell [1976] compares these two methods. Our presentation is based on the work of Daganzo & Sheffi [1977], Fisk [1980], Sheffi & Powell [1981, 1982], Powell & Sheffi [1982] and Daganzo [1982, 1983]. For a different approach to stochastic assignment, see Mirchandani & Soroush [1987].

Section 7. The spatial price equilibrium problem was introduced by Samuelson [1952] and developed in various directions by Takayama & Judge [1964, 1971]. MacKinnon [1975] gave a fixed point formulation and Kuhn & MacKinnon [1975] used this formulation in computing equilibrium solutions. The recent research in network equilibria and the connection made by Florian & Los [1982] between the network equilibrium and the spatial price equilibrium problems has rekindled interest in this problem and its extensions.

Florian & Los [1982] and Dafermos & Nagurney [1984a] addressed the variational inequality formulation of general models and Dafermos & Nagurney [1985] established the isomorphism with the traffic equilibrium problem. Our numerical example is based on that reference. The algorithms of Sections 4 and 5, as well as others, have been adapted for solving this problem. In particular, the ease of adapting the linear approximation method was pointed out by Florian & Los [1982], who have also suggested the use of relaxation methods for the asymmetric multiple product version of this problem. Dafermos [1983] and Friesz, Harker & Tobin [1984] have also suggested the use of projection and relaxation methods. This last reference includes computational comparisons as does the study by Nagurney [1987a]. Pang [1984] adapts Gauss–Seidel linearization methods to the spatial equilibrium problem. Nagurney [1989] and Dafermos & Nagurney [1989] developed algorithms for quadratic cost spatial price equilibrium problems. Nagurney & Kim [1989] consider such multicommodity problems and develop parallel and serial decomposition algorithms. More recently, Drissi-Kaïtouni & Florian [1991] proposed a Newton type algorithm for the spatial price equilibrium on a general network and Marcotte, Marquis & Zubieta [1992] have provided an efficient Newton-SOR method. For reviews and additional references see Nagurney [1987b] and Harker & Pang [1990].

Modeling and solving for currents and potentials in electrical networks by optimization techniques has been a standard application in nonlinear programming textbooks, such as Bazaraa & Shetty [1979], since the fundamental work of Dennis [1959]. For a particular application to large resistor networks which arise in the modeling of composite materials, see Ventura & Hearn [1988].

For the hydraulic network problem, Hall [1976] uses geometric programming theory in analyzing conditions for existence and uniqueness of solutions and suggests a gradient projection algorithm. Collins, Cooper, Helgason, Kennington & LeBlanc [1978] and Collins, Cooper, Helgason & Kennington [1978] provide a literature review and analyze the model using Karush–Kuhn–Tucker theory. They give four solution techniques based on the linear-approximation method with PARTAN, piece-wise linearization, the convex simplex algorithm, and the Newton–Raphson method. A computational comparison of these methods is provided. A hydraulic network is included in the computational study of restricted simplicial decomposition (Hearn, Lawphongpanich & Ventura, 1987). Our numerical example is from Collins, Cooper, Helgason & Kennington [1978].

Other topics

We provide reference notes for specialized topics, related to network equilibrium models and solution algorithms, which, while not covered in the text of this chapter, are important for a broader knowledge of this area.

Optimal toll patterns. A question that arises, due to the difference between user optimal and system optimal flows, is what toll to charge on the use of roads in order to induce travelers to choose routes that lead to system optimal flows. This question was studied by Tanner [1963], and then, in the context of NEMs by Dafermos & Sparrow [1971], Netter [1972a, b], Dafermos [1973] and Smith [1979a]. If it were possible to charge tolls on each arc that are the difference between the average cost, $s_a(v_a)$, and the marginal cost $s_a(v_a) + v_a s_a'(v_a)$, then the resulting flow pattern would be system optimal, even though the travelers choose user optimal routes.

Combined distribution-assignment. A special version of the variable demand NEM, called the combined distribution assignment model, where the demand is proportional to a negative exponential function of the O–D travel costs, and constraints on the total number of trips originating and terminating at origin and destination nodes are imposed as well, attracted the attention of many researchers. Florian, Nguyen & Ferland [1975], and Evans [1976] formulate the model and suggest as solution algorithms the linear approximation method and the partial linear approximation method, respectively. Florian & Nguyen [1978] extend the model to two modes, Erlander, Nguyen & Stewart [1979] discuss the calibration of the model from observed data and Jörnsten [1980] suggests a solution algorithm based on generalized Benders decomposition.

Bounded flows. The NEM as usually formulated (Section 1) does not have bounds on the arc flow variables. It is possible, however, to extend the notion of equilibrium to the model with bounds by including the multipliers associated with the bound constraints in the calculation of path costs (see Hearn & Ribera, 1980). Further, the multipliers may be interpreted as the tolls required to keep the arc flows within the bounds. The algorithm of section 4.4 addresses the symmetric cost version of this problem. Extension of the linear approximation method of Section 4.2 to the bounded variable problem is given in Daganzo [1977a, b] and Hearn & Ribera [1981].

Network aggregation. The large size of many network equilibrium problems has forced practitioners to consider aggregated versions of their models. Although called aggregation, this often takes the form of simply extracting the more important links from a large model and solving the NEM on the extracted subnetwork. See Hearn [1978, 1984] and Barton, Hearn & Ribera [1989] for discussion of both the practical applications and the theoretical development of link extraction as a form of generalized Benders decomposition. Also, Lawphongpanich & Hearn [1990] have extended these results to variational inequalities and applied the resulting algorithm to the asymmetric NEM.

Sensitivity and stability. Since the data used in applying network equilibrium models often is not very accurate, the sensitivity of the equilibrium flows to perturbations in the problem parameters, as well as their stability, are important issues. Hall [1978] and Smith [1979b] were the first to consider these problems. Dafermos & Nagurney [1984b, d] and Tobin & Friesz [1988] analyze the sensitivity of the symmetric and asymmetric NEMs. Dafermos & Nagurney [1984a], Chao & Friesz [1984] and Tobin [1987] consider the sensitivity analysis of spatial price equilibrium models. Kyparisis [1987], Dafermos [1988] and Qiu & Magnanti [1989, 1992] provide results for the sensitivity analysis of general variational inequality problems.

Application and validation. As remarked by Friesz [1985], there are only a handful of application and validation studies of NEMs in the literature. Florian & Nguyen [1976a, b], Dow & Van Vliet [1979] and Tobin [1979] are the only detailed descriptions of the goodness of fit of the predicted link flows against observed flows, as well as predicted travel times against observed travel times for fixed demand models. Florian, Chapleau, Nguyen, Achim, James-Lefebvre, and Fisk [1979] report on the calibration and validation of a two-mode asymmetric cost network equilibrium model with variable demand. Practitioners of transportation planning methods are not in the habit of publishing their results. If this were not the case, the literature would contain more articles of this type.

Extensions of network equilibrium models. In order to model the wide variety of traffic phenomena where NEMs may be applied, extensions of the fundamental model have been achieved for multiple classes of users (e.g., trucks, cars, cars that have access to high occupancy lanes, cars that pay tolls, etc.) by Dafermos [1972]. Also NEMs which consider multiple modes (e.g. cars and public transit) and integrate econometric mode choice functions have been developed by Florian [1977b], Abdulaal & LeBlanc [1979] and further studied by Fisk & Nguyen [1981] and Florian & Spiess [1983]. Some of these extensions are essential to the practical application of these models.

Books and surveys

There are relatively few books that cover the topics associated with NEM models and solution algorithms. The books of Potts & Oliver [1972] and Steenbrink [1974] appeared before most of the developments covered in this chapter. The book of Sheffi [1985] presents deterministic and stochastic NEMs. A different view is provided in the book of Newell [1980]. In the context of data communication networks, the books of Kleinrock [1964] and Bertsekas & Gallager [1987] contain valuable material. The later book of Bertsekas & Tsitsiklis [1989] has chapters dedicated to multicommodity convex cost flows, which relate to solution algorithms for the symmetric NEM as well as a chapter on VI algorithms. Nagurney's [1993] book emphasizes the modeling and solution of economic equilibria by using variational inequality formulations on appropriately defined networks. Such economic equilibria include spatial oligopolies, migration problems, Walrasian price equilibrium and others. Kennington & Helgason [1980] present algorithms for convex cost network flow optimization problems. The classical text of Kinderlehrer & Stampacchia [1980] is an excellent reference for the theoretical aspects of variational inequalities. As the study of symmetric and asymmetric NEMs was developing, survey articles were published periodically summarizing recent developments. We list some of these articles in chronological order: Van Vliet [1976a, b, c], Nguyen [1976], Magnanti & Golden [1978], Assad [1978], Gartner [1980a, b], Fernandez & Friesz [1983], Boyce [1984], Florian [1984, 1986], Magnanti [1984], Friesz [1985], Boyce, LeBlanc & Chou [1988], Patricksson [1991]. Lasdon & Warren [1980] survey nonlinear programming applications, which include the symmetric NEM. Harker & Pang [1990] survey developments in variational inequality theory and applications, which include network equilibrium models and spatial price equilibrium models.

Computer codes for NEMs

The literature contains several references to computer codes which may be used for fixed and variable demand NEMs, even though many more experimental codes have been written by researchers in this field. Nguyen & James [1975] made available the code TRAFFIC which implements the linear approximation method for fixed demand symmetric NEMs. Hall, Van Vliet & Williamson [1980] describe the fixed demand symmetric and asymmetric NEM algorithms of SATURN, a commercially available program. Babin, Florian, James-Lefebvre, and Spiess [1982] briefly describe EMME/2, another commercially available code, which uses the linear approximation method to solve fixed demand symmetric NEMs and the partial linear approximation method to solve variable demand symmetric NEMs. Bertsekas, Gendron & Tsai [1984] provide a code for solving symmetric NEMs which is a gradient projection method in the space of path flows embedded in a cyclical decomposition scheme. Hearn, Lawphongpanich, Ventura & Yang [1989] have a publicly available code RSDNET which uses restricted simplicial decomposition for solving one commodity network optimization problems. No doubt, commercial packages are available now to solve such problems; however, they are not reported in the literature.

Acknowledgement

The authors gratefully acknowledge the expert assistance of Luana Gibbons and Brenda Rayco for their many contributions to this chapter and express thanks to Luís Lopes for his expert preparation of this text. Appreciation is also expressed to the reviewers whose comments significantly improved the exposition.

References

Aashtiani, H.Z., and T.L. Magnanti (1981). Equilibria on a congested transportation network. *SIAM J. Algebraic Discr. Methods* 2, 213–226.

Aashtiani, H.Z., and T.L. Magnanti (1982). A linearization and decomposition algorithm for computing urban traffic equilibria, *Proc. 1982 IEEE Large Scale Systems Symposium*, 8–19.

Abdulaal, M., and L.J. LeBlanc (1979). Methods for combining modal split and equilibrium assignment models. *Transp. Sci.* 13, 292–314.

Ahn, B.-H. (1979). *Computation of Market Equilibria for Policy Analysis: the Project Independence Evaluation Systems (PIES) Approach.* Garland, New York, NY.

Ahn, B.-H., and W.W. Hogan (1982). On convergence of the PIES algorithm for computing equilibria. *Oper. Res.* 30, 281–300.

Arezki, Y., and D. Van Vliet (1990). A full analytical implementation of the PARTAN/Frank-Wolfe algorithm for equilibrium assignment. *Transp. Sci.* 24, 58–62.

Asmuth, R. (1978). Traffic network equilibria. Ph.D. dissertation, Stanford University, Stanford, CA.

Assad, A.A. (1978). Multicommodity network flows — a survey. *Networks* 8, 37–91.

Auchmuty, G. (1989). Variational principles for variational inequalities. *Numer. Funct. Anal. Optim.* 10, 863–874.

Auslender, A. (1976). *Optimisation. Méthodes Numériques.* Masson, Paris.

Babin, A., M. Florian, L. James-Lefebvre and H. Spiess (1982). EMME/2: Interactive graphic method for road and transit planning. *Transp. Res. Rec.* 866, 1–9.

Barton, R.R., D.W. Hearn and J. Ribera (1989). On the equivalence of transfer and generalized Benders decomposition. *Transp. Res.* 23B, 61–73.

Bazaraa, M.S., and C.M. Shetty (1979). *Nonlinear programming, theory and algorithms.* John Wiley & Sons, Inc., NY.

Beckmann, M.J. (1967). On the theory of traffic flow in networks. *Traffic Q.* 21, 109–117.

Beckmann, M., C.B. McGuire and C.B. Winsten (1956). *Studies in the Economics of Transportation*, Yale University Press, New Haven, CT.

Bertsekas, D.P. (1976). On the Goldstein–Levitin–Polyak gradient projection method. *IEEE Trans. Automatic Control* AC-21, pp. 174–184.

Bertsekas, D.P. (1979). Algorithms for nonlinear multicommodity network flow problems, in: A. Bensoussan and J.L. Lions (eds.), *Proceedings of the International Symposium on Systems Optimization and Analysis*, Springer-Verlag, NY, pp. 210–224.

Bertsekas, D.P. (1980). A class of optimal routing algorithms for communication networks, *Proc. 5th Int. Conf. on Computer Communications*, Atlanta, GA, pp. 71–76.

Bertsekas, D.P. (1982). Projected Newton methods for optimization problems with simple constraints. *SIAM J. Control Optim.* 20, pp. 221–246.

Bertsekas, D.P., and E.M. Gafni (1982). Projection methods for variational inequalities with application to the traffic assignment problem. *Math. Program. Study* 17, 139–159.

Bertsekas, D.P., and R.G. Gallager (1987). *Data Communication Networks.* Prentice-Hall, Englewood Cliffs, NJ.

Bertsekas, D.P., B. Gendron and W.K. Tsai (1984). Implementation of an optimal multicommodity network flow algorithm based on gradient projection and a path flow formulation. Technical Report LIDS-P-1364, Laboratory for Information and Decision Studies, Massachusetts Institute of Technology, Cambridge, MA.

Bertsekas, D.P., and J.N. Tsitsiklis (1989). *Parallel and Distributed Computation. Numerical Methods.* Prentice-Hall, Englewood Cliffs, NJ.

Boyce, D.E. (1984). Urban transportation network-equilibrium and design models: recent achievements and future prospects. *Environ. Plann.* 16A, 1445–1474.

Boyce, D.E., L.J. LeBlanc and K.S. Chou (1988). Network equilibrium models of urban location and travel choices: a retrospective survey. *J. Region. Sci.* 28, 159–183.

Braess, D. (1968). Über ein Paradoxen der Verkehrsplannung. *Unternehmenforschung* 12, 258–268.

Braess, D., and G. Koch (1979). On the existence of equilibria in asymmetrical multiclass-user transportation networks. *Transp. Sci.* 13, 56–63.

Bruynooghe, M., A. Gibert and M. Sakarovitch (1969). Une méthode d'affectation du trafic, in: W. Lentzback and P. Baron (eds.), *Fourth International Symposium on the Theory of Traffic Flow*, Karlsruhe, 1968, Beiträge zur Theorie des Verkehrsflusses Strassenbau und Strassenverkehrstechnik Heft 86, Herausgeben von Bundesminister fur Verkehr, Abteilung Strassenbau, Bonn, pp. 198–204.

Burrell, J.E. (1969). Multipath route assignment and its application to capacity restraint, in: W. Lentzback and P. Baron (eds.), *Fourth International Symposium on the Theory of Traffic Flow*, Karlsruhe, 1968, Beiträge zur Theorie des Verkehrsflusses Strassenbau und Strassenverkehrstechnik Heft 86, Herausgeben von Bundesminister fur Verkehr, Abteilung Strassenbau, Bonn.

Burrell, J.E. (1976). Multiple route assignment: A comparison of two methods, in: M.A. Florian, (ed.), *Traffic equilibrium methods, Proceedings of the International Symposium in Montreal*, Lecture Notes in Economics and Mathematical Systems 118, Springer-Verlag, Berlin, pp. 229–239.

Canon, M.D., and C.D. Cullum (1968). A tight upper bound on the rate of convergence of the Frank–Wolfe algorithm. *SIAM J. Control* 6, pp. 509–516.

Cantor, D.G., and M. Gerla (1974). Optimal routing in a packet switched network. *IEEE Trans. Comput.* C-23, 1062–1069.

Carey, M. (1985). The dual of the traffic assignment problem with elastic demands. *Transp. Res.* 19B, 227–237.

Chao, G.S., and T.L. Friesz (1984). Spatial price equilibrium sensitivity analysis. *Trans. Res.* 18B, 423–440.

Charnes, A., and W.W. Cooper (1958). Extremal principles for simulating traffic flow in a network, *Proc. National Academy of Science of the United States of America*, Washington, DC, pp. 201–204.

Charnes, A., and W.W. Cooper (1961). Multicopy traffic network models, in: R. Herman (ed.), *Theory of Traffic Flow*, Elsevier, Amsterdam, pp. 85–96.

Collins, M., L. Cooper, R. Helgason, J. Kennington and L. LeBlanc (1978). Solving the pipe network analysis problem using optimization techniques. *Manage. Sci.* 24, 747–760.

Collins, M., L. Cooper, R. Helgason and J. Kennington (1978). Solution of large scale pipe networks by improved mathematical approaches. IEOR Report 77016-WR77001, School of Engineering and Applied Science, Southern Methodist University.

Dafermos, S. (1971). An extended traffic assignment model with applications to two-way traffic. *Transp. Sci.* 5, 366–389.

Dafermos, S. (1972). The traffic assignment problem for multiclass-user transportation networks. *Transp. Sci.* 6, 73–87.

Dafermos, S. (1973). Toll patterns for multi-class user transportation networks. *Transp. Sci.* 7, 211–223.

Dafermos, S. (1980). Traffic equilibrium and variational inequalities. *Transp. Sci.* 14, 42–54.

Dafermos, S. (1982a). The general multimodal network equilibrium problem with elastic demand. *Networks* 12, 57–72.

Dafermos, S. (1982b). Relaxation algorithms for the general asymmetric traffic equilibrium problem. *Transp. Sci.* 16, 231–240.

Dafermos, S. (1983). An iterative scheme for variational inequalities. *Math. Program.* 26, 40–47.

Dafermos, S. (1988). Sensitivity analysis in variational inequalities. *Math. Oper. Res.* 13, 421–434.

Dafermos, S., and A. Nagurney (1984a). Sensitivity analysis for the general spatial economic equilibrium problem. *Oper. Res.* 32, pp. 1069–1086.

Dafermos, S., and A. Nagurney (1984b). Sensitivity analysis for the asymmetric network equilibrium problem. *Math. Program.* 28, 174–184.

Dafermos, S., and A. Nagurney (1984c). On some traffic equilibrium theory paradoxes. *Trans. Res.* 18B, 101–110.

Dafermos, S., and A. Nagurney (1984d). Stability and sensitivity analysis for the general network equilibrium-travel choice model, in: J. Volmuller and R. Hamerslag (eds.), *Proceedings of the*

Ninth International Symposium on Transportation and Traffic Theory, Delft, VNU Science Press, Utrecht, pp. 217–231.

Dafermos, S., and A. Nagurney (1985). Isomorphism between spatial price and traffic equilibrium models. Report LCDS #85-17, Lefschetz Center for Dynamical Systems, Brown University, Providence, RI.

Dafermos, S., and A. Nagurney (1989). Supply and demand equilibration algorithms for a class of market equilibrium problems. *Transp. Sci.* 23 , 118–124.

Dafermos, S., and F.T. Sparrow (1969). The traffic assignment problem for a general network. *J. Res. Nat. Bur. Stand.* 73B, 91–118.

Dafermos, S., and F.T. Sparrow (1971). Optimal resource allocation and toll patterns in user-optimised transport networks. *J. Trans. Econ. Policy* 5, 184–200.

Daganzo, C.F. (1977a). On the traffic assignment problem with flow dependent costs — 1. *Trans. Res.* 11, 433–437.

Daganzo, C.F. (1977b). On the traffic assignment problem with flow dependent costs — 2. *Trans. Res.* 11, 439–441.

Daganzo, C.F. (1982). Unconstrained extremal formulations of some transportation equilibrium problems. *Transp. Sci.* 16, 332–360.

Daganzo, C.F. (1983). Stochastic network equilibrium with multiple vehicle types and asymmetric, indefinite link cost Jacobians. *Transp. Sci.* 17, 282–300.

Daganzo, C.F., and Y. Sheffi (1977). On stochastic models of traffic assignment. *Transp. Sci.* 11, 253–274.

Dantzig, G.B., S.F. Maier and Z.F. Lansdowne (1976). The application of decomposition to transportation networks analysis. Report DOT-TSC-OST-76-26, U.S. Department of Transportation, Washington, D.C.

Dembo, R.S. (1987). A primal truncated Newton algorithm with application to large-scale nonlinear network optimization. *Math. Program. Study* 31, pp. 43–71.

Dembo, R.S., and J.G. Klincewicz (1981). A scale reduced gradient algorithm for network flow problems with convex separable costs. *Math. Program. Study* 15, 125–147.

Dembo, R.S., and U. Tulowitzki (1988). Computing equilibria on large multicommodity networks: an application of truncated quadratic programming algorithms. *Networks* 18, 273–284.

Dennis, J.B. (1959). *Mathematical programming and electrical networks*, MIT Press and John Wiley, NY.

Dial, R.B. (1971). A probabilistic multipath traffic assignment model which obviates path enumeration. *Trans. Res.* 5, 83–111.

Dow, P., and D. Van Vliet (1979). Capacity restrained road assignment. *Traffic Eng. Control* 20, 296–305.

Drissi-Kaïtouni, O., and M. Florian (1991). An algorithm for the spatial price equilibrium problem on a general network in the space of path flows. *J. Region. Sci.* 31, 171–190.

Dupuis, C., and J.-M. Darveau (1986). The convergence conditions of diagonalization and projection methods for fixed demand asymmetric network equilibrium problems. *Oper. Res. Lett.* 5, 149–155.

Erlander, S., S. Nguyen and N. Stewart (1979). On the calibration of the combined distribution-assignment model. *Trans. Res.* 13B, pp. 259–267.

Evans, S.P. (1976). Derivation and analysis of some models for combining trip distribution and assignment. *Trans. Res.* 10, pp. 37–57.

Fang, S.-C. (1982). Traffic equilibria on multiclass-user transportation networks analysed via variational inequalities. *Tamkang J. Math.* 13, 1–9.

Ferland, J.A. (1974). Minimum cost multicommodity circulation problems with convex arc-costs. *Transp. Sci.* 8, 355–360.

Fernandez, J.E., and T.L. Friesz (1983). Equilibrium predictions in transportation markets: the state of the art. *Trans. Res.* 17B, pp. 155–172.

Fisk, C. (1979). More paradoxes in the equilibrium assignment problem. *Trans. Res.* 13B, 305–309.

Fisk, C. (1980). Some developments in equilibrium traffic assignment. *Trans. Res.* 14B, 243–255.

Fisk, C.S., and D.E. Boyce (1983). Alternative variational inequality formulations of the network equilibrium-travel choice problem. *Transp. Sci.* 17, 454–463.

Fisk, C., and S. Nguyen (1981). Existence and uniqueness properties of an asymmetric two-mode equilibrium model. *Transp. Sci.* 15, pp. 318–328.

Fisk, C., and S. Nguyen (1982). Solution algorithms for network equilibrium models with asymmetric user costs. *Transp. Sci.* 16, pp. 361–381.

Fisk, C., and S. Pallottino (1981). Empirical evidence for equilibrium paradoxes with implications for optimal planning strategies. *Trans. Res.* 15A, 245–248.

Florian, M. (1977a). An improved linear approximation algorithm for the network equilibrium (packet switching) problem, *Proc. 1977 IEEE Conference on Decision and Control*, New Orleans, TX, pp. 812–818.

Florian, M. (1977b). A traffic equilibrium model of travel by car and public transit modes. *Transp. Sci.* 8, 168–179.

Florian, M., R. Chapleau, S. Nguyen, C. Achim, L. James-Lefebvre and C. Fisk (1979). Validation and application of EMME: an equilibrium based two-mode urban transportation planning method. *Trans. Res. Record* 728, 14–22.

Florian, M. (1984). An introduction to network models used in transportation planning, in: M. Florian, (ed.), *Transp. Plann. Models*, North-Holland, Amsterdam, pp. 137–152.

Florian, M. (1986). Nonlinear cost network models in transportation analysis. *Math. Program. Study* 26, 167–196.

Florian, M., J. Guélat and H. Spiess (1987). An efficient implementation of the PARTAN variant of the linear approximation method for the network equilibrium problem. *Networks* 17, 319–339.

Florian, M., and M. Los (1982). A new look at static spatial price equilibrium problems. *Region. Sci. Urban Econ.* 12, 579–597.

Florian, M., and S. Nguyen (1974). A method for computing network equilibrium with elastic demands. *Transp. Sci.* 8, 321–332.

Florian, M., and S. Nguyen (1976a). An application and validation of equilibrium trip assignment methods. *Transp. Sci.* 10, 374–390.

Florian, M., and S. Nguyen (1976b). Recent experience with equilibrium methods for the study of a congested urban area, in: M.A. Florian (ed.), *Traffic Equilibrium Methods, Proceedings of the International Symposium in Montreal*, Lecture Notes in Economics and Mathematical Systems 118, Springer-Verlag, New York, NY, pp. 382–395.

Florian, M., and S. Nguyen (1978). A combined trip distribution modal split and trip assignment model. *Trans. Res.* 12, 241–246.

Florian, M., S. Nguyen and J. Ferland (1975). On the combined distribution-assignment of traffic. *Transp. Sci.* 9, 43–53.

Florian, M., and H. Spiess (1982). The convergence of diagonalization algorithms for asymmetric network equilibrium problems. *Trans. Res.* 16B, 477–483.

Florian, M., and H. Spiess (1983). On binary mode choice/assignment models. *Transp. Sci.* 17, 32–47.

Frank, M. (1981). The Braess paradox. *Math. Program.* 20, pp. 283–302.

Frank, M. (1989). The equilibrium worth of a network link. *Transp. Sci.* 23, 125–138.

Frank, M., and P. Wolfe (1956). An algorithm for quadratic programming. *Nav. Res Logist. Q.* 3, 95–110.

Fratta, L., M. Gerla and L. Kleinrock (1973). The flow-deviation method: an approach to store-and-forward computer communication network design. *Networks* 3, 97–133.

Friesz, T.L. (1985). Transportation network equilibrium, design and aggregation: key developments and research opportunities. *Trans. Res.* 19A, 413–427.

Friesz, T.L., P.T. Harker and R.L. Tobin (1984). Alternative algorithms for the general network spatial price equilibrium problem. *J. Region. Sci.* 24, 475–507.

Fukushima, M. (1984a). A modified Frank–Wolfe algorithm for solving the traffic assignment problem. *Trans. Res.* 18B, pp. 169–177.

Fukushima, M. (1984b). On the dual approach to the traffic assignment problem. *Trans. Res.* 18B, 235–245.

Fukushima, M. (1986). A relaxed projection method for variational inequalities. *Math. Program.* 35, 58–70.

Fukushima, M. (1992). Equivalent differentiable optimization problems and descent methods for asymmetric variational inequalities. *Math. Program.* 53, 99–110.

Gartner, N.H. (1977). Analysis and control of transportation networks by Frank–Wolfe decomposition, *Proc. 7th Int. Symp. on Transportation and Traffic Theory*, Kyoto, Japan, pp. 591–623.

Gartner, N.H. (1980a). Optimal traffic assignment with elastic demands: a review. Part I. Analysis framework. *Transp. Sci.* 14, pp. 174–191.

Gartner, N.H. (1980b). Optimal traffic assignment with elastic demands: a review. Part II. Algorithmic approaches. *Transp. Sci.* 14, pp. 192–208.

Gibert, A. (1968a). A method for the traffic assignment problem when demand is elastic. Report LBS-TNT-85, Transportation Network Theory Unit, London Business School, London.

Gibert, A. (1968b). A method for the traffic assignment problem. Report LBS-TNT-95, Transportation Network Theory Unit, London Business School, London.

Goffin, J.L. (1987). Affine methods in nondifferentiable optimization. CORE Discussion paper No. 8744, Center for Operations Research and Econometrics, Université Catholique de Louvain, Louvain, 24 pp.

Guélat, J., and P. Marcotte (1986). Some comments on Wolfe's 'away step'. *Math. Program.* 35, 110–119.

Hall, M.A. (1976). Hydraulic network analysis using (generalized) geometric programming. *Networks* 6, 105–130.

Hall, M.A. (1978). Properties of the equilibrium state in transportation networks. *Transp. Sci.* 12, 208–216.

Hall, M.A., D. Van Vliet and L.G. Willumsen (1980). SATURN — A simulation assignment model for the evaluation of traffic management schemes. *Traffic Eng. Control* 21, 168–176.

Harker, P.T. (1986). A note on the existence of traffic equilibria. *Appl. Math. Comput.* 18, 277–283.

Harker, P.T. (1988). Accelerating the convergence of the diagonalization and projection algorithms for finite-dimensional variational inequalities. *Math. Program.* 41, 29–59.

Harker, P.T., and J.-S. Pang (1990). Finite-dimensional variational inequality and nonlinear complementarity problems: a survey of theory, algorithms and applications. *Math. Program.* 48, 161–220.

Hearn, D.W. (1978). Network aggregation in transportation planning models, Part I. Report DOT-TSC-RSPD-78-8, Mathtech, Princeton, NJ.

Hearn, D.W. (1979). Minimizing the gap function in certain convex programs, in: *Network Aggregation in Transportation Planning Models*, Report DOT-TSC-RSPA-79-18, Mathtec, Princeton, NJ, pp. 60–82.

Hearn, D.W. (1982). The gap function of a convex program. *Oper. Res. Lett.* 1, 67–71.

Hearn, D.W. (1984). Practical and theoretical aspects of aggregation problems in transportation planning models, in: M. Florian, (ed.), *Transportation Planning Models*, North-Holland, Amsterdam, pp. 257–287.

Hearn, D.W., and S. Lawphongpanich (1990). A dual ascent algorithm·for traffic assignment problems. *Trans. Res.* 24B, pp. 423–430.

Hearn, D.W., S. Lawphongpanich and S. Nguyen (1984). Convex programming formulations of the asymmetric traffic assignment problem. *Trans. Res.* 18B, 357–365.

Hearn, D.W., S. Lawphongpanich and J.A. Ventura (1985). Finiteness in restricted simplicial decomposition. *Operations Research Letters* 4, pp. 125–130.

Hearn, D.W., S. Lawphongpanich and J.A. Ventura (1987). Restricted simplicial decomposition: computation and extensions. *Math. Program. Study* 31, 99–118.

Hearn, D.W., S. Lawphongpanich, J.A. Ventura and K.C. Yang (1989). RSDNET — Restricted simplicial decomposition network code. *Eur. J. Oper. Res.* 38, 121–122.

Hearn, D.W., and S. Nguyen (1982). Dual and saddle functions related to the gap function. Research Report 82-4, Department of Industrial and Systems Engineering, University of Florida, Gainesville, FL.

Hearn, D.W., and J. Ribera (1980). Bounded flow equilibrium problems by penalty methods, *Proc. 1980 IEEE International Conference on Circuits and Computers*, pp. 162–166.

Hearn, D.W., and J. Ribera (1981). Convergence of the Frank–Wolfe method for certain bounded variable traffic assignment problems. *Trans. Res.* 15B, 437–442.

Helgason, R.V., and J.L. Kennington (1978). An efficient specialization of the convex simplex method for nonlinear network flow problems. Technical Report IEOR 77107, Department of Industrial Engineering and Operations Research, Southern Methodist University, Dallas, TX.

Holloway, C.A. (1974). An extension of the Frank and Wolfe method of feasible directions. *Math. Program.* 6, 14–27.

Jewell, W.S. (1967). Models for traffic assignment. *Trans. Res.* 1, 31–46.

Jörnsten, K.O. (1980). A maximum entropy combined distribution and assignment model solved by Benders decomposition. *Transp. Sci.* 14, 262–276.

Kennington, J.L., and R.V. Helgason (1980). *Algorithms for Network Programming* John Wiley & Sons, New York, NY.

Kinderlehrer, D., and G. Stampacchia (1980). *An Introduction to Variational Inequalities and Their Applications*, Academic Press, New York, NY.

Kleinrock, L. (1964). *Communication Nets: Stochastic Message Flow and Delay*, McGraw-Hill, New York, NY.

Klessig, R.W. (1974). An algorithm for nonlinear multicommodity flow problems. *Networks* 4, 343–355.

Klincewicz, J.G. (1983). A Newton method for convex separable network flow problems. *Networks* 13, 427–442.

Knight, F.H. (1924). Some fallacies in the interpretation of social costs. *Q. J. Econ.* 38, 582–606.

Kuhn, H.W., and J.G. MacKinnon (1975). Sandwich method for finding fixed points. *J. Optim. Theory Appl.* 17, 189–204.

Kyparisis, J. (1987). Sensitivity analysis framework for variational inequalities. *Math. Program.* 38, 203–213.

Larsson, T., and M. Patricksson (1991). A generalized gap function for variational inequalities. Department of Mathematics Research Report LITH-MAT-R-1990-18, Linköping University, Linköping.

Larsson, T., and M. Patricksson (1992). Simplicial decomposition with disaggregated representation for the traffic assignment problem. *Trans. Sci.* 26, 4–17.

Lasdon, L.S., and A.D. Waren (1980). Survey of nonlinear programming applications. *Oper. Res.* 28, 1029–1073.

Lawphongpanich, S., and D.W. Hearn (1984). Simplicial decomposition of the asymmetric traffic assignment problem. *Trans. Res.* 18B, 123–133.

Lawphongpanich, S., and D.W. Hearn (1986). Restricted simplicial decomposition with application to the traffic assignment problem. *Ric. Oper.* 38, 97–120.

Lawphongpanich, S., and D.W. Hearn (1990). Benders decomposition for variational inequalities. *Math. Program.* 48, 231–247.

LeBlanc, L.J., and K. Farhangian (1981). Efficient algorithms for solving elastic demand traffic assignment problems and mode split-assignment problems. *Transp. Sci.* 15, 306–317.

LeBlanc, L.J., R.V. Helgason and D.E. Boyce (1985). Improved efficiency of the Frank–Wolfe algorithm for convex network programs. *Transp. Sci.* 19, 445–462.

LeBlanc, L.J., E.K. Morlok and W.P. Pierskalla (1974). An accurate and efficient approach to equilibrium traffic assignment on congested networks. *Trans. Res. Rec.* 491, TRB — National Academy of Sciences, pp. 12–23.

LeBlanc, L.J., E.K. Morlok and W.P. Pierskalla (1975). An efficient approach to solving the road network equilibrium traffic assignment problem. *Trans. Res.* 9, 309–318.

Leventhal, T., G. Nemhauser and L. Trotter, Jr. (1973). A column generation algorithm for optimal traffic assignment. *Transp. Sci.* 7, pp. 168–176.

Lupi, M. (1986). Convergence of the Frank–Wolfe algorithm in transportation networks. *Civ. Eng. Systems* 3, 7–15.

MacKinnon, J.G. (1975). An algorithm for the generalized transportation problem. *Region. Sci. Urban Econ.* 5, 445–464.

Magnanti, T.L. (1984). Models and algorithms for predicting urban traffic equilibria, in: M. Florian, (ed.), *Transportation Planning Models*, North-Holland, Amsterdam, pp. 153–185.

Magnanti, T.L., and B.L. Golden (1978). Transportation planning: network models and their implementation, in: A.C. Hax (ed.), *Studies in Operations Management*, pp. 465–518.

Mahmassani, H.S., and K.C. Mouskos (1988). Some numerical results on the diagonalization algorithm for network assignment with asymmetric interactions between cars and trucks. *Trans. Res.* 22B, 275–290.

Marcotte, P. (1985). A new algorithm for solving variational inequalities with application to the traffic assignment problem. *Math. Program. Study* 33, 339–351.

Marcotte, P., and J.-P. Dussault (1985). A modified Newton method for solving variational inequalities, *Proc. 24th IEEE Conf. on Decision and Control*, Fort Lauderdale, FL, pp. 1433–1436.

Marcotte, P., and J.-P. Dussault (1987). A note on a globally convergent Newton method for solving monotone variational inequalities. *Oper. Res. Lett.* 6, 35–42.

Marcotte, P., and J. Guélat (1988). Adaptation of a modified Newton method for solving the asymmetric traffic equilibrium problem. *Transp. Sci.* 22, 112–124.

Marcotte, P., G. Marquis and L. Zubieta (1992). A Newton-SOR method for spatial price equilibrium. *Transp. Sci.* 26, 36–47.

Mirchandani, P., and H. Soroush (1987). Generalized traffic equilibrium with probabilistic travel times and perceptions. *Transp. Sci.* 21, 133–152.

Murchland, J.D. (1970). Road network traffic distribution in equilibrium, in: R. Henn, H.P. Künzi and H. Schubert (eds.), *Proceedings of Mathematical Models in the Social Sciences*, Matematisches Forschungsinstitut, Oberwolfach, October 1969, II. Oberwolfach-Tagung über Operations Research, Operations Research-Verfahren 8, Anton Hain Verlag, Meisenheim am Glan, pp. 145–183.

Murchland, J.D. (1970). Braess's paradox of traffic flow. *Transp. Res.* 4, 391–394.

Nagurney, A. (1987a). Computational comparisons of spatial price equilibrium models. *J. Region. Sci.* 27), 55–76.

Nagurney, A. (1987b). Competitive equilibrium problems, variational inequalities and regional science. *J. Region. Sci.* 27, 503–517.

Nagurney, A. (1988). An equilibration scheme for the traffic assignment problem with elastic demands. *Trans. Res.* 22B, 73–79.

Nagurney, A. (1989). An algorithm for the solution of a quadratic programming problem with application to constrained matrix and spatial price equilibrium problems. *Environ. Planning* A 21, 99–114.

Nagurney, A. (1993). *Network Economics: A Variational Inequality Approach*, Kluwer, Academic Publishers, Dordrecht.

Nagurney, A., and D.-S. Kim (1989). Parallel and serial variational inequality decomposition algorithms for multicommodity market equilibrium problems. *Int. J. Supercomput. Appl.* 3, 34–58.

Netter, M. (1972a). Equilibrium and marginal cost pricing on a road network with several flow types, in: G.F. Newell (ed.), *Traffic Flow and Transportation, Proceedings of the Fifth International Symposium on the Theory of Traffic Flow and Transportation*, American Elsevier, New York, NY, pp. 155–163.

Netter, M. (1972b). Affectations de trafic et tarification au coût marginal social: critique de quelques idées admises. *Trans. Res.* 6, pp. 411–429.

Newell, G.F. (1980). *Traffic Flow on Transportation Networks*, MIT Press, Cambridge, MA.

Nguyen, S. (1974). An algorithm for the traffic assignment problem. *Transp. Sci.* 8, 203–216.

Nguyen, S. (1976). A unified approach to equilibrium methods for traffic assignment, in: M.A. Florian (ed.), *Traffic Equilibrium Methods, Proceedings of the International Symposium in Montreal*, Lecture Notes in Economics and Mathematical Systems 118, Springer-Verlag, Berlin, pp. 148–182.

Nguyen, S., and C. Dupuis (1984). An efficient method for computing traffic equilibria in networks with asymmetric transportation costs. *Transp. Sci.* 18, 185–202.

Nguyen, S., and L. James (1975). TRAFFIC — an equilibrium traffic assignment program. Publication #17, Centre de Recherche sur les Transports, Université de Montréal.

Pang, J.-S. (1984). Solution of the general multicommodity spatial equilibrium problem by variational and complementarity methods. *J. Region. Sci.* 24, 403–414.

Pang, J.-S. (1985). Asymmetric variational inequality problems over product sets: applications and iterative methods. *Math. Program.* 31, pp. 206–219.

Pang, J.-S., and D. Chan (1982). Iterative methods for variational and complementarity problems. *Math. Program.* 24, 284–313.

Pang, J.-S., and C.-S. Yu (1984). Linearized simplicial decomposition methods for computing traffic equilibria on networks. *Networks* 14, pp. 427–438.

Patricksson, M. (1991). Algorithms for urban traffic network equilibria. Linköping Studies in Science and Technology, Department of Mathematics, Thesis No. 263, Linköping University, Linköping.

Patricksson, M. (1994). *The Traffic Assignment Problem — Models and Methods*, VSP, Utrecht.

Petersen, E.R. (1975). A primal-dual traffic assignment algorithm. *Manage. Sci.* 22, 87–95.

Potts, R.B., and R.M., Oliver *Flows in Transportation Networks*, Academic (1972). Press, New York, NY.

Powell, W.B., and Y. Sheffi (1982). The convergence of equilibrium algorithms with predetermined step sizes. *Transp. Sci.* 16, pp. 45–55.

Qiu, Y., and T.L. Magnanti (1989). Sensitivity analysis for variational inequalities defined on polyhedral sets. *Math. Oper. Res.* 14, pp. 410–432.

Qiu, Y., and T.L. Magnanti (1992). Sensitivity analysis for variational inequalities. *Math. Oper. Res.* 17, 61–76.

Rosenthal, R.W. (1973). The network equilibrium problem in integers. *Networks* 3, 53–59.

Ruiter, E.R. (1973). The prediction of network equilibrium: the state of the art. *Proc. Int. Conf. on Transportation Research*, Bruges. pp. 717–726.

Ruiter, E.R. (1974). Implementation of operational network equilibrium procedures. *Trans. Res. Record* 491, 40–51.

Samuelson, P.A. (1952). Spatial price equilibrium and linear programming. *Am. Econ. Rev.* 42, 283–303.

Serra, P., and A. Weintraub (1981). Convergence of decomposition algorithms for the traffic assignment problem, in: P. Hansen (ed.), *Studies on Graphs and Discrete Programming, Annals of Discrete Mathematics*, North-Holland Publishing Company, pp. 313–318.

Shah, B.V., R.J. Buehler and O. Kempthorne (1964). Some algorithms for minimizing a function of several variables. *SIAM J. Appl. Math.* 12, 74–92.

Sheffi, Y. (1985). *Urban Transportation Networks. Equilibrium Analysis with Mathematical Programming Methods*, Prentice-Hall, Englewood Cliffs, NJ.

Sheffi, Y., and W. Powell (1981). A comparison of stochastic and deterministic traffic assignment over congested networks. *Trans. Res.* 15B, 53–64.

Sheffi, Y., and W. Powell (1982). An algorithm for the equilibrium assignment problem with random link times. *Networks* 12, 191–207.

Smith, M.J. (1979a). The marginal cost taxation of a transportation network. *Trans. Res.* 13B, 237–242.

Smith, M.J. (1979b). Existence, uniqueness and stability of traffic equilibria. *Trans. Res.* 13B, 295–304.

Smith, M.J. (1983a). The existence and calculation of traffic equilibria. *Trans. Res.* 17B, 291–303.

Smith, M.J. (1983b). An algorithm for solving asymmetric equilibrium problems with a continuous cost-flow function. *Trans. Res.* 17B, pp. 365–371.

Smith, M.J. (1984). Two alternative definitions of traffic equilibrium. *Trans. Res.* 18B, 63–65.

Soumis, F. (1978). Planification d'une flotte d'avions. Publication #133, Centre for Research on Transportation, Université de Montréal.

Steenbrink, P.A. (1974). *Optimization of Transport Networks*, John Wiley & Sons, London.

Steinberg, R., and R.E. Stone (1988). The prevalence of paradoxes in transportation equilibrium problems. *Transp. Sci.* 22, 231–241.

Steinberg, R., and W.I. Zangwill (1983). The prevalence of Braess' paradox. *Transp. Sci.* 17, 301–318.

Stewart, N.F. (1980). Equilibrium vs. system-optimal flow: some examples. *Trans. Res.* 14A, 81–84.

Takayama, T., and G. Judge (1964). Equilibrium among spatially separated markets: A reformulation. *Econometrica* 32, 510–524.

Takayama, T., and G. Judge (1971). *Spatial and temporal price and allocation models*, North-Holland, Amsterdam.

Tanner, J.C. (1963). Pricing the use of the roads — a mathematical and numerical study, *Proc. 2nd Int. Symp. on the Theory of Traffic Flow*, London, pp. 317–345.

Tobin, R.L. (1979). Calculation of fuel consumption due to traffic congestion in a case-study metropolitan area. *Traffic Eng. Control* 20, pp. 101–113.

Tobin, R.L. (1987). Sensitivity analysis for general spatial price equilibria. *J. Region. Sci.* 27, 77–102.

Tobin, R.L., and T.L. Friesz (1988). Sensitivity analysis for equilibrium network flow. *Transp. Sci.* 22, 242–250.

Van Vliet, D. (1976a). Road assignment — I. Principles and parameters of model formulation. *Trans. Res.* 10, 137–143.

Van Vliet, D. (1976b). Road assignment — II. The GLTS model. *Transp. Res.* 10, 145–149.

Van Vliet, D. (1976c). Road assignment — III. Comparative tests of stochastic methods. *Trans. Res.* 10, 151–157.

Ventura, J., and D.W. Hearn (1988). Computing the effective resistance in a system of conducting sticks. *Comput. Ind. Eng.* 14, pp. 171–179.

von Hohenbalken, B. (1975). A finite algorithm to maximize certain pseudoconcave functions on polytopes. *Math. Program.* 9, 189–206.

von Hohenbalken, B. (1977). Simplicial decomposition in nonlinear programming algorithms. *Math. Program.* 13, 49–68.

Wardrop, J.G. (1952). Some theoretical aspects of road traffic research, *Proc. Institute of Civil Engineers, Part II* 1, pp. 325–378.

Weintraub, A. (1974). A primal algorithm to solve network flow problems with convex costs. *Manage. Sci.* 21, 87–97.

Weintraub, A., and J. Gonzáles (1980). An algorithm for the traffic assignment problem. *Networks* 10, 197–209.

Weintraub, A., C. Ortiz and J. González (1985). Accelerating convergence of the Frank-Wolfe algorithm. *Trans. Res.* 19B, 113–122.

Wolfe, P. (1970). Convergence theory in nonlinear programming, in: J. Abadie (ed.), *Integer and Nonlinear Programming*, North-Holland, Amsterdam, pp. 1–36.

Wu, J.H., M. Florian and P. Marcotte (1990). A general descent framework for the monotone variational inequality problem. Publication #723, Centre de Recherche sur les Transports, Université de Montréal, Montréal.

Wu, J.H., M. Florian and P. Marcotte (1991). A new optimization formulation of the variational inequality problem with application to the network equilibrium problem. Publication #722, Centre de Recherche sur les Transports, Université de Montréal, Montréal.

Zuhovickiĭ, S.I., R.A. Poljak and M.E. Primak (1969). Two methods of search for equilibrium points of n-person concave games. *Sov. Math. Dokl.* 10, 279–282.

M.O. Ball et al., Eds., *Handbooks in OR & MS, Vol. 8*

Chapter 7

Location on Networks

Martine Labbé

FNRS Research Associate, Université Libre de Bruxelles, Belgium

Dominique Peeters

Université Catholique de Louvain, Belgium

Jacques-François Thisse

Université de Paris I-Sorbonne and CERAS-ENPC, France

1. Introduction

Facility location analysis deals with the problem of locating one or several new facilities with regard to existing facilities and clients in order to optimize some economic criterion. Examples of facilities are plants, warehouses, schools, hospitals, administrative buildings, department stores, waste material dumps, ambulance or fire engine depots, etc. Alfred Weber is generally given credit for having introduced the first location model in a book published in 1909. Perhaps more importantly, he set up a paradigm for location based on the minimization of transportation costs which was to have a lasting influence on the subsequent developments. However, it was only after 1960 that significant progress occurred and that facility location analysis emerged as a field.

In this chapter, we consider the problem of selecting one or several points of a *network* in order to optimize a function which is *distance-dependent* with respect to given points of the network. The problem is motivated by a number of potential applications. For example: several plants are to be set up at some points of a transportation system to minimize production and shipment costs; an emergency service unit is located in a rural area to minimize the maximal intervention time to population centers; a computer is established at some point of a communication network to minimize transmission costs from and toward peripheral units.

There are points dispersed over the network where demand arises for the service supplied by facilities that are to be located on the network. The demand points typically correspond to existing facilities and clients, and are generally taken to be at the vertices of the network. As far as these points are finite in number, this entails no loss of generality since any demand point may be regarded as a vertex of the network. The problem is then to place the facilities in some optimal manner. The definition of optimality varies from problem to problem. A common feature, however, is that the objective function depends, at least in part, on the distances to the demand points; the distance between a facility and a demand point is defined by the length of the shortest path linking the corresponding points in the network.

Ever since the seminal paper by Hakimi [1964], a thread running through network location theory is the identification of a *finite* subset of the network that necessarily contains an optimal solution for all the instances of a particular location problem. Hakimi has considered the problem of minimizing the sum of the weighted distances to the vertices. He then showed that an optimal solution can always be found in the set of vertices, regardless of the lengths of the edges and of the weights associated with the vertices. This property still holds for a host of extensions of the basic problem, in which the objective is to minimize the sum of different functions of distances. This family of problems, called *minisum*, is reviewed extensively in Section 3.

Besides the sum of weighted distances, Hakimi has also proposed to retain the maximum distance to the vertices as an alternative objective to be minimized when locating a facility. It is no longer true that the set of vertices contains an optimal solution. However, it is possible to identify a larger set, containing also some points interior to edges, for which the desired property holds true. This alternative criterion has served as a basis for another major family of location problems, called *minimax*, which we study in Section 4.

The above two families of models have been built within the framework of operations research. Yet, location theory has also attracted the attention of other scholars, especially economists and regional scientists. For a long time, their analyses typically presumed that facilities were to be located in a continuous space (e.g., the plane). Recently, economists and regional scientists have also contributed to the development of network location theory. This has been done by integrating more economic variables, such as prices, and by putting more emphasis on the decision making context. In particular, a new subfield has emerged, *competitive location problems*, in which one or several facilities are to be located with respect to competing facilities under the control of independent decision makers. These problems are studied by means of tools borrowed from noncooperative game theory. The main economic models of location are discussed in Section 5.

In Section 6, it is assumed that the set of possible locations is a given and finite subset of the network. The models are then called *discrete location models*. Their solution methods generally belong to combinatorial and integer programming. They have been paid far and away the most attention by the practitioners. A drawback is that the list of possible sites must be carefully drawn up, omissions being sometimes disastrous and yet unnoticed. An advantage is their flexibility, which allows the practitioner to incorporate irregularities in the landscape and characteristics of the facilities and of the clients without changing the formal structure. Only the prototype of discrete location models — the uncapacitated facility location problem — is considered together with its main extensions.

The formal definitions of network and distance, as well as their properties, are presented in Section 2. Some new avenues for future research are briefly discussed in Section 7.

A bibliography devoted to facility location analysis, published in 1985 by Domschke and Drexl, contains over 1500 references. A glance at the papers reveals a large variance of problems. Therefore we had to perform a selection. This one reflects our idiosyncrasies. As a result, numerous location theorists who have contributed to the field might feel frustrated by our choice of menu: we owe our apologies to them for not having given full account of their work.

2. The general framework and its properties

2.1. The network

In this subsection, we describe the mathematical tools for modeling networks such as roads, river, transportation or communications systems. The following definitions are useful: an *edge* of length $\ell, \ell > 0$, is the image of $[0, \ell]$ by a continuous mapping f from $[0, \ell]$ to $\mathbb{R}^d$ such that $f(\theta) \neq f(\theta')$ for any $\theta \neq \theta'$ in $[0, \ell]$. A *network* is then defined as a subset N of $\mathbb{R}^d$ which satisfies the following conditions: (i) N is the union of a finite number of edges; (ii) any two edges intersect at most at their extremities; (iii) N is connected.

The set of *vertices* of the network is made of the extremities of the edges defining N; it is denoted by $V = \{v_1, v_2, \cdots, v_n\}$. Vertices v_i correspond to transportation nodes (crossroads, railway junctions, ...) and tips (demand points uniquely connected to the network) of the real space to model. The set of edges defining the network is denoted by E; an edge $[v_i, v_j] \in E$ iff in the real space there is a transportation line (road, railway, ...) linking the sites corresponding to v_i and v_j and passing through no other sites corresponding to vertices of N. The length of the edge $[v_i, v_j] \in E$ is given and denoted by ℓ_{ij}. Each point $x \in N$ belongs to some edge of E but x may not be a vertex. For any two points $x_1, x_2 \in [v_i, v_j]$, the subset of points of $[v_i, v_j]$ between and including x_1 and x_2 is a *subedge* $[x_1, x_2]$. Let f_{ij} be the mapping from $[0, \ell_{ij}]$ to $\mathbb{R}^d$ defining $[v_i, v_j]$ and θ_{ij} the inverse of f_{ij}; to each point $x \in [v_i, v_j]$ corresponds one and only one value $\theta_{ij}(x)$ in $[0, \ell_{ij}]$. Then the length of $[x_1, x_2]$ is $|\theta_{ij}(x_1) - \theta_{ij}(x_2)|$. A *path* $P(x_1, x_2)$ joining $x_1 \in N$ and $x_2 \in N$ is a minimal connected subset of N containing x_1 and x_2. The length of a path is equal to the sum of the lengths of all its constituent edges and subedges. The *distance* $d(x_1, x_2)$ between $x_1 \in N$ and $x_2 \in N$ is equal to the length of a shortest path joining x_1 and x_2. Clearly, d is a metric defined on N. We say that $x \in N$ is *between* x_1 and $x_2 \in N$ iff $d(x_1, x) + d(x, x_2) = d(x_1, x_2)$; the set of points of N between x_1 and x_2 is denoted by $B(x_1, x_2)$. Without loss of generality, we assume that there is no redundant edge, that is, $B(v_i, v_j) = [v_i, v_j]$ for all pairs of vertices v_i and v_j such that $[v_i, v_j] \in E$.

2.2. Basic properties of the distance on N

The distance between a given point of N and a variable point along an edge of E obeys the following property.

Theorem 2.1. *Let $\bar{x}$ be any given point of N and $x = f_{ij}(\theta) \in [v_i, v_j] \in E$. Then, $d(\bar{x}, f_{ij}(\theta))$ is a function of θ which is:*
(i) continuous and concave on $[0, \ell_{ij}]$;
(ii) linearly increasing with slope $+1$ on $[0, \bar{\theta}_{ij}[$ and linearly decreasing with slope -1 on $]\bar{\theta}_{ij}, \ell_{ij}]$, where

$$\bar{\theta}_{ij} = \tfrac{1}{2}[\ell_{ij} + d(\bar{x}, v_j) - d(\bar{x}, v_i)].$$

Proof. From the definition of the distance, $d(x, f_{ij}(\theta)) = \min\{d(x, v_i) + \theta, d(x, v_j) + \ell_{ij} - \theta\}$. Hence $d(x, f_{ij}(\theta))$ is the minimum of two linear functions in θ with slope $+1$ or -1 and it reaches its maximum where these two functions take the same value. $\square$

More specific properties of the distance are obtained when the network is a *tree* T, i.e., a network such that any pair of two distinct points is joined by a single path.

Theorem 2.2 [Dearing, Francis & Lowe, 1976]. *Let $\bar{x}, x_1$ and x_2 be any three points of a tree T. If $x_3 \in P(x_1, x_2)$, then along $P(x_1, x_2), d(\bar{x}, x_3)$ is*
(i) piecewise linear in $d(x_1, x_3)$ with at most two pieces, the slopes of which are -1 and $+1$;
(ii) convex in x_3, i.e. $d(\bar{x}, x_3) \le \alpha d(\bar{x}, x_1) + (1 - \alpha)d(\bar{x}, x_2)$ where $d(x_3, x_2) = \alpha d(x_1, x_2)$ for $0 \le \alpha \le 1$.

Given that the objective functions considered in location theory are often convex in distances, Theorem 2.2 allows one to obtain stronger properties and algorithms with lower complexity for trees than for general networks.

Let v_k be any vertex of N. If the point

$$x_{ij}(v_k) = f_{ij}\left[\tfrac{1}{2}(\ell_{ij} + d(v_k, v_j) - d(v_k, v_i))\right]$$

is interior to $[v_i, v_j]$, then it is called a *bottleneck point*. The distance between v_k and $x_{ij}(v_k)$ via v_i is equal to that via v_j; so there are two shortest paths linking v_k and $x_{ij}(v_k)$. Let B denote the set of bottleneck points associated with N. Since any vertex generates at most one bottleneck point on each edge, $|B| = O(|V||E|)$. Clearly, if N is a tree then $|B| = 0$. A subedge $[x_1, x_2]$ delimited by two successive vertices or bottleneck points of an edge is called a *segment*. Along a segment the distance from any vertex v_k is either linearly increasing or linearly decreasing.

Let Q be the set of points equidistant from any two distinct vertices v_k, v_ℓ via v_i and v_j in V (or v_j and v_i) respectively: $Q = \{x \in N : [v_i, v_j] \in E$ with $x \in [v_i, v_j]$ and $v_k, v_\ell \in V, v_k \neq v_\ell$, exist such that either $d(v_k, v_i) + d(v_i, x) = d(v_\ell, v_j) + d(v_j, x)$ or $d(v_\ell, v_i) + d(v_i, x) = d(v_k, v_j) + d(v_j, x)\}$. In a general network $|Q| = O(|V|^2|E|)$, while in a tree $|Q| = O(|V|^2)$.

We now consider the distance between a given edge of E and a variable point along an edge of E. The *remoteness* of $x \in [v_i, v_j] \in E$ relative to $[v_k, v_\ell] \in E$ is

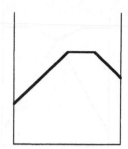

Fig. 1. The remoteness function at $(f_{ij}(\theta), [v_k, v_\ell])$ when $[v_k, v_\ell] \neq [v_i, v_j]$.

defined by

$$\bar{d}(x, [v_k, v_\ell]) = \max_{y \in [v_k, v_\ell]} d(x, y).$$

Theorem 2.1 is to be replaced by the following one.

Theorem 2.3 [Frank, 1967]. *Let $[v_k, v_\ell]$ be a given edge of E and $x = f_{ij}(\theta) \in [v_i, v_j]$. If $[v_k, v_\ell] \neq [v_i, v_j]$ then $\bar{d}(f_{ij}(\theta), [v_k, v_\ell])$ is:*
   (i) continuous and concave on $[0, \ell_{ij}]$;
   (ii) piecewise linear with at most three pieces having slopes $+1$, 0 and -1.
If $[v_k, v_\ell] = [v_i, v_j]$ then $\bar{d}(f_{ij}(\theta), [v_k, v_\ell])$ is:
   (i) continuous on $[0, \ell_{ij}]$;
   (ii) concave on $[0, \ell_{ij}/2]$ and $[\ell_{ij}/2, \ell_{ij}]$.

Proof. If $[v_k, v_\ell] \neq [v_i, v_j]$, then

$$\bar{d}(f_{ij}(\theta)), [v_k, v_\ell]) = \tfrac{1}{2}[\ell_{k\ell} + d(f_{ij}(\theta), v_k) + d(f_{ij}(\theta), v_\ell)],$$

i.e., it is the sum of two concave and piecewise linear functions in θ with slopes -1 or $+1$.
   If $[v_k, v_\ell] = [v_i, v_j]$, then

$$\bar{d}(f_{ij}(\theta), [v_k, v_\ell]) = \max\{d(f_{ij}(\theta), v_k), d(f_{ij}(\theta), v_\ell)\}$$

where $d(f_{ij}(\theta), v_k)$ and $d(f_{ij}(\theta), v_\ell)$ are linear with slope -1 or $+1$. $\quad\square$

   Typical examples of remoteness functions are given in Figures 1 and 2 for $[v_k, v_\ell] \neq [v_i, v_j]$ and $[v_k, v_\ell] = [v_i, v_j]$ respectively.

2.3. The demand structure

   The demand side of the model is as follows. Two cases are considered. In the first one, demand arises only at the vertices; let $w_i \geq 0$ be the demand associated with $v_i \in V$. More generally, w_i can be interpreted as the number of clients at v_i, the probability that a demand occurs at v_i, or the importance of a potential

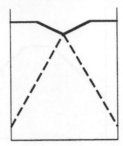

Fig. 2. The remoteness function at $(f_{ij}(\theta), [v_k, v_\ell])$ when $[v_k, v_\ell] = [v_i, v_j]$.

damage arising at v_i. For any subset S of V, let $w(S) = \sum_{v_i \in S} w_i$; of course, $w(N) = \sum_{i=1}^{n} w_i$ represents the total demand.

In the second case, demand is continuously distributed over the network N. To each edge $[v_i, v_j] \in E$ is associated a continuous density function $w_{ij}(y) \geq 0$ for $y \in [v_i, v_j]$. The total demand is now given by

$$w(N) = \sum_{[v_i,v_j] \in E} \int_{[v_i,v_j]} w_{ij}(y)\,\mathrm{d}y.$$

3. The median (minisum) problems

3.1. Introduction

Consider a finite number of points $v_1 \cdots v_n$ listed from left to right along the real line, as with settlements in a valley or along a coast; each point v_i is assigned a positive weight w_i that indicates its importance in some way, for example the population size. Suppose now that we want to place a facility in order to minimize the weighted sum of distances, $\sum_{i=1}^{n} w_i |v_i - x|$, where $|\cdot|$ denotes the absolute value function. It is then well known that the optimal solution is given by the (any of the) median(s) of the weight distribution $(w_1 \cdots w_n)$. This result, due to Edgeworth [1883] and rediscovered independently by several authors (see Rosenhead [1973] for a brief historical sketch) is known as the principle of median location.

The purpose of this section is to discuss the main properties of the median location on a network and to review several of the most relevant extensions. The basic problem — i.e., locating a single facility on a network with the aim of minimizing the weighted sum of distances — as well as the most straightforward extensions are considered in 3.2. In Subsection 3.3 we take a broader perspective and analyze further extensions of the objective function. We describe in 3.4 some techniques used in sensitivity analysis for the basic problem. Finally, we extend the framework in Subsection 3.5 to deal with the simultaneous location of several facilities.

Possible applications of the models discussed in this section are many, and range from the location of industrial plants or warehouses to the setting of some public facilities (schools, post offices, ...).

3.2. The median

A point $m \in N$ is a *median* iff

$$F(m) = \sum_{i=1}^{n} w_i d(v_i, m) \leq F(x) = \sum_{i=1}^{n} w_i d(v_i, x), \quad \forall x \in N.$$

A median is also called a centroid or a Launhardt–Weber point by reference to Launhardt [1882] and Weber [1909]'s use of the weighted sum of distances in continuous location theory. The reason for the terminology used in this section should be clear from the introduction above. The set of medians is never empty; it is called the *median set* and denoted by M. For simplicity, we assume in this subsection that the w_i are non-negative integers.

The main characterization of a median location on a network is given by the following result.

Theorem 3.1 [Hakimi, 1964]. *The set V of vertices contains a median. Furthermore, when $w(N)$ is odd, any median is a vertex.*

Proof. The first part of the theorem is a direct consequence of the concavity of $d(v_i, f_{k\ell}(\theta))$ over $[0, \ell_{k\ell}]$ for each topological edge $[v_k, v_\ell] \in A$ (cf. Theorem 2.1) and of the theorem of minimization of a concave function [Berge, 1966]. Consider now the second part and assume that m belongs to the interior of $[v_k, v_\ell]$. Then, the theorem of minimization of a concave function implies that $F(x)$ is constant over $[v_k, v_\ell]$. This can never hold when $w(N)$ is odd, which leads to a contradiction. □

When $w(N)$ is odd, finding the median set — which is included in V — can be done in $O(|V|^2)$ operations by the following method: (i) Set $F(m) = F(v_1)$ and $M = \{v_1\}$. (ii) Compute $F(v_k)$ for $k = 2 \cdots n$. If $F(v_k) > F(m)$, leave $F(m)$ and M unchanged; if $F(v_k) = F(m)$, add v_k to M; if $F(v_k) < F(m)$, set $F(m) = F(v_k)$ and $M = \{v_k\}$. On the other hand, when $w(N)$ is even, a median may belong to the interior of an edge; but then, $F(x) = F(m)$ for all x on this edge. Consequently, the median set can be determined in $O(|V||E|)$ operations as follows: (i) Find $M \cap V$ by the method described when $w(N)$ is odd. (ii) Consider in turn each $[v_k, v_\ell] \in E$ such that both v_k and v_ℓ belong to $M \cap V$. Compute $F(x_{k\ell})$ where $x_{k\ell}$ is the middle point of $[v_k, v_\ell]$; if $F(x_{k\ell}) = F(m)$, add $[v_k, v_\ell]$ to M.

The median set has been studied by Slater [1980] for a special class of networks.

Theorem 3.2 [Slater, 1980]. *Let N be a network whose edges have a unit length and vertices a unit weight. Then, there exists a network $N' \supseteq N$ such that the subnetwork induced by the median set of N' is equal to N.*

This result shows that we can always find a network, the median set of which is identical to any vertex set given a priori. Hence, we must add more structure to the model in order to derive specific properties of the medians. This is done in the following theorem where sufficient conditions are given for a subnetwork S of N to contain a median. Two types of restriction are imposed on S. The first is that S is a *dominant subnetwork* in the sense that

$$w(S) \geq w(N - S). \tag{3.1}$$

The second is that S is *gated*, which means that there exists a mapping g from $N - S$ to S such that

$$d(x_1, x_2) = d[x_1, g(x_1)] + d[g(x_1), x_2] \text{ for any } x_1 \in N - S \text{ and } x_2 \in S. \tag{3.2}$$

In words, this means that the shortest route between $x_1 \in N - S$ and $x_2 \in S$ necessarily passes through the point $g(x_1)$ – the 'gate' – which depends on the endpoint in $N - S$ but not on the endpoint in S.

Theorem 3.3 [Goldman & Witzgall, 1970]. *If S is a dominant and gated subnetwork, then $S \cap V$ contains a median.*

Proof. If $S \cap V$ contains a median, the theorem is proved. Assume now that $x \in N - S$ is a median. We then have:

$$F(x) = \sum_{v_i \in N-S} w_i d(v_i, x) + \sum_{v_i \in S} w_i d(v_i, x)$$

$$= \sum_{v_i \in N-S} w_i d(v_i, x) + \sum_{v_i \in S} w_i \{d[v_i, g(x)] + d[g(x), x]\}$$
$$\text{since } S \text{ is gated}$$

$$\geq \sum_{v_i \in N-S} w_i d(v_i, x) + \sum_{v_i \in N-S} w_i d[x, g(x)] + \sum_{v_i \in S} w_i d[v_i, g(x)]$$
$$\text{since } S \text{ is dominant}$$

$$\geq \sum_{v_i \in N-S} w_i d[v_i, g(x)] + \sum_{v_i \in S} w_i d[v_i, g(x)]$$
$$\text{by the triangle inequality}$$

$$= F[g(x)].$$

Consequently, if $x \in N - S$ is a median, then $g(x) \in S$ is also a median. Taking the endpoint in S of the edge containing $g(x)$ yields the desired point. □

This result has several implications which are discussed below.

(i) Clearly, a singleton is a gated subnetwork. Hence, if $w_k \geq w(N)/2$, i.e., v_k is a *dominant vertex*, then v_k is a median. Furthermore, when the inequality is strict, v_k is the only median [Witzgall, 1964].

(ii) Assume that N contains a *cut-vertex*, i.e., a vertex v_k such that the deletion of v_k and of the edges incident to v_k yields disjoint subnetworks $S_1 \cdots S_H$ (this property holds for a star-shaped network when v_k is the center of the network). Then, v_k is a gate of each of the subnetworks $S_1 \cdots S_H$. Consequently, if $w(S_h) \leq w(N)/2$ for $h = 1 \cdots H$, v_k is a median. Furthermore, when the inequality is strict for $h = 1 \cdots H$, the median is unique.

(iii) Suppose now that the network possesses a *cut-edge*, i.e., an edge $[v_k, v_\ell]$ such that $N - [v_k, v_\ell]$ is composed of two disjoint subnetworks S_k and S_ℓ (in this case, $[v_k, v_\ell]$ may correspond to a bridge connecting S_k and S_ℓ). Clearly, $v_k(v_\ell)$ is the only gate to enter $S_k(S_\ell)$. It is then straightforward to show that S_k contains a median if and only if S_k is a dominant subnetwork. Furthermore, if $w(S_k) > w(N)/2$, any median belongs to S_k [Goldman, 1971].

Goldman [1972a] has shown how the relaxation of (3.1) and (3.2) may affect Theorem 3.3. Specifically, for any $\mu > 0$, S is called a μ-*dominant subnetwork* if

$$w(S) \geq \mu w(N - S). \tag{3.3}$$

When $\mu = 1$, (3.3) is identical to (3.1). The subnetwork S is θ-*gated*, for $\theta \in]0, 1[$, if there exists a mapping g from $N - S$ to S such that

$$d(x_1, x_2) \geq \theta\{d[x_1, g(x_1)] + d[g(x_1), x_2]\}$$
$$\text{for any } x_1 \in N - S \text{ and } x_2 \in S. \tag{3.4}$$

When $\theta = 1$, (3.4) and the triangle inequality imply (3.2). Then, we have:

Theorem 3.4 [Goldman, 1972a]. *If S is a μ-dominant θ-gated subnetwork, then S contains at least a point $\overline{x}$ such that*

$$F(\overline{x}) \leq \left(\frac{1}{\mu\theta}\right) F(m). \tag{3.5}$$

Hence, if $\mu\theta$ is close to one, S always contains a nearly optimal location. In particular, if $w_k = \mu w(N - \{v_k\})$ with $\mu \in]0, 1[$, v_k is called a μ-*dominant vertex* and (3.5) implies that

$$F(v_k) \leq \frac{1}{\mu} F(m).$$

In this case, locating the facility at an almost dominant vertex yields, in the worst case, a value of the objective function close to the optimum. Note also that other extensions of Theorem 3.3 are presented in Chen, Francis, Lawrence, Lowe & Tufekci [1985] and Bandelt [1990].

The vertex optimality property (cf. Theorem 3.1) has been extended in various directions. Levy [1967] deals with concave functions of distance $C_i[d(v_i, x)]$, thus accounting for scale economies in transportation. The facility is now located in order to minimize the total cost $\sum_{i=1}^{n} C_i[d(v_i, x)]$. Clearly, for each edge $[v_k, v_\ell]$, the function $\sum_{i=1}^{n} C_i[d(v_i, f_{k\ell}(\theta))]$ is concave on $[0, \ell_{k\ell}]$ so that the proof of Theorem 3.1 still applies. Furthermore, if at least one function C_j is strictly concave, any minimizer of $\sum_{i=1}^{n} C_i[d(v_i, x)]$ belongs to V. In the special case where $C_i[d(v_i, x)]$ equals zero if $x = v_i$ and $k_i + w_i d(v_i, x)$ otherwise,

Louveaux, Thisse & Beguin [1982] have shown that any vertex with both positive k_i and w_i is a local minimizer of $\sum_{i=1}^{n} C_i[d(v_i, x)]$. Adding capacity constraints over edges, and hence congestion in transportation, or using topological arcs (instead of edges) to deal with asymmetries in transportation, does not affect the vertex optimality property [see Wendell and Hurter, 1973]. Mirchandani & Odoni [1979] consider another extension of the median in which both the weights and the network are random. Specifically, the weights and the network undergo probabilistic transitions among a finite number H of states. Each state $h = 1 \cdots H$ corresponds to a particular realization of the weights and of the edge lengths, and has a probability p_h which is independent of the previous state. Notwithstanding the latter assumption, this model allows for various forms of uncertainty in the median problem. Denoting by $d^h(x, y)$ the distance between x and y and by w_i^h the weight of vertex v_i in state h, the facility is to be located in order to minimize the expected cost given by $\sum_{h=1}^{n} p_h \sum_{i=1}^{n} w_i^h d^h(v_i, x)$. The above authors then show that the vertex set contains a location minimizing the expected cost. In the case of a multi-modal transportation system, the set of points containing an optimal solution is given by the vertices of each network and the junctions between networks [see Louveaux, Thisse & Beguin, 1982]. Finally, Batta and Palekar [1988] have considered the case of planar/network location problems in which some vertices are replaced by meganodes, representing cities where travel follows the Manhattan (ℓ_1) metric. Again, the set of candidate locations can be reduced to a finite easily identifiable set formed by the regular vertices and by some intersection points of the travel grids associated with the meganodes.

When the cost functions C_i are not concave, the vertex optimality property does not hold true. To illustrate, consider a network with two vertices connected by an edge of unit length; let also $C_i(x) = w_i x^\alpha$ with $\alpha > 1$. The objective function to be minimized is then given by $w_1 x^\alpha + w_2 (1 - x)^\alpha$. Since $\alpha > 1$, this function is strictly convex and differentiable. Applying the first order condition therefore yields the unique optimal location

$$x^* = \frac{(w_2/w_1)^{1/(\alpha-1)}}{1 + (w_2/w_1)^{1/(\alpha-1)}}$$

which is interior to the edge $[v_1, v_2]$.

Shier & Dearing [1983] have identified a necessary and sufficient condition for a vertex or a point interior of an edge to be a local minimizer of $\sum_{i=1}^{n} w_i[d(v_i, x)]^\alpha$, with $\alpha > 1$: a point $x \in N$ is a local minimizer of $\sum_{i=1}^{n} w_i[d(v_i, x)]^\alpha$ if and only if for every vertex v_k adjacent to x we have

$$\sum_{i \in I_k} w_i[d(v_i, x)]^{\alpha-1} \leq \sum_{i \notin I_k} w_i[d(v_i, x)]^{\alpha-1} \tag{3.6}$$

where $I_k = \{i = 1 \cdots n; v_k \text{ is between } v_i \text{ and } x\}$. Intuitively, this condition means that the derivative of the objective function at x in the direction of each adjacent vertex v_k is nonnegative. Clearly, a similar condition holds for any family of functions C_i which are strictly convex and continuously differentiable.

In this case, the global optimum can be determined by studying the behavior of the objective function along each segment of each edge. (i) If the right derivative at the left extremity and the left derivative at the right extremity of a segment have the same sign, the convexity of the objective function over the segment implies that there is no interior local minimizer. (ii) If the above derivatives have opposite signs, the objective function has an interior local minimizer that can be obtained by using any procedure devised for minimizing a one-dimensional convex function. This is so because distances are linear on each segment. (iii) The global optimum can be found by comparing the so-obtained values to those of the objective functions at the points of $B \cup V$ [see also Hooker, 1986].

The special case of a tree has attracted the attention of several authors, including Jordan [1869], Witzgall [1964], and Zelinka [1968]. Assuming that there is no vertex with at most two incident edges and zero weight, we can summarize their results as follows.

Theorem 3.5. *Let T be a tree. If $w(T)$ is odd, then $M = \{v_k\}$ for some vertex v_k of T; if $w(T)$ is even, then either $M = \{v_k\}$ for some vertex of T or $M = [v_k, v_\ell]$ where v_k and v_ℓ are two adjacent vertices of T.*

Hence the median set of a tree is also a tree. A generalization of this result has been provided by Barthélemy [1983] in the special case of networks whose edges have a unique length and vertices a unit weight: N is a tree if and only if the median set relative to any subset V' of vertices, with $|V'| \geq 3$, is a tree.

A nice characterization of a tree can be given in terms of local medians. For $x \in N$ and $\varepsilon > 0$, let $N(x, \varepsilon)$ denote the neighborhood of x defined as $N(x, \varepsilon) = \{y \in N | d(x, y) \leq \varepsilon\}$. A point $m_\ell \in N$ is a *local median* if there exists $\varepsilon > 0$ such that for all $x \in N(m_\ell; \varepsilon) \subset N$ we have $F(m_\ell) \leq F(x)$.

Theorem 3.6 [Bandelt, 1990]. *A network is a tree if and only if all the local medians are medians.*

The necessary condition is a straightforward consequence of Theorem 2.2. The sufficiency condition is more involved.

Goldman [1971] has shown how to find the median set of a tree in $O(|V|)$ operations. The algorithm is based on the following observation which can be derived from Theorem 3.3: given any partitioning of T into two connected subsets T_1 and T_2, T_i contains a median if and only if $w(T_i) \geq w(T)/2$. It proceeds as follows. (i) If $T = \{v_\ell\}$, then $M = \{v_\ell\}$ and stop. (ii) Select a pending vertex v_k of T, i.e., a vertex with a single adjacent vertex v_ℓ. (iii) If $w_k > w(T)/2$, then $M = \{v_k\}$ and stop; if $w_k = w(T)/2$, then $M = [v_k, v_\ell]$ and stop; otherwise delete v_k, set $w_\ell := w_\ell + w_k$ and return to (i).

Note that the edge lengths play no role in this algorithm.

Remark. Before studying further extensions of the median, it is worth mentioning that some of the results discussed in this subsection have been discovered independently by Gülicher in two papers written in German in 1964 and 1965, the diffusion of which has been very limited. Recently, Gülicher [1987] has made his work available to a larger audience by providing an English translation in a publication of the Institute for Econometrics and Statistics of Munster University.

In the foregoing, the objective function was given a priori. However, such a choice may appear as rather ad hoc in that it is justified by considerations which are not formally taken into account in the model itself. Holzman [1990] has recently proposed to tackle the problem of choosing an objective function, that depends only upon the distances to the clients, from a completely different perspective: the objective function is now obtained from a set of axioms which are deemed sensible. In this approach, a solution is defined as a mapping $\ell(\underline{z})$ which associates with the client configuration $\underline{z} = (z_1 \cdots z_{w(T)}) \in T^{w(T)}$ the 'best' location $z \in T$ for the facility (when $w_i > 0$, the vertex v_i is repeated w_i times in the configuration $\underline{z}$). More precisely, $\ell(\underline{z})$ is the minimizer of a particular objective function which is here determined endogenously. Of course, the form of the so-obtained objective function depends on the axioms chosen. Holzman considers the following three axioms.

($\mathcal{U}$) *Unanimity*: $\forall z \in T, \ell(z \cdots z) = z$.

If all clients are located at the same place, then the ideal location for the facility is this particular location. This hardly needs any further justification.

($\mathcal{I}$) *Invariance*: Consider two configurations $\underline{z}$ and $\underline{z}'$ such that (i) $z_j = z_j'$ for all $j \neq k$, and (ii) $z_k \neq z_k'$ satisfy $d[\ell(\underline{z}), z_k] = d[\ell(\underline{z}), z_k']$ and belong to the same subtree of T obtained after having deleted $\ell(\underline{z})$ and its incident edges. Then $\ell(\underline{z}) = \ell(\underline{z}')$.

In words, given that the locations of the other clients are unchanged, changing one client's location while preserving the distance and the direction from the solution $\ell(\underline{z})$ does not affect the solution. This seems to be a reasonable requirement.

($\mathcal{L}$) *Lipschitz*: Let $\underline{z}$ and $\underline{z}'$ be such that $z_j = z_j'$ for all $j \neq k$ and $z_k \neq z_k'$. Then, $d[\ell(\underline{z}), \ell(\underline{z}')] \leq d(z_k, z_k')/w(T)$.

This means that the facility location should not be too sensitive to small changes in the client configuration. The coefficient $1/w(T)$ expresses the following idea: no client is able to influence the solution by more than his share in the total weight $w(T)$. Clearly, this is the most demanding axiom in that it excludes a priori several possible locations (for example, the median).

Theorem 3.7 [Holzman, 1990]. *Let T be a tree. Then, there exists a unique solution which satisfies ($\mathcal{U}$), ($\mathcal{I}$) and ($\mathcal{L}$); this solution is given by*

$$\ell(\underline{z}) = \operatorname{argmin} \sum_{j=1}^{w(T)} [d(z_j, x)]^2 = \operatorname{argmin} \sum_{i=1}^{n} w_i [d(v_i, x)]^2.$$

Hence the solution is the *squared median* and is unique. This location can be obtained in $O(|V|^2)$ as follows. (i) For each edge, check if some interior point satisfies the optimality conditions (3.6). If such a point exists, it is the squared median. (ii) Otherwise the optimal location is at some vertex which can be found by evaluating the objective function at each of them. Note that Goldman (personal communication) has obtained a linear algorithm.

Recently, Hansen & Roberts [1991] have pursued this line of research in the case of networks containing at least one cycle. They show that there exists no mapping $\ell(\underline{z})$ satisfying the Lipschitz axiom $(\mathcal{L})$ as well as the standard conditions of Pareto optimality and of anonymity. This suggests that, for general networks, the choice of a 'good' objective function for placing a (public) facility may be very sensitive to changes in the clients' locations.

3.3. Extensions

3.3.1. The stochastic queue median

Consider a facility with one server and assume that demand for this facility's service, arising at vertices of a network, occurs in time as a random process. Then, the classical median location model becomes inappropriate. This is because it ignores the possibility for the server to be busy when some demand occurs, which implies that the response to this demand must be delayed. Recently, new location models have been proposed to deal with such *queueing* aspects (see Berman, Chiu, Larson, Odoni & Batta [1990] and Brandeau & Chiu [1990] for surveys).

Suppose that the vertices v_i of N generate demands for service according to independent Poisson processes with rates $w_i \geq 0$ per time unit. When a demand arises at v_i, and the server is available, the server is requested to travel to v_i; the corresponding travel time is $d(v_i, x)$. When the server is not available, the new demand enters a queue in a first-in-first-out manner. Each time the server finishes his service, he returns to the facility and is assigned to the first client in line. Finally, if the line is empty, the server waits at the facility for the next demand.

The *average travel time* from the facility at x to the vertices is given by

$$\overline{F}(x) = \frac{1}{w(N)} \sum_{i=1}^{n} w_i d(v_i, x).$$

Since we have an $M/G/1$ queue with Poisson arrival at a rate $w(N)$ and since the service time is defined by the round-trip travel time, it is readily verified [see, e.g. Kleinrock, 1975] that the *expected queue delay* is equal to

$$\overline{Q}(x) = \begin{cases} \dfrac{2\displaystyle\sum_{i=1}^{n} w_i[d(v_i, x)]^2}{1 - 2\displaystyle\sum_{i=1}^{n} w_i d(v_i, x)} & \text{if } 2\overline{F}(x) < w(N) \\ \infty & \text{otherwise} \end{cases}$$

Note that the denominator in $\overline{Q}(x)$ is linear in the expected service time, while the numerator is the second moment of the service time (up to a numerical factor).

A point $m^Q \in N$ is a *stochastic queue median* (*SQM*) if

$$\overline{TR}(m^Q) = \overline{F}(m^Q) + \overline{Q}(m^Q) \leq \overline{TR}(x) = \overline{F}(x) + \overline{Q}(x), \quad \forall x \in N$$

that is, the facility is to be located at a point minimizing the *average response time* (average travel time plus expected queue delay). Clearly, since a median m minimizes $\overline{F}(x)$, it follows from the definition of $\overline{Q}(x)$ that $0 < w(N) < 1/2\overline{F}(m)$ must hold for the *SQM* problem to have a solution.

When $w(N)$ is small enough, queuing rarely occurs and $\overline{F}(x)$ dominates $\overline{Q}(x)$ so that a median is an *SQM*. For values of $w(N)$ close to $1/2\overline{F}(m)$, the median m minimizing the numerator of $\overline{Q}(x)$ is also an *SQM* [Chiu, 1986a]. For intermediate values of $w(N)$, the *SQM* is no longer a median and can be an interior point of an edge. The trajectory of m^Q as a function of $w(N)$ has been studied by Brandeau & Chiu [1990].

Berman, Larson & Chiu [1985] have proposed an algorithm based on the following result to determine an *SQM*: the function $\overline{Q}[f_{k\ell}(\theta)]$ is strictly convex on any segment $[a, b]$ of an edge $[v_k, v_\ell] \in E$. Hence $\overline{TR}[f_{k\ell}(\theta)]$ is also strictly convex over $[a, b]$. The algorithm then proceeds as follows. (i) The set of bottleneck points is identified and N is decomposed into segments. (ii) For each segment $[a, b]$, it is first checked if $[a, b]$ contains some points where $\overline{TR}$ is finite (the linearity of $\overline{F}[f_{k\ell}(\theta)]$ on $[a, b]$ implies that this condition is satisfied when $\min\{\overline{F}(a), \overline{F}(b)\} \leq 1/w(N)$). If so, the point x minimizing $\overline{TR}(x)$ over $[a, b]$ is determined by solving $d\overline{TR}[f_{k\ell}(\theta)]/d\theta = 0$ which leads to a quadratic equation.

In the special case of a tree, the following results can be shown [Chiu, Berman & Larson, 1985]: (i) m^Q must lie on the path linking the squared median and the median closest to it; (ii) $\overline{TR}(x)$ is strictly convex along this path. Therefore, a descent (direction finding) procedure can be applied from the median to find the *SQM*.

In the model above, the service time was given by the round-trip travel time. This model can be generalized in order to account for on-scene and off-scene service time at each vertex, which are assumed to be i.i.d. random variables [see, e.g. Berman, Chiu, Larson, Odoni & Batta, 1990]. Furthermore, other service schemes have been considered: (i) demands arising when the server is busy are lost at some given cost, in which case the median with respect to weights w_i is an *SQM* [Berman, Larson & Chiu, 1985]; (ii) rejection of selected demands [Batta, 1988]; (iii) queueing which allows demands to be grouped in different priority classes; and (iv) queueing disciplines depending on expected service time [Batta, 1989].

3.3.2. The traveling salesman median

Minimizing total weighted distance to clients implicitly assumes that the server returns to the facility after servicing some client. In many cases (such as delivery, customer pick-up, repair and maintenance service), a server can visit sequentially several clients before returning to the facility. Hence, in such cases, the median does not constitute an optimal location anymore.

Suppose that, at the beginning of each time period, each vertex v_i of N generates (does not generate) a demand for service with probability w_i $(1 - w_i)$, where $0 \leq w_i \leq 1$. Demands from the various vertices occur independently of each other so that the probabilities that only the vertices from a given subset $S \subseteq V$ generate demand is given by

$$P(S) = \prod_{v_i \in S} w_i \prod_{v_i \in V - S} (1 - w_i).$$

At the beginning of each time period, a server is given a list of vertices generating demand for this period. He then starts from his location $x \in N$ an optimal traveling salesman tour passing through the vertices of S. Denote by $L(S \cup \{x\})$ the length of an optimal tour.

A point $m^{\text{TS}} \in N$ is a *traveling salesman median* (*TSM*) if

$$\overline{TL}(m^{\text{TS}}) = \sum_{S \subseteq V} P(S) L(S \cup \{m^{\text{TS}}\})$$

$$\leq \overline{TL}(x) = \sum_{S \subseteq V} P(S) L(S \cup \{x\}), \qquad \forall x \in N$$

that is, the facility is to be located at a point minimizing the expected length of an optimal traveling salesman tour connecting the vertices that generate demand.

It is readily verified that the length $L(S \cup \{x\})$ is concave in θ when x moves along an edge. Thus, the objective function $\overline{TL}(x)$ is also concave so that the vertex optimality property holds for the *TSM* [Berman & Simchi-Levi, 1986]. Unlike most single facility location problems for which this property holds, the *TSM* problem is very complex since the number of sets to be explored grows exponentially with $|V|$. Indeed, finding a *TSM* involves determining the length $L(S \cup \{v_i\})$ of the optimal tour for each subset $S \cup \{v_i\}$, with $S \subseteq V$ and $v_i \in V$.

Simchi-Levi & Berman [1988] propose the following heuristic. An approximation of $L(S \cup \{v_i\})$ is given by $L(S) + 2d(v_i, S)$. Since $L(S)$ is independent of v_i, the vertex v_i^* which minimizes $\sum_{S \subseteq V} P(S)[L(S) + 2d(v_i, S)]$ is also a minimizer of $\sum_{S \subseteq V} P(S) d(v_i, S)$. Then, the heuristic proceeds as follows. (i) For each vertex $v_i \in V$, sort the distances $d(v_i, v_j)$ for $v_j \in V$ such that $d(v_i, v_{j_1}) \leq \cdots \leq d(v_i, v_{j_n})$. (ii) Compute $\sum_{S \subseteq V} P(S) d(v_i, S) = \sum_{k=2}^{n} w_k \prod_{\ell=1}^{k-1} (1 - w_\ell) d(v_i, v_{j_k})$ for all $v_i \in V$ and select the vertex with the smallest value. The complexity of this heuristic is $O(|V|^2 \log |V|)$ and its relative worst-case error equals $1/2$. Note that Bertsimas [1989] has provided a refined, but data-dependent, bound of $(1 - \min_{i=1 \cdots n} w_i)/2$.

The above heuristic presents a major drawback. Though it often finds a relatively good solution, it does not provide an estimate for the value of $\overline{TL}$ at the chosen vertex. Furthermore, the heuristic does not address the question of constructing the optimal tours for the subsets S of V. This can be circumvented by using Christofides' heuristic for each subset S with complexity $O(|V|^3)$ [see Johnson & Papadimitriou, 1985]. In this case, the relative worst-case error becomes $(1 - (1/2) \min_{i=1 \cdots n} w_i)$.

An alternative approach is to select an a priori tour. Then, the vertices of a given subset $S \subseteq V$ are visited in the order they appear in the a priori tour. The problem is now to choose *simultaneously* the facility location and an a priori tour in order to minimize the expected distance traveled. Berman & Simchi-Levi [1988a] show that the vertex optimality property also holds for this problem. Given an a priori tour, they provide an $O(|V|^3)$ algorithm to find an optimal location. Bertsimas [1989] improves the method to obtain a complexity of $O(|V|^2)$ and proposes a heuristic to find both a location and an a priori tour which gives the optimal value of the objective function up to a factor of $O(\log|V|)$.

On a tree, the problem of finding a *TSM* becomes polynomial. Let S be a subset of vertices of a tree T and T_S be the smallest subtree including S. Then, the length $L(S \cup \{x\})$ of an optimal tour on $S \cup \{x\}$ is given by $L(S \cup \{x\}) = L(S) + 2d(x, T_S)$, so that a *TSM* on a tree is a vertex minimizing $\sum_{S \subseteq V} P(S)d(v_i, T_S)$. This function is convex along any path of T, a property used by Berman & Simchi-Levi [1986] to derive a linear time 'trimming' algorithm. Berman & Simchi-Levi [1988b] adapt their algorithm to networks in which each link belongs at most to one cycle.

Finally, many location problems with routing aspects often involve the choice of *several* facilities. Different models have been presented and the reader is refereed to Laporte [1988] for a detailed survey.

3.3.3. The continuous median

So far clients were located at the network vertices. We assume here that clients are continuously distributed over the network, with density $w_{ij}(y) \geq 0$ for all $y \in [v_i, v_j] \in E$.

There are two formulations of the continuous median. In the former, a point $m^S \in N$ is an *S-continuous median* iff

$$F^S(m^S) = \sum_{[v_i, v_j] \in E} \int_{[v_i, v_j]} w_{ij}(y) \, d(y, m^S) \, dy$$

$$\leq F^S(x) = \sum_{[v_i, v_j] \in E} \int_{[v_i, v_j]} w_{ij}(y) \, d(y, x) \, dy, \quad \forall x \in N.$$

In words, the facility is located in order to minimize the sum of the weighted distances along each edge. Suppose that $x \in [v_k, v_\ell]$. Then, for each edge $[v_i, v_j] \neq [v_k, v_\ell]$, the corresponding term in $F^S(x)$ can be rewritten as a concave function of $\theta = f_{k\ell}^{-1}(x)$ over $[0, \ell_{k\ell}]$, given by the parametric integral of a function concave in θ (to see it, apply Theorem 2.1 for $\bar{x}$ varying in $[v_k, v_\ell]$). However, when $[v_i, v_j] = [v_k, v_\ell]$ the corresponding term

$$\int_0^\theta w_{k\ell}(y)(\theta - y) \, dy + \int_\theta^{\ell_{k\ell}} w_{k\ell}(y)(y - \theta) \, dy$$

is convex in θ. Hence, $F^S[f_{k\ell}(\theta)]$ is no longer concave on $[0, \ell_{k\ell}]$ and the vertex optimality property ceases to hold. Chiu [1987] has shown that F^S is generally neither concave nor convex over a segment of an edge. Global optimization techniques must therefore be used to determine a local minimizer for each segment. In the special case of a uniform density on each edge, F^S is quadratic over each segment so that an S-continuous median can be easily obtained.

In the latter formulation, a point $m^M \in N$ is an *M-continuous median* iff

$$F^M(m^M) = \sum_{[v_i,v_j] \in E} \overline{d}(m^M, [v_i, v_j]) \leq F^M(x)$$

$$= \sum_{[v_i,v_j] \in E} \overline{d}(x, [v_i, v_j]), \qquad \forall x \in N$$

where $\overline{d}$ is the remoteness function defined in 2.2. The facility is now set up with the aim of minimizing the sum of the maximum distances to each edge.

Let Q_e denote the set of *middle points* of the edges in $E : Q_e = \{x_{ij} \in [v_i, v_j] \in E : d(v_i, x_{ij}) = d(x_{ij}, v_j)\}$. Hansen & Labbé [1989] have shown, using Theorem 2.3, that the set $V \cup Q_e$ contains an M-continuous median. Minieka [1977] has obtained a simple necessary condition for the middle point x_{ij} to be an M-continuous median: $|F^M(v_i) - F^M(v_j)| \leq d(v_i, v_j)$. Given these two results, finding an M-continuous median can be done in $O(|E|^2)$ operations by the following method: (i) Compute $F^M(v_i)$ for $i = 1 \cdots n$. (ii) For each edge $[v_i, v_j] \in E$, check if $|F^M(v_i) - F^M(v_j)| \leq d(v_i, v_j)$. If so, compute $F^M(x_{ij})$; otherwise proceed to the next edge. (iii) A point yielding the smallest of the values $F^M(v_i)$ and $F^M(x_{ij})$ computed in steps (i) and (ii) is an M-continuous median.

Finally, note that specialized algorithms are provided by Chiu and by Hansen and Labbé for the two types of continuous medians when the network is a tree (see also Chiu [1986b] for a continuous median model incorporating queueing delays).

3.4. Sensitivity analysis

In practice, accurate estimates of the parameters of a model are often difficult to obtain: relevant data may not be available or the values of the parameters may be based on forecasts. This motivates the need to study the sensitivity of a model's solution to variations in its parameters.

Labbé, Thisse & Wendell [1991] propose a framework to perform a sensitivity analysis for the median problem in which the weights w_i are subject to perturbations. Let $\hat{w}_i$ be an estimate of w_i and let $\hat{m}$ be a median corresponding to $\sum_{i=1}^n \hat{w}_i d(v_i, x)$. Consider the perturbed problem

$$\min_{x \in N} F(x, \gamma) = \sum_{i=1}^n (\hat{w}_i + \gamma_i w_i') d(v_i, x)$$

where w_i' is a given nonnegative number and γ_i is a multiplicative factor. Although results are developed for any specification of the w_i', the main focus is on multiplicative perturbations, i.e., $w_i' = \hat{w}_i$, for $i = 1 \cdots n$. In this case, γ_i can be interpreted as a percentage of deviation from the estimated value $\hat{w}_i$ of w_i.

Even when the values of w_i cannot be specified precisely, it is often possible to determine some region within which these parameters are known to vary. Let Γ be the set of n-dimensional vectors $\gamma = (\gamma_1, \cdots, \gamma_n)$ such that $(\hat{w}_1 + \gamma_1 w_1', \cdots, \hat{w}_n + \gamma_n w_n')$ belongs to this region. The set Γ is assumed to be a polytope. One important special case is when a given interval $[\underline{\gamma}_i, \overline{\gamma}_i]$ can be found for each γ_i, with $-\infty < \underline{\gamma}_i \leq 0$ and $0 \leq \overline{\gamma}_i < \infty$.

For a given $\gamma \in \Gamma$, the solution $\hat{m}$ may or may not minimize $F(x, \gamma)$. However, as long as the components of γ are not too large, $\hat{m}$ can be expected to be at least near-optimal. This justifies the concept of *degree of optimality for tolerance* $\tau, \alpha^*(\tau)$, which is defined as follows: for $\tau \geq 0$,

$$\alpha^*(\tau) = \max_{\substack{\gamma \in \Gamma \\ \|\gamma\|_\infty \leq \tau}} [F(\hat{m}, \gamma) - \phi(\gamma)]$$

where $\phi(\gamma) = \min_{x \in N} F(x, \gamma)$ and $\|\gamma\|_\infty = \max_{i=1\cdots n} |\gamma_i|$. For any given $\tau \geq 0, \alpha^*(\tau)$ gives the maximum divergence of the value of the objective function $F(\hat{m}, \gamma)$ at $\hat{m}$ from the optimal value function $\phi(\gamma)$ for $\gamma \in \Gamma$ and $|\gamma_i| \leq \tau$. Since τ determines the maximum variability in the coefficients w_i, the function $\alpha^*(\tau)$ expresses a trade-off between perturbation size and degree of optimality.

A characterization of $\alpha^*(\tau)$, which does not require the computation of $\phi(\gamma)$, can be obtained via the vertex optimality property of the median:

$$\begin{aligned}
\alpha^*(\tau) &= \max_{\substack{\gamma \in \Gamma \\ \|\gamma\|_\infty \leq \tau}} [F(\hat{m}, \gamma) - \min_{i=1\cdots n} F(v_i, \gamma)] \\
&= \max_{i=1\cdots n} \max_{\substack{\gamma \in \Gamma \\ \|\gamma\|_\infty \leq \tau}} [F(\hat{m}, \gamma) - F(v_i, \gamma)] \\
&\equiv \max_{i=1\cdots n} \alpha_i(\tau).
\end{aligned}$$

When the number of vertices is small, the determination of the function $\alpha^*(\tau)$ is relatively easy. For each $v_i \in V$, the function $\alpha_i(\tau)$ can be found by solving the linear parametric program

$$\max \ [F(\hat{m}, \gamma) - F(v_i, \gamma)]$$

$$\text{s. t.} \quad \gamma \in \Gamma \text{ and } \|\gamma\|_\infty \leq \tau$$

and $\alpha^*(\tau)$ is obtained by taking the maximum of the $\alpha_i(\tau)$ over the vertices. Note that $\alpha_i(\tau)$ is nondecreasing, piecewise linear and concave, which implies that $\alpha^*(\tau)$ is nondecreasing and piecewise linear; however, it need not be concave or convex.

In the case of a tree, a special property holds. Let v_i be a median for $F(x, \overline{\gamma})$ and $\{v_{j_k}\}_{k=0}^s$ be the sequence of vertices on the path linking $\overline{m}$ and v_i, with $v_{j_0} = \overline{m}$ and $v_{j_s} = v_i$. Then, there exists a sequence $\{\mu_k\}_{k=0}^s$ where $0 \leq \mu_0 \leq \mu_1 \leq \cdots \leq \mu_s = 1$ such that v_{j_k} is a median of $F(x, \mu\overline{\gamma})$ for $\mu \in [\mu_k, \mu_{k+1}[$.

This result allows Labbé, Thisse & Wendell [1991] to derive an iterative procedure to determine $\alpha^*(\tau)$. At the first iteration, $\alpha^*(\tau)$ is determined for τ

belonging to some interval $[0, \tau^1[$ by comparing $\alpha_i(\tau)$ at the vertices v_i adjacent to $\hat{m}$. At the next iteration, $\alpha^*(\tau)$ is obtained for a second interval $[\tau^1, \tau^2[$ by taking the maximum of the functions $\alpha_i(\tau)$ over the vertices considered in the previous iteration and those adjacent to them. The function $\alpha^*(\tau)$ is then constructed by repeating this iteration. At each iteration, only a few new functions $\alpha_i(\tau)$ are to be found.

Considering the scalar parametric case, i.e., $\Gamma = \{\gamma; \gamma_i = \mu$ for $i = 1 \cdots n\}$, Erkut & Tansel [1989] present a linear time algorithm for determining the sequence $\{\mu_k\}_{k=0}^s$.

3.5. The K-median

So far we have considered the location of a single facility. Now suppose that K facilities have to be located simultaneously, where K is an integer larger than one. The *K-median problem* consists in finding a set of K points of N so that the sum of the weighted distances to the vertices is minimum, i.e.,

$$\min \ F(S) = \sum_{i=1}^n w_i d(v_i, S)$$

s. t. $|S| = K$ and $S \subset N$

where $d(v_i, S) = \min\{d(v_i, s); s \in S\}$ is the distance between v_i and S. In the K-median, each vertex is therefore assigned to the nearest open facility.

3.5.1. Formulation of the problem

Hakimi [1965] has shown that the set of possible sites for the facilities can be restricted to the set of vertices. The argument is as follows. At the optimum, each facility must be located so as to minimize the sum of the weighted distances to the vertices assigned to it. Hence, by Theorem 3.1, there is an optimal configuration such that each facility is located at a vertex of the network. Notice that the vertex optimality property is kept when the weighted distances are replaced by concave functions of the distance [Levy, 1967].

Given the vertex optimality property, the K-median problem can be expressed as a mixed-integer linear program:

$$\min_{\underline{y}, X} \ F(\underline{y}, X) = \sum_{i=1}^n \sum_{j=1}^n w_i x_{ij} d_{ij} \tag{3.7}$$

$$\text{s. t.} \ \sum_{j=1}^n x_{ij} = 1, \qquad i = 1 \cdots n \tag{3.8}$$

$$0 \le x_{ij} \le y_j, \qquad i = 1 \cdots n, \ j = 1 \cdots n \tag{3.9}$$

$$\sum_{j=1}^n y_j = K \tag{3.10}$$

$$y_j \in \{0, 1\}, \qquad j = 1 \cdots n, \tag{3.11}$$

where d_{ij} is the distance between vertices v_i and v_j. The variable y_j is set equal to one if v_j belongs to the K-median and to zero otherwise; y denotes the vector of the y_j. The variable x_{ij} is the fraction of the demand arising at v_i satisfied by the facility at v_j (if any); X is the matrix of the x_{ij}. It is readily verified that the program above has an all-integer solution by setting x_{ij} equal to one for one of the indices such that $d_{ij} = \min\{d_{i\ell}; y_\ell = 1\}$. This corresponds to assigning the clients at v_i to the nearest vertex belonging to the K-median; all the other x_{ij} are set equal to zero.

The constraints (3.8) means that the demand at v_i must be met by some facility. The second group (3.9) prevents an assignment to a vertex not in the K-median. The constraint (3.10) ensures that exactly K vertices belong to the K-median.

The above formulation is not unique. The n^2 constraints $x_{ij} \leq y_j$ can be replaced by the n logically equivalent constraints $\sum_{j=1}^{n} x_{ij} \leq ny_j$. This reduction in the number of constraints may seem desirable if the problem is to be solved by standard linear programming techniques. However, the bound obtained by relaxing the integrability constraints $y_j \in \{0, 1\}$ by $y_j \in [0, 1]$ is generally worse than that obtained by using the original formulation, which reduces its practical usefulness [see Cornuéjols, Fisher & Nemhauser, 1977a].

It is well known that the linear programming relaxation often yields integer solutions, or at least provides a tight bound F_{LP} on the optimal value F^* [ReVelle & Swain, 1970]. A probabilistic analysis of this relaxation has been carried out by Ahn, Cooper, Cornuéjols & Frieze [1988]. When the vertices are drawn from a uniform distribution over the unit square and have unit weights, and when the distance is Euclidean, they show that $(F^* - F_{LP})/F^* \sim 0.00284$ almost surely when $K \to \infty$ and $n/(K \log n) \to \infty$. On a tree, the equality $F^* = F_{LP}$ holds when $K = 1$ or $K \geq \lfloor (n - 1)/2 \rfloor$. For the intermediate values of K, the equality does not necessarily hold and the conjecture $(F^* - F_{LP})/F^* \leq (K - 1)/2K$ is supported by extensive computational experiments.

3.5.2. Complexity

The following result shows that solving the K-median problem can be difficult in the worst-case.

Theorem 3.7 [Kariv & Hakimi, 1979b]. *The K-median problem is $\mathcal{NP}$-hard.*

For any *given* value of K, the K-median problem is polynomially solvable since it suffices to enumerate all sets S of K vertices, to assign every vertex of $V - S$ to the nearest one in S, to compute the corresponding value of F, and to keep the best value and solution found. Furthermore, in the case of a tree, Matula & Kolde [1978] have shown that the K-median can be solved in polynomial time. Kariv & Hakimi [1979b] have proposed an $O(K^2|V|^2)$ algorithm based on a dynamic programming formulation of the K-median problem on a tree, while Hassin & Tamir [1991] have refined this bound to $O(K|V|)$ when the tree is a path. These results imply that an $O(K|V|^2)$ algorithm can be obtained for a single cycle network: it suffices to discard one edge at a time, to apply Hassin and

Tamir's procedure to the resulting path, and to select the best of the solutions so obtained.

Even finding a good approximation of the K-median is difficult. The worst-case performance of a heuristic H is defined by

$$\tau_H = \sup_{I \in \mathcal{I}} \{\tau : F^*(I) \geq \tau F_H(I)\}$$

where I denotes an instance of K-median, $F^*(I)$ the corresponding optimal value of the objective function, $F_H(I)$ the value of F found by applying H to I, and $\mathcal{I}$ the set of all instances. Obviously $0 \leq \tau_H \leq 1$, and H is guaranteed to yield a solution within $1/\tau_H$ times the optimal value $F^*(I)$ for all $I \in \mathcal{I}$. A polynomial approximation scheme is a solution method with performance $\tau_H \geq \tau$ whose running time is polynomial in $|V|$ and K for any fixed $\tau, 0 < \tau < 1$.

Theorem 3.8 [Fisher, 1980]. *Unless* $\mathcal{P} = \mathcal{NP}$, *there exists no polynomial approximation scheme for the K-median.*

Note that this negative result holds only for the case where the d_{ij}'s do not satisfy the triangle inequality. To the best of our knowledge, the case of distances, for which the triangle inequality holds, is still open.

3.5.3. Exact algorithms

The large number of variables and constraints of the mixed-integer linear formulation, even for small-sized problems, deters from using standard LP packages and leads to the use of methods exploiting the special structure of the problem. Schrage [1975], for example, has proposed to consider the constraints $x_{ij} \leq y_j$ as 'variable upper bounds' and has tailored the simplex algorithm in consequence. Another approach, suggested by Morris [1978], consists in solving the formulation involving the constraints $\sum_i x_{ij} \leq n y_j, j = 1 \cdots n$, and generating the constraints $x_{ij} \leq y_j$ to cut off non-integer solutions only when necessary. However, the most successful approaches developed so far rely on the concept of Lagrangean relaxation and duality.

Cornuéjols, Fisher & Nemhauser [1977a] and Narula, Ogbu & Samuelsson [1977] have proposed to determine the K-median by means of a dual approach based on the Lagrangean relaxation of the constraints (3.8). Denoting λ_i the Lagrangean multiplier associated with the ith constraint, we obtain the relaxed problem:

$$F(\underline{\lambda}) = \min_{\underline{y}, X} \sum_{i=1}^{n} \sum_{j=1}^{n} w_i x_{ij} d_{ij} + \sum_{i=1}^{n} \lambda_i \left(1 - \sum_{j=1}^{n} x_{ij}\right)$$

$$\text{s. t.} \quad 0 \leq x_{ij} \leq y_j, \quad i = 1 \cdots n, \ j = 1 \cdots n$$

$$\sum_{j=1}^{n} y_j = K$$

$$y_j \in \{0, 1\}, \quad j = 1 \cdots n.$$

The objective function can be rewritten as

$$F(\underline{\lambda}) = \sum_{i=1}^{n} \lambda_i - \max_{\underline{y}, X} \sum_{i=1}^{n} \sum_{j=1}^{n} (\lambda_i - w_i d_{ij}) x_{ij}.$$

Clearly, x_{ij} must be set equal to y_j when $\lambda_i - w_i d_{ij} > 0$ and to zero if $\lambda_i - w_i d_{ij} < 0$. Defining $(\lambda_i - w_i d_{ij})^+ = \max(\lambda_i - w_i d_{ij}, 0)$ and

$$\rho_j(\underline{\lambda}) = \sum_{i=1}^{n} (\lambda_i - w_i d_{ij})^+,$$

the relaxed problem becomes

$$F(\underline{\lambda}) = \sum_{i=1}^{n} \lambda_i - \max_{\underline{y}} \sum_{j=1}^{n} \rho_j(\underline{\lambda}) y_j$$

$$\text{s. t.} \qquad \sum_{j=1}^{n} y_j = K$$

$$y_j \in \{0, 1\}, \qquad j = 1 \cdots n.$$

For a given vector $\underline{\lambda}$, this problem can be solved by inspection: the K facilities are located at the vertices corresponding to the K largest values of $\rho_j(\underline{\lambda})$. Since $F(\underline{\lambda})$ is a lower bound on F^* for any $\underline{\lambda}$, the best possible bound is given by

$$F_D = \max_{\underline{\lambda}} F(\underline{\lambda})$$

which corresponds to the Lagrangean dual problem. Solving this problem is done by means of a subgradient method.

It can be shown that $F_D = F_{LP}$: even if the bound F_D does not improve upon the bound given by the LP relaxation, having F_D often gives a very tight bound. In case of a genuine duality gap (the subgradient procedure has failed to reach the optimum), an implicit enumeration completes the algorithm. Its performance can be improved by adding tests. For example, let $\overline{F}$ be an upper bound on F^* and relabel the vertices such that $\rho_1(\underline{\lambda}) \geq \rho_2(\underline{\lambda}) \geq \cdots \geq \rho_n(\underline{\lambda})$. By definition of $F(\underline{\lambda})$, if

$$F(\underline{\lambda}) + \rho_j(\underline{\lambda}) - \rho_{K+1}(\underline{\lambda}) \geq \overline{F}$$

for $j = 1 \cdots K$, we must have $y_j = 1$ in any optimal solution. On the other hand,

$$F(\underline{\lambda}) + \rho_K(\underline{\lambda}) - \rho_j(\underline{\lambda}) \geq \overline{F}$$

for $j = K + 1 \cdots n$, we can set $y_j = 0$ in any optimal solution. These bounds were proposed by Christofides & Beasley [1982]. Hanjoul & Peeters [1985] have shown that dramatic reductions in the number of implicit enumerations can be reached by implementing these penalties.

A different relaxation, in which the constraint (3.10) is dualized, has been studied by Hanjoul & Peeters [1985]. The relaxed problem is now given by

$$F(\mu) = \min_{\underline{y},X} \sum_{i=1}^{n} \sum_{j=1}^{n} w_i x_{ij} d_{ij} + \sum_{j=1}^{n} \mu y_j - \mu K$$

s. t.
$$\sum_{j=1}^{n} x_{ij} = 1, \qquad i = 1 \cdots n$$

$$0 \le x_{ij} \le y_j, \qquad i = 1 \cdots n, \; j = 1 \cdots n$$

$$y_j \in \{0, 1\}, \qquad j = 1 \cdots n,$$

which is an uncapacitated facility location problem with equal fixed cost μ (see 6.2). The Lagrangean dual

$$F'_D = \max_{\mu} \; F(\mu)$$

yields a bound tighter than the linear programming relaxation. The relaxed problem is $\mathcal{NP}$-hard too, but is well solved in many practical instances (see 6.2). Furthermore, the procedure involves a single multiplier μ and F'_D can be found by considering only a finite number of values for μ. There is rarely a duality gap, in which case an implicit enumeration is required.

The above special-purpose algorithms have a major drawback: they cannot easily accommodate new constraints such as capacity constraints, obligation to open a certain number of facilities in some subset of vertices, or precedence constraints (a facility at some vertex cannot be open unless one or several facilities are open in prespecified vertices), etc. General purpose algorithms offer more flexibility in this respect, but need to be adapted to the particular structure of the K-median problem. A survey of decompositional methods, such as Dantzig–Wolfe and Benders decompositions, applied to the K-median can be found in Magnanti & Wong [1990].

3.5.4. Heuristics

Despite the good performance of exact algorithms for small-and average-sized problems, there is still a need for heuristic methods. First, heuristics are the only way to tackle very large problems. Second, most exact procedures converge more rapidly when good initial solutions are available. Several heuristics have been proposed to deal with the K-median problem and only the main ones will be reviewed here.

The first method is due to Maranzana [1964]. It proceeds as follows. (i) Choose an initial selection of K vertices defining the set S°; each vertex of V is then allocated to the nearest point in S°, which yields a partition $\{V_1^1 \cdots V_K^1\}$ of V. (ii) For each cluster $V_k^{t-1}, k = 1 \cdots K$, determine the corresponding median in V; let S^t be the set formed by these medians, which in turn yields a new partition $\{V_1^t \cdots V_K^t\}$ of V. This procedure is carried out until a stable solution is reached, i.e., $S^t = S^{t-1}$. Though intuitively appealing, this procedure suffers from two major defects: remote vertices incorporated in the initial selection S° often belong to the subsequent median sets S^t; vertices with a large weight which are cut-vertices, once included in S^t, tend to prevent the emergence of new medians. Therefore, the performance of this heuristic depends highly on the choice of S° so that several runs with different initial solutions are necessary. Furthermore, no polynomial bound on the running time of the heuristic is known.

Another method, known as the GREEDY heuristic, has been proposed by Kuehn & Hamburger [1963]. Let $F(S) = \sum_{i=1}^{n} \min_{v_j \in S} d_{ij}$, and $\rho_j(S) \equiv F(S) - F(S \cup \{v_j\})$ for $S \subseteq V$ and $v_j \in V - S$ which corresponds to the decrement in the objective function when adding vertex v_j to the median set S. In the initial step, select v_j such that $F(v_j) = \min_{v_k \in V} F(v_k)$ and set $S := \{v_j\}$. Next, choose $v_j \in V - S$ such that $\rho_j(S) = \max_{v_k \in V-S} \rho_k(S)$ and set $S := S \cup \{v_j\}$. The procedure is carried out until $|S| = K$. Clearly, the complexity is $O(K|V|^2)$.

The STINGY heuristic, proposed by Feldman, Lehrer & Ray [1966], proceeds by starting with an initial situation where a facility is located at each vertex; set $S := V$ and $F(S) = 0$. Then, if $\sigma_j(S) \equiv F(S - \{v_j\}) - F(S)$, for any $v_j \in S \subseteq V$, denotes the increase in the objective function when closing the facility at v_j, select v_j such that $\sigma_j(S) = \min_{k \in S} \sigma_k(S)$ and set $S := S - \{v_j\}$. At each step, a facility is dropped until $|S| = K$.

Another heuristic, due to Teitz & Bart [1968], starts with an arbitrary pattern of K facilities defining a set S. Let $v_j \in S$ and $v_k \in V - S$; define $\tau_{jk}(S) \equiv F(S) - F(S - \{v_j\} \cup \{v_k\})$ as the variation of the objective function when the facility located at v_j is moved to v_k, the other facilities being unchanged. INTERCHANGE selects a vertex v_j in S and a vertex v_k in $V - S$, computes $\tau_{jk}(S)$ and sets $S := S - \{v_j\} \cup \{v_k\}$ if this value is positive. The procedure stops when no further change decreases the objective function. The number of iterations required by this heuristic depends on the criterion chosen for selecting the vertex pairs $\{v_j, v_k\}$. Nemhauser, Wolsey & Fisher [1978] give an example where a poor choice leads to an exponential complexity. In practice, an efficient rule is to select, for a given vertex v_j, the vertex v_k such that $\tau_{jk}(S) = \max_{v_\ell \in V-S} \tau_{j\ell}(S)$.

The outcome of INTERCHANGE depends on the initial configuration of facilities. Empirically, very good results are often obtained by starting with the solution produced by GREEDY or STINGY; performing two consecutive runs initiated with these alternative candidates and retaining the better solution gives either the optimum or a tight approximation.

When INTERCHANGE fails to yield the optimal configuration, selecting a vertex which increases the objective function may lead to another local minimum. This idea lies at the root of the SIMULATED ANNEALING procedure [Kirkpatrick, Gelatt & Vecchi, 1983]. At the initialization stage, one chooses a configuration S° of K facilities and the parameters $T_\circ > 0$ and $0 < \beta < 1$. At step t, one randomly selects $v_j \in S^{t-1}$ and $v_k \in V - S^{t-1}$; if $\tau_{jk}(S^{t-1}) > 0$, set $S^t := S^{t-1} - \{v_j\} \cup \{v_k\}$; if $\tau_{jk}(S^{t-1}) < 0$, set $S^t := S^{t-1} - \{v_j\} \cup \{v_k\}$ with probability $\pi = \exp[\tau_{jk}(S^{t-1})/T_\circ \beta^t]$ and $S^t := S^{t-1}$ with probability $1 - \pi$. The probability of accepting an increase in the objective function is thus reduced at each step and the procedure stops after a predetermined number of iterations. Convergence to the optimum is guaranteed, though the rate of convergence is exponential.

SIMULATED ANNEALING explores many solutions and, therefore, takes a large amount of computational time. Nevertheless, the procedure can accommodate several types of additional constraints without affecting much the programming code.

Once a solution has been provided by a heuristic, it is useful to know something about its performance. A measure can be obtained via a duality relationship since we have

$$F(\underline{\lambda}) \le F_D \le F^* \le F_H$$

where, for example, $\underline{\lambda}$ is a vector of Lagrange multipliers associated with the constraints (3.8) (see 3.5.3) and $F(\underline{\lambda})$ the value of the relaxed problem. Hence, $F_H - F(\underline{\lambda})$ is an upper bound on $F_H - F^*$ and one may think of heuristics for determining $F(\underline{\lambda})$ as well. For example, one may set $\lambda_i = \min_{v_j \in S} w_i d_{ij}$ for each set S of open facilities generated at each iteration of a heuristic H; then compute the corresponding values $F(\underline{\lambda})$ and retain the largest of the values so obtained [Cornuéjols, Fisher & Nemhauser, 1977a]. The reader is referred to Nemhauser & Wolsey [1988, II.5.3] for more details on this topic.

3.5.5. Extensions

In some cases, one may want each client at $v_i, i = 1 \cdots n$, to be within a maximum distance $\overline{d}_i$ in order to guarantee a minimum standard of quality service (see also 4.1 for another interpretation based on equity considerations). This implies that the constraints $\sum_{i=1}^{n} x_{ij} = 1$ are to be replaced by $\sum_{v_j \in V_i} x_{ij} = 1$ where $V_i = \{v_j \in V; d_{ij} \le \overline{d}_i\}$. As pointed out by Feldman, Lehrer & Ray [1966], this new problem keeps the structure of the K-median problem provided that d_{ij} is replaced by an arbitrarily large number when $v_j \notin V_i$. The vertex optimality property does no longer hold for this problem. However, Church and Meadows [1977] have shown that the vertex set supplemented by the points situated on the edges at distance $\overline{d}_i$ from $v_i, i = 1 \cdots n$, always contains on optimal configuration.

When an infinite value is obtained for the objective function, this means there is no solution to the problem with the thresholds $\overline{d}_i$ and the number K of facilities. Church & ReVelle [1974] then suggest using a substitute objective which is the maximization of the number of clients served:

$$\max_{\underline{y}, X} \ F(\underline{y}, X) = \sum_{i=1}^{n} \sum_{j=1}^{n} w_{ij} x_{ij}$$

$$\text{s. t. } \sum_{i=1}^{n} x_{ij} \le 1, \quad j = 1 \cdots n$$

and (3.9)–(3.11), where w_{ij} is equal to w_i if $d_{ij} \le \overline{d}_i$ and to zero otherwise. The constraints are the same as those of the K-median problem except that it is not required to supply all clients any more. This problem, called the *maximal covering problem*, can be reduced to the K-median if a fictitious facility $n + 1$ is considered with $w_{i,n+1} = 0$ for $i = 1 \cdots n$.

In the K-median problem, each client is assigned to the nearest (open) facility. However, the corresponding server may be busy (see 3.3.1) so that the closest available facility must be used. This has led Weaver & Church [1985] to consider the generalized model in which demand at v_i is distributed among the first, second, ... nearest facilities. This new problem, called *vector assignment K-median problem*, can be efficiently solved by a Lagrangean relaxation of the assignment constraints. The vertex optimality property still holds when the nearer servers are used at least as frequently as the more distant ones [Hooker, Garfinkel & Chen, 1991].

4. The center (minimax) problems

4.1. Introduction

Some authors have questioned the pertinence of the median for placing a public facility (see Dear [1978] for a detailed discussion of this criterion). This is because total — or average — distance minimization tends to favor clients who are clustered in population centers to the detriment of clients who are spatially dispersed. Discrimination of this kind with regard to accessibility may have a severe impact on remote clients in the case of an emergency service (ambulances, fire brigades, police cars, ...). As a result, the decision maker may want to consider a criterion focusing more on clients who get poorly served. A simple approach consists of weighting long distances more than short distances. This can be done by applying a power $\alpha > 1$ to the distances in the minisum problem (see the model of Shier & Dearing [1983] discussed in 3.2). In this context, α is to be interpreted as a measure of aversion for inequality. The largest aversion occurs when α tends to infinity, which implies the choice of a location minimizing the maximal (weighted) distance to the facility (this corresponds to the application of Rawls' [1971] equity criterion to location decisions). This problem, called the center problem, has been known for a long time in Graph Theory in the case where the facility must be established at some vertex [see, e.g. Berge, 1967]. Hakimi [1964] was the first to formulate the problem when the location set is the whole network.

However, the center may yield a significant increase in the total distance. This has led Halpern [1976, 1978, 1980] to model the corresponding trade-off as a bicriterion problem in which a convex combination of total distance and maximum distance is minimized. The resulting solution is called a cent-dian. We consider a slightly modified version of Halpern's problem in which the total distance is replaced by the average distance, because average and maximum distances are more directly comparable in terms of magnitude [Hansen, Labbé & Thisse, 1991].

In this section, we discuss the main properties of the center and several of the most relevant extensions. The unweighted and weighted versions of the center problem are dealt with in 4.2 as well as the K-center. Extensions are analyzed in 4.3. Finally, in Subsection 4.4, we study the cent-dian.

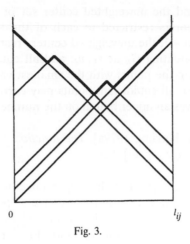

$$0 \qquad\qquad\qquad\qquad\qquad\qquad l_{ij}$$

Fig. 3.

4.2. The center

4.2.1. The unweighted center

A point $c \in N$ is an *unweighted center* iff

$$G_u(c) = \max_{v_i \in V} d(v_i, c) \le G_u(x) = \max_{v_i \in V} d(v_i, x), \qquad \forall x \in N.$$

The set of unweighted centers of N, called the *unweighted center set*, is denoted by C_u. The value of the objective function G_u at any point of C_u is called the *unweighted radius* r_u of the network N.

For a given edge $[v_k, v_\ell]$, Figure 3 describes the graphs of the distance functions $d(v_i, x)$, for all $v_i \in V$ and $x \in [v_k, v_\ell]$ in terms of $\theta = f_{k\ell}^{-1}(x)$. When the distance function $d(v_i, f_{k\ell}(\theta))$, has two pieces then its maximum corresponds to a bottleneck point associated with v_i. When the graphs of two different distance functions intersect, we obtain an equidistant point with respect to the corresponding vertices. Taking the upper envelope of all the distance functions over $[v_k, v_\ell]$ yields a piecewise linear function which is $G_u(f_{k\ell}(\theta))$. The concavity of $d(v_i, f_{k\ell}(\theta))$, for all $v_i \in V$, implies that any local maximizer of G_u is a bottleneck point, while any local minimizer is an equidistant point. As a consequence, we have the following result.

Theorem 4.1 [(Minieka, 1970]. *Any unweighted center belongs to $V \cup Q$.*

An *unweighted center restricted to the edge* $[v_k, v_\ell] \in E$ is a point $x^* \in [v_k, v_\ell]$ such that $G_u(x^*) = \min_{x^* \in [v_k, v_\ell]} G_u(x)$. The value $G_u(x^*)$ is called the *unweighted radius* of $[v_k, v_\ell]$ and is denoted by $r_u(v_k, v_\ell)$. It follows from the definition of the unweighted center that

$$r_u = \min_{[v_k, v_\ell] \in E} r_u(v_k, v_\ell).$$

Hence, in order to find the unweighted center set of N, it is sufficient to find all the unweighted centers restricted to each of the $|E|$ edges [Hakimi, 1964]. According to Theorem 4.1, the unweighted centers restricted to any edge $[v_k, v_\ell]$ are determined by evaluating G_u at v_k, v_ℓ, and all equidistant points interior to $[v_k, v_\ell]$ and by selecting the points with the smallest value. However, a glance at Figure 3 shows that not all equidistant points may correspond to local minima of G_u. The next result gives an upper bound on the number of such local minima.

Theorem 4.2 [Kariv & Hakimi, 1979a]. *The function $G_u(x)$ has at most $|V| + 1$ local minima on each edge $[v_k, v_\ell] \in E$.*

Proof. The function $G_u(f_{k\ell}(\theta))$ is the upper envelope of $|V|$ piecewise linear and concave functions. Since these functions have slope one, up to the sign, they contribute to G_u at most once each. The extremities of the interval along which a given distance function coincide with $G_u(x)$ are local minimizers of this function. This implies that $G_u(x)$ has at most $|V| - 1$ local minima in the interior of $[v_k, v_\ell]$. Since v_k and v_ℓ may also be minimizers of $G_u(x)$, we obtain the desired result. $\square$

Consider an edge $[v_k, v_\ell]$. Assume now that the vertices are ranked by nonincreasing order of their distances from v_k, i.e.,

$$d(v_1, v_k) \geq d(v_2, v_k) \geq \cdots \geq d(v_i, v_k) \geq \cdots \geq d(v_n, v_k) = 0$$

(v_k is the last vertex). Furthermore, if $d(v_i, v_k) = d(v_{i+1}, v_k)$, then the ranking is such that $d(v_i, v_\ell) \geq d(v_{i+1}, v_\ell)$.

Looking at Figure 4, it is easy to see that, if $\bar{x} = f_{k\ell}(\bar{\theta}) \in]v_k, v_\ell[$ minimizes locally $G_u(x)$, then there exists two vertices, say v_i and v_j, such that $d(v_i, \bar{x}) = d(v_j, \bar{x})$ and v_j is the first vertex with an index larger than i and for which $d(v_j, v_\ell) > d(v_i, v_\ell)$. Furthermore, by Theorem 2.1, none of the vertices v_p, with $i < p < j$, contributes to $G_u(x)$.

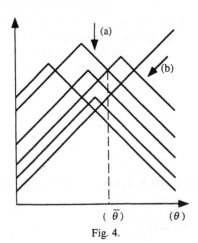

Fig. 4.

This property allows one to determine all the local unweighted centers belonging to an edge by exploring the sorted list of the vertices exactly once. This is the basis of Kariv & Hakimi's [1979a] algorithm: (i) A list of vertices arranged by nonincreasing order of their distance has to be established for each vertex v_k, which can be done in $O(|V|^2 \log |V|)$ operations. (ii) The local unweighted centers of each edge have to be determined, which requires $O(|V| |E|)$ operations. Hence, the complexity of the above algorithm is $O(|V|^2 \log |V| + |E| |V|)$. Other solution procedures have been proposed by Cuninghame-Green [1984], Hakimi, Schmeichel & Pierce [1978] and Minieka [1981], but all have higher complexity.

Similar to the median, when N is a network with unit edge lengths, it is possible to construct a network $N' \supseteq N$ such that the subnetwork induced by the unweighted center set of N' is equal to N [Buckley, Miller & Slater, 1981]. However, when the network N possesses a cut-edge $[v_k, v_\ell]$, the following theorem either identifies which subnetwork S_k or S_ℓ contains the unweighted center set C_u or allows to conclude that $C_u \subset [v_k, v_\ell]$.

Theorem 4.3 [Goldman, 1972b]. *Let $[v_k, v_\ell]$ be a cut-edge of N and S_k and S_ℓ be the two disjoint subnetworks of $N - [v_k, v_\ell]$.*
    (i) *If* $\max_{v_i \in S_k \cap V} d(v_i, v_k) - \max_{v_i \in S_\ell \cap V} d(v_i, v_\ell) \geq \ell_{k\ell}$, *then S_k contains C_u.*
    (ii) *If* $\max_{v_i \in S_\ell \cap V} d(v_i, v_\ell) - \max_{v_i \in S_k \cap V} d(v_i, v_k) \geq \ell_{k\ell}$, *then S_ℓ contains C_u.*
    (iii) *If* $\left| \max_{v_i \in S_\ell \cap V} d(v_i, v_\ell) - \max_{v_i \in S_k \cap V} d(v_i, v_k) \right| < \ell_{k\ell}$, *then $[v_k, v_\ell]$ contains the unique unweighted center.*

An extension of Theorem 4.3 is presented in Chen, Francis & Lowe [1988].

Since each edge of a tree T is a cut-edge, Goldman [1972b] exploits Theorem 4.3 to develop a 'trimming' algorithm for a tree. However, Handler [1973] proposes the simpler following algorithm: (i) Choose any point $x \in T$ and find a vertex v_k such that $G_u(x) = d(x, v_k)$. (ii) Find a vertex v_ℓ such that $G_u(v_k) = d(v_k, v_\ell)$. (iii) Let c be the middle point of the path $P(v_k, v_\ell)$. Then $C_u = \{c\}$.

This linear algorithm is justified by the following theorem, the proof of which is based on Theorem 2.2.

Theorem 4.4 [Handler, 1973]. *On a tree T,*
    (i) *the center c is unique;*
    (ii) $G_u(c) = \frac{1}{2} G_u(v_k)$ *for any $v_k \in V$ such that there exists $x \in T$ for which $G_u(x) = d(x, v_k)$;*
    (iii) $d(c, v_k) = \frac{1}{2} G(v_k)$ *and c is located on $P(v_k, v_\ell)$ with v_k such that $G(v_k) = d(v_k, v_\ell)$.*

A generalization of the unweighted center problem has been studied by Chen, Francis & Lowe [1988], Goldman [1972b], Halfin [1974], Hedetniemi, Cockayne & Hedetniemi [1981], and Lin [1975], by introducing *addends* a_i for each $v_i \in V$. In this case the function to be minimized is $\max_{v_i \in V} \{a_i + d(v_i, x)\}$. This new problem can be solved by straightforward modification of the algorithms for the unweighted center set. Specifically, the unweighted centers with addends of a network N are equal to the unweighted centers of the expanded network N' obtained from N by adding $|V|$ vertices v_i' and $|V|$ arcs $[v_i, v_{i'}]$ with length equal to a_i.

4.2.2. The weighted center

A point $c \in N$ is a *weighted center* iff

$$G_w(c) = \max_{v_i \in V} w_i d(v_i, c) \le G_w(x) = \max_{v_i \in V} w_i d(v_i, x), \quad \forall x \in N.$$

The set of weighted centers of N, called the *weighted center set*, is denoted by C_w. The value of the objective function G_w at any point of C_w is called the *weighted radius* r_w of N.

Clearly, Theorem 4.1 can be adapted to the weighted center: a weighted center c of a network N belongs to $V \cup Q_w$ where $Q_w = \{x \in N : [v_i, v_j] \in E$ with $x \in [v_i, v_j]$ and (v_k, v_ℓ) exist such that $w_k d(v_k, x) = w_k[d(v_k, v_i) + d(v_i, x)] = w_\ell d(v_\ell, x) = w_\ell[d(v_\ell, v_j) + d(v_j, x)]\}$.

To find C_w, it is again sufficient to determine all the weighted centers restricted to each of the $|E|$ edges of N. To do so for a particular edge $[v_k, v_\ell]$, the piecewise linear function $G_w(x)$ has to be constructed and evaluated at v_k, v_ℓ and all interior breakpoints which are local minimizers of G_w. The next result gives an upper bound on the number of interior breakpoints.

Theorem 4.5 [Kariv & Hakimi, 1979a]. *The function $G_w(x)$ has at most $3|V| - 2$ breakpoints on edge $[v_k, v_\ell] \in E$.*

Proof. The proof is by induction. Hence, notice, first, that this result holds true for $|V| \le 2$. Furthermore, assume that the vertices $v_i \in V$ are ordered in such a way that $w_1 \le w_2 \le \cdots \le w_n$ and let $G_w^q(x) = \max_{i=1\cdots q} w_i d(v_i, x)$. Hence, $G_w^{q+1}(x) = \max\{G_w^q(x), w_{q+1} d(v_{q+1}, x)\}$. Now, since the slope of $w_{q+1} d(v_{q+1}, x)$ is, in absolute value, larger than or equal to the slope of any segment of $G_w^q(x)$, it intersects $G_w^q(x)$ in at most two points. Hence, $G_w^{q+1}(x)$ has at most three more breakpoints (the two intersection points plus, possibly, a bottleneck point where $w_{q+1} d(x, v_{q+1})$ reaches its maximum) than $G_w^q(x)$. This implies the desired result. $\square$

The algorithm of Kariv & Hakimi [1979a] to find the weighted center set restricted to an edge is based on the proof of Theorem 4.5. The vertices $v_i \in V$ are first ranked by nondecreasing order of their weights w_i. Then, the functions $G_w^q(x)$ are iteratively constructed to finally obtain $G_w(x) = G_w^n(x)$. The weighted center set restricted to that edge is formed by the local minimizers yielding the smallest value of $G_w(x)$. This algorithm is of $O(|V| \log |V|)$ if a 2-3 tree is used to store the data when constructing the functions $G_w^k(x), k = 1 \cdots n$. (Since the number of breakpoints of the upper envelope of k weighted distance functions is in $O(k)$, a standard divide-and-conquer approach can be used to determine the weighted center set restricted to an edge in $O(|V| \log |V|)$ [see Tamir, 1988].)

To find the weighted center set of N, the weighted centers restricted to each edge are determined and those with smallest value constitute C_w. Therefore, one deduces that the complexity of the algorithm of Kariv and Hakimi [1979a] is $O(|E| \, |V| \log |V|)$.

A completely different approach has been proposed by Christofides [1975]. Let $N(v_i, R)$ be the set of points of N whose weighted distance to v_i is not larger than some given number R. The weighted absolute radius is the smallest R for which $\cap_{v_i \in V} N(v_i, R)$ is nonempty. Christofides' method is iterative: it starts with R equal to a lower bound on r_w, computes the intersection $\cap_{v_i \in V} N(v_i, R)$, increases R by a small amount if the intersection is empty and stops otherwise. Dearing & Francis [1974] prove that

$$\max_{v_i, v_j \in V} \left[\frac{w_i \, w_j}{(w_i + w_j)} \right] d(v_i, v_j)$$

is a lower bound on r_w.

Christofides' iterative procedure can also be easily adapted to find a good approximate solution when nonlinear functions of the distance are used. In that case, n strictly increasing and continuous functions $f_1, \cdots f_n$ are given and the optimal solution x, called a *nonlinear center*, is such that

$$G_n(x^*) = \max_{v_i \in V} f_i[d(v_i, x^*)]$$

$$\leq G_n(x) = \max_{v_i \in V} f_i[d(v_i, x)], \qquad \forall x \in N.$$

Dearing [1977] and Francis [1977] adopt Dearing & Francis' [1974] lower bound on r_w to this nonlinear case and also study the special case of a tree.

The convexity property of the distance along any path of a tree T allows to derive the following useful properties of the weighted center.

Theorem 4.6 [Kariv & Hakimi, 1979a].

(i) *Let v_i be a vertex of a tree T whose degree is H_i and let $T_{v_i,1}, \cdots T_{v_i,H_i}$ be the H_i subtrees rooted at v_i such that $\cup_{j=1}^{H_i} T_{v_i,j} = T$ and $T_{v_i,j} \cap T_{v_i,k} = \{v_i\}$ for j and $k = 1 \cdots H_i$. Then, the unique weighted center of T belongs to the subtree which contains the vertex v_j such that $G_w(v_i) = w_j d(v_j, v_i)$.*

(ii) *If there exists two vertices v_i and v_j such that $w_i d(v_i, x) = w_j d(v_j, x)$ holds for some $x \in T$, then x is the unique weighted center of T.*

Kariv & Hakimi [1979a] suggest a straightforward algorithm based on this theorem. (i) Let T_0 be the original tree T; choose some nonpending vertex v_0 of T_0 and determine the subtree T_{v_0,k_0} on which the weighted center c_w lies by using condition (i) of Theorem 4.6. Hence, the subtree $T_1 = T_0 \cap T_{v_0,k_0}$ contains c_w. (ii) Choose some nonpending vertex v_1 of T_1 and find the subtree T_{v_1,k_1} containing c_w. Thus c_w must be on $T_2 = T_1 \cap T_{v_1,k_1}$. (iii) The process is iterated until a subtree T_k which consists of a single edge is found. The point on this edge which satisfies condition (ii) of Theorem 4.6 is the weighted center.

Each iteration of this procedure requires $O(|V|)$ operations to determine T_{v_i, k_i} and T_{i+1}. The last step, in which the local center of an edge must be found, is in $O(|V| \log |V|)$. Hence, the total complexity is $O(k|V| + |V| \log |V|)$ where k is the number of iterations. Since each subtree rooted at an unweighted median of a tree T contains at most $|V|/2$ vertices, Kariv & Hakimi [1979a] obtain a total complexity of $O(|V| \log |V|)$ by choosing, at each iteration, the unweighted median of T_i for v_i.

Megiddo [1983] proposes a linear algorithm using the property that a linear number of vertices can be discarded at each iteration of the above procedure. Other algorithms which run in $O(|V|^2)$ have been proposed by Dearing & Francis [1974], and Hakimi, Schmeichel & Pierce [1978].

4.2.3. The K-center

The *K-center problem* consists in finding a set S of cardinality K in N such that the largest (weighted) distance between a vertex and S is minimum:

$$\min \ G(S) = \max_{v_i \in V} \ w_i d(v_i, S)$$

$$\text{s. t.} \ \ |S| = K \ \text{ and } \ S \subset N.$$

Kariv & Hakimi [1979a] have proved that the problem is $\mathcal{NP}$-hard, even for very simple networks (e.g., a planar graph with maximum vertex degree 3). Theorem 4.1 can easily be extended to the K-center problem. Hence, any K-center is included in $V \cup Q_w$. Most of the existing algorithms rely upon this property and are based on the ideas developed below [Halpern & Maimon, 1982].

Let $S \subseteq V \cup Q_w$ be a candidate solution. For each $s_k \in S$, set $V_k = \{v_i \in V; d(v_i, s_k) = d(v_i, S)\}$. Since $s_k \in V \cup Q_w$, there exist v_i and v_j such that $r(s_k) \equiv w_i d(v_i, s_k) = w_j d(v_j, s_k)$; furthermore, we notice that if v_i and v_j do not belong to V_k then $G(S)$ can either be reduced or be kept constant while replacing S by S' defined by the weighted centers of the sets V_k. Based on this observation, given an upperbound $\overline{G}$ on the optimal value G^*, any equidistant point s_k such that $r(s_k) \geq \overline{G}$ can be discarded from the search. Let also $r(S) = \max_{s_k \in S} r(s_k)$ and $R(S) \equiv \{v_i \in V; w_i d(v_i, S) \leq r(S)\}$. Algorithms typically start with an initial solution S, where $|S| \leq K$, $R(S) \neq V$ and $r(S) = r$ which is given a priori and called the range. Then, they aim at finding an optimal solution S^* by increasing the number of elements in S up to K, while keeping its maximum range r at its feasible minimum (i.e., there is no S' such that $|S'| = K$ and $r(S') < r$) and ensuring that $R(S^*) = V$. This can be done by selecting $S' \neq S$ in one or more of the following ways: (i) add facilities such that $R(S) \subset R(S')$ and/or $r(S) > r(S')$; (ii) extend the coverage $R(S') \supset R(S)$ while $|S| = |S'|$ and/or $r(S') \leq r(S)$; and (iii) reduce $r(S') < r(S)$ while $|S| = |S'|$ and/or $R(S') \supset R(S)$. The algorithms of Christofides & Viola [1971], Handler [1990], Kariv & Hakimi [1979a], Minieka [1970], and Tamir [1988] are all based on those principles. The best complexity known so far is by Tamir who reports $O(|E|^K |V|^K \log^2 |V|)$ in the weighted case, and $O(|E|^K |V|^{K-1} \log^3 |V|)$ in the unweighted case. In practice, the most efficient algorithm for the unweighted case seems to have been proposed by Handler

[1990]. To describe it, we need the following notation. Let v_i and v_j be a pair of vertices. We denote the set of equidistant points from v_i and v_j by $Q_w(v_i, v_j)$ and set $Q_w(v_i, v_i) = \{v_i\}$. Let $Q_w[(v_i, v_j); r] = \{s \in Q_w(v_i, v_j) : r(s) < r\}$ for a given r. Finally, for any subset of vertices $W \subseteq V$, we consider the set of equidistant points from all pairs of vertices in W whose radius $r(s)$ is strictly smaller than the range r: $Q_w(W, r) = \bigcup_{v_i, v_j \in W} Q_w[(v_i, v_j); r]$. Handler's procedure successively constructs optimal solutions for $p = 1 \cdots K$. It involves the following steps.

Step 0. Initialization. Set $p = 1$. Select any vertex v. Set $W = \{v\}$, $r^* = \infty$, $Q_w(W, r) = \{v\}$, $S_1 = \{v\}$, and $r(S_1) = 0$.

Step 1. Improvement. If $d(v, S_p) \le r^*$, $\forall v \in V - W$, then S_p is an improved solution: set $S_p^* = S_p$ and $r^* = r(S_p)$ and go to Step 3. Otherwise select $\bar{v}$ such that $d(\bar{v}, S_p) = \max_{v \in V} d(v, S_p)$. If $r(S_p) < G(S_p) < r^*$, then S_p is still an improved solution: set $r^* = G(S_p)$ and $S_p^* = S_p$ and go to Step 2.

Step 2. Generation of equidistant points. Add $\bar{v}$ to W and update $Q_w(W, r^*)$ using the relationships $Q_w(W \cup \{\bar{v}\}, r^*) = Q_w(W, r^*) \cup (\bigcup_{v \in W} Q_w[(v, \bar{v}), r^*]) \cup Q_w(\bar{v}, \bar{v})$.

Step 3. Covering problems. Construct the matrix B of size $(|W|, |Q_w(W, r^*)|)$ whose element $b_{vq} = 1$ if $d(v, q) < r^*$ and $b_{vq} = 0$ otherwise. Solve the covering problem $h = \min(|S| : \sum_{q \in S} b_{vq} \ge 1, \forall v \in W, S \subseteq Q_w(W, r^*))$. If $h \le p$, go to Step 1. Otherwise, go to Step 4.

Step 4. Optimality. S^* is an optimal p-center. Stop if $p = K$; otherwise, increase p by 1 and return to Step 3.

Note that Handler [1990] gives sufficient conditions for the existence of an equidistant point between v and $\bar{v}$ on a given edge, which can improve the search in Step 2. Similarly, different algorithms for solving the set covering problem in Step 3 are available in the literature [see, e.g., Balas, 1982; Balas & Ng, 1989; Cornuéjols & Sassano, 1989; Nobili & Sassano, 1989; Sassano, 1989]. Handler [1990] proves finite convergence of his algorithm and gives practical details about its implementation.

Polynomial algorithms have been obtained when the network is a tree. For the weighted case Megiddo and Tamir [1983] propose an $O(|V| \log^2 |V| \log\log |V|)$ algorithm.

It was proved in Theorem 3.8 that no polynomial approximation exists for the K-median problem. For the K-center problem, Dyer & Frieze [1985] have shown that the simple heuristic which consists in initially selecting the vertex set with maximum weight and subsequently augmenting the solution set S with the vertex corresponding to the largest weighted distance to S until the cardinality of S equals K, runs in $O(K|V|)$ and yields a value of the objective function which is bounded in the worst case by $\min(3, 1 + \alpha)$ times the optimal value, where α is the ratio of the maximum weight over the minimum weight. In the unweighted case, this gives a bound of 2, a result also obtained by Hochbaum & Schmoys [1985] for a complete graph with triangle inequality on the edge lengths. Moreover, Hochbaum and Schmoys prove that unless $\mathcal{P} = \mathcal{NP}$ no better polynomial algorithm can be obtained for the worst-case criterion.

4.3. Extensions

4.3.1. The stochastic queue center

When queueing aspects have to be taken into account in minimax location models, an optimal solution, called a *stochastic queue center* (SQC), is defined as a point c^Q such that

$$\overline{Q}(c^Q) + G_u(c^Q) \leq \overline{Q}(x) + G_u(x), \qquad \forall x \in N$$

where $\overline{Q}(x)$ is defined as in 3.3.1. Hence, the facility is to be located at a point minimizing the expected response time to the furthest vertex.

Brandeau & Chiu [1990] outline an algorithm for finding an SQC on a general network, based on the fact that the objective function is strictly convex on each segment $[a, b]$. They also study the trajectory of the SQC on a tree T in terms of $w(T)$.

4.3.2. The traveling salesman center

Suppose that, at the beginning of each time period, a server is given a list S of vertices to be visited via an optimal tour. To each $S \subseteq V$ is assigned a nonnegative weight $P(S)$ representing, for example, the relative importance of the vertices in S or the probability that only the vertices of S have to be visited during the time period considered.

A point $c^{TS} \in N$ is a *traveling salesman center* (TSC) if

$$TL^{\max}(c^{TS}) = \max_{S \subseteq V} \{P(S)L(S \cup \{c^{TS}\})\}$$

$$\leq TL^{\max}(x) = \max_{S \subseteq V} \{P(S)L(S \cup \{x\})\}, \qquad \forall x \in N,$$

that is, the facility is to be located at a point minimizing the maximum weighted length of a tour associated with each possible subset S of vertices.

This problem has been studied in the special case of a tree by Berman, Simchi-Levi & Tamir [1988]. Let T_S be the smallest subtree including S. The objective function can then be written as

$$TL^{\max}(x) = \max_{S \in \mathcal{F}} \{P(S)[L(S) + 2d(x, T_S)]\}$$

where $\mathcal{F} \equiv \{S \subseteq V; P(S) > 0\}$. Since T_S is connected, this function is convex and piecewise linear along any part of T. Furthermore, it is readily verified that if, for some $x \in N$, $TL^{\max}(x)$ is attained at some set S such $x \in T_S$, then x is a TSC.

If an explicit list of the sets S in $\mathcal{F}$ is given as well as the $P(S)$, for $S \in \mathcal{F}$, a polynomial time algorithm can be obtained by adapting Megiddo's [1983] procedure to find the weighted center of a tree.

Berman, Simchi-Levi & Tamir [1988] also provide results for the special case where the $P(S)$ depend on the weights w_i. In particular, if the w_i are all equal (so that all the sets S in $\mathcal{F}$ with the same cardinality have the same $P(S)$), then the unweighted center of a tree is a TSC.

4.3.3. The continuous center

When clients are uniformly distributed over the whole network, a point minimizing the distance to the furthest client is called a *continuous center* and is defined as a point $c^c \in N$ such that

$$G^c(c^c) = \max_{y \in N} d(y, c^c) \le G^c(x) = \max_{y \in N} d(y, x), \qquad \forall x \in N.$$

Using the remoteness function introduced in 2.2, the function $G^c(x)$ can be rewritten as

$$G^c(x) = \max_{[v_i, v_j] \in E} \overline{d}(x, [v_i, v_j]),$$

i.e., G^c is the upper envelope of the $|E|$ remoteness functions corresponding to the edges of N.

As in the cases of the center and unweighted center, a continuous center can be found by determining the continuous centers restricted to each edge. This means that for $[v_k, v_\ell] \in E$, the piecewise linear function $G^c(x)$ has to be evaluated at v_k, v_ℓ and at all interior breakpoints which are local minimizers of G^c. Hansen et al. [1991] propose an iterative method to construct $G^c(x)$ for $x \in [v_k, v_\ell]$. First, the remoteness functions $\overline{d}(x, [v_i, v_j])$ are ranked by nondecreasing order of their highest value. Then, at each subsequent iteration, the function $G^c(x)$ is updated by including the next remoteness function. The total number of breakpoints of $G^c(x)$ on $[v_k, v_\ell]$ is of $O(|E|)$ and determining the continuous centers restricted to $[v_k, v_\ell]$ takes $O(|E| \log |E|)$. A divide-and-conquer approach can also be used to find a continuous center; it has the same complexity [Tamir, 1988].

On a tree it can be shown that the continuous center is identical to the unweighted center [Hansen, Labbé & Nicolas, 1991].

When K points of N are to be chosen and the clients are uniformly distributed over N, the solution is called a *continuous K-center*. Tamir [1987] identifies a finite set of rationals containing the optimal solution value when the edge lengths of N are positive integers. Specifically, the optimal solution value is then of the form $T/2q$, where T is the length of an Euler tour of a subnetwork of N which belongs to one of four possible types ('racket', 'dumbbell', simple cycle, or simple path) and q is an integer bounded from above by 2K.

Given a positive real number r, the *continuous r-covering problem* is to find the minimum number $K = K(r)$ of points, say $y_1 \cdots y_{K(r)}$, of N such that $\min\{d(x, y_i), i = 1 \cdots K(r)\} \le r$ for all $x \in N$. Tamir's result implies that a continuous K-center can be found by solving a finite number of continuous r-covering problems.

4.4. The λ-cent-dian

A point $h_\lambda \in N$ is a *λ-cent-dian* iff

$$H_\lambda(h_\lambda) = \lambda G_u(h_\lambda) + (1 - \lambda)\frac{F(h_\lambda)}{w(N)} \le H_\lambda(x) =$$

$$= \lambda G_u(x) + (1 - \lambda)\frac{F(x)}{w(N)}, \qquad \forall x \in N.$$

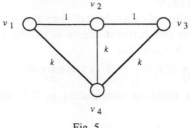

Fig. 5.

The set of λ-cent-dians is denoted $\lambda - CD$. When $\lambda = 0$, the λ-cent-dian is a center; when $\lambda = 1$, it is a median. For $0 < \lambda < 1$, the λ-cent-dian minimizes a convex combination of the average and maximum distances to the vertices. Hence, it can be viewed as a solution to a location problem where both efficiency and equity criteria are relevant. The value of λ reflects the importance attributed to the maximum distance compared to the average distance.

When choosing a center for placing the facility, one may still observe a large discrepancy in the distances separating the clients and the facility. This seems to contradict any intuitive idea of distributional justice in accessibility to the facility, thus justifying another solution, called the generalized center, which minimizes the difference between maximum and average distances. However, this may lead to an 'unreasonable' location. To illustrate, consider the network of Figure 5 with $w_1 = w_3 = 1$, $w_2 = n - 2$ and $w_4 = 0$, while $k > 1$. Clearly, v_4 is the only generalized center. Hence, if the objective is to reduce as much as possible the discrepancies in the distances to the vertices, the facility is to be located at v_4 which can be far away from all the clients (it suffices to choose k large enough). In this example, all clients would prefer to have the facility at v_2 since it is closer to all of them than v_4. This motivates the set of feasible locations to be the set PO of Pareto-optimal points with respect to distances. A point $x \in N$ is *Pareto-optimal with respect to distances* if there exists no other point $y \in N$ which satisfies the following two conditions:

(i) $d(v_i, y) \leq d(v_i, x), \forall v_i \in V$ such that $w_i > 0$;
(ii) $d(v_j, y) < d(v_j, x)$ for at least one vertex v_j such that $w_j > 0$.

Then, a point $c^g \in PO$ is a *generalized center* iff

$$G_u(c^g) - \frac{F(c^g)}{w(N)} \leq G_u(x) - \frac{F(x)}{w(N)}, \qquad \forall x \in PO.$$

Consider now the function $H_\lambda(x)$ for $\lambda > 1$. Since $H_\lambda(x) = F(x)/w(N) - \lambda(G_u(x) - F(x)/w(N))$, the limit of $H_\lambda(x)/\lambda$ for $\lambda \to \infty$ equals $G_u(x) - F(x)/w(N)$ and the λ-cent-dian is the generalized center. Hence, when $\lambda > 1$ but finite, the λ-cent-dian can be interpreted as the solution to a location problem where both efficiency and equalitarism are relevant. The value chosen for λ reflects the weight placed on the average distance with respect to the difference between the maximum and average distances. As in the case of generalized center, locations have to be restricted to the set PO.

4.4.1. The case $0 \leq \lambda \leq 1$

The following theorem identifies a finite set of points containing all the λ-cent-dian for any given λ.

Theorem 4.7 [Halpern, 1978]. *For each edge $[v_i, v_j] \in E$ and any given value of λ, $H_\lambda(x)$ is a piecewise linear function*

(i) with a finite number of breakpoints, all being either bottleneck points or local minimizers of $G_u(x)$, and

(ii) with a finite number of local minimizers of $H_\lambda(x)$, all being local minimizers of $G_u(x)$.

This result has been used by several authors to obtain algorithms of complexity $O(|V||E|\log|V||E|)$ to determine all the λ-cent-dian for a given network N and a given value of λ [Halpern, 1980; Hansen, Labbé & Thisse, 1991].

On a tree, $H_\lambda(x)$ is convex along any path so that a complete characterization of the set $\lambda - CD$ can be obtained [Halpern, 1976]. Let $P(m, c)$ be the path linking the center of T to the closest median m; let EF denote the set of efficient points with respect to $F(x)/w(N)$ and $G_u(x)$, i.e., the points $x \in T$ for which no point $y \in T$ satisfies $F(x) \leq F(y)$ and $G_u(y) \leq G_u(x)$ with at least one strict inequality; finally, a vertex $v_i \in P(m, c)$ is said to be active if either $w_i > 0$ or there exists some vertex $v_j \in V$ such that $w_j > 0$ and both $d(v_j, m) = d(v_j, v_i) + d(v_i, m)$ and $d(v_j, c) = d(v_j, v_i) + d(v_i, c)$ hold.

Theorem 4.8. *On a tree, $P(m, c) = EF = \cup\{\lambda - CD; 0 \leq \lambda \leq 1\}$. Furthermore, m is a λ-cent-dian for λ such that $0 \leq \lambda \leq 1/(w(N) + 1)$ and c is a λ-cent-dian for $(w(N) - 2)/2(w(N) - 1) \leq \lambda \leq 1$. Finally, let $P(v_i, v_j)$ be the path linking two consecutive active vertices; any interior point of $P(v_i, v_j)$ is a λ-cent-dian for $\lambda = 1 - w(N)/2w(V_i)$ where $V_i = \{v_k \in V; d(v_k, v_j) = d(v_k, v_i) + d(v_i, v_j)\}$.*

By locating a facility at a λ-cent-dian, one implicitly makes a compromise between efficiency and equity. The following result provides an upper bound on the increase in the value of the objective function when the median is chosen instead of the λ-cent-dian.

Theorem 4.9 [Hansen, Labbé & Thisse, 1991]. *On a tree, let h_λ be a λ-cent-dian and m be a median. Then,*

$$\frac{H_\lambda(m)}{H_\lambda(h_\lambda)} \leq \begin{cases} 1 & if \ 0 \leq \lambda \leq \dfrac{1}{w(N) + 1}, \\ \dfrac{2\lambda(w(N) - 1) + 2}{\lambda(w(N) - 3) + 3} & if \ \dfrac{1}{w(N) + 1} \leq \lambda \leq 1. \end{cases}$$

4.4.2. The case $\lambda > 1$

Recall that the set of locations is now restricted to the set PO of Pareto-optimal locations with respect to distances. This set can be determined by using Hansen et al.'s [1986a] algorithm with complexity $O(|V|^2|E|^2 \log |V|)$. Furthermore, it is constituted of several connected subnetworks; the set of endpoints of the subedges of each connected subnetwork is denoted by I.

Since the slope of $G_u(f_{k\ell}(\theta))$ on $[v_k, v_\ell]$ is one, up to the sign, and the slope of $F(f_{k\ell}(\theta))/w(N)$ belongs to $[-1, 1]$, the sign of $H_\lambda(f_{k\ell}(\theta))$ is the same as that of $G_u(f_{k\ell}(\theta))$. Consequently, a point $x \in N$ is a local minimizer of H_λ if and only if it is a local minimizer of G_u. Therefore, a point $x \in PO$ minimizes H_λ if either x is a local minimizer of G_u or $x \in I$. For any $\lambda > 1$, the $\lambda - CD$ can be found by determining both PO and I and by applying Kariv & Hakimi's [1979a] algorithm to identify the local minimizers of G_u.

On a tree, the convexity of the distances implies that the center (which belongs to PO), is the unique λ-cent-dian for all $\lambda > 1$ [Hansen, Labbé & Thisse, 1991].

5. Economic models of location

5.1. Introduction

The main distinctive features of the location models developed in economics deal with the following two aspects. First, *price*, as well as location, is a basic variable controlled by the decision-maker. Second, the location of a particular facility is often the outcome of a collective procedure involving *several* decision-makers.

Concerning the first aspect, price and location are together the decision variables and the objective function is now profit maximization. Besides cost considerations, the model must also account for a more realistic demand side in which clients' requirement are no longer fixed but price-sensitive. This is taken up in Subsection 5.2 where we consider the case of a *single* facility, the market of which is geographically separated from markets of other facilities selling the same commodity. This simplifying assumption allows us to discuss different price policies and to study their impact on the vertex optimality property.

As for the second aspect, we analyze two types of collective procedures. In the former, we consider locating (simultaneously) several competing facilities, each maximizing its own profits. Clearly, the 'optimal' location of a facility depends on the locations chosen by the others. The resulting interdependence is modeled and studied within the framework of *noncooperative game theory*. This theory deals with formal structures which take into account the *interaction* of the decision-makers' optimizing models. Location games have been very much studied in economics (see, e.g. Gabszewicz & Thisse [1992] for a recent survey, and Eiselt, Laporte & Thisse [1993] for an annotated bibliography). However, simple and particular models — for example, a uniform density of clients along a line or a circle — are usually assumed there. Network formulations are more complex and have only recently been studied. This is why only a few results are yet available. They are briefly reviewed in 5.3.

In the latter, the problem of locating a public facility is viewed as the result of a collective action in which clients pursue their own interest within the mutual dependence imposed by a voting rule. Therefore, the focus shifts from the conflict of interest among facilities to the *confrontation* of clients' preferences. Given the assumption that clients prefer to have the facility set up as close as possible to them, there is no way in general to meet the desiderata of all clients. Consequently, the facility location must necessarily be a compromise which reflects the underlying principles of the voting rule. Several solution concepts have been proposed in *spatial voting theory* to find such a compromise (see, e.g. Demange [1983] for a discussion of the main concepts and results). Their application to network location problems is recent. The main existing results are presented in Subsection 5.4.

5.2. The location of a facility under alternative price policies

Up to now, it was assumed that each client had a fixed, given demand (normalized to one). We suppose here that clients' demand depends on the unit price (inclusive of transportation costs) they have to pay in order to buy the commodity supplied by the facility. Let $x \in N$ be the facility's location and $w_i[p_i(x)]$ denote the demand arising at vertex v_i when the unit price there is $p_i(x)$.

If the volume of output is q, the corresponding production cost function is $C(q; p^1(x) \cdots p^m(x))$ where $p^j(x)$ is the unit price (gross of transportation costs) of input $j = 1 \cdots m$. Price $p^j(x)$ is an increasing and concave function of the distance between the facility at x and the source of input j (typically, $p^j(x)$ is equal to the price at the input source plus the corresponding transportation cost). A classical result in production theory is that, for any given q, C is nondecreasing and concave in the price vector $(p^1(x) \cdots p^m(x))$ [Varian, 1984]. This implies that C is a concave function of $\theta = f_{k\ell}^{-1}(x)$ on $[0, \ell_{k\ell}]$, for $x \in [v_k, v_\ell] \in E$.

5.2.1. Delivered pricing

When the facility supplies transportation, it is able to set a specific delivered price p_i at each vertex v_i. This price policy is called *discriminatory pricing*. In this case, the facility chooses a location $x \in N$ and a price vector $(p_1 \cdots p_n)$ is order to maximize its profits

$$\Pi(x, p_1, \cdots, p_n) = \sum_{i=1}^{n} [p_i - t_i(x)] w_i(p_i) - C(q; p^1(x) \cdots p^m(x)) \quad (5.1)$$

where $q = \sum_{i=1}^{n} w_i(p_i)$ is the sum of the quantities demanded, while $t_i(x)$, the unit transportation cost of the commodity, is an increasing and concave function of the distance between the facility at x and the clients at v_i.

Let $(x^*, p_1^* \cdots p_n^*)$ be a maximizer of (5.1) and set $q^* = \sum_{i=1}^{n} w_i(p_i^*)$. It is readily verified that x^* is a minimizer of production plus transportation costs $\Gamma(x) \equiv \sum_{i=1}^{n} w_i(p_i^*) t_i(x) + C(q^*; p^1(x) \cdots p^m(x))$. Clearly, this function is concave in $\theta = f_{k\ell}^{-1}(x)$ over $[0, \ell_{k\ell}]$. Hence, the proof of Theorem 3.1 can still be applied so that either $x^* \in V$ or $x^{**} \in V$ exists such that $\Gamma(x^*) = \Gamma(x^{**})$. In other words, V contains a profit-maximizing location.

A special case of the above price policy is when all clients pay the same price p_U regardless of their locations, given that the facility may refuse to supply when the serving cost exceeds the price p_U. This is called *uniform pricing*. This case can be dealt with as in the argument above provided that p_i^* be replaced throughout by p_U^*.

Theorem 5.1. *Under discriminatory or uniform pricing, the set V of vertices contains a profit-maximizing location.*

5.2.2. Mill pricing

When clients ships the commodity from the facility on their own, the resulting price policy is *mill pricing*. This means that the price paid by a client at v_i is equal to a mill (f.o.b.) price p_M, the same for all clients irrespective of their locations, plus the transportation cost: $p_i(x) = p_M + t_i(x)$. The facility now chooses a location $x \in N$ and a mill p_M so as to maximize profits given by

$$\Pi(x, p_M) = \sum_{i=1}^{n} p_M w_i[p_M + t_i(x)] - C(q; p^1(x) \cdots p^m(x)) \qquad (5.2)$$

where $q = \sum_{i=1}^{n} w_i[p_M + t_i(x)]$ — the total quantity sold — depends on the facility's location. This implies that the decomposition argument used in the delivered pricing case is no longer valid. Of course, the vertex optimality property might still hold. We will see below that it does not. However, it remains valid when the demands w_i are decreasing and convex, and when the marginal production cost is independent of the volume of production: $C(q; p^1(x) \cdots p^m(x)) = q \ c(p^1(x) \cdots p^m(x)) \equiv q \ c(x)$. The profit function (5.2) then becomes

$$\Pi(x, p_M) = [p_M - c(x)] \sum_{i=1}^{n} w_i[p_M + t_i(x)]. \qquad (5.3)$$

Let x^* and p_M^* be a maximizer of (5.3). Since $w_i(p_i)$ is convex in p_i, for all $p_i' \geq 0$ there exist $a_i(p_i') \geq 0$ and $b_i(p_i') > 0$ such that

$$w_i(p_i) \geq a_i(p_i')[b_i(p_i') - p_i]$$

and

$$w_i(p_i') = a_i(p_i')[b_i(p_i') - p_i']$$

for all $p_i > 0$. Hence,

$$w_i(p_i) = \max_{p_i' \geq 0} a_i(p_i')[b_i(p_i') - p_i]$$

so that (5.3) evaluated at (x^*, p_M^*) can be rewritten as

$$\Pi(x^*, p_M^*) = \max_{\underline{p}', p_M, x} [p_M - c(x)] \sum_{i=1}^{n} a_i(p_i')[b_i(p_i') - p_M - t_i(x)]$$

with $\underline{p}' = (p'_1, \cdots, p'_n)$. When $a(\underline{p}') \equiv \sum_{i=1}^{n} a_i(p'_i) > 0$, set

$$b(\underline{p}', x) = \sum_{i=1}^{n} a_i(p'_i)[b_i(p'_i) - t_i(x)]/a(\underline{p}');$$

otherwise, set $b(\underline{p}', x)$ equal to any constant. Therefore,

$$\Pi(x^*, p_M^*) = \max_{\underline{p}', p_M, x} a(\underline{p}') [p_M - c(x)] [b(\underline{p}', x) - p_M].$$

Maximizing the RHS of this expression with respect to p_M yields

$$\Pi(x^*, p_M^*) = \max_{\underline{p}', x} \tfrac{1}{4} a(\underline{p}')[b(\underline{p}', x) - c(x)]^2.$$

Since $b(\underline{p}', x)$ is convex and $c(x)$ is concave in $\theta = f_{k\ell}^{-1}(x)$ over $[0, \ell_{k\ell}]$, $[b(\underline{p}', x) - c(x)]^2$ is convex in θ. Taking the upper envelope of these functions with respect to $\underline{p}' \geq 0$ yields a convex function in θ. This implies that $x^* \in V$ or that $x^{**} \in V$ exists such that $\Pi(x^*, p_M^*) = \Pi(x^{**}, p_M^*)$. Consequently,

Theorem 5.2 [Hanjoul & Thisse, 1984]. *Assume that the demand functions are nonincreasing and convex and that the unit production cost is independent of the volume of production. Then, under mill pricing the set V of vertices contains a profit-maximizing location.*

The convexity assumption is essential for the result above. Indeed, using strictly concave demand functions is sufficient to invalidate the vertex optimality property. To see this, consider the following example. There are two clients located at the extremities of $[0, 1]$; the demand function of a client is given by $\max\{0, a - (p_M + |x - v_i|)^2\}$ where a is a positive constant and $v_1 = 0, v_2 = 1$; finally, there is no production cost. If the facility is located at $x \in [0, 1]$, the profit function is $p_M[a - (p_M + x)^2 + a - (p_M + (1 - x))^2]$ for $p_M < \sqrt{a} - 1$. Clearly, for a sufficiently large, the only profit-maximizing location is given by the middle point $x = 1/2$.

For a more detailed discussion of this class of models, the reader is referred to the survey by Beckmann & Thisse [1986].

5.3. The location of competing facilities

Consider K facilities competing with the aim of attracting clients from a given area. It is assumed that (mill) prices are fixed and equal, and that clients patronize the *nearest* facility; when a client is equidistant from two (or several) such facilities, its requirement is equally split between them.

Noncooperative game theory seeks to predict the behavior of rational, intelligent players (the facility owners, say) acting independently. Players are *rational* if they make decisions by maximizing their own objective function. Players are *intelligent* if they recognize that other players are rational. Intelligent players can put themselves in the other players' position and can reason from their points of view. In other words, players can form assumptions about the others' behavior, which are then treated as parameters in their own optimizing model. The outcome of this interactive decision-making process is a 'noncooperative equilibrium', i.e., a state in which players share a common view of how the game will be played and have, therefore, no regrets once all decisions have been made and observed by each of them (see Friedman [1990] for a detailed overview of noncooperative game theory, and Johansen [1982] for a nice, intuitive discussion of the concept of equilibrium).

For the moment, assume there are two competing facilities. In this section, two types of games are discussed. First, locations are assumed to be the only decision variables available to the facilities. Players move simultaneously so that no player can observe his competitor's decision before making his own. However, as mentioned above, each player is able to reproduce the other's reasoning. The outcome of the game is then a location pair such that no facility can unilaterally increase its number of clients by locating at another point on the network, given the location of the other facility. This is called a location equilibrium.

When prices are also decision variables, the competitive process is usually modeled as a *two-stage* game. The division into stages is motivated by the fact that prices are more easily adjusted than locations. Thus locations are simultaneously chosen in the first stage; once they are observed, prices are simultaneously chosen in the second stage. The solution of a location-then-price game proceeds backward. Given any location pair, facilities compete in (mill or delivered) prices and a second-stage equilibrium is such that no facility can increase its profits by unilaterally changing its price(s). Clearly, this equilibrium depends on the locations chosen in the first stage. After replacement, the profit functions depend only upon the facility locations so that these profits can be used as objective functions in the first stage game in which facilities choose locations. The equilibrium is called a subgame perfect Nash equilibrium; it captures the idea that, when facilities select their location, they both anticipate the consequences of their choice on the subsequent price competitive process. This concept is meaningful if, for any location pair, there exists one, and only one, second-stage equilibrium (otherwise, the first-stage payoffs would be undefined or multivalued). These conditions are very demanding. This is especially true under mill pricing for which no significant results are available yet. Thus we will concentrate below on the delivered pricing case only.

5.3.1. Competitive locations with fixed prices

Let K competing facilities be located at $\underline{x} = (x_1 \cdots x_K)$ along a network N. Since mill prices are assumed to be fixed and equal, it is reasonable to consider sales maximization as the players' objective functions (profit maximization reduces to sales maximization when marginal production costs are constant). If clients patronize the nearest facility and split equally their requirements when there are several such facilities, the sales of facility j are given by

$$A_j(\underline{x}) = \left\{ \sum_{v_i} \frac{w_i}{K_i(\underline{x})}; \ d(v_i, x_j) \leq d(v_i, x_k) \ \text{for} \ k = 1 \cdots K \right\}$$

where $K_i(\underline{x}) \geq 1$ is the number of facilities nearest from v_i in the configuration $\underline{x}$.

Assume first that the facilities simultaneously choose their locations on N. A *location equilibrium* is an K-tuple $\underline{x}^*$ of locations such that

$$A_j(\underline{x}^*) \geq A_j(x_j, \underline{x}^*_{-j}) \ \text{for all} \ x_j \in N \ \text{and} \ j = 1 \cdots K$$

where $\underline{x}^*_{-j}$ stands for $\underline{x}^*$ from which x_j^* has been deleted.

Consider the case of two facilities. Since each facility can secure half of the total sales by having the same location as its competitor, the sales of the two facilities must be equal in equilibrium. Therefore, if (x_1^*, x_2^*) is a location equilibrium, (x_1^*, x_1^*) and (x_2^*, x_2^*) are also location equilibria.

Theorem 5.3 [Slater, 1975]. *Let $m = 2$. On a tree the set of location equilibria is M^2, where M is the median set.*

This implies that both facilities located at a median of a tree is a location equilibrium [see also Wendell & McKelvey, 1981]. Furthermore, it follows from Theorem 3.5 that V^2 always contains a location equilibrium, and that this equilibrium is unique when $w(T)$ is odd. Hence, a property akin to the vertex optimality property holds for two competing facilities on a tree. When the two facilities must locate at distinct points of T, an equilibrium still exists but facilities do not have equal sales when the median is unique (see Eiselt & Laporte [1991] for further discussion of the 2-facility problem on a tree).

In the case of a general network, a location equilibrium may not exist. For example, there is no equilibrium when the network is a 3-vertex cycle with equal weights. Even when a location equilibrium exists, V^2 may not contain any such equilibrium. This is illustrated by the example of Figure 6 borrowed from Hakimi [1983], where $w_1 = \cdots w_4 = 1$ and $w_5 = w_6 = 0$. When one facility is located at v_i, it is readily verified that the other wants to be at the mid-point of $[v_5, v_6]$. Furthermore, having the two facilities located at this point is a location equilibrium.

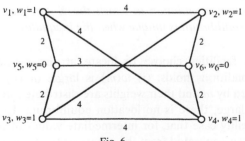

Fig. 6.

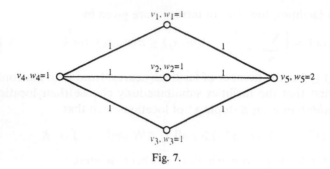

Fig. 7.

However, a partial generalization of Theorem 5.3 can be established for a general network when the client configuration obeys some kind of symmetry around a point of N. Let σ be a permutation defined on the client configuration $\underline{z} = (z_1 \cdots z_{w(N)}) \in N^{w(N)}$. Given some $x \in N$, we say that x has the betweeness property for σ if $x \in B[z_i, \sigma(z_i)]$ for all $i = 1 \cdots w(N)$.

Theorem 5.4 [Hansen, Thisse & Wendell, 1986b]. *Let σ be a permutation on the client configuration z. If $\hat{x} \in N$ has the betweeness property for σ, then $\hat{x}$ is a median and $(\hat{x}, \hat{x})$ a location equilibrium.*

This result is illustrated in Figure 7 where $w_1 = \cdots = w_4 = 1$ and $w_5 = 2$. The vertex v_5 satisfies the betweeness property under the permutation: $z_1 \to z_6, z_6 \to z_1, z_2 \to z_3, z_3 \to z_2, z_4 \to z_5, z_5 \to z_4$.

Note that, when Theorem 5.4 holds for some point of N, it also holds for some vertex [Hansen, Thisse & Wendell, 1986b]. Furthermore, the betweeness property reduces to a simple condition when $\hat{x}$ is a cut-vertex: if $w(S_h) \leq w(N)/2$ for $h = 1 \cdots H$, then $\hat{x}$ is a median (see 3.2) and $(\hat{x}, \hat{x})$ is a location equilibrium.

Suppose now K facilities. Consider a tree T and denote by T_m^* a subtree with the largest weight among the subtrees generated by deleting a median $m \in V$ and the edges incident to m. When the facilities are located at m, they equally share $w(T)$. If one of these facilities were to locate elsewhere, its sales would be at most equal to $w(T_m^*)$. This implies the following result:

Theorem 5.5. *Let $m \in V$ be a median of a tree T. If $w(T_m^*) \leq W(T)/K$, then there exists a location equilibrium in which the K facilities are at the median m. Furthermore, the equilibrium is unique when the inequality is strict.*

Intuitively, all facilities choose to locate at the median when one of the following two conditions holds: (i) $w(m)$ is large, or (ii) there are at least K subtrees generated by m and their weights are distributed in a fairly balanced way. When $w(T_m^*)$ is large, there is no location equilibrium. Eiselt & Laporte [1993] show in the 3-facility case that, for intermediate values of $w(T_m^*)$, equilibria exist where one facility is separated from the others or the three facilities are dispersed.

In either case, at least one facility is located at a median. The same authors also give a full characterization of the location equilibria when the three facilities must be at distinct points of T.

5.3.2. Competitive locations with variable prices

Consider two facilities producing and selling a homogeneous product to clients located in the vertex set. Facility $j = 1, 2$ chooses a location on the network and a vector of delivered prices, specifying a price p_{ji} at which the facility is willing to supply clients at $v_i \in V$. The competitive process is modeled as a two-stage game: first, facilities simultaneously choose locations and these decisions become known; second, facilities simultaneously choose delivered prices. This game is solved by backward induction.

Assume that x_1 and x_2 have been selected in the first stage and consider the second stage. Throughout this subsection the unit production cost, $c_j(x_j)$, is assumed to be independent of the facility's volume of production. Facility j's profit function is then given by

$$\Pi_j(p_{11} \cdots p_{1n}, p_{21} \cdots p_{2n}; x_1, x_2) =$$
$$= \sum_{i=1}^{n} [p_{ji} - c_j(x_j) - t_i(x_j)] S(p_{1i}, p_{2i}) w_i, \quad j = 1, 2 \qquad (5.4)$$

where

$$S(p_{1i}, p_{2i}) = \begin{cases} 1 & \text{if } p_{1i} < p_{2i} \text{ or } \left(p_{1i} = p_{2i} \text{ and } c_1(x_1) + t_i(x_1) \right. \\ & \qquad\qquad\qquad\qquad\qquad\qquad\left. < c_2(x_2) + t_i(x_2) \right) \\ 0 & \text{if } p_{2i} < p_{1i} \text{ or } \left(p_{1i} = p_{2i} \text{ and } c_2(x_2) + t_i(x_2) \right. \\ & \qquad\qquad\qquad\qquad\qquad\qquad\left. < c_1(x_1) + t_i(x_1) \right) \\ \frac{1}{2} & \text{if } p_{1i} = p_{2i} \text{ and } c_1(x_1) + t_i(x_1) = c_2(x_2) + t_i(x_2) \end{cases}$$

is a rule describing how clients at v_i choose to buy from facility 1 or 2: clients order from the low price firm, or from the low cost firm in the event of a price tie. Since unit production and transportation costs are independent of the quantity produced and hauled, the equilibrium prices p_{1i}^* and p_{2i}^* can be determined independently of the prices set at other vertices. It is then readily verified that there exists a unique equilibrium for the price subgame induced by x_1 and x_2, which is given by

$$p_{1i}^* = p_{2i}^* = \max \{c_1(x_1) + t_i(x_1), c_2(x_2) + t_i(x_2)\}, \quad i = 1 \cdots, n. \qquad (5.5)$$

This means that the equilibrium delivered price at v_i is set by the low cost facility at the unit production and transportation cost of the high cost facility. Introducing (5.5) into (5.4) yields the objective functions of the two players for the first stage game.

To determine an equilibrium of this game, define

$$\Gamma(x_1, x_2) \equiv \sum_{i=1}^{n} \min_{j=1,2} \{c_j(x_j) + t_i(x_j)\} w_i,$$

the total production and transportation costs associated with two facilities (clients are supposed to be supplied from the low cost facility). Under the usual assumptions of concavity of the transportation cost functions, the following result holds:

Theorem 5.6 [Lederer & Thisse, 1990]. *Assume that clients have fixed requirements and that two facilities compete in prices in the second stage. Then, the locations minimizing production and transportation costs are an equilibrium of the first stage game. Furthermore, V^2 always contains such an equilibrium.*

Thus, the search for a location equilibrium can be limited to the vertices. The theorem above can easily be extended to the case of K facilities, so that a location equilibrium can be found by solving a K-median problem (see Subsection 3.5).

An alternative approach has been taken by Labbé & Hakimi [1991] to analyze the case where the assumption of fixed requirements is replaced by that of linear demands. They assume that firms compete in quantities, instead of prices, at each vertex v_i: each facility sets the quantity of product it is willing to supply there. Price at v_i is now determined by the total quantity sold by the two facilities. In the second stage, facility i's profit function is therefore given by

$$\Pi_j(w_{11} \cdots w_{1n}, w_{21} \cdots w_{2n}; x_1, x_2) =$$
$$= \sum_{i=1}^{n} [p_i(w_{1i} + w_{2i}) - c_j(x_j) - t_i(x_j)] w_{ji}, \quad j = 1, 2$$

where $p_i(w_{1i} + w_{2i}) = \alpha_i - \beta_i(w_{1i} + w_{2i})$ and $w_{ji} \geq 0$ is the quantity sold by facility j at v_i. It is easy to show that there exists a unique equilibrium for the quantity subgame induced by x_1 and x_2.

Assuming that the unit costs are low enough for each facility to sell a positive quantity at each vertex, it can be shown:

Theorem 5.7 [Labbé & Hakimi, 1991]. *Assume that clients have linear demands and that two facilities compete in quantities in the second stage. Then, there exists a location equilibrium for the first stage game and V^2 contains such an equilibrium.*

An equilibrium can be found in $O(|V|^3)$ by computing and comparing the values of Π_1 and of Π_2 for each pair of vertices.

5.4. Voting for a facility location

Consider a public facility (e.g. a school or a park) to be located so as to satisfy the needs of clients distributed over the vertices of a transportation network. Since the population is dispersed, a collective rule is required to select a 'compromise' among the possible locations. In a democratic society, the location may result from a voting process in which each client casts a vote. A seemingly natural solution is then obtained when the facility is located in a way such that no other location is preferred by (i.e., closer to) a majority of clients. A location satisfying this property is called a Condorcet solution, after the Marquis de Condorcet who pioneered the mathematical theory of voting. Formally, a point $s \in N$ is a *Condorcet solution* iff

$$\left\{ \sum_{v_i} w_i; \ d(v_i, x) < d(v_i, s) \right\} \leq \frac{w(N)}{2}, \quad \forall x \in N.$$

The choice of a Condorcet solution can be described as follows. An initial site for the facility is selected. The clients are then asked to cast a vote for or against the site. If those rejecting the site are more than half the number of clients, another site is proposed and the process is repeated. If those rejecting the site do not get more than fifty percent of the votes, the process stops and the facility is set up at the site considered (see Rushton, McLafferty & Ghosh [1981] for an intuitive discussion of the voting process in locational analysis).

A Condorcet solution does not necessarily exist. Consider a 3-vertex cycle with equal weights. Any point interior to an edge is defeated by at least one of the adjacent vertices. So, only the vertices are candidate Condorcet solutions. But then, it is readily seen that any of them is defeated by any point belonging to the corresponding opposite edge. This is reminiscent of the nonexistence of a location equilibrium for two competing facilities (see 5.3.1.). There is indeed a relationship between the two problems: a Condorcet solution exists if and only if there is a location equilibrium for the 2-facility competitive location problem. The argument runs as follows. First, a Condorcet solution s is such that (s, s) is a location equilibrium since no alternative site can give a facility more than fifty percent of the clients. Conversely, if (x_1^*, x_2^*) is a location equilibrium, then we have seen that (x_1^*, x_1^*) and (x_2^*, x_2^*) are also location equilibria. Since ties in the vote are resolved in favor of those supporting the candidate point, the definition of a location equilibrium implies that x_1^* and x_2^* are both Condorcet solutions.

This connection between the two concepts suggests that a result similar to Theorem 5.3 holds in the voting framework.

Theorem 5.8 [Hansen & Thisse, 1981]. *On a tree, the set of Condorcet solutions is the median set.*

This result has two interesting implications. First, it says that a Condorcet solution always exists when the network is a tree. Second, it provides a simple condition under which planning according the minisum criterion and voting according to the one-man-one-vote system yield the same outcome. This equivalence is interesting from the social perspective since it reconciles both planning and voting procedures: a median is supported by a majority of clients and the location chosen through a vote of the clients is a median.

Bandelt [1985, Theorem 1] has generalized the result above and has provided a complete characterization of the class of networks for which the equivalence between the Condorcet solutions and the medians holds. A partial equivalence has also been obtained by Labbé [1985, Theorem 4] in the case of networks for

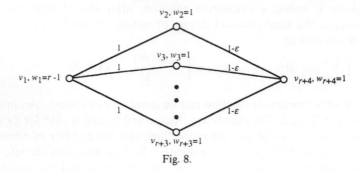

Fig. 8.

which any two cycles intersect at most at one vertex: either the set of Condorcet solutions is the median set, or there is no Condorcet solution.

For a general network, even when a Condorcet solution exists, the discrepancy between the two types of points can be large. The next theorem gives the maximum value of the bias evaluated in terms of the total distance to the facility.

Theorem 5.9 [Labbé, 1985]. *Let m be a median and s a Condorcet solution. Then*

$$\frac{F(s)}{F(m)} \le \frac{2w(N) - (\lceil w(N)/2\rceil + 1)}{\lceil w(N)/2\rceil + 1}.$$

Thus, in the worst case, a substantial loss in accessibility may result from selecting the facility location by means of a vote of the clients. This is reinforced by the fact that the above bound is the best possible. Consider the network represented in Figure 8. There are $w(N) = 2r + 1$ clients; clients $1 \cdots r - 1$ are located at v_1; clients $r \cdots 2r + 1$ are located one at each of the vertices $v_2 \cdots v_{r+3}$. The lengths of the edges are indicated at the figure, where $\varepsilon > 0$ is arbitrarily small. It is then easy to see that v_1 is the unique median and that v_{r+4} is the unique Condorcet solution. Hence

$$F(s) = 3r - \varepsilon(2r + 1) \text{ and } F(m) = r + 2, \text{ so that } \frac{F(s)}{F(m)} = \frac{3r}{r + 2}$$

minus a term linear in ε which can be made arbitrarily small.

Conversely, locating the facility at the median may dissatisfy a very large number of clients. This is shown by the example of Figure 9 due to Bandelt & Labbé [1986]. There are one client at v_1 and r clients at each of the other two vertices (v_2 and v_3). Clearly, v_1 is the only median for $\varepsilon < 1/r$, but $2r = w(N) - 1$ clients prefer the middle point on the edge $[v_2, v_3]$ to v_1.

Existence of a voting solution can be restored in the case of a general network by restricting the locations to be compared. Specifically, a *local Condorcet solution* is a point $s_\ell \in N$ for which there exists $\varepsilon > 0$ such that no point in $N(s_\ell; \varepsilon)$ is preferred to s_ℓ by a majority of clients.

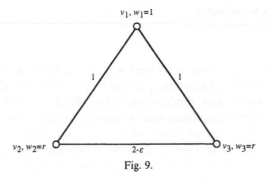

Fig. 9.

Theorem 5.10 [Hansen, Thisse & Wendell, 1986b]. *In a general network, the set of local Condorcet solutions is identical to the set of local medians.*

Hence, there is a local equivalence between the voting and planning procedures for general networks. This implies that the local Condorcet solution satisfies the vertex optimality property. In particular, when $w(N)$ is odd the set V contains all the local Condorcet solutions.

The equivalence property, especially in the case of trees, may generate some substantial discrimination in the accessibility to the facility (see 4.1). This has led Vohra [1989] to investigate a generalized voting process in which each client commands a number of votes given by a power α of his distance to the facility. More precisely, if the facility is located at $x \in N$, a client at v_i is given $[d(v_i, x)]^\alpha$ votes, where $\alpha > 1$ is an integer (the condition $\alpha > 1$ is required for the distant clients to have more influence in the vote than those close to the facility). When $\alpha = 0$, this reduces to the one-man-one-vote principle used above. An α-*Condorcet solution* is then a point $s_\alpha \in N$ which is never defeated in a vote where the clients at v_i are given each $[d(v_i, s_\alpha)]^\alpha$ votes and where all clients choose between s_α and any other point of the network. When the network is a tree, Vohra shows that *a point of T is an α-Condorcet solution if and only if this point minimizes $\sum_{i=1}^{n} w_i [d(v_i, x)]^{\alpha+1}$ over T* (the minimizer is unique since $\alpha > 1$). When α becomes arbitrarily large, the α-Condorcet solution converges to the weighted center (see 4.2.2) which may therefore be sustained as the outcome of a particular voting procedure.

For a tree, Theorem 5.8 implies the Condorcet solutions can be obtained by using Goldman's algorithm (see 3.2). In the case of a general network, the (possibly empty) set of Condorcet solutions can be determined in polynomial time by comparing pairs of points belonging to different segments [Hansen & Labbé, 1988]. When $w(N)$ is odd, V contains all the Condorcet solutions (if any), which simplifies the procedure. Hansen and Labbé also propose a method to find the point in N which is the least objectionable in that no other point has less clients wanting to move the facility there (see Bandelt & Labbé [1986] and Hakimi [1983] for properties of this solution).

6. Discrete location problems

6.1. Introduction

The purpose of this section is to deal with some discrete location problems, where the set of possible locations is defined by a *finite* subset of the network. The reasons for using such models are many. First, as seen in the previous sections, several location problems can be proved to have an optimal solution in an identifiable finite subset of points of the network. Second, land availability, zoning regulation and facility design considerations may be such that the set of eligible sites turns out to be finite. Third, and last, discrete models prove to be very flexible because they allow one to incorporate a number of geographical and economic features.

The prototype of discrete location theory is a close relative to the K-median problem studied in 3.5: the uncapacitated facility location problem (in short UFLP) which was originally formulated by Balinski [1966], Kuehn & Hamburger [1963], Manne [1964] and Stollsteimer [1963]. Properties and algorithms for the UFLP are discussed in 6.2. Various extensions, including the capacitated facility location problem, are considered in 6.3. In Subsection 6.4, we show how the UFLP can be generalized to cope with different spatial price policies (see 5.2).

6.2. The uncapacitated facility location problem

Though the set of possible locations in N may be arbitrary in the UFLP, we assume for simplicity that it is given by a subset W of V with $|W| = m$. Each client must be supplied from a facility (i.e., a plant, a warehouse, or a shop) where a commodity is produced (or stored, or sold). Facilities can be located at vertices $v_j \in W$ where the (site-specific) fixed set-up costs are given and denoted f_j. The cost of supplying clients at $v_i \in V$ from vertex $v_j \in W$ is known and denoted c_{ij} (this includes variable production costs and transportation costs). The problem is to find the number and locations of the facilities to be operated, as well as the allocation of clients to facilities, in order to minimize total costs. Formally, the *uncapacitated facility location problem* can be expressed as a mixed-integer linear program:

$$\min_{\underline{y}, X} F(\underline{y}, X) = \sum_{i=1}^{n} \sum_{j=1}^{m} c_{ij} x_{ij} + \sum_{j=1}^{m} f_j y_j \tag{6.1}$$

$$\text{s. t.} \sum_{j=1}^{m} x_{ij} = 1, \qquad i = 1 \cdots n \tag{6.2}$$

$$0 \le x_{ij} \le y_j, \qquad i = 1 \cdots n, \ j = 1 \cdots m \tag{6.3}$$

$$y_j \in \{0, 1\}, \qquad j = 1 \cdots m. \tag{6.4}$$

The variables y_j and x_{ij} are to be interpreted as in the K-median problem. The constraints (6.2) and (6.3) are similar to (3.8) and (3.9), respectively. The number of facilities is determined as the solution to the trade-off between fixed costs and supplying costs. The UFLP has always an integer solution in x_{ij}.

Not surprisingly, the UFLP and the K-median problem can be solved by similar techniques. In what follows, we briefly present the main properties and algorithms. The reader is referred to Cornuéjols, Nemhauser & Wolsey [1990] and Krarup & Pruzan [1983] for further details.

6.2.1. Properties of the UFLP

Solving the UFLP is theoretically difficult as shown by the following theorem.

Theorem 6.1 [Krarup & Pruzan, 1983]. *The UFLP is $\mathcal{NP}$-hard.*

When the network is a tree and c_{ij} is the distance between v_i and v_j, the UFLP can be solved in polynomial time using Kolen's [1983] $O(|V|^3)$ algorithm. Hassin & Tamir [1991] have refined this bound to $O(|V|)$ when the tree is a path. This implies that the UFLP is solvable in $O(|V|^2)$ for a single cycle network.

The polytope Π of the UFLP is defined as the convex hull in $\mathbb{R}^{m(n+1)}$ of the set of integral solutions, i.e., $\Pi = \text{Conv}\{(X, \underline{y}) : \sum_{j=1}^{m} x_{ij} = 1, i = 1 \cdots n; 0 \leq x_{ij} \leq y_j \leq 1, i = 1 \cdots n, j = 1 \cdots m; y_j$ integer, $j = 1 \cdots m\}$. When $n \leq 2$ or $m \leq 2$, the linear programming relaxation has always integral extreme points [Krarup & Pruzan, 1983]. Cho, Johnson, Padberg & Rao [1983] provide a complete description of Π in terms of inequalities when $n \leq 3$ or $m \leq 3$. So far, for larger problem, only a partial description of Π can be provided. The most significant results can be found in Cho, Johnson, Padberg & Rao [1983], Cho, Padberg & Rao [1983] and Cornuéjols & Thizy [1982].

Consider first the polytope Π^R of the linear programming relaxation of the UFLP. A complete characterization of the fractional extreme points of Π^R can be found in Cornuéjols, Fisher & Nemhauser [1977b]. Let $J_1 = \{j : 0 < y_j < 1\}$ and $I_1 = \{i : x_{ij}$ is noninteger for some j and $x_{ij} = 0$ or y_j for all $j \in J_1\}$, where (X, y) is a fractional solution of the relaxed problem. Let F be the $|I_1| \times |J_1|$ matrix whose elements are $f_{ij} = 1$ if $x_{ij} > 0$ and $f_{ij} = 0$ otherwise.

Theorem 6.2. *(X, y) is a fractional extreme point of Π^R if and only if*
   (i) $y_j = \max_i x_{ij}$, for all $j \in J_1$;
   (ii) for all $i = 1 \cdots n$, there is at most one j with $0 < x_{ij} < y_j$;
   (iii) $\text{rank}(F) = |J_1|$.

See also Guignard [1980] for complementary results.

Various classes of valid inequalities and facets for the polytope Π of the UFLP have been obtained in Cho, Johnson, Padberg & Rao [1983], Cho, Padberg & Rao [1983], Cornuéjols & Thizy [1982], and Guignard [1980]. To derive them, it is usual to consider an equivalent formulation of the UFLP in which the objective function is

$$\max \sum_{i=1}^{n} \sum_{j=1}^{m} \bar{c}_{ij} x_{ij} - \sum_{j=1}^{m} f_j y_j$$

with $\bar{c}_{ij} = \max_k c_{ik} - c_{ij}$ and the constraints (6.2) are replaced by the inequalities $\sum_{j=1}^{m} x_{ij} \leq 1$, $i = 1 \cdots n$. This new problem will be referred to as $P^{\leq}$. Note that Π is a face of the polytope of $P^{\leq}$, say $\Pi^{\leq}$, but polyhedral properties are more easily established with the latter since $\Pi^{\leq}$ is full dimensional in $\mathbb{R}^{m(n+1)}$. A further task will be to establish which facets of $\Pi^{\leq}$ are also facets of Π. To obtain facets of $\Pi^{\leq}$, a second transformation is applied using the variables $\bar{y}_j = 1 - y_j$, $j = 1 \cdots m$. The new problem is

$$\max \sum_{i=1}^{n} \sum_{j=1}^{m} \bar{c}_{ij} x_{ij} + \sum_{j=1}^{m} f_j \bar{y}_j$$

$$\text{s. t.} \quad \sum_{j=1}^{m} x_{ij} \leq 1, \qquad i = 1 \cdots n$$

$$x_{ij} + \bar{y}_j \leq 1, \qquad i = 1 \cdots n, \ j = 1 \cdots m$$

$$x_{ij}, \bar{y}_j \in \{0, 1\}, \qquad i = 1 \cdots n, \ j = 1 \cdots m.$$

and has the structure of a set packing problem.

As in Cornuéjols & Thizy [1982], one can show that the polytope Π' of the latter problem is the convex hull of the vertex packings in the associated *intersection graph* $G^I = (V^I, E^I)$, which is defined as follows. We associate a node with every variable x_{ij} and $\bar{y}_j$. For all i, there is an edge between every pair of nodes corresponding to x_{ij} and x_{ik} for $j \neq k$. There is also an edge between the nodes associated to x_{ij} and $\bar{y}_j$ for every pair (i, j). An example is provided in Figure 10 for the case $m = 4$ and $n = 3$.

Several results are known about facets of vertex packings polytopes; most of the theorems reported here use the special structure of the graph G^I to obtain facets of Π'. Let $G^S = (V^S, E^S)$ be a subgraph of G^I induced by a subset of vertices

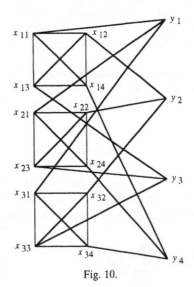

Fig. 10.

$V^S \subset V^I$, with $e \in E^S$ if $e \in E^I$ and its extremities belong to V^S. Let also $J^S = \{j : \overline{y}_j \in V^S\}$ and $I^S = \{i : x_{ij} \in V^S$ for some $j\}$. The matrix S of dimension $|I^S| \times |J^S|$ whose element $s_{ij} = 1$ if $(x_{ij}, \overline{y}_j) \in E^S$ and $s_{ij} = 0$ otherwise is an *adjacency matrix*. We denote $\alpha(G^S)$ the *independence number* of G^S, i.e., the maximum cardinality of a node packing in G^S, and $\beta(G^S)$ the *covering number* of G^S, i.e., the minimum number of plants in J^S to cover all destinations of I^S. S is a *maximal adjacency matrix* if changing a zero element of S to one increase $\alpha(G^S)$ by one or decreases $\beta(G^S)$ by one. Finally, S is called a *pd-adjacency matrix* and G^S a *pd-adjacency subgraph* if: (i) G^S is connected; (ii) there is at least one zero element in each column of S; (iii) $|I^S| \geq 3$ and $|J^S| \geq 3$. The following result can be shown to hold.

Theorem 6.3 [Cho, Padberg & Rao [1983]. *For any pd-subgraph G^S, the inequality*

$$\sum_{i \in I^S} \sum_{j \in J^S} s_{ij} x_{ij} + \sum_{j \in J^S} \overline{y}_j \leq |I^S| + |J^S| - k$$

is a valid inequality for $P^{\leq}$ iff $k \leq \beta(G^S)$. Furthermore, $\alpha(G^S) = |I^S| + |J^S| - \beta(G^S)$.

The following theorem gives valid inequalities for the UFLP that are violated by some fractional extreme points of Π^R as characterized in Theorem 6.2.

Theorem 6.4 [Cho, Johnson, Padberg & Rao, 1983]. *Let B be a $|I^B| \times |J^B|$ nonsingular adjacency matrix where $|I^B| = |J^B|$ and for which $Bz = e$ has a nonnegative solution, e being a column vector of ones. If $\beta(G^B) > \sum_{j \in J^B} z_j$, then the inequality*

$$\sum_{i \in I^B} \sum_{j \in J^B} b_{ij} x_{ij} + \sum_{j \in J^B} \overline{y}_j \leq |I^B| + |J^B| - \beta(G^B)$$

is a valid inequality for $P^{\leq}$ which cuts off some fractional extreme points of Π^R.

Consider the special case of this theorem for a $k \times k$ cyclic matrix Q, i.e., a matrix whose rows are $0 - 1$ vectors in which t consecutive ones are successively moved one position to the right. Assume also that k and t are relatively prime. Then Q is a nonsingular adjacency matrix and the valid inequality is now

$$\sum_{i \in I^Q} \sum_{j \in J^Q} q_{ij} x_{ij} + \sum_{j \in J^Q} \overline{y}_j \leq 2k - \left\lceil \frac{k}{t} \right\rceil. \tag{6.5}$$

This inequality has been obtained by Cornuéjols, Fisher & Nemhauser [1977b]. If $t = k - 1$, we have the inequality of Guignard [1980], which turns out to be a facet. Finally, when $t = 2$ and k is odd, we have the odd-cycle inequality of Padberg [1973].

We now turn to facet-defining inequalities for $P^{\leq}$. The first result is that the linear inequalities used in the formulation of the problem are facets.

Theorem 6.5 [Cornuéjols & Thizy, 1982]. *The following inequalities are facets of* $P^{\leq}$ *(known as 'trivial facets'):*

$$\sum_{j=1}^{m} x_{ij} \leq 1, \quad i = 1 \cdots n$$

$$x_{ij} \leq y_j, \quad i = 1 \cdots n, \ j = 1 \cdots m$$

$$x_{ij} \geq 0, \quad i = 1 \cdots n, \ j = 1 \cdots m$$

$$y_j \leq 1, \quad j = 1 \cdots m.$$

The trivial facets are very useful in practice since the linear programming relaxation of the UFLP often yields an integer solution. This connection is further illustrated by some results derived by Ahn, Cooper, Cornuéjols & Frieze [1988]. Assume $m = n$ and generate according to the uniform distribution n points $v_1 \cdots v_n$ in the unit square; let c_{ij} be the Euclidean distance between v_i and v_j; and suppose that the fixed costs f are the same across points $v_1 \cdots v_n$.

Theorem 6.6 [Ahn, Cooper, Cornuéjols & Frieze, 1988]. *Let* $n^{-1/2+\varepsilon} \leq f \leq n^{1/2+\varepsilon}$ *for some fixed* $\varepsilon > 0$. *Then, under the above conditions* $(F^* - F_{LP})/F^* \sim 0.00189$ *almost surely, where* F^* *is the optimal value of* F *and* F_{LP} *the optimal value corresponding to the linear relaxation.*

Although nontrivial facets have been little used so far in the practical implementation of algorithms for the UFLP, it is worth reporting the most important classes that have been discovered. The following theorems characterizes the facets $\gamma x + \delta y \leq \varepsilon$, where γ and δ are 0 or 1.

Theorem 6.7 [Cho, Padberg & Rao, 1983]. *Let* $I^S \subseteq \{1 \cdots n\}$ *and* $J^S \subseteq \{1 \cdots m\}$. *Then the inequalities with* $s_{ij} = 0$ *or* 1

$$\sum_{i \in I^S} \sum_{j \in J^S} s_{ij} x_{ij} + \sum_{j \in J^S} \overline{y}_j \leq \alpha(G^S)$$

is a nontrivial facet of $P^{\leq}$ *iff* S *is a* $|I^S| \times |J^S|$ *maximal pd-adjacency matrix*

Cornuéjols & Thizy [1982] have obtained a particular family of facets of $P^{\leq}$.

Theorem 6.8. *Consider any two integers* k *and* ℓ *such that* $2 \leq k < \ell \leq m$ *and* $\binom{\ell}{k} = m$, *and any two sets* $I \subseteq \{1 \cdots n\}$ *and* $J \subseteq \{1 \cdots m\}$ *such that* $|I| = \binom{\ell}{k}$ *and* $|J| = \ell$. *Let* $A^{\ell k} = (a_{ij}^{\ell k})_{i \in I, j \in J}$ *be a matrix with* $\binom{\ell}{k}$ *rows and* ℓ *columns whose rows are all the different* $0 - 1$ *vectors with* k *ones and* $\ell - k$ *zeros. Then, the inequality*

$$\sum_{i \in I} \sum_{j \in J} a_{ij}^{\ell k} x_{ij} + \sum_{j \in J} \overline{y}_j \leq \binom{\ell}{k} + k - 1$$

is a facet of $\Pi^{\leq}$.

Note that setting $l = m$ and $k = m - 1$, we get the facets of Guignard [1980].

A second family of facets can be obtained from (6.5). Cho, Padberg & Rao [1983] give a way of lifting this inequality to facet of $\Pi^\leq$. Let p and q denote respectively the dividend and the remainder of the division of k by t: $k = pt + q$. We define $Q_s, s = 1 \cdots p$ as the submatrix of Q obtained by deleting the first $t(s-1)$ rows and columns of Q. The following algorithm generates a new matrix $\tilde{Q}$ having the property that $\beta(G^Q) = \beta(G^{\tilde{Q}})$.

(1) Start with Q and Q_p.

(2) Transform matrix Q_p into a $(t+q) \times (t+q)$ cyclic matrix with $t+q-1$ consecutive ones per row by adding elements equal to one in the required places. Call this matrix $\tilde{Q}_p$.

(3) For $s = p - 1 \cdots 1$, generate $\tilde{Q}_s$ by appending additional t columns and rows to $\tilde{Q}_{s+1}$ so that the first t columns of $\tilde{Q}_s$ are replicates of the first columns of Q_s.

(4) Let $\tilde{Q}_1 = \tilde{Q}$.

Cho, Padberg & Rao [1983] prove that $\tilde{Q}$ is a maximal pd-adjacency matrix with $\beta(G^{\tilde{Q}}) = p + 1$ and hence obtain the following result.

Theorem 6.9. *The inequality*

$$\sum_{i \in I^{\tilde{Q}}} \sum_{j \in J^{\tilde{Q}}} \tilde{q}_{ij} x_{ij} + \sum_{j \in J^{\tilde{Q}}} \bar{y}_j \leq 2k - (p+1)$$

is a facet of $\Pi^\leq$.

Other lifting procedures are described in the above-mentioned papers.

Finally, it remains to determine which facets of $\Pi^\leq$ are also facets of Π. Cornuéjols & Thizy [1982] prove the following theorem.

Theorem 6.10. *Any facet of* $\Pi^\leq$

$$\sum_{i \in I} \sum_{j \in J} b_{ij} x_{ij} - \sum_{j \in J} y_j \leq r$$

where $I \subseteq \{1 \cdots n\}$ *and* $J \subseteq \{1 \cdots n\}$ *are nonempty subsets and* $b_{ij} = 0$ *or* 1 *for all* i *and* j *also defines a facet of* Π.

By consequence, all the inequalities presented above except $\sum_{j=1}^m x_{ij} \leq 1$ for $i = 1 \cdots n$ are facets of Π.

6.2.2. Algorithms for the UFLP

Given the similarity between the UFLP and the K-median problem, most of the algorithms described in Subsection 3.5 can readily be adapted.

The GREEDY heuristic can be described as in 3.5.4. with

$$\rho_j(S) = \sum_{i=1}^n (\min_{v_k \in S} c_{ik} - c_{ij})^+ - f_j, \tag{6.6}$$

while the procedure stops when $\rho_j(S) < 0$ for all $v_j \in W - S$. Similarly, STINGY proceeds with $\sigma_j(S) = -\sum_{i=1}^{n}(\min_{v_k \in S - \{v_j\}} c_{ik} - c_{ij})^+ + f_j$ and stops when $\sigma_j(S) > 0$ for all $v_j \in S$. Adapting INTERCHANGE may be slightly more difficult. For a given configuration S of facilities, let

$$\tau_{jk}(S) = \sum_{i=1}^{n} \max \left\{ \min_{v_\ell \in S} c_{i\ell} - c_{ik}, \min_{v_\ell \in S} c_{i\ell} - \min_{v_\ell \in S - \{v_j\}} c_{i\ell} \right\} - f_j + f_k$$

denote the variation of the objective function F when moving the facility at $v_j \in S$ to the vacant site $v_k \in W - S$. Define the neighborhood of S as being any configuration obtained by opening, closing or moving one facility in regard to S. The heuristic starts with an arbitrary configuration S of facilities and seeks a neighboring configuration that brings a decrease in F; iterations are performed until a local minimum is reached. Even if good solutions are generally obtained by using the above heuristics, one may also adapt SIMULATED ANNEALING to avoid being stuck in a poor local minimum.

It has been mentioned earlier that the linear programming relaxation of the UFLP often yields an integral optimal solution or, at least, a tight lower bound F_{LP} on F^*. Unfortunately, like in the K-median case, solving the relaxed problem by standard linear programming packages is a difficult task because of the large number of variables and constraints. Therefore, alternative strategies for getting F_{LP} are needed.

Consider first the relaxation obtained when replacing the constraints (6.4) by $y_j \geq 0$, $j = 1 \cdots m$. Let λ_i and μ_{ij} denote the dual variables associated with (6.2) and (6.3), respectively. The dual problem can then be written as follows (there are no restrictions on the λ_i):

$$\max \sum_{i=1}^{n} \lambda_i \tag{6.7}$$

$$\text{s. t.} \quad \sum_{i=1}^{n} \mu_{ij} \leq f_j, \qquad j = 1 \cdots m \tag{6.8}$$

$$\lambda_i - \mu_{ij} \leq c_{ij}, \qquad i = 1 \cdots n, \ j = 1 \cdots m \tag{6.9}$$

$$\mu_{ij} \geq 0, \qquad\qquad i = 1 \cdots n, \ j = 1 \cdots m. \tag{6.10}$$

Clearly, for any fixed values of the λ_i, μ_{ij} is equal to its minimum value $(\lambda_i - c_{ij})^+$ by (6.9) and (6.10). This leads to a condensed version of the dual by elimination of the μ_{ij}:

$$\max \sum_{i=1}^{n} \lambda_i$$

$$\text{s. t.} \quad \sum_{i=1}^{n} (\lambda_i - c_{ij})^+ \leq f_j, \qquad j = 1 \cdots m.$$

Denote $\rho_j(\underline{\lambda}) = \sum_{i=1}^{n}(\lambda_i - c_{ij})^+ - f_j$ (compare with (6.6)). Solving the dual up to optimality may be difficult and approximate solutions are often sufficient in practice. Bilde and Krarup [1977] and Erlenkotter [1978] have proposed a greedy-like heuristic called 'dual ascent'. It starts with $\lambda_i = \min_{v_j \in W} c_{ij}$ for $i = 1 \cdots n$. Cycling through the indices $i = 1 \cdots n$, one attempts to increase the variables λ_i one at a time while maintaining the constraints $\rho_j(\underline{\lambda}) \leq 0$. More precisely, if λ_i can be increased, one sets $\lambda_i := \min_{v_j \in W}\{c_{ij}; c_{ij} > \lambda_i\}$ if this entails no violation of the constraints $\rho_j(\underline{\lambda}) \leq 0$ or λ_i equal to the minimum value for which some $\rho_j(\underline{\lambda}) = 0$. The ascent is stopped when all the λ_i can no longer be increased.

The complementary slackness conditions allows one to construct a primal solution from a dual solution. Let $S(\underline{\lambda}) = \{v_j \in W : \rho_j(\underline{\lambda}) = 0\}$. Consider a subset $S^+(\underline{\lambda}) \subseteq S(\underline{\lambda})$ such that $|\{v_j \in S^+(\underline{\lambda}); c_{ij} \leq \lambda_i\}| \geq 1, i = 1 \cdots n$. Then, set $y_j = 1$ if $v_j \in S^+(\underline{\lambda})$ and $y_j = 0$ otherwise, and $x_{ij} = 1$ for some j such that $c_{ij} = \min_{v_k \in S^+(\underline{\lambda})} c_{ik}$ and $x_{ij} = 0$ otherwise. This choice of a primal solution is optimal if $k_i \equiv |\{v_j \in S^+(\underline{\lambda}); c_{ij} < \lambda_i\}| \leq 1$ for all $i = 1 \cdots n$. Otherwise, there is a duality gap whose value is given by

$$\overline{F} - \underline{F} = \sum_{i=1}^{m} \sum_{\{j; v_j \in S^+(\underline{\lambda}) \text{ and } x_{ij}=0\}} (\lambda_i - c_{ij})^+$$

where $\overline{F}$ and $\underline{F}$ denote respectively the values of the primal and dual solutions.

This gap can possibly be reduced by decreasing one λ_i, for which $k_i > 1$, to its previous value in dual ascent. This creates slacks on some constraints in $S(\underline{\lambda})$ and the dual ascent procedure can be applied again. If an improved dual solution is obtained, a new primal solution is then derived. The procedure is iterated until either an optimal solution has been found or no further improvement is possible. In the latter case, an implicit enumeration scheme on the variables y_j is undertaken. The whole procedure, known as DUALOC, has been proposed by Erlenkotter [1978] and seems to be the most efficient approach for solving the UFLP. Further refinements by Körkel [1989] makes it capable of dealing with large-sized problems.

Lagrangean relaxation offers another way to tackle the problem. Cornuejols, Fisher & Nemhauser [1977a] suggest dualizing the constraints (6.2), which yields the following relaxed problem:

$$F(\underline{\lambda}) = \min_{\underline{y}, X} \sum_{i=1}^{n} \sum_{j=1}^{m} c_{ij} x_{ij} + \sum_{j=1}^{m} f_j y_j + \sum_{i=1}^{n} \lambda_i (1 - \sum_{j=1}^{m} x_{ij})$$

$$= \sum_{i=1}^{n} \lambda_i + \sum_{i=1}^{n} \sum_{j=1}^{m} (c_{ij} - \lambda_i) x_{ij} + \sum_{j=1}^{m} f_j y_j$$

s. t. $\quad 0 \leq x_{ij} \leq y_j, \quad i = 1 \cdots n, \ j = 1 \cdots m$

$\qquad y_j \in \{0, 1\}, \qquad j = 1 \cdots m.$

The variables x_{ij} can be eliminated by simple inspection: $x_{ij} = 0$ when $c_{ij} - \lambda_i$ is positive and $x_{ij} = y_j$ when $c_{ij} - \lambda_i$ is negative. Therefrom, we obtain the condensed form of the relaxed problem:

$$F(\underline{\lambda}) = \sum_{i=1}^{n} \lambda_i - \sum_{j=1}^{m} \Big[\sum_{i=1}^{n} (\lambda_i - c_{ij})^+ - f_j \Big] y_j$$

$$= \sum_{i=1}^{n} \lambda_i - \sum_{j=1}^{m} \rho_j(\underline{\lambda}) y_j \tag{6.11}$$

s. t. $y_j \in \{0, 1\}, \quad j = 1 \cdots m.$

An optimal solution to this problem is obtained by setting $y_j = 1$ if $v_j \in S^+(\underline{\lambda})$ where $\{v_j \in W : \rho_j(\underline{\lambda}) > 0\} \subseteq S^+(\underline{\lambda}) \subseteq \{v_j \in W : \rho_j(\underline{\lambda}) \geq 0\}$ and $y_j = 0$ if $v_j \notin S^+(\underline{\lambda})$. Clearly,

$$F_D \equiv \max_{\underline{\lambda}} F(\underline{\lambda}) \tag{6.12}$$

is a lower bound on F^*. The problem (6.12) is the Lagrangean dual of the UFLP relative to the relaxation of the constraints (6.2). Since the constraints (6.4) can be replaced by $0 \leq y_j \leq 1$ in the relaxed problem, the Lagrangean dual does not improve upon the linear programming relaxation. Solving (6.12) is generally performed by means of subgradient optimization techniques.

Despite the better performance of DUALOC, the Lagrangean approach has the advantage that any vector $\underline{\lambda}$ of $\mathbb{R}^n$ is feasible and yields a lower bound on F^*. For any $S \subseteq W$, one may set $\lambda_i = \min_{v_j \in S} c_{ij}$ for $i = 1 \cdots n$ and thus derive a dual solution from a primal solution. Accordingly, given a solution to the UFLP obtained by applying a heuristic, one may compute (6.11) for this solution. By coupling these primal and dual heuristics, one may get an approximate solution of the dual. Alternative methods for solving the UFLP are reviewed in Cornuéjols, Nemhauser & Wolsey [1990].

The Lagrangean approach also suggests an interesting economic interpretation. Instead of having a central agency deciding on the location of the facilities and the allocation of the clients, it is assumed that each potential facility j acts as a profit-maximizing agent. Now one looks for a price system that would sustain the optimal solution to the UFLP via decentralized decision-making. In (6.11), λ_i is to be interpreted as the revenue generated when the demand at v_i is met. The amount $(\lambda_i - c_{ij})^+$ measures the profit earned by facility j in supplying clients at v_i, so that $\rho_j(\underline{\lambda})$ is the total profit made by this facility at the prices corresponding to the λ_i (remember that clients at v_i have a fixed requirement w_i, which implies that λ_i/w_i is the price prevailing there). Given $\underline{\lambda}$, the optimal decision for facility j is to operate when $\rho_j(\underline{\lambda})$ is positive and to supply those clients for whom $\lambda_i - c_{ij}$ is positive, and not to operate when $\rho_j(\underline{\lambda})$ is negative. When $\rho_j(\underline{\lambda}) = 0$ the question is left open. For a given $\underline{\lambda}$ and the corresponding set $S^+(\underline{\lambda})$ of open facilities, clients at v_i will raise λ_i if they are not supplied and lower λ_i when more than one facility is willing to supply them. Even if there is a single facility j that wishes to supply v_i, there is an incentive for the corresponding clients to reduce λ_i either to c_{ij} or to a value such that $\rho_j(\underline{\lambda}) = 0$. A spatial equilibrium is reached at $\underline{\lambda}^*$ when clients at each vertex are supplied by exactly one facility in $S^+(\underline{\lambda}^*)$ and when facilities in $S^+(\underline{\lambda}^*)$ make zero profits. In this case, $F(\underline{\lambda}^*) = F_D = F^*$ and locational decisions can be decentralized. When $F_D < F^*$, then the optimal solution to the UFLP cannot be sustained by a price system based on clients' locations. The empirical finding that $F^* - F_D$ is equal to zero in many practical instances suggests that indivisibilities in locational decisions do not often preclude the competitive mechanism to sustain the optimal configuration of facilities.

6.3. Extensions

6.3.1. The lock-box problem

The decision maker may face a constraint on the number of facilities to be open. In this case, the constraint

$$\sum_{j=1}^{m} y_j \leq K$$

must be appended to (6.1)–(6.4). The resulting problem, which combines the features of the UFLP and of the K-median problem, is known as the lock-box problem [Cornuéjols, Fisher & Nemhauser, 1977a]. Most of the heuristics and exact algorithms described above can be readily adapted.

6.3.2. Capacity constraints

In the UFLP, the capacity of a facility is endogenous and is equal to the sum of the demands allocated to that facility. When existing facilities are already operated, there are limitations on the demands that can be met at the corresponding sites. It may also happen that only standardized equipments are available, which implies capacity constraints on existing and new facilities, as in the case of power stations or computers. The following constraints must then be added to the UFLP:

$$\sum_{i=1}^{n} w_i x_{ij} \leq \overline{w}_j y_j, \quad j = 1 \cdots m \tag{6.13}$$

where $\overline{w}_j$ is the maximal capacity of the potential (or existing) facility at v_j. The resulting problem is known as the *capacitated facility location problem* (in short CFLP), which proves to be much more difficult to solve than the UFLP. The exact procedures are of the branch-and-bound type; they use different lower bounds on the objective function that are obtained by different relaxations. As the number of constraint sets increases, the possible relaxations become numerous: Cornuéjols, Sridharam & Thizy [1991] review 41 relaxations and find out that they yield only 7 genuinely different bounds.

The first approach, suggested by Geoffrion & McBride [1978] and Nauss [1978], consists of a Lagrangean relaxation of the constraints (6.2) with the multipliers λ_i.

The relaxed problem can be decomposed into m subproblems.

$$\rho_j(\underline{\lambda}) = \max_{\underline{x}_j} \sum_{i=1}^{n}(\lambda_i - c_{ij})x_{ij} - f_j$$

$$\text{s. t.} \quad \sum_{i=1}^{n} w_i x_{ij} \le \overline{w}_j$$

$$0 \le x_{ij} \le 1, \quad i = 1 \cdots n$$

which are continuous knapsack problems that can be solved in $O(|V|\log|V|)$ by ranking the vertices v_i by decreasing order of $(\lambda_i - c_{ij})/w_i$ and allocating them to v_j until the corresponding capacity constraint is tight. This decomposition also yields the master problem

$$F_1(\underline{\lambda}) = \min_{\underline{y}} \sum_{i=1}^{n} \lambda_i - \sum_{j=1}^{m} \rho_j(\underline{\lambda}) y_j \tag{6.14}$$

$$\text{s. t.} \quad y_j \in \{0, 1\}, \quad j = 1 \cdots m.$$

whose optimal solution can be found by inspection. The Lagrangean dual problem $F_1 = \max_\lambda F_1(\underline{\lambda})$, which can be solved by subgradient optimization techniques, does not improve upon the linear programming relaxation investigated by Guignard & Spielberg [1979]. Moreover, the solution to (6.14) does not always enable one to construct a primal-feasible solution since the sum of the capacities of the open facilities may be smaller than the total demand. To avoid this difficulty, Nauss [1978] has proposed to add a total demand constraint to the CFLP:

$$\sum_{j=1}^{m} \overline{w}_j y_j \ge \sum_{i=1}^{n} w_i. \tag{6.15}$$

Using the Lagrangean relaxation described above, one gets the same subproblems, but the master problem is now a knapsack problem. This implies that the solution F_2 of the Lagrangean dual is a tighter bound: $F_1 \le F_2 \le F^*$. Since knapsack problems can be solved efficiently [see, e.g. Martello & Toth, 1990], the bound F_2 can be obtained by subgradient optimization techniques.

Another approach, proposed by Van Roy [1986], consists in dualizing the capacity constraints (6.12). The relaxed problem is now an UFLP. The method developed by Van Roy for solving the Lagrangean dual and getting the bound F_3 is based on the observation that, for any subset of open facilities, the resulting subproblem is a transportation problem. The optimal dual solutions of this transportation problem are used as Lagrangean multipliers to generate a new UFLP. The algorithm incorporates these alternative steps. When it ceases to make progress, the solutions to both the UFLP and transportation problem allow one to generate new cuts for a Benders master problem. The solution to this problem gives a new seed for the alternating procedure.

The bound F_3 is tighter than the one given by the linear programming relaxation. It is not dominated by F_2, but Cornuéjols, Sridharam & Thizy [1991] report computational experiments where F_2 is consistently better than F_3. To improve upon F_3, Van Roy [1986] uses the CFLP augmented with the constraint (6.15). Using the same relaxation as above yields a UFLP in which (6.15) is in turned dualized. The bound F_4 obtained in this way requires a substantial computational effort but it dominates F_2 and is extremely tight. To sum-up, we have: $F_1 \leq \max\{F_2, F_3\} \leq F_4 \leq F^*$.

Heuristic methods for solving the CFLP can be found in Domschke & Drexl [1985] and Jacobsen [1983].

Results about the polyhedral structure of the CFLP are not many. Leung & Magnanti [1989] have investigated the special case where all capacities are equal, i.e., $\overline{w}_j = \overline{w}$ for all j, while Aardal [1992] has tackled the general case. She proves that the dimension of the polytope Π^C of the CFLP is $m \cdot n + n - m$. Next, she gives conditions under which the constraints of the CFLP are facet-defining. Let $w(V) = \sum_{i=1}^{n} w_i$.

Theorem 6.11.

   (i) *For every* $j = 1 \cdots m$, *if* $\overline{w}_j < w(V)$, *then* $\sum_{i=1}^{n} w_i x_{ij} \leq \overline{w}_j y_j$ *is a facet of* Π^C.

   (ii) *For every* $i = 1 \cdots n$ *and* $j = 1 \cdots m$, *if* $\sum_{k=1}^{m} \overline{w}_k - \max_{k=1 \cdots m} \overline{w}_k > w(V)$ *and* $w_i < \overline{w}_j$, *then* $x_{ij} \leq y_j$ *is a facet of* Π^C.

   (iii) *For every* $j = 1 \cdots m$, $y_j \leq 1$ *is a facet of* Π^C.

   (iv) *For every* $i = 1 \cdots n$ *and* $j = 1 \cdots m$, $x_{ij} \geq 0$ *is a facet of* Π^C.

Aardal [1992] also reports various classes of nontrivial facets and valid inequalities. In the special case where $\overline{w}_j = \overline{w}$ for all j, $\overline{w}$ being an integer, Leung & Magnanti [1989] have introduced the *residual capacity* (RC) inequalities

$$\sum_{i \in I} \sum_{j \in J} w_i x_{ij} - r \sum_{j \in J} \leq w(V) - r \left\lceil \frac{w(V)}{\overline{w}} \right\rceil$$

with $I \subseteq V, J \subseteq W$, and $r = w(V) \pmod{\overline{w}}$. They prove the following result:

Theorem 6.12.

   (i) *(RC) is a valid inequality for the CFLP.*

   (ii) *If* $|J| \geq \lceil w(V)/\overline{w} \rceil$ *and* $1 \leq r \leq \overline{w} - 1$, *(RC) is a facet of* Π^C.

6.3.3. Nonlinear cost functions

Assume that a facility operates at v_j. Let q_j be the corresponding output, $C_j(q_j)$ the production cost function, q_{ij} the quantity of output shipped to client i, and t_{ij} the unit transportation cost between v_i and v_j. The problem is to supply the clients at minimum production and transportation costs. This can be formulated as follows:

$$\min \sum_{i=1}^{n} \sum_{j=1}^{m} t_{ij} q_{ij} + \sum_{j=1}^{m} C_j(q_j) \tag{6.16}$$

$$\text{s. t.} \quad \sum_{j=1}^{m} q_{ij} = w_i, \quad i = 1 \cdots n \tag{6.17}$$

$$\sum_{i=1}^{n} q_{ij} = q_j, \quad j = 1 \cdots n \tag{6.18}$$

$$q_{ij} \geq 0, \quad i = 1 \cdots n, \ j = 1 \cdots m. \tag{6.19}$$

If $C_j(q_j) = f_j + c_j q_j$ when $q_j > 0$ and $C_j(0) = 0$, the above problem can be rewritten as a UFLP: for each site v_j, we introduce a variable y_j such that $y_j = 1$ when $q_j > 0$ and $y_j = 0$ when $q_j = 0$; we also set $x_{ij} = q_{ij}/w_i$; finally, q_j can be replaced by $\sum_{i=1}^{n} w_i x_{ij}$.

As observed by Feldman, Lehrer & Ray [1966], concave production cost functions can be easily dealt with. Specifically, $C_j(q_j)$ can be approximated by the lower envelope of a finite set of affine functions $C_{jh}(q_j) = f_{jh} + c_{jh} q_j$, with $h = 1 \cdots H_j$. The approximate problem can then be solved as a UFLP provided that each site v_j is split into H_j sites endowed with the production cost functions $C_{j1} \cdots C_{jH_j}$. It is readily verified that an optimal solution always exists in which a single function C_{jh} is chosen at each site where production takes place. In principle, $C_j(q_j)$ can be approximated to any required precision, but the approach is limited by the number of pseudo-sites generated in this way.

Non-concave production cost functions can also be considered. To this end, we approximate $C_j(q_j)$ by the affine function $C_{jh}(q_j)$ in the interval $[\underline{q}_h, \overline{q}_h]$. This can be formulated as follows:

$$C_j(q_j) = \sum_{h=1}^{H_j} C_{jh}(q_j) y_{jh}$$

$$\text{s. t.} \quad \sum_{h=1}^{H_j} y_{jh} \leq 1$$

$$\underline{q}_h y_{jh} \leq q_j \leq \overline{q}_h y_{jh}, \quad h = 1 \cdots H_j$$

$$y_{jh} \in \{0, 1\}, \quad h = 1 \cdots H_j.$$

Replacing now $C_j(q_j)$ by this expression in (6.16)–(6.19), we get a CFLP in which the capacity constraints are given by

$$\underline{q}_h y_{jh} \leq \sum_{i=1}^{n} w_i x_{ijh} \leq \overline{q}_h y_{jh}, \quad h = 1 \cdots H_j, \ j = 1 \cdots m$$

and which is augmented with the constraints $\sum_{h=1}^{H_j} y_{jh} \leq 1$.

6.3.4. The economic lot-sizing problem

The UFLP can also be useful in describing other problems. An example is provided by the *economic lot-sizing problem* in which demand arises and production occurs at the same location, but are distributed over a given number T of periods. It is assumed that the demand requirement w_t in period $t(= 1 \cdots T)$

is known beforehand and must be met at the beginning of the period. Production can take place at the beginning of each period and the production cost is $C_t(q_t) = f_t + c_t q_t$ if $q_t > 0$ and zero otherwise. Quantities produced but not sold are held in inventories; let s_t be the inventory at the end of period t and h_t the unit storage cost in period t. The problem of the firm is to establish a production schedule that minimizes the production and inventory costs subject to the restriction that it meets all the demand requirements on time. It is well known that this problem can be formulated as follows:

$$\min \sum_{t=1}^{T}(c_t q_t + f_t y_t) + \sum_{t=1}^{T} h_t s_t$$

$$\text{s. t. } s_{t-1} + q_t = w_t + s_t, \qquad t = 1 \cdots T$$
$$0 \le q_t \le M y_t, \qquad t = 1 \cdots T$$
$$y_t \in \{0, 1\}, \qquad t = 1 \cdots T$$
$$s_0 = s_T = 0$$

where $M \ge \sum_{t=1}^{T} w_t$. This problem can be solved in $O(T \log T)$ by dynamic programming (see Wagelmans, Van Hoesel & Kolen [1992] and Federgruen & Tzur [1991]). This bound can be refined to $O(T)$ when the variable production costs verify the conditions $c_t + h_t \ge c_{t+1}$, $t = 1 \cdots T$.

Now rewrite this problem as a UFLP whose underlying network is a path with $T + 1$ vertices. Let $q_{tt'}$ denote the quantity produced at t and sold at $t' \ge t$. Set $c_{tt'} = c_t + \sum_{\tau=t}^{t'} h_\tau$ if $t' > t$ and $c_{tt} = c_t$. An equivalent formulation of the lot-sizing problem is then as follows:

$$\min \sum_{t=1}^{T} \sum_{t'=t}^{T} c_{tt'} q_{tt'} + \sum_{t=1}^{T} f_t y_t$$

$$\text{s. t. } \sum_{t=1}^{t'} q_{tt'} = w_{t'}, \qquad t' = 1 \cdots T$$
$$0 \le q_{tt'} \le M y_t, \qquad t = 1 \cdots T, \ t' = t \cdots T$$
$$y_t \in \{0, 1\}, \qquad t = 1 \cdots T$$

which has the structure of the UFLP in which the path linking the vertices is oriented, i.e. $q_{tt'}$ is not defined for $t' < t$.

6.4. The UFLP under alternative spatial price policies

Consider a region containing n local markets with a demand function $w_i(p)$, $i = 1 \cdots n$, for a given commodity. It is assumed that $w_i(p)$ is decreasing and continuous in the delivered price charged at v_i. The commodity is made available within the region by a firm that may operate several plants to be located in W, with $C(q_j) = f_j + c_j q_j$ if $q_j > 0$ and $C_j(0) = 0$, where f_j and c_j denote respectively the fixed and marginal production costs at $v_j \in W$. Let t_{ij} be the unit transportation cost between v_i and v_j. The problem of the firm is to determine the price charged at each local market, the level of production at each open facility, and the distribution system that maximize its profit.

6.4.1. Delivered pricing

Let p_i be the delivered price at market v_i; $y_j = 1$ if some production is undertaken at v_j and $y_j = 0$ otherwise; x_{ij} is the fraction of the demand at v_i supplied from v_j. The profit of the firm is equal to its revenue, minus transportation costs and production costs:

$$
\begin{aligned}
\pi(\underline{p}, \underline{y}, X) &= \sum_{i=1}^{n} \sum_{j=1}^{m} p_i w_i(p_i) x_{ij} - \sum_{i=1}^{n} \sum_{j=1}^{m} t_{ij} w_i(p_i) x_{ij} - \\
&\quad - \sum_{j=1}^{n} \left[f_j y_j + c_j \sum_{i=1}^{n} w_i(p_i) x_{ij} \right] \\
&= \sum_{i=1}^{n} \sum_{j=1}^{m} (p_i - t_{ij} - c_j) w_i(p_i) x_{ij} - \sum_{j=1}^{m} f_j y_j \\
&= \sum_{i=1}^{n} \sum_{j=1}^{m} B_{ij}(p_i) x_{ij} - \sum_{j=1}^{m} f_j y_j,
\end{aligned}
$$

where $B_{ij}(p_i)$ can be interpreted as the revenue obtained when supplying the local market at v_i from v_j at the delivered price p_i. The problem of the firm can then be formulated as follows:

$$
\max_{\underline{p}, \underline{y}, X} \pi(\underline{p}, \underline{y}, X) \tag{6.20}
$$

$$
\text{s. t.} \quad \sum_{j=1}^{m} x_{ij} \le 1, \qquad i = 1 \cdots n \tag{6.21}
$$

$$
0 \le x_{ij} \le y_j, \qquad i = 1 \cdots n, \ j = 1 \cdots m \tag{6.22}
$$

$$
y_j \in \{0, 1\}, \qquad j = 1 \cdots m. \tag{6.23}
$$

Consider first the case of discriminatory pricing, where the firm sets market-specific prices. For each market and each potential location, we solve the problem

$$
B_{ij}^* = \max_{p_i} B_{ij}(p_i).
$$

Replacing in the above formulation, we obtain

$$
\pi_D = \max_{\underline{y}, X} \sum_{i=1}^{n} \sum_{j=1}^{m} B_{ij}^* x_{ij} - \sum_{j=1}^{m} f_j y_j
$$

subject to (6.21)–(6.23). By introducing a fictitious facility, this problem is amenable to the UFLP [see Erlenkotter, 1977, and Hansen & Thisse, 1977].

The case of uniform delivered pricing is obtained by appending the constraint

$$p_i = p, \quad i = 1 \cdots n$$

to (6.20)–(6.23). Observe that for any price level p, the problem results in a UFLP. Define $\pi(p) = \max_{y, X} \pi(p, y, X)$. The function $\pi(p)$ is neither concave nor convex, and generally exhibits local maxima and points of nondifferentiability; however, it is Lipschitzian. Therefore, the solution pertains to global optimization. An algorithm has been proposed by Hansen, Thisse & Hanjoul [1981] and further refined by Hanjoul, Hansen, Peeters & Thisse [1990]. This one is based on the following principles. Let $\alpha_i = \min\{p; w_i(p) = 0\}$. Clearly, the range of admissible prices can be restricted to a $[\min_{\substack{v_i \in V \\ v_j \in W}} c_j + t_{ij}, \max_{v_i \in V} \alpha_i]$. Let $[p^-, p^+]$ be a sub-interval. For any $i = 1 \cdots n$, consider a function $\psi(p)$ which dominates $\pi(p)$ over $[p^-, p^+]$ and which is concave over this domain; moreover, $\psi(p^-) = \pi(p^-)$ and $\psi(p^+) = \pi(p^+)$. The function ψ is then used in a branch-and-bound scheme in the following way. If $\psi(\tilde{p}) \equiv \max\{\psi(p); p \in [p^-, p^+]\} \leq \tilde{\pi}$, where $\tilde{\pi}$ is the best known value of π, then $[p^-, p^+]$ can be discarded. If $\psi(p)$ is monotonically increasing or decreasing, then $[p^-, p^+]$ possesses no interior solution and the interval can be discarded. Otherwise, compute $\pi(\tilde{p})$ and update $\tilde{\pi}$ if necessary. If $p^+ - p^- > \eta$, where η is some given tolerance, $[p^-, p^+]$ is subdivided into two subintervals $[p^-, \tilde{p}]$ and $[\tilde{p}, p^+]$.

More generally, the firm may partition the set of markets into K zones and set the same uniform delivered price $p_k, k = 1 \cdots K$, in each zone. (Note that the above two cases correspond to $K = n$ and $K = 1$.) The problem can now be written as follows:

$$\max \sum_{i=1}^{n} \sum_{j=1}^{m} \sum_{k=1}^{K} B_{ij}(p_k) x_{ijk} - \sum_{j=1}^{m} f_j y_j$$

$$\text{s. t.} \sum_{j=1}^{m} \sum_{k=1}^{K} x_{ijk} \leq 1, \quad i = 1 \cdots n$$

$$x_{ijk} \leq y_j, \qquad i = 1 \cdots n, \ j = 1 \cdots m, \ k = 1 \cdots K$$

$$x_{ijk} \in \{0, 1\} \text{ and } y_j \in \{0, 1\}$$

where $x_{ijk} = 1$ when a facility at v_j supplied the market v_i in the kth area. Clearly, for any given vector of prices one gets a UFLP. The algorithm for the uniform delivered pricing case can then be adapted [see Hansen, Peeters & Thisse, 1992].

6.4.2. Mill pricing

Assume first that all facilities charge the same mill price p_M so that

$$p_i = \min_{\{v_j; y_j = 1\}} (p_M + t_{ij}), \quad i = 1 \cdots n.$$

Under this price policy, the transportation costs are paid by the clients who decide which facility to patronize. Some discrepancies may arise between the clients' preferences (clients at v_i visit the facility j for which $t_{ij} = \min_{\{v_k; y_k=1\}} t_{ik}$)

and the firm's preferences (clients at v_i should visit the facility j for which $B_{ij}(p_M + t_{ij}) = \max_{\{v_k; y_k=1\}} B_{ik}(p_M + t_{ik})$). Therefore, we must add the following constraint:

$$\sum_{\{v_k; t_{ik} \leq t_{ij}\}} x_{ik} \geq y_j, \quad i = 1 \cdots n, \; j = 1 \cdots m \text{ and } w_i(p_M + t_{ij}) > 0. \quad (6.24)$$

For any given p, the problem is equivalent to a UFLP augmented with (6.24) (see Hanjoul & Peeters [1987] for a solution method). Again, the algorithm described for the uniform delivered pricing case can be adapted [Hanjoul, Hansen, Peeters & Thisse, 1990].

Consider now the case where each facility can set its own mill price p_{Mj}. The delivered price at the ith market is then

$$p_i = \min_{\{v_j; y_j=1\}} (p_{Mj} + t_{ij}), \quad i = 1 \cdots n,$$

and the additional constraints are

$$\sum_{\{v_k; \, p_{Mk}+t_{ik} \leq p_{Mj}+t_{ij}\}} x_{ik} \geq y_j,$$
$$i = 1 \cdots n, \; j = 1 \cdots m \text{ and } w_i(p_{Mj} + t_{ij}) > 0.$$

The objective function now exhibits sharp discontinuities and no solution method for this problem is known yet.

7. Conclusion

This chapter is far from covering everything that has been written in network location theory. There are three new lines of research we would like to mention. First, as seen in 4.1, equity considerations may lead the decision maker to elect the center for placing a public facility. However, such a location may still be characterized by large discrepancies in accessibility. Various indices have been proposed to measure inequality, including the variance, the Lorenz and Gini indices (see Mulligan [1991] for a discussion of these measures in a locational context). Any such index can be used to define a particular objective function capturing distributional justice aspects. A systematic comparison of the resulting locations would allow one to understand the implications of a specific objective function when locating a facility.

Second, in most multifacility location problems, the allocation side is specified in a way such that, at the optimum, the demand of a client is met by a single facility, typically the nearest one (see 3.5). This is consistent with the management of a delivered service where all the phases of the process are controlled by the decision maker. The nearest facility rule then follows from the minimization of the over-all cost of the system. On the other hand, when a fixed service is considered, clients often decide which facility to patronize. Empirical evidence collected by regional scientists suggests that a client distributes his trips among the facilities according to a gravity-like pattern. This is in accord with modern discrete choice theory in which a client's utility is modeled by a stochastic utility such as $- d_{ij} + \varepsilon_{ij}$, where ε_{ij} is a random variable with zero mean [see, e.g. Ben-Akiva & Lerman, 1985]. In an early contribution, Leonardi [1983] has studied a generalization of the K-median problem in which the allocation variables x_{ij} are given by $A_j e^{-\beta d_{ij}} / \sum_{v_k \in S} A_k e^{-\beta d_{ik}}$, where A_k is a measure of the attractiveness of a facility at v_k and β the coefficient of spatial friction. We thus obtain a nonlinear mixed program whose solutions can be determined by the heuristics used for solving the K-median problem. More work is called for here. An excellent introduction is provided by Fotheringham & O'Kelly [1989].

Third, and last, throughout this chapter we have considered only the case of desirable facilities in that some travel costs are to be minimized. However, some facilities may be undesirable or semi-desirable. In the former case, although necessary to society as a whole, such facilities generate nuisances to the individuals who are near them. In particular, they may have a negative impact on property values and/or may lower the quality of life through different sorts of pollution. The latter case is even more complex. As clients, individuals like to be close to the facilities; as residents, they like to be further away because of the local inconveniences associated with the facilities. Modeling the location such facilities gives rise to several conceptual problems [see Erkut & Neuman, 1989]. Clearly, there is here a whole set of problems that deserves the attention of location scientists.

Acknowledgement

We thank Arie Tamir and two referees for their comments and suggestions. The first author also thanks the European Institute for Advanced Studies in Management for financial support.

References

Several books and papers offer surveys of network location theory. An early reference is Handler & Mirchandani [1979]. Mirchandani & Francis [1990] is a collection of papers devoted to discrete location models, but a few chapters deal with network problems. Tansel, Francis & Lowe [1983] is a technically oriented survey, while Hansen, Labbé, Peeters & Thisse [1987a] take a broader perspective, but consider only the location of a single facility. Other domains of location theory, in particular continuous facility location, are covered in Love, Morris & Wesolowsky [1988] and Francis, McGinnis & White [1992]. Hansen, Peeters & Thisse [1983] and Hansen, Labbé, Peeters & Thisse [1987b] put forward the economic interpretation of the models. Finally, more specific problems are reviewed by Erkut & Neuman [1989] for obnoxious facility location and by Friesz, Miller & Tobin [1988] for competitive location problems; Brandeau & Chiu

[1989] give an overview of what appears to them as the canonical problems in the field.

Aardal, K. (1992). *On the Solution of One and Two-Level Capacitated Facility Location Problems by the Cutting Plane Approach*, Unpublished Ph.D. Thesis, Université Catholique de Louvain, Faculté des Sciences Appliquées.

Ahn, S., C. Cooper, G. Cornuéjols and A.M. Frieze (1988). Probabilistic analysis of a relaxation for the k-median problem, *Math. Oper. Res.* 13, 1–31.

Balas, E. (1982). A class of location, distribution and scheduling problems: modeling of solution methods, *Rev. Belge Statistique, Inf. Rech. Opér.*, 82, 36–69.

Balas, E., and S.M. Ng (1989). On the set covering polytope: II. Lifting the facets with coefficients in {0, 1, 2}, *Math. Program.*, 45, 1–20.

Balinski, M.L. (1966). On finding integer solutions to linear programs, in: *Proc. IBM Scientific Symp. on Combinatorial Problems*, White Plains, NY, IBM Data Processing Division.

Bandelt, H.-J. (1985). Networks with Condorcet solutions, *Eur. J. Oper. Res.* 20, 314–326.

Bandelt, H.-J. (1990). Centroids and medians of finite metric spaces, mimeo.

Bandelt, H.-J., and M. Labbé (1986). How bad can a voting location be?, *Social Choice and Welfare* 3, 125–145.

Barthélemy, J.-P. (1983). Caractérisation médiane des arbres, *Ann. Discrete Math.* 17, 39–46.

Batta, R. (1988). Single server queueing-location models with rejection, *Transp. Sci.* 22, 209–216.

Batta, R. (1989). A queueing-location model with expected service time dependent queueing disciplines, *Eur. J. Oper. Res.* 39, 192–205.

Batta, R., and U.S. Palekar (1988). Mixed planar/network facility location problems, *Comput. Oper. Res.* 15, 61–67.

Beckmann, M.J., and J.-F. Thisse (1986). The location of production activities, in: E.S. Mills and P. Nijkamp (eds.), *Handbook of Regional and Urban Economics*, North-Holland, Amsterdam, pp. 21–95.

Ben-Akiva, M., and S. Lerman (1985). *Discrete Choice Analysis: Theory and Applications to Predict Travel Demand*, MIT Press, Cambridge, MA.

Berge, C. (1966). *Espaces topologiques. Fonctions multivoques*, Dunod, Paris.

Berge, C. (1967). *Théorie des graphes et ses applications*, Dunod, Paris.

Berman, O., S.S. Chiu, R.C. Larson, A.R. Odoni and R. Batta (1990). Location of mobile units in a stochastic environment, in: P.B. Mirchandani and R.L. Francis (eds.), *Discrete Location Theory*, Wiley, New York, NY, pp. 503–549.

Berman, O., R.C. Larson and S.S. Chiu (1985). Optimal server location on a network operating as an M/G/1 queue, *Oper. Res.* 33, 746–771.

Berman, O., and D. Simchi-Levi (1986). Minisum location of a traveling salesman, *Networks* 16, 239–254.

Berman, O., and D. Simchi-Levi (1988a). Finding the optimal a priori tour and location of a traveling salesman with non homogeneous customers, *Transp. Sci.* 22, 148–154.

Berman, O., and D. Simchi-Levi (1988b). Minisum location of a travelling salesman on simple networks, *Eur. J. Oper. Res.* 36, 241–250.

Berman, O., D. Simchi-Levi and A. Tamir (1988). The minimax multistop location problem on a tree, *Networks* 18, 39–49.

Bertsimas, D.J. (1989). Traveling salesman facility location problems, *Transp. Sci.* 23, 184–191.

Bilde, O., and J. Krarup (1977). Sharp lower bounds and efficient algorithms for the simple plant location problem, *Ann. Discrete Math.* 1, 79–97.

Brandeau, M.L., and S.S. Chiu (1989). An overview of representative problems in location research, *Manage. Sci.* 35, 645–674.

Brandeau, M.L., and S.S. Chiu (1990). A unified family of queueing location models, *Oper. Res.* 38, 1034–1044.

Buckley, F., Z. Miller and P.J. Slater (1981). On graphs containing a given graph as center, *J. Graph Theory* 5, 427–434.

Chen, M.L., R.L. Francis and T.J. Lowe (1988). The 1-center problem: Exploiting block structure, *Transp. Sci.* 22, 259–269.

Chen, M.L., R.L. Francis, J.F. Lawrence, T.J. Lowe and S. Tufekci (1985). Block-vertex duality and the one-median problem, *Networks* 15, 395–412.

Chiu, S.S. (1986a). A dominance theorem for the stochastic queue median problem, *Oper. Res.* 34, 942–944.

Chiu, S.S. (1986b). Optimal M/G/1 server location on a tree network with continuous link demands, *Comput. Oper. Res.* 13, 653–669.

Chiu, S.S. (1987). The minisum location problem on an undirected network with continuous link demands, *Comput. Oper. Res.* 14, 369–383.

Chiu, S.S., O. Berman and R.C. Larson (1985). Locating a mobile server queueing facility on a tree network, *Manage. Sci.* 31, 764–772.

Cho, D.C., E.L. Johnson, M.W. Padberg and M.R. Rao (1983). On the uncapacitated plant location problem. I. Valid inequalities and facets, *Math. Oper. Res.* 8, 579–589.

Cho, D.C., M.W. Padberg and M.R. Rao (1983). On the uncapacitated plant location problem. II. Facets and lifting theorems, *Math. Oper. Res.* 8, 590–612.

Christofides, N. (1975). *Graph Theory: An Algorithmic Approach*, Academic Press, New York, NY.

Christofides, N., and P. Viola (1971). The optimal location of multi-centers on a graph, *Oper. Res. Q.* 22, 145–154.

Christofides, N., and J.E. Beasley (1982). A tree search algorithm for the *p*-median problem, *Eur. J. Oper. Res.* 10, 196–204.

Church, R.L., and M.E. Meadows (1979). Location modelling utilizing maximum service distance criteria, *Geogr. Anal.* 11, 358–373.

Church, R.L., and C.S. ReVelle (1974). The maximal covering location problem, *Pap. Reg. Sci. Assoc.* 32, 101–118.

Cornuéjols, G., M.L. Fisher and G.L. Nemhauser (1977a). Location of bank accounts to optimize float: An analytic study of exact and approximate algorithms, *Manage. Sci.* 23, 789–810.

Cornuéjols, G., M.L. Fisher and G.L. Nemhauser (1977b). On the uncapacitated location problem, *Ann. Discrete Math.* 1, 163–177.

Cornuéjols, G., G.L. Nemhauser and L.A. Wolsey (1990). The uncapacitated facility location problem, in: P.B. Mirchandani and R.L. Francis (eds.), *Discrete Location Theory*, Wiley, New York, NY, pp. 119–171.

Cornuéjols, G., and A. Sassano (1989). On the 0, 1 facets of the set covering polytope, *Math. Program.*, 43, 45–55.

Cornuéjols, G., R. Sridharam and J.M. Thizy (1991). A comparison of heuristics and relaxations for the capacitated plant location problem, *Eur. J. Oper. Res.*, 50, 280-297.

Cornuéjols, G., and J.M. Thizy (1982). Some facets of the simple plant location polytope, *Math. Program.* 23, 50–74.

Cuninghame-Green, R.A. (1984). The absolute center of a graph, *Discrete Appl. Math.* 7, 275–283.

Dear, M. (1978). Planning for mental health care: a reconsideration of public facility location theory, *Int. Reg. Sci. Rev.* 3, 93–111.

Dearing, P.M. (1977). Minimax location problems with nonlinear costs, *J. Res. Nat. Bur. Stand.* 82, 65–72.

Dearing, P.M., and R.L. Francis (1974). A minimax location problem on a network, *Transp. Sci.* 8, 333–343.

Dearing, P.M., R.L. Francis and T.J. Lowe (1976). Convex location problems on tree networks, *Oper. Res.* 24, 628–642.

Demange, G. (1983). Spatial models of collective choice, in: J.-F. Thisse and H.G. Zoller (eds.), *Locational Analysis of Public Facilities*, North-Holland, Amsterdam, pp. 153–182.

Domschke, W., and A. Drexl (1985). *Location and Layout Planning: An International Bibliography*, Springer Verlag, Berlin.

Domschke, W., and A. Drexl (1985). ADD- and DROP- procedures for capacitated plant location with different capacities, *Eur. J. Oper. Res.* 21, 47-53.

Dyer, M.E., and A.M. Frieze (1985). A simple heuristic for the p-centre problem, *Oper. Res. Lett.* 3, 285–288.

Edgeworth, F.Y. (1883). The method of least squares, *Philos. Mag.* 16, 360–375.

Eiselt, H.A., and G. Laporte (1991). Locational equilibrium of two facilities on a tree, *RAIRO Rech. Opér.* 25, 5–17.

Eiselt, H.A., and G. Laporte (1993). The existence of equilibria in the 3–facility Hotelling model in a tree, *Transp. Sci.* 27, 39–43.

Eiselt, H.A., G. Laporte and J.-F. Thisse (1993). Competitive location models: a framework and bibliography, *Transp. Sci.* 27, 44–54.

Erkut, E., and S. Neuman (1989). Analytical models for locating undesirable facilities, *Eur. J. Oper. Res.* 40, 275–291.

Erkut, E., and B.C. Tansel (1989). On parametric medians of trees, Department of Finance and Management Science, University of Alberta.

Erlenkotter, D. (1977). Facility location with price-sensitive demands: private, public, and quasi-public, *Manage. Sci.* 24, 378–386.

Erlenkotter, D. (1978). A dual-based procedure for uncapacitated facility location, *Oper. Res.* 26, 992–1009.

Federgruen, A., and M. Tzur (1991). A simple forward algorithm to solve general dynamic lot sizing models with n periods in $O(n \log n)$ or $O(n)$ time, *Manage. Sci.* 37, 909–925.

Feldman, E., F.A. Lehrer and T.L. Ray (1966). Warehouse locations under continuous economies of scale, *Manage. Sci.* 12, 670–684.

Fisher, M.L. (1980). Worst-case analysis of heuristic algorithms, *Manage. Sci.* 26, 1–18.

Fotheringham, A.S., and M.E. O'Kelly (1989). *Spatial Interaction Models: Formulations and Specifications*, Kluwer, Dordrecht.

Francis, R.L. (1977). A note on a nonlinear minimax location problem in a tree network, *J. Res. Nat. Bur. Stand.* 82, 73–80.

Francis, R.L., L.F. McGinnis and J.A. White (1992). *Facility Layout and Location: An Analytical Approach*, Prentice Hall, Englewood Cliffs, NJ.

Frank, H. (1967). A note on a graph theoretic game of Hakimi's, *Oper. Res.* 15, 567–570.

Friedman, J. (1990). *Game Theory with Applications to Economics*, Oxford University Press, Oxford.

Friesz, T.L., T. Miller and R.L. Tobin (1988). Competitive network facility location models, *Pap. Reg. Sci. Assoc.* 65, 47–57.

Gabszewicz, J., and J.-F. Thisse (1992). Location, in: R. Aumann and S. Hart (eds.), *Handbook of Game Theory with Economic Applications*, North-Holland, Amsterdam, pp. 281–304.

Geoffrion, A.M., and R. McBride (1978). Lagrangian relaxation applied to facility location problems, *AIIE Trans.* 10, 40–47.

Goldman, A.J. (1971). Optimal center location in simple networks, *Transp. Sci.* 5, 212–221.

Goldman, A.J. (1972a). Approximate localization theorems for optimal facility placement, *Transp. Sci.* 6, 195–201.

Goldman, A.J. (1972b). Minimax location of a facility in a network, *Transp. Sci.* 6, 407–418.

Goldman, A.J., and C.J. Witzgall (1970). A localization theorem for optimal facility placement, *Transp. Sci.* 4, 406–409.

Guignard, M. (1980). Fractional vertices, cuts and facets of the simple plant location problem, *Math. Program. Study* 12, 150–162.

Guignard, M., and K. Spielberg (1979). A direct dual method for the mixed plant location problem with some side constraints, *Math. Program.* 17, 198–228.

Gülicher, H. (1987). Some early contributions to minisum location theory, Institut für Ökonometrie und Wirtschaftsstatistik, Universität München.

Hakimi, S.L. (1964). Optimum locations of switching centers and the absolute centers and medians of a graph, *Oper. Res.* 12, 450–459.

Hakimi, S.L. (1965). Optimum distribution of switching centers in a communication network and some related graph theoretic problems, *Oper. Res.* 13, 462–475.

Hakimi, S.L. (1983). On locating new facilities in a competitive environment, *Eur. J. Oper. Res.* 12, 29–35.

Hakimi, S.L., E.F. Schmeichel and J.G. Pierce (1978). On *p*-centers in networks, *Transp. Sci.* 12, 1–15.

Halfin, S. (1974). On finding the absolute and vertex center of a tree with distances, *Transp. Sci.* 8, 75–77.

Halpern, J. (1976). The location of a cent-dian convex combination on an undirected tree, *J. Reg. Sci.* 16, 237–245.

Halpern, J. (1978). Finding minimal center-median convex combination (cent-dian) of a graph, *Manage. Sci.* 24, 534–544.

Halpern, J. (1980). Duality in the cent-dian of a graph, *Oper. Res.* 28, 722–735.

Halpern, J., and O. Maimon (1982). Algorithms for the *m*-center problems: A survey, *Eur. J. Oper. Res.* 10, 90-99.

Handler, G.Y. (1973). Minimax location of a facility in an undirected tree graph, *Transp. Sci.* 7, 297–293.

Handler, G.Y., and P.B. Mirchandani (1979). *Location on Networks: Theory and Algorithms*, MIT Press, Cambridge, MA.

Handler, G.Y. (1990). *p*-center problems, in: P.B. Mirchandani and R.L. Francis (eds.), *Discrete Location Theory*, Wiley, New York, NY, pp. 305–347.

Hanjoul, P., P. Hansen, D. Peeters and J.-F. Thisse (1990). Uncapacitated plant location under alternative spatial price policies, *Manage. Sci.* 36, 41–57.

Hanjoul, P., and D. Peeters (1985). Comparison of two dual-based procedures for solving the *p*-median problem, *Eur. J. Oper. Res.* 20, 387–396.

Hanjoul, P., and D. Peeters (1987). A facility location problem with client's preference orderings, *Reg. Sci. Urban Econ.* 17, 451–473.

Hanjoul, P., and J.-F. Thisse (1984). The location of a firm on a network, in: A.J. Hughes-Hallet (ed.), *Applied Decision Analysis and Economic Behaviour*. M. Nijhoff, Den Haag, pp. 289–326.

Hansen, P., and M. Labbé (1988). Algorithms for voting and competitive location on a network, *Transp. Sci.* 22, 278–288.

Hansen, P., and M. Labbé (1989). The continuous *p*-median of a network, *Networks* 19, 595–606.

Hansen, P., M. Labbé and B. Nicolas (1991). The continuous center set of a network, *Discrete Appl. Math.* 30, 181–195.

Hansen, P., M. Labbé and J.-F. Thisse (1991). From the median to the generalized center, *RAIRO Rech. Opér.* 25, 73–86.

Hansen, P., M. Labbé, D. Peeters and J.-F. Thisse (1987a). Single facility location on networks, *Ann. Discrete Math.* 31, 113–146.

Hansen, P., M. Labbé, D. Peeters and J.-F. Thisse (1987b). Facility location analysis, *Fundam. Pure Appl. Econ.* 22, 1–70.

Hansen, P., D. Peeters and J.-F. Thisse (1983). Public facility location models: A selective survey, in: J.-F. Thisse and H.G. Zoller (eds.), *Locational Analysis of Public Facilities*, North-Holland, Amsterdam, pp. 223–262.

Hansen, P., D. Peeters and J.-F. Thisse (1992). Facility location under zonal pricing, GERAD, Université de Montréal.

Hansen, P., and F.S. Roberts (1991). An impossibility result in axiomatic location theory, RUTCOR, Rutgers University.

Hansen, P., and J.-F. Thisse (1977). Multiplant location for profit maximisation, *Environment and Planning A* 9, 63–73.

Hansen, P., and J.-F. Thisse (1981). Outcomes of voting and planning: Condorcet, Weber, and Rawls locations, *J. Public Econ.* 16, 1–15.

Hansen, P., J.-F. Thisse and P. Hanjoul (1981). Simple plant location under uniform delivered pricing, *Eur. J. Oper. Res.* 6, 94–103.

Hansen, P., J.-F. Thisse and R.E. Wendell (1986a). Efficient points on a network, *Networks* 16, 358–367.

Hansen, P., J.-F. Thisse and R.E. Wendell (1986b). Equivalence of solutions to network location problems, *Math. Oper. Res.* 11, 672–678.

Hassin, R., and A. Tamir (1991). Improved complexity bounds for location problems on the real line, *Oper. Res. Letters* 10, 395–402.

Hedetniemi, S.M., E.J. Cockayne and S.T. Hedetniemi (1981). Linear algorithms for finding the Jordan center and path center of a tree, *Transp. Sci.* 15, 98–114.

Hochbaum, D.S., and D.B. Shmoys (1985). A best possible heuristic for the k-center problem, *Math. Oper. Res.* 10, 180–184.

Holzman, R. (1990). An axiomatic approach to location on networks, *Math. Oper. Res.* 15, 553–563.

Hooker, J.N. (1986). Solving nonlinear single-facility network location problems, *Oper. Res.* 34, 732–743.

Hooker, J.N., R.S. Garfinkel and C.K. Chen (1991). Finite dominating sets for network location problems, *Oper. Res.* 39, 100–118.

Jacobsen, S.K. (1983). Heuristics for the capacitated plant location model, *Eur. J. Oper. Res.* 12, 253–261.

Johansen, L. (1982). On the status of the Nash type of noncooperative equilibrium in economic theory, *Scand. J. Econ.* 84, 421–441.

Johnson, D.S., and C.H. Papadimitriou (1985). Performance guarantees for heuristics, in: E.L. Lawler, J.K. Lenstra, A.H.G. Rinnooy Kan and D.B. Shmoys (eds.), *The travelling Salesman Problem*, Wiley, New York, NY, pp. 87–143.

Jordan, C. (1869). Sur les assemblages de lignes, *Z. Reine Angew. Math.* 70, 185–190.

Kariv, O., and S.L. Hakimi (1979a). An algorithmic approach to network location problems, I: The p-centers, *SIAM J. Appl. Math.* 37, 513–538.

Kariv, O., and S.L. Hakimi (1979b). An algorithmic approach to network location problems, II: The p-medians, *SIAM J. Appl. Math.* 37, 539–560.

Kirkpatrick, S., C.D. Gelatt and M.P. Vecchi (1983). Optimization by simulated annealing, *Science* 220, 671–680.

Kleinrock, L. (1975). *Queueing Systems. Vol. 1*, Wiley, New York, NY.

Kolen, A. (1983). Solving covering problems and the uncapacitated plant location problem on trees, *Eur. J. Oper. Res.* 12, 266–278.

Körkel, M. (1989). On the exact solution of large-scale simple plant location problems, *Eur. J. Oper. Res.* 39, 157–173.

Krarup, J., and P.M. Pruzan (1983). The simple plant location problem: Survey and synthesis, *Eur. J. Oper. Res.* 12, 36–81.

Kuehn, A.A., and M.J. Hamburger (1963). A heuristic program for locating warehouses, *Manage. Sci.* 9, 643–666.

Labbé, M. (1985). Outcomes of voting and planning in single facility location problems, *Eur. J. Oper. Res.* 20, 299–313.

Labbé, M., and S.L. Hakimi (1991). Market and locational equilibrium for two competitors, *Oper. Res.* 39, 749–756.

Labbé, M., J.-F. Thisse and R.E. Wendell (1991). Sensitivity analysis in minisum facility location problems, *Oper. Res.* 39, 961–969.

Laporte, G. (1988). Location-routing problems, in: B.L. Golden and A.A. Assad (eds.), *Vehicle Routing: Methods and Studies*, North Holland, Amsterdam, pp. 163–198.

Launhardt, W. (1882). Die Bestimmung des Zwechkmässigsten Standortes einer Gewerblichten Anlage, *Z. Ver. Dtsch. Ing.* 26, columns 105–116.

Lederer, P.J., and J.-F. Thisse (1990). Competitive location on networks under delivered pricing, *Oper. Res. Lett.* 9, 147–153.

Leonardi, G. (1983). The use of random-utility theory in building location-allocation models, in: J.-F. Thisse and H.G. Zoller (eds.), *Locational Analysis of Public Facilities*, North-Holland, Amsterdam, pp. 357–383.

Leung, J.M.Y., and T.L. Magnanti (1989). Valid inequalities and facets of the capacitated plant location problem, *Math. Program.* 44, 271–291.

Levy, J. (1967). An extended theorem for location on a network, *Oper. Res. Q.* 18, 433–442.

Lin, C.C. (1975). On vertex addends in minimax location problems, *Transp. Sci.* 9, 165–168.

Louveaux, F., J.-F. Thisse and H. Beguin (1982). Location theory and transportation costs, *Reg. Sci. Urban Econ.* 12, 529–545.

Love, R.F., J.G. Morris and G.O. Wesolowsky (1988). *Facilities Location*, North-Holland, Amsterdam.

Magnanti, T.L., and R.T. Wong (1990). Decomposition methods for facility location problems, in: P.B. Mirchandani and R.L. Francis (eds.), *Discrete Location Theory*, Wiley, New York, NY, pp. 209–262.

Manne, A.S. (1964). Plant location under economies-of-scale — Decentralization and computation, *Manage. Sci.* 11, 213–235.

Maranzana, F.E. (1964). On the location of supply points to minimize transport costs, *Oper. Res. Q.* 15, 261–270.

Martello, S., and P. Toth (1990). *Knapsack Problems: Algorithms and Computer Implementations*. Wiley, New York, NY.

Matula, D.W., and R. Kolde (1976). Efficient multimedian location in a cyclic networks. ORSA/TIMS Meeting, Miami, FL.

Megiddo, N. (1983). Linear-time algorithms for linear programming in R^3 and related problems, *SIAM J. Comput.* 12, 759–776.

Megiddo, N., and A. Tamir (1983). New results on the complexity of p-center problems, *SIAM J. Comput.* 12, 751–758.

Minieka, E. (1970). The m-center problem, *SIAM Rev.* 12, 138–139.

Minieka, E. (1977). The centers and medians of a graph, *Oper. Res.* 25, 641–650.

Minieka, E. (1981). A polynomial time algorithm for finding the absolute center of a network, *Networks* 11, 351–355.

Mirchandani, P.B., and R.L. Francis (eds.) (1990). *Discrete Location Theory*, Wiley, New York, NY.

Mirchandani, P.B., and A.R. Odoni (1979). Location of medians on stochastic networks, *Transp. Sci.* 13, 85–95.

Morris, J.G. (1978). On the extent to which certain fixed-charge depot location problems can be solved by LP, *J. Oper. Res. Soc.* 29, 71–76.

Mulligan, G.F. (1991). Equality measures and facility location, *Pap. Reg. Sci.* 70, 345–365.

Narula, S.C., U.I. Ogbu and H.M. Samuelsson (1977). An algorithm for the p-median problem, *Oper. Res.* 25, 709–713.

Nauss, R.M. (1978). An improved algorithm for the capacitated facility location problem. *J. Oper. Res. Soc.* 29, 1195–1201.

Nemhauser, G.L., and L.A. Wolsey (1988). *Integer and Combinatorial Optimization*, Wiley, New York, NY.

Nemhauser, G.L., L.A. Wolsey and M.L. Fisher (1978). An analysis of approximations for maximising submodular set functions. I., *Math. Program.* 14, 265–294.

Nobili, P., and A. Sassano (1989). Facets and lifting procedures for the set covering polytope, *Math. Program.*, 45, 111–137.

Padberg, M.W. (1973). On the facial structure of set packing polyhedra, *Math. Program.* 5, 199–215.

Rawls, J. (1971). *A Theory of Justice*, Harvard University Press, Cambridge, MA.

ReVelle, C.S., and R. Swain (1970). Central facilities location, *Geogr. Anal.* 2, 30–42.

Rosenhead, J.V. (1973). The optimum location of transverse transport links, *Transp. Res.* 7, 107–123.

Rushton, G., S. McLafferty and A. Ghosh (1981). Equilibrium locations for public services: individual preferences and social choice, *Geogr. Anal.* 13, 196–202.

Sassano, A. (1989). On the facial structure of the set covering polytope, *Math. Program.*, 43, 45–55.

Schrage, L. (1975). Implicit representation of variable upper bounds in linear programming, *Math. Program. Studies* 44, 181–202.

Shier, D.R., and P.M. Dearing (1983). Optimal locations for a class of nonlinear, single-facility location problems on a network, *Oper. Res.* 31, 292–303.

Simchi-Levi, D., and O. Berman (1988). A heuristic algorithm for the traveling salesman location problem on networks, *Oper. Res.* 36, 478–484.

Slater, P.J. (1975). Maximin facility location, *J. Res. Nat. Bur. Stand.* B79, 107–115.

Slater, P.J. (1980). Medians of arbitrary graphs, *J. Graph Theory* 4, 389–392.

Stollsteimer, J.F. (1963). A working model for plant numbers and locations, *J. Farm Econ.* 45, 631–645.

Tamir, A. (1987). On the solution value of the continuous p-center location problem on a graph, *Math. Oper. Res.* 12, 340–349.

Tamir, A. (1988). Improved complexity bounds for center location problems on networks by using dynamic data structures, *SIAM J. Discrete Math.* 1, 377–396.

Tansel, B.C., R.L. Francis and T.J. Lowe (1983). Location on networks: A survey — Part I: The p-center and p-median problems, *Manage. Sci.* 29, 482–497.

Tansel, B.C., R.L. Francis and T.J. Lowe (1983). Location on networks: A survey — Part II: Exploiting the tree network structure, *Manage. Sci.* 29, 498–511.

Teitz, M.B., and P. Bart (1968). Heuristic methods for estimating the generalized vertex median of a weighted graph, *Oper. Res.* 16, 955–961.

Van Roy, T.J. (1986). A cross decomposition algorithm for capacitated facility location, *Oper. Res.* 34, 145–163.

Varian, H.R. (1984). *Microeconomic Analysis*, Norton, New York, NY.

Vohra, R.V. (1989). Distance weighted voting and a single facility location problem, *Eur. J. Oper. Res.* 41, 314–320.

Wagelmans, A.P.M., C.P.M. van Hoesel and A.W.J. Kolen (1992). Economic lot-sizing: an $O(n \log n)$-algorithm that runs in linear time in the Wagner–Whitin case, *Oper. Res.* 40, S1455–S156.

Weaver, J.R., and R.L. Church (1985). A median location problem with nonclosest facility service, *Transp. Sci.* 7, 18–23.

Weber, A. (1909). *Ueber den Standort der Industrien*, J.C.B. Mohr, Tübingen.

Wendell, R.E., and A.P. Hurter (1973). Optimal locations on a network, *Transp. Sci.* 7, 18–33.

Wendell, R.E., and R.D. McKelvey (1981). New perspectives in competitive location theory, *Eur. J. Oper. Res.* 6, 174–182.

Witzgall, C. (1964). Optimal location of a central facility: mathematical models and concepts, Report 8388, National Bureau of Standards.

Zelinka, B. (1968). Medians and the peripherians of trees, *Archivum Math.* 4, 87–95.

M.O. Ball et al., Eds., *Handbooks in OR & MS, Vol. 8*

Chapter 8

VLSI Network Design

Rolf H. Möhring

Fachbereich Mathematik, Technische Universität Berlin, Straße des 17. Juni 136, 10623 Berlin, Germany

Dorothea Wagner

Fakultät für Mathematik, Universität Konstanz, Postfach 5560, 78434 Konstanz, Germany

Frank Wagner

Institut für Informatik, Fachbereich Mathematik, Freie Universität Berlin, Takustraße 9, 14195 Berlin, Germany

Introduction

The design of integrated circuits is one of the broadest and most varied areas of application for combinatorial optimization methods and discrete algorithms. Many problems coming up during the layout process of chips or boards admit, for example, abstract formulations as graph problems. Over a period of more than one decade, mathematicians and theoretical computer scientists have intensively studied graph-theoretic and combinatorial aspects of circuit layout. In many cases, their investigations have led to exact and elegant models which help to understand the highly complex problems coming up in practice. Moreover, theoretical results have revealed that many of these problems may be attacked using classical optimization methods.

This article gives a survey on the interaction between integrated circuit layout and combinatorial optimization. The viewpoint taken is that of a combinatorialist, which means that main emphasis is given to aspects of circuit layout that are (by now) theoretically well understood and/or belong currently to the most prominent combinatorial problems in circuit layout.

On the other hand, the typical phases of the layout process have governed the organization of the layout material.

It starts in Part I with the *general layout problem*. In this overall approach, the physical components of the chip are considered as given together with their interconnection structure, the so-called *net lists*. It aims at *placing* components and *routing* the nets simultaneously. The main optimization goal is to minimize the layout area.

Typical combinatorial (optimization) problems occurring in this phase are the partitioning of graphs and hypergraphs such that the number of cut edges is small.

Part II is devoted to the *routing phase* which usually follows the placement. In this phase, the physical components have already been placed, and the wiring between these components has to be laid out. We discuss the graph-theoretic and combinatorial aspects of this problem.

The routing problem is usually divided into two subproblems, the *layout* problem and the *layer assignment* or *wiring* problem. During the layout phase the course of the wires in one single plane is determined. Then, in a second step, the wires are assigned in different layers to avoid the contact between different wires.

The layout phase is strongly related to the general layout problem discussed in Part I. In the most general case, it may be viewed as a Steiner tree problem in planar graphs. Actually, within the context of routing, the specified vertices that have to be connected in the layout are (in contrast to the general layout problem) assumed to lie on a small number of faces of the graph.

Depending on the model assumptions made for the layout phase, the wiring problem is a non-trivial task. It gives rise to some very attractive combinatorial problems. So, for example, the wiring of knock-knee layouts leads to a partitioning problem of the layout area.

Finally, Part III deals with special layout technologies that usually occur in *logic synthesis*, i.e. in the construction of the physical components that are given for the layout and routing phases. This so-called *linear layout methods* reveal some rich and surprising connections to independent and currently very active areas of graph theory such as graph searching, Robertson–Seymour theory, embedding graphs into interval graphs, and graph separation.

This survey is not intended as a complete overview or reference article on all aspects of combinatorial circuit layout. Rather, it aims at complementing other, recent monographs and articles (also in this volume) and at treating new developments, in particular in routing and linear layout technologies in greater detail. For more information on combinatorial methods in VLSI design we especially refer to Lengauer [1990] and Korte, Lovasz, Prömel & Schrijver [1990].

Part I. The General Layout Problem

In this part we discuss an overall approach to the layout process. Here, the physical components of the chip are considered as given together with their interconnection structure. The aim is to place the components and route the nets simultaneously. The main optimization goal is to minimize the layout area.

The typical algorithmic approach to solve this task is a *divide-and-conquer* approach. The hard combinatorial optimization problem occurring along these lines is the partitioning of hypergraphs such that the number of cut edges is small.

Most of this part is based on work done by Leighton [1982], unified by Bhatt & Leighton [1984] and presented in a much more readable and smooth form by Lengauer [1990]. The current version corrects some flaws and simplifies several proofs, in particular the combinatorial lemma, Lemma 3.6, and its use in balancing bifurcators. Whenever we give credits for a lemma or theorem we try to point out the source which comes nearest to the form given here.

1. Layout by graph partitioning

In its most general form, the layout of a circuit or network can be modeled as a hypergraph embedding problem [Thompson, 1980; Ullmann, 1983; Lengauer, 1990]:

Definition 1.1. *An* embedding *of a hypergraph* $H = (V, E)$ *into a grid-graph*

$$G_{n \times m} = (V_{n \times m}, E_{n \times m}),$$

where

$$V_{n \times m} := \{(i, j) \mid 1 \le i \le n, \ 1 \le j \le m\}$$

and

$$E_{n \times m} := \{\{v, w\} \mid v, w \in V, \ \|v - w\| = 1\},$$

is a mapping ℓ of the vertices and edges of H. The vertices are mapped to distinct vertices of $V_{n \times m}$; each edge $\{v_1, v_2, \ldots, v_k\}$ of H is mapped to a connected subgraph of $G_{n \times m}$ containing the vertices $\ell(v_1), \ldots, \ell(v_k)$. These subgraphs have to be pairwise edge-disjoint.

Typically the subgraphs realizing the hyperedges are minimal, i.e. trees. A necessary and sufficient condition for a hypergraph to be embeddable is that the maximum vertex degree, i.e. the number of edges containing a vertex, is at most four. An example is shown in Figure 1.

The *general layout problem* is then:

Given: A vertex- and edge-weighted hypergraph H.

Problem: Embed H into a rectangular grid, such that certain optimization criteria are fulfilled simultaneously or according to a certain priority order.

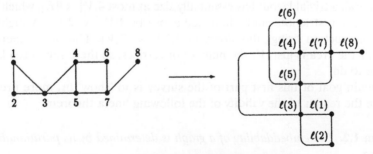

Fig. 1. A graph and an embedding in the grid.

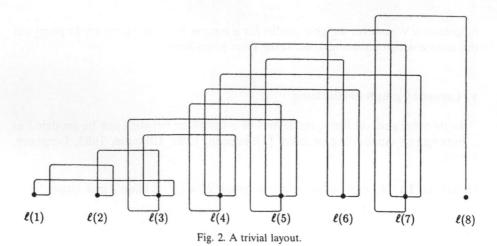

$\ell(1)$ $\ell(2)$ $\ell(3)$ $\ell(4)$ $\ell(5)$ $\ell(6)$ $\ell(7)$ $\ell(8)$

Fig. 2. A trivial layout.

Optimization criteria:
- The by far most important criterion is the *area* which is just $m \cdot n$, the number of vertices of the grid. We will essentially only discuss this goal.
- In addition, it is desirable to find, under the possible embeddings with small area, one with small *wire length*, i.e., with short paths, with few *bends*, and few *crossings*. Overall, the layout should have a low complexity resulting in good technological properties.

In the rest of this part we will concentrate on ordinary graphs without loops, parallel edges, or weights since, for most problems we discuss, the theory is not developed for hypergraphs. On the other hand, the results on graphs often seem to give the right flavor of what is true for hypergraphs.

It is easy to embed any given graph of maximum degree at most four by placing all vertices on the same horizontal line and realizing the edges one after the other by edge-disjoint paths, e.g. as shown in (Figure 2). We call such a layout *trivial*.

Minimizing the layout area is $\mathcal{NP}$-hard [Formann & Wagner, 1990; Kramer & van Leeuwen, 1984], even if the graph to be laid out is planar.

In general, a trivial layout has essentially size at most $4|V| \times |E|$, which because of the degree restriction can be bounded above by $4|V| \times 2|V|$. A slightly more clever construction brings this down to $2|V| \times 2|V|$. The only general lower bound for the area is just $|V|$ the number of vertices, so there remains a large gap and a lot to do.

The main goal of this first part of the survey is to 'show' or, more precisely to convince the reader of the validity of the following 'meta-theorem'

Theorem 1.2. *The embeddability of a graph is determined by its partitionability and vice versa.*

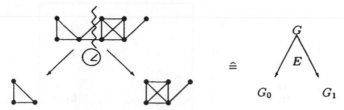

Fig. 3. A bipartition of G into G_0 and G_1.

2. The divide-and-conquer layout algorithm

Definition 2.1. *A graph* (V_G, E_G) *is bipartitioned by* $E \subseteq E_G$ *into two disjoint parts* $G_0 = (V_{G_0}, E_{G_0})$ *and* $G_1 = (V_{G_1}, E_{G_1})$, *if*

$$V_{G_0} \cap V_{G_1} = \emptyset, \; V_G = V_{G_0} \cup V_{G_1},$$

$$E_G = E_{G_0} \cup E_{G_1} \cup E,$$

and none of the edges in $E_{G_0} \cup E_{G_1}$ *connect vertices of* G_0 *and* G_1 *(see Figure 3).*

(G_0, E, G_1) *is a bipartition of* G *of size* $|E|$. *It is proper if* $V_{G_0} \neq \emptyset \neq V_{G_1}$. *We denote such a bipartition by* $G \rightarrow (G_0, E, G_1)$.

A layout algorithm based on this concept was introduced independently in Leiserson [1983] and Valiant [1981] and can be formulated on a very abstract level as follows:

Algorithm 2.2 *(divide-and-conquer layout).*

Step 1. Construct a proper bipartition (G_0, E, G_1) *of* G *and, recursively, find layouts for* G_0 *and* G_1.

Step 2. Arrange the two partial layouts side by side.

Step 3. Layout the edges of E.

Based on this concept, we now explain the details backwards, step by step:

Step 3: The missing edges are laid out one after the other. We start the realization of an edge connecting v and w in G, which means the construction of a conflict-free path from $\ell(v)$ to $\ell(w)$ in L, by adding at most two horizontal and at most two vertical grid lines.

To be precise, we need at most three and not four additional lines — at most two in one dimension and one in the other (Figure 4). If the two lines at the free 'valences' of $\ell(v)$ and $\ell(w)$, respectively, are perpendicular on each other, even two additional lines suffice, one in each direction (Figure 5).

Of course, the introduction of new grid lines interrupt the paths that cross these lines, this is avoided by stretching these paths by one unit across the new line.

Step 2: This is done by alternatingly placing partial layouts on top of the other and one beside each other, respectively. By that, the layouts obtained are square, respectively, undesirably long and thin layouts are avoided.

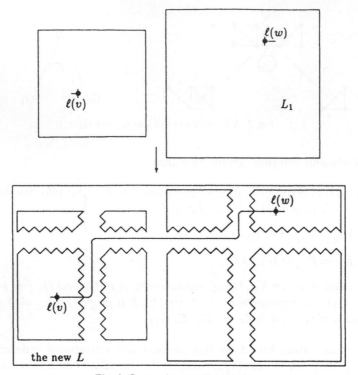

Fig. 4. Connecting parallel valencies.

Step 1: What we need in the course of the complete algorithm is not only a single graph bipartition, but, in order to apply the procedure recursively, a full *partition tree* (Figure 6)

In general, this will be a binary tree. Each forking corresponds to a bipartition and is labeled with the deleted edges. Each leaf corresponds to a single-vertex-graph. For every such partition tree, the algorithm produces a layout; of course of varying quality, depending on the concrete tree. Our main goal is to minimize the layout area. So it is natural to ask:

– What kind of partition tree induces a small area layout?

At first glance these questions sound by far too complex. So, instead we look backwards at the process and ask:

– Given a good layout, what kind of partition tree does the graph laid out have?

So we start with a hypothetical layout of area A, and, to circumvent any problems concerning the divisibility of numbers, we increase height and width of the layout to the next power of two. The new area A' is at most $4A$. We bisect the padded layout successively, always along the shorter side, thus producing a canonical bipartition tree (Figure 7).

In each partitioning step the area of the remaining layouts is halved, i.e., $A', A'/2, A'/4, \ldots, 1$. The shorter side always has length at most $\sqrt{A'}, \sqrt{A'/2}$,

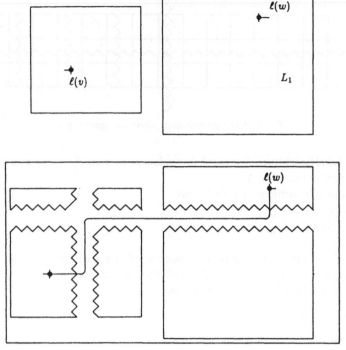

Fig. 5. Connecting perpendicular valencies.

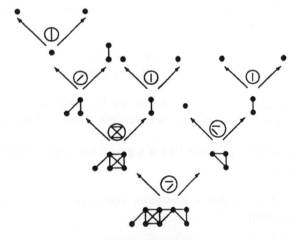

Fig. 6. A partition tree.

$\sqrt{A'/4}, \ldots, 1$. Each bisection at height i thus induces a bipartition of the embedded graph of size equal to this side length. The height of the induced partition tree is $\log_2 A'$.

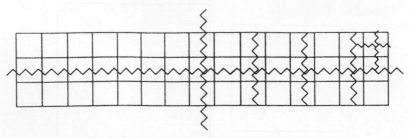

Fig. 7. Repeated bisections along the shorter side.

Theorem 2.3. [Leighton]. *If a graph has a layout of area A, it has a partition tree*
  – of height at most $\log A'$,
such that every bipartition at height i has
  – a size of at most $\sqrt{A'}/\sqrt{2}^{\,i}$,
where A' is at most 4A.

The factor 4 is just a technical consequence of the padding process, it does not reflect 'the real thing'. Theorem 2.3 and the last remark motivate the notion of bifurcator introduced in the following section.

3. Bifurcators and separators

The layout theory based on partition *trees* was developed in Leiserson [1983] and, in a sense, completed to its bifurcator-based version in Leighton [1982] and Bhatt & Leighton [1984].

Definition 3.1. *A bifurcator of size b is a partition tree of height at most* $\log b^2$, *such that every bipartition at height i has a size of at most* $b/\sqrt{2}^{\,i}$.

b must be at least $\sqrt{n}$, since there is no binary tree with n leaves of height less than $\log n$; the height restriction enforces a certain amount of balancedness of the bipartitions.
   Theorem 2.3 can now be restated as: If a graph has a layout of area A, it has a bifurcator of size $\sqrt{4A}$.

Corollary 3.2. *If a graph has a minimum layout area* $A_{\min}$ *and a minimum bifurcator size* $b_{\min}$, *then*

$$A_{\min} \geq \frac{b_{\min}^2}{4}$$

So the minimum bifurcator size of a graph yields a lower bound of order $\Omega\left(b_{\min}^2\right)$ on the layout area. The following theorem [Bhatt & Leighton, 1984] will give a similar upper bound and completes the 'proof' of our meta-theorem. It can be seen easily by an analysis of width and height of the layout obtained by Algorithm 2.2 using the bifurcator.

Theorem and Algorithm 3.3 [Bhatt & Leighton].

(i) $A_{\min}$ is in the order of $O\left(b_{\min}^2 \log^2 b_{\min}\right)$.

(ii) Given a bifurcator of size b, one can efficiently construct a layout of area $O(b^2 \log^2 b)$.

How good are the layouts produced by Algorithm 3.3? Since the graphs we deal with have at most $2n$ edges, $b_{\min}$ can be at most $O(n)$. Then Algorithm 3.3 gives a layout of area $O(n^2 \log^2 n)$ which is worse than the trivial $O(n^2)$ layout. On the other hand, if $b_{\min}$ is extremely small, i.e. $b_{min} \in O(\sqrt{n})$, we have a layout of size as small as $O(n \log^2 n)$, which is far off the bound given by the trivial method.

Of course, it would be nice to have a method which is always better than the trivial one. Such an algorithm can be constructed as a hybrid algorithm according to the following rule:

Stop applying the clever layout algorithm at a recursion level when the trivial one yields a better subresult on the remaining small graph and lay out the graph trivially.

Since the size of a trivial layout of a graph is a function of its number of vertices, we need a better control of this parameter during the bipartition process.

Definition 3.4. *A bifurcator is balanced, if for every single bipartition*
$G_\alpha \to (G_{\alpha 0}, E_\alpha, G_{\alpha 1})$,

$$\left| |V_{\alpha 0}| - |V_{\alpha 1}| \right| \leq 1.$$

Later on we will see how to construct balanced bifurcators from ordinary bifurcators. Observe that a subgraph at height i has at most $\lceil n/2^i \rceil$ vertices.

Given such a balanced bifurcator, we use Algorithm 2.2 only up to that height of the partition tree where the trivial layout becomes better. This is the case at height $\log(n/b)^2$. All subgraphs at this level are laid out trivially in one step. This procedure induces the following result [Bhatt & Leighton, 1984].

Theorem and Algorithm 3.5 [Bhatt & Leighton].

(i) $A_{\min}$ is in the order of $O(\beta_{\min}^2 \log^2 n/\beta_{\min})$, where $\beta_{\min}$ is the size of a minimum balanced bifurcator.

(ii) Given a balanced bifurcator of size β, we can efficiently construct a layout of area $O(\beta^2 \log^2 n/\beta)$.

This is always an improvement over Theorem 3.3 since $b \geq \sqrt{n}$ and thus $n/b \leq b$. If $b \in O(\sqrt{n})$, we have $A \in O(n \log^2 n)$ as before. If $b \in O(n)$, we have $A \in O(n^2)$. We will now balance a bifurcator at low cost. Let us start with the following combinatorial lemma (with a beautiful proof):

Lemma 3.6 [Lengauer]. *From every necklace N consisting of an even number of ruby and emerald pearls, we can efficiently obtain a subnecklace S of half the length containing about half of the ruby and half of the emerald pearls, i.e.*

$$|\text{\# ruby pearls in } S - \text{\# ruby pearls in } N \backslash S| \leq 1.$$

Example: ◯ ◯ ⟮◯ ◯ ● ●⟯ ◯ ●

$$S$$

Proof. Number the pearls from left to right $1, 2, \ldots, |N|$ and move a window of width $|N|/2$ from left to right. Stop, when the view is balanced. This will work, since

for Balance $(i) := \text{\# ruby pearls in } S_i := \{i + 1, \ i + |N|/2\}$
$\qquad\qquad\qquad - \text{\# ruby pearls in } N \backslash S_i$

we have

(i) Balance $(0) = -$Balance $(|N|/2)$.

(ii) Balance $(i + 1) - $Balance$(i) \in \{0, -2, +2\}$

Thus, there is an efficiently computable i_0 with |Balance $(i_0)| \leq 1$, so $S := S_{i_0}$ will do. □

Using the necklace lemma as the main tool, we balance the given bifurcator successively, starting at its root:

Step 1. Fill the partition tree up to a complete binary tree by adding dummy-nodes, -edges, and -labels. The dummy-nodes represent empty graphs, e.g. replace

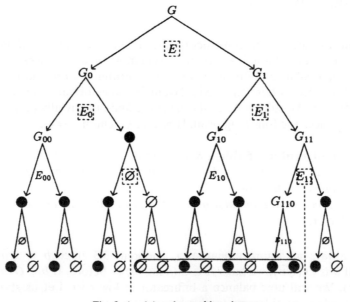

Fig. 8. Applying the necklace lemma.

a single-vertex leaf at non-maximum height. This yields a tree whose leafs are single-vertex graphs ($\hat{=}$ ruby) and empty ($\hat{=}$ emerald) graphs.

Step 2. According to Lemma 3.6, choose a subnecklace S containing about half of the ruby leaves. This induces (almost) a bisection of G by the edge sets along the paths from the left and right borders of S to the root (Figure 8) into two forests (sets of tree) representing the two parts G_0' and G_1'. (G_0', E', G_1') is the first bipartition of the balanced bifurcator we aim at, where E' is the union of the edge sets along the two paths. Observe that the height of the two forests is at least one smaller than the height of the given bifurcator (Figure 9).

Iterating Step 2 yields a balanced bifurcator of G (Figure 10). The number of cut edges in a bipartition at height i is at most

$$2 \sum_{j=1}^{\log b^2} \frac{b}{\sqrt{2}^i} \sum_{j=0}^{\infty} \left(\frac{1}{\sqrt{2}}\right)^j = \frac{2(2+\sqrt{2})b}{\sqrt{2}^i}$$

So we have constructed a balanced bifurcator of the desired size β. In summary, we have shown the following theorem [Bhatt & Leighton, 1984].

Theorem 3.7 [Bhatt & Leighton]. *Given a bifurcator of size b of a graph G, one can efficiently construct a balanced bifurcator of size*

$$\beta \le (4 + 2\sqrt{2})b$$

for the same graph G.

If the graph to be embedded has bipartitions of a very small size, i.e. in the range of $O(1)$ up to $O(\sqrt{n})$, the bifurcator concept is of no use. A partial substitute is then an approach based on *separators*. To this end, we introduce the following generalization of bifurcators.

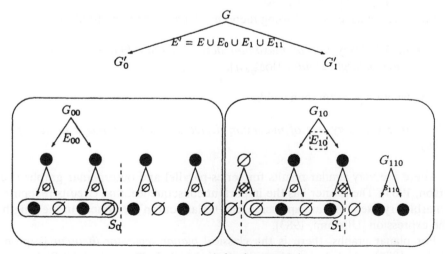

Fig. 9. Applying it for the second time.

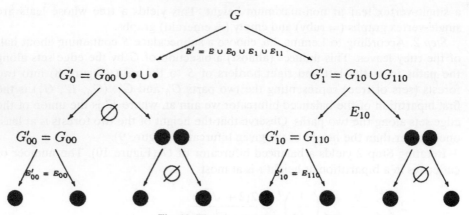

Fig. 10. The complete balancing.

Definition 3.8. *A partition tree is a* $(z_0, z_1, \ldots, z_h)$-*partition tree, if every bipartition at height* i *has size at most* z_i *and if all the leafs at height* h *represent empty or one-vertex-graphs.*

A bifurcator of size b is then just a

$$\left(b, \frac{b}{\sqrt{2}}, \frac{b}{\sqrt{2}^2}, \frac{b}{\sqrt{2}^3}, \ldots, \frac{b}{\sqrt{2}^h} \right) \text{-partition tree}$$

where h is at most $\log b^2$. Now we can proceed analogously as in the previous section by using such a partition tree as input for the divide-and-conquer algorithm, although the other direction of our meta-theorem is no longer valid.

The two most prominent examples of graph classes having non-trivial partition trees are trees and planar graphs:

For trees we have the following theorem which seems to be folklore.

Theorem 3.9. *Every tree of maximum degree at most four has a* $(1, 1, \ldots, 1)$-*partition tree of height at most* $\lceil \log_{5/4} n \rceil$.

The layout algorithm then yields:

Theorem 3.10. *Every tree of maximum degree at most four has a layout of area* $O(n)$.

There are very similar results for series-parallel and outerplanar graphs [Leiserson, 1983]. The former has the important consequence, that circuits recognizing an arbitrary fixed regular expression, can be laid out in area linear in the length of the expression [Ullmann, 1983].

For planar graphs, there is the classical *planar separator theorem* [Lipton & Tarjan, 1979], which has applications far beyond the context of VLSI layout. It essentially motivated the creation of the bifurcator concept and can be stated as follows:

Theorem 3.11. *Every planar graph of maximum degree at most four has a bipartition of size at most $8\sqrt{2}\sqrt{n}$ such that each of the two parts contains at most two thirds of the vertices.*

Formulated in our language, we thus obtain by repeated application of this theorem:

Theorem 3.12. *Every planar graph of maximum degree at most four has a bifurcator of size $O(\sqrt{n})$ and thus a layout of area $O(n\log^2 n)$.*

The partition tree approach provides a very good insight into the close connection of layout area and divisibility. However, except for planar graphs, no algorithms are known that are efficient and yield provably good partition trees as a whole. So, in practice one applies heuristics such as the *min-cut* heuristic which we sketch in the concluding section of this part.

4. Bipartitioning

The min-cut approach, which is legitimated by its success in practice, consists of doing just one single partition at a time, without having the rest of the recursive layout process in mind.

To ensure at least a certain amount of similarity between the subgraphs at one level, typically bisections, i.e. bipartitions into equal-sized parts, are considered. Driven by the intuition that edges which are cut in early stages of the recursion tend to be rather long in the complete layout, a bisection with minimum size is desired.

Unfortunately, this MINIMUM BISECTION problem is $\mathcal{NP}$-hard [Garey, Johnson & Stockmeyer, 1976].

There are two major approaches to overcome this hardness. A large and increasing number of papers [e.g. Leighton & Rao, 1988; Garg, Vazirani & Yannakakis, 1993; Klein, Agrawal, Ravi & Rao, 1990; Klein, Plotkin & Rao, to appear; Plotkin & E. Tardos, to appear; Leighton, Makedon, Plotkin, Stein, Tardos & Tragoudas, 1991; Shahrokhi & Matula, 1990] is devoted to attack the bisection problem by *uniform multicommodity flow techniques*. Given a graph with edge capacities, this problem consists in finding a maximal flow value that can be realized simultaneously between each pair of nodes of the network. There is a rather simple but unfortunately rather weak interdependence between this value and the bisection size. At least so far, the algorithms developed are very complicated. They need polynomial time but they are not really efficient, and their results fulfill only very weak quality guarantees (polylog-factors).

Another approach concentrates on planar graphs. The main drawback of the planar separator theorem is that it does not guarantee to find a partition that is good compared to the best possible in the given specific graph. It just gives a worst-case bound depending on the number of vertices. Empirically, this algorithm does not tend to find good or optimal partitions whenever the graph is sparse. Typically, graphs describing networks are sparse. So the approach in Rao [1987, 1992] leads far beyond the capabilities of the planar separator theorem as it gives quite good optimality guarantees. Its main theoretical contribution consists in a deeper insight into the shape of minimum cuts in planar graphs.

The by far most often used and very flexible graph partitioning algorithm is the one by Kernighan & Lin [1970]. It is 'only' a heuristic, nothing can be proved about its performance quality, but the idea is very interesting and occurs also in other heuristic approaches to graph problems, e.g. the traveling salesman problem.

Essentially, it is a very clever variant of greedy approach which in pure form would be as follows.

Start with a random bisection, look for a pair of vertices, one in each half of the bisection, whose interchange decreases the size of the cut, until you get stuck. This can be refined by taking the vertex pair, which decreases the size by the largest amount. But both of these approaches suffer from being absolutely blind with respect to *local* minima.

The one-eyed variant of Kernighan and Lin looks into the future. It, successively chooses the best pairs of vertices to be switched until all vertices have changed sides. Every vertex does this exactly once. All these $n/2$ interchanges are just tentative. In a second run, the best moment to stop interchanging is determined, and only all interchanges up to this moment are really performed. The best pair is that pair of vertices that, at a certain state, yields the largest decrease or, if no such pair exists, the smallest increase of the size of the cut. The main advantages of this definition of 'best' are that there is always such a best pair, and that the algorithm is able to overcome 'local minima' of the first kind.

Algorithm 4.1. *Given a graph with an even number of vertices:*

   Step 1. Start with a random bisection V_1, V_2; all vertices are unmarked.

   Step 2. $W_1 := V_1$; $W_2 := V_2$ (store the bisection found so far). Choose among all pairs $(v_1, v_2) \in V_1 \times V_2$ which consist of unmarked vertices, the one whose interchange decreases the cut size by the largest amount or increases it by the smallest amount (if there is no decreasing pair) and interchange it. Mark these two vertices.

   Step 3. Repeat step 2 until there is no unmarked vertex left, i.e. $n/2$ times. This yields a sequence $(v_1^1, v_2^1), \ldots, (v_1^{n/2}, v_2^{n/2})$. The amount of decrease in step i is called δ_i. This value is negative if it is an increase in size.

   Step 4. Choose j such that

$$\Delta_j := \sum_{i=1}^{j} \delta_j$$

is minimum.

   If $\Delta_j \geq 0$ then stop. Output W_1, W_2.

If $\Delta_j < 0$ then set

$$V_1 := (W_1 \setminus \{v_1^1, \ldots, v_1^j\} \cup \{v_2^1, \ldots, v_2^j\})$$

$$V_2 := (W_2 \setminus \{v_2^1, \ldots, v_2^j\} \cup \{v_1^1, \ldots, v_1^j\}).$$

Make all vertices unmarked again and goto 2.

In general the algorithm typically stops after 3-5 rounds. Every round costs $O(n^3)$ elementary operations. By rather elementary changes this can be brought down to a linear number of operations [Fiduccia & Mattheyses, 1982].

The algorithm is very flexible; it can be adapted to handle the following variants of the bisection problem:

1. Find a bipartition of small size such that each part has a fixed given number of vertices.

2. Find a bipartition of small size such that each part has at least a given number of vertices, e. g. one third of the total number of vertices.

3. Find a bisection of small size such that certain vertices lie in prescribed parts.

4. Find bipartitions of hypergraphs and weighted graphs satisfying analogous conditions.

So the Kernighan/Lin heuristic tries to avoid local minima by considering an increase of the cut size found so far at least as a hypothetical intermediate result.

In Johnson, Aragon, McGeoch & Schevon [1989], this algorithm is compared with the graph partitioning version of *simulated annealing*. This general concept leads to a partitioning algorithm that can be viewed as a randomized version of Kernighan/Lin.

It starts with a random bisection, chooses a random pair of vertices, one in each part, and interchanges them depending on the change in cut size. If the cut size decreases, the change is always performed; if it increases it is performed with a certain probability that depends on the amount of increase (the larger the increase, the smaller the probability of acceptance), and on the running time (the longer the algorithm runs, the smaller the probability). The starting values and the functional dependencies have to be tested and tuned according to a number of heuristic rules of thumb. The running times of Simulated Annealing are extremely large, so it is only useful if, in a given application, time is not the expensive thing.

To compare the two algorithms, the running time of Kernighan/Lin is equalized with that of annealing; i.e. it is run so often that the Annealing time is reached.

The overall conclusion of Johnson, Aragon, McGeoch & Schevon [1989] is that both algorithms are of comparable quality and that their performance is very good.

Currently, there are no really practical graph bisection algorithms, that have reasonable quality guarantees. So, as an outlook of this first part, we refer to a very interesting development with respect to the 'ordinary' minimum cut problem. Until recently, there was no way known to find even a unbalanced minimum size bipartition without using maximum flow techniques. The algorithms used to solve the minimum cut problems thus lacked of simplicity and were not intuitively adapted to the problem and, although polynomial, not that efficient. Now, there is a fundamental paper [Nagamochi & Ibaraki, 1992] which contains a 'flow-free' very simple minimum cut algorithm of greedy type of which we think that it has the potential to be really practical. It could be that the concentration on guaranteeing the balance is wrong, misleading and unnecessary in practice.

When we formulated the general layout problem, we started with hypergraphs. An elegant and general way to apply graph partitioning algorithms to hypergraphs would be to model hypergraphs by graphs and apply the graph algorithms to these models. Of course such models have to simulate the given hypergraphs with respect to their cut properties. In Ihler, Wagner & Wagner [1993] some 'negative' results are proved for this problem. An edge-weighted graph (V, E) is a *cut-model* for an edge-weighted hypergraph (V, H) if the weight of the edges cut by any bipartition of V in the graph is the same as the weight of the hyperedges cut by the same bipartition in the hypergraph. It is shown that there is no cut-model in general.

A natural question is if the introduction of 'dummy' vertices can help. An edge-weighted graph $(V \cup D, E)$ is a *min-cut-model* for an edge-weighted hypergraph (V, H) if the weight of the hyperedges cut by a bipartition of the hypergraphs vertices is the same as the weight of a minimum cut separating the two parts in the graph. In Ihler, Wagner & Wagner [1993] such models are constructed using positive *and* negative weights. On the other hand, it is shown that there is no min-cut-model in general if only positive weights are allowed.

Part II. Routing

In this part the routing problem is discussed. During the design process it occurs after the placement phase, and specifies the course of the wires connecting the cells.

Depending on the model assumptions, the routing phase is usually carried out in two steps. At first, the course of the wires in a single plane, the *layout*, is determined. The abstract formulation of this problem is essentially a path problem or Steiner tree problem in graphs. In a second step, the *wiring* step, the wires are assigned to different layers to avoid conflicts between different wires. Both phases not only lead to solvability problems, but must be considered under many different optimization criteria, e.g., the minimization of the layout area and a short realization of wires, or wirability in a small number of layers and with a small number of contacts which connect different layers. Thus, they result in various problems, which are usually computationally hard.

Typically, the problems are decomposed into many smaller subtasks. Depending on the model assumptions, each of the main subproblems of the routing problem, i.e. the layout problem, the layer assignment problem, and the contact minimization problem, may be essentially viewed as combinatorial or graph-theoretic problems.

In this part we give a survey of these problems under different routing models. We concentrate on results that are interesting with respect to the combinatorial methods applied. Sections 5–7 discuss the main results concerning the *knock-knee model*. From the mathematical point of view this model is maybe the most attractive one. The model assumptions require that the routing phase is carried out in two steps, the layout and the wiring phase. Both of them are well understood. The layout phase corresponds just to edge-disjoint Steiner trees or paths packing problems in planar graphs. Section 8 surveys the most important theoretical results on the more practical *Manhattan model*. The model corresponding to vertex-disjoint trees or paths packing problems in planar graphs, the *single layer model* is discussed in Section 10. Combinatorial methods for the *via minimization problem* are reviewed in Section 9. One important aspect of routing, *homotopic routing problem*, is not considered in this part. We refer instead to the excellent overview article [Schrijver, 1990].

5. From planar graphs to switchboxes

In general, the graph oriented version of the edge-disjoint routing problem can be viewed as the problem of packing edge-disjoint *Steiner trees*, i.e.,

Given: A routing graph $G = (V, E)$, $|V| = n$, and sets $\mathcal{N}_1, \ldots, \mathcal{N}_N \subseteq V$. In this context we call the $\mathcal{N}_i$ *nets* and the elements of $\mathcal{N}_i$ *terminals*.

Problem: Find N edge-disjoint Steiner trees $T - 1, \ldots, T_N$, each of which connecting the terminals of one net $\mathcal{N}_i$. The Steiner tree T_i is called *wire* for net $\mathcal{N}_i$. The solution is also called *layout*.

If the routing graph is a grid, the edge-disjoint version of the routing problem is often referred to as *knock-knee* routing. The name 'knock-knee' model comes from the form of two wires that bend at the same vertex in a grid.

The problem is $\mathcal{NP}$-complete even if G is planar and the $\mathcal{N}_i$ are all *two-terminal nets* [Kramer & van Leeuwen, 1984], or if G is planar and $N = 2$ [Korte, Prömel & Steger, 1990]. For planar problems, i.e. problems where G together with the 'demand' edges induced by the $\mathcal{N}_i$ is planar, which satisfy the *evenness condition* (see below), the two-terminal case again is polynomially solvable [Seymour, 1981]. By 'demand' edges, we mean edges added to the graph between the two terminals of the nets. The problem becomes $\mathcal{NP}$-complete again if planarity is dropped [Middendorf & Pfeiffer, 1993]. If $\sum_{i=1}^{N} |\mathcal{N}_i|$ is fixed the problem is polynomially solvable [Robertson & Seymour, 1990].

Typically, in VLSI-design the routing graph is planar and the terminals lie on a fixed (small) number of faces. A basic result is the theorem of Okamura & Seymour [1981] for the two-terminal case in a planar graph with all terminals lying on the outer face boundary.

Given: A planar embedded graph $G = (V, E)$ and a set of two-terminal nets $\{\{s_1, t_1\}, \ldots, \{s_N, t_N\}\}$, with $s_1, \ldots, s_N, t_1, \ldots, t_N \in V$ lying on the boundary of the infinite face.

Problem: Find edge-disjoint paths in G connecting s_i and t_i for $1 \leq i \leq N$.

The problem can be viewed as a *multicommodity flow problem*. The solvability strongly depends on the *cut condition* together with the *evenness condition*.

A subset $X \subseteq V$ is called a *cut* of G. For a cut X, the *capacity cap(X)* of X is the number of edges leaving X, and the *density dens(X)* of X is the number of nets leaving X, i.e.,

$$cap(X) := |\{\{u, v\} \in E : u \in X, v \in V \setminus X\}|$$
$$dens(X) := |\{\{s_i, t_i\} \in \mathcal{N} : s_i \in X, t_i \in V \setminus X\}|.$$

The *free capacity* of X is defined as

$$fcap(X) := cap(X) - dens(X).$$

Theorem 5.1 [Okamura & Seymour]. *Let* $G = (V, E)$ *be a planar graph, and* $\{\{s_1, t_1\}, \ldots, \{s_N, t_N\}\}$ *a set of two-terminal nets such that all* s_i, t_i *lie on the same face of* G. *For all* $X \subseteq V$ *let fcap(X) be even. Then there exist edge-disjoint paths connecting the* s_i *and* t_i *iff fcap$(X) \geq 0$ for all* $X \subseteq V$.

We call the condition '$fcap(X) \geq 0$' the *cut condition*, and '$fcap(X)$ even' the *evenness condition*.

The proof of Theorem 5.1 is constructive and yields an algorithm that preserves the cut condition and the evenness condition as invariants. We assume that G is embedded in the plane such that the terminals lie on the boundary of the outer face. Then the core of the algorithm can be formulated as follows.

Algorithm 5.2.

while $E \neq \emptyset$ **do**
    choose an edge $e = \{u, v\}$ on the outer face,
    if there is a cut X with $u \in X, v \notin X$ and $fcap(X) < 0$ **then** stop;
    (*cut condition is violated*)
    if there is a cut X with $u \in X, v \notin X$ and $fcap(X) = 0$ **then**
        choose an appropriate net $\{s, t\}$ with $s \in X, t \notin X$;
        reserve e for $\{s, t\}$;
        delete e;
        replace $\{s, t\}$ by nets $\{s, u\}, \{v, t\}$;
    else
        delete e;
        add a dummy net $\{u, v\}$.

Important for the efficiency of the algorithm are the following facts:

Fact 1. *Restriction to connected cuts.* For all cuts $X \subseteq V$, $fcap(X) \geq 0$ iff $fcap(X) \geq 0$ for all cuts $X \subseteq V$ with $G|X$ and $G|(V \setminus X)$ connected.

Fact 2. *Restriction to local evenness.* For all cuts $X \subseteq V$, $fcap(X)$ is even iff $fcap(v)$ is even for all vertices $v \in V$.

Becker & Mehlhorn [1986] relax the evenness condition and give an efficient version of the algorithm.

Theorem 5.3 [Becker & Mehlhorn]. *Let* $G = (V, E)$ *be a planar graph, and* $\{\{s_1, t_1\}, \ldots, \{s_N, t_N\}\}$ *a set of two-terminal nets. Assume* G *is embedded in the plane such that all* s_i, t_i *lie on the outer face. For all interior vertices* $v \in V$ *let* $fcap(v)$ *be even. Then the existence of edge-disjoint paths connecting the* s_i *and* t_i *can be decided and, in case they exist, determined in* $O(n^2)$ *time (where n denotes the number of vertices).*

The condition '$fcap(v)$ even for all interior $v \in V$' may be viewed as *weak evenness condition*. The basic idea of Becker & Mehlhorn's approach is to transform *weakly even* instances into *even* instances by introducing dummy nets between odd vertices on the outer face boundary. To get an $O(n^2)$ implementation they use the fact that minimum capacity cuts through two edges on the outer face boundary in G are equivalent to shortest paths between the corresponding vertices in the *multiple source dual* of G [Hassin, 1984; Matsumoto, Nishizeki & Saito, 1985; Suzuki, Nishizeki & Saito, 1989]. For the more restricted class of square grid graphs the running time reduces to $O(n^{3/2})$.

In Kaufmann & Klär [1991], the complexity of the algorithm has been improved to $O(n^{5/3} \log \log n^{1/3})$ for even planar problems by using Frederickson's decomposition method for planar graphs [Frederickson, 1987].

There are several faster algorithms, mostly based on the Okamura & Seymour approach for special grid graphs, where the improvement of the running time is based on the restricted shape of cuts to be considered. In Nishizeki, Saito & Suzuki [1985] Nishizeki, Saito & Suzuki handle convex grids and achieve a running time of $O(n)$. For more general grids, so-called *general switchboxes*, Kaufmann & Mehlhorn [1985] present a routing algorithm with a running time of $O(n \log^2 n)$. If the class of generalized switchboxes with every horizontal or vertical cut crossing the boundary at most twice is considered, which also contains convex grids, the running time can be reduced to $O(n)$ [Kaufmann, 1990]. Very recently, Wagner & Weihe introduced a completely new algorithm, which solves all even planar instances in time $O(n)$ [Wagner & Weihe, 1993]. We will explain the main ideas of this approach.

In the sequel, we consider an instance $(G, \mathcal{H})$ consisting of a planar graph $G = (V, E)$ and a set of two-terminal nets $\mathcal{H} = \{\{s_1, t_1\}, \ldots, \{s_N, t_N\}\}$, and an embedding of G such that all s_i, t_i lie on the outer face. We assume that all terminals have degree 1 and all other vertices have even degree. Obviously, a simple modification transforms every instance into a completely equivalent instance that fulfills this assumption. Moreover, for convenience we assume that, according to a reverse clockwise ordering starting with an arbitrary *start terminal* x, s_i precedes t_i for $i = 1, \ldots, N$, and t_i precedes t_{i+1} for $i = 1, \ldots, N-1$.

Before we determine a solution for the instance $(G, \mathcal{H})$, we will first consider an 'easier' instance $(G, \mathcal{H}^0)$ of 'parenthesis structure'. That is, consider the $2N$-string of s-terminals and t-terminals on the outer face in reverse clockwise ordering, starting with x. The ith terminal is assigned a left parenthesis if it is an s-terminal, and a right parenthesis otherwise. By that, we obtain a $2N$-string of parentheses that induce a correct pairing of left and right parentheses. The s-terminals and

t-terminals are now newly paired according to this pairing of parentheses, i.e., two terminals are paired if and only if the corresponding parentheses match. It is easy to see that $(G, \mathcal{H}^0)$ is solvable, if $(G, \mathcal{H})$ is.

The following Procedure 5.4 determines such a solution $(q_1, \ldots, q_N)$ for $(G, \mathcal{H}^0)$. In contrast to the original nets, we denote the nets in $\mathcal{H}^0$ by $\{s_1^0, t_1^0\}, \ldots, \{s_N^0, t_N^0\}$, and we assume w.l.o.g. that $t_i = t_i^0$ for $i = 1, \ldots, N$.

This procedure determines each path by a *right-first search*, i.e., a depth-first search where in each step all possibilities of going forward are searched 'from right to left'. In principle, it proceeds in the same way as the well-known *stack-algorithm* for a similar problem, where *vertex-disjoint* paths are to be drawn [Suzuki, Akama & Nishizeki, 1990].

Let $v \in V$, and let e be incident to v. We will say that the *next edge after e* w.r.t. v is the first edge to follow e in the adjacency list of v in reverse clockwise ordering.

Procedure 5.4. *'Auxiliary graph'.*

for $i := 1$ **to** N **do**

    let q_i initially consist of the unique edge incident to s_i^0;

    $v :=$ the unique vertex adjacent to s_i^0;

    while v is not a terminal **do**

        let $\{v, w\}$ be the next free edge after the leading edge of q_i w.r.t. v;

        add $\{v, w\}$ to q_i;

        $v := w$;

    if $v \neq t_i^0$ **then stop**: return 'unsolvable';

return $(q_1, \ldots, q_N)$.

The *auxiliary paths* $q_1, \ldots, q_N$ yield a directed *auxiliary graph* $A(G, \mathcal{H}, x)$ of instance $(G, \mathcal{H})$ w.r.t. start terminal x. Just orient all edges on the paths $q_1, \ldots, q_N$ according to the direction in which they are traversed in Procedure 5.4. Then $A(G, \mathcal{H}, x)$ consists of all vertices of G and of all oriented edges.

The following properties of the auxiliary paths, respectively auxiliary graph are easy to see.

Lemma 5.5. *The auxiliary paths $q_1, \ldots, q_N$ neither cross themselves nor each other. In particular, the left and the right sides of each of them are well defined. All edges to the right of an auxiliary path q_i are contained in $A(G, \mathcal{H}, x)$. The auxiliary graph $A(G, \mathcal{H}, x)$ does not contain a right-cycle, i.e., no directed cycle whose inner vertices lie to its right.*

The paths $p_1, \ldots, p_N$ for the original instance $(G, \mathcal{H})$ are now determined in the auxiliary graph. That is, edges that are not contained in the auxiliary graph will not be occupied by a path $p_1, \ldots, p_N$ of the final solution. Even more, the edges occupied by the final solution are exactly the edges of the auxiliary graph. The solution paths p_i are determined by a 'directed' right-first search. This means that edges belonging to $A(G, \mathcal{H}, x)$ are used according to their orientations in $A(G, \mathcal{H}, x)$.

Algorithm 5.6. *'Final paths'.*

Procedure 5.4

(∗determine $A(G, \mathcal{H}, x)$ for an arbitrary start terminal x∗)

for $i := 1$ **to** N **do**

  let p_i initially consist of the unique edge leaving s_i in $A(G, \mathcal{H}, x)$;

  $v :=$ the head of this edge;

  while v is not a terminal **do**

    let (v, w) be the next free edge leaving v after the leading edge of p_i w.r.t. v;

    add (v, w) to p_i;

    $v := w$;

  if $v \neq t_i$ **then stop**: return 'unsolvable';

return $(p_1, \ldots, p_N)$.

Procedure 5.4 is just a right-first search in an undirected graph and is easily implemented to run in linear time. The main part of Algorithm 5.6 is a right-first search in a directed graph. For a linear-time implementation, a special case of union-find is used which also runs in amortized linear time [Gabow & Tarjan, 1985].

The paths p_i determined by Algorithm 5.6 have some nice properties similar to those of the auxiliary paths stated in Lemma 5.5.

Lemma 5.7. *The paths $p_1, \ldots, p_N$ do not cross themselves. In particular, the left and the right sides of each of them are well defined. All edges to the right of a path p_i are either occupied by another path p_j, or contained in $A(G, \mathcal{H}, x)$ and directed towards p_i.*

Theorem 5.8 [Wagner & Weihe]. *Let $G = (V, E)$ be a planar graph, $\mathcal{H} = \{\{s_1, t_1\}, \ldots, \{s_N, t_N\}\}$ a set of two-terminal nets. Assume G is embedded in the plane such that all s_i, t_i lie on the outer face, and all $X \subseteq V$ fcap(X) are even. The Algorithm 5.6 determines edge disjoint paths in G connecting the s_i and t_i in time $O(n)$, provided such paths exist.*

Sketch of the proof. For an instance $(G, \mathcal{H})$ and a path p_i determined by Algorithm 5.6, consider the induced *residual instance*. This is the instance consisting of the subgraph of G induced by the edges that are not occupied by $p_1, \ldots, p_i$, and the set of nets $\{s_{i+1}, t_{i+1}\}, \ldots, \{s_N, t_N\}$. Then for a solvable instance $(G, \mathcal{H})$, Algorithm 5.6 maintains the following invariants.

I1. For any path p_i, the induced residual instance is again solvable.

I2. For any path p_i, the subgraph of $A(G, \mathcal{N}, x)$ induced by the edges that are not occupied by $p_1, \ldots, p_i$ is equal to the auxiliary graph that Procedure 5.4 determines for the induced residual instance (respectively the corresponding instance of parenthesis structure).

To prove these invariants, it suffices to prove that path p_1 determined by Algorithm 5.6 is correct, i.e. connects s_1 and t_1, and that I1 and I2 are correct for p_1. The main part of the correctness proof consists of the proof that I1 is satisfied for p_1. The proof does not use the Theorem of Okamura & Seymour. Instead, it is proved that there is a solution for the instance containing p_1. More precisely, it is shown that p_1 is just the rightmost path of all paths connecting s_1 and t_1 that are contained in a solution.

The correctness is based on the fact that the paths determined by the algorithm are in some sense 'extremal'. In fact, p_1 (respectively any path p_i) is the rightmost path connecting s_1 and t_1 (respectively s_i and t_i) with the property that the cut induced by the set of vertices lying to the right of p_1 (respectively p_i) is saturated (in the remaining problem). This means that the paths determined by the algorithm 'run along saturated cuts'.

The approach of Wagner & Weihe [1993] does not only lead to an algorithm with linear running time to find a solution for any solvable problem, but also yields a linear-time algorithm to find a connected oversaturated cut for a non-solvable instance. It thus gives an alternative proof for the Theorem of Okamura & Seymour and for Fact 1.

None of the algorithms mentioned so far takes care of optimization criteria such as e .g. the wire length or the number of bends. In fact, implementations show that they are bad with respect to both of these. Wagner & Wolfers modify the algorithm of Nishizeki, Saito & Suzuki such that the total wire length, the length of the longest wire and the number of knock-knees is substantially decreased [Wagner & Wolfers, 1991].

Notice, that for all these algorithms the complexity depends on the size of the routing region.

If the routing graph is a rectangle, the solvability depends only on cut conditions for the *horizontal* and *vertical* cuts [Frank, 1982].

Consider a rectangle graph $G = (V, E)$ consisting of vertical lines $1, \ldots, s$ and horizontal lines $1, \ldots, w$. The *vertical cuts* $V_1, \ldots, V_{s-1}$ are defined by

$$V_k := \{v \in V : v \text{ lies on one of the vertical lines } i, 1 \le i \le k\}.$$

The *horizontal cuts* $H_1, \ldots, H_{w-1}$ are defined analogously. Let $V_{i_1}, \ldots, V_{i_r}$ be the *saturated* vertical cuts , i.e. those vertical cuts with *fcap* equal to zero. Now, consider a horizontal cut H. The *parity* of a set $H \cap V_{i_j} \setminus V_{i_{j-1}}$ for a saturated vertical cut V_{i_j} is the parity of number of *odd* vertices (vertices of odd *fcap*) contained in $H \cap V_{i_j} \setminus V_{i_{j-1}}$. The *parity density* of H is the sum of the parities of the sets $H \cap V_{i_j} \setminus V_{i_{j-1}}, 1 \le j \le r$, plus the density of H. Then the *revised horizontal cut conditions* say that, for any horizontal cut H, the parity density must not exceed its capacity. The revised vertical cut conditions are defined analogously.

Theorem 5.9 [Frank]. *A two-terminal routing problem in a rectangle is solvable iff the revised vertical and horizontal cut conditions are satisfied.*

The proof of the theorem yields an $O(sw)$ algorithm. (Notice, that here the number of vertices is sw.)

Mehlhorn & Preparata [1986] present a routing algorithm for rectangles whose running time is sublinear in the number of grid points, i.e. $O(N \log N)$ where N

is the *number of nets*. The approach is completely different from the Okamura & Seymour based algorithms, but similar to standard channel routing methods, i.e. it processes the layout row by row. It guarantees a total of only $O(N)$ knock-knees.

For further references on routing in grid graphs we also refer to Kaufmann & Mehlhorn [1990].

6. Channel routing in the knock-knee mode

One of the mostly discussed routing problems is the *channel routing problem*, since most layout systems first place modules on a chip and divide the routing area between the modules into disjoint rectangles with terminals placed only on the upper and lower boundary.

Formally, a *channel* of *width* w and *spread* s is a rectilinear grid determined by horizontal lines $j, 1 \leq j \leq w$, called *tracks* and vertical lines $i, 1 \leq i \leq s$. A net $\mathcal{N} = (\{t_1, \ldots, t_l\}, \{b_1, \ldots, b_m\})$ is a collection of terminals, where $t_1, \ldots, t_l \in \{1, \ldots, s\}$ are the upper terminals or *input terminals* located on the crossing point of the top track and the corresponding vertical line and analogously $b_1, \ldots, b_m \in \{1, \ldots, s\}$ are the lower terminals or *exit terminals* on the bottom track. Normally, a channel routing problem (CRP) consists of a collection of nets, where no two nets share an input or exit.

Again, in knock-knee mode the solution consists of edge-disjoint Steiner trees through the grid connecting the terminals of the nets.

Clearly, the solvability of a CRP is based on the cut condition just as in the general case of planar graphs. But, because of the special structure of a channel routing problem, it can be formulated in simpler terms. First, the cut condition for the vertical cuts transforms to the notion of the *density* of a CRP, which leads to a lower bound for the number of tracks used for a layout. So, consider the segment between two vertical lines i and $i + 1$, also called a *column*. The *(local) density* of a column is the number of nets that have to cross the column, i.e. $|\{\mathcal{N}|\mathcal{N}$ is a net with $\min\{t_1, b_1\} \leq i$ and $\max\{t_l, b_m\} \geq i + 1\}|$. The *(global) density* d of a CRP is the maximum of all local densities. Obviously, the density of the CRP is a lower bound for the channel width needed for the layout. If the CRP contains k-terminal nets with $k \geq 5$, the problem of deciding if a layout exists in a channel of width d (independently from the spread of the channel) was proved to be $\mathcal{NP}$-complete by Sarrafzadeh [1987]. For the two-terminal case Frank gives a complete characterization of the solvable CRPs', and thus offers a polynomial algorithm that minimizes both width and spread of the channel [Frank, 1982]. But before we discuss the two-terminal case in Section 6.2, let us consider multiterminal problems.

6.1. Multiterminal nets

Approximation algorithms are obtained by Preparata & Sarrafzadeh that guarantee a layout within a channel of width $(3/2)d$ for problems containing only two-

or three-terminal nets [Preparata & Sarrafzadeh, 1985], and a layout of width $2d - 1$ for problems containing also k-terminal nets with $k > 3$ [Sarrafzadeh & Preparata, 1985]. Mehlhorn, Preparata & Sarrafzadeh [1986] have simplified these algorithms and obtain the following result.

Theorem 6.1. *For a CRP with N nets and density d, a layout can be constructed in time $O(N)$ that uses $2d - 1$ tracks. If all nets are two- or three-terminal nets, $(3/2)d$ tracks suffice.*

The algorithms process the channel in a single left-to-right sweep, where the sweep stops in every vertical line and performs a layout action to its right depending on the status of the terminals lying on that vertical line. For a vertical line i there are three possible cases: the density to its right is increased, decreased or preserved. On the other hand, in the current layout to the left of i only nets having at least one terminal to the left of i occupy a track. So, these nets may be divided in those having at least one terminal to the left of i and at least one terminal on i or to its right, the *active* nets, and those having all terminals to the left of i, the *extended* nets.

Now, considering all possible states of the current vertical line, the layout action may be performed always in such a way that the number of tracks occupied by active or extended nets never exceeds $(3/2)d$, respectively $2d - 1$.

The method can be viewed as a 'pushing' of the nets through the channel from left to right. Gao & Kaufmann [1987] improved the upper bound for the channel width to $(3/2)d + O(\sqrt{d \log d})$. The approach they offer is completely different from the algorithm of Mehlhorn, Preparata & Sarrafzadeh. It is based on decomposing a multiterminal net problem appropriately into two problems, a problem containing only one-sided nets, and a generalized switchbox problem containing only two-terminal nets.

The decomposition problem is formulated as an interval coloring problem and solved using an approximation for integer programming. For the subproblems containing only one-sided nets, the layout is performed in the upper, respectively lower part of the channel, and the layout for the middle part is constructed by using the algorithm of Kaufmann & Mehlhorn [1985] for generalized switchboxes. The method of Gao & Kaufmann is very complicated and the constant factor in the $O(\sqrt{d \log d})$-term is extremely large. Moreover, the running time is indeed polynomial, but not really efficient.

6.2. Two-terminal nets

The two-terminal channel routing problem treated as a solvability problem is considered in Frank [1982] for the case that all nets have exactly one upper and one lower terminal (i.e. no one-sided nets).

Theorem 6.2 [Frank]. *For a CRP of density d in a channel of width $m \geq d$, a correct layout exists if and only if one of the following properties is satisfied.*

(1) There is one vertical line not occupied by an upper terminal, such that the local density of any column to the right or to the left of it is less than m.

(2) There are two vertical lines not occupied by an upper terminal, such that the local density of any column between them is less than m.

Notice, that a special case of (1) is if one upper corner of the channel is 'free' (i.e. not occupied by a terminal).

The proof yields an algorithm that constructs for any solvable CRP an area-optimal layout. It determines the layout track by track by scanning the channel alternating from left-to-right or from right-to-left. Except in the first scan, the algorithm always starts at the leftmost, respectively rightmost vertical line of the CRP and performs a layout action if the current scanline is free. The starting position of the first scan depends on the criterion (1) or (2) of Theorem 6.2 satisfied by the CRP. A layout action, e.g. in a left-to-right scan may be described as a pulling of an appropriate net from the right to the current scan line. Actually, the algorithm works very 'carefully', i.e. in a layout action, a net is pulled only over a short internal. As a consequence, the layouts contain $O(N^2)$ bends, respectively knock-knees. Since the running time of a routing algorithm strongly depends on the number of bends performed by the nets, the running time of Frank's algorithm is $O(N^2)$ as well (Figure 11).

Area-optimal layouts containing only $O(N)$ bends are guaranteed by the algorithm of Kuchem, Wagner & Wagner [1989]. It runs in $O(N \log N)$. Since this algorithm is of special interest in connection with the layer assignment problem, we will discuss it in detail in Section 7.

Further channel routing algorithms for two-terminal net problems are due to Mehlhorn, Preparata & Sarrafzadeh [1986] and to Preparata & Lipski [1984]. These algorithms guarantee layouts within the optimum number of tracks (i.e. d tracks), but not within minimum spread. Both algorithms are 'pushing-algorithms' similar to the algorithm for multiterminal net problems of Mehlhorn, Preparata & Sarrafzadeh, i.e. they process the channel in a single left-to-right sweep. They run in $O(N)$ time, where an essential reason to achieve linear running time is that the layouts are not area-optimal.

It is easy to observe that a CRP of density d containing at least one one-sided net is always solvable in a channel of width $m \geq d$. Area-optimal layouts for these more general problems are guaranteed by the algorithm of Kuchem, Wagner & Wagner. But also the algorithms in Frank [1982], Mehlhorn, Preparata & Sarrafzadeh [1986] and Preparata & Lipski [1984] can be modified such that they can handle one-sided nets as well. For a CRP considered as an optimization problem, beside the number of layers needed for the wiring, the most important goal is to minimize the layout area. But, to ensure fast and reliable propagation of signals along the wires also the *length* of the wires is of central importance.

Sarrafzadeh [1987] gives an algorithm that finds a layout with bounded-length wires, where each of these can have up to twice its minimum length. This implies an upper bound on the total wire length of twice its optimum.

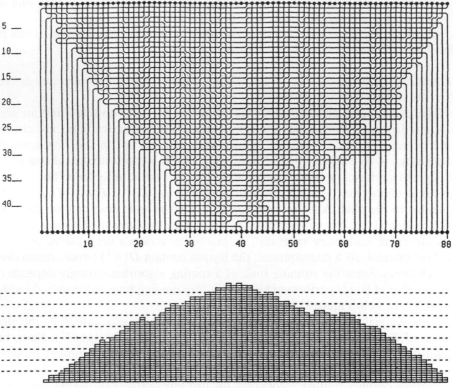

Fig. 11. A large layout determined by the algorithm of Frank and the density diagram of the problem.

The total wire length itself is considered recently by Formann, Wagner & Wagner [1993]. The algorithm they present constructs layouts with minimum total wire length for *dense* problems, i.e. problems in which all vertical lines are occupied by an upper and a lower terminal. The algorithm can be performed such that each wire has at most twice its minimal length and thus (in addition to all other optimality features) fulfills the optimality property of Sarrafzadeh [1987]. The 'key-result' to obtain this optimal algorithm is a tight lower bound for the wire length.

Consider a dense CRP of density d containing N nets, then, by Frank [1982], we know that at least one additional vertical line is needed for a layout. We assume that the channel has width d and spread $N \neq 1$ to achieve area-optimal layouts. Then a trivial lower bound for the total wire length is given by the number $d \cdot N$ of necessary vertical edges plus the sum of the local densities for the horizontal edges.

Now, to find a non-trivial lower bound, we must characterize the subproblems enforcing *detours* (i.e. extended nets). There are two types of detour-enforcing subproblems, *density valleys* and *autonomous intervals*.

Definition 6.3. *In a CRP a column is called a density valley iff there exists a column to its right and a column to its left whose local densities are greater than the local density of that column. In a layout a column is called a layout density valley iff there exists a column to its right and a column to its left whose number of horizontal edges occupied by a wire is greater than the number of horizontal edges of that column that are occupied by a wire.*

Lemma 6.4. *In every area-optimal layout for a dense CRP density valleys are filled up, i.e. there are no layout density valleys.*

An essential condition for the correctness of the lemma is that the CRP is dense.

Definition 6.5. *In a CRP an internal (of vertical lines) $[a, b]$ is called autonomous (AI) iff*
   (i) *every vertical line in $[a, b]$ is occupied by an input- and an exit terminal,*
   (ii) *no net leaves $[a, b]$, i.e. the input terminal of a net is in $[a, b]$ iff its exit terminal is in $[a, b]$;*
   (iii) *neither a nor b are straight lines, i.e. carry input- and exit terminal of the same net.*

Lemma 6.6. *Each AI enforces a detour. The number of detour edges is $2 \cdot (min\{l, r\} + 1)$ where l, respectively r is the number of straight lines entirely to the left, respectively right of the AI.*

The proof of the lemma follows directly from Frank [1982] since a *AI* is a dense CRP by itself. The naive approach to combine these lower bounds would add them together. But we cannot be sure (and indeed in general it is not the case) that the detour edges enforced by an *AI* are different from those that fill the density valleys. So, in addition to density valleys, we only have to consider those *AI*s that do not lie beside a density valley, the so-called *bad AI*s. We can conclude:

Theorem 6.7 [Formann, Wagner & Wagner]. *An area-optimal layout for a dense CRP of density d with N nets has to use $d \cdot N$ vertical edges and as many horizontal edges as:*
   (i) *the sum of all local densities plus*
   (ii) *the edges needed to fill the density valleys plus*
   (iii) *the detour edges caused by the bad AIs.*

The algorithm of Formann, Wagner & Wagner that produces area-optimal layouts with minimum total wire length for dense problems, is (as Frank's algorithm) a 'pulling method'. It also proceeds track by track, and the layout of a track is determined in alternating left-to-right, respectively right-to-left scans (Figures 12 and 13). During one scan the CRP is modified to a residual CRP whose density is decreased by one. In one scan, e.g. a left-to-right scan there are two possible

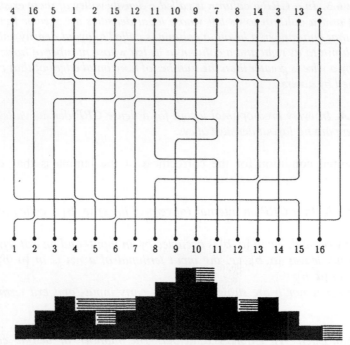

Fig. 12. A layout constructed by the algorithm of Formann, Wagner and Wagner, and its layout-density diagram. The density valleys are [4,7], [12,13]. The bad AIs are [9,10] end [1,16].

actions: either pull an 'appropriate' net from the right to the actual scan line or change to the next (right-to-left) scanning phase (using the track below). The choice of the appropriate net to pull is made such that the following invariants are maintained:

1. no unforced detours are made,

2. no additional detours are forced by creating new or deeper density valleys, new bad AIs or by adding new straight nets directly beside a bad AI.

Actually, each layout action depends on several different cases, which makes the algorithm somewhat complicated. Informally, the characteristic properties of the algorithm are 'laziness' and 'watchfulness'. A scanning phase stops as soon as possible after there is no column of maximal density left (laziness) and whenever the algorithm has the opportunity to fill a density valley or to get rid of a bad AI, then the algorithm seizes it (watchfulness). Similar as in Frank's algorithm, the nets are pulled only over a short interval per layout action. Therefore, again the layouts contain $O(N^2)$ bends, respectively knock-knees, and the running time is $O(N^2)$.

D. Wagner [1992] presents a new channel routing algorithm that produces for dense problems area-optimal layouts with minimum total wire length and only $O(N)$ bends, where the total number of bends is at most $d - 2$ more than the minimum. The running time is $O(N)$. The approach is completely different from the previously known algorithms, which all proceed track by track and on each

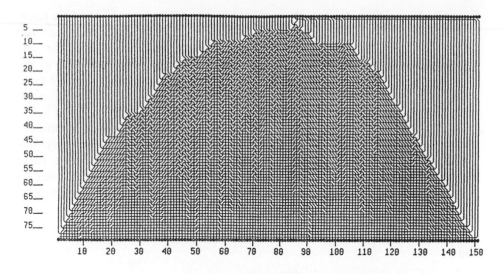

Fig. 13. A large layout determined by the algorithm of Formann, Wagner and Wagner, and the density diagram of the CRP.

track column by column performing the layout depending on the local state of the column. Instead, it is based on the 'cycle structure' of the CRP and thus may be viewed as a global approach. Again, the minimization of the number of bends relies on a good lower bound.

Obviously, every wire realizing a non-straight net contains an even number of bends, at least two. For two nets $\mathcal{N}_1 = (t_1, b_1), \mathcal{N}_2 = (t_2, b_2)$ with $t_1 = b_2$ (i.e. the input terminal of $\mathcal{N}_1$ lies above exit terminal of $\mathcal{N}_2$) we know that $\mathcal{N}_1$ must be moved before $\mathcal{N}_2$ to avoid vertical conflicts. This observation leads to the substructures enforcing additional bends, the cycles of a CRP. A sequence of nets $< (t_1, b_1), \ldots, (t_k, b_k) >$ is a *cycle* iff $t_i = b_{i+1}$ for $1 \le i < k$ and $t_k = b_1$. Since the cycles of a dense CRP induce a partition of the set of nets, we have

Lemma 6.8. *A layout for a dense CRP with m non-straight nets consisting of c cycles contains at least $2m + 2c$ bends.*

So, a good strategy to avoid bends is to lay out each net entirely on one track, i.e. to perform the layout cycle-wise. If the CRP consists of exactly one cycle, a layout with minimum total wire length and minimum number of bends is easily constructed by first routing the net, say with the leftmost exit terminal on the uppermost track, to the free vertical line to its left, and then routing the nets according to the cycle structure of the problem. Thereby, subsequent nets of the same 'type' i.e. both *left* ($t_i > b_i$), respectively both *right*, are routed on the same track, while a right net (left net) following a left net (right net) is laid out on the track below.

But, if the CRP consists of two or more cycles, the canonical procedure enforces unnecessarily long detours. Moreover, if the CRP contains a density valley, this procedure possibly does not even achieve minimum area. To overcome these problems, the algorithm first performs from the cycles one 'cycle arrangement' by applying a sequence of appropriate operations on the cycles (Figures 14 and 15).

Consider two cycles

$$C = \langle (t_1, b_1), \ldots, (t_k, b_k) \rangle$$

and

$$C' = \langle (t'_1, b'_1), \ldots, (t'_l, b'_l) \rangle.$$

By 'cutting' C and 'doubling' an appropriate net of C' we can combine C and C' to a 'cycle arrangement'

$$\langle (t_1, b_1), \ldots, (t_i, b_i), (t'_j, b'_j), (t'_{j+1}, b'_{j+1}), \ldots, (t'_j, b'_j),$$
$$(t_{i+1}, b_{i+1}), \ldots, (t_k, b_k) \rangle.$$

This cycle arrangement is transformed into a layout by the following procedure. First the nets of C are routed, each entirely on one track, until (t_i, b_i) is laid out. Then (t'_j, b'_j) is routed until the vertical line t_i followed by the subsequent nets of C' until (t'_j, b'_j) is routed again. Finally, the remaining nets of C are laid out. By moving (t'_j, b'_j) twice, the two additional bends enforced by C' are performed by either forming a detour or a stair. Essentially, there are three different interpretations of these operations on cycles: *patching* (in order to fill a density valley by a detour), *melting* overlapping cycles and *inserting* bad *AI*s into the outer problem.

Beside the optimal running time and the nearly minimum number of bends, the advantage of this approach is the restricted number and combinations of knock-knees per vertical line it guarantees, since the knock-knees contained in a layout are of fundamental importance for the wirability problem. In Wagner [1993], the approach from Wagner [1991] has been used to develop an algorithm that guarantees three-layer wirable layouts with nearly minimum total wire length and nearly minimum number of bends. (See Section 7).

7. Layer assignment

The design phase following the layout in the VLSI-design process is the *layer assignment*, i.e. the conversion of a layout in the plane to an actual

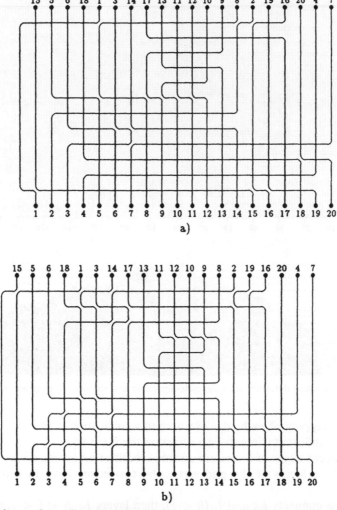

15 5 6 18 1 3 14 17 13 11 12 10 9 8 2 19 16 20 4 7

1 2 3 4 5 6 7 8 9 10 11 12 13 14 15 16 17 18 19 20

a)

15 5 6 18 1 3 14 17 13 11 12 10 9 8 2 19 16 20 4 7

1 2 3 4 5 6 7 8 9 10 11 12 13 14 15 16 17 18 19 20

b)

Fig. 14. a) A layout determined by the algorithm of Formann, Wagner and Wagner, and b) the layout determined by the cycle-algorithm for the same CRP. The bad AIs of the CRP are $[9,13]$ and $[10,12]$. Density valleys are between 6 and 7, 5 and 8, 15 and 16. The cycles of the CRP are $\langle 1,5,2,15\rangle$, $\langle 3,6\rangle$, $\langle 4,19,16,17,8,14,7,20,18\rangle$, $\langle 9,13\rangle$, $\langle 10,11,12\rangle$.

three-dimensional configuration of wires in order to avoid physical contacts between different wires. In the case of knock-knee layouts this is a non-trivial task.

A *conducting layer*, or simply *layer*, is a graph isomorphic to the routing graph. Conducting layers $L_1, \ldots, L_k$ are assumed to be stacked on top of each other, with L_1 on the bottom and L_k on the top. A contact between two layers, called a *via*, can be placed only at a vertex. A correct layer assignment or *wiring* of a given

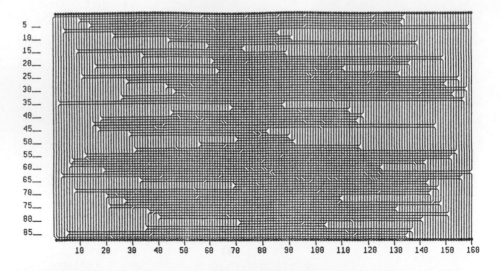

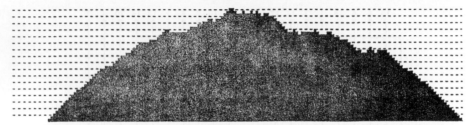

Fig. 15. A layout determined by the cycle-algorithm and the density diagram of the CRP.

layout is a mapping of each edge of a wire to a layer, such that:

1. No two different wires share a vertex on the same layer.

2. If adjacent edges of a wire are assigned to different layers, a via is established between these layers at the grid vertex the two edges have in common in the layout.

3. If a via connects L_h and L_j ($h < j$), then layers $L_i, h < i < j$, are not used at that vertex by another wire.

Given a routing problem, an important goal is to find, if it exists, a layout that is wirable with a minimum number of layers.

We review the combinatorial framework introduced by Lipski & Preparata, which applies to layouts in uniform grids [Preparata & Lipski, 1984; Lipski & Preparata, 1987]. Nearly nothing is known for layouts in more general graphs.

Consider a layout in a square grid. For the wiring problem only those grid vertices where two different wires intersect, i.e. cross or form a knock-knee are relevant. Denote the part of a layout induced by these grid vertices as the *core* of the layout. Then the following lemma holds.

Lemma 7.1. *A layout is wirable in k layers iff its core is.*

Obviously, the number of layers needed for a wiring of the core of a layout is at most the number of layers needed for the layout itself. On the other hand, if we have a wiring of the core of a layout in k layers, it induces a wiring of the layout in k layers by placing a via between arbitrary layers on any grid vertex outside the core if necessary. So, in the following we only consider the layout core.

The basic idea now is, that any correct wiring of a layout in a fixed number of layers induces a partition of the layout area into the following two types of regions:

V-region: the region where vertical wire edges lie above horizontal ones.

H-region: the region where horizontal wire edges lie above vertical ones.

We refer to this rule, namely that in a V-region vertical wire edges lie above horizontal ones, and in a H-region vice-versa, as the *H–V-rule*. By the partition in V-regions and H-regions, the layout can be considered as a *two-colorable map*. Obviously, the unit square around a crossing in the layout belongs to one color region. consider the unit square around a knock-knee. Two wires that form a knock-knee can not change their relative position through their common grid vertex in a correct wiring. Therefore, the unit square around a knock-knee must belong to both regions, say the 'triangle' above the imaginary diagonal through the knock-knee to the V-region and the triangle below to the H-region or vice-versa. Consequently, all diagonals through knock-knees belong to the boundary between two different regions. In addition, the boundary between two different regions contains 'appropriate' vertical and horizontal edges of the dual graph. The following properties of two-colorable maps give information how the boundary between different regions, i.e., a partition inducing a correct wiring has to look like.

Lemma 7.2. *A set P of diagonals and dual grid edges defines a two-colorable map iff*

(1) each interior vertex (of the dual of the routing graph) is incident with an even number of edges of P,

(2) each connected component of the boundary of the layout core is incident with an even number of edges of P.

We call the edges of P partition edges.

The main question now is: How must a set of partition edges look like in order to induce a wiring in a prescribed number of layers? For two layer wirings the answer is easy.

Lemma 7.3. *A set of partition edges defining a two-colorable map induces a two layer wiring iff it contains no horizontal or vertical edges.*

For, assume there is e.g. a vertical partition edge. Then a horizontal wire passing this edge must lie above all vertical wires on one side of the partition edge, and below all vertical wires on the other side of the partition edge. This enforces that the horizontal wire has to change layers, which is impossible inside the layout core with only two layers.

Consequently, two-layer wirable layouts are precisely those layouts in which the endpoints of the diagonals induced by the knock-knees are incident with one or three other endpoints of diagonals, or with the boundary of the layout core, and in which each connected component of the core of the boundary is incident with an even number of such vertices. Obviously, this is a very restrictive condition, which can be tested in time linear in the number of knock-knees.

To find the minimum number of layers needed for a layout in general, the notion of the layer graph is introduced in Lipski & Preparata [1987]. For a set of partition edges P inducing a two-colorable map and a fixed two-coloring the *layer graph* $G(P) = (V, E)$ is a directed graph defined by:

1. V is the set of grid vertices,
2. E is induced by the wire edges passing a horizontal or vertical edge from P, where the orientation of such a wire edge from a to b is (a, b) iff the wire edge must lie below some other wire edge through b by the H–V-rule. This means that all horizontal edges in E are oriented from the H-region to the V-region, and all vertical edges vice-versa.

The main result now is the following.

Theorem 7.4 [Lipski & Preparata]. *A partition P of a layout in a two-colorable map induces a wiring in k layers iff the length of the longest directed path in $G(P)$ is at most $k - 2$.*

Sketch of the proof. One direction follows from the fact that in any wiring a directed path of length l requires at most $l + 2$ layers.

Now, let $l(P)$ be the length of the longest path in $G(P)$. We get a correct wiring in $l(P) + 2$ layers $B \prec L_1 \prec \ldots \prec L_{l(P)} \prec T$ by the following assignment:

$$
W(e) := \begin{cases}
B & \text{if } e \text{ is horizontal in the V-region or} \\
  & e \text{ is vertical in the H-region, and } e \text{ is not in } G(P); \\
T & \text{if } e \text{ is vertical in the V-region or} \\
  & e \text{ is horizontal in the H-region, and } e \text{ is not in } G(P); \\
L_{t(e)} & \text{if } e \text{ is in } G(P),
\end{cases}
$$

where $t(e)$ denotes the length of the longest directed path in $G(P)$ terminating with edge e.

For fixed k, the decision problem if a layout is wirable in k layers is just the problem to decide if there exists a partition P in a two-colorable map that does not induce a directed path of length $k - 1$ in $G(P)$. That is, we have to characterize the forbidden patterns of partition edges inducing a directed path of length $k - 1$.

Corollary 7.5 [Preparata and Lipski]. *A layout is three-layer wirable iff there exists a partition in a two-colorable map containing none of the eight patterns shown in Figure 16.*

Using this fact, Lipski [1984] has proved that it is $\mathcal{NP}$-complete to decide if a layout is three-layer wirable (Figure 17).

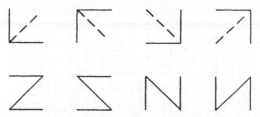

Fig. 16. Forbidden patterns for a three-layer partition. Broken lines denote the absence of edges.

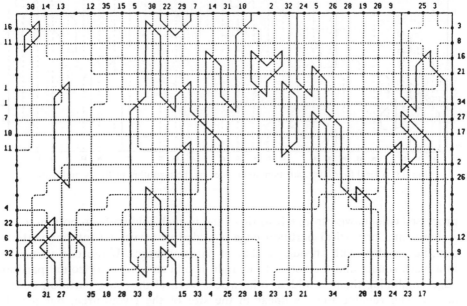

Fig. 17. A legal three-layer partition.

For four-layer wirability one obtains:

Corollary 7.6. *A layout is four-layer wirable iff there exists a partition in a two-colorable map containing no two 'neighboring' forbidden patterns for three layer wirability. (Neighboring means sharing a vertical or horizontal partition edge.)*

Brady & Brown [1984] present an algorithm that constructs for any layout a partition inducing a four-layer wiring.

Theorem 7.7 [Brady & Brown]. *Any layout can be wired with at most four layers.*

The partition algorithm of Brady & Brown is strikingly simple. The layout grid is considered as a chess-board pattern (the left-lower corner is white). It is

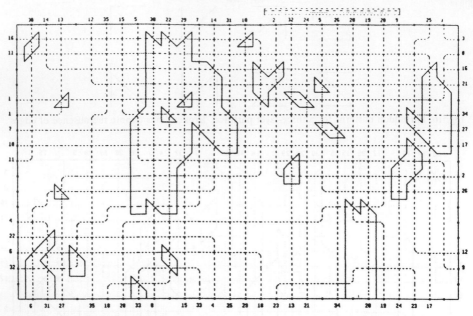

Fig. 18. A legal four layer partition determined by the algorithm of Brady and Brown.

processed column-wise from left-to-right, and in each column from bottom to top. A horizontal or vertical partition edge is added whenever an interior vertex is odd, i.e. incident with an odd number of partition edges. If the odd vertex lies in a black square and the vertex to its right is odd as well, a horizontal partition edge combining them is added, otherwise a vertical up-going partition edge is added. The running time is linear in the size of the layout-area, but may be implemented to run in time linear in the number of knock-knees contained in the layout (Figure 18).

Another approach that guarantees wirings within four layers for any layout is given in Tollis [1988]. The technique is equivalent to that of Brady & Brown. The notation and representation used is different, i.e. two-colorable maps are not introduced explicitly.

Since only very special layouts are wirable within two layers, the best one may expect in general is three-layer wirability. But to decide this is an $\mathcal{NP}$-complete problem [Lipski, 1984].

Therefore, a good approach to overcome this problem is to go one step back within the design process, and to consider the routing problem itself, i.e. to aim at layouts that are provably three-layer wirable.

Given: A routing problem.
Problem: Find a layout that is three-layer wirable.

The first result towards this goal is presented in Preparata & Lipski [1984].

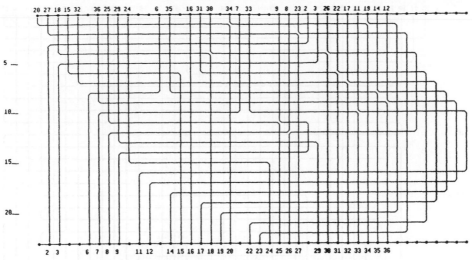

Fig. 19. Layout determined by the algorithm of Preparata & Lipski.

Theorem 7.8 [Preparata & Lipski]. *Any two-terminal net CRP containing only top-to-bottom nets admits a three-layer wirable layout within a channel of width d, where at most $d/2$ additional vertical lines to the right of the rightmost terminal are used (Figure 19).*

This result has been improved in Kuchem, Wagner & Wagner [1989].

Theorem 7.9 [Kuchem, Wagner & Wagner]. *Any two-terminal net CRP admits a three-layer wirable layout within a channel of minimum area (Figure 20).*

Both result are obtained by specifying an algorithm that constructs layouts with only a very small number of knock-knees. The possible combinations of knock-knees are very restricted, i.e. at most two knock-knees per vertical line and in case of two knock-knees they lie in opposite directions.

These very special layouts are obtained by routing the nets 'run-wise'. A *run* is a maximal sequence of nets $R = \langle (t_1, b_1), \ldots, (t_k, b_k) \rangle$ with $t_i = b_{i+1}$ for $1 \le i < k$, and either $t_i < b_i$ *(right run)* or $b_i < t_i$ *(left run)* for all $i, 1 \le i \le k$. The algorithm of Preparata & Lipski proceeds in a single left-to-right sweep, where runs are laid out completely on one track. As a consequence, nets are often extended to the right of the rightmost terminal. The algorithm may be viewed as a simultaneous 'right-pushing' on d tracks. In contrast, the algorithm of Kuchem, Wagner & Wagner proceeds in d 'pushing-phases', where right-pushing phases and left-pushing phases alternate, and where the layout on one track is determined in one phase. Again, runs are laid out completely on one track. To achieve minimum area, the algorithm produces at the beginning of each phase a free vertical line far enough to the right, respectively left of the channel, such

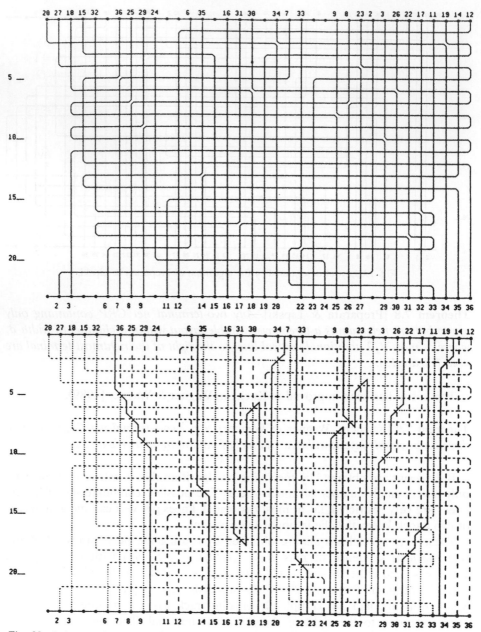

Fig. 20. A layout determined by the algorithm of Kuchem, Wagner and Wagner, and a legal
three-layer partition with the corresponding three-layer wiring.

that a net which is extended in the following phase may turn within the minimum
channel spread. (The generalization of runs in case of a CRP containing one-sided
nets is straight-forward.)

The run-oriented proceeding of these algorithms may be viewed as a first approach towards the global method developed in Wagner [1991].

The running time of the algorithm of Preparata & Lipski is $O(N)$, while the algorithm of Kuchem, Wagner & Wagner runs in $O(N \log N)$ using a special kind of segment trees as data structure. Obviously, both algorithms produce only $O(N)$ bends.

The three-layer wiring for a layout determined by one of these algorithms is now obtained by producing a 'legal partition' of the layout area. In both cases, the partition is constructed by adding only vertical edges. The layout is scanned column-wise from left to right and appropriate vertical edges are added in such a way that, after one scanning step, the following conditions hold:

1. the partition to the left of the scan line does not contain any forbidden pattern;

2. at most one diagonal (induced by a knock-knee) touching the scan line to the right forms an acute angle with any of the added edges.

To prescribe the appropriate way of adding vertical edges, all possible combinations of diagonals to the left and to the right of a scan line must be considered. It turns out, that in some cases the layout must be modified locally. This scan algorithm runs in time $O(N)$. The legal partition obtained induces a three-layer wiring with only $O(N)$ vias.

Using this approach, the layouts obtained by the algorithm of Mehlhorn, Preparata & Sarrafzadeh [1986] for multi-terminal net CRPs, can be wired within three layers as well. This result is also contained in Gonzales & Zheng [1988] in a more general setting. Recently, Wieners-Lummer [1991] modified the algorithm of Gao & Kaufmann for multi-terminal net CRPs to guarantee three-layer wirable layouts as well. But all these algorithms use additional vertical lines to the right of the rightmost terminal.

A different way to attack the wiring problem is to look for a *stretching* of the layout to derive two- or three-layer wirability. So, the general question is:

Given: A layout.

Problem: Find a set of tracks and/or vertical lines to be added such that the resulting layout is two-respectively three-layer wirable.

Stretching a layout increases the area. Obviously, to find a minimum stretching for three-layer wirability is $\mathcal{NP}$-hard. It remains $\mathcal{NP}$-hard for minimum stretching within only one dimension [Brady & Sarrafzadeh, 1990]. For the two-layer case, to find a minimum stretching at all is $\mathcal{NP}$-hard as well [Kaufmann & Molitor, 1991], while minimum stretching within one dimension is solvable in time linear in the layout area. Brady & Sarrafzadeh [1990] have shown that any layout may be transformed into a two-layer wirable layout by doubling the area, and into a three-layer wirable layout by increasing the area by $1/3$. Recently, Sarrafzadeh, Wagner, Wagner & Weihe [1994] have presented a global approach to attack the three-layer wiring problem. They consider the following question.

Given: A layout.

Problem: Does there exist a legal partition for three-layer wirability with only vertical (horizontal) additional partition edges.

This problem can be formulated as a path problem in a special graph, and may be decided in $O(N)$ time. Local layout modifications, as these applied in Preparata & Lipski [1984] and Kuchem, Wagner & Wagner [1989], and the local introduction of the orthogonal dimension (e.g. local use of horizontal edges within the vertical approach) are contained as well. Moreover, the method yields the minimum number of vertical lines (respectively tracks) to be added in order to transform a layout into a stretched layout that admits a legal partition for three-layer wirability with only vertical (respectively horizontal) additional partition edges. Using this method, one can obtain examples of layouts constructed by the algorithm from Preparata & Lipski [1984] or Kuchem, Wagner & Wagner [1989], where layout modifications are really necessary in order to derive a legal three-layer partition by adding only vertical partition edges. There also exist layouts containing only two knock-knees per column that admit no such legal partition, e.g. layouts computed by the algorithm from Wagner [1991]. For a survey on wiring, we also refer to Molitor [1991].

8. Manhattan routing

The most popular routing model for practitioners is the *Manhattan model*. It is in some sense intermediate between the edge-disjoint and the vertex-disjoint case. In a Manhattan layout, wires have to be edge-disjoint but may not form a knock-knee. From the algorithmical point of view the routing problem in Manhattan mode is very difficult. Even the most modest optimization goals seem to be intractable. Szymanski [1985] has proved that the channel width minimization problem for k-terminal net CRPs is $\mathcal{NP}$-hard for $k \geq 3$. In the same paper he claims, that it is even $\mathcal{NP}$-hard for the two-terminal case. But up to now, no proof of this claim has been published. The advantage of the Manhattan model is the fact that layer assignment is trivial. Obviously, any Manhattan layout is wirable within two layers, e.g. horizontal edges in the bottom layer and vertical edges in the top layer.

In this survey, we will concentrate on mathematically founded concepts for Manhattan routing, i.e., on the approximation algorithms given in Baker, Bhatt & Leighton [1984], Gao & Kaufmann [1987] and Wieners-Lummer [1990], and on the exact algorithms for special problems that are based on the vertical and horizontal *constraint graphs*.

Let us first consider lower bounds for the channel width for Manhattan CRPs. Again, a CRP of spread s consists of a set of nets, where a net $\mathcal{N} = (\{t_1, \ldots, t_l\}, \{b_1, \ldots, b_m\})$ is a collection of upper terminals $t_1, \ldots, t_m \in \{1, \ldots, s\}$ and lower terminals $b_1, \ldots, b_m \in \{1, \ldots, s\}$. Obviously, the density d is a lower bound for the channel width in Manhattan mode. This lower bound may be improved slightly. Let d_c denote the *closed density* of a CRP, i.e.

$$d_c = \max_{i \in \{1, \ldots, s\}} |\{\mathcal{N} \mid \mathcal{N} \text{ is a net with } \min\{t_1, b_1\} \leq i \text{ and } \max\{t_l, b_m\} \geq i\}|.$$

Since knock-knees are not allowed, d_c is a lower bound for the channel width. Obviously, $d_c \leq d + 1$ for all CRPs.

It is easy to see that the closed density is not a tight lower bound. For consider the so-called 'one shift problem' consisting of N two-terminal top-to-bottom nets $\mathcal{N}_1, \ldots, \mathcal{N}_N$ with $\mathcal{N}_i = (i, i+1)$. This problem has closed density two, but needs at least $\sqrt{2N+1} - 1$ tracks [Brown & Rivest, 1981]. The study of such pathological examples as the one shift problem leads to a second lower bound, the *flux* of a problem.

First, observe that in a layout each non-straight net must bend at least twice, once from the vertical to the horizontal direction and once vice-versa. Moreover, in a Manhattan layout a bend from the horizontal to the vertical direction requires a free vertical line. Now, consider a subproblem of a CRP consisting of r consecutive vertical lines. Let the subproblem contain t non-straight nets having at least one upper terminal on one of these r vertical lines, and a second terminal that is a lower terminal or outside the subproblem. We refer to such a subproblem of a CRP as $SCRP_{r,t}$. On the upper track, at most $r - t + 2$ such nets from $SCRP_{r,t}$ may be routed: at most $r - t$ to possibly $r - t$ free vertical lines inside $SCRP_{r,t}$, and two additional nets to the right, respectively left of $SCRP_{r,t}$. Accordingly, on the second track at most $r - t + 4$ such nets may be routed etc. Observe, that this approach does not take into account that a vertical line in $SCRP_{r,t}$ occupied by a second upper terminal of one of the t nets is *not free* at the beginning.

In general, the number of tracks necessary to route all t nets from $SCRP_{r,t}$ is at least

$$f(SCRP_{r,t}) := \min\{w \mid t \leq \sum_{i=1}^{w}(r - t + 2i)\} =$$

$$= \min\{w \mid t \leq (r - t + 1)w + w^2\},$$

which we call the *flux of the SCRP$_{r,t}$*.

Then, for a CRP the *flux* is defined as

$$f := \max_{SCRP_{r,t}} \{f(SCRP_{r,t})\}.$$

Lemma 8.1. *The minimum width for any Manhattan layout solving a CRP of closed density d_c and flux f is at least* $\max(d_c, f)$.

In the worst case, the flux can be in $O(\sqrt{N})$ for a CRP with N nets. Closed density and flux may be seen as completely 'unrelated'. Flux is a more 'local' phenomenon than density, and less likely to grow as the number of nets increases. There are problems where density dominates flux, say $d_c = N$ and $f = 1$. But on the other hand, there are also problems, as the one-shift problem, where flux dominates the density. Obviously, $\max(d_c, f)$ is not a tight lower bound for the channel width. It is easy to construct a CRP that requires more tracks than $max(d_c, f)$.

Baker, Bhatt & Leighton [1984] have used the concept of flux to design an approximation algorithm for Manhattan channel routing.

Theorem 8.2 [Baker, Bhatt & Leighton]. *For a CRP of closed density d_c and flux f, there exists a Manhattan layout using at most $d_c + \lceil 49,55 \cdot f \rceil + O(1)$ tracks if the CRP contains only two-terminal nets, respectively $2 \cdot d_c + \lceil 89,18 \cdot f \rceil + O(1)$ tracks if the CRP contains multi-terminal nets.*

This result has been improved by Wieners-Lummer [1990].

Theorem 8.3 [Wieners-Lummer]. *For a CRP of closed density d_c and flux f there exists a Manhattan layout using at most $d_c + \lceil 13,95 \cdot f \rceil + O(1)$ tracks in case the CRP contains only two-terminal nets, respectively $2 \cdot d_c + \lceil 17,65 \cdot f \rceil + O(1)$ tracks in case the CRP contains multi-terminal nets.*

We roughly describe the approximation algorithms from Baker, Bhatt & Leighton [1984], respectively Wieners-Lummer [1990] for two-terminal net CRPs, and outline their differences.

Both algorithms run in four phases.

Phase 1. Partitioning of the channel into *groups*. In Baker, Bhatt & Leighton [1984], the least b is determined such that the channel can be partitioned into groups of b^2 consecutive vertical lines, each group containing $3 \cdot b$ free upper and lower vertices. Then $b \leq 18 \cdot (f + 1)$.

In Wieners-Lummer [1990] integers $b_i, 1 \leq i \leq l$ for an appropriate l are chosen, such that the channel can be partitioned into groups of $c \cdot b_i^2$ consecutive vertical lines, each group containing $c \cdot b_i$ free upper and lower vertices. Then $b_i \leq \lceil (\sqrt{1/c + 1/4} + 1/2) \cdot f \rceil + 1$, where l and c must be calculated appropriately.

Phase 2. Uniform distribution of the free upper and lower vertices. In Baker, Bhatt & Leighton [1984] the groups are divided into b *blocks* of size b. Then nets are routed appropriately such that afterwards there exist three free upper and lower vertices for each block. The resulting problem has density at most $d_c + 6 \cdot b$.

In Wieners-Lummer [1990] the groups are divided into $c \cdot b_i$ blocks of size $b_i, 1 \leq i \leq l$. Nets are routed such that there exist two free upper and lower vertices for each block. The resulting problem then has density at most $d_c + 2 \cdot d \cdot \max\{b_i\}$.

Phase 3. Interblock routing, i.e. routing of the nets into the right blocks. This phase is the core of the algorithm.

In Baker, Bhatt & Leighton [1984] the nets are routed into their right block, but not necessarily to the right position, depending on a case analysis of the different types of nets. For this routing $d_c + 6 \cdot b$ tracks are used.

In Wieners-Lummer [1990] the interblock routing is a kind of 'right-pushing' algorithm, very similar to the knock-knee routing algorithm from Mehlhorn, Preparata & Sarrafzadeh [1986]. It is performed within $d_c + 2 \cdot c \cdot \max\{b_i\} + 8$ tracks, and leads to a more convenient starting problem for the last phase.

Phase 4. Routing of the right nets within each block to the correct position.

Both algorithms use an approach of Kawamoto & Kajitani [1979] to route the nets to their correct position. This method guarantees a layout within $(3/2)N$ tracks for a problem containing N nets: one track per net plus at most $(1/2)N$ tracks to break $(1/2)N$ possible cycles. Thus, it routes each block within at most $(3/2)b$ tracks.

In Wieners-Lummer [1990] the number of tracks used in this phase is improved to $(5/4)\max\{b_i\} + O(1)$ by taking into account that the input problems for phase 4 have a special structure.

The approximation methods for multi-terminal net CRPs are of the same flavor, but usually more complicated.

Both methods are not applicable to real world problems because of their 'terrible' approximation factors. But the algorithm from Wieners-Lummer [1990] is at least compatible with the *greedy strategy* from Rivest & Fiduccia [1982], which seems to work well in practice. The reason is, that the interblock routing method used is similar to the procedure of the greedy strategy, which is a kind of 'right-pushing' algorithm as the algorithm in Mehlhorn, Preparata & Sarrafzadeh [1986] for knock-knee routing.

The greedy router scans the channel from left to right and routes the nets from an upper terminal to the uppermost free track in the channel, respectively from a lower terminal to the lowermost free track. If a net occupies more than one track, the router tries to free up as many tracks as possible by bringing 'split-up' nets together, respectively closing extended nets. Moreover, if possible, nets are routed closer to their next top- or bottom terminal. So, the approximation algorithm of Wieners-Lummer can be included in the greedy method. Whenever the number of tracks required by the greedy strategy exceeds a prescribed value during the scanning, the current column is re-routed using the approximation algorithm.

An interesting variant of the Manhattan model is the *jog-free model*.

Given: A CRP.
Problem: Find a Manhattan layout, where each non-straight net is routed entirely on one track, i.e. no bends additional to the two trivial bends are created, especially no detours.

Not even every CRP is *solvable* under the jog-free model. The solvability problem leads to the consideration of *vertical* and *horizontal* constraints.

Consider a CRP consisting of N nets. The *vertical constraint graph* of the CRP is a directed graph $G_v = (V_v, E_v)$, where V_v corresponds to the set of nets, and where $(\mathcal{N}_i, \mathcal{N}_j) \in E_v$ for $1 \le i, j \le N$ iff net $\mathcal{N}_i$ has a lower terminal on the same vertical line where $\mathcal{N}_j$ has an upper terminal.

The vertical constraint graph reflects the fact that net $\mathcal{N}_i$ has to be routed on some track below the track used by $\mathcal{N}_j$ whenever $(\mathcal{N}_i, \mathcal{N}_j) \in E_v$. The *horizontal constraint graph* of a CRP is an undirected graph $G_h = (V_h, E_h)$, where V_h again corresponds to the set of nets, and where $\{\mathcal{N}_i, \mathcal{N}_j\} \in E_h$ for $1 \le i, j \le N$ iff the interval between the leftmost and the rightmost terminal of $\mathcal{N}_i$ and the interval between the leftmost and the rightmost terminal of $\mathcal{N}_j$ are not disjoint. The horizontal constraint graph reflects the closed density of a CRP, so d_c is just the maximum clique size of G_h.

Lemma 8.4. *A CRP is solvable in the jog-free Manhattan model iff G_V is acyclic.*

For acyclic CRPs, G_v and G_h provide lower bounds for the channel width. Let p_v denote the length of the longest directed path in G_v.

Lemma 8.5. *For a CRP solvable in the jog-free Manhattan model* $\max(d_c, p_v + 1)$ *is a lower bound for the channel width.*

The channel width minimization problem may be formulated as a special coloring problem on graphs. Consider the *constraint graph* $G = (V, E)$ of a CRP, whose vertex set V is the set of nets and whose edge set E is the set E_v together with those edges from E_h which have no corresponding arc in E_v. Then any correct jog-free Manhattan layout for the CRP corresponds to a vertex coloring of G with the following properties. The colors are numbered. Vertices joined by an edge from E have to be colored differently, and in addition net $\mathcal{N}_i$ is assigned a lower numbered color than net $\mathcal{N}_j$ iff $(\mathcal{N}_i, \mathcal{N}_j) \in E_v$.

Minimizing the number of colors and, equivalently, minimizing channel width in jog-free Manhattan mode is $\mathcal{NP}$-hard [LaPaugh, 1980]. But for special cases the problem is polynomially solvable.

Lemma 8.6. *For a CRP with N nets and $E_v = \emptyset$, a jog-free Manhattan layout can be determined within a channel of width d_c in $O(N)$ time.*

The channel width minimization problem is in that case equivalent to 'scheduling jobs with fixed starting and completion times' [Gupta, Lee & Leung, 1979]. The layout is determined by passing the nets ordered according to their leftmost terminals. The tracks are stored in a stack, where the uppermost track lies on top of the stack. Now, a net is assigned to the track on top of the stack, and a track is put back on the stack as soon as it is 'free' (not occupied by a net).

This optimal algorithm for vertical constraint free problems induces a $\sqrt{N}$-approximation for cycle-free CRPs with $E_v \neq \emptyset$ [LaPaugh, 1980]. The algorithm is often referred to as the 'left-edge algorithm'. Again the nets are assigned to the 'last-freed' track according to their leftmost terminal, now regarding also the partial order induced by E_v. Using a balanced search tree as data structure, the algorithm can be implemented to run in $O(N \log N)$ time [Hashimoto & Stevens, 1971].

Heuristics based on the concept of constraint graphs were presented by Yoshimura & Kuh [1982] and Yoshimura [1984].

There are a lot of heuristics that work quite well in practice. Most of them are based somehow on the left-edge algorithm, also heuristics routing with jogs, e.g. Reed, Sangiovanni-Vincentelli & Santomauro [1985]. For an overview on channel routing, especially in Manhattan mode, we also refer to LaPaugh & Pinter [1990].

9. Via minimization

An important problem coming up with the layer assignment problem is the via minimization problem.

Given: A layout, not necessarily grid based.
Problem: Find a correct layer assignment such that the number of vias is minimum.

Only heuristics are known for this problem if the layout requires three or more layers for a correct wiring. But the via minimization problem for two-layer wirable layouts is very well understood, for both the grid based and the non-grid based case.

This problem has been first formulated as a graph theoretic maximum cut problem by Hashimoto & Stevens [1971]. Based on this formulation, several heuristics [Chang & Du, 1987; Grötschel, Jünger & Reinelt, 1988], approximate methods [Stevens & VanCleemput, 1979], and exact methods using integer programming [Cielski & Kinnen, 1981] or a cutting plane algorithm [Barahona, Grötschel, Jünger & Reinelt, 1988] have been proposed.

Naclerio, Masuda & Nakajima [1989] have proved that the via minimization problem for two-layer assignment is $\mathcal{NP}$-hard, even if the layout is grid based or vias are restricted to lie at *junctions* (points, where two or more wire segments meet), or if the junction degree is limited to six. Choi, Nakajima & Rim [1989] improve this result by showing that the problem is $\mathcal{NP}$-hard even when restricted to a maximum junction degree of four.

Kajitani [1980] has shown that the problem transforms to a maximum cut problem in planar graphs if the maximum junction degree is three, or if vias are not allowed to be placed on junctions. Several polynomial time algorithms based on Hadlock's planar maximum cut algorithm [Hadlock, 1975], or related methods have been proposed in Kajitani [1980], Chen, Kajitani & Chan [1983], Molitor [1987], Naclerio, Masuda & Nakajima [1987]. They all have running time $O(n^3)$, where n is the number different *wire-segment clusters* (possibly larger than the number of nets). In Pinter [1984], the problem is also formulated as a maximum cut problem in planar graphs, but since also negative edge weights appear, Hadlock's algorithm does not work in this model (although the opposite is claimed in Pinter [1984]).

Recently, Kuo, Chern & Shih [1988] have presented a solution method for the via minimization problem with an improved time complexity of $O(n^{3/2} \log n)$. They use the model from Pinter [1984], but apply a new $O(n^{3/2} \log n)$ algorithm for planar real-weight maximum cut proposed by Shih, Wu & Kuo [1990].

We introduce this modelling here, and briefly describe the new maximum cut algorithm.

Consider a layout as a collection of paths in a planar graph, possibly in a grid graph. The following modelling is valid for any layout style, Manhattan, knock-knee or overlap.

The wires of the layout can be divided into different subpaths:

(i) *Via candidates*. These are maximal subpaths which can accommodate a via, i.e. do not cross or overlap any other wire and offer enough space for a via (e.g. contain at least one grid vertex in a grid based layout).

(ii) *Conflict segments*. These are the remaining subpaths, i.e. subpaths which can not accommodate a via or where vias are forbidden. Conflict segments connect two via candidates, or a via candidate and a terminal, or two terminals.

Two conflict segments belong to the same *conflict cluster* iff they cross each other or cross the same conflict segment. Obviously, this relation induces a partition of the set of conflict segments. We can now model the layout as a graph, the *cluster graph* $G_c = (V_c, E_c)$. The vertices in V_c correspond to the conflict clusters, and two vertices are connected by an edge of E_c iff there exist conflict segments in the corresponding conflict clusters which are connected by the same via candidate in the layout. If the layout is two-layer wirable, then the layer assignment for one cluster is uniquely determined by assigning one representative of a cluster to a fixed layer. Analogously, a layer assignment to two layers is induced by choosing from each conflict cluster one representative to be assigned to the bottom layer. Now, to solve the via minimization problem, we can associate edge-weights with every edge $e \in E_c$. Let $c(e)$ be the number of via candidates connecting elements of the two conflict clusters incident to e. For an arbitrary, but fixed two-layer wiring of the layout, let $v(e)$ be the number of vias introduced by via candidates associated with e. Define

$$\omega(e) := v(e) - (c(e) - v(e)) \text{ for } e \in E_c,$$

i.e. the *weight* of an edge e is the 'via reduction' that can be achieved along e by 'flipping' the layer assignment on one conflict cluster incident to e. Now, each two-layer wiring of the layout can be obtained from the fixed wiring associated with v by 'flipping' the layer assignment on some conflict clusters, i.e. on a subset $X \subseteq V_c$. Clearly, a two-layer wiring with minimum number of vias then is induced by the subset of V_c that maximizes the total via reduction. So, the via minimization problem is equivalent to finding a cut $X \subseteq V_c$ in G_c that maximizes the sum of the weights of all edges leaving X, i.e., to calculate

$$\max_{X \subseteq V_c} \sum_{e \in (X, V_c \setminus X)} w(e),$$

where $(X, V_c \setminus X)$ is the set of edges going from X to $V_c \setminus X$.

Observe, that the edge weights may be positive or negative, but that a maximum cut is always positive. If the maximum junction degree in the given layout is at most three, the cluster graph G_c is planar.

Theorem 9.1 [Pinter; Kuo, Chern & Shih]. *The via minimization problem for a two-layer wirable layout with maximum junction degree three is equivalent to the maximum cut problem in a planar graph.*

The maximum cut problem in arbitrary graphs is $\mathcal{NP}$-hard. If the underlying graph is planar it transforms to a minimum complete matching problem in a very special graph. W.l.o.g. we may assume that G_c is connected. The maximum cut algorithm consists of the following four steps.

Step 1. Construct the *triangulation* $G_t = (V_c, E_t)$ of $G_c = (V_c, E_c)$. G_t is a connected planar graph with

   (i) $E_c \subseteq E_t$,

   (ii) each vertex from G_t has degree at least two,

   (iii) each face of G_t is enclosed by a simple cycle of length three,

   (iv) every two faces of G_t share at most one edge.

All new edges are assigned a weight of zero. Then obviously, a maximum cut in G_t is equivalent to a maximum cut in G_c.

Step 2. Construct the *geometric dual* $G_d = (V_d, E_d)$ of $G_t = (V_c, E_t)$, G_d is the graph obtained by

   (i) associating a vertex in V_d with each face of G_t,

   (ii) introducing an edge in E_d between vertices in V_d for each edge shared by the corresponding faces in G_t,

   (iii) assigning to each edge of E_d the same weight as its corresponding edge in E_t.

Since G_t is a planar triangulation, G_d is a three-regular planar graph.

We call a subset $D \subseteq E_d$ an *even-degree edge set* iff each vertex of V_d is incident to an even number of edges in D.

Lemma 9.2. *Finding a maximum cut in G_t is equivalent to finding an even-degree edge set of maximum weight in G_d.*

The proof is contained in Shih, Wu & Kuo [1990].

Step 3. Construct from $G_d = (V_d, E_d)$ a graph $G^* = (V^*, E^*)$ by replacing each vertex $v \in V_d$ by a 'star-graph' consisting of seven vertices $\{v_1, \ldots v_4, u_1, u_2, u_3\}$, and edges $\{\{v_i, v_{i+1}\} : 1 \leq i < 4\} \cup \{\{v_i, u_j\} : 1 \leq i \leq 4, 1 \leq j \leq 3\}$. For each edge in E_d we introduce an edge in E^* between corresponding 'u-vertices', such that the u-vertices in G^* all have degree three. (Observe, that G_d is three-regular.) All star-edges are assigned a weight of zero, all other edges the weights of the corresponding edges in E_d.

Lemma 9.3. *A minimum complete matching $M \subseteq E^*$ in G^* induces a maximum even-degree edge set $E_d \setminus M$ in G_d.*

Step 4. Compute a minimum complete matching in G^*. This can be done by computing a maximum weight matching in the same graph with appropriately replaced edge-weights. Lipton & Tarjan [1980] present a maximum weight matching algorithm with running time $O(n^{3/2} \log n)$ for planar graphs by applying the planar separator theorem [Lipton & Tarjan, 1979] which we mentioned in Part I. Although G^* is not planar, the same method can be applied to G^* as well [Matsumoto, Nishizeki & Saito, 1986].

Theorem 9.4 [Shih, Wu & Kuo]. *A maximum cut in a planar graph can be found in* $O(n^{3/2} \log n)$ *time.*

10. Single layer routing

The most restrictive routing problem is the *single layer routing problem*, where only one layer is available for the layout of the nets. It comes up when e.g. critical nets are to be placed on a preferred layer. In the most general form it is just the *vertex-disjoint Steiner tree problem*, i.e.,

Given: A routing graph $G = (V, E)$, $|V| = n$, and nets $\mathcal{N}_1, \ldots, \mathcal{N}_N \subseteq V$.
Problem: Find N pairwise vertex-disjoint Steiner trees for $\mathcal{N}_1, \ldots, \mathcal{N}_N$.

The problem is $\mathcal{NP}$-complete, even for planar graphs [Lynch, 1975] or grids [Kramer & van Leeuwen, 1984]. However, Robertson & Seymour showed that the problem is solvable in polynomial time if the number of nets is fixed [Robertson & Seymour, 1990] or if G is planar and the terminals lie only on one or two face boundaries [Robertson & Seymour, 1986]. But the algorithm they give is far from being implementable. In Liao & Sarrafzadeh [1990] and in Suzuki, Akama & Nishizeki [1990] an $O(n)$ algorithm is given for the case that all terminals lie on one face boundary. The algorithm runs in two phases, first the topological solvability is tested and then the layout of the nets is determined, assuming enough capacity is available. *Topological solvability* can be decided by a simple *stack algorithm*. The terminals are scanned clockwise around the boundary and every new visited terminal is pushed onto the stack. If the pushed terminal is the last non-visited terminal of the corresponding net, it is tested if all terminals of the net lie on top of the stack. If this is not the case, the problem is not topological solvable. Otherwise, all terminals of the net are popped from the stack. When all the terminals of all nets have been visited and no conflict has been detected before, the instance is topological solvable iff the stack is empty.

To test *routability* for a topologically solvable problem the *density* and the *capacity* of appropriate cuts must be considered, as in the edge-disjoint case.

The relevant cuts again are just subsets X of the vertex-set of the routing graph, and the density of a cut X is the number of nets leaving X. The relevant capacity for vertex-disjoint routing is the maximum number of vertex-disjoint edges leaving X, which we refer to as the *strong capacity* $cap_s(X)$. Then the *free capacity* of X in this case is defined as

$$fcap_s(X) := cap_s(X) - dens(X).$$

Obviously, in any planar graph $G = (V, E)$, a necessary condition for routability of a topological solvable vertex-disjoint Steiner tree problem is $fcap_s(X) \geq 0$ for all $X \subseteq V$. For the vertex-disjoint case (in contrast to the edge-disjoint case considered in Section 5), this condition is also sufficient, even when restricted to only connected cuts.

Lemma 10.1. *Given a planar graph* $G = (V, E)$, *a set of nets, and an embedding of G with all terminals lying on the outer face such that the vertex-disjoint problem is topologically solvable then there exists a correct layout iff* $fcap_s(X) \geq 0$ *for all connected cuts* $X \subseteq V$.

To prove this lemma, we can apply a simple layout algorithm that is based on the stack algorithm for testing solvability. To route the nets correctly, they are considered in the order in which they have been deleted from the stack. They are routed counterclockwise along the boundary beginning from the last terminal pushed onto the stack. Then the boundary is corrected by deleting all used edges and vertices, and all edges incident to them. If there is enough capacity, this method leads to a correct layout. The running time is $O(n)$.

The problem is much more complicated if the terminals are allowed to lie on two face boundaries. Suzuki, Akama & Nishizeki [1990] have developed an algorithm for this problem with run time $O(n \log n)$. Recently, Ripphausen-Lipa, Wagner & Weihe [1993] have presented a *linear time* algorithm for the planar vertex-disjoint Menger-problem. Using this algorithm, the two-face Steiner-tree problem may be solved in linear time as well.

The case when the routing graph is a grid graph, especially a channel, has been considered e.g. in Cole & Siegel [1984], Leiserson & Pinter [1983], Leiserson & Maley [1985], Mirzaian [1987], Siegel & Dolev [1988]. It turns out that, for the vertex-disjoint CRP, the number of cuts that must be tested for routability can be reduced significantly.

At first it suffices to consider only connected cuts that cut one edge on the upper boundary and one edge on the lower boundary. Then the edges leaving such a cut induce a path in the dual graph starting with the edge induced by the upper boundary edge and terminating with the edge induced by the lower boundary edge. Obviously, all connected cuts induced by the same pair of upper and lower boundary edges have the same density but possibly different capacity. To test routability, clearly only such cuts of minimum capacity must be considered, the so-called *diagonal cuts* or *critical cuts*. Observe, that the critical cut for an upper boundary edge $\{u, u + 1\}$ and a lower boundary edge $\{l, l + 1\}$ (u and l indicate vertical lines) induces a path in the dual graph. It starts with the edge dual to $\{u, u + 1\}$, continues by alternatingly horizontal and vertical edges, possibly followed by a sequence of vertical or horizontal edges, and terminates with the edge dual to $\{l, l + 1\}$.

We call the critical cut induced by $\{u, u + 1\}$ and $\{l, l + 1\}$ an (u, l)-*cut*. Then the capacity of such a critical cut in a channel of width w is just $cap_s(u, l) = \max\{|l - u|, w\}$.

Lemma 10.2. *For a topologically solvable CRP, a correct vertex-disjoint layout exists iff* $fcap_s(u, l) \geq 0$ *for all critical* (u, l)-*cuts.*

Testing all critical cuts leads to an $\Omega(N^2)$ algorithm (where N denotes number of nets). But it turns out that testing $O(N)$ cuts at all is sufficient. At first, observe that we only have to consider (u, l)-cuts where u or l is incident with a leftmost or rightmost upper, respectively lower terminal of a net. For a fixed upper edge $\{u, u + 1\}$ consider all (u, l)-cuts for lower boundary edges $\{l, l + 1\}$. Obviously, many of these (u, l) cuts have $fcap_s(u, l) \geq 0$ independently from the width of the channel, i.e., all these (u, l)-cuts where l is far enough to the left or to the right of u to ensure a high capacity of vertical edges. In other words, the critical (u, l)-cuts for u that must be tested are precisely the cuts whose density is greater than the number of vertical edges leaving the cuts. We call these cuts *dense*. For a fixed u the dense cuts have the shape of a *cone*.

Now let (u_1, l_1) be the 'leftmost' dense cut for u_1 and (u_1, r_1) the 'rightmost' dense cut for u_1. For an upper edge $\{u_2, u_2 + 1\}$ with $u_1 < u_2$, the corresponding leftmost, respectively rightmost dense cuts, say (u_2, l_2) and (u_2, r_2), must satisfy $l_1 \leq l_2$ and $r_1 \leq r_2$, respectively.

Based on these considerations, it can be proved that the condition '$fcap_s(u, l) \geq 0$ for all dense (u, l)-cuts' is testable by a single left-to-right scan through the channel which requires only $O(N)$ time. E. g., for the most special case that the CRP contains only two-terminal top-to-bottom nets, say $\{t_1, b_1\}, \ldots, \{t_N, b_N\}$, the test if the problem is routable within a channel of width w is just the test

$$t_{i+w} - b_i \geq w$$

and

$$b_{i+w} - t_i \geq w \text{ for } 1 \leq i \leq N - w.$$

Obviously, this (N) procedure can be applied to also solve the channel width minimization problem. In this case, just start with $w = 0$ and increase w whenever a violation occurs. For more information about different routing models we also refer to Lengauer [1990].

Part III. Linear Layout Methods

We study a class of graph-theoretic problems that arise in certain linear VLSI layout styles such as Weinberger arrays, gate matrix layout and PLA-folding. These layout styles are typically used in *cell synthesis*. This is the part of the layout process in which the physical building blocks for the placement and routing phase are constructed.

The term *linear* refers to the fact that the most important degrees of freedom in the underlying VLSI architecture consist of linear (i.e. one- dimensional) arrangements of the relevant physical objects, *the gates*.

In the most general form, an instance of such a linear layout problem consists of an $m \times n$ $0, 1$ matrix $M = (m_{ij})$ (*the net-gate matrix*), whose rows and columns represent the nets $N_1, \ldots, N_m$ and the gates $G_1, \ldots, G_n$ of the circuit, respectively.

The gates may be thought of as the basic electronic devices that are arranged linearly in a row, and the nets as realizing connections between them (details are given in Section 11). Net N_i must connect all gates G_j with $m_{ij} = 1$. Connections are realized row-wise by reserving for a given permutation of the gates (columns) for every net N_i the part of the row from the leftmost to the rightmost gate to which a connection must be established.

This can be expressed more formally by considering for a permutation π of the gates the *augmented net-gate matrix* $M_\pi = (\bar{m}_{ij})$ with

$$\bar{m}_{ij} = \begin{cases} 1 & \text{if there are gates } G_r, G_s \text{ with } \pi(r) \leq \pi(s) \text{ and } m_{ir} = m_{is} = 1 \\ 0 & \text{otherwise} \end{cases}$$

Nets of the augmented net-gate matrix may share the same row (called *track*) if they have no gate in common. An assignment of augmented nets to tracks preserving this property is called a *feasible track assignment*.

The additional ones in M (with respect to the same column permutation of M) are called *fill-ins*. They are represented by a '$*$' to distinguish them from the given '1's in M (see Figure 21).

The result of a permutation of the gates (*gate arrangement*) and an associated feasible track assignment is called a *layout*. Its area is proportional to # gates $\times$ # tracks $= n \times$ # tracks.

So constructing an area-minimal layout (*optimal layout*) is equivalent to finding a gate arrangement and an associated feasible track assignment such that the number of tracks is minimum.

In matrix terminology, this leads to the following *matrix permutation problem (MPP)*:

Given: A 0, 1 matrix (net-gate matrix) M.

Problem: Find a permutation of the columns and an assignment of the augmented rows (nets) to tracks such that the number of tracks is minimum.

An example is given in Figure 21. We denote the minimum number of tracks by $t(M)$ and call a layout with $t(M)$ tracks an *optimal* layout. Due to the mentioned origin of this problem in VLSI- applications, there is an enormous body of papers on it.

Section 11 deals with the VLSI background and models the linear VLSI layout technologies 'Weinberger arrays', 'gate matrix layout' and 'PLA folding' as instances of the general MPP. Some of these applications lead to restricted MPP's in the sense that either the permutation of the gates (e.g. fixed first and last gate) or the assignment of nets to tracks (e.g. at most two per track) are restricted.

Section 12 gives several equivalent graph theoretic problems with different and partly independent background. Among them are: augmentation of a graph to an interval graph with small clique size, a node search problem, determining the path-width of a graph, matching problems with side constraints and others.

Section 13 is devoted to the complexity of the problem, and to reductions between some of the specialized versions. The general problem is already $\mathcal{NP}$-hard on chordal graphs, but solvable in polynomial time on trees and cographs and if the number of tracks is fixed. Sharper versions of $\mathcal{NP}$- completeness results are also obtained for certain variants of PLA folding. Finally, Section 14 deals with exact and approximation algorithms for solving MPP's. Due to the practical relevance of the problem, many heuristics have been proposed in the literature.

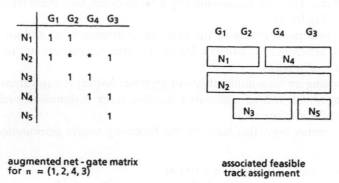

	G_1	G_2	G_3	G_4
N_1	1	1		
N_2	1			1
N_3		1		1
N_4			1	1
N_5		1		

net - gate matrix

	G_1	G_2	G_4	G_3
N_1	1	1		
N_2	1	*	*	1
N_3		1	1	
N_4			1	1
N_5		1		

augmented net - gate matrix
for n = (1, 2, 4, 3)

associated feasible
track assignment

Fig. 21. An example MPP.

Nevertheless, the problem of the existence of an approximation algorithm with constant relative performance bound has remained open.

11. The VLSI background

The matrix permutation problem is typical for a number of 'regular' layout styles for the generation of random logic modules in VLSI.

Such a random logic module may be seen as an irregular structure of basic components such as transistors, gates, flip-flops etc. It is given by some input description, e.g., by a transistor scheme, a logic scheme, or a set of Boolean functions. From this description, a concrete layout (a physical module) must be constructed according to some layout architecture style. A *regular* layout style is a style in which basic topological relationships between the physical components on the chip area are known in advance (e.g. restricted placement, predefined locations for the arrangements of gates etc.).

Typical such layout styles are *Weinberger arrays, gate matrix layout* and *programmable logic arrays* discussed below, see also Gajski & Lin [1988] and Brayton, McMullen, Hachtel & Sangiovanni-Vincentelli [1984].

11.1. Weinberger arrays

Weinberger arrays were introduced in Weinberger [1967] as a layout architecture for Boolean functions that are given by a circuit consisting only of NOR-gates (see Figure 22). Each NOR-gate is converted into an nMOS gate (i.e. a gate in the nMOS VLSI technology, see e.g. Mead & Conway [1980] for details); and these gates are arranged in a linear array that constitute the columns of an associated MPP (see Figure 23). Each column of this MPP consists of two vertical wires. One wire is connected to the pull-up transistor and serves as the output port, while the other wire is connected with the ground power line. (Usually,

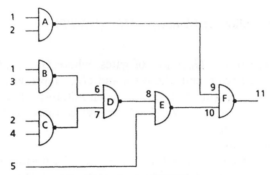

Fig. 22. A circuit of NOR gates.

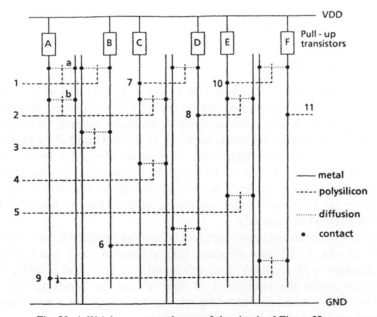

Fig. 23. A Weinberger array layout of the circuit of Figure 22.

two neighboring gates share a common ground wire.) The input signals to the gates are obtained from horizontal polysilicon intervals on a row. A transistor is formed by the intersection of an extension of such a polysilicon interval with a diffusion segment between the output and ground lines. For example, transistors a and b are formed by connecting row 1 and 2 to diffusion segments in gate A. The output of gate A is connected to the last row which serves as input to gate F, etc.

Note that connections (i.e. polysilicon intervals) may be placed on the same line if they do not overlap. Since the number of gates is fixed, minimization of the layout area is equivalent to reducing the number of rows, i.e. by finding a suitable permutation of the gates (which defines the length of the polysilicon intervals) and an associated row assignment of the intervals such that the number of rows is minimum.

So we obtain the following *Weinberger MPP (WMPP)*:

Given:
- A collection $G_0, G_1, \ldots, G_n, G_{n+1}$ of gates, where $G_1, \ldots, G_n$ represent the NOR gates of the circuit, and where G_0 and G_{n+1} represent the input (on the left) and output signals (on the right) of the circuit, respectively.
- A collection of nets $N_1, \ldots, N_m$. Each net N_i consists of those gates G_j, to which it is output or input.
- The net-gate matrix M.

Problem: Find a permutation of $G_1, \ldots, G_n$ (i.e. the positions of G_0 and G_{n+1} are fixed) and an associated feasible track assignment (layout) such that the number of tracks is minimum.

Figure 24 shows the matrix M, a layout corresponding to Figure 23 and an optimal layout for the above example.

11.2. Gate matrix layout

This architecture was introduced by Lopez & Law [1980] as a regular layout style for large scale transistor circuits in the CMOS technology. In such a layout, a vertical polysilicon wire corresponding to an input, internal or output signal is placed in every column (columns A, B, C, D, E, F, G and Z in Figure 25). All transistors using the same signal are constructed along the same column (e.g. transistors 1 and 7 in column A of Figure 25). Connections among transistors are made by horizontal metal lines, while connections to Vdd/Vss are in a second metal layer (and irrelevant to the underlying MPP). They are indicated by up and downward arrows in Figure 25.

A net is a collection of metal lines and transistors to which it must be connected. Net N_1 in row A of Figure 25 spans from column A to G and is connected to three transistors (1, 2, and 3) and one metal line (G). The (slightly simplified) assumption about the realization of nets is that the series-parallel transistor circuit of each net and its output signals can be realized in a row regardless the permutation of the metal lines and transistors.

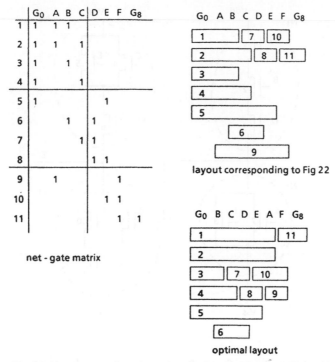

net - gate matrix

layout corresponding to Fig 22

optimal layout

Fig. 24. Net-gate matrix and two layouts for the WMPP of Figure 22.

As with Weinberger arrays, minimizing the layout area leads to the following *gate matrix permutation problem (GMPP)*.

Given:
- A collection $G_1, G_2, \ldots, G_n$ of gates representing metal lines or transistors of a gate matrix.
- A collection $N_1, \ldots, N_m$ of nets, where each N_i is a subset of $\{G_1, \ldots \ldots, G_n\}$.
- The net-gate matrix M.

Problem: Find a permutation of $G_1, G_2, \ldots, G_n$ and an associated feasible track assignment (layout) such that the number of tracks is minimum.

Figure 26 displays the net-gate matrix and two layouts for the above example.

The GMPP is the combinatorial core of the problem to construct an area minimal gate matrix layout. Additional (and usually neglected) features are 1) that one may distinguish two collections of rows (p-devices and n-devices) that permit independent column permutations, and 2) that a net may require more than one row depending on the permutation of its gates. While this second feature can be modeled within the MPP formulation by appropriate net splitting, incorporation of the first feature may lead to better layouts [Nakatani, Fujii, Kikuno & Yoshita, 1986]. For further technical information we refer to Schmidt & Mueller-Glaser [1983] and Wing, Huang & Wang [1985].

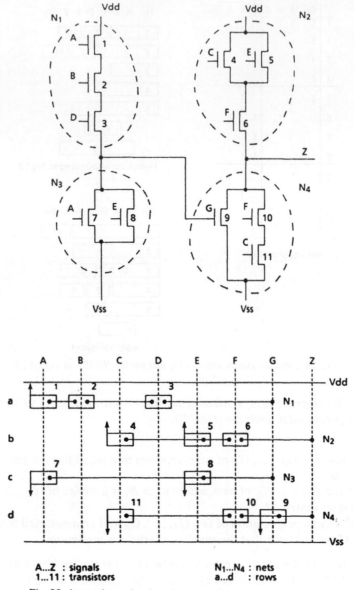

A...Z : signals   N₁...N₄ : nets
1...11 : transistors   a...d   : rows

Fig. 25. A transistor circuit and an associated gate matrix layout.

11.3. Programmable logic arrays

A programmable logic array (PLA) realizes a collection of Boolean functions given in disjunctive form (two-level sum of product form) on a two-dimensional array (see Figure 27).

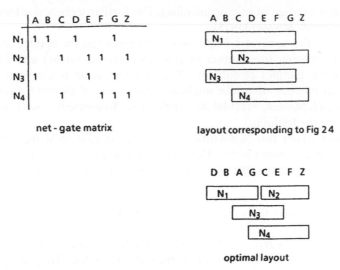

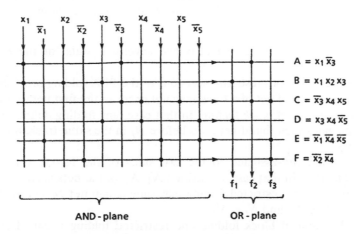

net - gate matrix

layout corresponding to Fig 2 4

optimal layout

Fig. 26. Net-gate matrix and two layouts for the GMPP of Figure 25.

$$f_1 = B + D, \; f_2 = A + C, \; f_3 = C + E + F$$

Fig. 27. A PLA layout of 3 Boolean functions.

This array consists of an AND-plane and an OR-plane. For every variable x_i of the Boolean functions, there is an input signal to the AND-plane (in fact, both inputs x_i and $\bar{x}_i$ are generated). Each row of the AND-plane produces a term that is an input to the OR-plane. The columns of the OR-plane correspond to the different Boolean functions and combine the appropriate product terms by an OR operation. By adding storage elements and simple feedback connections, a PLA can very easily be used to implement a sequential circuit. This application of PLA's is popular in the design of microcontrollers. For further technical information, we refer to Fleisher & Maissel [1975].

For reducing the area of a PLA, two techniques can be applied. *Logic minimization* for reducing the number of rows (= product terms) and *PLA folding* for reducing the number of columns. The first technique is the same as finding the minimum number of prime implicants for a set of Boolean functions [see e.g. Brayton, McMullen, Hachtel & Sangiovanni-Vincentelli, 1984]. It is usually applied before the folding.

The folding allows two (sometimes also more) signals to share a row (in the AND-plane) or a column (in the OR-plane). This leads to the same class of MPP's in both the AND- and the OR-plane, which we refer to as *PLAMPP*:

Given:
- A collection of gates $G_1, \ldots, G_n$ that correspond to the signals in one plane of a PLA.
- A collection of nets $N_1, \ldots, N_m$, where each net is a set of signals that have to be combined by AND (in the AND-plane) or OR (in the OR-plane).
- The net-gate matrix M.

Problem: Find a permutation of the gates $G_1, \ldots, G_n$ and a feasible assignment of at most two nets to a track (*PLA layout*) such that the number of tracks is minimum.

Figure 28 displays the net-gate matrix and several layouts for the example of Figure 27.

There are several more restrictive versions of the PLA folding problem. If gates occurring in the second net of a track may not occur in the first net of a track with two nets, one speaks of *block folding*. In that case, any folding defines a partition $\mathcal{G}_1, \mathcal{G}_2$ of the gates such that if net N_i is before N_j in the same track, then $N_i \subseteq \mathcal{G}_1$ and $N_j \subseteq \mathcal{G}_2$.

If the nets are pre-assigned to the two sides of the layout, one speaks of *constrained folding*. In that case, a partition $\mathcal{N}_1, \mathcal{N}_2$ of the nets is given as input to the problem, and any restricted folding may only assign net N_i before N_j on the same track if $N_i \in \mathcal{N}_1$ and $N_j \in \mathcal{N}_2$.

The combination of block folding and restricted folding is called *constrained block folding*. Figure 29 shows optimal layouts for these different folding problems for the example of Figure 28.

12. Graph-theoretic formulations and related problems

We will now consider several graph theoretic problems that are equivalent to the VLSI layout problems discussed in the previous sections. These problems have their own graph theoretic background and have to a large extent been investigated independently of each other.

		A	B	C	D	E	F
1	x_1	1	1				
2	$\overline{x}_1$				1		
3	x_2	1					
4	$\overline{x}_2$					1	
5	x_3		1	1			
6	$\overline{x}_3$	1	1				
7	x_4			1	1		
8	$\overline{x}_4$					1	1
9	x_5		1				
10	$\overline{x}_5$				1	1	

net - gate matrix

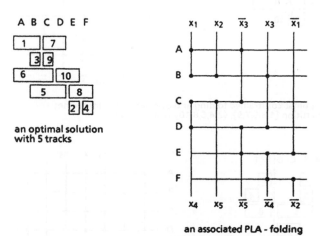

an optimal solution
with 5 tracks

an associated PLA - folding

Fig. 28. The net-gate matrix and an optimal layout for the AND-plane of Figure 27.

12.1. Interval graph augmentation

This formulation occurs already in the first papers on gate matrix layout [Wing, 1982, 1983]. For Weinberger arrays, similar considerations are made in Ohtsuki, Mori, Kuh, Kashiwabara & Fujisawa [1979].

Let V be a finite set and $(I_v)_{v \in V}$ be a collection of (not necessarily distinct) intervals I_v of a linearly ordered universe (such as the real line or a permutation of the gates). Such a collection of intervals $(I_v)_{v \in V}$ defines a partial order $P = (V, <)$ on V by putting

$$u < v \Leftrightarrow I_u \text{ is entirely to the left of } I_v. \tag{1}$$

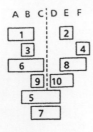
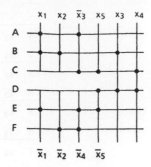

a) optimal block folding and associated PLA with 6 tracks

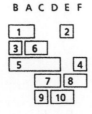
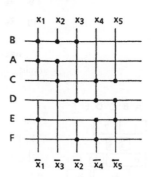

b) optimal constrained folding and associated PLA for the net
partition {1,3,5,7,9}, {2,4,6,8,10} with 5 tracks

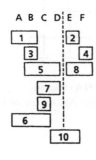
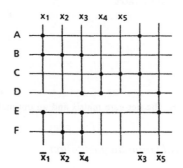

c) optimal constrained block folding and associated PLA for the
partition from b) with 7 tracks

Fig. 29. Restricted PLA-foldings.

It also defines an undirected graph $G = (V, E)$ on V by putting

$$(u, v) \in E \Leftrightarrow I_u \text{ and } I_v \text{ intersect (i.e. } I_u \cap I_v \neq \emptyset) \qquad (2)$$

A partial order P and a graph G obtained in that way are called an *interval order*
and an *interval graph*, respectively, and an associated collection of intervals $(I_v)_{v \in V}$

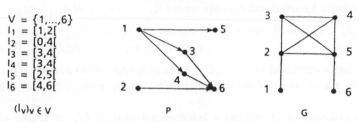

$$V = \{1,...,6\}$$
$$I_1 = [1,2[$$
$$I_2 = [0,4[$$
$$I_3 = [3,4[$$
$$I_4 = [3,4[$$
$$I_5 = [2,5[$$
$$I_6 = [4,6[$$

$$(I_v)_v \in V$$

Fig. 30. A collection of intervals with associated interval order P (as transitively reduced directed acyclic graph) and interval graph G.

is called an *interval representation* of P or G. An example is given in Figure 30.

Interval orders and interval graphs model the sequential and intersection structure of a set of intervals of the real line. This is why they have many applications dealing with intersection and consecutiveness such as the gene structure in molecular biology, seriation in archeology, preference and indifference relations in measurement theory, and consecutive retrieval, VLSI channel routing, and gate matrix layout in computer science. For more information on these applications, see Golumbic [1980, 1985] and Möhring [1985, 1989].

Note that different interval representations with the same intersection behavior define the same interval graph but possibly different interval orders. We call all interval orders related in this way to a fixed interval graph G the interval orders *associated* with G.

Consider now an MPP with net-gate matrix M. This matrix M defines a graph $G = ((V(G), E(G))$ by taking the *intersection graph* of the rows, i.e., $V(G)$ is the set of nets (rows), and two nets are connected by an edge if they share a gate (i.e. there is a column with 1's in both rows). G is called the *net adjacency graph* [Deo, Krishnamoorty & Langston, 1987], or *incompatibility graph* [Arbib, Lucertini & Nicoloso, 1988] since its edges (u, v) express that the nets u, v cannot be assigned the same track in a feasible layout.

For any gate permutation π of M, the associated augmented matrix M_π defines a collection of intervals (the augmented nets) of the linear order $G_{i_1}, \ldots, G_{i_n}$ defined by π on the gates. This collection of intervals defines an interval graph $H = (V(H), E(H))$ that *contains* G in the sense that $V(G) = V(H)$ and $E(G) \subseteq E(H)$, i.e. by *augmenting* the edge set of G. (This follows directly from the fact that two nets that share a gate in M share also a gate in M_π.) A feasible track assignment for M_π corresponds then obviously to a *coloring* of H, in which tracks correspond to color classes.

This shows that every feasible layout for M induces a coloring of an interval graph augmentation of G, the incompatibility graph of M. The converse is also true as the following lemma shows.

Lemma 12.1. *Let* $G = (V(G), E(G))$ *be the incompatibility graph of a net-gate matrix* M. *Let* $H = (V(H), E(H))$ *be an interval graph augmentation of* G *and let* $P = (V(H), <)$ *be any interval order associated with* H. *Then:*

(1) P induces a partial order on the gates $G_1, \ldots, G_n$ of M by putting

$$G_r < G_s \iff \exists \text{ nets } N_i \neq N_j \text{ incident to } G_r \text{ and } G_s, \text{ respectively}$$
$$(\text{i.e. } m_{ir} = m_{js} = 1), \text{ and } N_i < N_j \text{ in } P. \quad (3)$$

(2) Any linear extension $G_{j_1}, \ldots, G_{j_n}$ of the gate order of a) induces a permutation π such that the augmented net-gate matrix M_π is an interval representation of H and P.

(3) Any coloring of H induces a track assignment of M_π by taking each color class as a track and ordering the nets of the color class according to P.

The proof is straightforward and left to the reader. An example of these constructions is presented in Figure 31. An immediate consequence of these consideration is:

Theorem 12.2 [Wing]. *The minimum number of tracks of a feasible layout for a net-gate matrix M is equal to the smallest chromatic number of an interval graph augmentation H of the incompatibility graph G of M, i.e.*

$$t(M) = \min\{\chi(H) \mid H \text{ is an interval graph with } E(G) \subseteq E(H)\} \quad (4)$$

We briefly discuss the computational complexity of the constructions of Lemma 12.1. An adjacency matrix of G can be constructed from M in $O(nm^2)$ time by looking at each column separately and inserting the corresponding edge entries in the adjacency matrix.

An interval representation of an interval graph H and an associated interval order P (in interval representation) can be constructed in $O(|V(H)| + |E(H)|)$ time by PQ-tree techniques [Booth & Lueker, 1976; Korte & Möhring, 1989].

The gate order induced by P can be obtained efficiently by scanning an interval representation of P from left to right and constructing an ordered partition of the gates as follows:

For the ith right endpoint encountered in the scan, let the set $\mathcal{G}_i$ consist of all gates that are incident to nets ending (as intervals) at the current endpoint and do not belong to any previously constructed $\mathcal{G}_j$. The partition $\mathcal{G}_1, \ldots, \mathcal{G}_k$ thus constructed defines the gate ordering: all gates from $\mathcal{G}_i$ are incomparable among each other and precede all gates from $\mathcal{G}_{i+1}$, $(i = 1, \ldots, k-1)$. This follows from the fact that all gates not yet considered at the current endpoint must belong to nets that come later in the interval representation and are thus successors in P of the currently ending nets. So linear extensions of the gate ordering are just permutations within the classes $\mathcal{G}_i$ of the partition. It follows that the partition and a linear extension (gate permutation) can be constructed in $O(nm)$ time.

An optimal coloring of an interval graph H can be obtained in $O(|V(H)|)$ time from an interval representation by scanning the interval representation from left to right [Gupta, Lee & Leung, 1982]. When the left endpoint of an interval is encountered in the scan, the corresponding vertex of H is assigned the smallest color from the set $\{1, 2, \ldots, n\}$ of colors that has not been assigned to intervals containing the current endpoint.

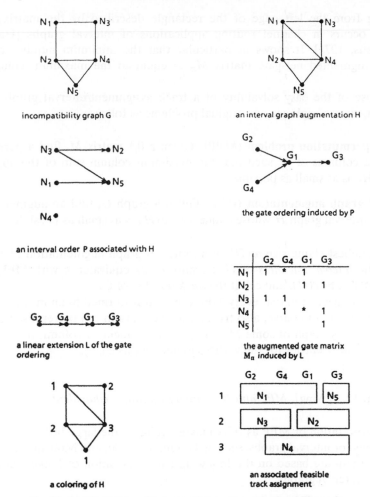

Fig. 31. An illustration of Lemma 13.1 on the example of Figure 21.

Obviously, the number of colors thus required is equal to the maximum number of intervals that intersect in a common point, i.e. the maximum size $\omega(H)$ of a clique of H. Since any coloring requires at least $\omega(H)$ colors, if follows that the coloring is optimal and $\chi(H) = \omega(H)$.

These arguments show that the hard core of the MPP is the construction of the right gate permutation, or, in terms of the interval graph augmentation, to find the right augmentation of the incompatibility graph to an interval graph. The track assignment problem or interval graph coloring problem can then be solved optimally by a simple linear-time algorithm.

Note that the interval graph coloring algorithm described here can of course also be directly carried out on augmented net-gate matrices (remember that they represent interval graphs). In this context, it is known as the *left-edge algorithm* (starting from the left edge of the rectangle described by the matrix), and it already occurs in channel routing applications of interval graphs [Hashimoto & Stevens, 1971]. It shows in particular that the minimum number of tracks for an augmented net-gate matrix M_π is equal to the maximum column sum of M_π.

Because of the easy solvability of a track assignment/interval graph coloring problem, we can rephrase our original problems as follows:

Matrix permutation problem (MPP): Given a 0,1 matrix M, find a permutation P of the columns of M such that the maximum column sum of the augmented matrix M_π is as small as possible.

Interval graph augmentation (IGA): Given a graph G, find an augmentation of G to an interval graph H whose clique size $\omega(H)$ is as small as possible.

The smallest clique size $\omega(H)$ of an interval graph augmentation of G is also called the *interval thickness* of G. Because of its equivalence with MPP, we will denote it also by $t(G)$, and call it the *track number* of G.

So far, we have seen that every MPP can be transformed to an instance of IGA. The converse is, of course, also true, since every graph can be represented as the incompatibility graph of some MPP (e.g. by introducing a column for every edge of G with two 1-entries for the vertices joined by this edge).

This gives:

Theorem 12.3 [Wing]. *MPP and IGA are polynomially equivalent.*

The interpretation of MPP as IGA makes available the large body of algorithmic techniques for interval graphs, see e.g. Golumbic [1980] and Möhring [1985, 1989]. Most of them are based on the following characterization of Fulkerson & Gross [1965] of interval graphs.

Theorem 12.4 [Fulkerson & Gross]. *A graph G is an interval graph iff its maximal cliques can be linearly ordered such that, for every vertex v, the maximal cliques containing v occur consecutively.*

Any such arrangement is called a *consecutive clique arrangement*. Such arrangements can be constructed and maintained by PQ-trees [Booth & Lueker, 1976] or their specialized versions for interval graphs, the MPQ-trees [Korte & Möhring, 1989]. These data structures will be useful for approximation algorithms discussed in Section 14.

Loosely speaking, the general idea of the interval graph augmentation approach to the MPP can be expressed as making the cliques of G consecutive (in the sense of Theorem 12.4) by extending or joining them to cliques of an interval graph while keeping the size of the new cliques small.

12.2. Path width

The path width of a graph was considered by Robertson & Seymour [1983] in the first part of their series of papers on graph minors.

They define a *path decomposition* of a graph G as a sequence $X_1, \ldots, X_r$ of subsets of $V(G)$ such that

$$\text{for every edge } e \text{ of } G, \text{ some } X_i \text{ contains both ends of } e \tag{5}$$

and

$$\text{for } 1 \leq i \leq j \leq l \leq r, \ X_i \cap X_l \subseteq X_j \tag{6}$$

hold. The *path width* of G (denoted by $pw(G)$) is the minimum value of $k \geq 0$ such that G has a path decomposition $X_1, \ldots, X_r$ with $|X_i| \leq k+1$ ($i = 1, \ldots, r$).

This notion is almost identical to interval graph augmentation. In fact, if G has a path decomposition $X_1, \ldots, X_r$ with $|X_i| \leq k + 1$, then the graph H defined by letting $X_1, \ldots, X_r$ be its maximal cliques is an interval graph because of (6) and Theorem 12.4, and fulfills $E(G) \subseteq E(H)$ because of (5), and $\omega(H) = \max_i |X_i| \leq k + 1$. Conversely, if H is an interval graph augmentation of G, then any consecutive clique arrangement $C_1, \ldots, C_r$ of H defines a path partition of G because of Theorem 12.4, and $|C_i| \leq k + 1$ with $k = \omega(G) - 1$. This gives:

Proposition 12.5. *Determining the path width of a graph G is polynomially equivalent to IGA. In particular,* $pw(G) = t(G) - 1$.

This equivalence permits a direct translation of deep results from the Robertson–Seymour theory to gate matrix layout. The most important of these is related to the notion of the minor of a graph.

H is a *minor* of G if H can be obtained from G by deleting some vertices and/or edges, and/or contracting some edges. It is easy to see that

$$t(H) \leq t(G) \text{ if } H \text{ is a minor of } G. \tag{7}$$

This implies directly that for any fixed k, the class of graphs G with $t(G) \leq k$ is *closed under taking minors* (i.e., if G belongs to this class and H is a minor of G then H belongs also to this class). This is the starting point for the application of the following results of Robertson & Seymour [1988].

Theorem 12.6 [Robertson & Seymour]. *Let $\mathcal{F}$ be any set of graphs closed under minors. Then there are finitely many graphs $H_1, \ldots, H_r$ such that $G \in \mathcal{F} \Leftrightarrow G$ does not contain H_i as a minor, $i = 1, \ldots, r$.*

This is a direct consequence of the proof of the Wagner conjecture (no class of graphs has infinite antichains under the minor ordering) in Robertson & Seymour [1988] and the closedness property under taking minors.

Theorem 12.7 [Robertson & Seymour]. *For any fixed graph H, it can be tested in polynomial time whether a graph G contains a minor isomorphic to H.*

The combination of these Theorems yields the existence of a polynomial time algorithm for testing membership for any minor-closed family of graphs, thus in particular for the class of graphs with bounded path width k. This is interesting in view of the $\mathcal{NP}$-hardness of the general problem when k is part of the input (see Section 13).

However, since no proof of Wagner's conjecture can be entirely constructive [Friedman, Robertson & Seymour, 1987], Theorem 12.6 is a pure *existence result* for the finite family of forbidden minors. Moreover, though the algorithms for minor recognition have low degree polynomials as worst case bounds, their constants of proportionality are enormous, rendering them impractical for practical problems [see e.g. Johnson, 1987]. The general bound is $O(n^3)$ for a graph with n vertices, and even $O(n^2)$ if the family $\mathcal{F}$ excludes a planar graph. This is the case for the path width (see Robertson & Seymour [1983] and Proposition 12.13 below), and thus:

Theorem 12.8 [Robertson & Seymour]. *Within the class of graphs with bounded pathwidth k, k fixed, the interval graph augmentation problem can be solved in $O(n^2)$ time for a graph with n vertices.*

Much research has been done to turn this existence result into a real (though maybe not practical) algorithm for calculating the pathwidth and an associated path decomposition for fixed k. An $O(n^{2k^2+4k+8})$ *dynamic* algorithm has been obtained in the context of graph searching [Ellis, Sudborough & Turner, 1987] (see also below). With help of the notion of *self-reduction*, it is possible to avoid the non-constructive aspects of Theorem 13.8, but at the cost of a further increase of the constant factors [Bodlaender, 1990]. Pure decision algorithms (is $pw(G) \leq k$?) of orders below $O(n^2)$ have e.g. be obtained in Lagergren [1990] and Fellows & Abrahamson [1990]. Based on Lagergren [1990], Bodlaender & Kloks [1991] obtain an $O(n \log^2 n)$ algorithm that, in addition to deciding whether the pathwidth is at most k, also constructs a path-decomposition of width at most k if there exists one. This algorithm is only based on constructive arguments, and its constant factor is *only* singly exponential in k.

The application of the Robertson–Seymour theory to gate matrix layout is discussed (among other problems) in a series of papers [Fellows & Langston, 1987, 1988]. For a survey about computational implications of the Robertson–Seymour theory, we refer to Johnson [1987].

12.3. Node searching

This problem formulation refers to a searching game on graphs introduced in Kirousis & Papadimitriou [1986] as a variant of the more investigated edge searching [Parsons, 1976].

In node searching, the edges of a graph represent a system of pipes or tunnels that are considered contaminated by a gas. The object of node searching is to clear all edges by a search. A *search* is a sequence of moves where a player places a *searcher* (also called *guard*) on a node of the graph that carries no searcher or deletes the searcher from a guarded node.

An edge is *cleared* if both its endpoints simultaneously carry a searcher. A cleared edge may be *recontaminated* if, at a later stage of the search, there is a path from an uncleared edge to the cleared edge without any searchers on it. So in order to avoid recontamination of cleared edges, the guarded nodes must after each move form a separating set that separates the still *unsearched part* of the graph (the not yet visited vertices) from the already *searched part* (all vertices that carried a searcher in the past).

A search is called *optimal* if the maximum number of searchers on the graph at any point is as small as possible. This number is called the *node-search number* of G, and denoted by $ns(G)$.

It was shown in Kirousis & Papadimitriou [1986] (a simpler proof is given in Bienstock & Seymour [1991]) that there always is an optimal search without recontamination of cleared edges. This was used in Kirousis & Papadimitriou [1985] to show the following unexpected relationships to interval graph augmentation:

Theorem 12.9 [Kirousis & Papadimitriou]. *For any graph G, $ns(G) = t(G)$.*

The proof of this theorem is based on the following ideas. If H is an interval graph augmentation of G, then any consecutive arrangement $C_1, \ldots, C_k$ of the maximum cliques of H defines a search by letting the searchers move in this order through $C_1, \ldots, C_n$. If C_i is guarded, then searchers from $C_i - C_{i+1}$ and possibly new searchers may be moved to $C_{i+1} - C_i$ until C_{i+1} is guarded. It is easy to see that this defines a search without recontamination with $\omega(H)$ searchers.

In the converse direction, any recontamination-free search of G assigns to every node v of G the interval $[i, j]$ whose endpoints are the first and last step in the search at which v is occupied by a searcher. (Note that this definition makes sense since the search is recontamination-free.) Since every edge is searched, the intervals assigned to its endpoints intersect. So the interval representation induces an interval graph H with $E(G) \subseteq E(H)$. The maximum number k of searchers in this search is obviously just the maximum number of pairwise intersecting intervals, i.e. $\omega(H)$.

Node searching is also closely related to other linear graph layout problems and to pebbling games on graphs [Kirousis & Papadimitriou, 1986]. In particular, the interpretation as progressive pebbling game gives the following useful result [Kirousis & Papadimitriou, 1986] for investigating or generating searches.

Consider an acyclic orientation of the edges of G and a dynamic assignment of searches to vertices that observes the rules:

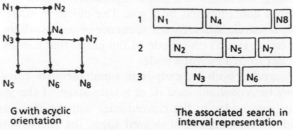

| | G with acyclic orientation | | The associated search in interval representation |

Fig. 32. An illustration of a search.

A vertex may accept a searcher only when all its immediate predecessors carry a searcher. (8)

Every vertex is assigned a searcher exactly once. (9)

Then:

Proposition 12.10. *Every assignment observing (8) and (9) defines a recontamination-free search of G, and every recontamination-free search of G can be obtained in that way.*

The first part follows easily by induction and the observation that the 'foremost' searchers (those that have an unvisited immediate successor) form a separating set in G that separates the searched part from the unsearched part. The other direction is obtained by considering the orientation of G defined by $u < v$ if u is visited by a searcher before v. An example is given in Figure 32.

We call such a search a *directed search*.

Another useful application is the combination of the search interpretation with a structural decomposition of graphs, the split decomposition [Cunningham, 1982].

A *split* in an undirected graph G is a partition $V(G) = V_1 \cup V_2$ of the vertex set of G such that $|V_i| \geq 2$ $(i = 1, 2)$ and

The edges of G from V_1 to V_2 induce a complete bipartite graph. (10)

Lemma 12.11. *Let $V(G) = V_1 \cup V_2$ be a split of G and let $A_i \subseteq V_i$ $(i = 1, 2)$ be the vertices of the associated complete bipartite graph. Then every recontamination-free search of G has a step at which all vertices from A_1 or all vertices from A_2 simultaneously carry a searcher. So in particular, $ns(G) \geq min \{|A_1|, |A_2|\}$.*

This can be seen as follows. If the statement is not true, then there is a first step of the search at which a searcher is deleted from the endpoint $(u$, say$)$ of an already cleared edge $(u, v) \in A_1 \times A_2$. Since neither A_1 nor A_2 are completely visited at that step, there is an uncleared edge $(x, y) \in A_1 \times A_2$. But then (u, v) is recontaminated via the path $(x, y), (y, u)$.

Still another application is the following argument from Kirousis & Papadimitriou [1986], which, in the context of edge searching, is due to Parsons [1976]. It also shows that the search number is unbounded on the class of trees. This gives also the missing argument for the $O(n^2)$ algorithm of Theorem 12.8 (the class of graphs with bounded tree width excludes some trees and thus a planar graph).

Lemma 12.12. *Let G contain a vertex v of degree 3 whose deletion separates G into three connected components G_1, G_2, G_3, each of which has node search number $ns(G_i) = t$. Then $ns(G) = t + 1$.*

It is easy to see that $t + 1$ searchers suffice. (Put a searcher on v and search G_1, G_2, G_3 with the remaining t searchers). To show that they are also necessary, assume that t searchers suffice for G.

Since t searchers are already required for each of the subgraphs G_1, G_2, G_3, there is a moment at which one of them (G_1, say) has already been searched, all searchers are on the second one (G_2, say), and the last of them is still unvisited. But then recontamination takes place between G_3 and G_1, a contradiction.

As a consequence, one obtains [Kirousis & Papadimitriou, 1986]:

Proposition 12.13. *The search number of a complete ternary tree T is equal to its height plus one.*

It was already mentioned that node searching is a variant of the more investigated edge searching. In edge searching, an edge is cleared by letting a searcher go through it (instead of by occupying both endpoints as in node searching). It is therefore possible [Kirousis & Papadimitriou, 1986] to obtain (optimal) node searches on G from (optimal) edge searches on a slight modification of G (replace each edge of G by three parallel edges). Exploitation of this transformation and known results for edge searching on trees [Megiddo, Hakimi, Garey, Johnson & Papadimitriou, 1988] and dynamic programming formulations [Ellis, Sudborough & Turner, 1987] yield a linear time search algorithm for trees (see also Section 13) and a polynomial time algorithm of order $O(n^{2k^2+4k+8})$ for graphs with search number at most k, k fixed. Finally, for a combination of node searching and edge searching with a similar flavor we refer to Bienstock & Seymour [1991].

12.4. Alternating paths

This final equivalence is considered in many papers on PLA folding. It views tracks in a layout of M as directed paths in the complement G of the incompatibility graph G of M. The directed edges of these paths may be considered as being added to G. This gives the following formulation.

Let G be the incompatibility graph of a MPP, and consider the edges of G as colored red. A *path partition* (or *multiple folding*) of G is a set of directed green edges F such that the following two conditions are satisfied.

> The subgraph defined by the green edges is a collection of directed paths (i.e., indegree and outdegree of every vertex is at most 1). This constraint is called the *degree constraint*. $\qquad(11)$

There exists no cycle of alternating directed green path segments
and red edges (*alternating cycle*). This constraint is called the   (12)
cycle constraint.

The *size* of a path partition is the number of green paths.

Theorem 12.14. *Determining a minimum size path partition is polynomially equivalent to interval graph augmentation.*

This can be seen as follows. From a path partition with t paths, one can construct
an interval representation of an interval augmentation H of G with $\omega(H) = t$
by a left to right scan through the t paths. Initially, the intervals corresponding
to the minimal vertices in the t paths are 'opened'. When an interval is closed,
the interval of the next vertex in the corresponding path is opened etc. Given a
collection of t opened intervals, the next interval to close corresponds to a vertex u
such that there are no red edges (u, w) to a vertex w that occurs after a currently
open interval v on a green path containing v. Note that there is such a vertex u
because of (12).

Since exactly t intervals are open at any moment, $\omega(H) = t$. To see that
H augments G, let $(u, v) \in E(G)$ and, w.l.o.g., let u be opened before v (as
intervals). Then the choice defined above ensures that u is only closed after
v is opened. Hence the corresponding intervals overlap. An example of this
construction is given in Figure 33.

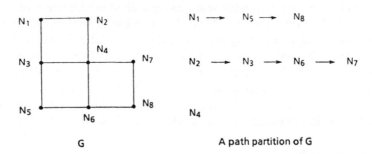

G   A path partition of G

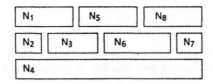

The associated interval representation
of Theorem 3.15

Fig. 33. An illustration of a path partition.

The converse direction is obvious since every optimal coloring of an optimal interval graph augmentation H of G defines a path partition of size $\chi(H) = \omega(H)$. This interpretation is particularly useful for the special cases of the MPP dealing with PLA folding.

There are several other equivalent or related graph-theoretic notions. We just mention vertex separation, min-cut linear arrangement, bandwidth and several modifications of these problems. While vertex separation is equivalent to node search [Kirousis & Papadimitriou, 1986], the other notions define bounds on $t(G)$ [Kirousis & Papadimitriou, 1986; Bodlaender, 1988].

12.5. Restrictions of the MPP

The restrictions discussed in Section 11 have natural formulations within several of the different representations of this section.

For instance, the WMPP can be modeled by fixing two cliques of G as belonging to the first and last maximal clique of the interval augmentation to construct, or by requiring the searchers to start and finish their search on specified cliques of G.

For PLA folding problems, at most two vertices may share a common track. This means in the path partition formulation that the directed green paths reduce to directed green edges. This gives the most common formulation of PLA-folding, which is due to Hachtel, Newton & Sangiovanni-Vincentelli [1982].

Proposition 12.15. *Let G be the incompatibility graph of a PLA folding problem. Then the minimum number of tracks for a PLA folding is equal to $|V(G)| - s$, where s is the maximum number of green directed arcs that can be added to G such that*

> *The green arcs form a matching in the complement of G (degree constraint).* (13)

> *There is no alternating cycle of directed green edges and undirected red edges of G (cycle constraint).* (14)

Such a set F of green arcs is called a *folding set* or simply *folding*. So Proposition 12.15 states that the PLA-folding problem is equivalent to finding a folding set of maximum size.

Another characterization can be obtained from the observation that, in an optimal layout of the associated augmented matrix M_π, the rows can be permuted in such a way that the rows with two nets appear on top and that the rightmost ones of the first nets in these rows define a 'decreasing staircase'.

This staircase pattern corresponds to the special subgraph $Z_{m,m}$ of $\overline{G}$ defined below.

A $Z_{m,m}$ (also called a *triangular clique* in Hu & Kuo [1987]) is a bipartite graph $G = (U, V, E)$ with $U = \{a_1, \ldots, a_m\}$, $V = \{b_1, \ldots, b_m\}$, and $E = \{(a_i, b_j) | 1 \le i \le j \le m\}$.

This gives [Hu & Kuo, 1987]:

Proposition 12.16. *Finding a maximum folding set in G is equivalent to finding a maximum size $Z_{m,m}$ as (partial, not induced) subgraph of $\overline{G}$.*

Both conditions can easily be sharpened for block folding and constrained folding problems. Call a folding set F resulting in a block folding a *block folding set*. Then one obtains the following characterization of block folding, see e.g. Ravi & Loyd [1988].

Proposition 12.17. *Let G be the incompatibility graph of a PLA folding problem. Then:*

(1) A set F of directed green arcs added to G is a block folding set if F satisfies the degree constraint (13) and

$$\text{There is no red edge } (u, v) \text{ from the head of a green edge to the tail of another green edge.} \tag{15}$$

(2) Finding a maximum block folding set in G is equivalent to finding a maximum size $K_{m,m}$ (the complete bipartite graph on 2m vertices) as (partial, not induced) subgraph of $\overline{G}$.

Remember that in constrained PLA folding, the nets are preassigned to the two sides of the layout. Thus only incompatibility relations between these two sides are of importance, i.e., the incompatibility graph G can be assumed to be bipartite. Therefore, constrained PLA-folding is sometimes also called *bipartite folding* [Egan & Liu, 1984; Hu & Kuo, 1987].

The above conditions then specialize further for block folding.

Proposition 12.18. *Finding a maximum block folding in a bipartite graph G is equivalent to finding a maximum size $K_{m,m}$ as induced subgraph of the (bipartite) complement of G.*

13. Complexity results

It was already mentioned several times that the general MPP is $\mathcal{NP}$-hard. In fact, this has been obtained independently in many of the equivalent formulations, e.g. for interval graph augmentation in Kashiwabara & Fujisawa [1979], for node search in Kirousis & Papadimitriou [1986], for directed node search in Lengauer [1981], and for path width in Arnborg, Corneil & Proskurowski [1987].

The $\mathcal{NP}$-hardness of the PLA folding problems (including the general MPP, but excluding block folding) is shown in Luby, Vazirani, Vazirani & Sangiovanni-Vincentelli [1982] by a series of reductions from *matrix upper triangulation* (given an $n \times n$ 0–1 matrix A, is there a permutation of the rows and another of the columns such that resulting matrix is upper triangular?).

We will here sketch a different series of reductions that starts from constrained block folding and contains block folding and a sharper version for interval graph augmentation (i.e. the general MPP), which turns out to be $\mathcal{N}P$-hard already on chordal graphs. Our starting point is [see e.g. Egan & Liu, 1984]:

Theorem 13.1 [Egan & Liu]. *Constrained block folding is $\mathcal{N}P$-hard.*

This follows in fact directly from Proposition 12.18 that gives the equivalence to 'balanced complete bipartite subgraph' which is stated to be $\mathcal{N}P$-complete in Garey & Johnson [1979] (the proof has appeared in Johnson [1987]).

Preassignment of nets to sides can easily be enforced by adding two gates G_0, G_{n+1} that are connected to the nets from the left and right side, respectively. Then any (block) folding of the augmented problem can only fold nets incident to G_0 with nets incident to G_{n+1}, and the sides are (up to reversal of the layout) fixed by the position of G_0 and G_{n+1}. In view of Theorem 14.1, this gives:

Theorem 13.2. *Block folding is $\mathcal{N}P$-hard.*

This result has been sharpened in Müller & Wagner [1991] by a different reduction from GRAPH BISECTION. Exploiting techniques from Bui, Chaudhuri, Leighton & Sipser [1987] and Wagner & Wagner [1993], they obtain:

Theorem 13.3 [Müller & Wagner]. *Block folding is $\mathcal{N}P$-hard already for graphs with degree at most k for any fixed $k \geq 3$.*

The reduction from block folding to IGA on chordal graphs is based on the following equivalent formulation of block folding.

Given: A graph G and an integer k.
Question: Is there a partition of $V(G)$ into three sets V_1, V_2, V_3 such that
(i) every path from V_1 to V_3 goes through a vertex of V_2,
(ii) min $\{|V_1|, |V_3|\} \geq k$?

The answer to such an instance is obviously yes iff there exists a block folding set of G with k green edges (viz. from vertices to V_1 to vertices of V_3). Based on this formulation, the following result is obtained in Gustedt [1993].

Theorem 13.4 [Gustedt]. *Gate matrix layout and pathwidth are $\mathcal{N}P$-hard already on the class of chordal graphs.*

Recall that a graph is *chordal (or triangulated)* if every elementary cycle $v_1, v_2, \ldots, v_l, v_1$ of length $l \geq 4$ possesses a chord, i.e. an edge (v_i, v_j) with $1 \leq i < j + 1 \leq l + 1$.

Chordal graphs form a natural generalization of interval graphs, see e.g. Golumbic [1980] for more information about chordal graphs. The chordal graphs needed in the proof are quite special. They consist of a set of maximal cliques that overlap in a central clique.

In more detail, let G, k be an instance of the above formulation of block folding such that G is w.l.o.g. connected. From G we construct such a special chordal graph H as follows. H contains a maximal clique C_0 (called the *central* clique) with vertex set $V(G)$. For each edge $(u, v) \in E(G)$, a maximal clique C_{uv} is added that consists of the vertices $u, v \in C_0$ and $|V(G)|$ additional vertices that are incident only to vertices in C_{uv}. Clearly, this graph is of the desired type.

From H one can construct a net-gate matrix M with incompatibility graph H by introducing a gate for each of the maximal cliques of H. Consider an augmented matrix M_π of M and let C^+ and C^- be the cliques (gates) before and after the central clique (gate) C_0 in M_π. Since every net corresponding to an original vertex of G that is incident to some gate from C^+ or C^- is also incident to C_0, the order of the gates in C^+ or C^- does not influence the number of tracks.

Let V^+ and V^- denote the nets of C_0 (vertices of G) incident to a gate from C^+ and C^-, respectively, and let $V_1 := V^+ - V^-$, $V_2 := V^+ \cup V^-$ and $V_3 := V^- - V^+$. It can then be shown that V_1, V_2, V_3 form a partition of $V(G)$ with the above disconnecting property, and that min $\{|V_1|, |V_3|\} \geq k$ is equivalent to $t(M) \leq 2|V(G)| - k$.

Conversely, every partition V_1, V_2, V_3 of G with the above properties can be transformed into an augmented matrix M_π by first taking all gates (cliques) C_{uv} with $u, v \in V_1 \cup V_2$ (in any order), followed by C_0 and all cliques C_{uv} with $u, v \in V_2 \cup V_3$. This proves the theorem. An example of this construction is given in Figure 34.

Note that if all maximal cliques that intersect the central clique are mutually disjoint, then the problem can be solved in $O(|V(G)|^2)$ time by a dynamic programming algorithm [Gustedt, 1993]. This confirms that the borderline between easy and hard problems for subclasses of chordal graphs G depends essentially on the overlapping behavior of the maximal cliques of G [see also Arnborg, Corneil & Proskurowski, 1987].

Another sharpening has been obtained in Monien & Sudborough [1988] in the context of vertex separation for the class of planar graphs.

Theorem 13.5 [Monien & Sudborough]. *Gate matrix layout and pathwidth are $\mathcal{NP}$-hard already on the class of planar graphs.*

A reduction from gate matrix layout to constrained PLA-folding can be obtained by turning an arbitrary graph G into a bipartite graph $H = (U, V, E)$ as follows: $U := V(G)$, $V := \{v' \mid v \in V(G)\}$ (a copy of $V(G)$), and $(u, v') \in E$ (with $u \in U$ and $v' \in V$) if $(u, v) \in E(G)$ or $u = v$.

Then any constrained folding set F for H corresponds uniquely to a collection of tracks for G by combining the edges of F to paths (tracks) (x_1, x_2'), (x_2, x_3'), $(x_3, x_4'), \ldots, (x_{l-1}', x_l)$. Then $t(G) = |V(G)| - |F|$, i.e. maximizing $|F|$ in H corresponds to minimizing $t(G)$ in G. Hence:

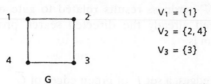

$V_1 = \{1\}$

$V_2 = \{2, 4\}$

$V_3 = \{3\}$

An associated partition
V_1, V_2, V_3 with $k = 1$

	C_0	C_{12}	C_{14}	C_{23}	C_{34}
1	1	1	1		
2	1	1		1	
3	1			1	1
4	1		1		1
5		1			
6		1			
7		1			
8		1			
9			1		
10			1		
11			1		
12			1		
13				1	
14				1	
15				1	
16				1	
17					1
18					1
19					1
20					1

	C_{12}	C_{14}	C_0	C_{23}	C_{34}
1	1	1	1		
2	1	*	1	1	
3			1	1	1
4		1	1	*	1
5	1				
6	1				
7	1				
8	1				
9		1			
10		1			
11		1			
12		1			
13				1	
14				1	
15				1	
16				1	
17					1
18					1
19					1
20					1

The net gate-matrix of H **The augmented matrix corresponding to V_1, V_2, V_3**

Fig. 34. An illustration of the proof of Theorem 13.4.

Theorem 13.6. *Constrained PLA folding is NP-hard.*

Finally, the reduction from constrained PLA folding to PLA folding is achieved in the same way as from constrained block folding to block folding.

Theorem 13.7. *PLA folding is NP-hard.*

The complexity status of PLA folding for special graph classes is investigated in Damaschke [1990]. With regard to $\mathcal{NP}$-completeness, the following holds:

Theorem 13.8 [Damaschke]. *Block folding and PLA folding are NP-hard both for bipartite graphs and for split graphs.*

There are several other $\mathcal{NP}$-hardness results related to gate matrix layout and PLA folding. We mentioned already the directed search problem [Lengauer, 1981]. Another such problem is

ORDERABILITY:
  Given: A graph G of red edges, a set F of green edges of $\overline{G}$.
  Question: Is there an orientation of the edges of F such that F is a folding?

Orderability is shown to be $\mathcal{NP}$-complete in Hachtel, Newton & Sangiovanni-Vincentelli [1982]. It is solvable in polynomial time for constrained PLA folding [Ravi, 1989].

  Other variants of the PLA-folding problem not discussed here are shown to be $\mathcal{NP}$-hard in Arbib, Lucertini & Nicoloso [1988].

  We consider now some special classes of graphs on which the gate matrix layout problem can be solved in polynomial time. Most of the arguments leading to polynomial algorithms come from node searching (in particular Lemma 12.11 and Lemma 12.12) and demonstrate again the usefulness of this interpretation.

  We will start with the class of trees. As mentioned before, the polynomial algorithm for edge searching on trees [Megiddo, Hakimi, Garey, Johnson & Papadimitriou, 1988] can be transformed by the principles of Kirousis & Papadimitriou [1986] to a polynomial algorithm for node searching on trees. This requires $O(n)$ time for determining $ns(G)$, and $O(n \log n)$ time for finding the associated search. Different and faster algorithms with much simpler correctness proofs have independently been obtained in Möhring [1990] and Scheffler [1989, see also 1990].

  The algorithms peel the tree, i.e. they start with the leaves and works its way towards the 'center' of the tree. So at a typical step of the algorithm, certain subtrees $T_1, \ldots, T_r$ of the tree T have already been investigated. Each of these trees T_i has a vertex v_i connecting it to the still unsearched part of T.

  The peeling is done in phases $t = 1, 2, \ldots, ns(T)$. At the beginning of phase t, the vertices v_i are the leaves of the remaining tree, and every T_i requires t searchers, while every $T_i - v_i$ can be searched with $t - 1$ searchers. Then the remaining tree is peeled starting from the v_i until one reaches new vertices u_j in which the search number must go up to $t + 1$ (or T has been searched completely). The current phase is completed when all v_i have been processed and are connected to some u_j in the already searched part of the tree.

  The argument for an increase of the search number in each phase is always Lemma 12.12. This is in fact due to the 'converse' of Lemma 12.12 [Parsons, 1976] stating that, for any tree T, $ns(T) \geq t + 1$ iff there is a vertex v at which there are three or more subtrees with search number t ore more.

  Altogether, this gives:

Theorem 13.9. *The track number $t(T)$ of a tree T and an optimal layout can be obtained in $O(|V(T)|)$ time.*

  It is open whether these techniques extend to other 'tree-structured' graphs such as (2-connected) outerplanar graphs, k-trees, and possibly partial k-trees.

Lemma 12.11 can be used to obtain a polynomial algorithm for cographs. Recall that a *cograph (or complement reducible graph*, see Corneil, Lerchs & Stewart Burlingham [1981] for details) can be defined recursively by:

The one-vertex graph is a cograph (16)

If $G_1 = (V_1, E_1)$ and $G_2 = (V_2, E_2)$ are cographs on disjoint sets V_1, V_2, then

$$G_1 + G_2 := (V_1 \cup V_2, E_1 \cup E_2) \text{ and}$$
$$G_1 * G_2 := (V_1 \cup V_2, E_1 \cup E_2 \cup (V_1 \times V_2))$$
(17)

are also cographs.

Then Lemma 12.11 gives:

$$ns(G_1 + G_2) = \max\{ns(G_1), ns(G_2)\}$$ (18)
$$ns(G_1 * G_2) = \min\{|V_1| + ns(G_2), |V_2| + ns(G_1)\}$$ (19)

An associated interval representation of an optimal search can be defined similarly as in the tree algorithm by defining offsets in each operation. So if a sequence of operations '*' and '+' producing the whole graph is given (a parse tree defining such a sequence for G can be obtained in $O(|V(G)| + |E(G)|)$ time [Corneil, Pearl & Stewart, 1985]) one obtains [see Bodlaender & Möhring, 1990]:

Theorem 13.10 [Bodlaender & Möhring]. *The track number of a cograph G (given in 'decomposed' form or by its parse tree) and an optimal layout can be obtained in $O(|V(G)| + |E(G)|)$ time.*

The operation '*' can be seen as a special split in which (with the terminology of Lemma 12.11) $A = V_1$ and $B = V_2$. It is still open whether there is a polynomial time search algorithm for the larger class of graphs that can be recursively decomposed by splits. They are known as *distance- hereditary graphs* [Bandelt & Mulder, 1986] or *completely separable* graphs [Hammer & Maffray, 1990], and contain both the class of trees and the class of cographs.

Finally, we mention the class of chordal graphs discussed in connection with Theorem 14.4. By using a dynamic programming approach similar to that for PARTITION in Garey & Johnson [1979] and Gustedt [1993] obtains:

Theorem 13.11 [Gustedt]. *If the maximal cliques $C_0, C_1, \ldots, C_m$ of G fulfill*

$$C_i \cap C_j \neq \emptyset \Rightarrow C_i \cap C_j = C_i \cap C_0,$$ (20)

then $t(G)$ and an optimal layout can be obtained in $O(|V(G)|^2)$ time.

Condition (13.3) means that the cliques $C_1, \ldots, C_n$ have a special overlap structure with the 'central' clique C_0. This class of graphs contains the split graphs [Golumbic, 1980], i.e., those chordal graphs whose complement is also chordal.

We close this section with a summary of polynomially solvable cases for PLA-folding problems. An algorithm for constrained PLA-folding on trees has been obtained in Hu & Kuo [1987]. Results of Bodlaender [1987] on the balanced complete bipartite subgraph problem (in particular the transformation applied there) show, when combined with Propositions 12.16 and 12.18, the polynomial solvability of constrained block folding and block folding on partial k-trees. It is easy to also obtain polynomial time dynamic programming algorithms for these cases.

Polynomial algorithms are given in Damaschke [1990] for a number of (mostly) perfect graph classes. It turns out that PLA folding is polynomially solvable for cographs, interval graphs and trees, while block folding is polynomially solvable for trapezoid graphs, circular arc graphs, and directed path graphs. Other graph classes are covered in Arbib [1991].

Moreover, it is shown in Damaschke [1991] that the class of cographs has a natural interpretation for PLA folding. It contains all incompatibility graphs arising from Boolean functions that are indicator functions of intervals in $\{0, 1\}^n$. This includes e. g. all threshold functions.

14. Algorithms

Due to the VLSI-background, many algorithms have been proposed and studied in the literature. The majority of them can be classified as (sometimes a combination of) heuristics, branch-and-bound algorithms, or dynamic programming algorithms.

Representatives for the different technologies discussed here are Yoshizawa, Kawanishi & Kami [1975], Asano [1982] for Weinberger arrays, Luby, Vazirani, Vazirani & Sangiovanni-Vincentelli [1982], Li [1983], Wing, Huang & Wang [1985], Leong [1986], Nakatani, Fujii, Kikuno & Yoshita [1986], Deo, Krishnamoorty & Langston [1987] for gate matrix layout, and Lewandowski & Liu [1984], Hwang, Dutton & Blank [1986], Hu & Kuo [1987], Kuo & Hu [1987], Arbib, Lucertini & Nicoloso [1988] for PLA folding (see Gajski & Lin [1988] for additional references). The problems studied in these papers are usually not larger than 50×60 (in terms of the net-gate matrix), with the exception of 100×80 in Arbib, Lucertini & Nicoloso [1988].

Only little is known about the performance of heuristics for these problems. By using standard arguments from Garey & Johnson [1979], the existence of an approximation algorithm for gate matrix layout with a constant absolute performance guarantee is ruled out in Deo, Krishnamoorty & Langston [1987]:

Theorem 14.1 [Deo, Krishnamoorty & Langston]. *Unless* $\mathcal{P} = \mathcal{NP}$, *there is no approximation algorithm A for gate matrix layout with*

$$A(I) \leq OPT(I) + K, \quad K \in \mathbf{N} \text{ fixed} \tag{21}$$

for all instances I.

It is well known [Garey & Johnson, 1979] that this also rules out the existence of a fully polynomial approximation scheme for gate matrix layout.

Concerning approximation algorithms with a relative performance guarantee, the following result has recently been obtained in Bodlaender, Gilbert, Hafsteinsson & Kloks [1991].

Theorem 14.2 [Bodlaender, Gilbert, Hafsteinsson & Kloks]. *There exists a polynomial time algorithm that, given a graph $G = (V, E)$ with $|V| = n$, finds a path-decomposition of G with pathwidth at most $O(k \log^2 n)$, where k is the pathwidth of G.*

This algorithm is based on a recursive decomposition of the given graph by relatively small separators that cut the graph into large parts. The polynomial construction of such separators is, however, only an announced result of T. Leighton and as yet not available in written form.

The existence of approximation algorithms with a *constant relative* performance guarantee is open. There are, however, some indications that they might not exist.

For PLA-folding problems, such an indication is given in Ravi & Loyd [1988]. Call two problems *equivalent with respect to approximation* if both or none are approximable with constant relative performance guarantee in polynomial time. Then:

Theorem 14.3 [Ravi & Loyd]. *PLA-folding, block folding, and constrained block folding are equivalent with respect to approximation, when viewed as maximization problems (maximize the size of a folding).*

The proof is based on transforming feasible solutions of one problem to feasible solutions of the other while preserving a constant relative performance guarantee. For instance, let algorithm A produce, for an instance I of PLA-folding, a solution with $A(I)$ folded pairs such that $A(I) \leq K \cdot OPT(I)$, where $OPT(I)$ denotes the size of an optimal folding. Then we can transform this solution to a solution of block folding with size $A'(I)$ by taking, in a decreasing staircase arrangement of the layout (cf. the remarks preceding Proposition 13.17), the first nets of rows $1, 2, \ldots, \lceil A(I)/2 \rceil$ and fold them with the last nets of rows $\lceil A(I)/2 \rceil + 1, \ldots, A(I)$. Clearly, this gives a block-folding with $A'(I) = \lceil A(I)/2 \rceil$ folded nets, and so the maximum size $OPT'(I)$ of a block folding fulfills $OPT'(I) \leq OPT(I) \leq K \cdot A(I) = 2K \cdot A'(I)$.

This equivalence result is combined in Ravi & Loyd [1988] with the following, unexpected result that provides the indication for non-approximatebility.

Theorem 14.4 [Ravi & Loyd]. *If there is an approximation algorithm A for block folding (in the maximization version) with relative performance guarantee K, $K \in \mathbf{N}$ fixed, then there is also one with relative performance guarantee $1 + \varepsilon$ for every fixed $\varepsilon > 0$.*

Note, however, that these results carry only through for the *maximization* version of the PLA-folding problems (maximize the size of a folding set). Both proofs fail for the — perhaps more natural — minimization of the number of tracks required in a layout.

For gate matrix layout, an indication for the non-existence of approximation algorithms with constant relative performance guarantee is obtained in Deo, Krishnamoorty & Langston [1987] by considering algorithms that are *on-line* with respect to the nets. This means that the nets are processed in an incremental fashion according to the following rules:

1. A partial layout for the nets processed so far has already been constructed and may not be changed when the next net is processed.

2. The next net N_i to be processed is chosen such that its addition to the partial layout causes the least increase in the number of tracks.

By designing a class of examples, it is shown in Deo, Krishnamoorty & Langston [1987] that such on-line algorithms cannot guarantee a constant relative performance error.

This definition of on-line seems, however, to be very restrictive, since the algorithms even fail to construct an optimal layout for interval graphs. (In fact, the class of examples of Deo, Krishnamoorty & Langston [1987] consists entirely of interval graphs.) The reason for this behavior is the rigidity of the already constructed partial layout.

We suggest here a less rigid approach that is still on-line and equally fast, but allows more flexibility in modifying the partial layout. The basic idea is to maintain not a partial layout, but the corresponding interval graph (see Section 12), and to represent it by its MPQ-tree.

The MPQ-tree (see Korte & Möhring [1989] for details) is a data structure that represents an interval graph H and all associated interval orders (cf. Lemma 12.1) in $O(|V(H)|)$ space. It permits also fast updating when a vertex is added to H. Such an update will always recognize when $H + v$ is again an interval graph and modify the MPQ-tree accordingly. (So the examples from Deo, Krishnamoorty & Langston [1987] are solved optimally.) If $H + v$ is not an interval graph, then there are several possibilities to augment $H + v$ to an interval graph H^* by considering different interval orders associated with H. So keeping the maximum clique size small then means to permute or invert the nodes of the MPQ-tree representing H in such a way that adding v increases $\omega(H)$ as little as possible.

Altogether, this leads to a class of on-line algorithms based on incremental interval graph generation by means of MPQ-trees, which are more powerful than the 'static' on-line algorithms suggested in Deo, Krishnamoorty & Langston [1987].

Acknowledgement

Part of this work was done while the authors were with the Leonardo Fibonacci Institute for the Foundations of Computer Science, Trento, Italy, in 1991.

References

Arbib, C. (1991). Two polynomial algorithms in PLA folding, in: R.H. Möhring (ed.), *Proceedings 16th International Workshop on Graph-Theoretic Concepts in Computer Science WG'90*, Springer-Verlag, Berlin, Lecture Notes in Computer Science, Vol. 484, pp. 119–129.

Arbib, C., M. Lucertini and S. Nicoloso (1988). Optimal design of programmed logic arrays. Preprint, Universita degli Studi di Roma 'La Sapienza'.

Arnborg, S., D.G. Corneil and A. Proskurowski (1987). Complexity of finding embeddings in a *k*-tree. *SIAM J. Alg. Discr. Meth.* 8, 277–284.

Asano, T. (1982). An optimum gate placement algorithm for MOS one-dimensional arrays. *J. Digital Systems*, IV, 1–27.

Baker, B.S., S.N. Bhatt and F.T. Leighton (1984). An approximation algorithm for Manhattan routing, in: F.P. Preparata (ed.), *Advances in Computer Research, VOL 2: VLSI Theory*, JAI Press Inc., pp. 205–229.

Bandelt, H.J., and H.M. Mulder (1986). Distance-hereditary graphs. *J. Combin. Theory Ser. B* 41, 182–208.

Barahona, F., M. Grötschel, M. Jünger and G. Reinelt (1988). An application of combinatorial optimization to physics and circuit layout design. *Oper. Res.* 36, 493–513.

Becker, M., and K. Mehlhorn (1986). Algorithms for routing in planar graphs. *Acta Inform.* 23, 163–176.

Bhatt, S.N., and F.T. Leighton (1984). A framework for solving VLSI graph layout problems. *J. Comput. System Sci.* 28, 300–343.

Bienstock, D., and P. Seymour (1991). Monotonicity in graph searching. *J. Algorithms* 12, 239–245.

Bodlaender, H.L. (1987). Dynamic programming on graphs with bounded tree width. Technical Report RUU-CS-87-22, Rijksuniversiteit Utrecht.

Bodlaender, H.L. (1988). Some classes of graphs with bounded treewidth. *Bull. EATCS* 36, 116–125.

Bodlaender, H.L. (1990). Improved self-reduction algorithms for graphs with bounded treewidth, in: M. Nagl (ed.), *Proceedings 15th International Workshop on Graph-Theoretic Concepts in Computer Science WG'89*, Springer-Verlag, Berlin, Lecture Notes in Computer Science, Vol. 411, pp. 232–244.

Bodlaender, H.L., J.R. Gilbert, H. Hafsteinsson and T. Kloks (1991). Approximating treewidth, pathwidth, and minimum elimination tree height, in: G. Schmidt and R. Berghammer (eds.), *Proceedings 17th International Workshop on Graph-Theoretic Concepts in Computer Science WG'91*, Springer-Verlag, Berlin, Lecture Notes in Computer Science, Vol. 570, pp. 1–12.

Bodlaender, H.L., and T. Kloks (1991). Better algorithms for the pathwidth and treewidth of graphs. *Proc. 18th Int. Colloquium on Automata, Languages and Programming ICALP'91*, Springer-Verlag, Berlin, Lecture Notes in Computer Science, Vol. 510, pp. 544–555.

Bodlaender, H.L., and R.H. Möhring (1990). The pathwidth and treewidth of cographs. *Proc. 2nd Scandinavian Workshop on Algorithm Theory*, Springer-Verlag, Berlin, Lecture Notes in Computer Science, Vol. 477, pp. 301–309.

Booth, S., and G.S. Lueker (1976). Testing for the consecutive ones property, interval graphs, and planarity using PQ-tree algorithms. *J. Comput. System Sci.* 13, 335–379.

Brady, M.L., and D.J. Brown (1984). VLSI routing: Four layers suffice, in: F.P. Preparata (ed.), *Advances in Computer Research, VOL 2: VLSI Theory*, JAI Press Inc., pp. 245–257.

Brady, M.L., and M. Sarrafzadeh (1990). Stretching a knock-knee layout for multilayer wiring. *IEEE Trans. Comp.*, C-39, 148–152.

Brayton, R.K., C. McMullen, G.D. Hachtel and A. Sangiovanni-Vincentelli (1984). *Logic Minimization Algorithms for VLSI Synthesis*. Kluwer Acad. Publ., Dordrecht.

Brown, D.J., and R.L. Rivest (1981). New lower bounds for channel width, in: H. Kung, S. Sproull, and G. Steele (eds.), *Proceedings of the CMU Conference on VLSI Systems and Computation*, Computer Science Press, Rockville, pp. 178–185.

Bui, T.N., S. Chaudhuri, F.T. Leighton and M. Sipser (1987). Graph bisection algorithms with good average case behavior. *Combinatorica* 7, 171–191.

Chang, K.C., and D.H.-C. Du (1987). Efficient algorithms for layer assignment problems. *IEEE Trans. Comp.-Aided Design*, CAD-6, 67–78.

Chen, R.W., Y. Kajitani and S.P. Chan (1983). A graph-theoretic via minimization algorithm for two-layer printed circuit board. *IEEE Trans. Circuits and Systems*, CAS-30, 284–299.

Choi, H., K. Nakajima and C.S. Rim (1989). Graph bipartization and via minimization. *SIAM J. Discr. Math.* 2, 38–47.

Cielski, M.J., and E. Kinnen (1981). An optimum layer assignment for routing in ICs and PCBs. *Proc. 18th Design Automation Conference DAC'81*, pp. 733–737.

Cole, R., and A. Siegel (1984). River routing every which way, but loose. *Proc. 24th Annu. Symp. on Foundations of Computer Science FOCS'83*, pp. 65–73.

Corneil, D.G., H. Lerchs and L. Stewart Burlingham (1981). Complement reducible graphs. *Ann. Discr. Math.* 1, 145–162.

Corneil, D.G., Y. Pearl and L. Stewart (1985). A linear recognition algorithm for cographs. *SIAM J. Comput.* 14, 926–934.

Cunningham, W.H. (1982). Decomposition of directed graphs. *SIAM J. Alg. Discr. Meth.* 3, 214–228.

Damaschke, P. (1990). PLA folding in special graph classes. *Discr. Appl. Math.* (1994), 63–74..

Damaschke, P. (1991). Logic arrays for interval indicator functions, in: G. Schmidt and R. Berghammer (eds.), *Proceedings 17th International Workshop on Graph-Theoretic Concepts in Computer Science WG'91*, Springer-Verlag, Berlin, Lecture Notes in Computer Science, Vol. 570, pp. 219–225.

Deo, N., M.S. Krishnamoorty and M.A. Langston (1987). Exact amd approximate solutions for the gate matrix layout problem. *IEEE Trans. Comp.-Aided Design*, CAD-6, 79–84.

Egan, J.R., and C.L. Liu (1984). Bipartite folding and partitioning of a PLA. *IEEE Trans. Comp.-Aided Design*, CAD-3, 191–199.

Ellis, J., I.H. Sudborough and J. Turner (1987). Graph separation and search number. Technical Report DCS-66-IR, University of Victoria.

Even, S. (1979). *Graph Algorithms*. Pitman, London.

Even, S., A. Itai and A. Shamir (1976). On the complexity of time table and multicommodity flow problems. *SIAM J. Comput.* 5, 691–703.

Fellows, M.R., and K. Abrahamson (1990). Cutset regularity beats well-quasi-ordering for bounded treewidth. preprint.

Fellows, M.R., and M.A. Langston (1987). Nonconstructive advances in polynomial-time complexity. *Inform. Process. Lett.* 26, 157–162.

Fellows, M.R., and M.A. Langston (1988). Layout permutation problems and well-partially-ordered sets, in: J.H. Reif (ed.), *Proceedings of the Aegean Workshop on Computing*, Springer-Verlag, Berlin, Lecture Notes in Computer Science, Vol. 319, pp. 315–327.

Fellows, M.R., and M.A. Langston (1988). Nonconstructive tools for proving polynomial-time decidability. *J. Assoc. Comp. Mach.* 35, 727–739.

Fiduccia, C.M., and R.M. Mattheyses (1982). A linear time heuristic for improving network partitions. *Proc. 19th Design Automation Conference DAC'82*, pp. 175–181.

Fleisher, H., and L.I. Maissel (1975). An introduction to array logic. *IBM J. Res. Develop.* 19, 98–109.

Formann, M., D. Wagner and F. Wagner (1993). Routing through a dense channel with minimum total wire length. *J. Algorithms* 15, 267–283.

Formann, M., and F. Wagner (1990). The VLSI layout problem on various embedding models, in: R.H. Möhring (ed.), *Proceedings 16th International Workshop on Graph-Theoretic Concepts in Computer Science WG'90*, Springer-Verlag, Berlin, Lecture Notes in Computer Science, Vol. 484, pp. 130–139.

Frank, A. (1982). Disjoint paths in a rectilinear grid. *Combinatorica* 2, 361–371.

Frederickson, G.N. (1987). Fast algorithms for shortest paths in planar graphs with applications. *SIAM J. Comput.* 16.

Friedman, H., N. Robertson and P.D. Seymour (1987). The metamathematics of the graph minor theorem. *Contemporary Mathematics* 65, 229–261.

Fulkerson, D.R., and O.A. Gross (1965). Incidence matrices and interval graphs. *Pacific J. Math.* 15, 835–855.

Gabow, H.N., and R.E. Tarjan (1985). A linear-time algorithm for a special case of disjoint set union. *J. Comput. System Sci.* 30, 209–221.

Gajski, D.G., and Y.-L.S. Lin (1988). Module generation and silicon compilation, in: B. Preas and M. Lorenzetti (eds.), *Physical Design Automation of VLSI Systems*, Benjamin/Cummings, Menlo Park, CA, pp. 283–345.

Gao, S., and M. Kaufmann (1987). Channel routing of multiterminal nets. *Proc. 28th Annu. Symp. on Foundations of Computer Science FOCS'87*, pp. 316–325.

Garey, M.J., and D.S. Johnson (1979). *Computers and Intractibility: A Guide to the Theory of NP-Completeness*. Freemann, San Francisco.

Garey, M.J., D.S. Johnson and L. Stockmeyer (1976). Some simplified NP-complete graph problems. *Theor. Comp. Sci.* 1, 237–267.

Garg, N., V.V. Vazirani and M. Yannakakis (1993). Approximate max-flow min-(multi)cut theorems and their applications. *Proc. 25th Annu. ACM Symp. on Theory of Computing STOC'93*, pp. 698–707.

Golumbic, M.C. (1980). *Algorithmic Graph Theory and Perfect Graphs*. Academic Press, New York, New York.

Golumbic, M.C. (1985). Interval graphs and related topics. *Discr. Math.* 55, 113–121.

Gonzales, T., and S. Zheng (1988). Simple three-layer channel routing algorithms, in: J.H. Reif (ed.), *Proceedings of the Aegean Workshop on Computing*, Springer-Verlag, Berlin, Lecture Notes in Computer Science, Vol. 319, pp. 237–246.

Gonzales, T., and S. Zheng (1989). On ensuring three-layer wirability by stretching planar layouts. *INTEGRATION VLSI J.* 8, 111–141.

Grötschel, M., M. Jünger and G. Reinelt (1988). Via minimization with pin preassignment and layer preference. *Angew. Math. Mech.* 69, 393–399.

Gupta, U.I., D.T. Lee and J.Y.T. Leung (1979). An optimal solution for the channel assignment problem. *IEEE Trans. Comp.*, C-28, 807–810.

Gupta, U.I., D.T. Lee and J.Y.T. Leung (1982). Efficient algorithms for interval graphs and circular arc graphs. *Networks* 12, 459–467.

Gustedt. , J., Path width for chordal graphs is NP-complete. *Discr. Appl. Math.* (1993), 233–248.

Hachtel, G.D., A.R. Newton and A.L. Sangiovanni-Vincentelli (1982). An algorithm for optimal PLA folding. *IEEE Trans. Comp.-Aided Design*, CAD-1, 63–77.

Hadlock, F. (1975). Finding a maximum cut of a planar graph in polynomial time. *SIAM J. Comput.* 4, 221–225.

Hammer, P.L., and F. Maffray (1990). Completely separable graphs. *Discr. Appl. Math.* 27, 85–99.

Hashimoto, A., and J. Stevens (1971). Wire routing by optimizing channel assignment within large apertures. *Proc. 8th Design Automation Conference DAC'71*, pp. 155–169.

Hassin, R. (1984). On multicommodity flows in planar graphs. *Networks* 14, 225–235.

Hu, T.C., and Y.S. Kuo (1987). Graph folding and programmable logic array. *Networks* 17, 19–37.

Hwang, S.Y., R.W. Dutton and T. Blank (1986). A best-first search algorithm for optimal PLA folding. *IEEE Trans. Comp.-Aided Design*, CAD-5, 433–442.

Ihler, E., D. Wagner and F. Wagner (1993). Modeling hypergraphs by graphs with the same min-cut properties. *Information Processing Letters* 45 171-175.

Johnson, D.S. (1987). The NP-completeness column: An ongoing guide. *J. Algorithms* 8, 285–203.

Johnson, D.S. (1987). The NP-completeness column: An ongoing guide. *J. Algorithms* 8, 438–448.

Johnson, D.S., C. Aragon, L.A. McGeoch and C. Schevon (1989). Optimization by simulated annealing: An experimental evatuation; Part I, Graph Partitioning. *Oper. Res.* 37, 865–892.

Kajitani, Y. (1980). On via hole minimization of routing on a 2-layer board. In *Proceedings IEEE International Conference on Circuits and Computers*, pp. 295–298.

Kashiwabara, T., and T. Fujisawa (1979). NP-completeness of the problem of finding a minimum-clique-number interval graph containing a given graph as a subgraph. In *Proceedings 1979 International Symp. on Circuits and Systems*, pp. 657–660.

Kaufmann, M. (1990). A linear time algorithm for routing in a convex grid. *IEEE Trans. Comp.-Aided Design*, CAD-9, 180–184.

Kaufmann, M., and G. Klär (1991). A faster algorithm for edge-disjoint paths in planar graphs, in: W.L. Hsu and R.C.T. Lee (eds.), *ISA'91 Algorithms, 2nd International Symposium on Algorithms*, Springer-Verlag, Berlin, Lecture Notes in Computer Science, Vol. 557, pp. 336–348.

Kaufmann, M., and K. Mehlhorn (1985). Generalized switchbox routing. *J. Algorithms* 7, 510–531.

Kaufmann, M., and K. Mehlhorn (1990). Routing problems in grid graphs, in: B. Korte, L. Lovasz, H.J. Prömel and A. Schrijver (eds.), *Paths, Flows and VSLI-Layout*, Springer-Verlag, Berlin, pp. 165–1884.

Kaufmann, M., and P. Molitor (1991). Minimal stretching of a layout to ensure 2-layer wirability. *INTEGRATION VLSI J.*, pp. 339–352.

Kaufmann, M., and I.G. Tollis (1988). Channel routing with short wires, in: J.H. Reif (ed.), *Proceedings of the Aegean Workshop on Computing*, Springer-Verlag, Berlin, Lecture Notes in Computer Science, Vol. 319, pp. 226–236.

Kawamoto, T., and Y. Kajitani (1979). The minimum width routing of a 2-row 2-layer polycell layout. *Proc. 16th Design Automation Conference DAC'79*, pp. 290–296.

Kernighan, B.W., and S. Lin (1970). An efficient heuristic procedure for partitioning graphs. *Bell System Tech. J.* 49, 291–307.

Kirousis, L.M., and C.H. Papadimitriou (1985). Interval graphs and searching. *Discr. Math.* 55, 181–184.

Kirousis, L.M., and C.H. Papadimitriou (1986). Searching and pebbling. *Theor. Comp. Sci.* 47, 205–218.

Klein, P., A. Agrawal, R. Ravi and S. Rao (1990). Approximation through multicommodity flow. *Proc. 31th Ann. Symp. on Foundations of Computer Science FOCS'90*, pp. 726–737.

Klein, P., S. Plotkin and S. Rao. Planar graphs, multicommodity flow, and network decomposition, to appear.

Korte, B., L. Lovasz, H.J. Prömel and A. Schrijver, eds. (1990). *Paths, Flows and VLSI-Layout*. Springer-Verlag, Berlin.

Korte, B., H.J. Prömel and A. Steger (1990). Steiner trees in VLSI-layout, in: B. Korte, L. Lovasz, H.J. Prömel and A. Schrijver (eds.), *Paths, Flows and VLSI-Layout*, Springer-Verlag, Berlin, pp. 185–214.

Korte, N., and R.H. Möhring (1989). An incremental linear-time algorithm to recognize interval graphs. *SIAM J. Comput.* 18, 68–81.

Kramer, M.R., and J. van Leeuwen (1984). The complexity of wire-routing and finding minimum area layouts for arbitrary VLSI-circuits, in: F.P. Preparata (ed.), *Advances in Computer Research, VOL 2: VLSI Theory*, JAI Press Inc., pp. 129–146.

Kuchem, R., D. Wagner and F. Wagner (1989). Area-optimal three-layer channel routing. *Proc. 30th Annu. Symp. on Foundations of Computer Science FOCS'89*, pp. 506–511.

Kuo, Y.S., T.C. Chen and W. Shih (1988). Fast algorithm for optimal layer assignment. *Proc. 25th Design Automation Conference DAC'88*, pp. 554–559.

Kuo, Y.S., and T.C. Hu (1987). An efffective algorithm for optimal PLA column folding. *INTEGRATION VLSI J.* 5, 217–230.

Lagergren, J. (1990). Efficient parallel algorithm for tree-decomposition and related problems. *Proc. 31th Annu. Symp. on Foundations of Computer Science FOCS'90*, pp. 173–182.

LaPaugh, A.S. (1980). *Algorithms for Integrated Circuit Layout: An Analytic Approach*. PhD thesis, MIT.

LaPaugh, A.S., and R.Y. Pinter (1990). Channel routing for integrated circuits. *Annual Reviews Comp. Science* 4, 307–363.

Leighton, F.T. (1982). A layout strategy for VLSI which is provably good. *Proc. 14th Annu. ACM Symp. on Theory of Computing STOC'82*, pp. 85–98.

Leighton, F.T., F. Makedon, S. Plotkin, C. Stein, E. Tardos and S. Tragoudas (1991). Fast approximation algorithms for multicommodity flow problems. *Proc. 23th Annu. ACM Symp. on Theory of Computing STOC'91*, pp. 101–111.

Leighton, F.T., and S. Rao (1988). An approximate max-flow min-cut theorem for uniform multicommodity flow problems with applications to approximation algorithms. *Proc. 29th Annu. Symp. on Foundations of Computer Science FOCS'88*, pp. 422–431.

Leiserson, C.E. (1983). *Area-Efficient VSLI-Computation*. MIT Press, Cambridge, Massachussetts.

Leiserson, C.E., and F.M. Maley (1985). Algorithms for routing and testing routability of planar VLSI layouts. *Proc. 17th Annu. ACM Symp. on Theory of Computing STOC'85*, pp. 69–78.

Leiserson, C.E., and R.Y. Pinter (1983). Optimal placement for river routing. *SIAM J. Comput.* 12, 447–462.

Lengauer, T. (1981). Black-white pebbles and graph separation. *Acta Inform.* 16, 465–475.

Lengauer, T. (1990). *Combinatorial Algorithms for Integrated Circuit Layout*. Wiley-Teubner.

Leong, H.W. (1986). A new algorithm for gate matrix layout. In *International Conference on Computer-Aided Design*, pp. 316–319.

Lewandowski, J.L., and C.L. Liu (1984). A branch and bound algorithm for optimal PLA folding. *Proc. 21th Design Automation Conference DAC'84*, pp. 425–433.

Li, J.T. (1983). Algorithms for gate matrix layout. In *Proceedings of IEEE International Symposium on Circuits and Systems*, pp. 1013–1016.

Liao, K.-F., and M. Sarrafzadeh (1990). Vertex-disjoint trees and boundary single-layer routing, in: R.H. Möhring (ed.), *Proceedings 16th International Workshop on Graph-Theoretic Concepts in Computer Science WG'90*, Springer-Verlag, Berlin, Lecture Notes in Computer Science, Vol. 484, pp. 99–108.

Lipski, W., Jr. (1984). On the structure of three-layer wirable layouts, in: F.P. Preparata (ed.), *Advances in Computer Research, VOL 2: VLSI Theory*, JAI Press Inc., pp. 231–243.

Lipski, W., Jr. and F.P. Preparata (1987). A unified approach to layout wirability. *Math. Systems Theory* 19, 189–203.

Lipton, R.J., and R.E. Tarjan (1979). A separator theorem for planar graphs. *SIAM J. Appl. Math.* 36, 177–189.

Lipton, R.J., and R.E. Tarjan (1980). Applications of a planar separator theorem. *SIAM J. Comput.* 9, 615–627.

Lopez, A., and H. Law (1980). A dense gate matrix layout method for MOS VLSI. *IEEE Trans. on Electronic Devices*, ED-27, 1671–1675.

Luby, M., U. Vazirani, V. Vazirani and A.L. Sangiovanni-Vincentelli (1982). Some theoretical results on the optimal PLA folding problem. In *Proceedings 1982 International Symp. on Circuits and Systems*, pp. 165–170.

Lynch, J.F. (1975). The equivalence of theorem proving and the interconnection problem. *ACM SIGDA Newsletter* 5, 31–65.

Makedon, F., and I.H. Sudborough (1989). On minimizing width in linear layouts. *Discr. Appl. Math.* 23, 201–298.

Matsumoto, K., T. Nishizeki and N. Saito (1985). An efficient algorithm for finding multicommodity flows in planar networks. *SIAM J. Comput.* 14, 289–302.

Matsumoto, K., T. Nishizeki and N. Saito (1986). Planar multicommodity, maximum matchings and negative cycles. *SIAM J. Comput.* 15, 495–510.

Mead, C., and L. Conway (1980). *Introduction to VLSI Systems*. Addison-Wesley, Reading, NY, Reading, MA.

Megiddo, N., S.L. Hakimi, M.R. Garey, D.S. Johnson and C.H. Papadimitriou (1988). The complexity of searching a graph. *J. Assoc. Comp. Mach.* 35, 18–44.

Mehlhorn, K., and F.P. Preparata (1986). Routing through a rectangle. *J. Assoc. Comp. Mach.* 33, 60–85.

Mehlhorn, K., F.P. Preparata and M. Sarrafzadeh (1986). Channel routing in knock-knee mode: Simplified algorithms and proofs. *Algorithmica* 1, 213–221.

Middendorf, M., and F. Pfeiffer (1993). On the complexity of the disjoint path problem. *Combinatorica* 13, 97–107.

Mirzaian, A. (1987). River routing in VLSI. *J. Comput. System Sci.* 34, 43–54.

Möhring, R.H. (1985). Algorithmic aspects of comparability graphs and interval graphs, in: I. Rival (ed.), *Graphs and Order*, Nato Advanced Study Institutes Series, D. Reidel Publishing Company, Dordrecht, pp. 41–101.

Möhring, R.H. (1989). Computationally tractable classes of ordered sets, in: I. Rival (ed.), *Algorithms and Order*, Nato Advanced Study Institutes Series, D. Reidel Publishing Company, Dordrecht, pp. 105–193.

Möhring, R.H. (1990). Graph problems related to gate matrix layout and PLA folding, in: G. Tinhofer, E. Mayr, H. Noltemeier and M.M. Sysło (eds.), *Computational Graph Theory*, Computing Supplementum 7, Springer-Verlag, Wien, pp. 17–51.

Molitor, P. (1987). On the contact minimization problem. *Proc. 4th Annu. Symp. on Theoretical Aspects of Computer Science STACS'87*, pp. 159–165.

Molitor, P. (1991). A survey on wiring. *J. Inf. Process. Cybern. EIK* 27, 3–19.

Monien, B., and I.H. Sudborough (1988). Min cut is NP-complete for edge weighted trees. *Theor. Comp. Sci.* 58, 209–229.

Müller, R., and D. Wagner (1991). α-vertex separatior is NP-hard even for 3-regular graphs. *Computing* 46, 343–353.

Naclerio, N.J., S. Masuda and K. Nakajima (1987). Via minimization for gridless layouts. *Proc. 24th Design Automation Conference DAC'87*, pp. 159–165.

Naclerio, N.J., S. Masuda and K. Nakajima (1989). The via minimization problem is NP-complete. *IEEE Trans. Comp.*, C-38, 1604–1608.

Nagamochi, H., and T. Ibaraki (1992). Computing edge-connectivity in multigraphs and capacitated graphs. *SIAM J. Discr. Math.* 5, 54–66.

Nakatani, K., T. Fujii, T. Kikuno and N. Yoshita (1986). A heuristic algorithm for gate matrix layout. In *Digest Internat. Conference on Computer-Aided Design*.

Nishizeki, T., N. Saito and K. Suzuki (1985). A linear time routing algorithm for convex grids. *IEEE Trans. Comp.-Aided Design*, CAD-4, 68–76.

Ohtsuki, T., H. Mori, E. Kuh, T. Kashiwabara and T. Fujisawa (1979). One-dimensional logic gate assignment and interval graphs. *IEEE Trans. Circuits and Systems*, pp. 675–684.

Okamura, K., and P.D. Seymour (1981). Multicommodity flows in planar graphs. *J. Combin. Theory Ser. B* 31, 75–81.

Parsons, T.D. (1976). Pursuit-evasion in a graph, in: Y. Alavi and D. Lick (eds.), *Theory and Applications of Graphs*, Springer-Verlag, Berlin, pp. 426–441.

Pinter, R.Y. (1984). Optimal layer assignment for interconnect. *J. VLSI Comp. Systems* 1, 123–137.

Plotkin, S., and E. Tardos. Improved bounds on the max-flow min-cut ratio for multicommodity flows, to appear.

Preparata, F.P., and W. Lipski, Jr. (1984). Optimal three-layer channel routing. *IEEE Trans. Comp.*, C-33, 427–437.

Preparata, F.P., and M. Sarrafzadeh (1985). Channel routing of nets of bounded degree, in: P. Bertolazzi and F. Luccio (eds.), *Algorithms and Architecture*, North-Holland, pp. 189–203.

Preparata, F.P., and M.I. Shamos (1985). *Computational Geometry*. Springer-Verlag, Berlin, New York.

Rao, S.B. (1987). Finding near optimal separators in planar graphs, in: *Proceedings of the 28th Annual Symposium on Foundations of Computer Science FOCS'87*, pp. 225–237.

Rao, S.B. (1992). Faster algorithms for finding small edge cuts in planar graphs. *Proc. 24th Annu. ACM Symp. on Theory of Computing STOC'92*, pp. 229–240.

Ravi, S.S. (1989). A note on the orderability problem for PLA folding. *Discr. Appl. Math.* 25, 317–320.

Ravi, S.S., and E.L. Loyd (1988). The complexity of near-optimal programmable logic array folding. *SIAM J. Comput.* 17, 696–710.

Reed, J., A. Sangiovanni-Vincentelli and M. Santomauro (1985). A new symbolic channel router: YACR2. *IEEE Trans. Comp.-Aided Design*, CAD-4, 208–219.

Ripphausen-Lipa, H., D. Wagner and K. Weihe (1993). The vertex-disjoint Menger-problem in planar graphs. *Proc. 4th Annu. ACM–SIAM Symp. on Discrete Algorithms*, pp. 112–119.

Rivest, R.L., and C.M. Fiduccia (1982). A greedy channel router. *Proc. 19th Design Automation Conference DAC'82*, pp. 418–424.

Robertson, N., and P. Seymour (1986). Graph minors XIII, the disjoint paths problem. *Manuscript*.

Robertson, N., and P.D. Seymour (1983). Graph minors I, Excluding a forest. *J. Combin. Theory Ser. B* 35, 39–61.

Robertson, N., and P.D. Seymour (1986). Graph minors VI, disjoint paths across a disc. *J. Combin. Theory Ser. B* 41, 115–138.

Robertson, N., and P.D. Seymour (1988). Graph minors XV, Wagner's conjecture. Preprint.

Robertson, N., and P.D. Seymour (1990). An outline of a disjoint paths algorithm, in: B. Korte, L. Lovasz, H.J. Prömel and A. Schrijver (eds.), *Paths, Flows and VLSI-Layout*, Springer-Verlag, Berlin, pp. 267–292.

Sarrafzadeh, M. (1987). Channel routing problem in the knock-knee mode is NP-complete. *IEEE Trans. Comp.-Aided Design*, CAD-6, 503–506.

Sarrafzadeh, M. (1987). Channel routing with provably short wires. *IEEE Trans. Circuits and Systems*, CAS-34, 1133–1135.

Sarrafzadeh, M., and F.P. Preparata (1985). Compact channel routing of multiterminal nets. *Ann. Discr. Math.* 25, 255–280.

Sarrafzadeh, M., D. Wagner, F. Wagner and K. Weihe (1994). Wiring knock-knee layouts: A global approach. *IEEE Trans. Comp.*, 43: 581–589.

Scheffler, P. (1989). *Die Baumweite von Graphen als ein Maß für die Kompliziertheit algorithmischer Probleme*. PhD thesis, Akademie der Wissenschaften der DDR, Berlin.

Scheffler, P. (1990). A linear algorithm for the pathwidth of trees, in: R. Bodendiek and R. Henn (eds.), *Topics in Combinatorics and Graph Theory*, Physica-Verlag, Heidelberg, pp. 613–620.

Schmidt, K.H., and K.D. Mueller-Glaser (1983). NMOS dense gate matrix VLSI design. *IEEE J. of Solid-State Circuits*, SC-18, 157–159.

Schrijver, A. (1990). Homotopic routing methods, in: B. Korte, L. Lovasz, H.J. Prömel and A. Schrijver (eds.), *Paths, Flows and VLSI-Layout*, Springer-Verlag, Berlin, pp. 329–371.

Seymour, P.D. (1981). On odd cuts and plane multicommodity flows. *Proc. London Math. Soc.* 42, 178–192.

Shahrokhi, F., and D.W. Matula (1990). The maximum concurrent flow problem. *J. Assoc. Comp. Mach.* 37, 318–334.

Shih, W.-K., S. Wu and Y.S. Kuo (1990). Unifying maximum cut and minimum cut of a planar graph. *IEEE Trans. Comp.*, C-39, 694–697.

Siegel, A., and D. Dolev (1988). Some geometry for general river routing. *SIAM J. Comput.* 17, 583–605.

Stevens, K.R., and W.M. VanCleemput (1979). Global via elimination in generalized routing environment. In *Proceedings IEEE International Conference on Circuits and Computers*, pp. 689–692.

Suzuki, H., T. Akama and T. Nishizeki (1990). Finding Steiner forests in planar graphs. *Proc. 1st Annu. ACM–SIAM Symp. on Discrete Algorithms*, pp. 444–453.

Suzuki, H., T. Nishizeki and N. Saito (1989). Algorithms for multicommodity flows in planar graphs. *Algorithmica* 4, 471–501.

Szymanski, T. (1985). Dogleg channel routing is NP-complete. *IEEE Trans. Comp.-Aided Design*, CAD-4, 31–41.

Thompson, C.D. (1980). *A Complexity Theory for VLSI*. PhD thesis, Department of Computer Science, Carnegie-Mellon University, Pittsburg, Pennsylvania, USA.

Tollis, I.G. (1988). A new algorithm for wiring layouts, in: J.H. Reif (ed.), *Proceedings of the Aegean Workshop on Computing*, Springer-Verlag, Berlin, Lecture Notes in Computer Science, Vol. 319, pp. 257–267.

Ullmann, J.D. (1983). *Computational Aspects of VLSI*. Computer Science Press, Rockville, Maryland, USA.

Valiant, L.G. (1981). Universality considerations in VLSI circuits. *IEEE Trans. Comp.*, C-30, 135–140.

Wagner, D. (1991). A new approach to knock-knee channel routing, in: W.L. Hsu and R.C.T. Lee (eds.), *ISA'91 Algorithms, 2nd International Symposium on Algorithms*, Springer-Verlag, Berlin, Lecture Notes in Computer Science, Vol. 557, pp. 83–93.

Wagner, D. (1993). Optimal routing through dense channels. *Int. J. Comp. Geom. Appl.*, 3: 269–289.

Wagner, D., and F. Wagner (1993). Between min-cut and graph bisection, in: A.M. Borzyszkowski and S. Sokolowski (eds.), *MFCS'93, 18th International Symposium on Mathematical Foundations of Computer Science*, Springer-Verlag, Berlin, Lecture Notes in Computer Science, Vol. 711, pp. 744-750.

Wagner, D., and K. Weihe (1993). A linear time algorithm for edge-disjoint paths in planar graphs, in: T. Lengauer (ed.), *ESA'93, First European Symposium on Algorithms*, LNCS 726, pp. 384–395.

Wagner, F., and B. Wolfers (1991). Short wire routing in convex grids, in: W.L. Hsu and R.C.T. Lee (eds.), *ISA'91 Algorithms, 2nd International Symposium on Algorithms*, Springer-Verlag, Berlin, Lecture Notes in Computer Science, Vol. 557, pp. 72–83.

Weinberger, W. (1967). Large scale integration of MOS complex logic: A layout method. *IEEE J. Solid-State Circuits*, SC-2, 182–190.

Wieners-Lummer, C. (1990). Manhattan routing with good theoretical and practical performance. *Proc. 1st Annu. ACM–SIAM Symp. on Discrete Algorithms*, pp. 465–475.

Wieners-Lummer, C. (1991). Three-layer channel routing in knock-knee mode. Technical report, Universität Paderborn.

Wing, O. (1982). Automated gate matrix layout. In *Proc. IEEE Int. Symp. on Circuits and Systems*, pp. 681–685.

Wing, O. (1983). Interval-graph-based circuit layout. In *Proc. IEEE Int. Conf. on CAD*, pp. 84–85.

Wing, O., S. Huang and R. Wang (1985). Gate matrix layout. *IEEE Trans. Comp.-Aided Design*, CAD-4, 220–231.

Yoshimura, T. (1984). An efficient channel router. *Proc. 21th Design Automation Conference DAC'84*, pp. 38–44.

Yoshimura, T., and E.S. Kuh (1982). Efficient algorithms for channel routing. *IEEE Trans. Comp.-Aided Design*, CAD-1, 25–35.

Yoshizawa, H., H. Kawanishi and K. Kami (1975). A heuristic procedure for ordering MOS arrays. *Proc. 12th Design Automation Conference DAC'75*, pp. 384–389.

M.O. Ball et al., Eds., *Handbooks in OR & MS, Vol. 8*

Chapter 9

Network Models in Economics

William W. Sharkey

Institut d'Economie Industrielle, Toulouse, France

The discipline of economics is broadly concerned with the allocation of scarce resources. The fundamental economic hypothesis is that human beings behave as rational and self interested agents in the pursuit of their objectives, and that aggregate resources are limited. Economic behavior is therefore modeled as the solution of a constrained optimization problem. Typically both the objective function and the constraint set are assumed to possess technical properties such as continuity and convexity that ensure a well behaved solution to the optimization problem. Models of general economic equilibrium [Debreu, 1959] represent elegant mathematical structures which describe, at an abstract level, the workings of an entire economy. In models of this type, both the existence of equilibrium prices, and the fundamental optimality properties associated with equilibria, depend on the underlying convexity of the production possibility set. In recent years economists have become increasingly concerned with the restrictiveness of these assumptions, since they preclude the study of commonly observed phenomena such as increasing returns to scale and indivisibilities in the production process.[1] A particularly rich class of problems in this tradition is the study of discrete optimization problems on networks. The purpose of this chapter is to survey the literature in economics, and the closely related literature of cooperative and non-cooperative game theory, in which networks and related graph theoretic concepts play a prominent role.

Clearly there is a substantial interdisciplinary component to the subject matter of this chapter. However, while the specialist in operations research will typically be interested in issues such as computational complexity and the design of efficient algorithms, the economist addresses a different set of issues. Three fundamental economic issues will be considered in this chapter in the context of discrete optimization on networks. First, there is the issue of allocation, either of costs, or of surplus, resulting from a cooperative endeavor, such as the operation of a communications network. For this purpose the tools of cooperative game theory will prove to be particularly useful. Typically the question of proper

---

[1] See Sharkey [1989] for a survey of some of this issues.

allocation of cost or surplus is studied in the context of complete information. If there is a central planner, he or she is assumed to have access to all relevant information regarding the costs and benefits of the economic activity under consideration. If the allocation question is modeled as a cooperative game, then each player in the game is assumed to possess full information about the relevant characteristics of all players in the game, and about the technical structure of the game itself.

A second fundamental issue is the question of incentives. In this case it is assumed that economic agents have private information about some of the relevant variables in the optimization problem. For example, the users of a network service may have a private valuation of the service that cannot be easily observed by a central planner. In order to arrive at a solution to the optimization problem, the planner must be aware of the fact that agents may not truthfully reveal their actual preferences if it is not in their interest to do so. Incentive questions are typically modeled as non-cooperative games in which players are allowed to independently choose a strategy. Equilibrium outcomes correspond to joint strategy choices that no player wishes to deviate from, given the choices of every other player in the game.

The third fundamental economic issue concerns the role of prices in guiding resource allocation in problems involving discrete optimization. For this purpose the dual solution of the optimization problem itself typically contains the relevant information that determines whether such prices exist.

While the economic literature on discrete optimization and networks is not large, it is broad, and it covers a wide range of topics. Inherently any survey such as this is highly idiosyncratic. In particular, no attempt has been made to survey the literature on network externalities (since this literature is technically unrelated to most of the work covered here) or the literature on location models on networks (since this work is covered elsewhere in this volume). The chapter will consist of three sections.

Section 1 will be concerned with cost allocation on minimal cost spanning tree networks. Network optimization problems of this type have an interesting combinatorial structure that has been a subject of active research in recent years. In addition, there are a number of important economic applications involving fair allocation of cost among customers of communications, transportation and other networks.

Section 2 will survey the literature on matching models as they have been applied in economic theory. Matching models also possess a rich combinatorial structure that, while different from the structure of spanning tree networks, has many interesting parallels. Furthermore, like the subject matter of the previous section, matching models have important economic applications, such as the assignment of workers to jobs, production facilities to locations, and students to colleges.

Section 3 will consider the application of economic methodologies to the design and operation of stochastic service systems. In these papers, networking aspects are not emphasized; a service center can be thought of as a single link of a larger computer or communications network. The economic problem is the design of service disciplines, through pricing or flow control restrictions, which promote efficient utilization of the service center In each of the sections, it is hoped that students of operations research will become aware of some interesting and important economic issues, while economists will become aware of a class of problems in which further economic modeling can be productively applied.

1. Cost allocation on networks

This section will describe the way in which game theoretic tools have been applied to define reasonable procedures for allocating costs on a network. The need for cost allocation arises when network services are provided by regulated or public enterprises in which markets do not fully determine the prices that must be charged to individual users of the network. Cooperative game theory suggests two possible ways in which reasonable allocations may be defined. One is the axiomatic approach, in which a particular allocation is selected on the basis of reasonable properties that any allocation ought to possess. Leading examples of axiomatic allocation methods are the Shapley Value and the nucleolus. An alternative method of defining reasonable allocations is based on the requirement that allocations should in some circumstances be immune to disruption by coalitions of players. The coalition rationality requirement is particularly appropriate when allocations are used to determine prices of a regulated supplier of network services in the presence of competition by rival producers.

This issue of cost allocation on a spanning tree network was first raised by Claus & Kleitman [1973]. The spanning tree model corresponds to a variety of cost allocation problems in transportation and communications industries, "in which users located at a number of geographically distinct points are connected to a center by means of a minimal length (or cost) spanning tree. Such an arrangement can arise, for example, when several users are connected to a central computer or to a cable TV network; or when rental of a multipoint communications circuit is to be cost allocated."[2] A number of more or less arbitrary allocation methods were described by the above authors, and a fundamental incentive issue was raised. A naive, or improperly chosen allocation method might require some individual users to bear so large a share of the total cost that they would prefer to secede from the network, or to try to exclude others from joining.

Consider the following example. There are three customers represented by nodes 1,2, and 3 in the graph of Figure 1, and a central supplier represented by node 0. The minimal length tree has total cost $\alpha + \beta + \gamma$. One natural method of cost allocation would be to assign the cost of each link in the minimal cost network equally among those customers who make use of it. In this allocation, customer 1 would pay $\alpha/3$; customer 2 would pay $\alpha/3 + \beta/2$ and customer 3

---

[2] Claus & Kleitman [1973], p. 289.

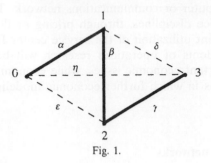

Fig. 1.

would pay $\alpha/3 + \beta/2 + \gamma$. Customer 3 would be allocated a cost greater than her direct connection cost if $\eta < \alpha/3 + \beta/2 + \gamma$, as is geometrically possible. If link costs are not proportional to length, then it is also possible for customer 2 or the coalition of customers 2 and 3 together to be overcharged in a similar way. Other plausible allocation methods based on the incremental cost of providing service lead to similar potential difficulties.

Bird [1976] and Granot & Huberman [1981] were the first to discover a satisfactory resolution of the cost allocation problem on spanning tree networks by applying the techniques of cooperative game theory. Following these authors, a substantial literature has developed using cooperative game theoretic methods in the study of spanning tree problems and related allocation problems on networks.[3] The next section develops the notation and defines some of the standard solution concepts from cooperative game theory. Readers familiar with these concepts should proceed to Section 1.2.

1.1. Game theoretic preliminaries

A cooperative game with 'transferable utility' (TU game) consists of a set $N = \{1, \ldots, n\}$ of players and a function $v : 2^N \to \Re_+$, which assigns a value to every coalition of players. Side payments among players in terms of a transferable medium of exchange are permitted, and the usual interpretation of v is that it describes the best outcome that the coalition S can achieve, given the most unfavorable actions on the part of the complementarity coalition $N \setminus S$. It is required that $v(\emptyset) = 0$. No other restrictions need to be placed on v, although it is typically assumed that the function v is *subadditive*, so that $v(S) + v(T) \leq v(S \cup T)$ whenever $S \cap T = \emptyset$. As a notational convention, for any vector $a = (a_1, \ldots, a_n)$ define the additive function $a(S) = \sum_{i \in S} a_i$ for every $S \subseteq N$.

A solution of a game (N, v) consists of a set of vectors x which assign a payoff x_i to every player $i \in N$. While there are a large number of solution concepts that have found an application in economic modeling, this chapter will be concerned with four solutions. The most elementary, but also most fundamental

---

[3] Granot & Huberman [1981, 1982, 1984]; Granot [1984, 1986]; Granot & Hojati [1990]; Megiddo [1978a, b]; Sharkey [1991]; and Tamir [1991].

solution concept for most economic problems is the core. The core is defined as the set of payoff vectors x that are feasible for the grand coalition N, and which cannot be improved upon by any coalition $S \subseteq N$. Formally there is the following:

Definition 1.1. *The core of a TU game (N,v) is the set*

$$\{x : x(N) = v(N) \text{ and } x(S) \geq v(S) \text{ for all } S \subseteq N\}$$

The core offers a precise answer to the incentive question raised by Claus and Kleitman.[4] Any payoff vector in the core has the property that no coalition S of players would wish to secede from the grand coalition N. As long as v is superadditive, this is a desirable property, since the fragmentation of N implies an inefficient outcome in the game.

The core is the solution of a set of linear inequalities, for which there may be no feasible solution. The question of existence of a non-empty core is resolved in the following result, due to Bondareva [1962] and Shapley [1967].

Proposition 1.2. *The core of a game (N, v) is non-empty if and only if*

$$v(N) \geq \sum_{S \in \mathbf{B}} \delta_S v(S) \tag{1.1.1}$$

for every collection $\mathbf{B}$ for which

$$\sum_{S \in \mathbf{B}: i \in S} \delta_S = 1 \text{ for every } i \in N \tag{1.1.2}$$

A collection $\mathbf{B}$ which satisfies conditions (1.1.2) is said to be a 'balanced' collection, and a game (N, v) for which v satisfies (1.1.1) is said to be a balanced game.

If (N, v) is a game and $T \subset N$, then a 'subgame' (T, v^T) is the game with player set T and valuation function $v^T(S) = v(S)$ for every $S \subseteq T$. A game (N, v) is totally balanced, if and only if, for every subset $T \subseteq N$, the subgame (T, v^T) is balanced.

The core describes a set of outcomes that can in some sense survive a process of competition in which every player seeks to join a coalition offering the highest payoff. An alternative approach to solution theory postulates that the members of the grand coalition have agreed to cooperate, but that they nevertheless will bargain among themselves over the distribution of the total payoff. One model of such a bargaining process is defined by the kernel.[5] Given a game (N, v)

---

[4] At this point the core has been introduced as a general game theoretic solution. In a cost allocation framework, it is costs rather than values that must be allocated among the players, so the inequalities defining the core must be reversed. After these introductory remarks, most of the discussion in the present chapter will be in terms of the cost allocation problem.

[5] The kernel and relating bargaining solutions are more generally defined for coalition structures consisting of partitions of players. Throughout the chapter it will be assumed that the grand coalition is the only coalition structure under consideration.

and a payoff vector x, for every coalition $S \subseteq N$, define the excess function $e(S, x) = v(S) - x(S)$. For any two players i and j define the surplus of i against j by

$$s_{ij}(x) = \text{Max}\{e(S, x) : i \in S, \ j \notin S\} \tag{1.1.3}$$

One can say that player i 'outweighs' player j at an allocation x, and write $i \succ\succ j$, if and only if $s_{ij}(x) > s_{ji}(x)$ and $x_j > v(\{j\})$. Intuitively, if $i \succ\succ j$, then i can make a demand on j which j cannot contest, and so the allocation x is, in this sense, unstable. Formally there is the following: @@@

Definition 1.3. *The kernel of a game* (N, v) *is the set of allocations* x *such that for no* $i, j \in N$ *is it true that* $i \succ\succ j$.

The definition of the kernel is due to Davis & Maschler [1965], who proved that the kernel is non-empty for every coalition structure. Maschler & Peleg [1966] provided an algebraic existence proof, and demonstrated that the intersection of the kernel and the core is non-empty whenever the core itself is non-empty. Maschler, Peleg & Shapley [1979] proved that any outcome x contained in both the kernel and the core must lie at the midpoint of a line segment $R_{ij}(x)$, which defines the bargaining range between i and j, for every pair of players i and j. Thus, the kernel represents a completely symmetric fair division scheme, when fairness is defined with respect to pairwise bargaining between players.

The core is the solution to a set of 2^n inequality constraints in n variables. As such, the core of a game is frequently the empty set. On the other hand, in some games, both the core and the kernel may consist of a large number of payoff vectors, so that additional criteria may be necessary in order to choose a unique allocation. Two single valued solutions that have been widely studied are the Shapley value and the nucleolus.[6] In order to define the Shapley value, let us consider the set Ω of permutations of the player set N. For a particular permutation $\omega \in \Omega$, let $P^i(\omega) = \{j : \omega(j) < \omega(i)\}$ be the set of players which precede i in ω, where $P^i(\omega) = \emptyset$ if player i is first in the ordering ω. The marginal contribution of player i in permutation ω is given by

$$v^i(\omega) = v[P^i(\omega) \cup \{i\}] - v[P^i(\omega)] \tag{1.1.4}$$

The Shapley value is then defined as follows:

Definition 1.4. *The Shapley value of a game* (N, v) *is the payoff vector* $\phi(v)$ *defined by*

$$\phi_i(v) = \sum_{\omega \in \Omega} \left(\frac{1}{n!}\right) v^i(\omega).$$

[6] The Shapley value was defined by Shapley [1953]. The nucleolus is due to Schmeidler [1969]. Since both solutions have been widely studied, this chapter will not attempt to motivate the definitions of either the Shapley value or the nucleolus. The reader for whom both concepts are new is referred instead to a textbook on game theory, such as Owen [1982].

For each permutation $\omega \in \Omega$ the marginal contribution of player i is precisely the incremental value which that player brings into the game, assuming that players arrive according to ω. The Shapley value may therefore be thought of as the average contribution of a player to the game, under the assumption that each of the $n!$ permutations is equally likely. Clearly the Shapley value exists, and is single valued for all games. In addition, it satisfies a number of desirable properties, such as efficiency, so that $\sum_{i \in N} \phi_i(v) = v(N)$. Shapley [1953] characterized the value as the unique operator on the set of all possible operators, which satisfies a set of reasonable a priori axioms including efficiency.

An alternative single valued solution is the nucleolus. Given a game (N, v), one may define the 2^n dimensional vector $\theta(x)$ whose components are the values of the excess function for each of the coalitions $S \subseteq N$ arranged in decreasing order. That is, if the subsets $S \subseteq N$ are enumerated as $S_1, \ldots, S_{2^n}$ and $e(S_k, x) \geq e(S_k + 1, x)$ for $k = 1, \ldots, 2^n - 1$, then $\theta_k(x) = e(S_k, x)$. Given any two vectors q and r of dimension m, q is lexicographically larger than r, $q \succ r$, if for some $k < m$, $q_i = r_i$ for all $i < k$, and $q_k > r_k$. A payoff vector is efficient if $x(N) = v(N)$. The nucleolus is defined as the set of efficient payoff vectors that lexicographically minimize the vector θ. That is, the nucleolus is the set of payoff vectors such that the maximum excess is as small as possible, and the maximum is attained on as few subsets as possible; the next largest excess is as small as possible and is attained on as few subsets as possible; and so on. Formerly,

Definition 1.5. *The nucleolus of a game (N, v) is the set*

$$\eta(v) = \{x : x \text{ is efficient, and there is no efficient } y \text{ such that } \theta(x) \succ \theta(y)\}$$

The definition of the nucleolus is due to Schmeidler [1969], who also demonstrated that the set $\eta(v)$ consists of a single point. Notice that, if the core is nonempty, then there exists an allocation x such that $e(S, x) \leq 0$ for every S. It follows that the nucleolus is necessarily contained in the core, when the core is non-empty. Schmeidler also demonstrated that the nucleolus is contained in the kernel.

The above solution concepts for arbitrary TU games can be directly applied to the problem of cost allocation. However, since it is costs rather than benefits that are allocated, it is necessary to reinterpret the definitions of the core, Shapley value, and nucleolus. Formally, let $c : 2_N \rightarrow \Re_+$ be a cost function which satisfies $c(\emptyset) = 0$. The function c is subadditive if $c(S) + c(T) \geq c(S \cup T)$ whenever $S \cap T = \emptyset$. Every subadditive cost function is the negative of a superadditive value function. The core, kernel, Shapley value and nucleolus of the cost game (N, c) are therefore defined in terms of the corresponding solutions on $(N, -c)$ or on the dual game (N, c^*) where $c^*(S) = c(N) - c(N \setminus S)$. For example, the core of the cost game (N, c) is the set $\{x : x(N) = c(N), \text{ and } x(S) \leq c(S) \text{ for all } S \subset N\}$ or equivalently $\{x : x(N) = c^*(N), \text{ and } x(S) \geq c * (S)\}$. In a cost game, with excess function $c(S) - x(S)$, the surplus of i against j is reinterpreted as the minimum of $e(S, x)$, taken over all coalitions S containing i and not j, and one says that player i outweighs player j, if $s_{ij}(x) < s_{ji}(x)$ and $x_j < c(\{j\})$. In a cost game, the nucleolus is the payoff vector that lexicographically maximizes the vector θ of excesses arranged in increasing order. The nucleolus of a cost game (N, c) is equal to the (traditionally defined) nucleolus of the game (N, c^*). In the case of the Shapley value, the marginal cost functions are given by $c^i(\omega) = c[P^i(\omega) \cup \{i\}] - c[P^i(\omega)]$. The Shapley value is then defined by applying (1.1.5) to these functions instead of the marginal contribution functions $v^i(\omega)$. Since the Shapley value satisfies an additivity property, it is true that the value of a cost game (N, c) is equal to the value of the dual game (N, c^*).

1.2. Minimal cost spanning tree games

Consider now the application of the cooperative solution concepts to a class of minimal cost spanning tree games. Let $N = \{1, \ldots, n\}$ represent a set of 'customer' nodes, and let $\{0\}$ represent a distinguished supplier node, called the 'root.' For any subset $S \subseteq N$, define $V_S = S \cup \{0\}$. Consider a graph $G_N = (V_N, E)$, where $E = \{e_{ij}\}$ is a set of 'edges' (also called 'links'). The graph G_N is assumed to be complete and undirected, and it is assumed that G_N contains no loops or multiple edges connecting two given nodes. A symmetric matrix $C = \{c_{ij}\}$, for which $c_{ii} = 0$ and $c_{ij} \geq 0$, defines the 'cost' of edge e_{ij} if $i \neq j$. It is assumed that the service may be provided to any coalition S of customers by providing a connected graph containing the nodes V_S. For any $S \subseteq N$, let $T_S = (V_S, E_S)$ represent a minimal cost connected graph such that $E_S \subset E$. Since $c_{ij} \geq 0$ for all i and j, T_S is necessarily a tree, called the minimal cost spanning tree (m.c.s.t.) for coalition S of customers.

A minimal cost spanning tree can be efficiently constructed by means of a 'greedy algorithm' [see, e.g. Kruskal, 1956, or Prim, 1957]. For the grand coalition N, the algorithm is initialized by setting $\overline{V}$ equal to an arbitrary node, e.g. the root, and $\overline{E} = \emptyset$. Then choose a node $j \in V \setminus \overline{V}$ in order to minimize c_{ij} over all $i \in \overline{V}$, and $j \in V \setminus \overline{V}$, and update $\overline{V}$ and $\overline{E}$ by adding node j, and edge e_{ij}, respectively. The algorithm terminates when $\overline{V} = V$, and the minimal cost spanning tree is given by the edge set $\overline{E}$.

Let $G_N = (V, E)$ be any graph with m.c.s.t T_N. For every $i \in N$, let $p(i)$ denote the 'predecessor' of i in T, which is the unique node adjacent to i on the path in T from i to 0. Node j is an 'ancestor' of i if j lies on the path from i to 0. (Thus every node is an ancestor of itself.) A set $S \subset V$ is said to be 'closed' if for every $i \in S$, every ancestor j of i, $j \neq 0$ is contained in S. A set S is said to be a 'branch' of V of the tree T, if for some $j \in S$, it is true that $i \in S$, if and only if, j is an ancestor of i. For every node $j \in V$, let $B(j)$ be the branch consisting of all nodes for which j is an ancestor.

Definition 1.6. *Given a complete graph $G_N = (V_N, E)$ with cost matrix $C = \{c_{ij}\}$, a minimal cost spanning tree (m.c.s.t.) game, (N, c), is defined by the player set N and the cost function c for which*

$$c(\emptyset) = 0 \qquad\qquad (1.1.5)$$

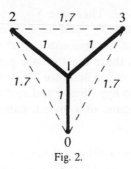

Fig. 2.

and

$$c(S) = \sum_{e_{ij} \in E_S} c_{ij} \text{ for every } S \subseteq N.$$
(1.1.6)

A monotone minimal cost spanning tree game $(N, \bar{c})$ is defined by the cost function $\bar{c}$ for which

$$\bar{c}(S) = \min_{T \supseteq S} c(T).$$
(1.1.7)

A m.c.s.t. game allows a coalition S of customers to use only nodes in V_S in order to construct a minimal cost network, while a monotone m.c.s.t. game allows a coalition S to use nodes from $N \setminus S$ in constructing a minimal cost network. Both c and $\bar{c}$ are well defined, subadditive cost functions. Moreover $\bar{c}(N) = c(N)$ and $\bar{c}(S) \leq c(S)$. The following example, due to Granot & Huberman [1981], demonstrates that the above inequalities may be strict.

Example 1.7. Consider the network with minimal cost spanning tree consisting of edges $\{e_{10}, e_{12}, e_{13}\}$ illustrated in Figure 2, in which $c_{20} = c_{30} = c_{23} = 1.7$ and $c_{ij} = 1$ otherwise. Then it is easily seen that $c(\{2, 3\}) = 3.4 > c(\{1, 2, 3\}) = c(\{2, 3\}) = 3$.

From the definition of the core it follows immediately that the core of a monotone m.c.s.t. game (N, c) is contained in the core of the corresponding m.c.s.t. game (N, c).

The most important property of m.c.s.t. games is the following result, due to Bird [1976] and Granot & Huberman [1981].

Proposition 1.8. *Every minimal cost spanning tree game is totally balanced.*

Proof. Two distinct proofs of this result have appeared in the literature. The first proof is due to Granot & Huberman [1981]. Let T_N be the minimal cost spanning tree for V_N, and let c_i be the cost of the edge $e_{i,p(i)}$ which is uniquely associated with i in G_N. It will be demonstrated that the allocation x for which $x_i = c_i$ satisfies the core constraints. Clearly $x(N) = \sum_{i=1}^{n} c_i = c(N)$. Suppose that $x(S) > c(S)$ for some $S \subset N$. Let T_S represent the minimal cost spanning tree for

V_S with associated edge costs c_i^S. Therefore $\sum_{i \in S} c_i > \sum_{i \in S} c_i^S$. Let E_S represent the set of edges associated with T_S, and let E_{-S} represent the edges associated with $N \setminus S$ in T_N. Then it can be demonstrated that the graph $(V, E_S \cup E_{-S})$ is a tree, since for each $i \in N$, there must exist a path consisting of edges in E_{-S} connecting i to some node $j \in S$. Therefore T_N could not have been the minimal cost spanning tree, contrary to hypothesis, and so it must be true that $x(S) \leq c(S)$ for every S. Since every subgame of a m.c.s.t. game is also a m.c.s.t. game, total balancedness is immediate. □

An alternative proof of Proposition 1.8 may be given, based upon a result of Edmonds [1967], as noted by Curiel [1988]. The cost function $c(N)$ may be written as the value of an integer program as follows.

$$\text{Min} \sum_{i,j \in N: i < j} y_{ij} c_{ij} \tag{1.1.8}$$

subject to

$$\sum_{i \in S: j \notin S} y_{ij} \geq 1 \quad \text{for all } S \subset V_N \tag{1.1.9}$$

$$y_{ij} \in \{0, 1\} \quad \text{for all } i, j \tag{1.1.10}$$

Here, $y_{ij} = 1$ corresponds to the inclusion of the edge e_{ij} in T_N. Condition (1.1.9) guarantees that the graph is connected and the objective function guarantees that there are no cycles. Edmonds demonstrated that (1.1.8)–(1.1.10) are equivalent to a corresponding linear program in which (1.1.10) is replaced by

$$y_{ij} \geq 0 \quad \text{for all } i, j \tag{1.1.10'}$$

The dual of this linear program is given by

$$\text{Max} \sum_{S \subseteq N} \delta_S \tag{1.1.11}$$

subject to

$$\sum_{S \subseteq N: i \in S, j \notin S} \delta_S \leq c_{ij} \quad \text{for all } i, j \in V_N \tag{1.1.12}$$

$$\delta_S \geq 0 \quad \text{for all } S \subseteq N \tag{1.1.13}$$

Let δ^* be optimal in (1.1.11)–(1.1.12), and define an allocation x such that

$$x_i = \sum_{T \subseteq N, i \in T} \delta_T^* |T|^{-1}$$

Then by the duality theorem $x(N) = \sum_{T \subseteq N} \delta_T^* = c(N)$. For any coalition S with m.c.s.t. T_S, and associated link costs c_i^S, let $p^S(i)$ denote the predecessor of i in T_S. It then follows that

$$\sum_{i \in S} x_i = \sum_{T \cap S \neq 0} \delta_T^* |T \cap S| |T|^{-1} \leq$$

$$\leq \sum_{T \cap S \neq 0} \delta_T^* \leq \sum_{i \in S} \left[\sum_{T: i \in T, p^S(i) \notin T} \delta_T^* \right] \leq \sum_{i \in S} c_{i, p^S(i)} = c(S).$$

Therefore x is in the core of (N, c). The second inequality follows since T cannot contain both i and its predecessor $p^S(i)$ for all $i \in T \cap S$ if T_S is a spanning tree for V_S.

1.3. The structure of the core of a m.c.s.t. game

Proposition 1.8 is an important result on cost allocation on networks. It also establishes the class of m.c.s.t. games as an interesting class of cooperative games. This section will explore the properties of this class. Note that while the first proof of Proposition 1.8 guarantees that a non-empty core exists, it reveals very little about the structure of the core. Indeed, the particular core allocation, that is constructed in the course of the proof, is an extreme point in the core that is systematically unfair to some players. In particular, given a minimal cost tree T_N, recall that a closed subset is one that includes every ancestor of every player, and a branch is a coalition of all players having a common ancestor. Then, any closed coalition S is allocated a cost equal to $c(S)$, the 'stand alone cost' of serving S. This is the maximum that S would be forced to pay in any core allocation. On the other hand, any branch S' is allocated a cost equal to $c^*(S) = c(N) - c(N \setminus S)$, the incremental cost of serving S, which is the minimum that S could expect to pay in any core allocation. In other words, subsets of customers close to the central supplier pay the minimum prices consistent with the core constraints, while those farthest from the supplier pay the maximum 'subsidy free prices.' Thus, one of the purposes of this section is to determine whether it is possible to systematically find other core allocations that are in some sense more fair to all of the set of all players.

Let us begin with the observation that m.c.s.t. games are a strict subset of the class of cost games having non-empty cores. This is easily seen in the case of three player games. Consider an arbitrary symmetric cost function for which $c(\{i\}) = 2$ and $c(\{i, j\}) = 3$ for $i, j \in \{1, 2, 3\}$. If this cost function were derived from a m.c.s.t. game it would have to be the case that the link costs satisfy $c_{i0} = c(\{i\}) = 2$ and $c_{ij} = c(\{i, j\}) - c(\{i\}) = 1$ for all i, j. It therefore follows that $c(\{1, 2, 3\}) = 4$. However, an arbitrary cost function would be totally balanced whenever $c(\{1, 2, 3\}) \leq 4.5$.

One interesting class of totally balanced cost functions is the class of submodular cost games.[7] A cost function is submodular (or concave) if

$$c(S) + c(T) \geq c(S \cup T) + c(S \cap T) \text{ for all } S \text{ and } T \tag{1.3.1}$$

An alternative characterization of submodular cost functions suggests their importance in a cost allocation framework. It is easily demonstrated that a function c is

---

[7] A characteristic function of a game would satisfy a corresponding supermodularity property. Shapley [1971] defined the class of 'convex' games which satisfy the property that $v(S) + v(T) \leq v(S \cup T) + v(S \cap T)$ for all $S, T \subseteq N$. Edmonds [1970] independently developed these concepts in his study of polymatroids.

submodular, if and only if,

$$c(S \cup \{i\}) - c(S) \geq c(S' \cup \{i\}) - c(S')$$

$$\text{whenever } S \subseteq S' \subseteq N \setminus \{i\} \tag{1.3.2}$$

Thus, submodular cost functions satisfy a property of non-increasing incremental cost as the size of a coalition grows. For any permutation of players ω, let $m(\omega) = [c^1(\omega), \ldots, c^n(\omega)]$ be the vector of incremental costs which arise as players arrive according to the ordering ω. Shapley [1971] demonstrated that $m(\omega)$ is contained in the core of every submodular game. In fact, the core of a submodular game is precisely the convex hull of the set of all marginal worth vectors. Ichiishi [1981] noted that the converse result also holds. If every marginal worth vector is in the core, then the game must be submodular. He also observed the close relationship of the problem of finding core allocations in submodular games, with the study of the greedy algorithm for finding optimal solutions to linear programs.

The following example demonstrates that m.c.s.t. games are not, in general, submodular.

Example 1.9. Let $N = \{1, 2, 3, 4\}$, and let link costs be such that $c_{20} = c_{30} = c_{24} = 2$ and $c_{ij} = 1$ for all other $i, j \in V_N$. A minimal cost spanning tree is the edge set $\{e_{40}, e_{14}, e_{24}, e_{34}\}$ and it is easily seen that $c(\{2\}) = c(\{1, 2\}) = 2$, $c(\{2, 3, 4\}) = 3$, and $C(N) = 4$, so that $c(\{1, 2\}) + c(\{2, 3, 4\}) < c(N) + c(\{2\})$.

Granot & Huberman [1982] defined a class of games based on the permutational submodularity of the cost function. A function is permutationally submodular if there exists a permutation ω such that

$$c[P^i(w) \cup S] - c[P^i(\omega)] \geq c[P^j(\omega) \cup S] - c[P^j(\omega)] \tag{1.3.3}$$

whenever j follows i in the ordering ω and $S \subseteq N \setminus P^j$. They then demonstrated that the particular marginal worth vector $m(\omega)$ is contained in the core of a game that is permutationally submodular with the ordering ω. As might be expected, m.c.s.t. games are permutationally submodular for any permutation which is consistent with the partial order imposed by the minimal cost tree T_N. That is, for any ω such that 'j is an ancestor of i' implies that '$\omega_j \leq \omega_i$,' the marginal worth vector $m(\omega)$ is contained in the core.

While m.c.s.t. games are not themselves submodular, there are two special cases of submodular m.c.s.t. game, both involving complete graphs derived from trees. Let T be any tree with node set V_N, which one can think of as the minimal cost tree for some arbitrary graph G. One can then define a another complete graph $G'(T)$ as follows.

Definition 1.10. *Given a tree T with edge set $E = \{e_{ij}\}$ and link costs c_{ij}, let $G'(T)$ be the complete graph, with link costs c_{ij} if $e_{ij} \in T$, and $c_{kl} = \max\{c_{ij}$ such that e_{ij} is in the cycle formed by the addition of e_{kl} to $T\}$ if $e_{kl} \notin T$. Let c' be the cost function derived from $G'(T)$.*

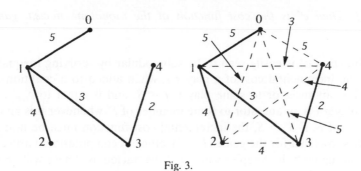

Fig. 3.

This definition is due to Bird [1976], who referred to the graph $G'(T)$ as a 'minimal network.' Given an arbitrary graph G with m.c.s.t. T, and cost function c, Bird also defined the 'irreducible core' of (N, c) as the set of allocations which are in the core of every game having T as a m.c.s.t. For example, the minimal network associated with the tree illustrated in Figure 3a is illustrated in Figure 3b.

Bird demonstrated that the irreducible core of (N, c) is precisely equal to the core of (N, c'). Notice that for any graph G with minimal cost spanning tree T and typical link cost c_{ij}, it must be true that $c'_{ij} \leq c_{ij}$, where c'_{ij} are the link costs associated with $G(T)$. Otherwise T would not have been a minimal tree. Therefore $c'(S) \leq c(S)$ for every S. Since $c'(N) = c(N)$ it follows that the irreducible core is a subset of the core of any m.c.s.t. game. In fact, the associated cost function c' is submodular [Granot & Huberman, 1982]. This result follows from the fact that for every permutation of players ω, it is possible to construct a minimal cost tree for the minimal network that is consistent with ω. Submodularity then follows by the result of Ichiishi [1981].[8] This result, and the observation that the irreducible core is a subset of the core, provides a third independent proof of Proposition 1.8. It also leads to a procedure for computing a large subset of core allocation for any m.c.s.t. game. In the example of Figure 3, the core of any game having the minimal cost tree of Figure 3a must contain the following core allocations: (5,4,3,2), (5,4,2,3), (4,5,3,2), (4,5,2,3), (3,5,4,2), (3,5,2,4), (3,4,5,2), and (3,4,2,5). The irreducible core is precisely the convex hull of these allocations.

Megiddo [1978a] defined a class of games related to m.c.s.t. games that are also submodular.

Definition 1.11. *Given a tree T on node set V_N, define a cost function c'', where $c''(S)$ represents the minimal cost of connecting every node $i \in S$ to the source node $\{0\}$ along a path in T. Equivalently, given a tree T, let $G''(T)$ be the complete graph with link costs c''_{ij}, where c''_{ij} is the length of the shortest path in T connecting*

---

[8] Bird proved that the irreducible core can be written as the convex hull of the set of core allocation computed in the usual way for every minimal cost spanning tree of the minimal network.

i and j. Then c″ is the cost function of the monotone m.c.s.t. game derived from G″.

The function $c″$ is easily seen to be submodular by verifying inequality (1.3.2). Clearly the incremental cost of a player i, when added to a coalition S, is equal to 0, if i is an ancestor to some player $j \in S$, and is equal to $c_{i,p(i)}$ otherwise. Since a player i is an ancestor to some member of S' whenever it is an ancestor to a member of S, if $S' \supseteq S$, the incremental cost function must be non-increasing. Megiddo shows that for the game $(N, c″)$, efficient computational procedures exist for computing both the Shapley value and the nucleolus. These will be considered later in this section.

Suppose that G is an arbitrary graph, and c is the cost function of the derived monotone m.c.s.t. game. Then it is easily seen that $c″(N) = c(N)$ and $c″(S) \geq c(S)$. Therefore, the core of (N, c) is a subset of the core of $(N, c″)$. If the cost function of an arbitrary m.c.s.t. game (N, c) is monotone, it therefore follows that there exist submodular cost functions c' and $c″$ such that $c'(N) = c(N) = c″(N)$, and $c'(S) \leq c(S) \leq c″(S)$ for every $S \subset N$.

Aarts & Driessen [1992] consider a modification of the link cost matrix $C = \{c_{ij}\}$ that is related to Bird's minimal network, but which preserves the core of the original game. Specifically, given a m.c.s.t. T_N for an arbitrary game G_N, consider a set of nodes i, j, k and l such that i and j are on the path from j to k in T_N, and $c_{ij} > c_{kl}$. From the optimality of T_N, it follows that $e_{ij} \notin T_N$ (i.e. i and j are not adjacent in T_N). Aarts and Driessen demonstrate that the cost function c''' derived from c by setting $c_{ij} = c_{kl}$ in all possible situations meeting the above criteria, results in a new game in which the core of (N, c''') is identically equal to the core of (N, c). Moreover, when the m.c.s.t. T_N is a 'chain', it is possible to use this technique to easily compute additional extreme points of the core. Further results on chain games are contained in Aarts [1992].

Let us now consider some of the other cooperative solution concepts defined in Section 1.1, beginning with the Shapley value. Shapley [1971] demonstrated that the value is contained in the core of every submodular game. In fact, the value can be expressed as $\phi(c) = (1/n!) \sum_\omega m(\omega)$. Since the core of a submodular game is precisely the convex hull of the incremental cost vectors $m(\omega)$ the value occupies a central location in the core of such a game. Thus, for games derived from minimal networks, and for the game defined by Megiddo, the value is in the core. For arbitrary m.c.s.t. games, however, the value may not be contained in the core. The value of the game in Example 1.9 is $\phi(c) = (2/3, 4/3, 7/6, 5/6)$ but, as already noted, the core consists of the unique allocation $x = (1, 1, 1, 1)$.

In general, there are no efficient algorithms for the computation of the Shapley value for m.c.s.t. games. However, Megiddo [1978a] demonstrated that the value may be easily computed in the game which he defined. This result follows from the fact that the cost function of this game can be written as $c(S) = \sum_{i \in N} h^i(S)$, where $h^i(S) = c_i$ if $S \cap B(i) \neq \emptyset$ and $h^i(S) = 0$ otherwise. (Here c_i represents the cost of the link associated with i in the tree T.) The value of the game h_i is given by $\phi_j(h^i) = c_i/|B(i)|$ if $j \in B(i)$, and is equal to 0 otherwise. Given the additivity property of the value, it follows that $\phi(c) = \sum_{i \in N} \phi(h^i)$. An algorithm for this computation requires only $O(n)$ operations.[9]

While the nucleolus is considerably more complicated to work with than the Shapley value, more is known about the properties of the nucleolus for general m.c.s.t. games than for the value. Two papers by Granot & Huberman [1981, 1984] describe many of the known results. The first paper contains a decomposition result. Let $P = \{1, \ldots, p\}$ be the set of nodes adjacent to the common supplier, and let $B(i)$ be the branch associated with i for each $i \in P$. Then one can demonstrate that $c(N) = \sum_1^p c(B(i))$. Moreover, for each $i \in P$, one can derive a cost game c^i on $B(i)$ from the original game in a straightforward manner by defining $c_{j0}^i = \min_{k \in \{0\} \cup N \setminus B(i)} c_{jk}$

Granot & Huberman [1981] then demonstrate the following result:

Proposition 1.12. *If, for any partition $P = \{N_1, \ldots, N_p\}$ of N, it is true that $c(N) = \sum_1^p c(N_i)$ then both the core and the nucleolus can be written as the Cartesian product of the cores and nuclei of games (N_i, c_i) defined on the components $\{N_1, \ldots, N_p\}$.*

Since the cost function can be decomposed in the above manner, if and only if there exists a minimal cost tree T_N in which there are exactly p nodes for which the source $\{0\}$ is the predecessor, Proposition 1.11 implies that one can effectively restrict attention, for purposes of computing the nucleolus, to games in which there is a single customer directly connected to the source.

In another paper, Granot & Huberman [1984] characterized the nucleolus of a m.c.s.t. game in more detail. Kopelowitz [1967] and Maschler, Peleg & Shapley [1979] demonstrated that for any TU game, an algorithm exists which computes the nucleolus by solving a sequence of at most 2^n linear programming problems in which a constraint containing the excess function $e(S, x)$ appears for every $S \subseteq N$. Given the special structure of m.c.s.t. games, it is possible to substantially reduce the number of steps required in such a computation. Granot and Huberman demonstrated that, in this algorithm, only subsets S for which the complement $N \setminus S$ is a connected component of the T_N need to be considered. While many of their other results are notationally complex, and will not be repeated here, the most important property of the nucleolus is contained in the following result.

Proposition 1.13. *In a minimal cost spanning tree game, the nucleolus is the unique point in the intersection of the kernel and the core.*

Since Maschler and coworkers have characterized geometrically the intersection of the kernel and core, Proposition 1.12 offers a geometric characterization of the nucleolus of a m.c.s.t. game.

---

[9] As noted by a referee, the game defined by Megiddo is a straightforward extension of the 'airport game' of Littlechild & Owen [1973], and the approach toward computing the Shapley value is derived from a suggestion of those authors.

Megiddo [1978a] also describes an algorithm for computing the nucleolus for the class of problems which he defined (see Definition 1.11). In these games the nucleolus satisfies the property that if j is an ancestor of i, then $\eta_j \leq \eta_i$. The algorithm which computes terminates within $O(n^3)$ operations. Further computational improvements for this class of games have been obtained by Galil [1980], Granot & Granot [1993] and Granot, Maschler & Owen [1993].

1.4. Steiner networks and multiple supplier networks

The remainder of Section 1 will consider various generalizations and extensions of m.c.s.t. games. One path for generalization is to allow coalitions of customer nodes to connect to the supplier by adding arbitrary junction nodes, if such nodes allow the coalition to reduce its total cost. The so-called 'Steiner network problem' addresses precisely the following issue: what is the ratio of the minimum length Steiner network (with junction nodes allowed), to the length of the ordinary spanning tree network. In the two dimensional Euclidean plane, a network consisting of the vertices of an equilateral triangle, and a node at the center of gravity gives the result that this ratio can be as low as $\sqrt{3/2} \cong 0.867$. The network of Example 1.7, which was used to demonstrate the possible non-monotonicity of the cost function, roughly approximates this result. Du & Hwang [1990] verified that this number is in fact a lower bound on the set of 2 dimensional networks.

Since it is known that a monotone m.c.s.t. game (which allows a coalition S of customers to make use of any nodes outside of S) has a non-empty core, one would expect the core of a Steiner network game to also be non-empty. However, this is not the case, as the following example, due to Megiddo [1978b], illustrates.

Example 1.14. Consider a network of 5 customers located symmetrically around a central supplier, as illustrated in Figure 4. The minimal cost Steiner network is illustrated, with three junction nodes. This network has the property that $c(\{1, 2, 3, 4, 5\}) = c(\{1, 2\}) + c(\{3, 4, 5\})$. If the distance between two adjacent

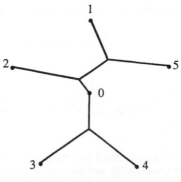

Fig. 4.

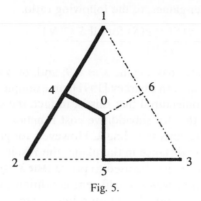

Fig. 5.

customers is equal to 1, then geometric computations reveal that $c(\{1, 2\}) = 1.5542$ and $c(\{3, 4, 5\}) = 2.3928$. Since $c(\{1, 2\})/2 = 0.7771 < c(N)/5 = 0.7894$, it follows from the symmetry of the example that the core is empty.

A more tractable way to present the Steiner network game is to assume a fixed set of possible junction nodes, M, in addition to the set of customer nodes N, and the supplier node $\{0\}$. Clearly all Euclidean Steiner networks can be described in this framework, since the maximum number of junction nodes is finite (less than or equal to $n - 2$). However, more general results are possible, since link costs c_{ij} are not constrained by the Euclidean metric. Sharkey [1982a, b] described a simple example illustrating the possible emptiness of the core in a framework such as this.[10]

Example 1.15. There are three customer nodes $\{1,2,3\}$, a source node $\{0\}$, and three junction nodes $\{4,5,6\}$ as illustrated in Figure 5. Assume that the cost of every link in the diagram is equal to 1, and that every other link has a cost of 2 or more. Then it follows that $c(\{i, j\}) = 3$ for any pair i, j and $c(\{1, 2, 3\}) = 5$. The symmetry of the problem then implies the emptiness of the core.

Bird [1976] made a similar observation regarding a network with relocatable sources. In fact, one can immediately interpret the set M of junction nodes as potential locations for the supplier, with the understanding that the link costs c_{j0} for $j \in M$ represent location specific fixed costs for each supplier.[11]

Examples, such as Example 4, raise the issue of whether there may be bounds on the 'corelessness' of a Steiner game. Let λ^* represent the greatest lower bound,

[10] See also Sharkey [1991] and Tamir [1991]

[11] If suppliers could costlessly locate at arbitrary nodes in the network, the cost of every coalition would clearly be equal to 0. Rosenthal [1987] made the observation that allowing costless supplier entry at a fixed set of non-customer nodes converts a minimum cost spanning tree game to a minimum cost spanning forest game in which the minimal cost network need not be connected. The two games have identical properties. Granot & Granot [1993] generalized Rosenthal's model.

over the set of all Steiner games, of the following ratio.

$$\frac{\max\{x(N) \text{ s.t. } x(S) \leq c(S) \text{ for all } S \subseteq N\}}{c(N)} \tag{1.4.1}$$

In Example 1.15, this ratio has a value $\lambda = 0.9$, and, of course, if the core is non-empty, it follows that $\lambda = 1$. In Sharkey [1991], an example was presented in which $\lambda = 0.875$, and it was conjectured that this is, in fact, the greatest lower bound. It was also demonstrated that for subadditive cost function games, in general, there is no lower bound on the corresponding λ. However, for games in which link costs are defined by Euclidean distance in the plane, the result on minimal cost Steiner networks mentioned above may be used to prove that $\lambda^* \geq \sqrt{3/2}$.

Skorin-Kapov [1993a] provides a sufficient condition for non-emptiness of the core of a Steiner network game, based on a linear programming relaxation of an integer programming formulation of the minimal cost (directed) Steiner network. Specifically, if the incidence vector of a minimum cost Steiner tree is an optimal solution to the corresponding linear program, this condition is a sufficient (and in some cases a necessary) condition for a non-empty core is. (See also Sharkey [1990], where a similar result was obtained for m.c.s.t. games.) Feasible solutions to the associated dual linear program can then be used to generate a lower bound on the cost that can be allocated among users, while satisfying the core constraints. (See the above discussion of Sharkey [1991].)

Skorin-Kapov [1993b, c] defines a capacitated network design game that includes, or is closely related to both the Steiner network games and network synthesis games, which will be discussed in the following section. These network design games are equivalent to capacitated Steiner network games, since each customer node $i \in N$, is assumed to have a non-negative demand for service d_i, a capacity, u_i limiting the number of customers who can be served at that point, and a cost c_i, which can be related to the cost of connecting node i to a fictitious root node $\{0\}$. While problems in this class are $\mathcal{NP}$-hard, it is nevertheless possible to provide an efficient representation of the core, for which a polynomial time algorithm can be used to generate core points (if any exist). Furthermore, when the core is non-empty, the nucleolus is characterized by, and can be efficiently computed from, the same set of constraints.

Granot & Maschler [1991] defined a further generalization of the m.c.s.t. game, called spanning network games. In a spanning network game, there can be costs associated with nodes, as well as edges, and these costs can be negative (in which case they correspond to profits rather than costs). Each player in the game is associated with a node in the network, although multiple players may inhabit the same node, and, as in the Steiner network, there can be junction nodes that are not inhabited by any player. Any coalition of players is assumed to be interested in building a network connecting every member of the coalition to the root in the cheapest possible way, given access to the entire set of nodes (whether or not they are inhabited). Granot and Maschler prove that every spanning network game is monotonic — i.e. if $c(R)$ represents the minimum cost (or maximum profit) attainable by a coalition R, and S and T are coalitions of players with $S \subset T$, then

$c(S) \leq c(T)$. In addition, they demonstrate that if all costs are non-negative, then a spanning network game is subadditive, although in general, spanning network games need not be subadditive. In a later paper, Van den Nouweland, Tijs & Maschler [1993] demonstrated that every monotonic game is also a spanning network game.

1.5. Models of communications and transportation networks

While most of the literature on cost allocation on networks has been concerned with spanning tree networks, because of their simple structure, many economic allocation problems involving communications and transportation networks require a more general framework for analysis. The network design (or synthesis) problem is one such generalization. In a network synthesis game, one assumes the existence of a graph $G = (N, E)$ where the node set is the set of players and there is no longer a distinguished source node. For every ordered set (i, j) of player nodes it is assumed that there is a demand for r_{ij} units of commodity to flow from i to j along the arcs of G. The problem of network synthesis is the discrete optimization problem which seeks to minimize the cost of the graph G that can accommodate the flow demands r_{ij}. It is possible to view the flow demands as entering the network either simultaneously, or non-simultaneously, and both versions of the problem have been studied.

Now let $c(S)$ represent the minimal cost of a network designed to accommodate the flow demands of coalition S. In either the simultaneous or the non-simultaneous case there is no ambiguity in the definition of $c(N)$. However, two possible assumptions can be made in the definition of $c(S)$ for $S \subset N$. In the first case, one can define $c'(S)$ as the minimum cost of a network designed to carry all flows r_{ij} with $i \in S$ and $j \in N$. Thus c' corresponds to a situation in which the players are exporters of a commodity. If players are viewed as both exporters and importers, one could define the cost function $c''(S)$ as the minimal cost of a network to carry all flows r_{ij} for which $i \in S$ or $j \in S$.[12] Since $c'(N) = c''(N)$ and $c'(S) \leq c''(S)$ for every S, it is clear that the core of (N, c') is contained in the core of (N, c'').

Granot & Hojati [1990] consider a network synthesis game in which the cost of constructing capacity on each link is assumed to be linear, and flow demands are assumed to be symmetric, so that $r_{ij} = r_{ji} > 0$.[13] Let d_{ij} represent the unit cost of capacity on the lowest cost path connecting nodes i and j. (If unit costs on each link satisfy the triangle inequality, then d_{ij} is just the marginal cost of capacity on edge e_{ij}). It is then a straightforward exercise to demonstrate that in the case of simultaneous flows, the cost function c'' has a simple additive form $c''(S) = \sum_{i<j} a_{ij}(S)$, where $a_{ij}(S) = r_{ij}d_{ij}$ if $i \in S$ or $j \in S$, and $a_{ij}(S) = 0$

---

[12] There is a third possibility, in which a coalition is only interested in flows r_{ij} for which $i \in S$ and $j \in N$. While this case clearly does not make sense in a cost allocation framework, it could be addressed in the surplus allocation game in which benefits are attached to flows, and each coalition seeks to maximize its net surplus.

[13] A closely related game is considered in Tamir [1991].

otherwise. The function c'' is therefore submodular, and it is possible to derive simple expressions for the Shapley value and nucleolus, which in this case coincide.

Granot and Hojati also consider the more difficult non-simultaneous network synthesis game with a linear cost structure. In this case, they demonstrate that the cost function c'' is submodular in the special case in which all link costs d_{ij} are equal. They also derive expressions for the value and nucleolus whenever the graph corresponding to the flow requirements r_{ij} is a tree. In the non-simultaneous game with unequal unit link costs d_{ij}, the cost function need not be submodular. However, it was independently proven in Granot [1986] that the core is non-empty for this case. This result will be described more fully in Section 1.6.

Tamir [1991] defined a network synthesis game which includes the examples of Granot & Hojati [1990] as special cases. In this formulation, the model satisfies the assumptions of Owen's [1975] linear production game, so that a subset of core allocations can be directly computed from solutions to the dual linear program. Kubo & Kasugai [1992] consider the general network design game, in which the cost function is defined in terms of Lagrangian relaxation of the original problem. Since core allocations of the modified game can be easily computed, this approach leads to an efficient method for computing approximate solutions to the original problem.

Sharkey [1991] defined a game closely related to a simultaneous network synthesis game, in which the set of players consists of the flow demands r_{ij} rather than the set of nodes N. This analysis is based on the concept of 'supportability', as defined in Sharkey & Telser [1978]. If $x \in \Re_+^n$ is a vector of outputs and $c : \Re_+^n \to \Re_+$ is a cost function, then a price vector p is a supporting price at an output α (and c is supportable at α) if $p \cdot \alpha = c(\alpha)$ and $p \cdot x \le c(x)$ for every $x \le \alpha$. In the network game, the cost of constructing q_{ij} units of capacity on edge e_{ij} is assumed to be a concave function $c_{ij}(q_{ij})$. It was demonstrated that whenever the minimal cost network for the grand coalition, consisting of all flow demands, is a complete network (in which every link is utilized) the cost function is supportable. In the special case in which link costs have the form $c_{ij}(q_{ij}) = f_{ij} + c_{ij}q_{ij}$, it follows that the minimal cost network is complete, and the cost function is supportable, if the vector of flow demands is sufficiently large. If the network required for the grand coalition is not complete, it was demonstrated, by example, that the cost function may fail to be supportable.

Another cost allocation problem on a network is the traveling salesman allocation problem, which is analyzed by Tamir [1989]. In this game, a service provider (e.g. a repairman visiting customers or a lecturer visiting universities) is located at a source node {0}, and is interested in visiting various subsets $S \subseteq N$ of cities, where the cost of traveling between every pair of cities is given by c_{ij} as in the m.c.s.t. game. The cost function is defined as the minimal cost of completing a tour, starting at node {0} which visits every node in S at least once and returns to {0}. It is obvious that this cost function is subadditive, so that there is an incentive for the grand coalition N to form, and to allocate the cost $c(N)$ among the members of N. Tamir demonstrates that the core may be empty for this class of games in the following example.

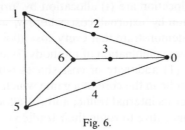

Fig. 6.

Example 1.16. Consider the network illustrated in Figure 6, in which every link is assumed to have cost $c_{ij} = 1$. It can be verified that $c(N) = 8$ and $c(\{1, 2, 4, 5\}) = 5$. Given the symmetry of the problem, it also follows that at any core allocation x, it must be true that $x_1 = x_5 = x_6$ and $x_2 = x_3 = x_4$. Therefore it must be true that $3x_1 + 3x_2 = 8$, and simultaneously $2x_1 + 2x_2 \leq 5$. Since it cannot be true that $8 = 3x_1 + 3x_2 = (3/2)(2x_1 + 2x_2) \leq 7.5$, it must be true that no cost allocation x can satisfy the core constraints.

In an interesting application of some of the results discussed in this section, Bittlingmayer [1990] considered a simple airline network consisting of three nodes, and consequently three city-pair routes. Declining average cost of capacity in each city pair market translates into a subadditive aggregate cost function. The minimal cost network could involve either direct flights on each link or a hub and spoke system (i.e. the minimum cost network is either a complete graph or a tree in a three node example) depending on the degree of scale economies. Moreover, the aggregate cost function is necessarily supportable.[14] However, if prices are set to maximize an aggregate welfare function, the resulting Ramsey price vector may fail to be supportable.

Woroch [1990] considers a related model of a triangular network in which the cost of constructing link capacity is not necessarily additive across links. Demand functions are assumed to be inelastic, but service providers are assumed to be unable to distinguish directly routed traffic from indirectly routed traffic. Finally, directly routed flows and indirectly routed flows are assumed to be valued differently by consumers. In all cases, the underlying cost function is subadditive. In this framework, the core of the cost allocation game need not exist. Various sufficient conditions for its existence are considered in the paper.

Henriet & Moulin [1993] describe an interesting axiomatic characterization of a set of cost allocation rules for a communications network. In their approach, each user has a fixed cost c_i of connection to the network, and total costs are simply $\sum_{i \in N} c_i$, where N is the set of subscribers. Let r_{ij} represent the traffic flows between subscribers i and j (with $r_{ij} = r_{ji}$ and $r_{ii} = 0$). Let y_i be the revenue collected from subscriber i, where it is required that $\sum_{i \in N} y_i = \sum_{i \in N} c_i$. Two

---

[14] In networks with more than three nodes, Sharkey [1991] demonstrated that even concave link capacity costs do not guarantee supportability of the cost function.

basic methods of cost allocation are (i) allocation by 'private' cost, where $y_i = c_i$ for all i, and (ii) allocation by 'external' cost, where $y_i = \sum_{j \in N} c_j [r_{ij} / \sum_{k \in N} r_{jk}]$. The authors are able to demonstrate that every cost allocation rule satisfying a set of axioms must be a convex combination of methods (i) and (ii). The axioms which establish this result are (1) additivity of cost shares with respect to connecting costs; (2) allocations must be in the core of a game which allows coalitions of users to construct a network for its internal traffic; and (3) allocations must not give any pair of users i and j an incentive to route their traffic via a third party k.

1.6. Other network models

This section will consider several papers in which networks are used in a more abstract way to model economic situations. Many combinatorial optimization problems, such as the transportation problem and the optimal assignment problem, can be formally modeled as network flow problems. In such problems the duality theory of linear programming becomes equivalent to the maximum flow, minimum cut theorem of Ford & Fulkerson [1962]. Many of the models of this section generalize not only the m.c.s.t. games considered earlier in this section, but also the matching games, which will be discussed in the following section.

Shapley [1961] defined the basic network flow model as follows. There is a directed graph $G = (V, E)$, consisting of v nodes and m links, in which each link $i \in E$ has non-negative capacity c_i. One or more nodes have only outward pointing links, and are designated as sources; one or more other nodes with only inward pointing links are designated as sinks. A flow pattern is a non-negative vector $f = (f_1, \ldots, f_m) \le (c_1, \ldots, c_m)$ such that for every node which is not a source or sink, the total incoming flow equals the total outgoing flow. The 'capacity' of such a network is defined as the maximum $\sum_1^m f_i$ over all flow patterns. Let $F(c)$ denote the capacity of a network with link capacities given by the vector c. Similarly, given a vector of initial capacities $\bar{c}$, let $F_{ij}(c_i, c_j)$ represent the capacity of a network with link capacities c_i, c_j, and $c_k = \bar{c}_k$ if $k \ne i, j$.

Shapley demonstrated that the functions F_{ij} are always either supermodular or submodular. When F_{ij} is supermodular the links i and j are complements; when F_{ij} is submodular links i and j are substitutes. Shapley further demonstrated the following results. If the terminal node of link i is the initial node of link j, or if i is a source and j is a sink, then i and j are complements. If links i and j have the same initial node, or the same terminal node, or if both are sources, or both sinks, then i and j are substitutes. In all other cases i and j can be either complements or substitutes, depending on the values of link capacities $\bar{c}_k$ for $k \ne i, j$. Shapley [1962] demonstrated that analogous results hold for the optimal assignment problem, in which case two players of the same type are substitutes and players of opposite types are complements. These results were generalized by Gale & Politof [1981], Topkis [1983, 1984, 1987] and Granot & Veinott [1985].

Kalai & Zemel [1982a] consider a network flow problem with a single source and a single sink. They define a cooperative game in which sets of links are owned by players, and the value of a coalition S is the capacity of the network defined by the links controlled by S. They then prove that the class of network flow games corresponds exactly to the class of totally balanced games.[15]

In a related paper, Kalai & Zemel [1982b] define a generalized network problem as follows. There are n players, and each player $i \in N$ controls a set of d_i variables in a general optimization problem defined by the maximization of a common objective function f over $d = \sum_{i \in N} d_i$ variables. Every coalition has a feasible set Y^S constraining the variables under the control of its members. For each coalition S, the value $v(S)$ is given by the maximum of f over Y^S. The collection $\{Y^S\}_{S \subseteq N}$ is said to be 'balanced' if for every balanced collection $\mathbf{B}$, with balancing weights δ_S, and every collection of feasible points $\{y^S : y^S \in Y^S, S \in \mathbf{B}\}$, it is true that $\sum_{S \in \mathbf{B}} \delta_S y^S \in Y^N$. The collection is totally balanced if this property holds for all subcoalitions $T \subset N$.

The objective function f is said to be 'super balanced' if for every balanced collection $\mathbf{B}$ with weights δ_S, and every collection of feasible points y^S, it is true that $f(\sum_{S \in \mathbf{B}} \delta_S y^S) \geq \sum_{S \in \mathbf{B}} \delta_S f(y^S)$; f is totally super balanced if this property holds for all subcoalitions. Finally f is totally bounded if the supremum of $f(y)$ over $y \in Y^S$ exists for every S. Kalai and Zemel prove that if $\{Y^S\}$ is totally balanced, and f is totally bounded and totally super balanced, then the game (N, v) defined above is totally balanced. The property of superbalancedness can be shown to hold for the class of 'market games' defined by Shapley & Shubik [1969].

Owen [1975] defined an important class of linear production games, for which he demonstrated that certain allocations in the core can be readily computed from optimal solutions to the linear program, which is the dual of the program which defines the optimal value of $v(N)$. (These core allocations are known as the dual set.) For some games, including some network optimization games, and the optimal assignment games, considered in Section 2.2, the core and the dual set coincide. Samet & Zemel [1984] and Dubey & Shapley [1984] describe related results on the core and dual set of linear programming games and more general convex programming games. Engelbrecht-Wiggans & Granot [1985] demonstrate that the core and dual set coincide in a class of linear production games that includes several of the above classes as special cases. Tamir [1980] defined a class of location games on tree networks in which the core and dual set also coincide. Sharkey [1990] considered a class of 'shared facility' games in which some related results on the core and dual set were obtained.

As noted above, the assignment game can be treated as a special case either of a network flow game, or of a linear production game (since it is known that an optimal solution to the linear program can be attained at integer values on the constraint set). Granot [1984] describes an optimization problem which generalizes both the m.c.s.t. game considered earlier in this section, and a variation of the assignment game, known as the roommates problem. Given a set N of players, it is assumed that every player $i \in N$ has preferences over the set of players in $N \setminus \{i\}$ with whom he could be matched. A feasible set of matchings,

---

[15] See the discussion following Proposition 1.2 for a definition of total balancedness.

known as a 'consistent organizational structure,' is a linkage structure defined by a graph $G = (N, E)$ in which there are no cycles. It should be clear that the m.c.s.t. games, and the games considered in Section 1.5 can be modeled in this framework. Any one to one, or many to one matching problem can also be described in this framework (see Section 3 for details). In particular, the roommates problem corresponds to a graph G in which the degree of every node is at most one. The optimal assignment problem is a special case of the roommate problem, in which G is bipartite.

For every edge $e_{ij} \in E$, let $d_{ij} \geq 0$ represent the benefit of linking i and j. Then $v(S)$ can be defined in a straightforward way, by maximizing the summation of d_{ij} over the set of all graphs G_S without cycles, which use only nodes and edges associated with S. Granot proves that for any integer r, one can construct a cooperative game in this manner having three or more players, such that the degree of every node in G_N is at most r, and such that the core is empty. However, for a fixed set of players N, one can show that there always exist values of r such that the core is non-empty. In particular, the core is always non-empty for $r = n - 1$, since the resulting game is then equivalent to a m.c.s.t. game.

Granot [1986] also considers a generalization of Owen's linear production game. In the linear production game, there are n players, m resources, and p final outputs. Each player is assumed to own a vector $b^i = (b^i_1, \ldots, b^i_m)$ of resources. There is a linear production technology, defined by the matrix A, where a_{ij} is the amount of resource i required to produce one unit of output j, which can be sold for a price of c_j. For any coalition S, let $b(S)$ be the m-vector defined by $b_j(S) = \sum_{i \in S} b^i_j$ for $j = 1, \ldots, m$. Then v is given by

$$v(S) = \max \ c \cdot x \tag{1.6.1}$$

subject to

$$Ax \leq b(S), \quad x \geq 0 \tag{1.6.2}$$

In Granot's generalized linear production model, the resources available to coalitions are assumed to be arbitrary set functions $b(S)$, rather than additive functions as assumed above. Granot proves that if each of the 'resource games' (N, b_j) is totally balanced, then the game (N, v) defined in (1.6.1)–(1.6.2) is totally balanced. This result follows from the work of Kalai & Zemel [1982b] on generalized network games. Granot also demonstrates that an appropriate generalization of the dual set is contained in the core of a generalized linear production game. Let y be an optimal solution to the linear program dual to (1.6.1)–(1.6.2) for $S = N$. Assume that the resource game (N, b_j) are balanced and that t^j is in the core of (N, b_j) for $j = 1, \ldots, m$. Then the vector $u = (u_1, \ldots, u_n)$ with $u_i = \sum_{j=1}^{m} y_j t^j_i$ is in the core of (N, v).

Given the above result, Granot provides an alternative proof of the non-emptiness of the core of a m.c.s.t. game, which closely follows the second proof of Proposition 1.8. The same proof technique can also be applied to certain generalizations of the m.c.s.t. model, in which directed links are allowed, and it is not assumed that link costs necessarily satisfy $c_{ij} = c_{ji}$. Finally, one can show that a non-simultaneous network synthesis game (see Section 1.5) satisfies the conditions of the generalized linear production model. Hence, such games necessarily have non-empty cores.

Granot & Granot [1993] consider another general class of games derived from network optimization problems, which include as special cases the maximum flow games of Kalai and Zemel and the assignment games. In this paper particular attention is directed to the characterization of the nucleolus and the intersection of the kernel and the core.

The remainder of this section will consider a class of network models which are not derived from an underlying optimization problem. In this literature, initiated by Myerson [1977], players in a cooperative game correspond to nodes in an undirected graph G. A function v defines the value of every coalition independently of the structure of G. However, G, is assumed to define the communication possibilities available to coalitions in the game. Specifically, players i and j in a coalition S are connected in S by G if there is a path in G from i to j which stays within S. Every graph G partitions every coalition S into a set of maximal connected subcoalitions, P_S in a natural way. The graph restricted game, v_G is then defined by

$$v_G(S) = \sum_{T \in P_S} v(T) \tag{1.6.3}$$

Myerson suggested the use of the Shapley value (see Definition 1.4) as a reasonable allocation method for $v_G(N)$. Owen [1986] examined some of the computational issues in computing the Shapley value. In particular, it was demonstrated that the value could be computed relatively easily if G happens to be a tree. Owen [1986] and Grofman & Owen [1982] also consider applications of the Shapley value, and the Banzhaf power index, which has been defined in political theory, as measures of the degree of 'centrality' in social networks. Specifically, they assume that v is symmetric, so that any differences $\phi_i(v_G) - \phi_j(v_G)$ can be attributed directly to differences in the centrality of i and j in G.

Rosenthal [1988] modeled communication within a game using a network flow model in which weights w_{ij} on individual links provide a quantitative measure of the degree of trust between two players. In this framework, one might interpret the maximum flow between any two players as the information transfer between them.

2. Matching models in economics

The literature on matching models in economics is quite extensive—more so than the literatures surveyed in the other sections of this chapter. There is a good reason for this. Matching models provide a natural framework for the study of economic markets with indivisibilities, such as the market for houses, and certain labor markets. Much of the technical literature on matching models has been motivated by concrete issues concerning the assignment of medical interns and residents to hospitals. (See Roth [1984c] and Roth [1990, 1991] for accounts of these markets in the U.S and U.K., respectively.) In some cases technical results obtained in this literature have a direct practical application. For example, a high percentage of American medical students and hospitals voluntarily participate in a centralized job matching clearinghouse, now known as the National Resident Matching Program. Under this program, rural hospitals typically fail to fill all of their positions, it has been suggested that an appropriate modification of the clearinghouse procedures might redress the imbalance. However, Roth [1986] demonstrated that as long as 'stable matchings', which will be defined in Section 2.1, are desired, than there are no possible modifications of the procedure which will increase the number of residents assigned to rural hospitals.

Matching models have been the subject of a number of recent surveys. See, for example, the monographs of Gusfield & Irving [1989] and Roth & Sotomayor [1990]. Accordingly, the survey of these models in this section will be relatively brief, and will focus primarily on the interconnections of the results on matching models with the results obtained for other network models that have been considered. In particular, matching models can be viewed as special cases of network games, which are related to some of the network games discussed in Section 1 (see, e.g. Gale & Politof, 1981].

2.1. Ordinal matching

The most fundamental unit of economic activity is the process of exchange, where the pairing of a buyer and a seller results in an exchange of one commodity for another such that the utility of both parties increases. In a market, the value of trade is maximized when the set of all buyers and sellers is matched in such a way that no further trading is possible. In a production context, there is a similar benefit which results when plants are matched with locations, and in a more general context, when firms are matched with sets of workers. Similar issues arise in social contexts, as when men and women are matched in marriage. Each of these optimization problems is an example of a matching problem.

As the above examples suggest, there can be different forms of matching problem, depending on the complexity of the bipartite graph from which feasible matchings are allowed. The simplest problem is that of one to one matching, where one agent of each type is matched to at most one agent of the opposite type. Gale & Shapley [1962] introduced this model in the context of the marriage problem. Suppose that a community consists of a set of women, $W = \{1, \ldots, n\}$, and men, $M = \{1, \ldots, m\}$. Each person has an ordinal ranking of each person of the opposite sex as a marriage partner. For any $x \in M \cup W$, write $y >_x y'$ whenever x prefers y to y', and $y \geq_x y'$ whenever x likes y at least as well as y'. It is assumed that every preference ordering $>_x$ is a complete and transitive ordering on the relevant set of feasible partners. A matching is then a set of monogamous marriages. Formally, a matching μ is a one-to-one correspondence from $M \cup W$ onto itself, such that for every $x \in M$, $\mu(x) \in \{x\} \cup W$; for every $x \in W$, $\mu(x) \in \{x\} \cup M$; and for every $x \in M \cup W$, $\mu(\mu(x)) = x$. If $\mu(x) = x$, then x remains single in the matching μ. Gale and Shapley propose the following definitions.[16]

Definition 2.1. *If $m \in M$ and $w \in W$, then the pair (m, w) is said to block a matching μ if $w >_m \mu(m)$, and $m >_w \mu(w)$. A matching μ is said to be stable if it is not blocked by any such pair.*

Definition 2.2. *A stable matching μ is said to be M-optimal if, for every $m \in M$, and every stable matching μ', $\mu(m) \geq_m \mu'(m)$. Similarly a stable matching v is W-optimal if, for every $w \in W$, and every stable matching v', $v(w) \geq_w v'(w)$.*

It is not obvious that optimal matchings, or even stable matchings exist. Gale and Shapley, however, propose a procedure, known as the deferred acceptance algorithm, which can be shown to always lead to stable matchings, and in the case of strict preferences, to lead to either an M-optimal or a W-optimal matching. The M-optimal version (with $m \leq n$) of the algorithm operates as follows. First every man proposes to his most preferred woman (or chooses randomly from the set of most preferred, in case of ties). Women who receive multiple proposals reject all but the most preferred man (or randomly choose from the set of most preferred in case of ties). Every man who is rejected at the first step, and every subsequent step, proposes to his next most preferred women, from the set of women who have not already rejected him. At each step, every women rejects all but the most preferred man from the set of current proposers and the man held over from the previous step. The procedure terminates when no man is rejected. This can occur only if every man is matched with some woman. In a generalized deferred acceptance algorithm, no man or woman x is required to make or accept a proposal with x' if $x >_x x'$ (i.e. if x prefers to remain single rather than be matched with x').

Suppose that the matching μ produced by the deferred acceptance algorithm is not stable. Then there must exist m and w such that $w >_m \mu(m)$ and $m >_w \mu(w)$. But the algorithm requires that m must have proposed to w, and been rejected by her, before m could have proposed to $\mu(m)$. Therefore, by transitivity of preferences, it must be true that $\mu(w) \geq_w m$, and so (m, w) cannot block μ.

The M-optimality of the deferred acceptance algorithm, when preferences are strict, follows by induction. A woman w is 'achievable' for a man m, if there is some stable matching μ with $\mu(m) = w$. Suppose that at some step in the algorithm it is true that no man has yet been rejected by an achievable woman. If at this step a woman w rejects a man m, there are two possibilities. Either $w >_w m$, in which case w is not achievable for m, or for some man m', $m' >_w m$. By the induction hypothesis m' prefers w to every woman who is achievable for him. Therefore, any matching μ which matches m and w could be blocked by (m', w), and so could not be stable. Therefore w must not be achievable for m. It therefore follows that

[16] Whenever possible the notation and terminology of this section will follow that in Roth & Sotomayor [1990].

every man m in this version of the deferred acceptance procedure is matched with his most preferred achievable woman, and by definition, the resulting matching is M-optimal. Since men and women are completely symmetric in the marriage model, it is clear that the deferred acceptance procedure, with women proposing, leads to a W-optimal matching.

These results are summarized in the following:

Proposition 2.3. *There exist stable matchings in the marriage problem for all preference rankings of men and women. When preferences are strict, there exist both M-optimal and W-optimal matchings. Furthermore, these matchings are selected by the two versions of the deferred acceptance algorithm in which men and women are allowed to propose.*

Roth [1982] proved that the matchings chosen by the deferred acceptance procedure have an even stronger optimality property. He demonstrated that there is no feasible (not necessarily stable) matching that is strictly preferred by all men (women) to the M-optimal (W-optimal) matching chosen by the algorithm. Gale & Sotomayor [1985b] establish two additional properties of stable matchings when preferences are strict. They prove that the set of individuals who remain single is the same in every stable matching. They also demonstrate that whenever an additional man (woman) enters the market, then no woman (man) is harmed in the W-optimal (M-optimal) matching.

Gale and Shapley also consider the more complicated many to one matching situation, exemplified by the college admissions problem. In a many to one matching, there is a set $C = \{C_1, \ldots, C_n\}$ of colleges, and a set $S = \{s_1, \ldots, s_m\}$ of students. Each college i has a quota q_i of positions, and a preference ordering over the set S of students, defining preferences over individual students. Each student is assumed to have a preference ordering over the set of colleges. A pair (C, s) then blocks a given matching μ if $s \notin \mu(C)$, C prefers s to some member of $\mu(C)$, and s prefers C to $\mu(s)$. The definitions of stable matchings, and optimal matchings are exactly as in Definitions 2.1 and 2.2 above.

For many purposes, the many to one matching problem can be reduced to the marriage problem, by the following device, as noted in Gale & Sotomayor [1985b]. Each college C_i can be replaced by q_i copies of itself, $C_{i1}, \ldots, C_{iq_i}$, each having a quota of 1, and preferences identical to C_i. Each student then inserts the string $C_{i1}, \ldots, C_{iq_i}$ in place of C_i in his original preference ordering. It is easy to see that stable matchings for the derived, one to one matching, are also stable in the many to one matching problem. Furthermore, it remains true that the deferred acceptance algorithm applied to the one to one matching problem generates both college optimal and student optimal stable matchings. Therefore, Proposition 2.3 remains true for the class of many to one matching problems.

As Proposition 2.3 suggests, the set of stable matchings has an interesting structure, which will be considered more fully in Section 2.4. One should note that an essential property of the marriage and college admission problems is the assumption that the set of economic agents can be partitioned into two distinct groups, such that every agent prefers being matched with himself to any matching from within the same group. Without this assumption, it is easily shown that stable matchings may not exist. Consider, for example, the situation of three potential roommates, 1, 2, 3, such that 1 prefers 2 to 3, 2 prefers 3 to 1, and 3 prefers 1 to 2, and everyone prefers any roommate to being assigned to a single room. If only pairs are allowed, it is clear that any matching is unstable, since the singleton is the most preferred roommate to one of the pairs in any match.[17]

In the original definition of the college admission problem, as outlined above, colleges are assumed to have preferences over individual students, but not over sets of students. When applied to a model of the labor market, this assumption precludes firms from having preferences over sets of workers—something likely to occur if individual workers are substitutes or complements for one another. Roth & Sotomayor [1990] present the following example which illustrates that stable matchings may fail to exist under more general preference structures.

Example 2.4. There are two firms and three workers, which are described by the sets $F = \{f_1, f_2\}$ and $W = \{w_1, w_2, w_3\}$. Preferences are represented as lists of the form $x_1, x_2, \ldots, x_t$ where x_i is preferred to x_{i+1} for $i = 1, \ldots, t - 1$.

$$f_1: \quad \{w_1, w_3\}, \{w_1, w_2\}, \{w_2, w_3\}, \{w_1\}, \{w_2\}, \emptyset, \{w_3\}$$
$$f_2: \quad \{w_1, w_3\}, \{w_2, w_3\}, \{w_1, w_2\}, \{w_3\}, \{w_1\}, \{w_2\}, \emptyset$$
$$w_1: \quad f_2, f_1, \emptyset$$
$$w_2: \quad f_2, f_1, \emptyset$$
$$w_3: \quad f_1, f_2, \emptyset$$

It is straightforward to verify that every matching without unemployment is blocked by some coalition of a firm and one or more workers. For example, when w_2 is matched with f_2, and the other workers with f_1, both f_2 and w_1 would prefer the matching in which they are assigned together. Similarly, any outcome in which there is unemployment can be blocked.

Roth [1985] proves that in a many to one matching situation, a simple restriction on preferences is sufficient to guarantee existence of stable matchings. Preferences of firms for workers (or colleges for students) are said to be 'responsive,' to preferences over individuals, if for any two assignments S and S' of workers to a given firm j, such that $S' = S \cup \{w_i\} \setminus \{w_j\}$, firm j prefers S', if and only if it also prefers w_i to w_j. He then proved that when all firms have responsive preferences, a stable matching always exists. However, even with responsive preferences, many of the other properties of stable matchings in the marriage model fail to generalize. These results will be considered in Section 2.3.

Crawford & Knoer [1981], Kelso & Crawford [1982], and Roth [1984a] all presented models of matching firms and workers in which different assumptions

---

[17] Gale & Shapley describe this example in the case of a fourth potential roommate who ranks last in each of the other three rankings.

on firm preferences were used to guarantee the existence of stable matchings. In addition, the models of Crawford and Knoer, and Kelso and Crawford allowed firms to earn profits, and workers to receive salaries, which are determined as part of the equilibrium. Roth also considered a many to many matching situation in which workers could be hired by multiple firms, so that the role of firms and workers again becomes completely symmetric.

In both one to one, and many to one matching models, there is a relationship of the set of stable matchings to the core of a game without side payments. An arbitrary vector $u = (u_1, \ldots, u_n)$ is said to be strictly dominated via a coalition S if there exists an alternative u', feasible for S, such that $u_i' > u_i$ for all $i \in S$. A vector u is said to be weakly dominated via S if there exists u' such that $u_i' \geq u_i$ for all $i \in S$, with strict inequality for some i. The core of a matching game is the set of utility outcomes which are attainable by the grand coalition of all agents, on both sides of the market, and which cannot be strictly dominated by any other coalition of agents. The weak domination core is the set of attainable allocations for the grand coalition that cannot be weakly dominated. When all preferences are strict, it is a simple matter to demonstrate that the core, the weak domination core, and the set of stable matchings are precisely the same set in the case of a one to one matching model. Since domination is more easily accomplished in the weak domination core, it is always a subset of the regular core. Roth & Sotomayor [1990] demonstrate that when preferences are strict and responsive, then the weak domination core is precisely the set of stable matchings in a many to one matching situation.

In the case of many to many matchings, the correspondence between the core and the set of stable matchings disappears entirely. Example 2.5, due to Blair [1988], describes a many to many matching with a unique (pairwise) stable matching, for which the core is empty.

Example 2.5. There are three firms and three workers. Using the notation of Example 2.4, suppose that preferences are given by

$$
\begin{aligned}
f_1\colon &\quad \{w_1, w_2\},\ \{w_2, w_3\},\ \{w_1\},\ \{w_2\},\ \{w_3\},\ \emptyset \\
f_2\colon &\quad \{w_2, w_3\},\ \{w_1, w_3\},\ \{w_2\},\ \{w_1\},\ \{w_3\},\ \emptyset \\
f_3\colon &\quad \{w_1, w_3\},\ \{w_1, w_2\},\ \{w_3\},\ \{w_1\},\ \{w_2\},\ \emptyset \\
w_1\colon &\quad \{f_1, f_2\},\ \{f_2, f_3\},\ \{f_1\},\ \{f_2\},\ \{f_3\},\ \emptyset \\
w_2\colon &\quad \{f_2, f_3\},\ \{f_1, f_3\},\ \{f_2\},\ \{f_1\},\ \{f_3\},\ \emptyset \\
w_3\colon &\quad \{f_1, f_3\},\ \{f_1, f_2\},\ \{f_3\},\ \{f_1\},\ \{f_2\},\ \emptyset
\end{aligned}
$$

In this example, preferences of both firms and workers are responsive. It is also possible to verify that the matching μ in which $\mu(f_i) = \{w_i\}$ for $i = 1, 2, 3$ is pairwise stable. However, the outcome associated with this matching is not in the core. This follows since the unstable matching in which firm 1 hires workers 2 and 3, firm 2 hires workers 1 and 3, and firm 3 hires workers 1 and 2, makes every agent better off.

2.2. *Matching models with transferable utility*

While the literature on matching models begins with the ordinal approach of Gale and Shapley, it is also possible to model matching situations as cooperative games with side payments and transferable utility.[18] In this framework, both one to one and many to one matching situations have been modeled. Koopmans & Beckmann [1957] studied the linear program associated with the optimal assignments in a one to one matching situation. This section will follow the approach of Shapley & Shubik [1972], in which the assignment problem is modeled as a cooperative game with transferable utility.

In a general assignment game there are two finite disjoint sets of players P and Q. For every $i \in P$ and $j \in Q$ the real number α_{ij} represents the monetary value of the matching of i and j. A 'matched coalition' is a coalition $S = (i, j)$ consisting of exactly one player from P and one player from Q. The value of any matched coalition S is $v(S) = \alpha_{ij}$. The value of any other coalition $S \in P \cup Q$ is given by

$$v(S) = \max \sum [v(S_1) + \ldots + v(S_t)] \qquad (2.2.1)$$

where the maximum in (2.2.1) is taken over all disjoint subsets of matched coalitions contained in S. The game (N, v) defined in (2.2.1) is referred to as the 'optimal assignment game.'

For the grand coalition $N = P \cup Q$ (or any other coalition), $v(N)$ is the optimal solution of an integer programming problem as follows.

$$v(P \cup Q) = \max \sum_{i \in P} \sum_{j \in Q} a_{ij} x_{ij} \qquad (2.2.2)$$

subject to

$$\sum_{i \in P} x_{ij} \leq 1 \qquad (2.2.3)$$

$$\sum_{j \in Q} x_{ij} \leq 1 \qquad (2.2.4)$$

$$x_{ij} \in \{0, 1\} \qquad (2.2.5)$$

Consider the linear program corresponding to (2.2.2)–(2.2.5) in which (2.2.5) is replaced by the non-negativity constraint

$$x_{ij} \geq 0 \qquad (2.2.5')$$

It is well known that problems of the form (2.2.2)–(2.2.5') have integral optimal solutions. That is, the optimal value of the program (2.2.2)–(2.2.5') is the same as the value of the integer program (2.2.2)–(2.2.5) defining $v(P \cup Q)$.

---

[18] Shapley [1962] is an early reference for this approach.

By the duality theorem of linear programming, the value $v(P \cup Q)$ can also be expressed as the optimal solution to the dual program

$$\min \left[\sum_{i \in P} u_i + \sum_{j \in Q} v_j \right] \tag{2.2.6}$$

subject to

$$u_i + v_j \geq \alpha_{ij} \quad \text{for } i \in P \text{ and } j \in Q \tag{2.2.7}$$

$$u_i, \, v_j \geq 0 \tag{2.2.8}$$

It follows that an optimal solution (u, v) to (2.2.6)–(2.2.8) satisfies $\sum_{i \in P} u_i + \sum_{j \in Q} v_j = v(P \cup Q)$. Furthermore, the feasibility condition (2.2.7) and the definition of $v(S)$ in (2.2.1) guarantee that

$$\sum_{S \cap P} u_i + \sum_{S \cap Q} v_j \geq v(S)$$

The following result can therefore be established.

Proposition 2.6. *The core of the optimal assignment game is precisely the set of optimal dual solutions to the linear program defined by (2.2.6)–(2.2.8).*

Proposition 2.6 states that the set of optimal assignments is precisely the set of equilibrium assignments. Furthermore an optimal assignment is generically unique in the following sense. If it happens that more than one optimal assignment exists, then there exist small perturbations in the values α_{ij} which would result in a unique solution.

Shapley and Shubik also prove the following important result concerning the structure of the core.

Proposition 2.7. *If (u, v) and (u', v') are two imputations in the core of the optimal assignment game, then $(u \wedge u', v \vee v')$ and $(u \vee u', v \wedge v')$ are also contained in the core, where*

$$x_i \wedge x_i' = \min\{x_i, x_i'\} \text{ and } x_i \vee x_i' = \max\{x_i, x_i'\}$$

It follows that the core of the optimal assignment game satisfies the properties of a complete lattice under the natural partial orderings defined on the payoffs to members of P, or to members of Q. Geometrically, the core is a closed, convex polyhedron, whose dimension is no greater than the maximum size of P and Q. The lattice property guarantees that the core tends to be 'elongated,' with a long axis connecting the points at which the returns to P and Q are maximized.

There have been a number of generalizations of the optimal assignment game. Kaneko [1976] considered an assignment game consisting of a set of buyers and sellers, in which each seller owns a collection of indivisible goods. Kaneko [1982] extends this model in a game without side payments. Kaneko & Wooders [1982] and Curiel & Tijs [1986] describe other combinatorial games which generalize the assignment games. Finally, Granot & Granot [1993] describe a general class of network flow games which include the optimal assignment game as a special case. In particular, they provide necessary and sufficient conditions for the core to be contained in the kernel of an assignment game, and provide a characterization of the nucleolus for such games.

In the assignment game, preferences are assumed to be expressed in terms of common units of transferable utility. In this framework, a linear programming representation of the set of stable matchings is completely natural. In an ordinal matching model, linear programming techniques can also be applied. Vande Vate [1989] demonstrated that the set of stable matchings of the Gale–Shapley marriage problem can be characterized as the solution of a linear program. That is, a linear programming formulation of the problem was demonstrated to have integer valued extreme points, even though the constraint matrix, in general, fails to be unimodular. Rothblum [1992] extended Vande Vate's results in the case of the generalized Gale–Shapley algorithm in which members of either sex are allowed to remain single rather than accept marriage to an 'unacceptable' partner. (For further discussion of this generalized algorithm see the discussion following Proposition 2.13 below.) Roth, Rothblum & Vande Vate [1990] explore the close relationship of the solutions of the dual linear program and the primal solutions. These results further clarify the structure of the set of stable assignments in the ordinal matching problem, and the relationship between the ordinal and cardinal versions of the matching problem. Additional results on the structure of the core of both ordinal and cardinal matching games are presented in Section 2.4.

2.3. Strategic issues in matching problems

This section will consider further some of the incentive issues introduced in Sections 2.1 and 2.2. In the case of a matching problem, the issue is whether individual economic agents, whether they are buyers being assigned objects at an auction, or students being assigned to a college, have the incentive to truthfully reveal their true preferences to the central planner in charge of making the optimal assignment. All of the discussion of the previous two sections was concerned only with the issue of attaining an optimal assignment, assuming these true preferences were known. Now assume that the economic agents on one, or both, sides of the matching problem are rational, and calculating individuals, who have private knowledge about their own true preferences, and who are interested in maximizing their own welfare. In situations such as this, the central planner must take account of the possibility that economic agents may misrepresent their preferences whenever it is in their interest to do so.

Formally the current section is an application of 'implementation theory' to the matching problem.[19] Agents on one, or both, sides of the matching model are

---

[19] See Maskin [1985] for a survey of implementation theory.

viewed as players in a noncooperative game. There is a central planner, who is responsible for selecting a matching of agents, but who in addition is responsible for specifying the rules of the game. Thus, while individual players have private knowledge about their preferences, the planner, in general, has the freedom to control the manner in which those preferences are reported. This section, however, will be concerned only with so-called 'direct revelation games' in which agents are simply asked to report on their true preferences. As in the previous sections, it will be assumed that the planner is interested only in achieving stable, and if possible optimal matchings, and not in his or her own welfare.

A strategy of a player consists of a reported preference ordering. Thus, in the marriage model, if $P(x)$ represents the true preferences of a man or woman, then a strategy consists of a report $Q(x)$ which may, or may not, coincide with $P(x)$. Two concepts of equilibrium will be considered. The most satisfactory equilibrium, from the point of view of the central planner, is one in which every player, in every matching situation, has the incentive to choose the same strategy, no matter what strategic moves the other players take. Formally, $Q(x)$ is a best response to any set of strategies $Q(i)$ for $i \neq x$. Such an equilibrium is called an equilibrium in dominant strategies. If it happens that truthful reporting of preferences is a dominant strategy equilibrium for every player, then the incentive problem is resolved in a very satisfactory manner. A weaker concept of equilibrium is one in which, for every player x, $Q(x)$ is a best response to the particular strategies $Q(i)$ for $i \neq x$. Such an equilibrium is known as a Nash equilibrium. Clearly the Nash equilibrium is less satisfactory from the social planners point of view, since it requires some implicit coordination among the agents as to which equilibrium is being played. A Nash equilibrium may also be subject to manipulation by coalitions of players. A partial answer to the manipulation issue is given if a Nash equilibrium is also a 'strong' equilibrium, which is the case if no coalition of agents can benefit by changing their strategy choice, given the choices made by every player outside of the coalition. The three concepts of equilibrium are summarized in the following definition.

Definition 2.8. *Given a set of players $N = \{1, \dots, n\}$ with strategy sets $Q(i)$, an equilibrium in dominant strategies is a vector $\bar{q} = (\bar{q}_1, \dots, \bar{q}_n)$, such that for every i, $\bar{q}_i$ is a best response to $(q_1, \dots, \bar{q}_i, \dots, q_n)$ for any set of $q_j \in Q(j)$ for $j \neq i$. A vector $\bar{q}$ is a Nash equilibrium if $\bar{q}_i$ is a best response to $\bar{q}$. A vector $\bar{q}$ is a strong equilibrium if there is no coalition S, and no alternative strategies q'_i for $i \in S$, such that every member of S prefers the outcome selected when S plays according to q' and $N \setminus S$ plays according to $\bar{q}$.*

A matching mechanism in which truthful reporting of preferences is a dominant strategy for all players (hence truth telling is a dominant strategy equilibrium) is called a 'strategy proof' mechanism. Roth [1982] demonstrated that in the marriage model there can be no strategy proof mechanism. Since the marriage model is a special case of the many to one and many to many matching problems, the impossibility result applies to all matching problems.

Proposition 2.9. *No stable matching mechanism in the marriage model exists in which truthful reporting of preferences is a dominant strategy for every man and every woman.*

The proof of this result is an example consisting of two men and two women, all of whom prefer marriage to either partner of the opposite sex to remaining single. m_1 prefers w_1; m_2 prefers w_2; w_1 prefers m_2; and w_2 prefers m_1. There are two stable matchings μ and ν, given by $\mu(m_i) = w_i$ and $\nu(m_i) = w_j$ for $j \neq i$. If the M-optimal matching μ is chosen, then w_2 can benefit by falsely reporting that she prefers to remain single, rather than marry m_2. If all others report truthfully, then the W-optimal matching ν is the only stable matching.

Proposition 2.9 was strengthened by Gale & Sotomayor [1985b] in the following sense. They demonstrated that whenever the M-optimal stable matching is chosen, and at least one other stable matching exists, then some woman will have an incentive to misrepresent her preferences. Zhou [1990] proves a related result for the case of a one-sided matching problem. In this case, it can be demonstrated that whenever n objects are to be assigned to n positions, with $n \geq 3$, then no strategy proof mechanism exists which also satisfies Pareto optimality and symmetry. The proof of Proposition 2.9 may also be used to prove the following corollary.

Corollary 2.10. *No stable matching mechanism in the marriage model exists in which truthful reporting of preferences is a Nash equilibrium strategy for every man and every woman.*

Despite the impossibility results of Proposition 2.9 and its corollary, there are positive results to be obtained if strategic behavior is likely to occur on only one side of the matching situation. Dubins & Freedman [1981] and Roth [1982] demonstrate the following result.

Proposition 2.11. *In the marriage model, the mechanism that selects the M-optimal stable matching, given the reported preferences, makes it a dominant strategy for every man to report his true preferences.*

Leonard [1983] proved a similar result in the case of assignment games with transferable utility. Dubins and Freedman proved the following stronger result.

Proposition 2.12. *In the marriage model, whenever the M-optimal matching is used, no coalition of men can benefit by misreporting their true preferences.*

Proposition 2.9 and its extensions demonstrate that whenever the M-optimal version of the deferred acceptance algorithm is used, it is not in the interest of women to report their true preferences, although Proposition 2.11 guarantees that men will report truthfully. This point does not completely determine whether or not a stable matching will be chosen by the algorithm, when it is applied to the presumably false preferences of one or more women. Roth [1984b] partially resolved this issue with the following result.

Proposition 2.13. *Suppose that, in the marriage model, in which the M-optimal deferred acceptance procedure is used, every man reports his true preferences $P(m_i)$, and the set of women choose any set of equilibrium strategies $Q(w_j)$. Then the stable matching associated with $[P(m_i), Q(w_j)]$ is also a stable matching for the game $[P(m_i), Q(w_j)]$ in which women report their true preferences.*

Proposition 2.13 states that when the M-optimal matching procedure is used, and women respond strategically, then any equilibrium attained is also a stable matching, given actual preferences of both men and women. The existence of an equilibrium, however, was not established. Gale & Sotomayor [1985a] demonstrate that when preferences are strict such an equilibrium does exist for an extended version of the Gale–Shapley deferred acceptance algorithm in which each woman can announce a list of unacceptable partners. (For any matching μ every woman w can list as unacceptable every man except $\mu(w)$.) Zhou [1991] demonstrates that a Nash equilibrium also exists for the original Gale–Shapley algorithm. Specifically, he proved that if μ is a matching that is stable, given true preferences $P(i)$ for all i, and which is an outcome of the M-optimal stable matching procedure, given reported preferences $Q(i)$, then there exist Nash equilibrium strategies $Q'(i)$ that also support μ.

Up to this point this survey has considered only the incentive issues in one to one matching situations. Since every one to one matching is formally a many to one matching, Proposition 2.9 applies in the more general case. Roth [1985], however, proved that Proposition 2.11 does not apply to many to one matchings, such as the college admissions problem. That is, there may be no stable matching procedure in which it is a dominant strategy for every college to state its true preferences.

Incentive issues in matching problems arise whenever the central planner, or mechanism designer, does not have complete information about every agents preferences. The analysis of the strategic (direct revelation) game, in which players report preferences has, so far, made an implicit assumption that the players do have complete information about other players preferences, as well as their own. (Of course, players do not know the strategies chosen by other players, except in equilibrium.) Roth [1989] examines whether the results obtained in the complete information strategic game also carry over to a game of incomplete information, in which players know only the probability distribution of other players preferences. He finds that Propositions 2.9 and 2.11 remain valid in the incomplete information setting, but that Proposition 2.12 and 2.13 are no longer valid.

2.4. The structure of the core of a matching game

This section will continue the analysis of Sections 2.1 and 2.2 by focusing on the structural properties of the core of a matching game. Matching games are in one respect similar to minimum cost spanning tree games in that the existence of a core is typically demonstrated by means of an algorithm which generates an extreme point in the core. The motivation of this section is similar to that of

Section 1.3, in that its goal is to characterize additional points in the core, that may have a greater appeal as fair allocations of the total surplus. The analysis will reveal that cores of matching games have an entirely different structure than cores of m.c.s.t. games. Whereas the main results of Section 1.3 demonstrated that, in many respects, m.c.s.t. games resemble submodular games, which have large and full dimensional cores, this section will attempt to make precise the sense in which the core of a matching game is 'narrow' and 'elongated.'[20]

Propositions 2.3 and 2.7 have already defined the basic structure of the core in the case of ordinal matching games and linear assignment games respectively. Formally these results demonstrate that the core of a one to one matching game has the structure of a complete distributive lattice.[21] For a one to one matching game, there are two partial ordering which define a lattice structure. In the marriage model these orderings are defined by $\geq_m$ and $\geq_w$ as in Section 2.1. In the case of a many to one matching game, Blair [1988] demonstrates that a comparable partial ordering in the case of many to one matching does not generate a lattice structure. However, in the college admissions problem, if one defines $m \geq n$ to mean that every college wishes to keep all of the students assigned by m, and not add any new students, even if all the students assigned by n were made available to it, then the resulting partial ordering is a lattice.[22]

In the assignment game of Shapley and Shubik, preferences were defined by linear utility functions giving the value of feasible matches in terms of a common transferable utility measure. Demange & Gale [1985] consider a generalization of this approach in which agent's preferences are represented by quasi-linear utility functions, expressing the utility of being matched with another agent, and receiving a particular monetary payment. As in Section 2.2, agents are of type P and Q and the matching is one to one. Side payments, but only within a matching, are allowed. Roth & Sotomayor [1990] describe a similar model in which side payments among arbitrary coalitions are feasible.

In both models, most of the lattice properties, and most of the incentive results described in Sections 2.1–2.3 are generalized. Thus, there exist both largest and smallest elements, $\bar{u}$ and $\underline{u}$, representing the highest and lowest payoffs to type P agents. Similarly $\bar{v}$ and $\underline{v}$ represent the highest and lowest payoffs to type Q agents. Demange and Gale then establish the following result.

Proposition 2.14. *If additional players of type Q enter the game, then $\bar{u}$ does not decrease and $\underline{v}$ does not increase. If additional players of type P enter the game, then $\bar{u}$ does not increase and $\underline{v}$ does not decrease.*

The effects of entry of new players is also studied by Mo [1988] in the context of the linear assignment game. When a new player j of type Q enters, it is interesting

---

[20] Note, however, that a one-side assignment game is naturally subadditive, and resembles in many respects a m.c.s.t. game.

[21] See Roth [1985] and Blair [1988] for details.

[22] It is not, however, a distributive lattice.

to compare the cores of $[P \cup Q, v]$ and $[P \cup Q \cup \{j\}, v]$, where v is defined in (2.2.1). For any sets T and T', in $\mathfrak{R}^n$, T is said to 'weakly dominate' T', if for every point $x' \in T'$ there exists $x \in T$ such that $x \geq x'$. T is said to 'strongly dominate' T', if for every $x \in T$ and $x' \in T'$, $x \geq x'$. Mo demonstrates that whenever a player j of type Q enters the game, the returns to type P in the core of $[P \cup Q \cup \{j\}, v]$ weakly dominate the returns to type P in the core of $[P \cup Q, v]$. Furthermore, the returns to type Q in the core of $[P \cup Q, v]$ weakly dominate the returns to type Q in the core of $[P \cup Q \cup \{j\}, v]$. Mo also demonstrates that for certain subsets of players who directly compete with the entrant j, the above results hold in terms of strong domination.[23]

Rochford [1984] defines a subset of the core of the optimal assignment game, which is defined by the equilibrium of a bargaining process between every pair of players. Given an optimal matching μ, every player $i \in P$ is assumed to calculate an optimal 'threat' point, as $t_i = \max_{j \in Q \setminus \mu(i)} \alpha_{ij} - v_j$. Similarly every $j \in Q$ computes $t_j = \max_{i \in P \setminus \mu(j)} \alpha_{ij} - u_i$. One can verify that $t_i \geq 0, t_j \geq 0$, and $t_i + t_j \leq a_{i,\mu(i)}$. Every pair of players i and $\mu(i)$ in an optimal matching then has a surplus defined by $a_{i,\mu(i)}$ which can be allocated between them. A bargaining solution, called the symmetrically pairwise-bargained (SPB) allocation, is a core allocation such that this surplus is evenly divided. Rochford then proves the following:

Proposition 2.15. *For every assignment game the set of symmetrically pairwise-bargained allocations is nonempty. Moreover, this set is precisely the intersection of the kernel and the core of the game* $(P \cup Q, v)$.[24]

Clearly the SPB allocations occupy a more central location in the core than the extreme allocations defined by the deferred acceptance proposal, which correspond to the allocations $(\overline{u}, \underline{v})$ and $(\underline{u}, \overline{v})$. Roth & Sotomayor [1988], however, demonstrate that this set is itself a lattice. That is, there is a SPB allocation that is unambiguously best for type P players and simultaneously worst for type Q players, and another SPB allocation that is best for type Q and worst for type P.

Roth & Vande Vate [1990] consider a different property of the core of an ordinal, one to one matching game. They ask whether it is possible to arrive at a stable (core) matching by starting with an arbitrary feasible matching, and allowing blocking pairs to generate new matchings in a particular order. Specifically, if the pair (i, j) blocks the matching μ, then the matching μ' is created, in which i is matched with j, and $\mu(i)$ and $\mu(j)$ are unmatched, and all other agents are matched in the same way in both matchings. Knuth [1976] demonstrated that cycles are possible in this framework, suggesting that stable matches may never be reached. Roth & Vande Vate, however, demonstrate that there is always some finite sequence of blocking pairs that terminates in a stable matching. It therefore

---

[23] Roughly speaking, a player directly competes with an entrant if that player is reassigned in any new optimal assignment in the enlarged game.

[24] The kernel is formally defined in Section 1.1.

follows that if blocking pairs are chosen randomly, with positive probability, then no matter what the initial position, the recontracting procedure will ultimately reach a core allocation. The authors suggest that this finding supports the idea that decentralized allocation procedures may, in some circumstances, work as well as centralized procedures, such as the deferred acceptance algorithm.

3. Optimal pricing and allocation of waiting time in stochastic service systems

This section will consider the allocation of waiting time in a queuing system. Imagine a set of users who send streams of jobs to a queue. Each job is characterized by its expected service time, and possibly by other attributes, and it is the responsibility of the network manager to impose a service discipline which determines the order in which jobs are completed. Each user imposes an externality on every other user, and the economic question is how to allocate the externality in a fair manner, and in a way that promotes the efficient utilization of the center. In the case of a finite queue, the externality arises from the fact that additional usage by one customer increases the probability that other customers will be dropped from the system. In the case of an infinite queue, the externality arises because increased usage by any one customer increases the waiting time for all customers. Situations such as this arise naturally in computer and telecommunications networks, but virtually any allocation question in which congestion and waiting time are important factors can be considered in this framework.

As in the previous section, it may also be true that users have private information concerning the benefits of completing their jobs, and their costs of delay. That is, utility functions may be unobservable to the system manager. The test of a desirable allocation mechanism may therefore depend on the way in which it elicits truthful reporting of individual preferences.

The papers which will be surveyed in this section assume a basic familiarity with queuing theory. A queuing system consists of a service facility, a process of arrival of customers who wish to be served, and a process of service. When a customer arrives at the facility, the server may be busy with another customer, in which case the new arrival is either dropped from the system, or placed in a queue to wait until the server is free. Generally both the arrival rates and service times are modeled as Poisson process with exponential distributions. In this case a system with s servers and maximum queue length of r is described as a $M/M/s/r$ queuing system.

A queuing discipline is a rule which determines the formation of the queue, and the behavior of customers while waiting. The simplest discipline is the first in, first out (FIFO) discipline in which customers are served in the order in which they arrive. Other disciplines are possible in which customers are assigned a priority within the queue. Customers may also have the option of leaving the queue (balking) if the expected waiting time is too long. The system administrator has various possible options for managing the queue. It may be possible to charge a price for entry into the system, or to directly control the flow rates of entering customers. Also it may be possible to determine the specific service discipline used. In the case of incomplete information about customer utility functions, the administrator also has the ability to design a mechanism to obtain information about customer's utility functions. Customer utility functions are assumed to depend on the value of completed service and the waiting time in the queue. The system administrator's objective is to maximize a function, generally the simple summation, of customer utility functions.

Section 3.1 will consider these issues in the simplest possible framework—in a model with complete information and deterministic arrival and service times. Section 3.2 considers stochastic queuing systems in both full information and incomplete information situations. Each of the models of Section 3.2 assumes that system capacity is fixed, and that the manager is restricted to the use of the FIFO service discipline. Section 3.3 considers priority pricing mechanisms, flow control mechanisms, and the optimal choice of service discipline when explicit pricing and side payments are not feasible.

3.1. Sequencing games

This section will describe a simple example in which some queueing theoretic issues can be analyzed in a framework of cooperative game theory. Consider a queue consisting of customers $N = \{1, \ldots, n\}$ waiting to be served. Each customer i is characterized by a (deterministic) service time $s_i > 0$, and a cost function c_i where $c_i(t)$ represents the delay cost of customer i, if the total waiting time plus service time is equal to t. Customers are assumed to arrive in an order described by a permutation $\omega^o \in \Omega$, where $\omega^o(i) = j$ indicates that customer i is initially in the jth position in the queue. For an arbitrary permutation ω, let $P_i(\omega)$ represent the predecessors of i. The total cost of serving customers according to ω is therefore given by

$$C_\omega = c_1 \left[s_1 + \sum_{i \in P^1(\omega)} s_i \right] + \ldots + c_n \left[s_n + \sum_{i \in P^n(\omega)} s_i \right] \tag{3.1.1}$$

In general, it is possible to reduce the total waiting cost by rearranging the order in which customers are served. Thus, the objective of the manager is to minimize C_ω with respect to $\omega \in \Omega$. If it is known that the waiting cost for every customer is a linear function $c_i(t) = \alpha_i t + \beta_i$, there is a simple rule which may be used to find the minimal cost permutation of customers. Define the urgency index of customer i by $u_i = \alpha_i / s_i$. Suppose that i is an immediate predecessor of j in the initial permutation ω^o. If i and j switch places, leading to a new permutation ω', the difference in cost is $C_{\omega'} - C_{\omega^o} = \alpha_i s_j - \alpha_j s_i$ regardless of the absolute positions of i and j in the queue. If $u_j > u_i$ this difference will be negative, and so the total cost decreases. Smith [1956] demonstrated that C_ω achieves a minimum total waiting cost if and only if $u_{\omega^{-1}(1)} \geq \ldots \geq u_{w^{-1}(n)}$.

Curiel, Pederzoli & Tijs [1989] describe the above model as a sequencing situation. Given an initial permutation ω^o they consider various ways in which the cost savings $C_{\omega^o} - C_{\omega^*}$ can be allocated among the customers. For example, a naive allocation rule might divide the cost savings equally among all of the customers. This rule has the disadvantage that the savings allocation does not depend in any way on the identities of the customers who contribute to the common good by agreeing to switch places.

There is a rule, however, that has a desirable incentive property. Define $g_{ij} = \max[0, \alpha_j s_i - \alpha_i s_j]$. Thus g_{ij} represents the gain which could be achieved by i and j in a situation in which i directly precedes j in a permutation ω. If $u_j > u_i$, the gain is positive, and can be achieved if i and j switch places. If $u_j \le u_i$ there is no achievable gain. A natural rule is therefore the equal gain splitting rule (EGS), which divides g_{ij} equally between i and j. More precisely the EGS rule allocates the total savings $C_{\omega^o} - C_{\omega^*}$ among the n customers according to the rule:

$$\text{EGS}_i = \tfrac{1}{2} \left[\sum_{k \in P^i(\omega^o)} g_{ki} + \sum_{j:i \in P^j(\omega^o)} g_{ij} \right] \tag{3.1.2}$$

With the EGS rule in place, every cost reducing switch between two neighbors in a queue can take place without coercion from the system manager.[25] Some additional incentive properties associated with the EGS rule can be determined by considering a cooperative game which arises naturally in a sequencing situation. In this game, coalitions other than N are allowed to make certain rearrangements in the queuing order in order to achieve reductions in waiting time. Consider a particular coalition S. If players i and j are neighbors in a queue who are both members of S, and $g_{ij} > 0$, then i and j can achieve a cost reduction for S by switching places. Furthermore, this switch does not adversely effect any member outside of S.

In a sequencing game with an initial permutation ω^o, a coalition S is said to be connected if for all $i, j \in S$, and $k \in N$ such that $\omega^o(i) < \omega^o(k) < \omega^o(j)$, it is true that $k \in S$. A connected coalition is allowed to rearrange its members arbitrarily. Thus the value of a connected coalition S can be defined as

$$v(S) = \sum_{i \in S} \left[\sum_{j \in P^i(\omega^o) \cap S} g_{ji} \right] \tag{3.1.3}$$

If $|S| = 1$ then S is connected, but no switches can be made, so that $v(\{i\}) = 0$ for every i. By definition of g_{ij}, $v(S) \ge 0$ for every connected coalition S. If S is not a connected coalition, its value is given by the maximum gain which can be achieved on any of its connected components. Formally, for any S

$$v(S) = \max \sum_{k=1}^{t} v(S_k) \tag{3.1.4}$$

---

[25] This rule, however, assumes that waiting costs are fully observable

where the maximum is taken over all partitions $\{S_1, \ldots, S_t\}$ of S for which each S_k is connected.

The most important result about sequencing games is the following.

Proposition 3.1. *Every sequencing game is supermodular.*

This result is easily proven by examining the marginal contributions $v(S \cup \{i\}) - v(S)$. A player i can increase the value of a coalition S if i is a neighbor of exactly one or exactly two connected subsets of S. Therefore, whenever $S_1 \subseteq S_2 \subseteq N \setminus \{i\}$, there exist coalitions $T_1 \subseteq T_2$ and $U_1 \subseteq U_2$, possibly empty, such that for $p = 1, 2$

$$v(S_p \cup \{i\}) - v(S_p) = \sum_{k \in T_p} g_{ki} + \sum_{k \in U_p} g_{ik} + \sum_{j \in T_p : k \in U_p} g_{jk} \qquad (3.1.5)$$

The supermodularity property $v(S_1 \cup \{i\}) - v(S_1) \leq v(S_2 \cup \{i\}) - v(S_2)$ then follows immediately.

Notice that for the equal gain splitting rule it is true that

$$\sum_{i \in S} \text{EGS}_i = \tfrac{1}{2} \sum_{i \in S} \left[\sum_{k \in P^i(\omega^o)} g_{ki} + \sum_{j : i \in P^j(\omega^o)} g_{ij} \right] \geq$$

$$\geq \tfrac{1}{2} \left[\sum_{k \in P^i(\omega^o) \cap S} g_{ki} + \sum_{j \in S : i \in P^j(\omega^o)} g_{ij} \right] =$$

$$= \sum_{i \in S} \sum_{k \in P^i(\omega^o) \cap S} g_{ki} \geq v(S)$$

Therefore the equal gain splitting rule is contained in the core of a sequencing game. It is also possible to derive a simple expression for the Shapley value of a sequencing game. From the supermodularity of v, it follows that the value is also contained in the core.

Curiel, Pederzoli & Tijs [1989] also note that if all customers have the same service time, so that $s_i = s_j$ for every i and j, then a sequencing game is a special case of the class of permutation games, defined in Tijs, Parthasarathy, Potters & Rajendra Prasad [1984], when customers have arbitrary non-linear cost functions $c_i(t)$. Permutation games are generalizations of the assignment games. Since both assignment games and permutation games are totally balanced, the nonemptiness of the core of the equal service time sequencing game, with arbitrary cost functions follows.

3.2. Pricing and allocation of waiting times in stochastic service systems with fixed capacity

In the model of the previous section, a fixed set of users placed a demand for a service, and the objective of the manager of that service was to minimize total delay cost summed over all users. This section extends that analysis in two ways. First it models the demand for service probabilistically using standard queuing theory techniques. Second, this section introduces the assumption that users of the facility have utility functions that are unobservable to the system administrator. Since the capacity of the system is limited, the randomness of demand raises the possibility that aggregate demand may exceed that capacity, unless active measures are taken to limit the flow demands of the users. As in the previous section, the objective function is assumed to be maximization of aggregate utility of the user population.

Naor [1969] was one of the first to consider the use of prices to improve system performance in a queuing system. Naor's model consisted of an $M/M/1/n$ queuing model — i.e. a single server with n positions in a queue, with a Poisson arrival rate λ and service rate μ. Each customer is assumed to have a benefit of R if he or she is served by the facility, and a cost per unit of waiting plus service time of c, measured in the same units as R. Customers are assumed to balk if their expected waiting cost exceeds R. If the system imposes a price equal to θ on arriving customers, then arriving customers are assumed to balk whenever $R - \theta$ is less than expected waiting cost. For every price θ, there is therefore a critical queue size $n(\theta) \leq n$ such that customers will choose to balk whenever the size of the queue exceeds $n(\theta)$. Whenever a customer arrives at the facility, and there are q customers waiting in line, the expected waiting plus service time is equal to $(q + 1)/\mu$. The critical size $n(\theta)$ is therefore defined as the solution to

$$\frac{c[n(\theta)]}{\mu} \leq R - \theta < \frac{c[n(\theta) + 1]}{\mu} \tag{3.2.1}$$

The socially optimal user fee is determined by computing the socially optimal queue length as follows. The steady state probability that there are k customers in the system, when the maximum queue length is n, is given by

$$p_k(n) = \frac{(\lambda/\mu)^k}{\sum_{i=0}^{n} (\lambda/\mu)^i} \tag{3.2.2}$$

The socially optimal system is given by the maximization of total expected net benefit per unit of time in the following expression.

$$\text{Expected benefit} = \lambda \sum_{k=0}^{n-1} p_k(n) \left[R - \theta - \frac{c(k + 1)}{\mu} \right] \tag{3.2.3}$$

Now suppose that the system administrator is interested in choosing n in order to maximize revenue per unit of time. Clearly the maximum price that can be charged is $R - cn/\mu$, since, at any higher price, customers would balk whenever there were $n - 1$ customers in the system. The revenue maximizing queue length is therefore given by the maximization of the following:

$$\text{Revenue} = \lambda \left[R - \frac{cn}{\mu} \right] \sum_{k=0}^{n-1} p_k(n) \tag{3.2.4}$$

Naor demonstrated that the revenue maximizing queue length is less than the socially optimal queue length. Naor's results were generalized by Knudsen [1972] to the case of a multiserver facility facing more general cost and benefit functions. Recall that $n(\theta)$ represents the size of queue at which a customer will balk if the price is θ, and let θ^* represent the socially optimal price and $\bar{\theta}$ the revenue maximizing price. Then Knudsen's results are summarized in the following:

Proposition 3.2. *If the cost of delay is a convex function of waiting plus service time for every customer, then $\bar{\theta} \geq \theta^*$ and $n(\bar{\theta}) \leq n(\theta^*) \leq n(0)$.*

Edelson & Hildebrand [1975] demonstrate that in Naor's model, if customers are required to make an irrevocable decision whether or not to join the queue before observing the system state, then the optimal price and revenue maximizing price coincide. However, this result no longer holds when customers have heterogeneous waiting times. Other generalizations of Naor's model may be found in Yechiali [1971], Lippman & Stidham [1977], Mendelson & Yechiali [1981], Bell & Stidham [1983], Whang [1989] and Dewan & Mendelson [1990].

3.3. Analysis of priority pricing mechanisms and gateway service disciplines

Models of the previous were concerned with the use of prices to achieve optimal outcomes in stochastic systems with a fixed capacity and fixed service discipline. This section will address several different allocation issues, in which the waiting times of individual users are under the control of a central administrator, through the choice of a queuing service discipline or through direct control of user throughputs. These results build upon the deterministic problem discussed in Section 3.1, in which the value of the coalition of all users was maximized by reordering the place of individual users within the queue. All of the models of this section maintain the assumption that individual users have private knowledge about their utility functions, and that each user is motivated to maximize his or her own utility, while taking the actions of other users as given.

As in Section 3.2, there are n users who generate independent Poisson input streams with arrival rate λ_i. It is assumed throughout the section that service rates are homogeneous, so that, with an appropriate normalization, $\mu_i = 1$ for all i. Each customer has a utility function $U_i(\lambda_i)$, and a waiting cost, per unit of time, equal to c_i. The average waiting time of each customer i, W_i, is partially under the control of the administrator, through the choice of service discipline. However, customers have private information about the functions U_i and costs c_i.

First consider a priority pricing scheme introduced by Dolan [1978] and generalized by Mendelson & Whang [1990]. It is assumed that there are n customers with waiting costs which satisfy $c_1 > \ldots > c_n$. As in Section 3.1, the priority scheme that minimizes expected waiting cost per unit of time is the one which assigns highest priority $\bar{\theta}$ to customer class 1, second highest priority to class 2, and so on, down to class n. Given this priority scheme, standard methods of queuing theory establish that waiting times are given by

$$W_i(\lambda) = 1 + \frac{\sum_{k=1}^{n} \lambda_k}{\left(1 - \sum_{k=1}^{i-1} \lambda_k\right)\left(1 - \sum_{k=1}^{i} \lambda_k\right)} \qquad (3.3.1)$$

Aggregate net benefits are given by

$$\sum_{j=1}^{n}[U_j(\lambda_j) - c_j\lambda_j W_j(\lambda)] \qquad (3.3.2)$$

It is assumed that the administrator seeks to maximize (3.3.2) by setting a price p_i for a customer who claims to be of type i. The administrator knows the set of possible types (U_i, c_i), but cannot identify the type of a particular customer. Given a price p_i, customer i's demand is given by

$$\frac{\mathrm{d}U_i}{\mathrm{d}\lambda_i} - p_i = c_i W_i(\lambda) \qquad (3.3.3)$$

This expression equates expected benefit and expected marginal cost. Solving the first order conditions for the maximization of (3.3.2), given (3.3.1) and (3.3.3) gives an optimal price vector p^* such that

$$p_i^* = \sum_{j=1}^{n} c_j\lambda_j^* \left(\frac{\partial W_j}{\partial \lambda_i}\right) \quad \text{for } i = 1, \ldots, n \qquad (3.3.4)$$

where λ^* maximizes (3.3.2).

Given the above priority pricing mechanism, each customer is allowed to announce a priority class, $r_i(p)$ and choose a level of demand $\lambda_i(p)$. A Nash equilibrium is a vector of strategies such that no customer can gain by deviating, taking as given the strategies of the other customers. In this model, equilibrium strategies must satisfy the following two conditions.

$$p_{r_i(p)} + c_i W_{r_i(p)}\big(\lambda(p), r(p)\big) = \min_{j}\{p_j + c_j W_j[\lambda(p), r(p)]\} \qquad (3.3.5)$$

$$\frac{\mathrm{d}U_i}{\mathrm{d}\lambda_i}[\lambda_i(p)] = p_{r_i(p)} + c_i W_{r_i(p)}[\lambda(p), r(p)] \qquad (3.3.6)$$

An equilibrium price vector p^e which satisfies (3.3.5)–(3.3.6) is said to be 'incentive compatible' if $r_i(p^e) = i$. Mendelson and Whang prove the following result.

Proposition 3.4. *The vector of optimal prices p^* which satisfy (3.3.4) are incentive compatible.*

Examination of (3.3.4) reveals that each customer is charged a price equal to the marginal delay cost that increased usage by that customer imposes on the system as a whole. In a related result, Dolan [1978] considers a model in which customers report their delay costs c_i, and the system manager determines a service discipline which serves customers in decreasing order of reported costs. Each customer is charged a price equal to the marginal delay costs that his position in the queue imposes on all customers with lower reported delay costs. It is proven that this pricing mechanism induces truthful reporting, not only as a Nash equilibrium, but also as a dominant strategy equilibrium. That is, each customer has the incentive to report true waiting costs, whatever strategies the other customers choose.

Sanders [1985] considers a model in which a system administrator can directly control the throughputs of each user in a single server queuing system. Customers have private information, unobservable to the system administrator, about the benefit of completing a job, and about the cost of waiting in a queue. (Sanders [1988a, b] applies this algorithm to more complex network situations.) Each of n customers generates a flow demand, which can be described as an independent Poisson process with arrival rate λ_i. The demands are served by a single server with service rate μ. It is a standard result that the average delay, consisting of queuing time plus service time is given by

$$D = \left(\mu - \sum_{1}^{n} \lambda_i \right)^{-1} \tag{3.3.7}$$

Each user has a utility function $U_i(\lambda_i, D)$, which is assumed to be differentiable in both arguments, concave, nondecreasing in λ_i and nonincreasing in D. Letting $\lambda = (\lambda_1, \ldots, \lambda_n)$, the manager is assumed to have the following objective function.

$$\max_{\lambda} W(\lambda) = \sum_{1}^{n} U_i(\lambda_i, D) \text{ s.t. } \lambda \geq 0 \text{ and } \sum_{1}^{n} \lambda_i < \mu \tag{3.3.8}$$

Given the concavity of user utility functions, a unique solution to (3.3.8) exists, assuming that utility functions are known. Note that it is implicit from (3.3.8) that the manager is able to set each user's arrival rate at any desired level.

If users were to report truthfully, an optimal solution to (3.3.8) could be implemented by means of a 'hill climbing' algorithm, in which each user is asked to supply information about the partial derivatives $\partial U_i / \partial \lambda_i$ and $\partial U_i / \partial D$. Given these reports the manager could then update the rate allocation to each user so as to ultimately arrive at an optimal solution.

As is well known in the incentives literature, however, users do not have the incentive to report truthfully. However, as in the implementation literature for public good economies [Green & Laffont, 1979; Maskin, 1985], it is known that mechanisms exist in which truth telling is a dominant strategy for every agent. That is, regardless of the reports of any other agent, a given agent can do no better than to reveal his or her true preferences. All such mechanisms have the property that the central planner allocates to each agent a share of the total cost that is independent of that agent's report, and for which untruthful reporting can only harm the agent by leading to a suboptimal level of public good provision. Following this approach, Sanders proposed an algorithm which iteratively adjusted arrival rates based on reports from users about the above partial derivatives. A side-payment was used to guarantee that truthful reporting is a dominant strategy for each user. More formally, Sanders established to following result:

Proposition 3.3. *There exists a rate allocation algorithm consisting of incremental adjustments which converges to an optimal solution to (3.3.8) and for which truth telling maximizes each user's objective function regardless of the strategies adopted by other users.*

Chakravorti [1994] noted that the mechanism used to induce truthful reporting has the following shortcomings: (i) it cannot be practically implemented in most situations due to the size of the budget deficit which the system administrator is required to incur, (ii) these budget deficits do not enter the administrator's objective function, (iii) the administrator's objective function is not well defined, and, in particular there may be no solution to (3.3.8) if user utility functions include the side payments which the mechanism requires. Chakravorti defines an alternative flow control mechanism in which side payments are not required (so that the system administrator's budget is balanced), individually rational and Pareto optimal outcomes are achieved, and truth-telling is again a dominant strategy. In a subsequent paper Chakravorti [1993] extends these results to the case of far sighted users who maximize a discounted sum of utilities.

Consider now a model, due to Shenker [1990a, b] in which explicit prices or side payments to customers are not feasible, but in which the system administrator wishes to choose a service discipline that is both fair to customers and efficient. Users have utility functions $U_i(\lambda_i, w_i)$ with exactly the same properties as above, except that w_i is now a private, rather than a public good. The system is again modeled as a single server queue with service rate $\mu = 1$. Given a vector of flow demands λ, the administrator is free to choose any allocation of waiting times W such that

$$\sum_{i=1}^{n} w_i = g\left(\sum_{i=1}^{n} \lambda_i\right) = \frac{\sum_{i=1}^{n} \lambda_i}{1 - \sum_{i=1}^{n} \lambda_i} \tag{3.3.9}$$

Furthermore, if users are numbered so that w_i/λ_i are in increasing order, it must also be true that $\sum_{i=1}^{k} w_i \geq g(\sum_{i=1}^{k} \lambda_i)$, for $1 \leq k < n$.

As an example, the FIFO discipline allocates to each user a waiting time $w_i(\lambda) = \lambda_i/(1 - \sum_{j=1}^{n} \lambda_j)$, which is proportional to the user's input demand. In a noncooperative game theoretic setting, the FIFO discipline can be shown to have particularly undesirable properties, since users do not take account of the increased waiting times of other users when sending jobs to a service center. Shenker defines an alternative 'fair share' allocation function as follows. Assume that the users are ordered in terms of increasing arrival rate, so that $\lambda_1 \leq \ldots \leq \lambda_n$. Let $D(\lambda_1, \ldots, \lambda_n) = \sum_{i=1}^{n} \lambda_i / (1 - \sum_{i=1}^{n} \lambda_i)$, and define the numbers $d_1, \ldots, d_n$ such that

$$d_0 = 0 \text{ and } d_i = D[\lambda_1 + \ldots + \lambda_{i-1} + (n - i + 1)\lambda_i] \text{ for } i = 1, \ldots, n$$

$$(3.3.10)$$

Then the fair share allocation is given by

$$w_1 = \frac{d_1}{n}$$

$$w_2 = \frac{d_1}{n} + \frac{d_2 - d_1}{n - 1}$$

$$(3.3.11)$$

$$w_i = \frac{\sum_{k=1}^{i}(w_k - w_{k-1})}{n - k + 1} \text{ for } i = 1, \ldots, n$$

The fair share allocation protects small users from arbitrarily long waiting times due to the demands of larger users. The allocation can be implemented by means of a scheduling algorithm in which the first λ_1 units of every user's demand is given highest priority; the next $\lambda_2 - \lambda_1$ are assigned second priority; and so on. Shenker also demonstrates that service disciplines associated with the fair share allocation also have desirable incentive properties as well. These follow from the fact that the fair share method, unlike the FIFO method, does not reward users for imposing large demands on the system.[26]

4. Concluding comments

This chapter has surveyed several distinct areas of economic theory in which combinatorial optimization is an inherent part of the framework or analysis of the underlying economic issue. In the analysis of cost allocation on networks, in Section 1, the combinatorial issues are particularly complex, and the literature up to this point has been concerned with deriving properties of various game theoretic solutions on networks. The literature on matching models in Section 2 is at a different stage of development. Here the combinatorial issues are well understood, and the economic literature has been concerned with extending and generalizing the fundamental matching paradigm in models in which more fundamental economic issues can be addressed. The final section of the chapter

---

[26] It is important to note, however, that the fair share allocation offers no protection against malicious users who could benefit by subdividing their demands into arbitrarily small units associated with fictitious users.

contains a brief review of economic models of the allocation of waiting time in queuing networks.

Acknowledgement

I am grateful to B. Chakravorti, D. Granot, R. McLean, Y. Speigel, S. Wilkie and three anonymous referees for helpful comments regarding the content of this survey.

References

Aarts, H.F.M. (1992). Marginal allocations in the core of minimal chain-games. Memorandum No. 1103, University of Twente.

Aarts, H.F.M., and T. Driessen (1992). On the core structure of minimum cost spanning tree games. Memorandum No. 1085, University of Twente.

Bell, C.E., and S. Stidham, Jr. (1983). Individual versus social optimization in the allocation of customers to alternative servers. *Manage. Sci.* 29, 831–839.

Bird, C.G. (1976). On cost allocation for a spanning tree: a game theoretic approach. *Networks* 6, 335–350.

Bittlingmayer, G. (1990). The economics of a simple airline network. *Int. J. Ind. Organ.* 8, 245–257.

Blair, C. (1988). The lattice structure of the set of stable matchings with multiple partners. *Math. Oper. Res.* 13, 619–628.

Bondareva, O. (1962). The core of an *n*-person game. *Vestnik Leningrad Univ.* 17, 141–142.

Chakravorti, B. (1993). A rate allocation algorithm for computer networks with non-myopic and manipulative privately informed users. Unpublished manuscript.

Chakravorti, B. (1994). Optimal flow control of an $M/M/1$ queue with a balanced budget, forthcoming.

Claus, A., and D.J. Kleitman (1973). Cost allocation for a spanning tree. *Networks* 3, 289–304.

Crawford, V.P., and E.M. Knoer (1981). Job matching with heterogeneous firms and workers. *Econometrica* 49, 437–450.

Curiel, I. (1988). Cooperative game theory and applications, Doctoral dissertation, Katholieke Universiteit van Nijmegen.

Curiel, I., G. Pederzoli and S. Tijs (1989). Sequencing games. *Eur. J. Oper. Res.* 40, 344–351.

Curiel, I.J., and S.H. Tijs (1986). Assignment games and permutation games. *Methods Oper. Res.* 54, 323–334.

Davis, M., and M. Maschler (1965). The kernel of a cooperative game. *Nav. Res. Logist. Q.* 12, 223–259.

Debreu, G. (1959). *Theory of Value*, John Wiley & Sons, New York, N.Y.

Demange, G., and D. Gale (1985). The strategy structure of two-sided matching markets. *Econometrica* 53, 873–888.

Dewan, S., and H. Mendelson (1990). User delay costs and internal pricing for a service facility. *Manage. Sci.* 36, 1502–1517.

Dolan, R.J. (1978). Incentive mechanisms for priority queuing problems. *Bell J. Econ.* 9, 421–436.

Du, D.-Z., and F.K. Hwang (1990). An approach for proving lower bounds: solution of Gilbert–Pollak's conjecture on Steiner ratio. *Proc. 31st Annu. Symp. on Foundations of Computer Science, Vol. 1.* Los Alamitos, CA, IEEE Computer Society Press, pp. 76–85.

Dubey, P., and L.S. Shapley (1984). Totally balanced games arising from controlled programming problems. *Math. Program.* 29, 245–267.

Dubins, L.E., and D.L. Freedman (1981). Machiavelli and the Gale–Shapley algorithm. *Am. Math. Monthly* 88, 485–494.

Edelson, N.M., and D.K. Hildebrand (1975). Congestion tolls for Poisson queuing processes. *Econometrica* 43, 81–92.

Edmonds, J. (1967). Optimum branchings. *Nat. Bur. Stand. J. Res.* 71B, 233–240.

Edmonds, J. (1970). Submodular functions, matroids, and certain polyhedra, in: R. Guy, H. Hanani, N. Sauer and J. Schönheim (eds.), *Proceedings of the Calgary International Conference on Combinatorial Structures and Their Applications*, Gordon and Breach, New York, N.Y., pp. 69–87.

Engelbrecht-Wiggans, R., and D. Granot (1985). On market prices in linear production games. *Math. Program.* 32, 366–370.

Ford, L.R., and D.R. Fulkerson (1962). *Flows in Networks*. Princeton University Press, Princeton, N.J.

Gale, D., and L.S. Shapley (1962). College admissions and the stability of marriage. *Am. Math. Monthly* 69, 9–15.

Gale, D., and M. Sotomayor (1985a). Ms. Machiavelli and the stable matching problem. *Am. Math. Monthly* 92, 261–268.

Gale, D., and M. Sotomayor (1985b). Some remarks on the stable matching problem. *Discr. Appl. Math.* 11, 223–232.

Gale, D., and T. Politof (1981). Substitutes and complements in network flow problems. *Discr. Appl. Math.* 3, 175–186.

Galil, Z. (1980). Application of efficient mergeable heaps for optimization problems on trees. *Acta Inf.* 13, 53–58.

Granot, D. (1984). A note on the room-mates problem and a related revenue allocation problem. *Manage. Sci.* 30, 633–643.

Granot, D. (1986). A generalized linear production model: a unifying model, *Math. Program.* 34, 212–222.

Granot, D., and F. Granot (1993). On the computational complexity of a cost allocation approach to a fixed cost spanning forest problem. *Math. Oper. Res.* 17, 765–780.

Granot, D., and G. Huberman (1981). Minimum cost spanning tree games, *Math. Program.* 21, 1–18.

Granot, D., and G. Huberman (1982). The relationship between convex games and minimal cost spanning tree games: a case for permutationally convex games. *SIAM J. Algebra Discr. Math.* 3, 288–292.

Granot, D., and G. Huberman (1984). On the core and nucleolus of minimum cost spanning tree games. *Math. Program.* 29, 323–347.

Granot, D., and M. Hojati (1990). On cost allocation in communication networks. *Networks* 20, 209–229.

Granot, D., and M. Maschler (1991). Network cost games and the reduced game property. Working Paper, Faculty of Commerce and Business Administration, University of British Columbia and Department of Mathematics, The Hebrew University of Jerusalem.

Granot, D., M. Maschler and G. Owen, (1993) The kernel/nucleolus of a standard tree game. Working Paper, Faculty of Commerce and Business Administration, University of British Columbia and Department of Mathematics, The Hebrew University of Jerusalem.

Granot, F., and A.F. Veinott Jr. (1985). Substitutes, complements and ripples in network flows. *Math. Oper. Res.* 10, 471–497.

Green, J., and J.-J. Laffont (1979). *Incentives in Public Decision-Making*. North-Holland, Amsterdam.

Grofman, G., and G. Owen (1982). A game-theoretic approach to measuring degree of centrality in social networks. *Networks* 4, 213–224.

Gusfield, D., and R.W. Irving (1989). *The Stable Marriage Problem: Structure and Algorithms*. MIT Press, Cambridge.

Henriet, D., and H. Moulin (1993). Traffic based cost allocation in a network. Unpublished manuscript.

Ichiishi, T. (1981). Super-modularity: Applications to convex games and to the greedy algorithm for LP. *J. Econ. Theor.* 25, 283–286.

Kalai, E., and E. Zemel (1982a). Generalized network problems yielding totally balanced games. *Oper. Res.* 30, 998–1008.

Kalai, E., and E. Zemel (1982a). Totally balanced games and games of flow. *Math. Oper. Res.* 7, 476–478.

Kaneko, M. (1976). On the core and competitive equilibria of a market with indivisible goods. *Nav. Res. Logist. Q.* 23, 321–337.

Kaneko, M. (1982). The central assignment game and the assignment markets. *J. Math. Econ.* 10, 205–232.

Kaneko, M., and M.H. Wooders (1982). Cores of partitioning games. *Math. Social Sci.* 3, 313–327.

Kelso, A.S., and V.P. Crawford (1982). Job matching, coalition formation, and gross substitutes. *Econometrica* 50, 1483–1504.

Knudsen, N.C. (1972). Individual and social optimization in a multi-server queue with a general cost-benefit structure. *Econometrica* 40, 515–528.

Knuth, D.E. (1976). *Marriages Stables*. Les Presses de l'Université de Montreal, Montreal.

Koopmans, T.C., and M.J. Beckmann (1957). Assignment problems and the location of economic activities. *Econometrica* 25, 53–76.

Kopelowitz, A. (1967). Computation of the kernels of simple games and the nucleolus of n-person games. Research Memorandum No. 31, Department of Mathematics, The Hebrew University of Jerusalem.

Kruskal, J.B. Jr., (1956). On the shortest spanning subtree of a graph and the traveling salesman problem. *Prod. Am. Math. Soc.* 7, 48–50.

Kubo, M., and H. Kasugai (1992). On the core of the network design game. *J. Oper. Res. Soc. Japan* 35, 250–255.

Leonard, H.B. (1983). Elicitation of honest preferences for the assignment of individuals to positions. *J. Polit. Econ.* 91, 461–479.

Lippman, S.A., and S. Stidham (1977). Individual versus social optimization in exponential congestion systems. *Oper. Res.* 25, 233–247.

Littlechild, S.C., and G. Owen (1973). A simple expression for the Shapley value in a special case. *Manage. Sci.* 20, 370–372.

Maschler, M., and B. Peleg (1966). A characterization, existence proof and dimension bounds for the kernel of a game. *Pacific J. Math.* 18, 289–328.

Maschler, M., B. Peleg and L.S. Shapley (1979). Geometric properties of the kernel, nucleolus, and related solution concepts. *Math. Oper. Res.* 4, 303–338.

Maskin, E.S. (1985). The theory of implementation in Nash equilibrium: a survey, in: L. Hurwicz, D. Schmeidler and H. Sonnenschein (eds.), *Social Goals and Organization: Essays in Memory of Elisha Pazner*, Cambridge University Press, Cambridge, U.K.

Megiddo, N. (1978a). Computational complexity of the game theory approach to cost allocation for a tree. *Math. Oper. Res.* 3, 189–196.

Megiddo, N. (1978b). Cost allocation for Steiner trees. *Networks* 8, 1–6.

Mendelson, H., and S. Whang (1990). Optimal incentive-compatible priority pricing for the M/m/1 queue. *Oper. Res.* 38, 870–883.

Mendelson, H., and U. Yechiali (1981). Controlling the G1/M/1 queue by conditional acceptance of customers. *Eur. J. Oper. Res.* 7, 77–85.

Mo, J.P. (1988). Entry and structures of interest groups in assignment games. *J. Econ. Theor.* 46, 66–96.

Mongell, S., and A.E. Roth (1991). Sorority rush as a two-sided matching mechanism. *Am. Econ. Rev.* 81, 441–464.

Myerson, R.B. (1977). Graphs and cooperation in games. *Math. Oper. Res.* 2, 225–229.

Naor, P. (1969). On the regulation of queue size by levying tolls. *Econometrica* 37, 15–24.

Owen, G. (1975). On the core of linear production games. *Math. Program.* 9, 358–370.

Owen, G. (1982). *Game Theory*. Academic Press, New York, N.Y.

Owen, G. (1986). Values of graph-restricted games. *SIAM J. Algebraic Discr. Methods* 7, 210–220.

Prim, R.C. (1957). Shortest connection networks and some generalizations. *Bell System Tech. J.* 36, 1389–1401.

Rochford, S.C. (1984). Symmetrically pairwise-bargained allocations in an assignment market. *J. Econ. Theor.* 34, 262–281.

Rosenthal, E.C. (1987). The minimum cost spanning forest game. *Econ. Lett.* 23, 355–357.

Rosenthal, E.C. (1988). Communication networks and their role in cooperative games. *Social Networks* 10, 255–263.

Roth, A.E. (1982). The economics of matching: stability and incentives. *Math. Oper. Res.* 7, 617–628.

Roth, A.E. (1984a). Stability and polarization of interests in job matching. *Econometrica* 52, 47–57.

Roth, A.E. (1984b). Misrepresentation and stability in the marriage problem. *J. Econ. Theor.* 34, 383–387.

Roth, A.E. (1984c). The evolution of the labor market for medical interns and residents: a case study in game-theory. *J. Polit. Econ.* 92, 991–1016.

Roth, A.E. (1985). The college admissions problem is not equivalent to the marriage problem. *J. Econ. Theor.* 36, 277–288.

Roth, A.E. (1986). On the allocation of residents to rural hospitals: a general property of two-sided matching markets. *Econometrica* 54, 425–427.

Roth, A.E. (1989). Two-sided matching with incomplete information about others' preferences. *Games Econ. Behavior* 1, 191–209.

Roth, A.E. (1990). New physicians: a natural experiment in market organization. *Science* 250, 1524–1528.

Roth, A.E. (1991). A natural experiment in the organization of entry level labor markets: regional markets for new physicians and surgeons in the U.K. *Am. Econ. Rev.* 81, 415–440.

Roth, A.E., and M. Sotomayor (1988). Interior points in the core of two-sided matching markets. *J. Econ. Theor.* 45, 85–101.

Roth, A.E., and M. Sotomayor (1990). *Two-sided Matching: A Study in Game-Theoretic Modeling and Analysis*. Cambridge University Press, Cambridge, U.K.

Roth, A.E., U.G. Rothblum and J.H. Vande Vate (1990). Stable matchings, optimal assignments, and linear programming, forthcoming.

Roth, A.E., and J.H. Vande Vate (1990). Random paths to stability in two-sided matching. *Econometrica* 58, 1475–1480.

Rothblum, U.G. (1992). Characterization of stable matchings as extreme points of a polytope. *Math. Program.* 54, 57–67.

Samet, D., and E. Zemel (1984). On the core and dual set of linear programming games. *Math. Oper. Res.* 9, 309–316.

Sanders, B.A. (1985). A private good/public good decomposition for optimal flow control of an M/M/1 queue. *IEEE Trans. Autom. Control* AC-30, 1143–1145.

Sanders, B.A. (1988a). An asynchronous, distributed flow control algorithm for rate allocation in computer networks. *IEEE Trans. Comput.* 37, 779–787.

Sanders, B.A. (1988b). An incentive compatible flow control algorithm for rate allocation in computer networks. *IEEE Trans. Comput.* 37, 1067–1072.

Schmeidler, D. (1969). The nucleolus of a characteristic function game. *SIAM J. Appl. Math.* 17, 1163–1170.

Shapley, L.S. (1953). A value for n-person games, in: H.W. Kuhn and A.W. Tucker (eds.), *Contributions to the Theory of Games*, Vol. II, Annals of Mathematics Studies No. 28, Princeton University Press, Princeton, N.J., pp. 307–317.

Shapley, L.S. (1961). On network flow functions. *Nav. Res. Logist. Q.* 8, 151–158.

Shapley, L.S. (1962). Complements and substitutes in the optimal assignment problem. *Nav. Res. Logist. Q.* 9, 45–48.

Shapley, L.S. (1967). On balanced sets and cores. *Nav. Res. Logist. Q.* 14, 453–460.

Shapley, L.S. (1971). Cores of convex games. *Int. J. Game Theor.* 1, 11–26.

Shapley, L.S., and M. Shubik (1969). On market games. *J. Econ. Theor.* 1, 9–25.

Shapley, L.S., and M. Shubik (1972). The assignment game I: The core. *Int. J. Game Theor.* 1, 111–130.

Sharkey, W.W. (1982a). *The Theory of Natural Monopoly*. Cambridge University Press, Cambridge, U.K.

Sharkey, W.W. (1982b). Suggestions for a game-theoretic approach to public utility pricing and cost allocation. *Bell J. Econ.* 13, 57–68.

Sharkey, W.W. (1989). Game theoretic modeling of increasing returns to scale. *Games Econ. Behavior* 1, 370–431.

Sharkey, W.W. (1990). Cores of games with fixed costs and shared facilities. *Int. Econ. Rev.* 30, 245–262.

Sharkey, W.W. (1991). Supportability of network cost functions, forthcoming.

Sharkey, W.W., and L.G. Telser (1978). Supportable cost functions for the multiproduct firm. *J. Econ. Theor.* 18, 23–37.

Shenker, S. (1990a). Efficient network allocations with selfish users. Unpublished manuscript, Xerox Palo Alto Research Center.

Shenker, S. (1990b). Making greed work in networks: a game theoretic analysis of gateway service disciplines. Unpublished manuscript, Xerox Palo Alto Research Center.

Skorin-Kapov, D. (1993a). On the core of the minimum Steiner tree game in networks. Technical Report, Harriman School for Management, SUNY at Stony Brook.

Skorin-Kapov, D. (1993b). On a cost allocation problem arising from a capacitated concentrator covering problem. *Oper. Res. Lett.* 13, 315–323.

Skorin-Kapov, D., and H.F. Beltran (1993c). A unifying efficient characterization of some cost allocation solutions associated with capacitated network design problems. Technical Report, Harriman School for Management, SUNY at Stony Brook.

Smith, W.E. (1956). Various optimizers for single-stage production. *Nav. Res. Logist. Q.* 3, 59–66.

Tamir, A. (1989). On the core of a traveling salesman cost allocation game. *Oper. Res. Lett.* 8, 31–34.

Tamir, A. (1991). On the core of network synthesis games. *Math. Program.* 50, 123–135.

Tijs, S.H., T. Parthasarathy, J.A.M. Potters and V. Rajendra Prasad (1984). Permutation games: another class of totally balanced games. *OR Spektrum* 6, 119–123.

Topkis, D.M. (1983). Activity selection games and the minimum-cut problem. *Networks* 13, 93–105.

Topkis, D.M. (1984). Complements and substitutes among locations in the two-stage transshipment problem. *Eur. J. Oper. Res.* 15, 89–92.

Topkis, D.M. (1987). Activity optimization games with complementarity. *Eur. J. Oper. Res.* 28, 358–368.

van den Nouweland, A., S. Tijs and M. Maschler (1993). Monotonic games are spanning network games. *Int. J. Game Theor.* 21, 419–427.

Vande Vate, J. (1989). Linear programming brings marital bliss. *Oper. Res. Lett.* 8, 147–153.

Whang, S. (1989). Cost allocation revisited: an optimality result. *Manage. Sci.* 35, 1264–1273.

Woroch, G.A. (1990). On the stability of efficient networks: integration and fragmentation in communication and transportation networks, presented at ORSA/TIMS Conference, Denver, 1988.

Yechiali, U. (1971). On optimal balking rules and toll charges in the G1/M/1 queue process. *Oper. Res.* 19, 348–370.

Zhou, L. (1990). On a conjecture by Gale about one-sided matching problems. *J. Econ. Theory* 52, 123–35.

Zhou, L. (1991). Stable matchings and equilibrium outcomes of the Gale–Shapley algorithm for the marriage problem. *Econ. Lett.* 36, 25–9.

Biographical Information

Arjang A. ASSAD holds a Ph.D. from the Sloan School of Management at MIT, where he also received a B.S. in mathematics and masters degrees in chemical engineering and operations research. He joined the College Business and Management of the University of Maryland in 1978, where he is currently Professor of Management Science and director of the IBM Total Quality Project. Dr. Assad's research and publications include transportation modeling, vehicle routing, mathematical programming, and the implementation of OR/MS models. His recent publications include two books: *Vehicle Routing: Methods and Studies* and *Excellence in Management Science Practice: A Readings Book.* He is serving or has served on the Editorial Boards of *ORSA Journal on Computing; Production and Operations Management; Transportation Science; Operations Research;* and the *American Journal of Mathematical and Managerial Sciences.* He has also acted as a consultant in the areas of distribution and manufacturing for several organizations in both the public and the private sectors. (Chapter 5.)

Jacques DESROSIERS is Professor at Ecole des Hautes Etudes Commerciales (HEC) de Montréal. His research interests include large-scale linear and integer programming optimization. His work and that of his colleagues at GERAD, the joint Research Center between several Canadian universities, has lead to the development of advanced optimization approaches for time-constrained vehicle routing and crew scheduling. A major achievement of the group has been the development of the GENCOL optimizer, a column generation based software whose application in different areas such as school busing, transportation of the handicapped, urban transit crew scheduling, as well as airline crew scheduling, crew rostering and aircraft fleet planning has permitted the solution of large instances of these problems. (Chapter 2.)

Yvan DUMAS, following his Ph.D. and development work at GERAD, is now Vice-President for R&D at AD OPT Technologies Inc., Montreal. He is heavily involved in the current implementation of GENCOL-related software systems. (Chapter 2.)

Awi FEDERGRUEN is the Charles E.-Exley Professor of Management at the Graduate School of Business at Columbia University. (Chapter 4.)

Marshall L. FISHER is a Stephen J. Heyman Professor, Professor of Operations and Information Management, Co-director of the Center for Manufacturing and Logistics Research of The Warton School, University of Pennsylvania.

Dr. Fisher is highly regarded for his pioneering work in logistics and supply chain coordination. His research has been successfully applied in both industry and government, and he has gained international recognition for the impact of his work in developing the operational capabilities required to support a customer responsive strategy.

Dr. Fisher has been a consultant to many Fortune 500 companies, including Air Products and Chemicals, Campbell Soup, Dupont, Exxon, Frito Lay and Scott Paper. He has served as principal investigator for several projects funded by the National Science Foundation and the Office of Naval Research, including studies to measure the impact of product variety on productivity and to develop optimization tools for logistics planning.

In 1994, Dr. Fisher was elected a member of the National Academy of Engineering. He also served as President of the Institute of Management Science during 1988–1989 and as departmental editor of *Management Science* from 1979–1983. He is a recipient of the 1977 Lanchester prize for his writings on operations research, the 1983 Edelman Prize from the Institute of Management Science for his development of a large-scale logistics planning model for a major industrial gas firm, and the E. Grosvenor Plowman Award from the Council of Logistics Management for contributions to logistics. In 1981, he co-founded Distribution Analysis, Research and Technology, Inc. and served as Chairman of the Board of Directors until 1990.

Dr. Fisher earned his S.B. in electrical engineering (1965), S.M. in management (1969), and Ph.D. in operations research (1970) from the Massachusetts Institute of Technology. Prior to coming to Wharton, he taught at Cornell University and the University of Chicago. (Chapter 1.)

Michael A. FLORIAN is Professor of Computer Science and Operations Research and carries out research at the Center for Research on Transportation (CRT) of the Université de Montréal. He holds a B.Eng. degree in Mechanical Engineering (McGill University, 1962) and M.Sc. and Dr.Eng.Sc. degrees in Operations Research (Columbia University, 1966, 1969). His current research interests include mathematical programming as it is applied to network analysis, especially transportation networks, interactive computing, parallel computing implementations of network optimization algorithms and transportation systems modelling. From 1973–1979 he served as Director of the CRT and was, at various time periods, Editor of *INFOR*, Associate Editor of *Operations Research*, *Transportation Science* and *Regional Planning and Urban Economics*. He is currently on the Editorial Advisory Board of *Transportation Science*, *International Transactions in Operations Research*, the *Logistics and Transportation Review* and *Sistemi Urbani*. He received the Canadian Operational Research Society Merit Award, the CRDT prize for R&D in transportation, the ACFAS Jacques Rousseau prize for interdisciplinary research and is a Fellow of the Royal Society of Canada. He has edited two books and authored approximately 70 refereed articles. (Chapter 6.)

Bruce L. GOLDEN is a Professor in the College of Business and Management at the University of Maryland. He has served as Chairman of the Department of Management Science and Statistics since 1980. His research interests include network optimization, distribution management, natural resource management, and applied operations research, and he has published many technical articles in these and related fields. In addition, he has edited numerous volumes on vehicle routing, computing in operations research, and public sector OR. Some of Dr. Golden's recent research has been funded by the Maryland Department of Natural Resources, the U.S. Agency for International Development, Westinghouse, and the U.S. Census Bureau. He has consulted for a wide variety of organizations including the American Red Cross, the National Soft Drink Association, MidAtlantic Toyota, Airco Industrial Gases, the Military Airlift Command, and Amerigas Corporation. Dr. Golden is Editor of the *American Journal of Mathematical and Management Sciences*, Editor of the *ORSA Journal on Computing*, and Departmental Editor of *Management Science*, and he has served as Area Editor of *Operations Research*. He received his bachelor's degree in mathematics from the University of Pennsylvavnia and his masters and doctoral degrees in operations research from M.I.T. (Chapter 5.)

Donald W. HEARN is Professor of Industrial and Systems Engineering and Co-Director of the Centre for Applied Optimization at the University of Florida. He received an undergraduate degree in physics at the University of North Carolina as a Morehead Scholar and received Masters and Ph.D. degrees from Johns Hopkins University in management science and operations research. His research interests include mathematical programming, transportation science and location theory. Recent and current research has concerned the development of efficient algorithms in large-scale optimization, with an emphasis on decomposition and dual methods. Applications of the methods have included optimization and equilibrium problems in urban traffic assignment, water pipe network design and electrical networks. He is editor of OPTIMA, the newsletter of the Mathematical Programming Society and an Associate Editor of the *Journal Computational Optimization and Applications*. He is Co-Editor of the recent book *Large-Scale Optimization: State of the Art* and author of approximately 50 refereed articles. (Chapter 6.)

Patrick JAILLET is Professor in the Department of Management Science and Information Systems and has a courtesy appointment in the Department of Civil Engineering at The University of Texas at Austin. He is also Professor at the Laboratory for Applied Mathematics, ENPC, Paris, France.

After his undergraduate studies in France (majors in Mathematics and Computer Science), he received a M.S. in transportation in 1982, and a Ph.D. in operations research in 1985, both from MIT. His Ph.D. thesis was the recipient of the 1985 ORSA/TSS Dissertation Prize.

His research interests include topics such as probabilistic combinatorial optimization problems, online graph problems, optimization on random graphs, asymptotic properties of Euclidean functionals, network design problems for airline industry and truck dispatching under real-time information.

Professor Jaillet has been invited as a Visiting Scientist in various universities and was a Fulbright Scholar in 1990. He currently holds a CBA Foundation Advisory Council Fellowship at The University of Texas at Austin. He is a member of the Institute for Operations Research and Management Science Society (INFORMS), the Mathematical Programming Society (MPS), and the Society of Industrial and Applied Mathematics (SIAM). He is an Associate Editor for *Operations Research* and an Appointed Reviewer for *Mathematics Abstracts*. (Chapter 3.)

Martine LABBÉ is research associate of the National Fund for Scientific Research of Belgium and is affiliated with the Service de Mathématiques de la Gestion of the Université Libre de Bruxelles. In this context, she heads a research group working on traffic and telecommunication models. Her research interests include transportation and telecommunication networks, location theory and combinatorial optimization. (Chapter 7.)

Rolf H. MÖHRING is currently Full Professor in the Department of Mathematics at the Technische Universität Berlin. He obtained his Diplom (1973), Ph.D. (1975) and Habilitation (1982) in Mathematics at the Rheinisch-Westfälische Technische Hochschule Aachen. He has held earlier positions in Bonn, Hildesheim and Aachen. He is member of the Editorial Board of several journals, among them *Discrete Applied Mathematics, Operations Research*, and the *SIAM Journal on Discrete Mathematics*. His research interests center around graph algorithms, combinatorial optimization, VLSI layout and scheduling. (Chapter 8.)

Amedeo R. ODONI is Professor of Aeronautics and Astronautics and of Civil and Environmental Engineering at the Massachusetts Institute of Technology (MIT). He holds S.B. and S.M. degrees in Electrical Engineering (MIT 1965, 1967) and a Ph.D. in Operations Research (MIT 1969). His research interests are in applied probability, risk analysis and the modeling of large scale systems, especially in air transportation. He has served as Co-Director of the Operations Research Center and of the Statistics Center at MIT, as well as Editor-in-Chief of *Transportation Science*, and is currently Head of the Systems Division of the Aeronautics and Astronautics Department at MIT. He has also been a consultant to many Airport Authorities and government air traffic control organizations around the world, playing a key role in several very large aviation infrastructure projects in the United States, Europe, East Asia and Australia. In 1991 he received the FAA Administrator's National Award for Excellence in Aviation Education. Dr. Odoni has authored or co-authored four books and approximately 50 refereed articles and book chapters on stochastic processes, dynamic queueing networks, probabilistic combinatorial optimization, location theory, airport planning and air traffic control. (Chapter 3.)

Dominique PEETERS, a member of CORE, is Assistant Professor of Geography at the Université Catholique de Louvain. He has published over 40 papers in refereed journals in geography and operations research. His main topics are in location theory and quantitative methods in geography. (Chapter 7.)

Warren POWELL is a Professor in the Department of Civil Engineering and Operations Research at Princeton University, and Director of the CASTLE Laboratory, specializing in the development of dynamic models for transportation and logistics. Since joining the faculty at Princeton in 1981 from MIT, he has specialized in the development and implementation of advanced optimization models for real-time planning of motor carriers, railroads and other large-scale logistics operations. Optimization models developed under his supervision are now in use by a number of major transportation companies. His research focus is in the development of stochastic models for dynamic settings, and the design of efficient solution algorithms that will work in a real-time setting. Tools used in these systems encompass stochastic programming, dynamic programming, stochastic control, large-scale optimization, integer programming, and queueing.

Professor Powell is a past recipient of a Presidential Young Investigator Award and has twice been a finalist in the Franz Edelman Management Science Achievement competition (1987 and 1991). He has been an Area Editor for *Operations Research*, and serves on the Editorial Advisory Board for *Transportation Science*. He is a past president of the Transportation Science Section, and co-founder of TRISTAN, a triennial conference on transportation models and algorithms. (Chapter 3.)

William SHARKEY is currently a visiting professor at the Institut d'Economie Industrielle in Toulouse. Previously he has been a member of the economics research group at Bellcore in Morristown, New Jersey and at Bell Laboratories in Murray Hill, New Jersey. His research interests include the economics of regulation, the economics of telecommunications, cooperative game theory, cost allocation, and the economics of networks. He is the author of *The Theory of Natural Monopoly* and has recently completed a series of articles on axiomatic methods of pricing and cost allocation. He holds a Ph.D. in economics from the University of Chicago and a B.S. from the University of Michigan. (Chapter 9.)

David SIMCHI-LEVI is an Associate Professor of industrial engineering and management sciences at Northwestern University. He received his B.Sc. in Aeronautical Engineering in the Technion, Israel Institute of Technology, and M.Sc. and Ph.D. in Operations Research from Tel-Aviv University, Israel. His research currently focuses on the analysis, development and implementation of robust and efficient techniques for the design, control and operation of logistics systems and telecommunication networks. He was involved in the development of a Computerized System for School Bus Routing in New York City, developed together with the NYC Board of Education/Office of pupil transportation, the Fund for the City of NY and Julien Bramel. The system won the *first place prize* in the government/public sector category of the Windows World Open Competition, Atlanta May 1994. The competition was sponsored by, among others, Microsoft, Borland, AT&T, the Computer World magazine and the Windows World Conference. Dr. Simchi-Levi is an Associate Editor for *Operations Research, Transportation Science* and Telecommunication Systems. (Chapter 4.)

Marius M. SOLOMON is Associate Professor in the Management Science Group at Northeastern University and a Visiting Member of GERAD. This article is part of his ongoing research on time-based mathematical programming models of logistics and production management problems. He is also Coordinator of the Vehicle Routing Special Interest Group of INFORMS. (Chapter 2.)

François SOUMIS is Professor at Ecole Polytechnique de Montréal and the current director of GERAD. His research interests span across the production, transportation and telecommunication fields. He is also director of a SYNERGIE project in air transportation. SYNERGIE, a large program undertaken jointly by Canadian government, business and academic institutions, is aimed at sponsoring research with immediate commercial application. Under his direction, the 25 researcher team has developed a number of planning and operations management software systems, several of which have been commercialized. (Chapter 2.)

Jacques-François THISSE, a Fellow of the Econometric Society, is Professor of Economics at the Sorbonne and Director of CERAS. He has published over 100 papers in refereed journals in economics and operations research. His main topics of interest are in economic geography and industrial economics. (Chapter 7.)

Dorothea WAGNER is a full Professor at the Department of Mathematics and Computer Science of the Universität Konstanz, Germany. She received her Diplom and Ph.D. degrees in mathematics from the Rheinisch-Westfälische Technische Hochschule Aachen in 1983 and 1986, respectively. In 1992 she received the Habilitation degree from the Department for Mathematics of the Technische Universität Berlin. Her research interests are in the areas of discrete optimization and graph algorithms particularly applied to VLSI-design and transportation systems. (Chapter 8.)

Frank WAGNER is currently an Assistant Professor at the Institut for Computer Science of the Freie Universität Berlin. He received his Diplom (1985) and Ph.D. (1989) degrees in mathematics from the Rheinisch-Westfälische Technische Hochschule Aachen, and his Habilitation (1995) degree in Computer Science from the Department for Mathematics and Computer Science of the Freie Universität Berlin. His research interest is the area of discrete applied algorithms. He works on graph algorithms with applications in VLSI layout, and on geometric algorithms with applications in cartography. (Chapter 8.)

Subject Index

Aggregation, 44, 47
Air transportation, 43–44, 124–127, 205–206
Aircraft fleet assignment, 43–44
A priori optimization, 142, 145, 147
A priori solutions, 158, 159, 162
Arc elimination, 57
Arc-path incidence matrix, 487
Artificial intelligence, 15
Asymmetric cost models, 490
Asymmetric NEM, 499
Asymptotic behavior, 147, 156, 157
Asymptotic properties, 152, 166
Asymptotic results, 164
    asymptotically optimal, 298
Augmented net-gate matrix, 675
Average performance, 297
Average reward, 188
    criterion, 187
Average-case bounds, 147
'Away' step variant, 536

Balanced bifurcator, 633
Balanced collection, 717
Balanced game, 717
Best lower bound, 507
Bifurcator, 632, 633
Bin-packing, 298
    perfect packing, 325
Bipartition, 629
Bisection, 637
Block, 122
    folding, 682, 696
Booking strategies, 144
Boundary conditions, 239, 240
Bounded flows, 539
Bounds, 158, 280
Braess' paradox, 489
Branch-and-bound strategies, 53, 109, 127
Branch-and-cut, 125
Branching, 95
Bucket, 75
Bus driver scheduling problem, 122

Capacitated multicommodity flow problems, 231
Capacitated network design game, 730
Cell synthesis, 674
Chain games, 726
Chance constrained programming, 190, 257
Channel, 647
Chordal graphs, 697
Circulation problem, 532
Clarke & Wright method, 7
Cograph, 701
College admissions problem, 740
Column generation methods, 90, 117, 127, 206, 501, 535
Comb inequalities, 20
Combinatorial optimization, 148, 152
Combined distribution-assignment, 539
Communication, 147
Complexity of a priori optimization, 158
Complexity of the PSPP, 174
Computer codes for NEMs, 541
Concave costs networks, 189, 190
Consecutive clique arrangement, 688
Constant-guarantee bound, 171
Constant-guarantee heuristic, 158, 160, 166, 170
Constrained folding, 682
Constraint graphs, 664
Container shipping, 205
Convex simplex method, 535, 536
Convexity constraint, 50
Cooperative game, 716
Core, 656, 717, 742
Cost allocation, 715
Crew pairing, 124–127
Crew scheduling, 116, 230
Critical cuts, 673
Critical path approach, 59
CRP, 647
Cut condition, 642
Cuts, 53, 95, 128
Cutting plane, 250
Cutting-stock problem, 95
2-Cycle elimination, 80, 91, 97
Cycle structure, 653

Handbooks in Operations Research and Management Science
Contents of Previous Volumes

Volume 1. Optimization
Edited by G.L. Nemhauser, A.H.G. Rinnooy Kan and M.J. Todd
1989. xiv + 709 pp. ISBN 0-444-87284-1

Volume 2. Stochastic Models
Edited by D.P. Heyman and M.J. Sobel
1990. xv + 725 pp. ISBN 0-444-87473-9

Volume 5. Marketing
Edited by J. Eliashberg and G.L. Lilien
1993. xiv + 894 pp. ISBN 0-444-88957-4

Printed and bound by CPI Group (UK) Ltd, Croydon, CR0 4YY

03/10/2024

01040330-0004